P9-APN-262

The Oxford English Minidictionary

Fourth Edition

Edited by
HELEN LIEBECK
ELAINE POLLARD

Clarendon Press · Oxford

Oxford University Press, Walton Street, Oxford OX2 6DP

Oxford New York
Athens Auckland Bangkok Bombay
Calcutta Cape Town Dar es Salaam Delhi
Florence Hong Kong Istanbul Karachi
Kuala Lumpur Madras Madrid Melbourne
Mexico City Nairobi Paris Singapore
Taipei Tokyo Toronto

and associated companies in
Berlin Ibadan

Oxford is a trade mark of Oxford University Press

Published in the United States by
Oxford University Press Inc., New York

ᶜ *Oxford University Press 1981, 1988, 1991, 1994, 1995*

First edition 1981
Second edition 1988
Third edition 1991
Revised edition 1994
Fourth edition 1995

All rights reserved. No part of this publication may be reproduced,
stored in a retrieval system, or transmitted, in any form or by any means,
without the prior permission in writing of Oxford University Press. Within
the UK, exceptions are allowed in respect of any fair dealing for the purpose
of research or private study, or criticism or review, as permitted under the
Copyright, Designs and Patents Act, 1988, or in the case of reprographic
reproduction in accordance with the terms of the licences issued by the
Copyright Licensing Agency. Enquiries concerning reproduction outside
these terms and in other countries should be sent to the Rights Department,
Oxford University Press, at the address above

This book is sold subject to the condition that it shall not, by way
of trade or otherwise, be lent, re-sold, hired out or otherwise circulated
without the publisher's prior consent in any form of binding or cover
other than that in which it is published and without a similar condition
including this condition being imposed on the subsequent purchaser

British Library Cataloguing in Publication Data
Data available

Library of Congress Cataloging in Publication Data
Data available
ISBN 0-19-861324-5

10 9 8 7 6 5 4 3

Printed in Great Britain by
Charles Letts (Scotland) Ltd.

Contents

Preface

The *Oxford English Minidictionary* is the smallest member of the Oxford family of dictionaries, and is written for those who need a compact guide to the spelling and meaning of words in the English language of today.

This new edition has been extensively updated in the light of evidence collected by the Oxford Dictionaries Department since the previous edition was published. Many new words and phrases have been added, and new features have been introduced giving extra guidance on the language. The overall layout of the entries has also been improved, making all the information provided clearer and more accessible.

In spite of its small format the dictionary gives many alternative spellings (a great many more than in previous editions, including American English variants). It also offers clear help in cases where there might be doubt about the formation or spelling of inflected forms: difficult plural forms of nouns and those inflections of verbs and comparatives of adjectives which are not wholly straightforward are spelt out in full.

Another new feature of the dictionary is easy-to-read guidance on the pronunciation of difficult words, using a system that is intended to be self-explanatory.

More prominence has been given in this edition to assistance on disputed and controversial usage, and special usage notes have been introduced to this end. There are over a hundred such notes giving advice on correct English, focusing on areas such as spelling, style, and commonly confused words.

H.F.L., E.B.P.

Pronunciation

A guide to pronunciation is given for any word that is difficult to pronounce, or difficult to recognize when read, or spelt the same as another word but pronounced differently. The pronunciation given represents the standard speech of southern England. It is shown in brackets, usually just after the word itself.

Words are broken up into small units, usually of one syllable. The syllable that is spoken with most stress in a word of two or more syllables is shown in heavy letters, like **this**.

The sounds represented are as follows:

a *as in* cat	i *as in* pin	r *as in* red
á *as in* ago	I *as in* pencil	s *as in* sit
ah *as in* calm	I *as in* eye	sh *as in* shop
air *as in* hair	j *as in* jam	t *as in* top
ar *as in* bar	k *as in* king	th *as in* thin
aw *as in* law	l *as in* leg	th *as in* this
ay *as in* say	m *as in* man	u *as in* cup
b *as in* bat	n *as in* not	ù *as in* circus
ch *as in* chin	ng *as in* sing,	uu *as in* book
d *as in* day	finger	v *as in* van
e *as in* bed	nk *as in* thank	w *as in* will
ě *as in* taken	o *as in* top	y *as in* yes
ee *as in* meet	õ *as in* lemon	or when preceded
eer *as in* beer	oh *as in* most	by a consonant = I
er *as in* her	oi *as in* join	*as in* cry, realize
ew *as in* few	oo *as in* soon	yoo *as in* unit
ewr *as in* pure	oor *as in* poor	yoor *as in* Europe
f *as in* fat	or *as in* corn	yr *as in* fire
g *as in* get	ow *as in* cow	z *as in* zebra
h *as in* hat	p *as in* pen	zh *as in* vision

A consonant is sometimes doubled to show that the vowel just before it is short (like the vowels in *cat, bed, pin, top, cup*). The pronunciation of a word is sometimes indicated by giving a well-known word that rhymes with it.

Abbreviations

adj. adjective	*poss.pron.* possessive pronoun
abbr. abbreviation	*pref.* prefix
adv. adverb	*prep.* preposition
Amer. American	*pron.* pronoun
Austral. Australian	*rel.pron.* relative pronoun
Brit. British	*S. Afr.* South African
comb. form combining form	*Scot.* Scottish
conj. conjunction	*sing.* singular
derog. derogatory	sp. spelling
esp. especially	*symb.* symbol
hist. historical	usu. usually
int. interjection	*v.* verb
n. noun	var. variant
N. Engl. northern England	vars. variants
n.pl. noun plural	*v.aux.* auxiliary verb
pl. plural	

Abbreviations that are in general use (such as cm, RC, and USA) appear in the dictionary itself.

Proprietary terms (trade marks)

This book includes some words which are or are asserted to be proprietary names or trade marks. Their inclusion does not imply that they have acquired for legal purposes a non-proprietary or general significance, nor is any other judgement implied concerning their legal status. In cases where the editor has some evidence that a word is used as a proprietary name or trade mark this is indicated by the label (*trade mark*), but no judgement concerning the legal status of such words is made or implied thereby.

Aa

A *abbr.* **1** ampere(s). **2** answer. □ **A1** (*colloquial*) in perfect condition; first-rate. **A1, A2, A3,** etc. the standard paper size, each half the previous one, e.g. A4 = 297 x 210mm.

a *adj.* (called the *indefinite article*) one, any; in, to, or for each; per.

Å *abbr.* angstrom(s).

AA *abbr.* **1** Automobile Association. **2** Alcoholics Anonymous.

aardvark *n.* an African animal with a long snout.

aback *adv.* □ **taken aback** disconcerted.

abacus *n.* a frame with balls sliding on rods, used for counting.

abalone (abă-loh-ni) *n.* an edible mollusc with a shell lined with mother-of-pearl.

abandon *v.* leave without intending to return; give up. ● *n.* a careless freedom of manner. □ **abandonment** *n.*

abandoned *adj.* (of manner etc.) showing abandon; depraved.

abase *v.* humiliate, degrade. □ **abasement** *n.*

abashed *adj.* embarrassed, ashamed.

abate *v.* make or become less intense. □ **abatement** *n.*

abattoir (ab-ă-twar) *n.* a slaughterhouse.

abbey *n.* a building occupied by a community of monks or nuns; a church belonging to this.

abbot *n.* a man who is the head of a community of monks.

abbess *n.* a woman who is the head of a community of nuns.

abbreviate *v.* shorten.

abbreviation *n.* a shortened form of a word or phrase.

ABC *n.* **1** the alphabet; an alphabetical guide. **2** the basic facts of a subject.

abdicate *v.* renounce the throne. □ **abdication** *n.*

abdomen *n.* the part of the body containing the digestive organs. □ **abdominal** *adj.*

abduct *v.* kidnap. □ **abduction** *n.*, **abductor** *n.*

aberrant *adj.* showing aberration. □ **aberrance** *n.*

aberration *n.* a deviation from what is normal; distortion.

abet *v.* (**abetted**) encourage or assist in wrongdoing. □ **abettor** *n.*

abeyance *n.* □ **in abeyance** not being used for a time.

abhor *v.* (**abhorred**) detest. □ **abhorrence** *n.*

abhorrent *adj.* detestable.

abide *v.* bear, tolerate. □ **abide by** keep (a promise); accept (consequences etc.).

abiding *adj.* lasting, permanent.

ability *n.* the power to do something; cleverness.

ab initio (ab in-ish-i-oh) *adv.* from the beginning.

abject *adj.* wretched; lacking all pride. □ **abjectly** *adv.*

abjure *v.* renounce; repudiate.

ablaze *adj.* blazing.

able *adj.* having power or ability. □ **ably** *adv.*

ablutions *n.pl.* the process of washing oneself.

abnegate *v.* renounce.

abnormal *adj.* not normal. □ **abnormally** *adv.*, **abnormality** *n.*

aboard *adv.* & *prep.* on board.

abode *n.* a home, a dwelling place.

abolish v. put an end to. □ **abolition** n.

abominable adj. very bad or unpleasant. □ **abominably** adv.

abominate v. detest. □ **abomination** n.

aboriginal adj. existing in a country from its earliest times. ● n. an aboriginal inhabitant, esp. (**Aboriginal**) of Australia.

aborigine (ab-er-**ij**-in-ee) n. an aboriginal inhabitant, esp. (**Aborigine**) of Australia.

abort v. (cause to) expel a foetus prematurely; end prematurely and unsuccessfully.

abortion n. the premature expulsion of a foetus from the womb; an operation to cause this.

abortionist n. a person who performs abortions.

abortive adj. 1 unsuccessful. 2 causing an abortion. □ **abortively** adv.

abound v. be plentiful.

about adv. & prep. 1 near; here and there; in circulation. 2 approximately. 3 in connection with. 4 so as to face in the opposite direction. □ **be about to** be on the point of (doing).

about-face n. (also **about-turn**) a reversal of direction or policy.

above adv. & prep. at or to a higher point (than); beyond the level or understanding of. □ **above board** without deception.

abracadabra n. a magic formula.

abrasion n. rubbing or scraping away; an injury caused by this.

abrasive adj. causing abrasion; harsh. ● n. a substance used for grinding or polishing.

abreast adv. side by side. □ **keep abreast of** keep up to date with.

abridge v. shorten by using fewer words. □ **abridgement** n.

abroad adv. away from one's home country.

abrogate v. repeal, abolish. □ **abrogation** n.

abrupt adj. 1 sudden. 2 curt. 3 steep. □ **abruptly** adv., **abruptness** n.

abscess n. a collection of pus formed in the body.

abscond v. go away secretly or illegally.

abseil v. descend a steep rock face using a rope fixed at a higher point.

absence n. being absent; lack.

absent (ab-sènt) adj. not present; lacking, non-existent. □ **absent oneself** (ab-sent) stay away.

absentee n. a person who is absent from work etc. □ **absenteeism** n.

absent-minded adj. with one's mind on other things; forgetful.

absinthe n. a green liqueur.

absolute adj. complete; unrestricted. □ **absolutely** adv.

absolution n. a priest's formal declaration of forgiveness of sins.

absolutism n. a principle of government with unrestricted powers. □ **absolutist** n.

absolve v. clear of blame or guilt.

absorb v. take in, combine into itself or oneself; occupy the attention or interest of. □ **absorption** n.

absorbent adj. able to absorb moisture etc.

abstain v. 1 refrain, esp. from drinking alcohol. 2 decide not to use one's vote. □ **abstainer** n., **abstention** n.

abstemious adj. not self-indulgent, esp. in eating and drinking. □ **abstemiously** adv., **abstemiousness** n.

abstinence n. abstaining, esp. from food or alcohol.

abstract *adj.* (ab-strakt) **1** having no material existence; theoretical. **2** (of art) not representing things pictorially. ● *n.* (ab-strakt) **1** an abstract quality or idea. **2** a summary. **3** a piece of abstract art. ● *v.* (åb-strakt) **1** take out, remove. **2** make a summary of. □ **abstraction** *n.*

abstruse *adj.* hard to understand, profound.

absurd *adj.* not in accordance with common sense, ridiculous. □ **absurdly** *adv.*, **absurdity** *n.*

ABTA *abbr.* Association of British Travel Agents.

abundant *adj.* plentiful; having plenty of something. □ **abundantly** *adv.*, **abundance** *n.*

abuse *v.* (å-bewz) ill-treat; attack with abusive language. ● *n.* (å-bewss) ill-treatment; abusive language.

abusive *adj.* using harsh words or insults. □ **abusively** *adv.*

abut *v.* (abutted) border (upon), end or lean (against); have a common boundary.

abysmal *adj.* very bad.

abyss *n.* a bottomless chasm.

AC *abbr.* (also **ac**) alternating current.

Ac *symb.* actinium.

a/c *abbr.* account.

acacia (å-kay-shå) *n.* a flowering tree or shrub.

academic *adj.* **1** of a college or university; scholarly. **2** of theoretical interest only. ● *n.* an academic person. □ **academically** *adv.*

academician *n.* a member of an Academy.

Academy *n.* **1** a society of scholars or artists. **2** (**academy**) a school, esp. for specialized training.

acanthus *n.* a plant with large spiny leaves.

ACAS *abbr.* Advisory, Conciliation, and Arbitration Service.

accede (ak-seed) *v.* agree (to).

accelerate *v.* increase the speed (of). □ **acceleration** *n.*

accelerator *n.* a pedal on a vehicle for increasing speed.

accent *n.* (ak-sènt) a mark showing how a vowel is pronounced; a particular regional, national, or other way of pronouncing words; emphasis on a word. ● *v.* (ak-sent) pronounce with an accent; emphasize.

accentuate *v.* emphasize; make prominent. □ **accentuation** *n.*

accept *v.* say yes (to); take as true. □ **acceptance** *n.*

acceptable *adj.* worth accepting; tolerable. □ **acceptably** *adv.*, **acceptability** *n.*

access *n.* a way in; the right to enter or visit.

accessible *adj.* able to be reached or obtained. □ **accessibly** *adv.*, **accessibility** *n.*

accession *n.* **1** reaching a rank or position. **2** an addition; being added.

accessory *adj.* additional. ● *n.* **1** something that is extra or decorative. **2** a person who helps in or knows about a crime but does not take part in it.

accident *n.* **1** an unexpected event, esp. one causing damage. **2** chance.

accidental *adj.* happening by accident. □ **accidentally** *adv.*

acclaim *v.* welcome or applaud enthusiastically. ● *n.* a shout of welcome; applause; approval. □ **acclamation** *n.*

acclimatize *v.* (also **-ise**) make or become used to a new climate. □ **acclimatization** *n.*

accolade *n.* bestowal of a knighthood or other honour; praise.

accommodate *v.* **1** provide lodging or room for. **2** adapt to.

■ **Usage** Accommodate, accommodation, etc. are spelt with two *m*s, not one.

accommodating *adj.* willing to do as asked.

accommodation *n.* a place to live.

accompany *v.* **1** go with. **2** play an instrumental part supporting (a singer or instrument). □ **accompaniment** *n.*, **accompanist** *n.*

accomplice *n.* a partner in crime.

accomplish *v.* succeed in doing or achieving.

accomplished *adj.* skilled; having many accomplishments.

accomplishment *n.* a useful ability.

accord *v.* be consistent. ● *n.* consent, agreement. □ **of one's own accord** without being asked.

accordance *n.* conformity.

according *adv.* □ **according to** as stated by; in proportion to. □ **accordingly** *adv.*

accordion *n.* a portable musical instrument with bellows and keys or buttons.

accost *v.* approach and speak to.

account *n.* **1** a statement of money paid or owed; a credit arrangement with a bank or firm. **2** a description, a report. □ **account for** give a reckoning of; explain; kill, overcome. **on account of** because of.

accountable *adj.* obliged to account for one's actions. □ **accountability** *n.*

accountant *n.* a person who keeps or inspects business accounts. □ **accountancy** *n.*

accoutrements *n.pl.* (*Amer.* also **accouterments**) equipment, trappings.

accredited *adj.* officially recognized or certified.

accretion *n.* growth; matter added.

accrue *v.* accumulate. □ **accrual** *n.*

accumulate *v.* acquire more and more of; increase in amount. □ **accumulation** *n.*

accumulator *n.* **1** a rechargeable electric battery. **2** a bet on a series of events with winnings restaked.

accurate *adj.* free from error. □ **accurately** *adv.*, **accuracy** *n.*

accusative *n.* the grammatical case expressing the direct object.

accuse *v.* lay the blame for a crime or fault on. □ **accusation** *n.*, **accuser** *n.*

accustom *v.* make used (to).

ace *n.* **1** a playing card with a single spot. **2** an expert. **3** an unreturnable stroke in tennis.

acerbic *adj.* (of manner) harsh and sharp. □ **acerbity** *n.*

acetate *n.* a synthetic textile fibre.

acetic acid *n.* a colourless liquid (also called *ethanoic acid*), an essential ingredient of vinegar.

acetone *n.* a colourless liquid used as a solvent.

acetylene *n.* a colourless gas burning with a bright flame.

ache *n.* a dull continuous pain. ● *v.* suffer an ache.

achieve *v.* accomplish; reach or gain by effort. □ **achievable** *adj.*, **achievement** *n.*, **achiever** *n.*

Achilles heel *n.* a person's vulnerable point.

Achilles tendon *n.* the tendon attaching the calf muscles to the heel.

achromatic *adj.* free from colour.

acid *adj.* sour. ● *n.* any of a class of substances that contain hydrogen and neutralize alkalis. □ **acidly** *adv.*, **acidity** *n.*, **acidic** *adj.*

acid rain *n.* rain made acid by pollution.

acknowledge v. **1** admit the truth of. **2** confirm receipt of. □ **acknowledgement** n.

acme (ak-mi) n. the peak of perfection.

acne (ak-ni) n. an eruption of pimples.

acolyte n. a person assisting a priest in a church service.

acorn n. the oval nut of the oak tree.

acoustic adj. of sound. ● n.pl. (**acoustics**) the qualities of a room that affect the way sound carries in it.

acquaint v. make known to. □ **be acquainted with** know slightly.

acquaintance n. a slight knowledge; a person one knows slightly.

acquiesce v. assent, agree. □ **acquiescent** adj., **acquiescence** n.

acquire v. get possession of.

acquired immune deficiency syndrome see Aids.

acquisition n. acquiring; something acquired.

acquisitive adj. eager to acquire things. □ **acquisitiveness** n.

acquit v. (acquitted) declare to be not guilty. □ **acquit oneself** conduct oneself; perform. □ **acquittal** n.

acre (ay-ker) n. a measure of land, 4,840 sq. yds (0.405 hectares).

acreage n. a number of acres.

acrid adj. bitter.

acrimonious adj. angry and bitter. □ **acrimony** n.

acrobat n. a performer of acrobatics.

acrobatic adj. involving spectacular gymnastic feats. ● n.pl. (**acrobatics**) acrobatic feats.

acronym n. a word formed from the initial letters of others.

acrophobia n. an abnormal fear of heights.

acropolis n. the upper fortified part of an ancient Greek city.

across prep. & adv. from side to side (of); on the other side (of).

acrostic n. a poem in which the first and/or last letters of lines form a word or words.

acrylic n. a synthetic fibre made from an organic substance.

act n. **1** a thing done. **2** a law made by parliament. **3** a section of a play. **4** an item in a variety show. ● v. **1** perform actions; behave. **2** play the part of; be an actor.

acting adj. serving temporarily.

action n. **1** the process of doing something or functioning; something done. **2** a lawsuit. **3** a battle.

actionable adj. giving cause for a lawsuit.

activate v. make active. □ **activation** n., **activator** n.

active adj. doing things; energetic. □ **actively** adv.

activist n. a person adopting a policy of vigorous action in politics etc. □ **activism** n.

activity n. action; a particular pursuit.

actor n. a performer in a play or film.

actress n. a woman who acts in a play or film.

actual adj. existing in fact; current.

actuality n. reality.

actually adv. in fact, really.

actuary n. an insurance expert who calculates risks and premiums. □ **actuarial** adj.

actuate v. activate; be a motive for. □ **actuation** n.

acumen n. shrewdness.

acupressure n. medical treatment involving pressing the body at specific points. □ **acupressurist** n.

acupuncture n. medical treatment involving pricking the

skin with needles. □ **acupuncturist** n.

acute adj. **1** sharp. **2** intense; (of illness) severe for a time. **3** quick at understanding. □ **acutely** adv., **acuteness** n.

acute accent n. the accent ´.

AD abbr. (in dates) after the supposed date of Christ's birth (from Latin *Anno Domini* = in the year of the Lord).

adage n. a saying.

adagio adv. (*Music*) in slow time.

adamant adj. not yielding to requests.

Adam's apple n. the lump of cartilage at the front of the neck.

adapt v. make or become suitable for new use or conditions. □ **adaptation** n., **adaptor** n.

adaptable adj. able to be adapted or to adapt oneself. □ **adaptability** n.

add v. join as an increase or supplement; say further; put together to get a total.

addendum n. (pl. **addenda**) section added to a book.

adder n. a small poisonous snake.

addict n. a person addicted to a habit, esp. to a drug.

addicted adj. doing or using something as a habit or compulsively. □ **addiction** n., **addictive** adj.

addition n. adding; something added.

additional adj. added, extra. □ **additionally** adv.

additive n. a substance added.

addle v. **1** muddle, confuse. **2** make (an egg) rotten.

address n. **1** particulars of where a person lives or where mail should be delivered. **2** a speech. ● v. **1** write the address on. **2** speak to. **3** apply (oneself) to a task.

addressee n. a person to whom a letter etc. is addressed.

adduce v. cite as proof.

adenoids n.pl. the enlarged tissue between the back of the nose and the throat. □ **adenoidal** adj.

adept adj. very skilful.

adequate adj. enough; satisfactory but not excellent. □ **adequately** adv., **adequacy** n.

adhere v. **1** stick. **2** continue to give one's support. □ **adherence** n., **adherent** adj. & n.

adhesion n. the process or fact of sticking to something.

adhesive adj. sticking, sticky. ● n. an adhesive substance.

ad hoc adv. & adj. for a specific purpose.

adieu (ă-dew) int. & n. (pl. **adieus** or **adieux**) (a) goodbye.

ad infinitum adv. for ever.

adipose adj. fatty.

adjacent adj. lying near; adjoining.

adjective n. a descriptive word adding information about a noun. □ **adjectival** adj.

adjourn v. move (a meeting etc.) to another place or time. □ **adjournment** n.

adjudge v. decide judicially.

adjudicate v. act as judge (of); adjudge. □ **adjudication** n., **adjudicator** n.

adjunct n. a thing that is subordinate to another.

adjure v. command or urge strongly.

adjust v. alter slightly so as to be correct or in the proper position; adapt (oneself) to new conditions. □ **adjustable** adj., **adjustment** n.

adjutant n. an army officer assisting in administrative work.

ad lib adv. as one pleases. ● adj. improvised. ● v. (**ad libbed**) improvise, speak or act without preparation.

administer v. **1** manage (business affairs). **2** give or hand out.

administrate v. act as manager (of). □ **administrator** n.

administration n. administering; management of public or business affairs. □ **administrative** adj.

admirable adj. worthy of admiration; excellent. □ **admirably** adv.

admiral n. a naval officer of the highest rank.

admire v. regard with pleasure; think highly of. □ **admiration** n.

admissible adj. able to be admitted or allowed. □ **admissibility** n.

admission n. admitting; a statement admitting something.

admit v. (**admitted**) **1** allow to enter. **2** accept as valid. **3** state reluctantly.

admittance n. admitting, esp. to a private place.

admittedly adv. as must be admitted.

admixture n. something added as an ingredient.

admonish v. exhort; reprove. □ **admonition** n.

ad nauseam adv. to a sickening extent.

ado n. fuss, trouble.

adobe (ă-doh-bi) n. a sun-dried brick.

adolescent adj. & n. (a person) between childhood and adulthood. □ **adolescence** n.

adopt v. take as one's own; accept responsibility for the maintenance of; accept, approve (a report etc.). □ **adoption** n.

adoptive adj. related by adoption.

adorable adj. very lovable.

adore v. love deeply. □ **adoration** n.

adorn v. decorate with ornaments; be an ornament to. □ **adornment** n.

adrenal (ă-dree-năl) adj. close to the kidneys.

adrenalin (ă-dren-ă-lin) n. (also **adrenaline**) a stimulant hormone produced by the adrenal glands.

adrift adj. & adv. drifting; loose.

adroit adj. skilful, ingenious.

adsorb v. attract and hold (a gas or liquid) to a surface.

adulation n. excessive flattery. □ **adulatory** adj.

adult adj. fully grown. ● n. an adult person. □ **adulthood** n.

adulterate v. make impure by adding a substance. □ **adulteration** n.

adulterer n. a person who commits adultery.

adultery n. sexual infidelity to one's wife or husband. □ **adulterous** adj.

advance v. **1** move or put forward. **2** lend (money). ● n. **1** forward movement; progress. **2** a loan; an increase in price. **3** (**advances**) attempts to establish a friendly relationship. □ **advancement** n.

advanced adj. **1** far on in time or progress. **2** not elementary.

advantage n. a favourable circumstance; a benefit. □ **take advantage of** make use of; exploit.

advantageous adj. profitable, beneficial.

Advent n. **1** the season before Christmas. **2** (**advent**) an arrival.

adventure n. an exciting or daring experience. □ **adventurer** n., **adventurous** adj.

adverb n. a word qualifying a verb, adjective, or other adverb. □ **adverbial** adj., **adverbially** adv.

adversary n. an opponent; an enemy. □ **adversarial** adj.

adverse adj. unfavourable; bringing harm. □ **adversely** adv., **adversity** n.

advertise v. make publicly known, esp. to encourage sales.

advertisement n. advertising; a public notice about something.

advice n. an opinion given about what should be done.

advisable adj. worth recommending as a course of action. □ **advisability** n.

advise v. give advice to; recommend; inform. □ **adviser** n.

advisory adj. giving advice.

advocaat n. a liqueur of eggs, sugar, and brandy.

advocacy n. speaking in support.

advocate n. (ad-vŏ-kăt) a person who recommends a policy; a person who speaks in court on behalf of another. ● v. (ad-vŏ-kayt) recommend.

adze n. (Amer. also **adz**) an axe with a blade at right angles to the handle.

aegis (ee-jiss) n. protection, sponsorship.

aeolian (ee-oh-li-ăn) adj. windborne.

aeon (ee-ŏn) n. (also **eon**) an immense time.

aerate v. 1 expose to the action of air. 2 add carbon dioxide to.

aerial adj. of or like air; existing or moving in the air; by or from aircraft. ● n. a wire for transmitting or receiving radio waves. □ **aerially** adv.

aero- comb. form air; aircraft.

aerobatics n.pl. spectacular feats by aircraft in flight. □ **aerobatic** adj.

aerobics n.pl. vigorous exercises designed to increase oxygen intake. □ **aerobic** adj.

aerodynamic adj. of the airflow around objects moving through air. ● n. (**aerodynamics**) the study of this. □ **aerodynamically** adv.

aerofoil n. an aircraft wing, fin, or tailplane giving lift in flight.

aeronautics n. the study of the flight of aircraft. □ **aeronautical** adj.

aeroplane n. a power-driven aircraft with wings.

aerosol n. a container holding a substance for release as a fine spray.

aerospace n. the earth's atmosphere and space beyond this.

aesthete (ees-theet) n. (Amer. also **esthete**) a person claiming to understand and appreciate beauty.

aesthetic (ess-the-tik) adj. (Amer. also **esthetic**) of or showing appreciation of beauty; artistic, tasteful. □ **aesthetically** adv.

aetiology (ee-ti-ol-ŏji) n. (Amer. **etiology**) the study of causes, esp. of diseases.

affable adj. polite and friendly. □ **affably** adv., **affability** n.

afar adv. far away.

affair n. 1 a thing to be done; business. 2 a temporary sexual relationship.

affect v. 1 pretend to have or feel or be. 2 have an effect on.

■ **Usage** Affect is often confused with effect, which means 'to bring about'.

affectation n. pretence, esp. in behaviour.

affected adj. full of affectation.

affection n. love, liking.

affectionate adj. loving. □ **affectionately** adv.

affidavit n. a written statement sworn on oath to be true.

affiliate v. connect as a subordinate member or branch. □ **affiliation** n.

affinity n. a close resemblance or attraction.

affirm v. state as a fact; declare formally. □ **affirmation** n.

affirmative adj. affirming; saying 'yes'. ● n. an affirmative statement. ● *int.* yes. □ **affirmatively** *adv.*

affix *v.* (ǎ-**fiks**) attach; add (a signature etc.). ● *n.* (**a**-fiks) a thing affixed; a prefix, a suffix.

afflict *v.* distress physically or mentally.

affliction *n.* distress.

affluence *n.* wealth. □ **affluent** *adj.*

afford *v.* have enough money or time for.

afforest *v.* convert into forest; plant with trees. □ **afforestation** *n.*

affray *n.* a public fight or riot.

affront *n.* an open insult. ● *v.* insult, offend.

afloat *adv. & adj.* floating; on the sea.

afoot *adv. & adj.* going on.

aforementioned, aforesaid *adj.* mentioned previously.

afraid *adj.* frightened; regretful.

afresh *adv.* anew, with a fresh start.

Afrikaans *n.* a language of South Africa, developed from Dutch.

Afrikaner *n.* an Afrikaans-speaking white person in South Africa.

aft *adv.* at or towards the rear of a ship or aircraft.

after *prep.* **1** behind; later (than); in pursuit of. **2** concerning; according to. ● *conj. & adv.* at a time later (than). ● *n.pl.* (**afters**) (*colloquial*) dessert.

afterbirth *n.* the placenta discharged from the womb after childbirth.

after-effect *n.* an effect persisting after its cause has gone.

aftermath *n.* the after-effects.

afternoon *n.* the time between noon or lunchtime and evening or sunset.

aftershave *n.* a lotion used after shaving.

afterthought *n.* something thought of or added later.

afterwards *adv.* at a later time.

Ag *symb.* silver.

Aga *n.* (*trade mark*) a large cooking stove that can also run a heating system.

again *adv.* another time, once more; besides.

against *prep.* in opposition or contrast to; in preparation or return for; into collision or contact with.

agar (**ay**-gar) *n.* (also **agar-agar**) a substance obtained from seaweed, used to thicken food or make it set like jelly.

agate *n.* a hard stone with patches or usu. bands of colour.

age *n.* the length of life or existence; the latter part of life; a historical period; (*colloquial*, usu. **ages**) a very long time. ● *v.* (**aged, ageing**) grow old, show signs of age; cause to do this.

aged *adj.* (**ayjd**) of the age of; (**ay**-jid) old.

ageism *n.* prejudice on grounds of age.

ageless *adj.* not growing old; not seeming old.

agency *n.* **1** the business or office of an agent. **2** a means of action by which something is done.

agenda *n.* a list of things to be dealt with, esp. at a meeting.

agent *n.* a person who does something, esp. on behalf of another; a thing producing an effect.

agent provocateur (azh-ahn prŏ-vok-ǎ-**ter**) *n.* a person employed to detect suspected offenders by tempting them to do something illegal.

agglomerate *v.* collect into a mass.

aggrandize *v.* (also **-ise**) cause to seem greater. □ **aggrandizement** *n.*

aggravate *v.* **1** make worse. **2** (*colloquial*) annoy. □ **aggravation** *n.*

■ **Usage** The use of *aggravate* to mean 'annoy' is considered incorrect by some people, but it is common in informal use.

aggregate *adj.* (ag-ri-găt) combined, total. ● *n.* (ag-ri-găt) a collected mass; broken stone etc. used in making concrete. ● *v.* (ag-ri-gayt) collect into an aggregate, unite; (*colloquial*) amount to. □ **aggregation** *n.*

aggression *n.* unprovoked attacking; hostile acts or behaviour.

aggressive *adj.* showing aggression; forceful. □ **aggressively** *adv.*

aggressor *n.* a person who begins hostilities.

aggrieved *adj.* having a grievance.

aghast *adj.* filled with horror.

agile *adj.* nimble, quick-moving. □ **agility** *n.*

agitate *v.* shake briskly; cause anxiety to; stir up concern. □ **agitation** *n.*, **agitator** *n.*

agitprop *n.* the spreading of political propaganda.

AGM *abbr.* annual general meeting.

agnostic *adj. & n.* (a person) believing that nothing can be known about the existence of God. □ **agnosticism** *n.*

ago *adv.* in the past.

agog *adj.* eager, expectant.

agonize *v.* (also **-ise**) cause agony to; suffer agony; worry intensely.

agony *n.* extreme suffering.

agoraphobia *n.* an abnormal fear of crossing open spaces.

AGR *abbr.* advanced gas-cooled (nuclear) reactor.

agrarian *adj.* of land or agriculture.

agree *v.* **1** approve as correct or acceptable. **2** hold or reach a similar opinion. □ **agree with** suit the health or digestion of.

agreeable *adj.* pleasing; willing to agree. □ **agreeably** *adv.*

agreement *n.* agreeing; an arrangement agreed between people.

agribusiness *n.* the industries manufacturing and distributing agricultural machinery and produce.

agriculture *n.* large-scale cultivation of land. □ **agricultural** *adj.*

agronomy *n.* soil management and crop production.

aground *adv. & adj.* (of a ship) on the bottom of shallow water.

ague (ay-gew) *n.* a fever; a fit of shivering.

ahead *adv.* further forward in position or time.

ahoy *int.* a seaman's shout for attention.

AI *abbr.* **1** artificial intelligence. **2** artificial insemination.

aid *v. & n.* help.

aide *n.* an aide-de-camp; an assistant.

aide-de-camp *n.* an officer assisting a senior officer.

Aids *abbr.* (also **AIDS**) acquired immune deficiency syndrome, a condition developing after infection with the HIV virus, breaking down a person's natural defences against illness.

aikido *n.* a Japanese form of self-defence.

ail *v.* make or become ill.

aileron *n.* a hinged flap on an aircraft wing.

ailment *n.* a slight illness.

aim *v.* point, send, or direct towards a target. ● *n.* aiming; intention.

aimless *adj.* without a purpose. □ **aimlessly** *adv.*, **aimlessness** *n.*

ain't (*colloquial*) am not, is not, are not; has not, have not.

■ **Usage** The use of *ain't* is incorrect in standard English.

air *n.* **1** a mixture of oxygen, nitrogen, etc., surrounding the earth; the atmosphere overhead. **2** an impression given. **3** a melody. ● *v.* **1** expose to air; dry off. **2** express publicly. □ **on the air** broadcasting by radio or television.

air bag *n.* a car safety device that fills with air in a collision to protect the driver.

air-bed *n.* an inflatable mattress.

airborne *adj.* carried by air or aircraft; (of aircraft) in flight.

airbrick *n.* a perforated brick for ventilation.

air-conditioning *n.* a system controlling the humidity and temperature of air in a building or vehicle. □ **air-conditioned** *adj.*

aircraft *n.* (*pl.* **aircraft**) a machine capable of flight in air.

aircraft carrier *n.* a ship carrying and acting as a base for aircraft.

aircrew *n.* the crew of an aircraft.

airfield *n.* an area with runways etc. for aircraft.

air force *n.* a branch of the armed forces using aircraft.

airgun *n.* a gun with a missile propelled by compressed air.

airlift *n.* the large-scale transport of supplies by aircraft. ● *v.* transport in this way.

airline *n.* a company providing an air transport service.

airliner *n.* a passenger aircraft.

airlock *n.* **1** a stoppage of the flow in a pipe, caused by an air bubble. **2** an airtight compartment giving access to a pressurized chamber.

airmail *n.* mail carried by aircraft. ● *v.* send by airmail.

airman *n.* (*pl.* **-men**) a member of an air force, esp. below the rank of officer.

airplane *n.* (*Amer.*) = **aeroplane**.

airplay *n.* the playing of a recording on radio.

airport *n.* an airfield with facilities for passengers and goods.

air raid *n.* an attack by aircraft.

airship *n.* a power-driven aircraft that is lighter than air.

airspace *n.* the air and skies above a country and subject to its control.

airstrip *n.* a strip of ground for take-off and landing of aircraft.

airtight *adj.* not allowing air to enter or escape.

airway *n.* **1** a regular route of aircraft. **2** a ventilating passage; a passage for air into the lungs.

airworthy *adj.* (of aircraft) fit to fly.

airy *adj.* (**airier**) **1** well-ventilated; light as air. **2** careless and light-hearted. □ **airily** *adv.*, **airiness** *n.*

aisle (*rhymes with* mile) *n.* the side part of a church; a gangway between rows of seats.

ajar *adv.* & *adj.* (of a door) slightly open.

aka *abbr.* also known as.

akimbo *adv.* with hands on hips and elbows pointed outwards.

akin *adj.* related, similar.

Al *symb.* aluminium.

à la *prep.* in the style of.

alabaster *n.* a translucent usu. white form of gypsum.

à la carte *adj.* & *adv.* ordered as separate items from a menu rather than a set meal.

alacrity *n.* eager readiness.

à la mode *adj.* & *adv.* in fashion.

alarm *n.* a warning sound or signal; fear caused by expectation of danger. ● *v.* cause alarm to.

alarmist *n.* a person who raises unnecessary or excessive alarm.

alas *int.* an exclamation of sorrow.

albatross *n.* a seabird with long wings.

albino n. (*pl.* **albinos**) a person or animal with no natural colouring in the hair or skin.

album n. **1** a blank book for holding photographs, stamps, etc. **2** a long-playing record.

albumen n. egg white.

albumin n. a protein found in egg white, milk, blood, etc.

alchemy n. a medieval form of chemistry, seeking to turn other metals into gold. □ **alchemist** n.

alcohol n. a colourless inflammable liquid, the intoxicant in wine, beer, etc.; drink containing this.

alcoholic adj. of alcohol. ● n. a person addicted to drinking alcohol. □ **alcoholism** n.

alcove n. a recess in a wall.

al dente (al den-tay) adj. (of pasta) cooked so as to be still firm.

alder n. a tree related to birch.

alderman n. (*pl.* **-men**) (*hist.*) a co-opted senior member of an English council. **2** an elected member of the municipal governing body in Australian and some American cities.

ale n. beer.

alert adj. watchful, observant. ● v. make alert, warn.

A level n. Advanced level, an examination of a higher level than ordinary level or GCSE.

alfalfa n. lucerne.

alfresco adv. & adj. in the open air.

algae n.pl. water plants with no true stems or leaves.

algebra n. a branch of mathematics using letters etc. to represent quantities. □ **algebraic** adj., **algebraically** adv.

algorithm n. a step by step procedure for calculation.

alias n. a false name. ● adv. also called.

alibi n. **1** evidence that an accused person was elsewhere when a crime was committed. **2** an excuse.

alien n. **1** a person who is not a citizen of the country where he or she lives. **2** a being from another world. ● adj. foreign; unfamiliar.

alienate v. cause to become unfriendly or hostile. □ **alienation** n.

alight v. get down from (a vehicle etc.); descend and settle. ● adj. on fire.

align v. place or bring into line; join as an ally. □ **alignment** n.

alike adj. like one another. ● adv. in the same way.

alimentary adj. of nourishment.

alive adj. living; alert; lively.

alkali n. any of a class of substances that neutralize acids. □ **alkaline** adj.

alkaloid n. an organic compound containing nitrogen.

all adj. the whole amount or number or extent of. ● n. all those concerned. □ **all but** almost. **all in** exhausted; including everything. **all out** using maximum effort. **all right** satisfactory, satisfactorily; in good condition.

allay v. lessen (fears).

all-clear n. a signal that danger is over.

allegation n. something alleged.

allege v. declare without proof.

allegedly adv. according to allegation.

allegiance n. support given to a government, sovereign, or cause.

allegory n. a story symbolizing an underlying meaning. □ **allegorical** adj., **allegorically** adv.

allegro adv. (*Music*) briskly.

alleluia int. & n. (also **hallelujah**) praise to God.

allergen n. a substance causing an allergic reaction.

allergic *adj.* having or caused by an allergy; having a strong dislike.

allergy *n.* a condition causing an unfavourable reaction to certain foods, pollens, etc.

alleviate *v.* lessen (pain or distress). □ **alleviation** *n.*

alley *n.* (*pl.* **alleys**) a narrow street; a long enclosure for tenpin bowling.

alliance *n.* an association formed for mutual benefit.

allied *adj.* of an alliance; related.

alligator *n.* a reptile of the crocodile family.

alliteration *n.* the occurrence of the same sound at the start of words. □ **alliterative** *adj.*

allocate *v.* allot, assign. □ **allocation** *n.*

allot *v.* (**allotted**) distribute officially; give as a share.

allotment *n.* **1** a share allotted. **2** a small area of land for cultivation.

allotrope *n.* one of the forms of an element that exists in different physical forms.

allow *v.* **1** permit. **2** give a limited quantity or sum; add or deduct in estimating. **3** admit.

allowance *n.* allowing; an amount or sum allowed. □ **make allowances for** be lenient towards or because of.

alloy *n.* a mixture of chemical elements at least one of which is a metal. ● *v.* **1** mix (with another metal). **2** spoil or weaken (pleasure etc.).

allspice *n.* a spice made from dried pimento berries.

allude *v.* refer briefly or indirectly.

allure *v.* entice, attract. ● *n.* attractiveness.

allusion *n.* a statement alluding to something. □ **allusive** *adj.*

alluvium *n.* a deposit left by a flood. □ **alluvial** *adj.*

ally *n.* (al-I) a country or person in alliance with another. ● *v.* (ăl-I) join as an ally.

almanac *n.* (also **almanack**) an annual publication containing a calendar, times of sunrise and sunset, astronomical data, anniversaries, and other information; a yearbook of sport, theatre, etc.

almighty *adj.* all-powerful; (*colloquial*) very great.

almond *n.* the kernel of a fruit related to the peach; the tree bearing this.

almost *adv.* very nearly.

aims *n.pl.* (*old use*) money given to the poor.

almshouse *n.* (*hist.*) a house built by charity for poor people.

aloe *n.* a plant with bitter juice.

aloft *adv.* high up; upwards.

alone *adj.* not with others; without company or help. ● *adv.* only.

along *adv.* **1** through part or all of a thing's length; onward. **2** in company with others. ● *prep.* beside the length of.

alongside *adv.* close to the side of a ship or wharf etc.

aloof *adv.* apart. ● *adj.* showing no interest; unfriendly.

alopecia *n.* loss of hair; baldness.

aloud *adv.* in a voice that can be heard, not in a whisper.

alpaca *n.* a llama with long wool; its wool; cloth made from this.

alpha *n.* the first letter of the Greek alphabet (A, α).

alphabet *n.* the letters used in writing a language. □ **alphabetical** *adj.*, **alphabetically** *adv.*

alphabetize *v.* (also **-ise**) put into alphabetical order. □ **alphabetization** *n.*

alpine *adj.* of high mountains. ● *n.* a plant growing on mountains or in rock gardens.

already *adv.* before this time; as early as this.

alright = all right.

■ **Usage** Although *alright* is widely used, it is not correct in standard English.

Alsatian *n.* a dog of a large strong smooth-haired breed.

also *adv.* in addition, besides.

altar *n.* a table used in religious service.

altarpiece *n.* a painting behind an altar.

alter *v.* make or become different. □ **alteration** *n.*

altercation *n.* a noisy dispute.

alternate *adj.* (awl-ter-nǎt) first one then the other successively. ● *v.* (awl-ter-nayt) place or occur alternately. □ **alternately** *adv.*, **alternation** *n.*

■ **Usage** See note at **alternative**.

alternative *adj.* **1** available as another choice. **2** different, unconventional. ● *n.* an alternative thing. □ **alternatively** *adv.*

■ **Usage** The adjective *alternative* is often confused with *alternate*, which is correctly used in 'there will be a dance on alternate Saturdays'.

alternator *n.* a dynamo giving alternating current.

although *conj.* though.

altimeter *n.* an instrument in an aircraft showing altitude.

altitude *n.* height above sea level or above the horizon.

alto *n.* (*pl.* **altos**) the highest adult male voice; a musical instrument with the second or third highest pitch in its group.

altogether *adv.* entirely; on the whole.

altruism *n.* unselfishness. □ **altruist** *n.*, **altruistic** *adj.*, **altruistically** *adv.*

alum *n.* a white mineral salt used in medicine and dyeing.

aluminium *n.* a metallic element (symbol Al), a lightweight metal.

always *adv.* at all times; whatever the circumstances.

alyssum *n.* a plant with small yellow or white flowers.

Alzheimer's disease (alts-hy-merz) *n.* a brain disorder causing senility.

AM *abbr.* amplitude modulation.

Am *symb.* americium.

a.m. *abbr.* (Latin *ante meridiem*) before noon.

amalgam *n.* an alloy of mercury, used esp. in dentistry; a mixture or blend.

amalgamate *v.* mix, combine. □ **amalgamation** *n.*

amaryllis *n.* a lily-like plant.

amass *v.* heap up, collect.

amateur *n.* a person who does something as a pastime rather than as a profession.

amateurish *adj.* lacking professional skill.

amatory *adj.* relating to love.

amaze *v.* overwhelm with wonder. □ **amazement** *n.*

amazon *n.* a fierce strong woman.

ambassador *n.* a diplomat representing his or her country abroad.

amber *n.* a hardened brownish-yellow resin; its colour.

ambergris *n.* a waxy substance found in tropical seas and obtained from sperm whales, used in perfume manufacture.

ambidextrous *adj.* able to use either hand equally well.

ambience *n.* surroundings. □ **ambient** *adj.*

ambiguous *adj.* having two or more possible meanings. □ **ambiguously** *adv.*, **ambiguity** *n.*

ambit *n.* scope, bounds.

ambition *n.* a strong desire to achieve something.

ambitious *adj.* full of ambition. □ **ambitiously** *adv.*

ambivalent *adj.* with mixed feelings towards something. □ **ambivalence** *n.,* **ambivalently** *adv.*

amble *v. & n.* (walk at) a leisurely pace.

ambrosia *n.* something delicious.

ambulance *n.* a vehicle equipped to carry sick or injured people.

ambuscade *n.* an ambush.

ambush *n.* troops etc. lying concealed to make a surprise attack; this attack. ● *v.* attack in this way.

ameba Amer. sp. of **amoeba**.

ameliorate *v.* make or become better. □ **amelioration** *n.*

amen *int.* (in prayers) so be it.

amenable *adj.* responsive, cooperative. □ **amenably** *adv.,* **amenability** *n.*

amend *v.* make minor alterations in (a text etc.). □ **make amends** compensate for something. □ **amendment** *n.*

amenity *n.* a pleasant or useful feature of a place.

American *adj.* of America; of the USA. ● *n.* an American person.

American football *n.* a form of football played with an oval ball and an H-shaped goal on a field marked as a gridiron.

Americanism *n.* an American word or phrase.

Americanize *v.* (also **-ise**) make American in character. □ **Americanization** *n.*

americium *n.* an artificially made radioactive metallic element (symbol Am).

amethyst *n.* a precious stone, a purple or violet quartz; its colour.

amiable *adj.* likeable; friendly. □ **amiably** *adv.,* **amiability** *n.*

amicable *adj.* friendly. □ **amicably** *adv.*

amid, amidst *prep.* in the middle of; during.

amino acid *n.* an organic acid found in proteins.

amiss *adv.* wrongly, badly. ● *adj.* wrong, faulty.

amity *n.* friendly feeling.

ammeter *n.* an instrument that measures electric current.

ammonia *n.* a strong-smelling gas; a solution of this in water.

ammonite *n.* a fossil of a spiral shell.

ammunition *n.* a supply of bullets, shells, etc.

amnesia *n.* loss of memory. □ **amnesiac** *adj. & n.*

amnesty *n.* a general pardon.

amniocentesis *n.* (amni-oh-sen-tee-sis) *n.* (*pl.* **amniocenteses**) a prenatal test for foetal abnormality involving the withdrawal of a sample of amniotic fluid.

amniotic fluid *n.* the fluid surrounding the foetus in the womb.

amoeba (ă-mee-bă) *n.* (*Amer.* also **ameba**) (*pl.* **amoebae** or **amoebas**) a simple microscopic organism capable of changing shape.

amok *adv.* (also **amuck**) □ **run amok** be out of control and do much damage.

among, amongst *prep.* surrounded by; in the number of; between.

amoral *adj.* not based on moral standards.

amorous *adj.* showing sexual love.

amorphous *adj.* shapeless.

amortize *v.* (also **-ise**) pay off (a debt) gradually.

amount *n.* a total of anything; a quantity. □ **amount to** add up to; be equivalent to.

amour propre *n.* self-respect.

amp *n.* (*colloquial*) **1** an ampere. **2** an amplifier.

ampere *n.* a unit of electric current (symbol A).

ampersand n. the sign & (= and).

amphetamine n. a stimulant drug.

amphibian n. an animal able to live on both land and water; an amphibious vehicle.

amphibious adj. able to live or operate both on land and in water.

amphitheatre n. (Amer. **amphitheater**) a semicircular unroofed building with tiers of seats round a central arena.

ample adj. plentiful, quite enough; large. □ **amply** adv.

amplify v. increase the strength of, make louder; add details to (a statement). □ **amplification** n., **amplifier** n.

amplitude n. breadth; abundance.

ampoule n. a small sealed container holding a liquid.

amuck var. of **amok**.

amputate v. cut off by surgical operation. □ **amputation** n.

amulet n. something worn as a charm against evil.

amuse v. cause to laugh or smile; make time pass pleasantly for. □ **amusement** n., **amusing** adj.

an adj. the form of a used before vowel sounds other than long 'u'.

anabolic steroid n. a steroid hormone used to build up bone and muscle.

anachronism n. something that does not belong in the period in which it is placed. □ **anachronistic** adj.

anaconda n. a large snake of South America.

anaemia (ă-nee-miă) n. (Amer. **anemia**) lack of haemoglobin in the blood.

anaemic (ă-nee-mik) adj. (Amer. **anemic**) adj. suffering from anaemia; pale and lacking in vitality.

anaerobic adj. not requiring air or oxygen.

anaesthesia (an-is-thee-ziă) n. (Amer. **anesthesia**) loss of sensation, esp. induced by anaesthetics.

anaesthetic (an-is-the-tik) adj. & n. (Amer. **anesthetic**) (a substance) causing loss of sensation.

anaesthetist (ă-neess-thě-tist) n. (Amer. **anesthetist**) a person who administers anaesthetics. □ **anaesthetize** v.

anaglypta n. thick wallpaper with a raised pattern.

anagram n. a word or phrase formed from the rearranged letters of another word or phrase.

anal adj. of the anus.

analgesic adj. & n. (a drug) relieving pain. □ **analgesia** n.

analogous adj. similar in certain respects.

analogue n. (Amer. also **analog**) an analogous thing.

analogy n. a partial likeness between things.

analyse v. (Amer. **analyze**) examine in detail; psychoanalyse. □ **analyst** n.

analysis n. (pl. **analyses**) a detailed examination or study of something.

analytical adj. (also **analytic**) of or using analysis. □ **analytically** adv.

anarchist n. a person who believes that government is undesirable and should be abolished. □ **anarchism** n.

anarchy n. total lack of organized control; lawlessness. □ **anarchical** adj., **anarchically** adv.

anathema (ă-nath-ěmă) n. a formal curse; a detested thing.

anathematize v. (also **-ise**) curse.

anatomist n. an expert in anatomy.

anatomize v. (also **-ise**) examine the anatomy or structure of.

anatomy n. bodily structure; the study of this. □ **anatomical** adj., **anatomically** adv.

anatto var. of **annatto**.

ANC abbr. African National Congress.

ancestor n. a person from whom one's father or mother is descended. □ **ancestral** adj.

ancestry n. a line of ancestors.

anchor n. a heavy metal structure for mooring a ship to the sea bottom. ● v. moor with an anchor; fix firmly.

anchorage n. a place where ships may anchor; lying at anchor.

anchorman n. (pl. **-men**) a strong member of a sports team playing a vital part; a coordinator; the compère of a radio or TV programme.

anchovy n. a small strong-tasting fish.

ancient adj. very old.

ancillary adj. helping in a subsidiary way.

and conj. together with; added to; then; afterwards.

andante adv. (Music) in moderately slow time.

andiron n. a metal stand for supporting wood in a fireplace.

androgynous adj. having both male and female parts.

anecdote n. a short amusing or interesting true story.

anemia etc. Amer. sp. of **anaemia** etc.

anemone n. a plant with white, red, or purple flowers.

aneroid barometer n. a barometer measuring air pressure by the action of air on a box containing a vacuum.

anesthesia etc. Amer. sp. of **anaesthesia** etc.

aneurysm n. excessive swelling of an artery.

anew adv. again; in a new way.

angel n. **1** a messenger of God. **2** a kind person. □ **angelic** adj.

angelica n. candied stalks of a fragrant plant; this plant.

angelus n. a Roman Catholic prayer to the Virgin Mary said at morning, noon, and sunset; the bell announcing this.

anger n. extreme displeasure. ● v. make angry.

angina n. (in full **angina pectoris**) pain in the chest caused by overexertion when the heart is diseased.

angle n. **1** the space between two lines or surfaces that meet. **2** a point of view. ● v. **1** place obliquely. **2** present from a particular point of view. **3** fish with hook and bait. **4** try to obtain by hinting. □ **angler** n.

Anglican adj. & n. (a member) of the Church of England. □ **Anglicanism** n.

Anglicism n. an English word or custom.

Anglicize v. (also **-ise**) make English in character.

angling n. the sport of fishing.

Anglo- comb. form English, British.

Anglo-Saxon n. & adj. (of) an English person or language before the Norman Conquest; (of) a person of English descent.

angora n. a long-haired variety of cat, goat, or rabbit; yarn or fabric made from the hair of these animals.

angostura n. an aromatic bitter bark of a South American tree.

angry adj. (**angrier**) feeling or showing anger. □ **angrily** adv.

angst n. anxiety; remorse.

angstrom n. a unit of measurement for wavelengths.

anguish n. severe physical or mental pain. □ **anguished** adj.

angular adj. having angles or sharp corners; forming an angle.

anhydrous adj. without water.

aniline n. a colourless oily liquid used in making dyes and plastics.

animal n. & adj. (of) a living thing with sense organs, able to move voluntarily.

animate adj. (an-i-măt) living. ● v. (an-i-mayt) give life or movement to; produce as an animated cartoon. □ **animation** n., **animator** n.

animism n. the belief that all natural things have a living soul.

animosity n. hostility.

animus n. animosity.

anion n. an ion with a negative charge.

aniseed n. the fragrant seed of a plant (**anise**), used for flavouring.

ankle n. the joint connecting the foot with the leg.

anklet n. a chain or band worn round the ankle.

annals n.pl. a narrative of events year by year; historical records.

annatto n. (also **anatto**) an orange-red dye from a tropical fruit, used as food colouring.

anneal v. toughen (metal or glass) by heat and slow cooling.

annex v. 1 take possession of. 2 add as a subordinate part. □ **annexation** n.

annexe n. an additional building.

annihilate v. destroy completely. □ **annihilation** n.

anniversary n. the yearly return of the date of an event.

annotate v. add explanatory notes to. □ **annotation** n.

announce v. make known publicly; make known the presence or arrival of. □ **announcement** n.

announcer n. a person who announces items in a broadcast.

annoy v. cause slight anger to; be troublesome to. □ **annoyed** adj., **annoyance** n.

annual adj. yearly. ● n. 1 a plant that lives for one year or one season. 2 a book published in yearly issues. □ **annually** adv.

annuity n. a yearly allowance provided by an investment.

annul v. (**annulled**) make null and void, cancel. □ **annulment** n.

annular adj. ring-shaped.

Annunciation n. the announcement by the angel Gabriel to the Virgin Mary that she was to be the mother of Christ.

anode n. an electrode by which current enters a device.

anodize v. (also **-ise**) coat (metal) with a protective layer by electrolysis.

anodyne n. something that relieves pain or distress. ● adj. relieving pain or distress.

anoint v. apply ointment or oil etc. to, esp. in religious consecration.

anomaly n. something irregular or inconsistent. □ **anomalous** adj.

anon adv. (old use) soon.

anon. abbr. anonymous.

anonymous adj. of unknown or undisclosed name or authorship. □ **anonymously** adv., **anonymity** n.

anorak n. a waterproof jacket with a hood.

anorexia n. reluctance to eat; (in full **anorexia nervosa**) a condition characterized by an obsessive desire to lose weight and refusal to eat normally. □ **anorexic** adj. & n.

another adj. one more; a different; any other. ● pron. another one.

answer n. something said, written, needed, or done to deal with a question etc.; the solution to a problem. ● *v.* **1** make or be an answer (to); act in response to. **2** take responsibility. **3** correspond (to a description).

answerable *adj.* having to account for something.

answering machine *n.* a machine that answers telephone calls and records messages.

ant *n.* a small insect that lives in highly organized groups.

antacid *n. & adj.* (a substance) preventing or correcting acidity esp. in the stomach.

antagonism *n.* active opposition, hostility. □ **antagonistic** *adj.*

antagonist *n.* an opponent.

antagonize *v.* (also **-ise**) rouse antagonism in.

Antarctic *adj. & n.* (of) regions round the South Pole.

ante *n.* a stake put up by a poker-player before drawing new cards.

ante- *pref.* before.

anteater *n.* a mammal that eats ants.

antecedent *n.* a preceding thing or circumstance. ● *adj.* previous.

antechamber *n.* an ante-room.

antedate *v.* put an earlier date on; precede in time.

antediluvian *adj.* of the time before Noah's Flood; (*colloquial*) antiquated.

antelope *n.* an animal resembling a deer.

antenatal *adj.* before birth; of or during pregnancy.

antenna *n.* **1** (*pl.* **antennae**) an insect's feeler. **2** (*pl.* **antennas**) a radio or TV aerial.

anterior *adj.* coming before in position or time.

ante-room *n.* a small room leading to a main one.

anthem *n.* a piece of music to be sung in a religious service.

anther *n.* a part of a flower's stamen containing pollen.

anthology *n.* a collection of passages from literature, esp. poems.

anthracite *n.* a form of coal burning with little flame or smoke.

anthrax *n.* a disease of sheep and cattle, transmissible to people.

anthropoid *adj.* resembling a human in form. ● *n.* an anthropoid ape.

anthropology *n.* the study of the origin and customs of mankind. □ **anthropological** *adj.*, **anthropologist** *n.*

anthropomorphic *adj.* attributing human form to a god or animal. □ **anthropomorphism** *n.*

anti- *pref.* opposed to; counteracting.

anti-aircraft *adj.* used against enemy aircraft.

antibiotic *n.* a substance that destroys bacteria.

antibody *n.* a protein formed in the blood in reaction to a substance which it then destroys.

antics *n.pl.* absurd often amusing behaviour.

anticipate *v.* deal with or use in advance; look forward to; expect. □ **anticipation** *n.*

anticlimax *n.* a dull ending where a climax was expected.

anticlockwise *adj. & adv.* in the direction opposite to clockwise.

anticyclone *n.* an outward flow of air from an area of high pressure, producing fine weather.

antidepressant *n.* a drug counteracting mental depression.

antidote *n.* a substance that counteracts the effects of poison.

antifreeze *n.* a substance added to water to prevent freezing.

antigen *n.* a foreign substance stimulating the production of antibodies.

anti-hero n. (pl. **anti-heroes**) a central character lacking conventional heroic qualities.

antihistamine n. a substance counteracting the effect of histamine.

anti-lock adj. (of brakes) designed to prevent locking and skidding.

antimacassar n. a protective covering for a chair-back.

antimony n. a metallic element (symbol Sb).

antipasto n. (pl. **antipasti**) an Italian starter or appetizer.

antipathy n. strong dislike.

antiperspirant n. a substance that prevents or reduces sweating.

antipodes (an-tip-ŏdeez) n.pl. places on opposite sides of the earth, esp. Australia and New Zealand (opposite Europe).

antiquarian adj. of the study of antiques. ● n. a person who studies antiques.

antiquated adj. very old; very old-fashioned.

antique adj. belonging to the distant past. ● n. an antique object.

antiquity n. ancient times; an object dating from ancient times.

antirrhinum n. a snapdragon.

anti-Semitic adj. hostile to Jews.

antiseptic adj. & n. (a substance) preventing things from becoming septic. □ **antiseptically** adv.

antisocial adj. destructive or hostile to other members of society.

antistatic adj. counteracting the effects of static electricity.

antithesis n. (pl. **antitheses**) an opposite; a contrast. □ **antithetical** adj.

antitoxin n. a substance that neutralizes a toxin. ● **antitoxic** adj.

antivivisectionist n. a person opposed to making experiments on live animals.

antler n. a branched horn of a deer.

antonym n. a word opposite to another in meaning.

anus n. the excretory opening at the end of the alimentary canal.

anvil n. an iron block on which a smith hammers metal into shape.

anxiety n. the state of being troubled and uneasy.

anxious adj. **1** troubled and uneasy in mind. **2** eager. □ **anxiously** adv.

any adj. one or some from a quantity; every.

anybody pron. any person.

anyhow adv. **1** anyway. **2** not in an orderly manner.

anyone pron. anybody.

anything pron. any item.

anyway adv. in any case.

anywhere adv. & pron. (in or to) any place.

AOB abbr. any other business.

aorta n. the main artery carrying blood from the heart.

apace adv. swiftly.

apart adv. separately, so as to become separated; to or at a distance; into pieces.

apartheid (ă-par-tayt) n. a former policy of racial segregation (esp. in South Africa).

apartment n. a set of rooms; (Amer.) a flat.

apathy n. lack of interest or concern. □ **apathetic** adj.

ape n. a tailless monkey. ● v. imitate, mimic.

aperitif n. an alcoholic drink taken as an appetizer.

aperture n. an opening, esp. one that admits light.

Apex n. (also **APEX**) a system of reduced fares for scheduled flights paid for in advance.

apex n. the tip; the highest point; a pointed end.

aphasia *n*. loss of the ability to use language as a result of brain damage.

aphid *n*. a small insect destructive to plants.

aphorism *n*. a short wise saying.

aphrodisiac *adj.* & *n*. (a substance) arousing sexual desire.

apiary *n*. a place where bees are kept. □ **apiarist** *n*.

apiece *adv.* to, for, or by each.

aplomb (à-plom) *n*. dignity and confidence.

apocalypse *n*. the end of the world; a violent event as described in **the Apocalypse** (the last book of the New Testament).

apocalyptic *adj.* prophesying great and dramatic events like those in the Apocalypse.

Apocrypha *n.pl.* the books of the Old Testament not accepted as part of the Hebrew scriptures.

apocryphal *adj.* of doubtful authenticity; invented.

apogee *n*. the point in the moon's orbit furthest from the earth; the highest point.

apologetic *adj.* making an apology. □ **apologetically** *adv.*

apologize *v*. (also **-ise**) make an apology.

apology *n*. **1** a statement of regret for having done wrong or hurt. **2** an explanation or defence.

apoplectic *adj.* **1** of or liable to suffer apoplexy. **2** (*colloquial*) enraged.

apoplexy *n*. a stroke; sudden loss of the ability to feel and move, caused by rupture or blockage of the brain artery.

apostasy *n*. abandonment of one's former religious belief.

apostate *n*. a person who renounces a former belief.

a posteriori *adj.* (of reasoning) proceeding from effect to cause.

Apostle *n*. any of the twelve men sent by Christ to preach the gospel. □ **apostolic** *adj.*

apostrophe (à-pos-trŏ-fi) *n*. the sign ' used to show the possessive case or omission of a letter.

apothecary *n*. (*old use*) a pharmaceutical chemist.

apotheosis *n*. (*pl.* **apotheoses**) elevation to godly status; a thing's highest development; a perfect example.

appal *v*. (*Amer.* **appall**) (**appalled**) fill with horror or dismay. □ **appalling** *adj.*

apparatus *n*. equipment for scientific or other work.

apparel *n*. clothing.

apparent *adj.* clearly seen or understood; seeming but not real. □ **apparently** *adv.*

apparition *n*. a thing appearing, esp. of a startling or remarkable kind; a ghost.

appeal *v*. **1** make an earnest or formal request. **2** refer to a higher court. **3** seem attractive. ● *n*. **1** an act of appealing. **2** attractiveness.

appear *v*. be or become visible; seem. □ **appearance** *n*.

appease *v*. soothe or conciliate, esp. by giving what was asked. □ **appeasement** *n*.

appellant *n*. a person who appeals to a higher court.

append *v*. add at the end.

appendage *n*. a thing appended.

appendectomy *n*. the surgical removal of the appendix.

appendicitis *n*. inflammation of the intestinal appendix.

appendix *n*. (*pl.* **appendices** or **appendixes**) **1** a section at the end of a book, giving extra information. **2** the small closed tube of tissue attached to the large intestine.

appertain *v*. be relevant.

appetite *n*. desire, esp. for food.

appetizer *n*. (also **-iser**) something eaten or drunk to stimulate the appetite.

appetizing *adj.* (also **-ising**) stimulating the appetite.

applaud v. express approval (of), esp. by clapping; praise. ● **applause** n.

apple n. a fruit with firm flesh.

appliance n. a device, an instrument.

applicable adj. appropriate; relevant. □ **applicability** n.

applicant n. a person who applies for a job.

application n. a thing applied; the process or capability of working hard.

applicator n. a device for applying something.

applied adj. for put to practical use.

appliqué (ap-lee-kay) n. a piece of fabric attached ornamentally.

apply v. **1** make a formal request. **2** bring into use or action. **3** be relevant. ● **apply oneself** give one's attention and energy.

appoint v. choose (a person) for a job, committee, etc.

appointee n. a person appointed.

appointment n. **1** an arrangement to meet or visit at a specified time. **2** a job.

apportion v. divide into shares.

apposite adj. appropriate.

apposition n. placing side by side; a grammatical relationship in which a word or phrase is placed with another which it describes, e.g. *Elizabeth, our Queen.*

appraise v. estimate the value or quality of. □ **appraisal** n.

appreciable adj. perceptible; considerable. □ **appreciably** adv.

appreciate v. **1** enjoy intelligently; understand; value. **2** increase in value. □ **appreciation** n., **appreciative** adj.

apprehend v. seize, arrest; grasp the meaning of. □ **apprehension** n.

apprehensive adj. feeling apprehension, anxious. □ **apprehensively** adv.

apprentice n. a person learning a craft. □ **apprenticeship** n.

apprise v. inform.

approach v. **1** come nearer (to). **2** set about doing. **3** go to with a request or offer. ● n. approaching; a way or means of this.

approachable adj. easy to talk to.

approbation n. approval.

appropriate adj. (ǎ-proh-pri-ǎt) suitable, proper. ● v. (ǎ-proh-pri-ayt) take and use; set aside for a special purpose. □ **appropriately** adv., **appropriation** n.

approval n. approving. □ **on approval** (of goods) supplied without obligation to buy if not satisfactory.

approve v. say that (a thing) is good or suitable; agree to.

approx. abbr. approximate(ly).

approximate adj. (ǎ-prok-sim-ǎt) almost but not quite exact. ● v. (ǎ-prok-sim-ayt) be almost the same; make approximate. □ **approximately** adv., **approximation** n.

APR abbr. annual percentage rate (of interest).

Apr. abbr. April.

après-ski (ap-ray-skee) n. social activities following a day's skiing.

apricot n. a stone fruit related to the peach; its orange-pink colour.

April n. the fourth month.

a priori adj. (of reasoning) from cause to effect; not derived from experience, assumed. ● adv. logically; as far as one knows.

apron n. **1** a garment worn over the front of the body to protect clothes. **2** part of a theatre stage in front of the curtain. **3** an area where aircraft are manoeuvred, loaded, etc.

apropos (aprŏ-poh) adv. concerning.

apse n. a recess with an arched or domed roof in a church.

apt adj. **1** suitable. **2** having a certain tendency. **3** quick to learn. □ **aptly** adv., **aptness** n.

aptitude n. natural ability.

aqualung n. a portable underwater breathing apparatus.

aquamarine n. a bluish-green gemstone; its colour.

aquaplane n. a board on which a person can be towed by a speedboat. ●v. (of a vehicle) glide uncontrollably on a wet road surface.

aquarium n. (pl. **aquariums** or **aquaria**) a tank for keeping living fish etc.; a building containing such tanks.

aquatic adj. living or taking place in or on water.

aquatint n. a kind of etching.

aqua vitae n. alcoholic spirit, esp. brandy.

aqueduct n. an artificial water channel on a raised structure.

aqueous (ay-kwi-ŭs) adj. of or like water.

aquifer n. a layer of water-bearing rock or soil.

aquiline adj. like an eagle; (of a nose) curved like an eagle's beak.

Ar symb. Argon.

Arab n. & adj. (a member of) a Semitic people of the Middle East.

arabesque n. **1** a dancer's posture with the body bent forward and leg and arm extended in line. **2** a decoration with intertwined lines etc.

Arabian adj. of Arabia.

Arabic adj. & n. (of) the language of the Arabs.

Arabic numerals n.pl. the symbols 1, 2, 3 etc.

arable adj. & n. (land) suitable for growing crops.

arachnid n. a member of the class to which spiders belong.

arachnophobia n. fear of spiders.

arbiter n. a person with power to judge in a dispute; an arbitrator.

arbitrary adj. based on random choice. □ **arbitrarily** adv.

arbitrate v. act as arbitrator. □ **arbitration** n.

arbitrator n. an impartial person chosen to settle a dispute.

arboreal adj. of or living in trees.

arboretum n. (pl. **arboretums** or **arboreta**) a place where trees are grown for study and display.

arbour n. (Amer. **arbor**) a shady shelter under trees or a framework with climbing plants.

arc n. **1** part of a curve, esp. of the circumference of a circle. **2** a luminous electric current crossing a gap between terminals.

arcade n. a covered walk between shops; a place with games machines etc.

arcane adj. mysterious.

arch[1] n. a curved structure, esp. as a support. ●v. form into an arch.

arch[2] adj. consciously or affectedly playful. □ **archly** adv., **archness** n.

archaeology n. (Amer. **archeology**) the study of civilizations through their material remains □ **archaeological** adj., **archaeologist** n.

archaic adj. belonging to former or ancient times.

archaism n. an archaic word or phrase.

archangel n. an angel of the highest rank.

archbishop n. a chief bishop.

archdeacon n. a priest ranking next below bishop.

archduke n. (hist.) a chief duke, esp the son of an Austrian Emperor.

archeology Amer. sp. of **archaeology**.

archer *n.* a person who shoots with bow and arrows. □ **archery** *n.*

arch-enemy *n.* a chief enemy.

archetype (ar-ki-typ) *n.* an original model from which others are copied. □ **archetypal** *adj.*

archipelago *n.* (*pl.* **archipelagos** or **archipelagoes**) a group of islands; the sea round this.

architect *n.* a designer of buildings.

architecture *n.* the designing of buildings; the style of a building. □ **architectural** *adj.*

architrave *n.* a moulded frame round a doorway or window.

archive (ar-kyv) *n.* (usu. **archives**) a collection of historical documents.

archivist (ar-kiv-ist) *n.* a person trained to deal with archives.

archway *n.* an arched entrance or passage.

Arctic *adj.* **1** of regions round the North Pole. **2** (**arctic**) (*colloquial*) very cold.

ardent *adj.* passionate, enthusiastic. □ **ardently** *adv.*

ardour *n.* (*Amer.* **ardor**) great warmth of feeling; enthusiasm.

arduous *adj.* needing much effort.

area *n.* **1** the extent or measure of a surface; a region; a sunken courtyard. **2** the range of a subject etc.

arena *n.* a level area in the centre of an amphitheatre or sports stadium; the scene of a conflict.

arête *n.* a sharp mountain ridge.

argon (ar-gon) *n.* a chemical element (symbol Ar), an inert gas.

argot (ar-goh) *n.* jargon.

arguable *adj.* open to argument, debatable. □ **arguably** *adv.*

argue *v.* **1** express disagreement; exchange angry words. **2** give as reasons.

argument *n.* **1** a discussion involving disagreement; a quarrel.

2 a reason put forward; a chain of reasoning.

argumentative *adj.* fond of arguing.

aria *n.* a solo in opera.

arid *adj.* dry, parched. □ **aridly** *adv.*, **aridity** *n.*

arise *v.* (**arose**, **arisen**, **arising**) **1** come into existence or to people's notice. **2** (*old use*) rise.

aristocracy *n.* the hereditary upper classes. □ **aristocratic** *adj.*

aristocrat *n.* a member of the aristocracy.

arithmetic *n.* calculating by means of numbers.

ark *n.* **1** Noah's boat. **2** (in full **Ark of the Covenant**) a wooden chest in which the writings of Jewish Law were kept.

arm *n.* **1** an upper limb of the human body; a raised side part of a chair. **2** (**arms**) weapons. ●*v.* equip with weapons; make (a bomb) ready to explode.

armada *n.* a fleet of warships.

armadillo *n.* (*pl.* **armadillos**) a mammal of South America with a body encased in bony plates.

Armageddon *n.* a final disastrous conflict; in the Bible, the last battle between good and evil before the Day of Judgement.

armament *n.* military weapons; the process of equipping for war.

armature *n.* the wire-wound core of a dynamo or electric motor.

armchair *n.* a chair with raised sides.

armistice *n.* an agreement to stop fighting temporarily.

armlet *n.* a band worn round an arm.

armorial *adj.* of heraldry.

armour *n.* (*Amer.* **armor**) a protective metal covering, esp. that formerly worn in fighting.

armoured *adj.* (*Amer.* **armored**) protected by armour.

armoury n. (*Amer.* **armory**) a place where weapons are kept.

armpit n. the hollow under the arm at the shoulder.

army n. an organized force for fighting on land; a vast group.

arnica n. a plant substance used to treat bruises.

aroma n. a smell, esp. a pleasant one. □ **aromatic** adj.

aromatherapy n. the use of essential plant oils in massage.

arose see **arise**.

around adv. & prep. **1** all round, on every side (of). **2** approximately.

arouse v. rouse.

arpeggio n. (*pl.* **arpeggios**) the notes of a musical chord played in succession.

arraign v. accuse; find fault with. □ **arraignment** n.

arrange v. put into order; form plans, settle the details of; adapt. □ **arrangement** n.

arrant adj. downright. □ **arrantly** adv.

array v. arrange in order; adorn. ●n. a display; an ordered arrangement.

arrears n.pl. money owed and overdue for repayment; work overdue for being finished.

arrest v. stop (a movement or moving thing); seize by authority of law. ●n. the legal seizure of an offender.

arrival n. arriving; a person or thing that has arrived.

arrive v. reach the end of a journey; (of time) come; be recognized as having achieved success.

arrogant adj. proud and overbearing. □ **arrogantly** adv., **arrogance** n.

arrogate n. claim, seize, or attribute without the right to do so.

arrow n. a straight shaft with a sharp point, shot from a bow; a line with a V at the end, indicating direction.

arrowroot n. an edible starch made from the root of a West Indian plant.

arse n. (*Amer.* **ass**) (*vulgar*) the buttocks.

arsenal n. a place where weapons are stored or made.

arsenic n. a semi-metallic element (symbol As); a strongly poisonous compound of this. □ **arsenical** adj.

arson n. the intentional and unlawful setting on fire of a building. □ **arsonist** n.

art n. **1** the production of something beautiful; paintings and sculptures. **2** (**arts**) subjects other than sciences, requiring sensitive understanding rather than use of measurement; creative activities (e.g. painting, music, writing).

artefact n. (*Amer.* **artifact**) a man-made object; a simple prehistoric tool or weapon.

arterial adj. of an artery.

arterial road n. a major road.

arteriosclerosis n. a thickening of the artery walls, hindering blood circulation.

artery n. a large blood vessel carrying blood away from the heart.

artesian well n. a well that is bored vertically into oblique strata so that water rises naturally with little or no pumping.

artful adj. crafty. □ **artfully** adv.

arthritis n. a condition in which there is pain and stiffness in the joints. □ **arthritic** adj.

arthropod n. an animal with a segmented body and jointed limbs (e.g. an insect or crustacean).

artichoke n. (also **globe artichoke**) a plant with a flower of leaflike scales used as a vegetable. □ **Jerusalem artichoke** a type of sunflower with a root eaten as a vegetable.

article n. a particular or separate thing; a piece of writing in a newspaper etc.; a clause in an agreement. □ **definite article** the word 'the'. **indefinite article** the word 'a' or 'an'.

articled clerk n. a trainee solicitor.

articulate adj. (ar-tik-yoo-lăt) spoken distinctly; able to express ideas clearly. ● v. (ar-tik-yoo-layt) **1** speak or express clearly. **2** form a joint, connect by joints. □ **articulation** n.

articulated lorry n. one with sections connected by a flexible joint.

artifact Amer. sp. of **artefact**.

artifice n. trickery; a clever deception.

artificial adj. not originating naturally; man-made. □ **artificially** adv., **artificiality** n.

artificial intelligence n. the development of computers to do things normally requiring human intelligence.

artillery n. large guns used in fighting on land; a branch of an army using these.

artisan n. a skilled manual worker, a craftsperson.

artist n. **1** a person who produces works of art, esp. paintings. **2** a person who does something with exceptional skill. **3** a professional entertainer. □ **artistry** n.

artiste (ar-teest) n. a professional entertainer, esp. a singer or dancer.

artistic adj. **1** of art or artists. **2** showing or done with good taste. □ **artistically** adv.

artless adj. free from artfulness, simple and natural. □ **artlessly** adv., **artlessness** n.

artwork n. pictures and diagrams in books, advertisements, etc.

arty adj. (**artier**) (colloquial) with an exaggerated or affected display of artistic style or interests.

As symb. arsenic.

as adv. & conj. in the same degree, similarly; while, when; because. ● prep. in the form or function of. **as for, as to** with regard to. **as well** in addition.

asafoetida n. (also **asafetida**) a strong-smelling spice used in Indian cooking.

a.s.a.p. abbr. as soon as possible.

asbestos n. a soft fibrous mineral substance; a fireproof material made from this.

asbestosis n. a lung disease caused by inhaling asbestos particles.

ascend v. go or come up.

ascendant adj. rising.

ascension n. an ascent, esp. (**Ascension**) that of Christ to heaven.

ascent n. ascending; a way up.

ascertain v. find out. □ **ascertainable** adj.

ascetic adj. not allowing oneself pleasures and luxuries. ● n. a person who is ascetic. □ **asceticism** n.

ASCII abbr. (Computing) American Standard Code for Information Exchange; a code assigning a different number to each letter and character.

ascorbic acid n. vitamin C.

ascribe v. attribute. □ **ascription** n.

asepsis n. an aseptic condition.

aseptic adj. free from harmful bacteria. □ **aseptically** adv.

asexual adj. without sex. □ **asexually** adv.

ash n. **1** a tree with silver-grey bark. **2** powder that remains after something has burnt.

ashamed adj. feeling shame.

ashcan n. (Amer.) a dustbin.

ashen adj. pale as ashes; grey.

ashlar n. square-cut stones; masonry made of this.

ashore adv. to or on shore.

ashram n. (originally in India) a place of religious retreat.

Asian adj. of Asia or its people. ● n. an Asian person.

Asiatic adj. of Asia.

aside adv. to or on one side, away from the main part or group. ● n. words spoken so that only certain people will hear.

asinine adj. silly.

ask v. call for an answer to or about; address a question to; seek to obtain; invite.

askance □ **look askance at** look at with distrust or disapproval.

askew adv. & adj. crooked(ly).

asleep adv. & adj. in or into a state of sleep.

asp n. a small poisonous snake.

asparagus n. a plant whose shoots are used as a vegetable.

aspect n. the look or appearance of something; a feature of a complex matter; the direction in which a thing faces.

aspen n. a poplar tree.

asperity n. harshness.

aspersion n. a derogatory remark.

asphalt n. a black substance like coal tar; a mixture of this with gravel etc. for paving.

asphyxia n. suffocation.

asphyxiate v. suffocate. □ **asphyxiation** n.

aspic n. a savoury jelly for coating cooked meat, eggs, etc.

aspidistra n. an ornamental plant with broad tapering leaves.

aspirant n. a person who aspires to something.

aspirate n. (ass-pi-răt) the sound of h. ● v. (ass-pi-rayt) pronounce with an h.

aspiration n. 1 an earnest desire or ambition. 2 aspirating.

aspire v. have an earnest ambition.

aspirin n. a drug that relieves pain and reduces fever; a tablet of this.

ass n. 1 a donkey. 2 a stupid person. 3 (Amer.) = **arse**.

assail v. attack violently.

assailant n. an attacker.

assassin n. a killer of an important person.

assassinate v. kill (an important person) by violent means. □ **assassination** n.

assault n. & v. attack.

assay n. a test of metal for quality. ● v. make an assay of.

assemble v. bring or come together; put or fit together.

assembly n. an assembled group; assembling.

assent v. consent; express agreement. ● n. consent, permission.

assert v. 1 state, declare to be true. 2 use (power etc.) effectively. □ **assertion** n., **assertive** adj., **assertiveness** n.

assess v. decide the amount or value of; estimate the worth or likelihood of. □ **assessment** n., **assessor** n.

asset n. 1 a property with money value. 2 a useful quality, a person or thing having this.

assiduous adj. diligent and persevering. □ **assiduously** adv., **assiduousness** n., **assiduity** n.

assign v. allot; designate to perform a task.

assignation n. an arrangement to meet.

assignment n. a task assigned.

assimilate v. absorb or be absorbed into the body, into a larger group, or into the mind as knowledge. □ **assimilation** n.

assist v. help. □ **assistance** n.

assistant n. a helper; a person who serves customers in a s

● *adj.* assisting, esp. as a subordinate.

assizes *n.pl.* formerly, a county court.

associate *v.* (ă-**soh**-si-ayt) **1** connect in one's mind. **2** socially. **3** join as a companion or supporter. ● *n.* (ă-**soh**-si-ăt) **1** a companion, a partner. **2** a subordinate member.

association *n.* associating; a group organized for a common purpose; a connection between ideas.

assonance *n.* resemblance of sound in syllables; a rhyme of vowel sounds. □ **assonant** *adj.*

assorted *adj.* of different sorts.

assortment *n.* a collection composed of several sorts.

assuage (ă-**swayj**) *v.* soothe, allay.

assume *v.* **1** accept as true without proof. **2** take on (a responsibility, quality, etc.).

assumption *n.* assuming; something assumed to be true.

assurance *n.* **1** a positive assertion. **2** self-confidence. **3** life insurance.

assure *v.* tell confidently, promise.

assured *adj.* **1** sure, confident. **2** insured.

assuredly *adv.* certainly.

astatine (**ass**-tă-teen) *n.* a radioactive element (symbol At).

aster *n.* a plant of the daisy family.

asterisk *n.* a star-shaped symbol (*).

astern *adv.* at or towards the stern; backwards.

asteroid *n.* any of the tiny planets revolving round the sun.

asthma *n.* a chronic condition causing difficulty in breathing. □ **asthmatic** *adj.* & *n.*

astigmatism *n.* a defect in an eye, preventing proper focusing. □ **astigmatic** *adj.*

astonish *v.* surprise very greatly. □ **astonishment** *n.*

astound *v.* shock with surprise.

astrakhan *n.* the dark curly fleece of lambs from Russia.

astral *adj.* of or from the stars.

astray *adv.* & *adj.* away from the proper path.

astride *adv.* & *prep.* with one leg on each side (of).

astringent *adj.* causing tissue to contract; harsh, severe. ● *n.* an astringent substance or drug. □ **astringency** *n.*, **astringently** *adv.*

astro- *comb. form* of the stars; relating to outer space.

astrology *n.* the study of the supposed influence of stars on human affairs. □ **astrologer** *n.*, **astrological** *adj.*

astronaut *n.* a person trained to travel in a spacecraft.

astronautics *n.* the study of space travel and its technology.

astronomer *n.* a person skilled in astronomy.

astronomical *adj.* **1** of astronomy. **2** enormous in amount. □ **astronomically** *adv.*

astronomy *n.* the study of stars and planets and their movements.

astrophysics *n.* the study of the physics and chemistry of heavenly bodies.

AstroTurf *n.* (*trade mark*) an artificial grass surface, esp. for sports fields.

astute *adj.* shrewd, quick at seeing how to gain an advantage. □ **astutely** *adv.*, **astuteness** *n.*

asunder *adv.* (*literary*) apart.

asylum *n.* **1** refuge, protection. **2** (*old use*) a mental institution.

asymmetrical *adj.* not symmetrical. □ **asymmetrically** *adv.*, **asymmetry** *n.*

At *symb.* astatine.

at *prep.* having as a position, time of day, condition, or price.

atavism *n.* resemblance to remote ancestors. □ **atavistic** *adj.*

ataxia *n.* difficulty in controlling body movements.

ate *see* **eat.**

atheist *n.* a person who does not believe in God. □ **atheism** *n.*

atherosclerosis *n.* damage to the arteries caused by a build-up of fatty deposits.

athlete *n.* a person who is good at athletics.

athlete's foot *n.* an infectious fungal condition of the foot.

athletic *adj.* of athletes; muscular and physically active. ● *n.* **(athletics)** track and field sports, e.g. running, jumping, and throwing. □ **athletically** *adv.*, **athleticism** *n.*

atlas *n.* a book of maps.

atmosphere *n.* **1** the mixture of gases surrounding a planet; air in any place; a unit of pressure. **2** a feeling conveyed by an environment or group. □ **atmospheric** *adj.*

atoll *n.* a ring-shaped coral reef enclosing a lagoon.

atom *n.* the smallest particle of a chemical element; a very small quantity.

atomic *adj.* concerning atoms or atomic energy.

atomic bomb *n.* a bomb deriving its power from atomic energy.

atomic energy *n.* energy obtained from nuclear fission.

atomize *v.* (also **-ise**) reduce to atoms or fine particles. □ **atomization** *n.*, **atomizer** *n.*

atonal (ay-toh-nàl) *adj.* (of music) not written in any key. □ **atonality** *n.*

atone *v.* make amends for an error or deficiency. □ **atonement** *n.*

atrium *n.* (*pl.* **atria** or **atriums**) **1** either of the two upper heart cavities. **2** the central court of an ancient Roman house.

atrocious *adj.* extremely wicked; very bad. □ **atrociously** *adv.*

atrocity *n.* wickedness; a cruel act.

atrophy *n.* wasting away through lack of nourishment or use. ● *v.* cause atrophy in; suffer atrophy.

attach *v.* fix to something else; join; attribute. be attributable. □ **attachment** *n.*

attaché (à-tash-ay) *n.* a person attached to an ambassador's staff.

attaché case *n.* a small rectangular case for carrying documents.

attached *adj.* fastened on. □ **attached to** very fond of or devoted to.

attack *n.* a violent attempt to hurt or defeat a person; strong criticism; a sudden onset of illness. ● *v.* make an attack (on). □ **attacker** *n.*

attain *v.* achieve. □ **attainable** *adj.*, **attainment** *n.*

attar *n.* a fragrant oil from rosepetals.

attempt *v.* try, make an effort to do. ● *n.* an effort.

attend *v.* give attention to; be present at. □ **attendance** *n.*

attendant *adj.* accompanying. ● *n.* a person present to provide service.

attention *n.* **1** applying one's mind; consideration, care. **2** an erect attitude in military drill.

attentive *adj.* giving attention. □ **attentively** *adv.*, **attentiveness** *n.*

attenuate *v.* make thin or weaker. □ **attenuation** *n.*

attest *v.* provide proof of; declare true or genuine. □ **attestation** *n.*

attic *n.* a room in the top storey of a house.

attire *n.* clothes. ● *v.* clothe.

attitude n. a position of the body; a way of thinking or behaving.

attorney n. (pl. **attorneys**) (Amer.) a lawyer.

Attorney-General n. the chief legal officer in England and some other countries.

attract v. draw towards itself by unseen force; arouse the interest or pleasure of. □ **attraction** n.

attractive adj. attracting; pleasing in appearance. □ **attractively** adv., **attractiveness** n.

attribute n. (at-ri-bewt) a characteristic quality. □ **attribute to** (ǎ-**tri**-bewt) regard as belonging to or caused by. □ **attributable** adj., **attribution** n.

attributive adj. (Grammar) (of a word) placed before the word it describes.

attrition n. wearing away.

attune v. adapt; tune.

atypical adj. not typical. □ **atypically** adv.

Au symb. gold.

aubergine n. (oh-ber-zheen) (also **eggplant**) a dark purple tropical vegetable; its colour.

aubrietia n. a perennial rock plant having purple or pink flowers in spring.

auburn adj. (of hair) reddish brown.

auction n. a public sale where articles are sold to the highest bidder. ● v. sell by auction.

auctioneer n. a person who conducts an auction.

audacious adj. bold, daring. □ **audaciously** adv., **audacity** n.

audible adj. loud enough to be heard. □ **audibly** adv.

audience n. **1** a group of listeners or spectators. **2** a formal interview.

audio n. sound or the reproduction of sound.

audiotape n. (also **audio tape**) a magnetic tape for recording sound; a recording on this.

audio-visual adj. using both sight and sound.

audit n. an official examination of accounts. ● v. make an audit of.

audition n. a test of a prospective performer's ability. ● v. test or be tested in an audition.

auditor n. a person who audits accounts.

auditorium n. (**auditoriums** or **auditoria**) the part of a building where the audience sits.

auditory adj. of hearing.

au fait (oh fay) adj. □ **au fait with** acquainted with (a subject).

Aug. abbr. August.

auger n. a boring tool with a spiral point.

augment v. increase. □ **augmentation** n., **augmentative** adj.

au gratin (oh gra-tan) adj. cooked with a topping of breadcrumbs or grated cheese.

augur v. be a sign of, bode.

augury n. an omen, interpretation of omens.

August n. the eighth month.

august (aw-gust) adj. majestic.

auk n. a northern seabird.

aunt n. the sister or sister-in-law of one's father or mother.

au pair n. a young person from overseas helping with housework in return for board and lodging.

aura n. the atmosphere surrounding a person or thing.

aural adj. of the ear. □ **aurally** adv.

aureole n. (also **aureola**) a halo.

au revoir (oh rě-vwahr) int. goodbye (until we meet again).

auscultation n. listening to the sound of the heart etc. for medical diagnosis.

auspice n. a forecast. □ **under the auspices of** with the patronage of.

auspicious adj. showing signs of success. □ **auspiciously** adv., **auspiciousness** n.

austere adj. severely simple and plain. □ **austerity** n.

Australasian adj. & n. (a native or inhabitant) of Australia, New Zealand, and neighbouring islands.

Australian adj. & n. (a native or inhabitant) of Australia.

authentic adj. genuine, known to be true. □ **authentically** adv., **authenticity** n.

authenticate v. prove the truth or authenticity of. □ **authentication** n.

author n. the writer of a book etc.; the originator. □ **authorship** n.

authoritarian adj. favouring complete obedience to authority. □ **authoritarianism** n.

authoritative adj. having or using authority. □ **authoritatively** adv.

authority n. **1** power to enforce obedience; a person with this. **2** a person with specialized knowledge.

authorize v. (also **-ise**) give permission for. □ **authorization** n.

autistic adj. suffering from a mental disorder that prevents a normal response to one's environment. □ **autism** n.

auto- comb. form self; own.

autobiography n. the story of a person's life written by that person. □ **autobiographical** adj.

autocracy n. rule by an autocrat.

autocrat n. a person with unrestricted power. □ **autocratic** adj., **autocratically** adv.

autocross n. motor racing across country or on dirt tracks.

autocue n. (trade mark) a device showing a speaker's script on a television screen unseen by the audience.

autograph n. a person's signature. ● v. write one's name in or on.

autoimmune adj. (of diseases) caused by antibodies produced against substances naturally present in the body.

automat n. (Amer.) a slot machine.

automate v. control by automation.

automatic adj. mechanical, self-regulating; done without thinking. ● n. an automatic machine or firearm. □ **automatically** adv.

automation n. the use of automatic equipment in industry.

automaton n. (pl. **automatons** or **automata**) a robot.

automobile n. (Amer.) a car.

automotive adj. concerned with motor vehicles.

autonomous adj. self-governing. □ **autonomy** n.

autopilot n. a device for keeping an aircraft or ship on a set course automatically.

autopsy n. a post-mortem.

autumn n. the season between summer and winter. □ **autumnal** adj.

auxiliary adj. giving help or support. ● n. a helper.

auxiliary verb n. (Grammar) a verb used in forming the tenses of other verbs.

avail v. be of use or help. ● n. effectiveness, advantage. □ **avail oneself of** make use of.

available adj. ready to be used; obtainable. □ **availability** n.

avalanche n. a mass of snow pouring down a mountain.

avant-garde (av-ahn-gard) adj. progressive, new. ● n. a group of innovators, esp. in the arts.

avarice n. greed for gain. □ **avaricious** adj., **avariciously** adv.

Ave. abbr. avenue.

avenge v. take vengeance for. □ **avenger** n.

avenue n. a wide street or road; a method of approach.

aver v. (**averred**) state as true.

average n. a value arrived at by adding several quantities together and dividing by the number of these; a standard regarded as usual. ● adj. found by making an average; of ordinary standard.

averse adj. unwilling, disinclined.

aversion n. a strong dislike.

avert v. turn away; ward off.

aviary n. a large cage or building for keeping birds.

aviation n. the practice or science of flying an aircraft.

avid adj. eager; greedy. □ **avidly** adv., **avidity** n.

avionics n. electronics in aviation.

avocado n. (pl. **avocados**) a pear-shaped tropical fruit with smooth oily flesh.

avocet n. a wading bird with a long upturned bill.

avoid v. keep oneself away from; refrain from. □ **avoidable** adj., **avoidance** n.

avoirdupois n. a system of weights based on a pound of 16 ounces.

avow v. declare. □ **avowal** n.

avuncular adj. of or like a kindly uncle.

AWACS abbr. airborne warning and control system; a long-range radar system.

await v. wait for.

awake v. (**awoke**, **awoken**, **awaking**) wake. ● adj. not asleep; alert.

awaken v. awake.

award v. give by official decision as a prize or penalty. ● n. a thing awarded.

aware adj. having knowledge or realization. □ **awareness** n.

awash adj. washed over by water.

away adv. **1** to or at a distance; into non-existence. **2** persistently. ● v. (of a match) played on an opponent's ground.

awe n. respect combined with fear or wonder. ● v. fill with awe.

aweigh adj. (of an anchor) raised clear of the sea bottom.

awesome adj. causing awe.

awful adj. (colloquial) extremely bad or unpleasant; excessive. □ **awfully** adv.

awhile adv. for a short time.

awkward adj. difficult to use or handle; clumsy, having little skill; inconvenient; embarrassed. □ **awkwardly** adv., **awkwardness** n.

awning n. a rooflike canvas shelter.

awl n. a tool for making holes in leather or wood.

awoke, awoken see **awake**.

AWOL abbr. absent without leave.

awry (ă-ry) adv. & adj. twisted to one side; amiss.

axe (Amer. usu. **ax**) n. a chopping tool with a sharp blade. ● v. (**axed, axing**) remove by abolishing or dismissing.

axiom n. an accepted general truth or principle. □ **axiomatic** adj.

axis n. (pl. **axes**) a line through the centre of an object, round which it rotates if spinning. □ **axial** adj.

axle n. a rod on which wheels turn.

ayatollah n. a Muslim religious leader in Iran.

aye (also **ay**) adv. (old use or dialect) yes. ● n. a vote in favour.

azalea n. a garden shrub with large flowers.

Aztec n. a member of a former Indian people ruling Mexico before the Spanish conquest in the 16th century.

azure adj. & n. (of) a deep sky-blue colour.

....

Bb

....

B n. (B, 2B, 3B, etc.) (of a pencil lead) soft; softer than H and HB. ● symb. boron.

b. abbr. born.

BA abbr. Bachelor of Arts.

ba symb. barium.

baa n. bleat.

babble v. chatter indistinctly or foolishly; (of a stream) murmur. ● n. babbling talk or sound.

babe n. a baby.

babel n. a confused noise.

baboon n. a large monkey.

baby n. a very young child or animal; (slang) a young woman, a sweetheart. □ **babyish** adj.

Babygro n. (pl. **Babygros**) (trade mark) a stretchy all-in-one baby suit.

babysit v. (babysat, babysitting) look after a child while its parents are out. □ **babysitter** n.

baccalaureate (bak-ă-lor-iăt) n. the final school examination in France.

baccarat (bak-er-ah) n. a gambling card game.

bachelor n. **1** an unmarried man. **2** a person with a university degree.

bacillus n. (pl. **bacilli**) a rod-shaped bacterium.

back n. the surface or part furthest from the front; the rear part of the human body from shoulders to hips; the corresponding part of an animal's body; a defensive player positioned near the goal in football etc. ● adj. **1** situated behind. **2** of or relating to the past. ● adv. **1** at or towards the rear; in or into a previous time, position, or state. **2** in return. ● v. **1** move backwards. **2** help, support. **3** lay a bet on. □ **back down** withdraw a claim or argument. **back up** support. □ **backer** n.

backache n. a pain in one's back.

backbencher n. an ordinary MP not holding a senior office.

backbiting n. spiteful talk.

backbone n. the column of small bones down the centre of the back.

backchat n. (colloquial) answering back rudely.

backcloth n. a painted cloth at the back of a stage or scene.

backdate v. declare to be valid from an earlier date.

backdrop n. a backcloth; a background.

backfire v. **1** make an explosion in an exhaust pipe. **2** produce an undesired effect.

backgammon n. a game played on a board with draughts and dice.

background n. the back part of a scene or picture; the conditions surrounding something.

backhand n. a backhanded stroke.

backhanded adj. **1** performed with the back of the hand turned forwards. **2** said with underlying sarcasm.

backhander n. **1** a backhanded stroke. **2** (slang) a bribe, a reward for services.

backlash n. a violent hostile reaction.

backlist n. a publisher's list of books available.

backlog n. arrears of work.

backpack n. a rucksack.

back-pedal v. (back-pedalled; Amer. back-pedaled) reverse one's previous action or opinion.

back seat n. an inferior position or status.

backside n. (colloquial) the buttocks.

backslide v. slip back from good behaviour into bad.

backspace v. move a computer cursor or typewriter carriage one space back.

backspin n. a backward spinning movement of a ball.

backstage adj. & adv. behind a theatre stage.

backstreet n. a side street. ● adj. illegal, illicit.

backstroke n. a swimming stroke performed on the back.

backtrack v. retrace one's route; reverse one's opinion.

backup n. a support, a reserve; (Computing) a copy of data for safety.

backward adj. directed backwards; having made less than normal progress; diffident. ● adv. backwards. □ **backwardness** n.

backwards adv. towards the back; with the back foremost.

backwash n. 1 receding waves created by a ship etc. 2 a reaction.

backwater n. 1 stagnant water joining a stream. 2 a place unaffected by new ideas or progress.

backwoods n.pl. a remote or backward region.

backyard n. a yard behind a house; the area near where one lives.

bacon n. salted or smoked meat from a pig.

bacteriology n. the study of bacteria. □ **bacteriological** adj., **bacteriologist** n.

bacterium n. (pl. **bacteria**) a microscopic organism. □ **bacterial** adj.

■ **Usage** It is a mistake to use the plural form bacteria when only one bacterium is meant.

bad adj. (**worse**, **worst**) having undesirable qualities; wicked, evil; harmful; decayed. □ **badly** adv., **badness** n.

bade see bid².

badge n. something worn to show membership, rank, etc.

badger n. a large burrowing animal with a black and white striped head. ● v. pester.

badminton n. a game played with rackets and a shuttlecock over a high net.

baffle v. be too difficult for; frustrate. □ **bafflement** n.

bag n. 1 a flexible container; a handbag. 2 (**bags**) (colloquial) a large amount. ● v. (**bagged**) (colloquial) take for oneself.

bagatelle n. 1 a board game in which balls are struck into holes. 2 something trivial. 3 a short piece of music.

baggage n. luggage.

baggy adj. (**baggier**) hanging in loose folds.

bagpipes n.pl. a wind instrument with air stored in a bag and pressed out through pipes.

baguette n. a long thin French loaf.

bail see also **bale**. n. 1 money pledged as security that an accused person will return for trial. 2 each of two crosspieces resting on the stumps in cricket. ● v. (also **bale**) scoop water out of. □ **bail out** obtain or allow the release of (a person) on bail; relieve by financial help.

bailey n. the outer wall of a castle.

bailiff n. a law officer empowered to seize goods for non-payment of fines or debts.

bailiwick n. an area of authority or interest.

bain-marie (ban mă-ree) n. (pl. **bains-marie**) a vessel of hot water in which a dish of food is placed for slow cooking.

bairn n. (Scot. & N. Engl.) a child.

bait n. food etc. placed to attract prey. ● v. **1** place bait on or in. **2** torment by jeers.

baize n. a thick woollen green cloth used for covering billiard tables.

bake v. cook or harden by dry heat.

baker n. a person who bakes or sells bread. □ **bakery** n.

baker's dozen n. thirteen.

baking powder n. a mixture used to make cakes rise.

baksheesh n. a tip, a gratuity.

balaclava (helmet) n. a woollen cap covering the head and neck.

balalaika n. a Russian guitar-like instrument with a triangular body.

balance n. **1** a weighing apparatus. **2** the regulating apparatus of a clock. **3** an even distribution of weight or amount. **4** difference between credits and debits; remainder. ● v. **1** consider by comparing. **2** be, put, or keep in a state of balance.

balcony n. a projecting platform with a rail or parapet; the upper floor of seats in a theatre etc.

bald adj. **1** with the scalp wholly or partly hairless; (of tyres) with the tread worn away. **2** without details. □ **baldly** adv., **baldness** n.

balderdash n. nonsense.

balding adj. becoming bald.

bale see also **bail**. n. a large bound bundle of straw etc.; a large package of goods. ● v. make into a bale or bales. □ **bale out** (or **bail out**) make an emergency parachute jump from an aircraft etc.

baleful adj. menacing, destructive. □ **balefully** adv.

balk var. of **baulk**.

ball n. **1** a spherical object used in games; a rounded part or mass; a single delivery of a ball by a bowler. **2** a formal social gathering for dancing.

ballad n. a song telling a story.

ballast n. heavy material placed in a ship's hold to steady it; coarse stones as the base of a railway or road.

ball-bearing n. a bearing using small steel balls; one of these balls.

ballcock n. a device with a floating ball controlling the water level in a cistern.

ballerina n. a female ballet dancer.

ballet n. a performance of dancing and mime to music.

ballistic missile n. a powered missile, directed at its launch, then falling by gravity to its target.

ballistics n.pl. the study of projectiles and firearms. □ **ballistic** adj.

balloon n. a bag inflated with air or lighter gas. ● v. swell like this.

ballot n. a vote recorded on a slip of paper; voting by this. ● v. (**balloted**) (cause to) vote by ballot.

ballpark n. **1** a baseball ground. **2** (colloquial) an area of activity or responsibility. ● adj. (colloquial) approximate.

ballpoint n. a pen with a tiny ball as its writing point.

ballroom n. a large room where dances are held.

ballyhoo n. a fuss; extravagant publicity.

balm n. **1** a soothing influence; (old use) an ointment. **2** a fragrant herb.

balmy adj. (**balmier**) fragrant; (of air) soft and warm.

baloney n. (also **boloney**) (*slang*) nonsense.

balsa n. a tropical American tree; its lightweight wood.

balsam n. **1** a soothing oil. **2** a kind of flowering plant.

baluster n. a short post or pillar in a balustrade.

balustrade n. a row of short pillars supporting a rail or coping.

bamboo n. a giant tropical grass with hollow stems.

bamboo shoot n. the young shoot of a bamboo, eaten as a vegetable.

bamboozle v. (*colloquial*) mystify, trick.

ban v. (**banned**) forbid officially. ● n. an order banning something.

banal adj. commonplace, uninteresting. □ **banality** n.

banana n. a curved yellow fruit; a tropical tree bearing this.

band n. **1** a strip, a hoop, a loop. **2** a range of values or wavelengths. **3** an organized group of people; a set of musicians. □ **bandleader** n., **bandmaster** n., **bandsman** n.

bandage n. a strip of material for binding a wound. ● v. bind with this.

bandanna n. a large coloured neckerchief.

b. & b. abbr. bed and breakfast.

bandit n. a member of a band of robbers.

bandstand n. a covered outdoor platform for a band playing music.

bandwagon n. **climb on the bandwagon** join a movement heading for success.

bandy v. pass to and fro; exchange (words). ● adj. (**bandier**) curving apart at the knees.

bane n. a cause of trouble or anxiety. □ **baneful** adj.

bang n. a noise of or like an explosion; a sharp blow. ● v. make this noise; strike; shut noisily. ● adv. abruptly; exactly.

banger n. **1** a firework that explodes noisily; (*slang*) a noisy old car. **2** (*slang*) a sausage.

bangle n. a bracelet of rigid material.

banish v. condemn to exile; dismiss from one's presence or thoughts. □ **banishment** n.

banisters n.pl. (also **bannisters**) the uprights and handrail of a staircase.

banjo n. (*pl.* **banjos**) a guitar-like musical instrument with a circular body.

bank n. **1** a slope, esp. at the side of a river; a raised mass of earth etc. **2** a row of lights, switches, etc. **3** an establishment for safe keeping of money; a place storing a reserve supply. ● v. **1** build up into a mound or bank; tilt sideways in rounding a curve. **2** place money in a bank. **3** base one's hopes.

banker's card = **cheque card**.

banker's order n. an instruction to a bank to pay money or deliver property.

banknote n. a printed strip of paper issued as currency.

bankrupt adj. unable to pay one's debts. ● n. a bankrupt person. ● v. make bankrupt. □ **bankruptcy** n.

banner n. a flag; a piece of cloth bearing a slogan.

bannisters var. of **banisters**.

banns n.pl. an announcement in church about a forthcoming marriage.

banquet n. an elaborate ceremonial meal.

banquette n. a long upholstered seat attached to a wall.

banshee n. a spirit whose wail is said to foretell a death.

bantam n. a small fowl.

bantamweight n. a weight between featherweight and flyweight (esp. in boxing).

banter *n.* good-humoured joking.
● *v.* joke in this way.

Bantu *adj. & n.* (*pl.* **Bantu** or **Bantus**) (a member) of a group of African Negroid peoples.

bap *n.* a large soft bread roll.

baptism *n.* a religious rite of sprinkling with water as a sign of purification, usu. with name-giving. □ **baptismal** *adj.*

Baptist *n.* a member of a Protestant sect believing in adult baptism by total immersion in water.

baptistery *n.* a place where baptism is performed.

baptize *v.* (also **-ise**) perform baptism on; name, nickname.

bar *n.* **1** a long piece of solid material; a strip. **2** a barrier. **3** a counter or room where alcohol is served. **4** a vertical line dividing music into units; this unit. **5** barristers, their profession. **6** a unit of atmospheric pressure. ● *v.* (**barred**) fasten or keep in or out with bars; obstruct; prohibit. ● *prep.* except.

barb *n.* **1** a backward-pointing part of an arrow, fish-hook, etc. **2** a wounding remark.

barbarian *n.* an uncivilized person.

barbaric *adj.* suitable for barbarians, rough and wild.

barbarity *n.* savage cruelty.

barbarous *adj.* uncivilized, cruel. □ **barbarously** *adv.*, **barbarism** *n.*

barbecue *n.* a frame for grilling food above an open fire; this food; an open-air party where such food is served. ● *v.* cook on a barbecue.

barbed *adj.* having barbs; (of a remark) hurtful.

barbed wire *n.* wire with many sharp points.

barber *n.* a men's hairdresser.

barber-shop *n.* close harmony singing for four male voices.

barbican *n.* an outer defence to a city or castle; a double tower over a gate or bridge.

barbiturate *n.* a sedative drug.

bar code *n.* a pattern of printed stripes as a machine-readable code identifying a commodity, its price, etc.

bard *n.* a Celtic minstrel; a poet. □ **bardic** *adj.*

bare *adj.* not clothed or covered; not adorned; scanty. ● *v.* reveal. □ **barely** *adv.*

bareback *adv.* on horseback without a saddle.

barefaced *adj.* shameless, undisguised.

bargain *n.* **1** an agreement with obligations on both sides. **2** something obtained cheaply. ● *v.* discuss the terms of an agreement. □ **bargain on** rely on, expect.

barge *n.* a large flat-bottomed boat used on rivers and canals. ● *v.* move clumsily. □ **barge in** intrude.

baritone *n.* a male voice between tenor and bass.

barium *n.* a white metallic element (symbol Ba).

bark *n.* **1** a sharp harsh sound made by a dog. **2** the outer layer of a tree. ● *v.* **1** make the sharp harsh sound of a dog; utter in a sharp commanding voice. **2** scrape skin off accidentally.

barley *n.* a cereal plant; its grain.

barley sugar *n.* a sweet made of boiled sugar.

barley water *n.* a drink made from pearl barley.

barmaid *n.* a woman serving in a pub etc.

barman *n.* (*pl.* **-men**) a man serving in a pub etc.

bar mitzvah *n.* a Jewish ceremony in which a boy of 13 takes on the responsibilities of an adult.

barmy adj. (**barmier**) (slang) crazy.

barn n. a simple roofed farm building for storing grain, hay, etc.

barnacle n. a shellfish that attaches itself to objects under water.

barney n. (pl. **barneys**) (colloquial) an argument.

barometer n. an instrument measuring atmospheric pressure, used in forecasting weather. □ **barometric** adj.

baron n. a member of the lowest rank of nobility. □ **baronial** adj.

baroness n. a woman of the rank of baron; a baron's wife or widow.

baronet n. the holder of a hereditary title below a baron but above a knight. □ **baronetcy** n.

baroque (bǎ-rok) adj. of the architectural or musical style of the 17th-18th centuries. ● n. this style.

barque n. a sailing ship.

barrack v. shout protests; jeer at.

barracks n.pl. buildings for soldiers to live in.

barracuda n. a large voracious tropical sea fish.

barrage n. 1 a heavy bombardment. 2 an artificial barrier.

barre n. a horizontal bar used by dancers when exercising.

barrel n. 1 a large round container with flat ends. 2 a tubelike part esp. of a gun.

barrel organ n. a mechanical instrument producing music by a pin-studded cylinder acting on pipes or keys.

barren adj. not fertile, unable to bear fruit or young. □ **barrenness** n.

barricade n. a barrier. ● v. block or defend with a barricade.

barrier n. something that prevents or controls advance or access.

barring prep. except.

barrister n. a lawyer representing clients in court.

barrow n. 1 a wheelbarrow; a cart pushed or pulled by hand. 2 a prehistoric burial mound.

barter n. & v. trade by exchange of goods for other goods.

basal adj. of or at the base of something.

basalt (ba-sawlt) n. a dark rock of volcanic origin.

base n. 1 the lowest part; a part on which a thing rests or is supported; a basis; headquarters. 2 a substance capable of combining with an acid to form a salt. 3 each of four stations to be reached by a batter in baseball. 4 the number on which a system of counting is based. ● v. use as a base or foundation or evidence for a forecast. ● adj. dishonourable; of inferior value.

baseball n. a team game in which the batter has to hit the ball and run round a circuit.

baseless adj. without foundation. □ **baselessly** adv.

basement n. a storey below ground level.

bash v. strike violently; attack. ● n. a violent blow or knock.

bashful adj. shy and self-conscious. □ **bashfully** adv., **bashfulness** n.

BASIC n. a computer programming language using familiar English words.

basic adj. forming a basis; fundamental. □ **basically** adv.

basil n. a sweet-smelling herb.

basilica n. an oblong hall or church with an apse at one end.

basilisk n. a small American lizard; a mythical reptile said to cause death by its glance or breath.

basin *n.* a deep open container for liquids; a washbasin; a sunken place where water collects, an area drained by a river. □ **basinful** *n.*

basis *n.* (*pl.* **bases**) a foundation or support; a main principle.

bask *v.* sit or lie comfortably exposed to pleasant warmth.

basket *n.* a container for holding or carrying things, made of interwoven cane or wire.

basketball *n.* a team game in which the aim is to throw the ball through a high hooped net.

basketwork *n.* material woven in the style of a basket.

Basque *n. & adj.* (a member, the language) of a people living in the western Pyrenees.

bas-relief *n.* a sculpture or carving in low relief.

bass[1] (bass) *n.* (*pl.* **bass**) a fish of the perch family.

bass[2] (bayss) *adj.* deep-sounding, of the lowest pitch in music. ● *n.* the lowest male voice; bass pitch; a double bass.

basset *n.* a short-legged hound.

bassoon *n.* a woodwind instrument with a deep tone.

bast *n.* the inner bark of the lime tree; a similar fibre.

bastard *n.* **1** (*offensive*) an illegitimate child. **2** (*vulgar slang*) an unpleasant or difficult person or thing. □ **bastardy** *n.*

baste *v.* **1** moisten with fat during cooking. **2** sew together temporarily with loose stitches. **3** thrash.

bastion *n.* a projecting part of a fortified place; a stronghold.

bat *n.* **1** a wooden implement for hitting a ball in games; a batsman. **2** a flying animal with a mouselike body. ● *v.* (**batted**) perform or strike with the bat in cricket etc.

batch *n.* a set of people or things dealt with as a group.

bated *adj.* □ **with bated breath** very anxiously.

bath *n.* a container used for washing the body; a wash in this. ● *v.* wash in a bath.

bathe *v.* immerse in liquid; swim for pleasure. ● *n.* a swim. □ **bather** *n.*

bathos *n.* an anticlimax, a descent from an important thing to a trivial one.

bathroom *n.* a room containing a bath, shower, washbasin, etc.

batik *n.* a method of printing textiles by waxing parts not to be dyed; fabric printed in this way.

batman *n.* (*pl.* **-men**) a soldier acting as an officer's personal servant.

baton *n.* a short stick, esp. used by a conductor.

batrachian (bă-tray-kiăn) *n.* a frog or toad.

batsman *n.* (*pl.* **-men**) a player batting in cricket etc.

battalion *n.* an army unit of several companies.

batten *n.* a bar of wood or metal, esp. holding something in place. ● *v.* fasten with battens. □ **batten on** thrive at another's expense.

batter *v.* hit hard and often. ● *n.* **1** a beaten mixture of flour, eggs, and milk, used in cooking. **2** a player batting in baseball.

battering ram *n.* an iron-headed beam formerly used for breaking through walls or gates.

battery *n.* **1** a group of big guns; an artillery unit. **2** a set of similar or connected units of equipment, poultry cages, etc. **3** an electric cell or cells supplying current. **4** an unlawful blow or touch.

battle *n.* a fight between large organized forces; a contest. ● *v.* engage in a battle, struggle.

battleaxe n. a heavy axe used as a weapon in ancient times; (*colloquial*) a domineering woman.

battlefield n. the scene of a battle.

battlements n.pl. a parapet with gaps for firing from.

battleship n. a warship of the most heavily armed kind.

batty adj. (**battier**) (*slang*) crazy.

bauble n. a valueless ornament.

baulk (also **balk**) v. shirk; frustrate. ● n. **1** a hindrance. **2** the starting area on a billiard table.

bauxite n. a mineral from which aluminium is obtained.

bawdy adj. (**bawdier**) humorously indecent. □ **bawdiness** n.

bawl v. shout; weep noisily. □ **bawl out** (*colloquial*) reprimand.

bay n. **1** part of the sea within a wide curve of the shore. **2** a recess. **3** a laurel, esp. a type used as a herb. **4** the deep cry of a large dog or of hounds in pursuit. ● v. make the deep cry of a dog. ● adj. (of a horse) reddish brown. □ **at bay** forced to face attackers.

bayonet n. a stabbing blade fixed to a rifle.

bay window n. one projecting from an outside wall.

bazaar n. **1** a market in an oriental country. **2** a sale of goods to raise funds.

bazooka n. a portable weapon firing anti-tank rockets.

BBC abbr. British Broadcasting Corporation.

BC abbr. (of a date) before Christ.

Be symb. beryllium.

be v. exist, occur; have a certain position, quality, or condition. ● v.aux. (used to form tenses of other verbs).

beach n. the shore. ● v. bring on shore from water.

beachcomber n. a person who salvages things on a beach.

beachhead n. a fortified position set up on a beach by an invading army.

beacon n. a signal fire on a hill.

bead n. a small shaped piece of hard material pierced for threading with others on a string; a drop or bubble of liquid.

beading n. a strip of trimming for wood.

beadle n. (formerly) a minor parish official.

beady adj. (**beadier**) (of eyes) small and bright.

beagle n. a small hound.

beak n. **1** a bird's horny projecting jaws; any similar projection. **2** (*slang*) a magistrate.

beaker n. a tall cup; a tumbler.

beam n. **1** a long piece of timber or metal carrying the weight of part of a building. **2** a ray of light or other radiation. **3** a bright smile. **4** a ship's breadth. ● v. **1** send out light etc. **2** smile radiantly.

bean n. a plant with kidney-shaped seeds in long pods; a seed of this or of coffee.

beanfeast n. (*colloquial*) a celebration.

bear n. a large heavy animal with thick fur; a child's toy like this. ● v. (**bore**, **borne**) **1** carry, support; have in one's heart or mind. **2** endure. **3** be fit for. **4** produce, give birth to. **5** take (a specified direction). **6** exert pressure.

bearable adj. endurable.

beard n. the hair around a man's chin. ● v. confront boldly.

bearing n. **1** deportment, behaviour. **2** relevance. **3** a compass direction. **4** a device reducing friction where a part turns.

bearskin n. a guardsman's tall furry cap.

beast n. a large animal; an unpleasant person or thing.

beastly *adj.* (**beastlier**) (*colloquial*) very unpleasant. □ **beastliness** *n.*

beat *v.* (**beat, beaten, beating**) **1** hit repeatedly; mix vigorously; (of the heart) pump rhythmically. **2** do better than; defeat. ● *n.* **1** a regular repeated stroke; a recurring emphasis marking rhythm. **2** an appointed course of a policeman or sentinel. □ **beat up** assault violently.

beatific *adj.* showing great happiness. □ **beatifically** *adv.*

beatify *v.* (in the RC Church) declare (a dead person) blessed, as the first step in canonization. □ **beatification** *n.*

beatitude *n.* blessedness.

beauteous *adj.* (*poetic*) beautiful.

beautician *n.* a person who gives beauty treatment.

beautiful *adj.* having beauty; pleasant; excellent. □ **beautifully** *adv.*

beautify *v.* make beautiful. □ **beautification** *n.*

beauty *n.* a combination of qualities giving pleasure to the sight, mind, etc.; a beautiful person or thing.

beaver *n.* an amphibious rodent that builds dams. ● *v.* work hard.

becalmed *adj.* unable to move because there is no wind.

because *conj.* for the reason that. □ **because of** by reason of.

beck *n.* a stream. □ **at the beck and call of** ready and waiting to obey.

beckon *v.* summon by a gesture.

become *v.* (**became, become, becoming**) **1** come or grow to be, begin to be. **2** give a pleasing appearance or effect upon; befit.

becquerel (bek-er-el) *n.* a unit of radioactivity.

bed *n.* **1** a thing to sleep or rest on; a framework with a mattress and coverings. **2** a flat base, a

foundation; the bottom of a sea or river etc. **3** a garden plot.

B.Ed. *abbr.* Bachelor of Education.

bedclothes *n.pl.* sheets, blankets, etc.

bedding *n.* beds and bedclothes.

bedding plant *n.* a plant grown to be planted when in flower and discarded at the end of the season.

bedevil *v.* (**bedevilled**; *Amer.* **bedeviled**) afflict with difficulties.

bedfellow *n.* **1** a person sharing one's bed. **2** an associate.

bedlam *n.* a scene of uproar.

Bedouin (bed-oo-in) *n.* (also **Beduin**) (*pl.* **Bedouin**) a member of an Arab people living in tents in the desert.

bedpan *n.* a pan for use as a lavatory by a person confined to bed.

bedraggled *adj.* limp and untidy.

bedridden *adj.* permanently confined to bed through illness.

bedrock *n.* **1** solid rock beneath loose soil. **2** basic facts.

bedroom *n.* a room for sleeping in.

bedside *n.* a position by a bed.

bedsitter *n.* (also **bedsit, bedsitting room**) a room used for both living and sleeping in.

bedsore *n.* a sore developed by lying in bed for a long time.

bedspread *n.* a covering for a bed.

bedstead *n.* the framework of a bed.

Beduin var. of **Bedouin**.

bee *n.* an insect that produces honey.

beech *n.* a tree with smooth bark and glossy leaves.

beef *n.* **1** meat from an ox, bull, or cow; muscular strength. **2** (*slang*) a grumble. ● *v.* (*slang*) grumble.

beefburger *n.* a fried cake of minced beef.

beefeater *n.* a warder in the Tower of London, wearing Tudor dress.

beefy *adj.* (**beefier**) having a solid muscular body.

beehive *n.* a structure in which bees live.

beeline *n.* □ **make a beeline for** go straight or rapidly towards.

beep *n.* a high-pitched sound like that of a car horn. ● *v.* make a beep. □ **beeper** *n.*

beer *n.* an alcoholic drink made from malt and hops. □ **beery** *adj.*

beeswax *n.* a yellow substance secreted by bees, used as polish.

beet *n.* a plant with a fleshy root used as a vegetable (*beetroot*) or for making sugar (*sugar beet*).

beetle *n.* **1** an insect with hard wing-covers. **2** a tool for ramming or crushing things. ● *v.* overhang, project.

beetroot *n.* the dark red root of a beet as a vegetable.

befall *v.* (**befell, befallen, befalling**) (*poetic*) happen; happen to.

befit *v.* (**befitted**) be suitable for.

before *adv., prep., & conj.* at an earlier time (than); ahead, in front of; in preference to.

beforehand *adv.* in advance.

befriend *v.* show kindness towards.

befuddle *v.* confuse.

beg *v.* (**begged**) **1** ask for as a gift or charity; request earnestly or humbly. **2** (of a dog) sit up expectantly with forepaws off the ground. □ **beg the question** assume the truth of a thing to be proved.

■ **Usage** The expression *beg the question* is often used incorrectly to mean 'to invite the obvious inference (that...)'.

beget *v.* (**begot, begotten, begetting**) (*literary*) be the father of; give rise to.

beggar *n.* a person who lives by begging. ● *v.* reduce to poverty. □ **beggary** *n.*

beggarly *adj.* mean and insufficient.

begin *v.* (**began, begun, beginning**) perform the first or earliest part of (an activity); be the first to do a thing; come into existence.

beginner *n.* a person beginning to learn a skill.

beginning *n.* a first part; a starting point, a source or origin.

begonia *n.* a garden with bright leaves and flowers.

begrudge *v.* be unwilling to give or allow.

beguile *v.* **1** deceive. **2** entertain pleasantly.

begum *n.* the title of a married Muslim woman.

behalf *n.* □ **on behalf of** as the representative of.

behave *v.* act or react in a specified way; (also **behave oneself**) show good manners.

behaviour *n.* (*Amer.* **behavior**) a way of behaving.

behead *v.* cut the head off.

beheld *see* **behold.**

behind *adv. & prep.* in or to the rear (of); in arrears; remaining after others' departure. ● *n.* the buttocks.

behold *v.* (**beheld, beholding**) (*old use*) see, observe. □ **beholder** *n.*

beholden *adj.* owing thanks.

behove *v.* (*formal*) be incumbent on; befit.

beige *n.* a light fawn colour.

being *n.* existence; a thing that exists and has life, a person.

belabour *v.* (*Amer.* **belabor**) beat; attack; labour (a subject).

belated *adj.* coming very late or too late. □ **belatedly** *adv.*

belay v. secure (a rope) by winding it round something.

belch v. send out wind noisily from the stomach through the mouth. ● n. an act or sound of belching.

beleaguer v. besiege; harass.

belfry n. a bell tower; a space for bells in a tower.

belie v. (**belied, belying**) contradict, fail to confirm.

belief n. believing; something believed.

believe v. accept as true or as speaking truth; think, suppose. □ **believe in** have faith in the existence of; feel sure of the worth of. □ **believer** n.

Belisha beacon n. a flashing amber globe marking a pedestrian crossing.

belittle v. disparage.

bell n. a cup-shaped metal instrument that makes a ringing sound when struck; its sound.

belladonna n. deadly nightshade; a medicinal drug made from this.

belle n. a beautiful woman.

belles-lettres (bel-letr) n.pl. literary writings or studies.

bellicose adj. eager to fight.

belligerent adj. waging war; aggressive. □ **belligerently** adv., **belligerence** n.

bellow n. a loud deep sound made by a bull; a deep shout. ● v. make this sound.

bellows n.pl. an apparatus for driving air into something.

belly n. the abdomen; the stomach.

bellyful n. (colloquial) enough or more than enough.

belong v. be rightly assigned as property, part, duty, etc.; have a rightful place. □ **belong to** be a member of.

belongings n.pl. personal possessions.

beloved adj. dearly loved.

below adv. & prep. at or to a lower position or amount (than).

belt n. a strip of cloth or leather etc. worn round the waist; a long narrow region. ● v. **1** put a belt round. **2** (slang) hit. **3** (slang) rush.

bemoan v. complain about.

bemused adj. bewildered; lost in thought. □ **bemusement** n.

bench n. **1** a long seat of wood or stone; a long working-table. **2** judges or magistrates hearing a case.

benchmark n. a surveyor's mark; a point of reference.

bend v. (**bent, bending**) make or become curved; turn downwards, stoop; turn in a new direction. ● n. curve, turn.

bender n. (slang) a wild drinking spree.

beneath adv. & prep. below, underneath; not worthy of.

benediction n. a spoken blessing.

benefactor n. one who gives financial or other help. □ **benefaction** n., **benefactress** n.

beneficent adj. doing good; actively kind. □ **beneficence** n.

beneficial adj. having a helpful or useful effect. □ **beneficially** adv.

beneficiary n. one who receives a benefit or legacy.

benefit n. something helpful, favourable, or profitable. ● v. (**benefited**; Amer. **benefitted**) do good to; receive benefit.

benevolent adj. kindly and helpful. □ **benevolently** adv., **benevolence** n.

benighted adj. in darkness; ignorant.

benign adj. kindly; mild and gentle; not malignant. □ **benignly** adv.

bent see **bend**. n. a natural skill or liking. □ **bent on** seeking or determined to do.

benzene | better

benzene n. a liquid obtained from petroleum and coal tar, used as a solvent, fuel, etc.

benzine n. a liquid mixture of hydrocarbons used in dry-cleaning.

benzol n. (unrefined) benzene.

bequeath v. leave as a legacy.

bequest n. a legacy.

berate v. scold.

bereave v. deprive, esp. of a relative, by death. □ **bereavement** n.

bereft adj. deprived.

beret (be-ray) n. a round flat cap with no peak.

beriberi n. a disease caused by lack of vitamin B.

berk n. a stupid person.

berkelium n. a man-made radioactive metallic element (symbol Bk).

berry n. a small round juicy fruit with no stone.

berserk adj. □ **go berserk** go into an uncontrollable destructive rage.

berth n. a bunk or sleeping place in a ship or train; a place for a ship to tie up at a wharf. ● v. moor at a berth. □ **give a wide berth** to keep a safe distance from.

beryl n. a transparent green gem.

beryllium n. a light metallic element (symbol Be).

beseech v. (besought, beseeching) implore.

beset v. (beset, besetting) hem in, surround; habitually affect or trouble.

beside prep. at the side of, close to; compared with. □ **be beside oneself** be at the end of one's self-control. **beside the point** irrelevant.

besides prep. in addition to, other than. ● adv. also.

besiege v. lay siege to.

besmirch v. dirty; dishonour.

besom n. a broom made from a bundle of twigs tied to a handle.

besotted adj. infatuated.

besought see **beseech**.

bespeak v. (bespoke, bespoken, bespeaking) 1 engage beforehand. 2 be evidence of.

bespoke adj. made to a customer's order.

best adj. of the most excellent kind. ● adv. in the best way; most usefully. ● n. the best thing; a victory. □ **best part of** most of.

bestial adj. of or like a beast, savage. □ **bestiality** n.

bestir v. (bestirred) □ **bestir oneself** exert oneself.

best man n. a bridegroom's chief attendant.

bestow v. present as a gift. □ **bestowal** n.

bestride v. (bestrode, bestridden, bestriding) stand astride over.

bet n. a pledge that will be forfeited if one's prediction is wrong. ● v. (bet or betted, betting) make a bet; (colloquial) predict.

beta n. the second letter of the Greek alphabet (Β, β).

beta blocker n. a drug used to prevent increased cardiac activity.

betake v. (betook, betaken, betaking) □ **betake oneself** go.

bête noire (bet nwar) n. (pl. **bêtes noires**) something greatly disliked

betide v. happen to.

betimes adv. (literary) in good time, early.

betoken v. be a sign of.

betray v. hand over to an enemy; be disloyal to. □ **betrayal** n., **betrayer** n.

betroth v. cause to be engaged to marry. □ **betrothal** n.

better adj. 1 of a more excellent kind. 2 recovered from illness. ● adv. in a better manner; more usefully. ● v. improve; do better than. ● n. a person who bets.

□ **better part of** more than half. **get the better of** overcome; outwit.

betting shop n. a bookmaker's office.

between prep. in the space, time, or quality bounded by (two limits); separating; to and from; connecting; shared by. ● adv. between points or limits etc.

bevel n. a sloping edge. ● v. (**bevelled**; Amer. **beveled**) give a sloping edge to.

beverage n. a drink.

bevvy n. (slang) an alcoholic drink.

bevy n. a company, a large group.

bewail v. wail over.

beware v. be on one's guard.

bewilder v. puzzle, confuse. □ **bewilderment** n.

bewitch v. put under a magic spell; delight very much.

beyond adv. & prep. at or to the further side (of); outside the range (of).

bezel n. the sloped edge of a chisel.

bhaji n. an Indian dish of vegetables fried in batter.

b.h.p. abbr. brake horsepower.

Bi symb. bismuth.

bi- comb. form two; twice.

biannual adj. happening twice a year. □ **biannually** adv.

bias n. **1** an influence favouring one of a group unfairly. **2** a tendency of a bowl to swerve because of its lopsided form. ● v. (**biased** or **biassed**) give a bias to, influence.

bib n. a covering put under a young child's chin to protect its clothes while feeding.

Bible n. the Christian or Jewish scriptures; (colloquial) any authoritative book.

biblical adj. of or in the Bible.

bibliography n. a list of books about a subject or by a specified author. □ **bibliographer** n., **bibliographical** adj.

bibliophile n. a book-lover.

bicarbonate of soda n. sodium bicarbonate, used in baking powder or as an antacid.

bicentenary n. a 200th anniversary.

bicentennial adj. happening every 200 years. ● n. a bicentenary.

biceps n. the large muscle at the front of the upper arm.

bicker v. quarrel about unimportant things.

bicycle n. a two-wheeled vehicle driven by pedals. ● v. ride a bicycle.

bid[1] n. an offer of a price, esp. at an auction; a statement of the number of tricks a player proposes to win in a card game; an attempt. ● v. (**bid**, **bidding**) make a bid (of), offer. □ **bidder** n.

bid[2] v. (**bid** or **bade**, **bidden**, **bidding**) command; say as a greeting.

biddable adj. willing to obey.

bide v. await (one's time).

bidet (bee-day) n. a low washbasin that one can sit astride to wash the genital and anal regions.

biennial adj. lasting for two years; happening every second year. ● n. a plant that flowers and dies in its second year. □ **biennially** adv.

bier (beer) n. a movable stand for a coffin.

biff (slang) n. a sharp blow. ● v. hit (a person).

bifocals n.pl. spectacles with lenses that have two segments, assisting both distant and close focusing.

bifurcate v. fork. □ **bifurcation** n.

big adj. (**bigger**) large in size, amount, or intensity.

bigamy *n.* the crime of going through a form of marriage while a previous marriage is still valid. □ **bigamist** *adj.*, **bigamous** *adj.*

bigot *n.* a person who is prejudiced and is intolerant towards those who disagree. □ **bigoted** *adj.*, **bigotry** *n.*

bijou *adj.* (bee-zhoo) ● *adj.* small and elegant.

bike *n.* (*colloquial*) a bicycle or motorcycle. □ **biker** *n.*

bikini *n.* a woman's two-piece swimming costume.

bilateral *adj.* having two sides; existing between two groups. □ **bilaterally** *adv.*

bilberry *n.* a small round dark blue fruit; a shrub producing this.

bile *n.* a bitter yellowish liquid produced by the liver.

bilge *n.* **1** a ship's bottom; water collecting there. **2** (*slang*) worthless talk, nonsense.

bilharzia *n.* a disease caused by a tropical parasitic flatworm.

bilingual *adj.* written in or able to speak two languages.

bilious *adj.* sick, esp. from trouble with the bile or liver. □ **biliousness** *n.*

bill *n.* **1** a written statement of charges to be paid. **2** a poster. **3** a programme. **4** a certificate. **5** (*Amer.*) a banknote. **6** a bird's beak. □ **bill and coo** exchange caresses.

billabong *n.* (*Austral.*) a backwater.

billboard *n.* a hoarding for advertisements.

billet *n.* a lodging for troops. ● *v.* (**billeted**) place in a billet.

billhook *n.* a tool with a broad curved blade for lopping trees.

billiards *n.* a game played with cues and three balls on a table.

billion *n.* a thousand million or (less commonly) a million million.

billow *n.* a great wave. ● *v.* rise or move like waves; swell out.

bimbo *n.* (*pl.* **bimbos**) (*slang*) an attractive but unintelligent young woman.

bimetallic *adj.* made of two metals.

bin *n.* a large rigid container or receptacle.

binary *adj.* of two.

binary digit *n.* either of the two digits (0 and 1) of the **binary system**, a number system used in computing.

bind *v.* (**bound, binding**) **1** tie, fasten together; cover the edge of so as to strengthen or decorate; fasten into a cover. **2** place under an obligation or legal agreement. ● *n.* (*colloquial*) a nuisance.

binding *n.* **1** a book cover. **2** braid etc. used to bind an edge.

bindweed *n.* wild convolvulus.

binge *n.* (*slang*) a bout of excessive eating and drinking.

bingo *n.* a gambling game using cards marked with numbered squares.

binocular *adj.* using two eyes. ● *n.pl.* (**binoculars**) an instrument with lenses for both eyes, making distant objects seem larger.

binomial *adj.* & *n.* (an expression or name) consisting of two terms.

bio- *comb. form* of living things.

biochemistry *n.* the chemistry of living organisms. □ **biochemical** *adj.*, **biochemist** *n.*

biodegradable *adj.* able to be decomposed by bacteria.

bioengineering *n.* the application of engineering techniques to biological processes.

biographer *n.* the writer of a biography.

biography n. the story of a person's life. □ **biographical** adj.

biology n. the study of the life and structure of living things. □ **biological** adj., **biologist** n.

biomass n. the total quantity or weight of organisms in a given area.

bionic adj. (of a person or faculties) operated electronically.

biopsy n. an examination of tissue cut from a living body.

biorhythm n. any of the recurring cycles of activity in a person's life.

biotechnology n. the use of micro-organisms and biological processes in industrial production.

bipartisan adj. involving two parties.

bipartite adj. consisting of two parts; involving two groups.

biped n. a two-footed animal such as man.

biplane n. an aeroplane with two pairs of wings.

birch n. a tree with smooth bark.

bird n. a feathered animal.

birdie n. (Golf) a score of one stroke under par for a hole.

biro n. (pl. **biros**) (trade mark) a ballpoint pen.

birth n. the emergence of young from its mother's body; parentage, origin.

birth control n. contraception.

birthday n. the anniversary of the day of one's birth.

birthmark n. an unusual coloured mark on the skin at birth.

birthright n. something that is one's right through being born into a certain family or country.

biscuit n. a small flat piece of pastry or cake-like substance baked crisp.

bisect v. divide into two equal parts. □ **bisection** n., **bisector** n.

bisexual adj. sexually attracted to members of both sexes. □ **bisexuality** n.

bishop n. **1** a clergyman of high rank. **2** a mitre-shaped chess piece.

bishopric n. the diocese of a bishop.

bismuth n. a metallic element (symbol Bi); a compound of this used in medicines.

bison n. (pl. **bison**) a wild ox; a buffalo.

bistro n. (pl. **bistros**) a small informal restaurant.

bit see **bite**. n. **1** a small piece or quantity; a short time or distance. **2** the mouthpiece of a bridle. **3** part of a tool that cuts, bores, or grips when twisted. **4** a small coin. **5** (Computing) a binary digit.

bitch n. **1** a female dog. **2** (slang) a spiteful woman, a difficult thing. ● v. (colloquial) speak spitefully or sourly. □ **bitchy** adj., **bitchiness** n.

bite v. (**bit, bitten, biting**) cut with the teeth; penetrate; grip or act effectively. ● n. **1** an act of biting; a wound made by this. **2** a small meal.

biting adj. causing a smarting pain; sharply critical.

bitter adj. tasting sharp, not sweet or mild; with mental pain or resentment; piercingly cold. ● n. beer flavoured with hops and slightly bitter. □ **bitterly** adv., **bitterness** n.

bittern n. a marsh bird.

bitty adj. (**bittier**) made up of unrelated bits. □ **bittiness** n.

bitumen n. a black substance made from petroleum. □ **bituminous** adj.

bivalve n. a shellfish with a hinged double shell.

bivouac n. a temporary camp without tents or other cover. ● v. (**bivouacked, bivouacking**) camp in a bivouac.

bizarre *adj.* strikingly odd in appearance or effect.

Bk *symb.* berkelium.

blab *v.* (**blabbed**) talk indiscreetly.

black *adj.* **1** of the very darkest colour, like coal; having a dark skin. **2** dismal, gloomy; hostile; evil. ● *n.* a black colour or thing; a member of a dark-skinned race. □ **in the black** with a credit balance, not in debt. **black out** cover windows so that no light can penetrate.

blackberry *n.* a bramble; its edible dark berry.

blackbird *n.* a European songbird, the male of which is black.

blackboard *n.* a board for writing on with chalk in front of a class.

blacken *v.* **1** make or become black. **2** say evil things about.

black eye *n.* a bruised eye.

blackguard (**blag-ard**) *n.* a scoundrel.

blackhead *n.* a small dark lump blocking a pore in the skin.

black hole *n.* a region in outer space from which matter and radiation cannot escape.

blackleg *n.* a person who works while fellow workers are on strike.

blacklist *n.* a list of people who are disapproved of or under suspicion.

blackmail *v.* demand payment or action from (a person) by threats. ● *n.* demands made in this way. □ **blackmailer** *n.*

black market *n.* illegal buying and selling.

blackout *n.* a temporary loss of consciousness or memory.

black pudding *n.* a sausage of blood and suet.

black sheep *n.* a member of a family etc. regarded as a disgrace.

blacksmith *n.* a smith who works in iron.

blackthorn *n.* a thorny shrub bearing white flowers and sloes.

bladder *n.* the sac in which urine collects in the body; an inflatable bag.

blade *n.* the flattened cutting part of a knife or sword; the flat part of an oar or propeller; a flat narrow leaf of grass; a broad bone.

blame *v.* hold responsible for a fault. ● *n.* responsibility for a fault. □ **blameless** *adj.*, **blameworthy** *adj.*

blanch *v.* **1** make or become white or pale. **2** immerse (vegetables) briefly in boiling water, peel by this method.

blancmange (blǎ-**monj**) *n.* a flavoured jelly-like dessert.

bland *adj.* mild; gentle and casual, not irritating or stimulating; dull and uninteresting. □ **blandly** *adv.*

blandishments *n.pl.* flattering or coaxing words.

blank *adj.* not written or printed on; without interest or expression. ● *n.* a blank space; a cartridge containing no bullet.

blank cheque *n.* one with the amount left blank for the payee to fill in.

blanket *n.* a warm covering made of woollen or similar material; a thick covering mass. ● *adj.* covering everything; inclusive.

blank verse *n.* verse without rhyme.

blare *v.* sound loudly and harshly. ● *n.* this sound.

blarney *n.* smooth talk that flatters and deceives.

blasé (blah-**zay**) *adj.* bored or unimpressed.

blaspheme *v.* utter blasphemies. □ **blasphemer** *n.*

blasphemy *n.* irreverent talk about sacred things. □ **blasphemous** *adj.*, **blasphemously** *adv.*

blast *n.* **1** a strong gust; a wave of air from an explosion; a sharp sound from a whistle etc. **2** a severe reprimand. ● *v.* **1** blow up with explosives; cause to wither, destroy. **2** (*colloquial*) reprimand severely. □ **blast off** (of a rocket etc.) take off from a launching site.

blatant *adj.* very obvious; shameless. □ **blatantly** *adv.*

blather *v.* (also **blether**) chatter foolishly.

blaze *n.* **1** a bright flame or fire; a bright light or display; an outburst. **2** a white mark on an animal's face; a mark chipped in the bark of a tree to mark a route. ● *v.* burn or shine brightly. □ **blaze a trail** mark out a route; pioneer.

blazer *n.* a loose-fitting jacket, esp. in the colours or bearing the badge of a school, team, etc.

blazon *n.* a heraldic shield, a coat of arms. ● *v.* proclaim; ornament with heraldic or other devices.

bleach *v.* whiten by sunlight or chemicals. ● *n.* a bleaching substance or process.

bleak *adj.* cold and cheerless. □ **bleakly** *adv.*, **bleakness** *n.*

bleary *adj.* (**blearier**) (of eyes) watery and seeing indistinctly. □ **blearily** *adv.*, **bleariness** *n.*

bleat *n.* the cry of a sheep or goat. ● *v.* utter this cry; speak or say plaintively.

bleed *v.* (**bled**, **bleeding**) leak blood or other fluid; draw blood or fluid from; extort money from.

bleep *n.* a short high-pitched sound. ● *v.* make this sound, esp. as a signal. □ **bleeper** *n.*

blemish *n.* a flaw or defect. ● *v.* spoil with a blemish.

blench *v.* flinch.

blend *v.* mix smoothly; mingle. ● *n.* a mixture.

blender *n.* an appliance for puréeing food.

bless *v.* call God's favour upon; make sacred or holy; praise (God). □ **be blessed with** be fortunate in having.

blessed (bles-id) *adj.* **1** holy, sacred; in paradise. **2** (*slang*) damned, cursed. □ **blessedly** *adv.*

blessing *n.* **1** God's favour; a prayer for this. **2** something one is glad of.

blether var. of **blather**.

blight *n.* a disease or fungus that withers plants; a malignant influence. ● *v.* affect with blight; spoil.

blimey *int.* (*slang*) an exclamation of surprise

blimp *n.* a small non-rigid airship.

blind *adj.* **1** without sight; without foresight, understanding, or adequate information. **2** (in cookery) without filling. ● *v.* make blind; rob of judgement. ● *n.* **1** a screen, esp. on a roller, for a window. **2** a pretext. □ **blindly** *adv.*, **blindness** *n.*

blindfold *n.* a cloth used to cover the eyes and block the sight. ● *v.* cover the eyes of (a person) with this.

blink *v.* open and shut one's eyes rapidly; shine unsteadily. ● *n.* an act of blinking; a quick gleam.

blinker *n.* a leather piece fixed to a bridle to prevent a horse from seeing sideways. ● *v.* obstruct the sight or understanding of.

blip *n.* a quick sound or movement; a small image on a radar screen; a minor problem or change.

bliss *n.* perfect happiness. □ **blissful** *adj.*, **blissfully** *adv.*

blister *n.* a bubble-like swelling on the skin; a raised swelling on a surface. ● *v.* raise a blister on; be affected with blister(s).

blithe adj. casual and carefree. □ **blithely** adv.

blitz n. a violent attack, esp. from aircraft. ● v. attack in a blitz.

blitzkrieg n. an intense military campaign aimed at a swift victory.

blizzard n. a severe snowstorm.

bloat v. swell with fat, gas, or liquid.

bloater n. a salted smoked herring.

blob n. a drop of liquid; a round mass.

bloc n. a group of parties or countries who combine for a purpose.

block n. **1** a solid piece of hard substance; (slang) the head. **2** a compact mass of buildings; a large building divided into flats or offices. **3** a large quantity treated as a unit. **4** a pad of paper for drawing or writing on. **5** an obstruction. **6** a pulley mounted in a case. ● v. obstruct, prevent the movement or use of.

blockade n. the blocking of access to a place, to prevent entry of goods. ● v. set up a blockade of.

blockage n. blocking; something that blocks.

blockbuster n. something very powerful or successful.

blockhead n. a stupid person.

block letters n.pl. plain capital letters.

bloke n. (colloquial) a man.

blond n. a fair-haired man. ● adj. (of hair) fair.

blonde n. a fair-haired woman. ● adj. (of a woman's hair) fair.

blood n. the red liquid circulating in the bodies of animals; race, descent, parentage, kindred; temper, courage. ● v. give a first taste of blood to (a hound); initiate (a person).

blood-curdling adj. horrifying.

bloodhound n. a large keen-scented dog, formerly used in tracking.

bloodless adj. without bloodshed. □ **bloodlessly** adv.

bloodshed n. killing or wounding.

bloodshot adj. (of eyes) red from dilated veins.

blood sports n.pl. sports involving killing.

bloodstained adj. stained with blood.

bloodstock n. thoroughbred horses.

bloodstream n. blood circulating in the body.

bloodsucker n. a creature that sucks blood; a person who extorts money.

bloodthirsty adj. eager for bloodshed.

blood vessel n. a tubular structure conveying blood within the body.

bloody adj. (**bloodier**) **1** bloodstained; with much bloodshed. **2** (vulgar slang) cursed. ● adv. (vulgar slang) extremely. ● v. stain with blood.

bloody-minded adj. (colloquial) deliberately uncooperative.

bloom n. a flower; beauty, perfection. ● v. bear flowers; be in full beauty.

bloomer n. **1** a blunder. **2** a long loaf with diagonal marks. **3** (**bloomers**) knickers with legs.

blossom n. flowers, esp. of a fruit tree. ● v. open into flowers; develop and flourish.

blot n. a spot of ink etc.; something ugly or disgraceful. ● v. (**blotted**) make a blot on.

blotch n. a large irregular mark. □ **blotchy** adj.

blotto adj. (slang) very drunk.

blouse n. a shirt-like garment worn by women.

blouson n. a short full jacket gathered at the waist.

blow v. (**blew, blown, blowing**) **1** send out a current of air or breath; move or flow as a current of air; move, shape, or sound by this; puff and pant. **2** (of a fuse) melt. ● n. **1** an act of blowing. **2** a hard stroke with a hand, tool, or weapon; a shock, a disaster. □ **blow up** explode, shatter by an explosion; inflate; exaggerate; enlarge (a photograph); lose one's temper.

blowfly n. a fly that lays its eggs on meat.

blowlamp n. (also **blowtorch**) a portable burner for directing a very hot flame.

blow-out n. **1** a burst tyre. **2** a melted fuse. **3** an uncontrolled rush of oil from a well. **4** (colloquial) a huge meal.

blowpipe n. a tube through which air etc. is blown, e.g. to heat a flame or send out a missile.

blowy adj. (**blowier**) windy.

blowzy adj. (**blowzier**) red-faced and coarse-looking.

blub v. (**blubbed**) (slang) sob.

blubber n. whale fat. ● v. sob noisily.

bludgeon n. a heavy stick used as a weapon. ● v. strike with a bludgeon; compel forcefully.

blue adj. **1** of a colour like the cloudless sky. **2** unhappy. **3** indecent. ● n. **1** a blue colour or thing. **2** (**blues**) melancholy jazz melodies, a state of depression. □ **out of the blue** unexpectedly.

bluebell n. a plant with blue bell-shaped flowers.

blueberry n. an edible blue berry; a shrub bearing this.

blue-blooded adj. of aristocratic descent.

bluebottle n. a large bluish fly.

blueprint n. a blue photographic print of building plans; a detailed scheme.

bluestocking n. a learned woman.

bluff v. deceive by a pretence. ● n. **1** bluffing. **2** a broad steep cliff or headland. ● adj. **1** with a broad steep front. **2** abrupt, frank, and hearty.

bluish adj. rather blue.

blunder v. move clumsily and uncertainly; make a bad mistake. ● n. a bad mistake.

blunderbuss n. an old type of gun firing many balls at one shot.

blunt adj. **1** without a sharp edge or point. **2** speaking or expressed plainly. ● v. make or become blunt. □ **bluntly** adv., **bluntness** n.

blur n. a smear; an indistinct appearance. ● v. (**blurred**) smear; make or become indistinct.

blurb n. a written description praising something.

blurt v. utter abruptly or tactlessly.

blush v. become red-faced from shame or embarrassment. ● n. blushing; a pink tinge.

blusher n. a cosmetic giving a rosy colour to cheeks.

bluster v. **1** blow in gusts. **2** talk aggressively, with empty threats. ● n. blustering talk. □ **blustery** adj.

BMA abbr. British Medical Association.

BMX n. bicycle racing on a dirt track; a bicycle for this.

BO abbr. body odour.

boa (**boh-ă**) n. a large South American snake that crushes its prey.

boar n. a wild pig; a male pig.

board n. **1** a long piece of sawn wood; a flat piece of wood or stiff material. **2** daily meals supplied in return for payment or services. **3** a committee. ● v. **1** cover or block with boards. **2** enter (a ship, aircraft, or vehicle). **3** provide with or receive meals and accommodation for payment. □ **on board** on or in a ship, aircraft, or vehicle.

boarder *n.* a person who boards with someone; a resident pupil.

boarding house *n.* a house at which board and lodging can be obtained for payment.

boarding school *n.* one where pupils receive board and lodging.

boardroom *n.* a room where a board of directors meets.

boast *v.* speak with great pride, trying to impress people; be the proud possessor of. ● *n.* a boastful statement; a thing one is proud of.

boastful *adj.* boasting frequently. □ **boastfully** *adv.*, **boastfulness** *n.*

boat *n.* a vessel for travelling on water.

boater *n.* a flat-topped straw hat.

boathouse *n.* a shed at the water's edge for boats.

boating *n.* rowing or sailing for pleasure.

boatman *n.* (*pl.* **-men**) a man who rows, sails, or rents out boats.

boatswain (boh-sŭn) (also **bosun**, **bo'sun**) *n.* a ship's officer in charge of rigging, boats, etc.

bob *v.* (**bobbed**) **1** move quickly up and down. **2** cut (hair) in a bob. ● *n.* **1** a bobbing movement. **2** a hairstyle with the hair at the same length just above the shoulders.

bobbin *n.* a small spool holding thread or wire in a machine.

bobble *n.* a small woolly ball as an ornament.

bobsleigh *n.* a sledge with two sets of runners with mechanical steering.

bode *v.* be a sign of, promise.

bodice *n.* part of a dress from shoulder to waist; an under-

garment for this part of the body.

bodily *adj.* of the human body or physical nature. ● *adv.* in person, physically; as a whole.

body *n.* **1** the structure of bones and flesh etc. of man or animal; a corpse. **2** the main part. **3** a group regarded as a unit. **4** a separate piece of matter, a distinct object. **5** a strong texture or quality.

body blow *n.* a severe setback.

bodyguard *n.* an escort or personal guard of an important person.

bodysuit *n.* a close-fitting one-piece stretch garment for women.

bodywork *n.* the outer shell of a motor vehicle.

Boer *n.* a South African of Dutch descent.

boffin *n.* (*colloquial*) a person engaged in scientific research.

bog *n.* permanently wet spongy ground. ● *v.* (**bogged**) make or become stuck and unable to progress. □ **bogginess** *n.*, **boggy** *adj.*

bogey *n.* (*pl.* **bogeys**) **1** (*Golf*) a score of one stroke over par at a hole. **2** (also **bogy**) something causing fear.

boggle *v.* be bewildered.

bogie *n.* an undercarriage on wheels, pivoted at each end.

bogus *adj.* false.

bohemian *adj.* socially unconventional.

boil *v.* bubble up with heat; boil so that liquid does this. ● *n.* an inflamed swelling producing pus.

boiler *n.* **1** a container in which water is heated. **2** a fowl too tough to roast.

boiler suit *n.* a one-piece garment for rough work.

boisterous *adj.* noisy and cheerful. □ **boisterously** *adv.*

bold *adj.* confident and courageous; (of colours) strong and vivid. □ **boldly** *adv.*, **boldness** *n.*

bole *n.* the trunk of a tree.

bolero *n.* (*pl.* **boleros**) **1** a Spanish dance. **2** a woman's short jacket with no fastening.

boll *n.* a round seed vessel of cotton, flax, etc.

bollard *n.* a short thick post.

bollocks (*vulgar slang*) *n.pl.* testicles. ● *int.* rubbish, nonsense.

boloney var. of **baloney**.

bolster *n.* a long pad placed under a pillow. ● *v.* support, prop.

bolt *n.* **1** a sliding bar for fastening a door; a strong metal pin used with a nut to hold things together. **2** a shaft of lightning. **3** a roll of cloth. **4** an arrow from a crossbow. ● *v.* **1** fasten with a bolt. **2** run away. **3** gulp (food) hastily.

bolt-hole *n.* a place into which one can escape.

bomb *n.* a case of explosive or incendiary material to be set off by impact or a timing device. ● *v.* attack with bombs.

bombard *v.* attack with artillery; attack with questions etc. □ **bombardment** *n.*

bombardier *n.* an artillery non-commissioned officer.

bombastic *adj.* using pompous words.

bomber *n.* an aircraft that carries and drops bombs; a person who places bombs.

bombshell *n.* a great shock.

bona fide (boh-nă fy-di) *adj.* genuine.

bonanza *n.* sudden great wealth or luck.

bond *n.* **1** something that unites or restrains; a binding agreement; an emotional link. **2** a document issued by a government or public company ac-knowledging that money has been lent to it and will be repaid with interest. **3** high-quality writing paper. ● *v.* unite with a bond. □ **in bond** stored in a customs warehouse until duties are paid.

bondage *n.* slavery, captivity.

bone *n.* each of the hard parts making up the vertebrate skeleton. ● *v.* remove bones from.

bone china *n.* fine china made of clay and bone ash.

bonfire *n.* a fire built in the open air.

bongo *n.* (*pl.* **bongos** or **bongoes**) each of a pair of small drums played with the fingers.

bonhomie (bon-ŏmi) *n.* geniality.

bonk *v.* **1** make an abrupt thudding sound; bump. **2** (*slang*) have sexual intercourse (with). ● *n.* a thudding sound.

bonnet *n.* **1** a hat with strings that tie under the chin. **2** a hinged cover over the engine of a motor vehicle.

bonny *adj.* (**bonnier**) healthy-looking; (*Scot.* & *N. Engl.*) good-looking.

bonsai *n.* (*pl.* **bonsai**) an ornamental miniature tree or shrub; the art of growing these.

bonus *n.* an extra payment or benefit.

bon voyage (bon vwa-**yah**zh) *int.* an expression of good wishes to a person starting a journey.

bony *adj.* (**bonier**) like bones; having bones with little flesh.

boo *int.* an exclamation of disapproval; an exclamation to startle someone. ● *v.* shout 'boo' (at).

boob *n.* (*slang*) **1** a blunder. **2** a breast.

booby *n.* a foolish person.

booby prize n. one given to the competitor with the lowest score.

booby trap n. a hidden trap as a practical joke; a hidden bomb.

boogie n. jazz piano music with a persistent bass rhythm. ● v. dance to this.

book n. 1 a set of sheets of paper bound in a cover; a literary work filling this; a main division of a literary work. 2 a record of bets made. ● v. 1 reserve or buy in advance. 2 engage a performer. 3 take details of (an offender). 4 enter in a book or list.

bookcase n. a piece of furniture with shelves for books.

bookie n. (colloquial) a bookmaker.

bookkeeping n. the systematic recording of business transactions.

booklet n. a small thin book.

bookmaker n. a person whose business is the taking of bets.

bookmark n. a strip of paper etc. to mark a place in a book.

bookworm n. 1 a person fond of reading. 2 a grub that eats holes in books.

boom v. 1 make a deep resonant sound. 2 have a period of prosperity. ● n. 1 a booming sound. 2 prosperity. 3 a long pole; a floating barrier.

boomerang n. an Australian missile of curved wood that can be thrown so as to return to the thrower.

boon n. a benefit.

boon companion n. a favourite companion.

boor n. an ill-mannered person. □ **boorish** adj., **boorishness** n.

boost v. push upwards; increase the strength or reputation of. ● n. an upward thrust; an increase. □ **booster** n.

boot n. 1 a shoe covering both foot and ankle. 2 the luggage compartment in a car. 3 (**the boot**) (colloquial) dismissal. ● v. kick.

bootee n. a baby's woollen boot.

booth n. a small shelter.

bootleg adj. smuggled, illicit. □ **bootlegger** n., **bootlegging** n.

booty n. loot.

booze (colloquial) v. drink alcohol. ● n. alcoholic drink; a drinking spree. □ **boozer** n., **boozy** adj.

borage n. a blue-flowered plant.

borax n. a compound of boron used in detergents.

border n. an edge, a boundary; a flower bed round part of a garden. ● v. put or be a border to. □ **border on** come close to being.

borderline n. a line marking a boundary.

bore see **bear**. v. 1 weary by dullness. 2 make (a hole) with a revolving tool. ● n. 1 a boring person or thing. 2 a hole bored; the hollow inside of a cylinder; its diameter. 3 a tidal wave in an estuary. □ **boredom** n.

boric adj. of boron.

born adj. brought forth by birth; having a specified natural quality.

born-again adj. converted to a religion (esp. Christianity).

borne see **bear**.

boron n. a chemical element (symbol B) very resistant to high temperatures.

borough n. a town or district with rights of local government.

borrow v. get temporary use of (a thing or money). □ **borrower** n.

Borstal n. the former name of an institution for young offenders.

bortsch n. (also **borsch**) beetroot soup.

borzoi n. a large Russian wolfhound.

bosom n. the breast.

bosom friend | bow

bosom friend n. a very dear friend.

boss n. **1** (colloquial) a person in charge of workers. **2** a projecting knob. ● v. (colloquial) be the boss of; give orders to.

boss-eyed adj. blind in one eye; cross-eyed.

bosun, **bo'sun** vars. of **boatswain**.

bossy adj. (**bossier**) fond of giving orders to people. □ **bossily** adv., **bossiness** n.

botany n. the study of plants. □ **botanical** adj., **botanist** n.

botch v. spoil by poor work.

both adj., pron., & adv. the two.

bother v. **1** cause trouble, worry, or annoyance to; pester. **2** take trouble, feel concern. ● n. worry, minor trouble. □ **bothersome** adj.

bottle n. a narrow-necked container for liquid. ● v. store in bottles; preserve in jars.

bottleneck n. an obstruction to an even flow of work, traffic, etc.

bottom n. the lowest part or place; the buttocks; the ground under a stretch of water. ● adj. lowest in position, rank, or degree.

bottomless adj. extremely deep.

botulism n. poisoning by bacteria in food.

bougainvillea n. a tropical shrub with large red or purple bracts.

bough n. a large branch coming from the trunk of a tree.

bought see **buy**.

bouillon (boo-yawn) n. thin clear soup.

boulder n. a large rounded stone.

boulevard n. a wide street.

bounce v. **1** rebound; move in a lively manner, spring. **2** (colloquial) (of a cheque) be sent back by a bank as worthless. ● n. a bouncing movement or power; a lively manner.

bouncer n. (colloquial) a person employed to eject troublemakers.

bound see **bind**. v. **1** run with a jumping movement. **2** limit, be a boundary of. ● n. **1** a bounding movement. **2** (usu. **bounds**) the limit of something. ● adj. going in a specified direction. □ **bound to** certain to. **out of bounds** beyond the permitted area.

boundary n. a line that marks a limit; a hit to the boundary in cricket.

bounden duty n. duty dictated by conscience.

boundless adj. without limits.

bountiful adj. giving generously; abundant. □ **bountifully** adv.

bounty n. generosity; a generous gift. □ **bounteous** adj.

bouquet (boo-kay) n. **1** a bunch of flowers. **2** the perfume of wine.

bouquet garni (boo-kay gar-ni) n. (pl. **bouquets garnis**) a bunch of herbs for flavouring stews etc.

bourbon (ber-bŏn) n. an American whisky made mainly from maize.

bourgeois (boor-zhwah) adj. conventionally middle-class.

bourgeoisie (boor-zhwah-zi) n. the bourgeois class.

bout n. **1** a period of exercise, work, or illness. **2** a boxing contest.

boutique n. a small shop selling fashionable clothes etc.

bovine adj. **1** of oxen or cattle. **2** dull and stupid.

bow¹ (boh) n. **1** a weapon for shooting arrows. **2** a rod with horsehair stretched between its ends, for playing a violin etc. **3** a knot with loops in a ribbon or string.

bow² (bow) n. bending of the head or body in greeting, respect, etc. ● v. bend in this way;

bend downwards under weight; submit.

bow² (bow) *n.* the front end of a boat or ship.

bowdlerize *v.* (also **-ise**) remove sections considered improper from (a book etc.). □ **bowdlerization** *n.*

bowel *n.* the intestine; (**bowels**) the innermost parts.

bower *n.* a leafy shelter.

bowie knife *n.* a hunting knife with a long curved blade.

bowl¹ *n.* a basin; the hollow rounded part of a spoon etc. **2** a heavy ball weighted to roll in a curve; (**bowls**) a game played with such balls; a ball used in skittles. ● *v.* **1** send rolling along the ground; go fast and smoothly. **2** send a ball to a batsman, dismiss by knocking bails off with the ball. □ **bowl over** knock down; overwhelm with surprise or emotion.

bowler *n.* **1** a person who bowls in cricket; one who plays at bowls. **2** (in full **bowler hat**) a hard felt hat with a rounded top.

bowling *n.* playing bowls, skittles, or a similar game.

box *n.* **1** a container or receptacle with a flat base. **2** a numbered receptacle at a newspaper office for holding replies to an advertisement. **3** a compartment in a theatre, stable, etc. **4** a small evergreen shrub; its wood. ● *v.* **1** put into a box. **2** fight with the fists as a sport. □ **boxing** *n.*

boxer *n.* **1** a person who engages in the sport of boxing. **2** a dog of a breed resembling a bulldog.

boxer shorts *n.pl.* men's loose underpants like shorts.

box office *n.* an office for booking seats at a theatre etc.

boxroom *n.* a small spare room.

boy *n.* a male child. □ **boyhood** *n.*, **boyish** *adj.*

boycott *v.* refuse to deal with or trade with. ● *n.* boycotting.

boyfriend *n.* a person's regular male companion or lover.

Bq *abbr.* becquerel.

Br *symb.* bromine.

bra *n.* a woman's undergarment worn to support the breasts.

brace *n.* **1** a device that holds things together or in position; (**braces**) straps to keep trousers up, passing over the shoulders. **2** a pair. ● *v.* give support or firmness to.

bracelet *n.* an ornamental band worn on the arm.

bracing *adj.* invigorating.

bracken *n.* a large fern that grows on open land; a mass of such ferns.

bracket *n.* **1** a projecting support. **2** any of the marks used in pairs for enclosing words or figures, (), [], { }. ● *v.* enclose by brackets; put together as similar.

brackish *adj.* slightly salt.

bract *n.* a brightly coloured leaflike part of a plant.

brag *v.* (**bragged**) boast.

braggart *n.* a person who brags.

Brahman *n.* (also **Brahmin**) a member of the highest Hindu caste, the priestly caste.

braid *n.* a woven ornamental trimming; a plait of hair. ● *v.* trim with braid; plait.

Braille *n.* a system of representing letters etc. by raised dots which blind people read by touch.

brain *n.* the mass of soft grey matter in the skull, the centre of the nervous system in animals; (also **brains**) the mind, intelligence.

brainchild *n.* a person's invention or plan.

brainstorm *n.* a violent mental disturbance; a sudden mental lapse; (*Amer.*) a bright idea; a

brainwash | break

spontaneous discussion in search of new ideas.

brainwash v. force (a person) to change his or her views by exerting great mental pressure.

brainwave n. a bright idea.

brainy adj. (**brainier**) clever.

braise v. cook slowly with little liquid in a closed container.

brake n. a device for reducing speed or stopping motion. ● v. slow by the use of this.

bramble n. a shrub with long prickly shoots, a blackberry.

bran n. the ground inner husks of grain, sifted from flour.

branch n. 1 an arm-like part of a tree; a similar part of a road, river, etc. 2 a subdivision of a subject. 3 a local shop or office belonging to a large organization. ● v. send out or divide into branches.

brand n. goods of a particular make; a mark of identification made with hot metal. ● v. mark with a brand.

brandish v. wave, flourish.

brand new adj. completely new, unused.

brandy n. a strong alcoholic spirit distilled from wine or fermented fruit juice.

brash adj. vulgarly self-assertive. □ **brashly** adv., **brashness** n.

brass n. a yellow alloy of copper and zinc; musical instruments made of this. ● adj. made of brass.

brasserie n. an informal licensed restaurant.

brassica n. a plant of the cabbage family.

brassière (bras-i-air) n. a bra.

brassy adj. (**brassier**) 1 like brass. 2 bold and vulgar. □ **brassiness** n.

brat n. (derog.) a child.

bravado n. a show of boldness.

brave adj. able to face and endure danger or pain; daring, spectacular. ● v. face and en-

dure bravely. □ **bravely** adv., **bravery** n.

bravo int. well done!

bravura n. a brilliant performance; a style of music requiring brilliant technique.

brawl n. a noisy quarrel or fight. ● v. take part in a brawl.

brawn n. 1 muscular strength. 2 pressed meat from a pig's or calf's head.

brawny adj. (**brawnier**) muscular.

bray n. a donkey's cry; a similar sound. ● v. make this cry or sound.

braze v. solder with an alloy of brass.

brazen adj. 1 like or made of brass. 2 shameless, impudent. □ **brazen it out** behave (after doing wrong) as if one has no need to be ashamed. □ **brazenly** adv.

brazier n. a basket-like stand for holding burning coals.

breach n. breaking or neglect of a rule or contract; estrangement; a broken place, a gap. ● v. break through, make a breach in.

bread n. food made of baked dough of flour and liquid, usu. leavened by yeast.

breadline n. □ **on the breadline** living in extreme poverty.

breadth n. width, broadness.

breadwinner n. the member of a family who earns money to support the others.

break v. (**broke**, **broken**, **breaking**) 1 divide or separate otherwise than by cutting; fall into pieces. 2 damage; become unusable. 3 fail to keep (a promise or law). 4 make or become discontinuous. 5 make a way suddenly or violently; appear suddenly. 6 reveal (news). 7 surpass (a record). 8 (of a ball) change direction after touching the ground. ● n. 1 breaking. 2 a gap; an interval. 3 a sudden dash. **4** points scored continuously in snooker. **5** (*colloquial*) an opportunity, a piece of luck. □ **break down** fail, collapse; give way to emotion; analyse. **break even** make gains and losses that balance exactly.

breakable *adj.* able to be broken.

breakage *n.* breaking.

breakdown *n.* **1** a mechanical failure; a collapse of health or mental stability. **2** an analysis.

breaker *n.* a heavy ocean wave that breaks on the coast.

breakfast *n.* the first meal of the day.

breakneck *adj.* dangerously fast.

breakthrough *n.* breaking through; a major advance in knowledge or negotiation.

breakwater *n.* a wall built out into the sea to break the force of waves.

bream *n.* a fish of the carp family.

breast *n.* the upper front part of the body; either of the two milk-producing organs on a woman's chest.

breastbone *n.* the bone down the upper front of the body.

breaststroke *n.* a swimming stroke performed face downwards.

breath *n.* air drawn into and sent out of the lungs in breathing; breathing in; gentle blowing. □ **out of breath** panting after exercise. **under one's breath** in a whisper. □ **breathy** *adj.*

breathalyse *v.* (*Amer.* **breathalyze**) test (a person) with a breathalyser.

breathalyser *n.* (*Amer.* **breathalyzer**) (*trade mark*) a device measuring the alcohol in a person's breath.

breathe *v.* **1** draw (air) into the lungs and send it out again. **2** utter.

breather *n.* a pause for rest; a short period in fresh air.

breathless *adj.* out of breath.

breathtaking *adj.* amazing.

bred *see* **breed**.

breech *n.* the back part of a gun barrel, where it opens.

breeches *n.pl.* trousers reaching to just below the knees.

breed *v.* (**bred**, **breeding**) produce offspring; train, bring up; give rise to. ● *n.* a variety of animals etc. within a species; a sort. □ **breeder** *n.*

breeding *n.* good manners resulting from training or background.

breeze *n.* a light wind. □ **breezy** *adj.*

breeze-block *n.* a lightweight building block.

brethren *n.pl.* (*old use*) brothers.

Breton *adj.* & *n.* (a native) of Brittany.

breve *n.* **1** a mark (ˇ) over a short vowel. **2** (in music) a long note.

breviary *n.* a book of prayers to be said by Roman Catholic priests.

brevity *n.* briefness.

brew *v.* make (beer) by boiling and fermentation; make (tea) by infusion; bring about, develop. ● *n.* a liquid or amount brewed.

brewer *n.* a person whose trade is brewing beer.

brewery *n.* a business that brews beer.

briar *var.* of **brier**.

bribe *n.* something offered to influence a person to act in favour of the giver. ● *v.* persuade by this. □ **bribery** *n.*

bric-a-brac *n.* odd items of ornaments, furniture, etc.

brick *n.* a block of baked or dried clay used to build walls; a rectangular block. ● v. block with a brick structure.

brickbat n. a missile hurled at someone; a criticism.

bricklayer n. a workman who builds with bricks.

bridal adj. of a bride or wedding.

bride n. a woman on her wedding day or when newly married.

bridegroom n. a man on his wedding day or when newly married.

bridesmaid n. a girl or unmarried woman attending a bride.

bridge n. **1** a structure providing a way across or joining something. **2** the captain's platform on a ship. **3** the bony upper part of the nose. **4** a card game developed from whist. ● v. make or be a bridge over; span as if with a bridge.

bridgehead n. a fortified area established in enemy territory, esp. on the far side of a river.

bridgework n. a dental structure covering a gap.

bridle n. a harness on a horse's head. ● v. put a bridle on; restrain; draw up one's head in pride or scorn.

bridle path (also **bridleway**) a path for riders or walkers.

brief adj. lasting only for a short time; concise; short. ● n. a set of instructions and information, esp. to a barrister about a case. ● v. employ (a barrister); inform or instruct in advance. □ **briefly** adv., **briefness** n.

briefcase n. a case for carrying documents.

briefs n.pl. short pants or knickers.

brier n. (also **briar**) a thorny bush, a wild rose.

brig n. a square-rigged sailing vessel with two masts.

brigade n. an army unit forming part of a division.

brigadier n. an officer commanding a brigade or of similar status.

brigand n. a member of a band of robbers.

bright adj. **1** giving out or reflecting much light, shining. **2** cheerful. **3** quick-witted, clever. □ **brightly** adv., **brightness** n.

brighten v. make or become brighter.

brilliant adj. **1** very bright, sparkling. **2** very clever. ● n. a cut diamond with many facets. □ **brilliantly** adv., **brilliance** n.

brim n. the edge of a cup or hollow; the projecting edge of a hat. ● v. (**brimmed**) be full to the brim.

brimstone n. (old use) sulphur.

brindled adj. brown with streaks of another colour.

brine n. salt water; sea water.

bring v. (**brought, bringing**) convey; cause to come. □ **bring about** cause to happen. **bring off** do successfully. **bring out** show clearly; publish. **bring up** look after and train (growing children).

brink n. the edge of a steep place or of a stretch of water; the point just before a change.

brinkmanship n. a policy of pursuing a dangerous course to the brink of catastrophe.

briny adj. of brine or sea water; salty.

briquette n. a block of compressed coal dust.

brisk adj. lively, moving quickly. □ **briskly** adv.

brisket n. a joint of beef from the breast.

bristle n. a short stiff hair; one of the stiff pieces of hair or wire in a brush. ● v. raise bristles in anger or fear; show indignation; be thickly set with bristles.

Britannic adj. of Britain.

British adj. of Britain or its people.

Briton n. a British person.

brittle adj. hard but easily broken. □ **brittleness** n.

broach v. open and start using; begin discussion of.

broad adj. **1** large across; wide; full and complete. **2** in general terms. **3** (of humour) rather coarse. **4** tolerant, liberal. □ **broadly** adv.

broad bean n. an edible bean with flat seeds.

broad-minded adj. having tolerant views.

broadcast v. (**broadcast, broadcasting**) send out by radio or television; make generally known; sow (seed) by scattering. ●n. a broadcast programme. □ **broadcaster** n.

broaden v. make or become broader.

broadsheet n. a large-sized newspaper.

broadside n. the firing of all guns on one side of a ship.

brocade n. a fabric woven with raised patterns.

broccoli n. a vegetable with tightly-packed green or purple flower heads.

brochure (broh-sher) n. a booklet or leaflet giving information.

broderie anglaise n. open embroidery on white cotton or other fabric.

brogue n. **1** a strong shoe with ornamental perforated bands. **2** a strong regional accent, esp. Irish.

broil v. grill; make or become very hot.

broiler n. a chicken suitable for broiling.

broke see **break**. adj. (colloquial) having spent all one's money; bankrupt.

broken see **break**. adj. (of a language) badly spoken by a foreigner.

broken-hearted adj. overwhelmed with grief.

broker n. an agent who buys and sells on behalf of others; a member of the Stock Exchange dealing in stocks and shares.

brokerage n. a broker's fee.

brolly n. (colloquial) an umbrella.

bromide n. a compound used to calm nerves.

bromine n. a dark red poisonous liquid element (symbol Br).

bronchial adj. of the branched tubes into which the windpipe divides.

bronchitis n. inflammation of the bronchial tubes.

bronco n. (pl. **broncos**) a wild or half-tamed horse of western North America.

brontosaurus n. a large planteating dinosaur.

bronze n. a brown alloy of copper and tin; something made of this; its colour. ●v. make or become suntanned.

brooch (brohch) n. an ornamental hinged pin fastened with a clasp.

brood n. young produced at one hatching or birth. ●v. **1** sit on (eggs) and hatch them. **2** think long and deeply.

broody adj. **1** (of a hen) wanting to brood. **2** thoughtful and depressed.

brook n. a small stream. ●v. tolerate, allow.

broom n. **1** a long-handled brush. **2** a shrub with white, yellow, or red flowers.

broomstick n. a broom-handle.

Bros. abbr. Brothers.

broth n. a thin meat or fish soup.

brothel n. a house where women work as prostitutes.

brother n. **1** the son of the same parents as another person. **2** a man who is a fellow member of a group, trade union, or Church. **3** a monk who is not a priest. □ **brotherly** adj.

brotherhood n. the relationship of brothers; comradeship.

brother-in-law n. (pl. **brothers-in-law**) the brother of one's husband or wife; the husband of one's sister.

brought see **bring**.

brow n. an eyebrow; a forehead; a projecting or overhanging part.

browbeat v. (**browbeat, browbeaten, browbeating**) intimidate.

brown adj. of a colour between orange and black. ● v. make or become brown.

browned off adj. (colloquial) bored, fed up.

browse v. **1** read or look around casually. **2** feed on leaves or grass.

bruise n. an injury that discolours skin without breaking it. ● v. cause a bruise on.

bruiser n. a tough brutal person.

brunch n. a meal combining breakfast and lunch.

brunette n. a woman with brown hair.

brunt n. the chief stress or strain.

brush n. **1** an implement with bristles. **2** a fox's tail. **3** a skirmish. **4** brushing. **5** undergrowth. ● v. use a brush on; touch lightly in passing. □ **brush off** reject curtly; snub. **brush up** smarten; study and revive one's knowledge of.

brushwood n. undergrowth; cut or broken twigs.

brusque (broosk) adj. curt and offhand. □ **brusquely** adv.

Brussels sprout n. the edible bud of a kind of cabbage.

brutal adj. cruel, without mercy. □ **brutally** adv., **brutality** n.

brutalize v. (also **-ise**) make brutal; treat brutally. □ **brutalization** n.

brute n. an animal other than man; a brutal person; (colloquial) an unpleasant person or thing. ● adj. unable to reason; unreasoning. □ **brutish** adj.

BS abbr. British Standard(s).

BSE abbr. bovine spongiform encephalopathy; mad cow disease (a fatal disease of cattle).

BST abbr. British Summer Time.

Bt. abbr. Baronet.

bubble n. a thin ball of liquid enclosing air or gas; an air-filled cavity. ● v. **1** rise in bubbles. **2** show great liveliness. □ **bubbly** adj.

bubonic adj. (of plague) characterized by swellings.

buccaneer n. a pirate; an adventurer.

buck n. **1** the male of a deer, hare, or rabbit. **2** an article placed before the dealer in a game of poker. **3** (Amer. & Austral. slang) a dollar. ● v. (of a horse) jump with the back arched. □ **buck up** (colloquial) hurry up; cheer up. **pass the buck** shift the responsibility (and possible blame).

bucket n. an open container with a handle, for carrying or holding liquid. ● v. pour heavily.

buckle n. a device through which a belt or strap is threaded to secure it. ● v. **1** fasten with a buckle. **2** crumple under pressure. □ **buckle down to** set about doing.

buckram n. stiffened cloth for binding books.

buckwheat n. a cereal plant; its seed.

bucolic adj. rustic.

bud n. a leaf or flower not fully open. ● v. (**budded**) put forth buds; begin to develop.

Buddhism n. an Asian religion based on the teachings of Buddha. □ **Buddhist** adj. & n.

budding adj. beginning to develop or be successful.

buddleia n. a tree or shrub with purple or yellow flowers.

buddy n. (colloquial) a friend.

budge v. move slightly.

budgerigar n. an Australian parakeet.

budget n. a plan of income and expenditure; an amount allowed. ● v. allow or arrange in a budget.

budgie n. (colloquial) a budgerigar.

buff n. **1** a fawn colour. **2** bare skin. **3** (colloquial) an enthusiast. ● v. polish with soft material.

buffalo n. (pl. **buffaloes** or **buffalo**) a wild ox; a North American bison.

buffer n. **1** something that lessens the effect of impact. **2** (slang) a person. ● v. act as a buffer to.

buffet¹ (buuf-ay) n. a counter where food and drink are served; a meal where guests serve themselves.

buffet² (buff-it) n. a blow, esp. with a hand. ● v. (**buffeted**) deal blows to.

buffoon n. a person who plays the fool. □ **buffoonery** n.

bug n. **1** a small unpleasant insect; (colloquial) a microorganism causing illness. **2** (slang) a secret microphone. **3** a defect. ● v. (**bugged**) (slang) **1** install a secret microphone in. **2** annoy.

bugbear n. something feared or disliked.

bugger n. (vulgar slang) int. damn. ● n. something unpleasant or difficult.

buggery n. sodomy.

buggy n. a light carriage; a small sturdy vehicle; a lightweight folding pushchair.

bugle n. a brass instrument like a small trumpet. □ **bugler** n.

build v. (**built**, **building**) construct by putting parts or material together. ● n. bodily shape. □ **build up** establish gradually; increase. □ **builder** n.

building n. a house or similar structure.

building society n. an organization that accepts deposits and lends money, esp. to people buying houses.

built-in adj. forming part of a structure.

built-up adj. covered with buildings.

bulb n. **1** the rounded base of the stem of certain plants. **2** something (esp. an electric lamp) shaped like this. □ **bulbous** adj.

bulge n. a rounded swelling. ● v. form a bulge, swell.

bulimia n. (in full **bulimia nervosa**) an eating disorder in which overeating alternates with self-induced vomiting, dieting, or purging. □ **bulimic** adj.

bulk n. size; mass; the greater part; a bulky thing. ● v. increase the size or thickness of.

bulkhead n. a partition in a ship etc.

bulky adj. (**bulkier**) taking up much space.

bull n. **1** the male of the ox, whale, elephant, etc. **2** the bull's-eye of a target. **3** a pope's official edict. **4** (slang) an absurd statement.

bulldog n. a powerful dog with a short thick neck.

bulldozer n. a powerful tractor with a device for clearing ground.

bullet n. a small missile fired from a rifle or revolver.

bulletin n. a short official statement of news.

bullfight n. the baiting and killing of bulls as an entertainment.

bullfinch n. a songbird with a strong beak and pinkish breast.

bullion n. gold or silver in bulk or bars, before manufacture.

bullock n. a castrated bull.

bull's-eye n. the centre of a target.

bullshit n. (*vulgar slang*) rubbish, nonsense.

bull terrier n. a terrier resembling a bulldog.

bully n. a person who hurts or intimidates others. ● v. behave as a bully towards. □ **bully off** put the ball into play in hockey by two opponents striking sticks together.

bulrush n. a tall rush.

bulwark n. **1** a wall of earth built as a defence. **2** a ship's side above the deck.

bum n. (*slang*) **1** the buttocks. **2** a beggar, a loafer.

bumble v. move or act in a blundering way.

bumble-bee n. a large bee.

bumf n. (also **bumph**) (*colloquial*) documents, papers.

bump v. knock with a dull-sounding blow; travel with a jolting movement. ● n. a bumping sound or knock; a swelling, esp. left by a blow. □ **bumpy** adj.

bumper n. **1** something unusually large. **2** a horizontal bar at the front or back of a motor vehicle to lessen the effect of collision.

bumpkin n. a country person with awkward manners.

bumptious adj. conceited.

bun n. a small round sweet cake; hair twisted into a bun shape at the back of the head.

bunch n. a cluster; a number of small things fastened together. ● v. make into a bunch; form a group.

bundle n. a collection of things loosely fastened or wrapped together. ● v. **1** make into a bundle. **2** push hurriedly.

bung n. a stopper for the hole in a barrel or jar. ● v. (*slang*) throw, put.

bungalow n. a one-storeyed house.

bungee jumping n. the sport of jumping from a height attached to an elasticated rope (a **bungee**).

bungle v. spoil by lack of skill, mismanage. ● n. a bungled attempt. □ **bungler** n.

bunion n. a swelling at the base of the big toe, with thickened skin.

bunk n. a shelflike bed. □ **do a bunk** (*slang*) run away.

bunker n. **1** a container for fuel. **2** a sandy hollow forming a hazard on a golf course. **3** a reinforced underground shelter.

bunkum n. nonsense, humbug.

Bunsen burner n. a small adjustable gas burner used in laboratories.

bunting n. **1** a bird related to the finches. **2** decorative flags.

buoy (boy) n. an anchored floating object serving as a navigation mark. ● v. □ **buoy up** keep afloat; sustain, hearten.

buoyant (boy-ănt) adj. **1** able to float. **2** cheerful. □ **buoyancy** n.

bur n. (also **burr**) a plant's seed case that clings to clothing etc.

burble v. make a gentle murmuring sound; speak lengthily.

burden n. something carried; a heavy load or obligation; trouble. ● v. put a burden on.

bureau n. (bew-roh) (*pl.* **bureaux** or **bureaus**) **1** a writing desk with drawers. **2** an office, a department.

bureaucracy (bew-rok-ră-si) n. government by unelected officials; excessive administration. □ **bureaucratic** adj.

bureaucrat n. a government official.

burgeon v. begin to grow rapidly.

burger n. (*colloquial*) a hamburger.

burglar n. a person who breaks into a building in order to steal. □ **burglary** n.

burgle v. rob as a burglar.

burgundy n. a dark purplish red colour.

burial n. burying.

burlesque n. a mocking imitation.

burly adj. (**burlier**) with a strong heavy body. □ **burliness** n.

burn v. (**burned** or **burnt**, **burning**) be on fire; damage, destroy, or mark by fire, heat, or acid; use as fuel; produce heat or light; feel a sensation (as) of heat. ● n. 1 a mark or sore made by burning. 2 (Scot.) a stream.

burner n. a part that shapes the flame in a lamp, cooker, etc.

burning adj. intense; hotly discussed.

burnish v. polish by rubbing.

burnt see burn.

burp n. & v. (colloquial) (make) a belch.

burr n. 1 a whirring sound; the strong pronunciation of 'r'; a country accent using this. 2 var. of bur.

burrow n. a hole dug by a fox or rabbit as a dwelling. ● v. dig a burrow, tunnel; form by tunnelling; search deeply, delve.

bursar n. a person who manages the finances and other business of a college.

bursary n. a scholarship or grant given to a student; a bursar's office.

burst v. (**burst**, **bursting**) force or be forced open; fly violently apart; begin or appear or come suddenly. ● n. bursting; an outbreak; a brief violent effort, a spurt.

bury v. place (a dead body) in the earth or a tomb; put or hide underground; cover up; involve (oneself) deeply.

bus n. (pl. **buses**; Amer. **busses**) a long-bodied public passenger vehicle. ● v. (**buses** or **busses**, **bussed**, **bussing**) travel by bus; transport by bus.

busby n. a tall fur hat.

bush n. 1 a shrub; thick growth; wild uncultivated land. 2 a metal lining for a hole in which something fits; an insulating sleeve.

bushy adj. (**bushier**) covered with bushes; growing thickly.

business n. an occupation, a trade; a task, a duty; something to be dealt with; buying and selling, trade; a commercial establishment.

businesslike adj. practical, systematic.

businessman n. (pl. **-men**) a man engaged in trade or commerce.

businesswoman n. (pl. **-women**) a woman engaged in trade or commerce.

busk v. perform as a busker.

busker n. an entertainer performing in the street.

busman's holiday n. leisure time spent doing something similar to one's work.

bust n. a sculptured head, shoulders, and chest; the bosom. ● v. (**busted** or **bust**, **busting**) (colloquial) burst, break. □ **go bust** (colloquial) become bankrupt.

bustle v. make a show of activity or hurry. ● n. 1 excited activity. 2 (old use) padding to puff out the top of a skirt at the back.

bust-up n. (colloquial) a quarrel.

busy adj. (**busier**) working, occupied; having much to do; full of activity. ● **busily** adv.

busybody n. a meddlesome person.

but adv. only. ● conj. however. ● prep. except.

butane n. an inflammable gas used in liquid form as fuel.

butch n. (slang) strongly masculine.

butcher n. a person who cuts up and sells animal flesh for food; one who butchers people. ● v. kill needlessly or brutally. □ **butchery** n.

butler n. a chief manservant, in charge of the wine cellar.

butt n. 1 a large cask or barrel. 2 the thicker end of a tool or weapon; a short remnant, a stub. 3 a mound behind a target; (**butts**) a shooting range. 4 a target for ridicule or teasing. ● v. 1 push with the head. 2 meet or place edge to edge. □ **butt in** interrupt; meddle.

butter n. a fatty food substance made from cream. ● v. spread with butter. ● **butter up** flatter.

buttercup n. a wild plant with yellow cup-shaped flowers.

butterfly n. an insect with four large often brightly coloured wings; a swimming stroke with both arms lifted at the same time.

buttermilk n. liquid left after butter is churned from milk.

butterscotch n. a hard toffee-like sweet.

buttock n. either of the two fleshy rounded parts at the lower end of the back of the body.

button n. a disc or knob sewn to a garment as a fastener or ornament; a small rounded object; a knob etc. pressed to operate a device. ● v. fasten with button(s).

buttonhole n. a slit through which a button is passed to fasten clothing; a flower worn in the buttonhole of a lapel. ● v. accost and talk to.

buttress n. a support built against a wall; something that supports. ● v. reinforce, prop up.

butty n. (colloquial) a sandwich.

buxom adj. plump and healthy.

buy v. (**bought, buying**) obtain in exchange for money. □ a purchase. ● n. a purchase. ● **buyer** n.

buyout n. the purchase of a controlling share in a company; the buying of a company by people who work for it.

buzz n. 1 a vibrating humming sound. 2 a rumour. 3 a thrill. ● v. 1 make or be filled with a buzz. 2 go about busily. 3 threaten (an aircraft) by flying close to it.

buzzard n. a large hawk.

buzzer n. a device that produces a buzzing sound as a signal.

buzzword n. (colloquial) a fashionable technical word.

by prep. & adv. near, beside, in reserve; along, via, past; during; through the agency or means of; not later than. ● **by and by** before long. **by and large** on the whole. **by oneself** alone, without help.

bye n. 1 a run scored from a ball not hit by the batsman. 2 having no opponent for one round of a tournament.

by-election n. an election of an MP to replace one who has died or resigned.

bygone adj. belonging to the past. ● n.pl. (**bygones**) bygone things.

by-law n. a regulation made by a local authority or corporation.

byline n. a line naming the writer of a newspaper article.

bypass n. a road taking traffic round a town. ● v. provide with a bypass; use a bypass round; avoid.

by-product n. something produced while making something else.

byre n. a cowshed.

byroad n. a minor road.

bystander n. a person standing near when something happens.

byte n. (Computing) a fixed number of bits.

byway n. a minor road.

byword n. a notable example; a familiar saying.

....

Cc

C abbr. **1** Celsius; centigrade. **2** coulomb(s). ● symb. carbon. ● n. (as a Roman numeral) 100.

c. abbr. **1** century. **2** cent(s).

c. abbr. circa, about.

Ca symb. calcium.

cab n. **1** a taxi. **2** a compartment for the driver of a train, lorry, etc.

cabal n. (the people involved in) a secret plot.

cabaret (kab-áray) n. entertainment provided in a nightclub etc.

cabbage n. a vegetable with a round head of green or purple leaves.

cabby n. (colloquial) a taxi driver.

caber n. a roughly trimmed tree trunk.

cabin n. **1** a small hut. **2** a compartment in a ship or aircraft.

cabinet n. **1** a cupboard with drawers or shelves. **2** (**the Cabinet**) the group of ministers chosen to be responsible for government policy.

cabinetmaker n. a maker of high-quality furniture.

cable n. **1** a thick rope of fibre or wire; a set of insulated wires for carrying electricity or signals. **2** a telegram.

cable car n. a small cabin on a loop of cable for carrying passengers up and down mountains etc.

cable television n. television transmission by cable to subscribers.

cabriolet n. a car with a folding top; a light two-wheeled carriage with a hood.

cacao n. the seed from which cocoa and chocolate are made; the tree producing this.

cache (kash) n. a hiding place for treasure or stores; things in this. ● v. put into a cache.

cachet (kash-ay) n. prestige; a distinctive mark or characteristic.

cackle n. the clucking of hens; chattering talk; a loud silly laugh. ● v. utter a cackle.

cacophony n. a harsh discordant sound. □ **cacophonous** adj.

cactus n. (pl. **cacti** or **cactuses**) a fleshy plant, often with prickles, from a hot dry climate.

cad n. (old use) a dishonourable man.

cadaver n. a corpse.

cadaverous adj. gaunt and pale.

caddie n. (also **caddy**) a golfer's attendant carrying clubs. ● v. act as caddie.

caddis-fly n. a four-winged insect living near water.

caddy n. a small box for tea.

cadence n. rhythm in sound; the rise and fall of the voice in speech; the end of a musical phrase.

cadenza n. an elaborate passage for a solo instrument or singer. □

cadet n. a young trainee in the armed forces or police.

cadge v. ask for as a gift, beg.

cadmium n. a metallic element (symbol Cd).

cadre n. a small group forming a nucleus that can be expanded.

caecum (see-kŭm) n. (Amer. **cecum**) (pl. **caeca**) a small pouch at the first part of the large intestine.

Caesarean section n. (Amer. **Cesarean, Cesarian**) an operation to deliver a child by an incision through the walls of the mother's abdomen and womb.

caesium (see-zi-ŭm) n. (Amer. **cesium**) a soft metallic element (symbol Cs).

café n. a small informal tea shop or restaurant.

cafeteria n. a self-service restaurant.

cafetière (ka-fě-tyair) n. a coffee pot with a plunger to keep the ground coffee separate from the liquid.

caffeine n. a stimulant found in tea and coffee.

caftan n. (also **kaftan**) a long loose robe or dress.

cage n. an enclosure of wire or with bars, esp. for birds or animals.

cagey adj. (**cagier**) (colloquial) secretive; shrewd; wary. □ **cagily** adv., **caginess** n.

cagoule n. a light hooded waterproof jacket.

cahoots n. (slang) □ **in cahoots with** in league with.

caiman var. of **cayman**.

cairn n. a mound of stones as a memorial or landmark.

caisson n. a watertight chamber used in underwater construction work.

cajole v. coax. □ **cajolery** n.

Cajun adj. in the style of French Louisiana.

cake n. **1** a baked sweet bread-like food. **2** a small flattened mass.

calabrese n. a variety of broccoli.

calamine n. a skin lotion containing zinc carbonate.

calamity n. a disaster. ● **calamitous** adj., **calamitously** adv.

calcify v. harden by a deposit of calcium salts. □ **calcification** n.

calcium n. a whitish metallic element (symbol Ca).

calculate v. reckon mathematically; estimate; plan deliberately. □ **calculation** n.

calculating adj. shrewd; scheming.

calculator n. an electronic device for making calculations.

calculus n. (pl. **calculi** or **calculuses**) **1** a branch of mathematics dealing with rates of variation. **2** a stone formed in the body.

Caledonian adj. of Scotland.

calendar n. a chart showing dates of days of the year.

calendar year n. 1 January to 31 December inclusive.

calf n. (pl. **calves**) **1** the fleshy part of the human leg below the knee. **2** the young of cattle, also of the elephant, whale, and seal.

calibrate v. mark or correct the units of measurement on (a gauge); find the calibre of. □ **calibration** n.

calibre n. (Amer. **caliber**) **1** a level of ability or importance. **2** the diameter of a gun, tube, or bullet.

calico n. a cotton cloth.

californium n. a radioactive metallic element (symbol Cf).

caliper var. of **calliper**.

caliph n. (formerly) a Muslim ruler.

calisthenics var. of **callisthenics**.

calk var. of **caulk**.

call v. **1** shout to attract attention; utter a characteristic cry. **2** summon; command, invite; rouse from sleep. **3** communicate (with) by telephone or radio. **4** make a brief visit. **5** name; describe or address as. ● n. **1** a shout; a bird's cry. **2** an invitation, a demand; a need; a vocation. **3** a telephone communication. **4** a short visit. □ **call off** cancel. □ **caller** n.

call box n. a telephone kiosk.

calligraphy n. (beautiful) handwriting. □ **calligrapher** n.

calliper n. (also **caliper**) a metal support for a weak leg.

callisthenics *n.pl.* (also **calisthenics**) exercises to develop strength and grace.

callous *adj.* feeling no pity or sympathy. □ **callously** *adv.*, **callousness** *n.*

callow *adj.* immature and inexperienced.

callus *n.* a patch of hardened skin.

calm *adj.* still, not windy; not excited or agitated. ● *n.* a calm condition. ● *v.* make calm. □ **calmly** *adv.*, **calmness** *n.*

Calor gas *n.* (*trade mark*) liquefied butane stored under pressure in containers.

calorie *n.* a unit of heat; a unit of the energy-producing value of food.

calorific *adj.* heat-producing.

calumniate *v.* slander.

calumny *n.* slander.

calve *v.* give birth to a calf.

Calvinism *n.* the teachings of the Protestant reformer John Calvin or his followers. □ **Calvinist** *n.*

calypso *n.* (*pl.* **calypsos**) a topical West Indian song.

calyx *n.* (*pl.* **calyxes** or **calyces**) a ring of leaves (sepals) covering a flower bud.

cam *n.* a projecting part on a wheel or shaft changing rotary to to-and-fro motion.

camaraderie *n.* comradeship.

camber *n.* a slight convex curve given to a surface esp. of a road.

cambric *n.* thin linen or cotton cloth.

camcorder *n.* a combined video and sound recorder.

came *see* **come**.

camel *n.* a large animal with one hump or two; its fawn colour.

camellia *n.* an evergreen flowering shrub.

cameo *n.* (*pl.* **cameos**) **1** a stone in a ring or brooch with coloured layers carved in a raised design. **2** a small part in a play

or film taken by a famous actor or actress.

camera *n.* an apparatus for taking photographs or film pictures. □ **in camera** in private. □ **cameraman** *n.*

camiknickers *n.pl.* a woman's one-piece undergarment combining camisole and knickers.

camisole *n.* a woman's bodice-like garment or undergarment with shoulder straps.

camomile *n.* (also **chamomile**) an aromatic herb.

camouflage *n.* disguise, concealment, by colouring or covering. ● *v.* disguise or conceal in this way.

camp *n.* **1** a place with temporary accommodation in tents; a place where troops are lodged or trained. **2** a group of people with the same ideals. **3** affected or exaggerated behaviour. ● *v.* sleep in a tent; live in a camp. ● *adj.* **1** affected, exaggerated. **2** homosexual.

campaign *n.* an organized course of action; a series of military operations. ● *v.* conduct or take part in a campaign. □ **campaigner** *n.*

campanology *n.* the study of bells, bell-ringing. □ **campanologist** *n.*

campanula *n.* a plant with bell-shaped flowers.

camp bed *n.* a portable folding bed.

camper *n.* a person who is camping; a large vehicle with beds, cooking facilities, etc.

camphor *n.* a strong-smelling white substance used in medicine and mothballs.

campion *n.* a wild plant with pink or white flowers.

campsite *n.* a place for camping.

campus *n.* the grounds of a university or college.

CAMRA *abbr.* Campaign for Real Ale.

camshaft n. a shaft carrying cams.

can¹ n. a container in which food etc. is sealed and preserved. ● v. (**canned**) put or preserve in a can.

can² v.aux. (**can, could**) be able or allowed to.

Canadian adj. & n. (a native or inhabitant) of Canada.

canal n. an artificial water-course; a duct.

canalize v. (also **-ise**) convert into a canal; channel. □ **canalization** n.

canapé (kan-ăpay) n. a small piece of bread etc. with savoury topping.

canard n. a false rumour.

canary n. a small yellow songbird.

cancan n. a lively high-kicking dance performed by women.

cancel v. (**cancelled**; Amer. **canceled**) declare that (something arranged) will not take place; order to be discontinued; cross out; neutralize. □ **cancellation** n.

cancer n. a malignant tumour; a disease in which these form; an evil influence. □ **cancerous** adj.

candela n. a unit measuring the brightness of light.

candelabrum n. (also **candelabra**; pl. **candelabra**; Amer. **candelabras** or **candelabrums**) a large branched candlestick or stand for lights.

candid adj. frank. □ **candidly** adv., **candidness** n.

candidate n. a person applying for a job or taking an examination. □ **candidacy** n., **candidature** n.

candied adj. encrusted or preserved in sugar.

candle n. a stick of wax enclosing a wick which is burnt to give light.

candlestick n. a holder for a candle.

candlewick n. fabric with a tufted pattern.

candour n. (Amer. **candor**) frankness.

candy n. (Amer.) sweets, a sweet.

candyfloss n. a fluffy mass of spun sugar.

candy stripe n. alternate stripes of white and colour. □ **candy-striped** adj.

candytuft n. a plant with flowers in flat clusters.

cane n. a stem of a tall reed or grass or slender palm; a light walking stick.

canine (kay-nyn) adj. of dogs. ● n. (in full **canine tooth**) a pointed tooth between the incisors and molars.

canister n. a small metal container.

canker n. a disease of animals or plants; an influence that corrupts.

cannabis n. the hemp plant; a drug made from this.

canned see **can¹**.

cannelloni n.pl. rolls of pasta with a savoury filling.

cannibal n. a person who eats human flesh. □ **cannibalism** n.

cannibalize v. (also **-ise**) use parts from (a machine) to repair another.

cannon n. **1** (pl. usu. **cannon**) a large gun. **2** the hitting of two balls in one shot in billiards. ● v. bump heavily (into).

cannonade n. continuous gunfire. ● v. bombard with this.

cannot the negative form of **can²**.

canny adj. (**cannier**) shrewd; good, clever. □ **cannily** adv.

canoe n. a light boat propelled by paddle(s). ● v. (**canoed**, **canoeing**) go in a canoe. □ **canoeist** n.

canon n. **1** a member of cathedral clergy. **2** a general rule or principle. **3** a set of writings accepted as genuine. □ **canonical** adj.

canonize v. (also **-ise**) declare officially to be a saint. □ **canonization** n.

canoodle v. (colloquial) kiss and cuddle.

canopy n. a covering hung up over a throne, bed, person, etc.

cant n. insincere talk; jargon.

cantabile (kan-tah-bi-lay) adv. (Music) smooth and flowing.

cantaloupe n. (also **cantaloup**) a small melon with orange-coloured flesh.

cantankerous adj. perverse, peevish. □ **cantankerously** adv.

cantata n. a choral composition.

canteen n. 1 a restaurant for employees. 2 a case of cutlery.

canter n. a gentle gallop. ● v. go at a canter.

cantilever n. a projecting beam or girder supporting a structure.

canto n. a division of a long poem.

canton n. a division of Switzerland.

canvas n. a strong coarse cloth; a painting on this.

canvass v. ask for political support; propose (a plan).

canyon n. a deep gorge.

CAP abbr. Common Agricultural Policy.

cap n. 1 a soft brimless hat, often with a peak; a headdress worn as part of a uniform; a cover or top. 2 an explosive device for a toy pistol. ● v. (**capped**) put a cap on; form the top of; surpass.

capable adj. having a certain ability or capacity; competent. □ **capably** adv., **capability** n.

capacious adj. roomy.

capacitance n. the ability to store an electric charge.

capacitor n. a device storing a charge of electricity.

capacity n. 1 the ability to contain or accommodate; an amount that can be contained or produced. 2 mental power. 3 a position or function.

caparison v. deck out. ● n. finery.

cape n. 1 a cloak; a short similar part. 2 a coastal promontory.

caper v. jump about friskily. ● n. 1 a frisky movement. 2 (slang) an activity. 3 a bramble-like shrub; one of its buds, pickled for use in sauces.

capercaillie n. a large grouse.

capillarity n. (also **capillary action**) the rise or depression of a liquid in a narrow tube.

capillary n. a very fine hairline tube or blood vessel.

capital adj. 1 chief, very important. 2 (of a letter of the alphabet) of the kind used to begin a name or sentence. 3 involving the death penalty. ● n. 1 the chief town of a country etc. 2 a capital letter. 3 money with which a business is started. 4 the top part of a pillar.

capitalism n. a system in which trade and industry are controlled by private owners.

capitalist n. a person who has much money invested; a supporter of capitalism.

capitalize v. (also **-ise**) 1 convert into or provide with capital. 2 write as or with a capital letter. □ **capitalize on** make advantageous use of. □ **capitalization** n.

capitation n. a fee paid per person.

capitulate v. surrender, yield. □ **capitulation** n.

capo n. (pl. **capos**) a device fitted across the strings of an instrument to raise their pitch.

capon n. a domestic cock castrated and fattened.

cappuccino (ka-poo-**chee**-noh) n. (pl. **cappuccinos**) coffee made with frothy steamed milk.

caprice (kă-*prees*) *n.* a whim; a piece of music in a lively fanciful style.

capricious *adj.* guided by caprice, impulsive; unpredictable. □ **capriciously** *adv.*, **capriciousness** *n.*

capsicum *n.* a tropical plant with hollow edible fruits, esp. varieties of the sweet pepper.

capsize *v.* overturn.

capstan *n.* a revolving post or spindle on which a cable etc. winds.

capsule *n.* **1** a small soluble gelatin case enclosing medicine for swallowing. **2** a detachable compartment of a spacecraft. **3** a plant's seed case.

captain *n.* a leader of a group or sports team; a person commanding a ship or civil aircraft; a naval officer next below rear admiral; an army officer next below major. ● *v.* be captain of. □ **captaincy** *n.*

caption *n.* a short title or heading; an explanation on an illustration.

captious *adj.* fond of finding fault, esp. about trivial matters.

captivate *v.* capture the fancy of, charm. □ **captivation** *n.*

captive *adj.* taken prisoner, unable to escape. ● *n.* a captive person or animal. □ **captivity** *n.*

captor *n.* one who takes a captive.

capture *v.* **1** take prisoner; take or obtain by force or skill. **2** portray; record on film. **3** cause (data) to be stored in a computer. ● *n.* capturing.

car *n.* a motor vehicle for a small number of passengers; a compartment in a cable railway, lift, etc.

carafe (kă-*raf*) *n.* a glass bottle for serving wine or water.

caramel *n.* brown syrup made from heated sugar; toffee tasting like this.

caramelize *v.* (also **-ise**) become caramel. □ **caramelization** *n.*

carapace *n.* the upper shell of a tortoise.

carat *n.* a unit of purity of gold.

caravan *n.* **1** a vehicle equipped for living in, able to be towed by a horse or car. **2** a company travelling together across desert. □ **caravanning** *n.*

caraway *n.* a plant with spicy seeds used for flavouring cakes etc.

carbine *n.* an automatic rifle.

carbohydrate *n.* an energy-producing compound (e.g. starch) in food.

carbolic *n.* a disinfectant.

carbon *n.* a chemical element (symbol C) occurring as diamond, graphite, and charcoal, and in all living matter.

carbonate *n.* a compound releasing carbon dioxide when mixed with acid. ● *v.* impregnate with carbon dioxide.

carbon copy *n.* a copy made with carbon paper; an exact copy.

carbon dating *n.* a method of deciding the age of something by measuring the decay of radiocarbon in it.

carbon paper *n.* paper coated with pigment for making a copy as something is typed or written.

carboniferous *adj.* producing coal.

carborundum *n.* a compound of carbon and silicon used for grinding and polishing things.

carboy *n.* a large round bottle surrounded by a protective framework.

carbuncle *n.* **1** a severe abscess. **2** a garnet cut in a round knob shape.

carburettor *n.* (*Amer.* **carburetor**) an apparatus mixing air and petrol in a motor engine.

carcass n. (also **carcase**) the dead body of an animal; the framework or basic structure of something.

carcinogen n. a cancer-producing substance. □ **carcinogenic** adj.

carcinoma n. (pl. **carcinomata** or **carcinomas**) a cancerous tumour.

card n. a piece of cardboard or thick paper; this printed with a greeting or invitation; a post-card; a playing card; a credit card; (**cards**) any card game. ● v. clean or comb (wool) with a wire brush or toothed instrument.

cardamom n. a cooking spice.

cardboard n. pasteboard or stiff paper.

cardiac adj. of the heart.

cardigan n. a knitted jacket.

cardinal adj. chief, most important. ● n. a member of the Sacred College of the RC Church which elects the pope.

cardinal number n. a whole number, 1, 2, 3, etc.

cardiogram n. a record of heart movements.

cardiograph n. an instrument recording heart movements.

cardiology n. the study of diseases of the heart. □ **cardiological** adj., **cardiologist** n.

cardphone n. a public telephone operated by a plastic machine-readable card.

card-sharp n. (also **card-sharper**) a swindler at card games.

care n. **1** serious attention and thought; caution to avoid damage or loss; protection, supervision. **2** worry, anxiety. ● v. feel concern, interest, affection, or liking.

careen v. tilt or keel over.

career n. a way of making one's living, a profession; a course through life. ● v. move swiftly or wildly.

careerist n. a person intent on advancement in a career.

carefree adj. light-hearted and free from worry.

careful adj. acting or done with care. □ **carefully** adv.

careless adj. not careful. □ **carelessly** adv., **carelessness** n.

carer n. a person who looks after a sick or disabled person at home.

caress n. a loving touch, a kiss. ● v. give a caress to.

caret n. a mark (∧) indicating an insertion in text.

caretaker n. a person employed to look after a building.

careworn adj. showing signs of prolonged worry.

cargo n. (pl. **cargoes** or **cargos**) goods carried by ship or aircraft.

Caribbean adj. of the West Indies or their inhabitants.

caribou n. (pl. **caribou**) a North American reindeer.

caricature n. an exaggerated portrayal of a person for comic effect. ● v. make a caricature of.

caries n. decay of a tooth or bone.

carillon (kă-ril-yŏn) n. a set of bells sounded mechanically; a tune played on these.

Carmelite n. a member of an order of white-cloaked friars or nuns.

carmine adj. & n. vivid crimson.

carnage n. great slaughter.

carnal adj. of the body or flesh, not spiritual. □ **carnally** adv.

carnation n. a cultivated fragrant pink.

carnelian var. of **cornelian**.

carnet (kar-nay) n. a permit.

carnival n. public festivities, usu. with a procession.

carnivore n. a carnivorous animal.

carnivorous *adj.* feeding on flesh.

carob *n.* the pod of a Mediterranean evergreen tree, similar in taste to chocolate.

carol *n.* a Christmas hymn. ● *v.* (**carolled**; *Amer.* **caroled**) sing carols; sing joyfully.

carotene *n.* an orange-coloured pigment found in carrots, tomatoes, etc.

carotid *n.* an artery carrying blood to the head.

carouse *v.* drink and be merry. □ **carousal** *n.*, **carouser** *n.*

carousel *n.* **1** (*Amer.*) a merry-go-round. **2** a rotating conveyor, esp. for luggage at an airport.

carp *n.* a freshwater fish. ● *v.* keep finding fault.

carpel *n.* the part of a flower in which the seeds develop.

carpenter *n.* a person who makes or repairs wooden objects and structures. □ **carpentry** *n.*

carpet *n.* a textile fabric for covering a floor. ● *v.* (**carpeted**) cover with a carpet. □ **on the carpet** (*colloquial*) being reprimanded.

carport *n.* a roofed open-sided shelter for a car.

carpus *n.* (*pl.* **carpi**) the set of small bones forming the wrist.

carriage *n.* a wheeled vehicle or support; a moving part; the conveying of goods etc., the cost of this.

carriage clock *n.* a small portable clock with a handle on top.

carriageway *n.* the part of the road on which vehicles travel.

carrier *n.* a person or thing carrying something; a paper or plastic bag with handles, for holding shopping.

carrion *n.* dead decaying flesh.

carrot *n.* **1** a tapering orange root vegetable. **2** an incentive.

carry *v.* **1** transport, convey; support; be the bearer of. **2** involve, entail. **3** take (a process

etc.) to a specified point. **4** get the support of (a motion etc.). **5** stock (goods for sale). **6** be audible at a distance. □ **carry on** continue; (*colloquial*) behave excitedly. **carry out** put into practice.

cart *n.* a wheeled structure for carrying loads. ● *v.* carry, transport.

carte blanche (kart **blahnsh**) *n.* full power to do as one thinks best.

cartel *n.* a manufacturers' or producers' union to control prices.

carthorse *n.* a horse of heavy build.

cartilage *n.* the firm elastic tissue in skeletons of vertebrates, gristle.

cartography *n.* map-drawing. □ **cartographer** *n.*, **cartographic** *adj.*

carton *n.* a cardboard or plastic container.

cartoon *n.* a humorous drawing; a film consisting of an animated sequence of drawings; a sketch for a painting. □ **cartoonist** *n.*

cartridge *n.* a case containing explosive for firearms; a sealed cassette.

cartridge paper *n.* thick strong paper.

cartwheel *n.* a handspring with limbs spread like the spokes of a wheel.

carve *v.* make, inscribe, or decorate by cutting; cut (meat) into slices for eating.

carvery *n.* a restaurant where meat is served from a joint as required.

Casanova *n.* a man noted for his love affairs.

casbah var. of **kasbah**.

cascade *n.* a waterfall; something falling or hanging like this. ● *v.* fall in this way.

case *n.* **1** an instance of something occurring; a situation. **2** a lawsuit. **3** a set of facts or arguments supporting something. **4** a container or protective covering; this with its contents; a suitcase. ● v. **1** enclose in a case. **2** (slang) examine (a building etc.) in preparation for a crime. □ **in case** lest.

casement n. a window opening on vertical hinges.

cash n. money in the form of coins or banknotes. ● v. give or obtain cash for (a cheque etc.). □ **cash in** (**on**) get profit or advantage (from).

cash card n. a plastic card with magnetic code for drawing money from a machine.

cashew n. an edible nut.

cashier n. a person employed to handle money. ● v. dismiss from military service in disgrace.

cashmere n. very fine soft wool; fabric made from this.

cashpoint n. (also **cash machine, cash dispenser**) a machine dispensing cash.

casino n. (pl. **casinos**) a public building or room for gambling.

cask n. a barrel for liquids.

casket n. a small usu. ornamental box for valuables; (Amer.) a coffin.

cassava n. a tropical plant; flour made from its roots.

casserole n. a covered dish in which meat etc. is cooked and served; food cooked in this. ● v. cook in a casserole.

cassette n. a small case containing a reel of magnetic tape or film.

cassis n. a blackcurrant-flavoured usu. alcoholic syrup.

cassock n. a long robe worn by clergy and choristers.

cassowary n. a large flightless bird related to the emu.

cast v. (**cast, casting**) **1** throw; shed; direct (a glance). **2** register (one's vote). **3** shape (molten metal) in a mould. **4** select

actors for a play or film; assign a role to. **5** calculate (a horoscope). ● n. **1** a throw of dice, a fishing line, etc. **2** a moulded mass of solidified material. **3** a set of actors in a play etc. **4** a type, a quality. **5** a slight squint.

castanets n.pl. a pair of shell-shaped pieces of wood clicked in the hand to accompany dancing.

castaway n. a shipwrecked person.

caste n. a social class, esp. in the Hindu system.

castellated adj. having turrets or battlements.

caster var. of castor.

castigate v. punish, rebuke, or criticize severely. □ **castigation** n.

casting vote n. a deciding vote when those on each side are equal.

cast iron n. a hard alloy of iron cast in a mould. ● adj. (**cast-iron**) made of cast iron; very strong.

castle n. a large fortified residence.

cast-off n. a discarded thing.

castor n. (also **caster**) **1** a small swivelling wheel on a leg of furniture. **2** a small container with a perforated top for sprinkling sugar etc.

castor oil n. a purgative and lubricant oil from the seeds of a tropical plant.

castor sugar n. finely granulated white sugar.

castrate v. remove the testicles of. □ **castration** n.

castrato n. (pl. **castrati**) a castrated male soprano or alto singer.

casual adj. happening by chance; not serious, formal, or methodical; not permanent. □ **casually** adv., **casualness** n.

casualty n. **1** a person killed or injured; something lost or destroyed. **2** (in full **casualty de-**

partment) part of a hospital treating accident victims.

casuist n. a theologian who studies moral problems; a sophist, a quibbler. □ **casuistry** n.

cat n. a small furry domesticated animal; a wild animal related to this.

cataclysm n. a violent upheaval or disaster. □ **cataclysmic** adj.

catacomb (kat-ăkoom) n. an underground gallery with recesses for tombs.

catafalque n. a platform for the coffin of a distinguished person before or during a funeral.

catalepsy n. a seizure or trance with rigidity of the body. □ **cataleptic** adj.

catalogue n. (Amer. also **catalog**) a systematic list of items. ●v. (**catalogued, cataloguing**) list in a catalogue.

catalyse v. (Amer. **catalyze**) subject to the action of a catalyst. □ **catalysis** n.

catalyst n. a substance that aids a chemical reaction while remaining unchanged.

catalytic converter n. part of an exhaust system that reduces the harmful effects of pollutant gases.

catamaran n. a boat with twin hulls.

catapult n. a device with elastic for shooting small stones. ●v. hurl from or as if from a catapult.

cataract n. **1** a large waterfall. **2** an opaque area clouding the lens of the eye.

catarrh n. inflammation of a mucous membrane, esp. of the nose, with a watery discharge.

catastrophe (kă-tas-trŏ-fi) n. a sudden great disaster. □ **catastrophic** adj., **catastrophically** adv.

catcall n. a whistle of disapproval.

catch v. (**caught, catching**) **1** capture, seize; grasp and hold. **2** detect; surprise, trick. **3** overtake. **4** be in time for. **5** become infected with. **6** hit. ●n. **1** an act of catching; something caught or worth catching. **2** a concealed difficulty. **3** a fastener. □ **catch on** (colloquial) become popular; understand what is meant. **catch out** detect in a mistake etc. **catch up** come abreast with; do arrears of work.

catching adj. infectious.

catchment area n. an area from which rainfall drains into a river; an area from which a hospital draws patients or a school draws pupils.

catchphrase n. a phrase in frequent current use, a slogan.

catch-22 n. a dilemma in which either choice will cause suffering.

catchword n. a catchphrase.

catchy adj. (**catchier**) (of a tune) pleasant and easy to remember.

catechism n. a series of questions and answers, esp. on the principles of a religion.

catechize v. (also **-ise**) put a series of questions to.

categorical adj. unconditional, absolute. □ **categorically** adv.

categorize v. (also **-ise**) place in a category. □ **categorization** n.

category n. a class of things.

cater v. supply food; provide what is needed or wanted. □ **caterer** n.

caterpillar n. **1** the larva of butterfly or moth. **2** (in full **caterpillar track** or **tread**) (trade mark) a steel band passing round the wheels of a tractor etc. for travel on rough ground.

caterwaul v. make a cat's howling cry.

catfish n. a large fish with whisker-like feelers round the mouth.

catgut n. gut as thread.

catharsis n. (pl. **catharses**) a release of strong feeling or tension. □ **cathartic** adj.

cathedral n. the principal church of a diocese.

Catherine wheel n. a rotating firework.

catheter n. a tube inserted into the bladder to extract urine.

cathode n. an electrode by which current leaves a device.

cathode ray tube n. a vacuum tube in which beams of electrons produce a luminous image on a fluorescent screen.

Catholic adj. 1 of all Churches or all Christians; Roman Catholic. 2 (**catholic**) universal. □ **Catholicism** n.

cation (kat-I-ŏn) n. a positively charged ion.

catkin n. a hanging flower of willow, hazel, etc.

catmint n. a strong-smelling plant attractive to cats.

catnap n. a short nap.

catnip n. catmint.

cat's cradle n. a child's game with string wound round the hands.

Catseye n. (trade mark) a reflector stud on a road.

cat's-paw n. a person used as a tool by another.

catsup (Amer.) = ketchup.

cattery n. a place where cats are boarded.

cattle n.pl. large animals with horns and cloven hoofs.

catty adj. (**cattier**) slightly spiteful. □ **cattily** adv., **cattiness** n.

catwalk n. a narrow strip for walking on as a pathway or platform.

caucus n. a local committee of a political party; (Amer.) a meeting of party leaders.

caught see **catch**.

caul n. a membrane enclosing a foetus in the womb.

cauldron n. (also **caldron**) a large deep cooking pot.

cauliflower n. a cabbage with a large white flower head.

caulk v. (also **calk**) stop up (a ship's seams) with waterproof material.

causal adj. of or forming a cause; of cause and effect. □ **causality** n.

causation n. causality.

cause n. 1 what produces an effect; the reason or motive for an action. 2 a principle supported. ● v. be the cause of, make happen.

cause célèbre (kohz se-lebr) n. (pl. **causes célèbres**) an issue arousing great interest.

causeway n. a raised road across low or wet ground.

caustic adj. burning by chemical action; sarcastic. ● n. a caustic substance. □ **caustically** adv.

caustic soda n. sodium hydroxide.

cauterize v. (also **-ise**) burn (tissue) to destroy infection or stop bleeding. □ **cauterization** n.

caution n. avoidance of rashness; a warning. ● v. warn; reprimand.

cautionary adj. conveying a warning.

cautious adj. having or showing caution. □ **cautiously** adv.

cavalcade n. a procession.

cavalier adj. arrogant, offhand. ● n. (**Cavalier**) (hist.) a supporter of Charles I in the English Civil War.

cavalry n. troops who fight on horseback.

cave n. a natural hollow. □ **cave in** collapse; yield.

caveat (kav-iat) n. a warning.

caveat emptor *n.* the principle that the buyer alone is responsible if dissatisfied.

caveman *n.* (*pl.* **-men**) a person of prehistoric times living in a cave.

cavern *n.* a large cave; a hollow part.

cavernous *adj.* like a cavern, large and hollow.

caviar *n.* the pickled roe of sturgeon or other large fish.

cavil *v.* (**cavilled**; *Amer.* **caviled**) raise petty objections. ● *n.* a petty objection.

caving *n.* the sport of exploring caves.

cavity *n.* a hollow within a solid body.

cavort *v.* leap about excitedly.

caw *n.* the harsh cry of a rook etc. ● *v.* utter a caw.

cayenne *n.* a hot red pepper.

cayman *n.* (*pl.* **caymans**) (also **caiman**) a South American alligator.

CB *abbr.* citizens' band.

CBE *abbr.* Commander of the Order of the British Empire.

CBI *abbr.* Confederation of British Industry.

cc *abbr.* (also **c.c.**) **1** cubic centimetre(s). **2** carbon copy or copies.

CD *abbr.* compact disc.

Cd *symb.* cadmium.

CD-ROM *n.* a compact disc holding data for display on a computer screen.

Ce *symb.* cerium.

cease *v.* come to an end, discontinue, stop.

ceasefire *n.* a signal to stop firing guns; a truce.

ceaseless *adj.* not ceasing.

cecum *Amer.* sp. of **caecum.**

cedar *n.* an evergreen tree; its hard fragrant wood.

cede *v.* surrender (territory etc.).

cedilla *n.* a mark written under c (ç) to show that it is pronounced as s.

ceilidh (kay-li) *n.* (*Scot. & Irish*) an informal gathering for traditional music and dancing.

ceiling *n.* the surface of the top of a room; an upper limit or level.

celandine *n.* a small wild plant with yellow flowers.

celebrant *n.* an officiating priest.

celebrate *v.* mark or honour with festivities. □ **celebration** *n.*

celebrated *adj.* famous.

celebrity *n.* a famous person; fame.

celeriac *n.* a variety of celery with a turnip-like root.

celerity *n.* (*literary*) swiftness.

celery *n.* a plant with edible crisp stems.

celestial *adj.* of the sky; of heaven.

celiac *Amer.* sp. of **coeliac.**

celibate *adj.* abstaining from sexual intercourse. □ **celibacy** *n.*

cell *n.* **1** a small room for a monk or prisoner; a compartment in a honeycomb. **2** a device for producing electric current chemically. **3** a microscopic unit of living matter. **4** a small group as a nucleus of political activities.

cellar *n.* an underground room; a stock of wine.

cello (che-loh) *n.* (*pl.* **cellos**) a bass instrument of the violin family. □ **cellist** *n.*

cellophane *n.* (*trade mark*) a thin transparent wrapping material.

cellphone *n.* a small portable radio-telephone.

cellular *adj.* **1** of cells. **2** woven with an open mesh.

cellular phone (or **radio**) *n.* a system of mobile communication with an area divided into cells, each served by a small transmitter.

cellulite n. a lumpy form of fat producing puckering of the skin.

celluloid n. plastic made from cellulose nitrate and camphor.

cellulose n. a substance in plant tissues, used in making plastics.

Celsius adj. of a centigrade scale with 0° as the freezing point and 100° as the boiling point of water.

Celt n. a member of an ancient European people or their descendants. □ **Celtic** adj.

cement n. a substance of lime and clay setting like stone; an adhesive; a substance for filling cavities in teeth. ● v. join with cement; unite firmly.

cemetery n. a burial ground other than a churchyard.

cenobite Amer. sp. of **coenobite.**

cenotaph n. a tomblike monument to people buried elsewhere.

censer n. a container for burning incense.

censor n. a person authorized to examine letters, books, films, etc., and remove or ban anything regarded as harmful. ● v. remove or ban in this way. □ **censorial** adj., **censorship** n.

■ **Usage** As a verb, censor is often confused with censure, which means 'to criticize harshly'.

censorious adj. severely critical.

censure n. hostile criticism and rebuke. ● v. criticize harshly.

census n. an official count of the population.

cent n. a 100th part of a dollar or other currency; a coin worth this.

centaur n. a mythical creature, half man, half horse.

centenarian n. a person 100 years old or more.

centenary n. a 100th anniversary.

centennial adj. of a centenary. ● n. a centenary.

center etc. Amer. sp. of **centre** etc.

centigrade adj. using a temperature scale of 100°; = Celsius.

centigram n. (also **centigramme**) a 100th of a gram.

centilitre n. (Amer. **centiliter**) 100th of a litre.

centime n. one 100th of a franc.

centimetre n. (Amer. **centimeter**) 100th of a metre, about 0.4 inch.

centipede n. a small crawling creature with many legs.

central adj. of, at, or forming a centre; most important. □ **centrally** adv., **centrality** n.

central heating n. heating of a building from one source.

centralism n. a system that centralizes an administration.

centralize v. (also **-ise**) bring under the control of a central authority. □ **centralization** n.

central nervous system n. the brain and the spinal cord.

centre n. (Amer. **center**) the middle point or part. ● v. (**centred**, **centring**) place in or at a centre.

centreboard n. (Amer. **centerboard**) the retractable keel of a sailing boat or sailboard.

centrefold n. (Amer. **centerfold**) a centre spread of a magazine, newspaper, etc.

centrifugal adj. moving away from the centre.

centrifuge n. a machine using centrifugal force for separating substances.

centripetal adj. moving towards the centre.

centurion n. a commander in the ancient Roman army.

century *n.* a period of 100 years; 100 runs at cricket.

■ **Usage** Strictly speaking, because the 1st century ran from the year 1 to the year 100, the first year of a century ends in 1. However, a century is commonly regarded as starting with a year ending in 00, the 20th century thus running from 1900 to 1999.

cephalic *adj.* of the head.

cephalopod *n.* a mollusc with tentacles (e.g. an octopus).

ceramic *adj.* of pottery or a similar substance. ●*n.* (**ceramics**) the art of making pottery.

cereal *n.* a grass plant with edible grain; this grain; breakfast food made from it.

cerebellum *n.* (*pl.* **cerebellums** or **cerebella**) a small part of the brain at the back of the skull.

cerebral *adj.* of the brain; intellectual. □ **cerebrally** *adv.*

cerebral palsy *n.* paralysis resulting from brain damage before or at birth.

cerebrum *n.* (*pl.* **cerebra**) the main part of the brain.

ceremonial *adj.* of or used in ceremonies, formal. ●*n.* ceremony; rules for this. □ **ceremonially** *adv.*

ceremonious *adj.* full of ceremony. □ **ceremoniously** *adv.*

ceremony *n.* a set of formal acts.

cerise *adj. & n.* light clear red.

cerium *n.* a metallic element (symbol Ce).

certain *adj.* **1** feeling sure; believed firmly. **2** specific but not named; some.

certainly *adv.* without doubt; yes.

certainty *n.* being certain; something that is certain.

certifiable *adj.* **1** able or needing to be certified. **2** (*colloquial*) mad.

certificate *n.* an official document attesting certain facts.

certify *v.* declare formally.

certitude *n.* a feeling of certainty.

cerulean *adj.* (*literary*) deep blue like a clear sky.

cervix *n.* (*pl.* **cervices**) the neck; a necklike structure, esp. of the womb. □ **cervical** *adj.*

Cesarean, Cesarian Amer. sp. of **Caesarean**.

cesium Amer. sp. of **caesium**.

cessation *n.* ceasing.

cession *n.* ceding, giving up.

cesspit *n.* (also **cesspool**) a covered pit designed to receive liquid waste or sewage.

cetacean (si-**tay**-shăn) *adj. & n.* (a member) of the whale family. *Cf.* **cetology**.

Cf. *symb.* californium.

cf. *abbr.* compare.

CFC *abbr.* chlorofluorocarbon, a gaseous compound thought to harm the ozone layer.

chafe *v.* **1** warm by rubbing; make or become sore by rubbing. **2** become irritated or impatient.

chafer *n.* a large beetle.

chaff *n.* **1** corn husks separated from seed; chopped hay and straw. **2** banter. ●*v.* banter, tease.

chaffinch *n.* a European finch.

chafing dish *n.* a heated pan for keeping food warm at the table.

chagrin *n.* annoyance and embarrassment.

chain *n.* a series of connected metal links; a connected series or sequence. ●*v.* fasten with chain(s).

chain reaction *n.* a change causing further changes.

chainsaw *n.* a saw with teeth set on a circular chain.

chair *n.* **1** a movable seat for one person, usu. with a back and four legs. **2** (the position of) a chairperson; the position of a professor. ● *v.* act as chairman of.

chairlift *n.* a series of chairs on a cable for carrying people up a mountain.

chairman *n.* (*pl.* **-men**) a person who presides over a meeting or board of directors.

chairperson *n.* a chairman or chairwoman.

chairwoman *n.* (*pl.* **-women**) a woman who presides over a meeting or board of directors.

chaise longue (shayz lawng) *n.* (**chaises longues** or **chaise longues**) a sofa with a backrest at only one end.

chalcedony *n.* a type of quartz.

chalet (sha-lay) *n.* a Swiss hut or cottage; a small villa; a small hut in a holiday camp etc.

chalice *n.* a large goblet.

chalk *n.* white soft limestone; a piece of this or similar coloured substance used for drawing. □ **chalky** *adj.*

challenge *n.* a call to try one's skill or strength; a demand to respond or identify oneself; a formal objection; a demanding task. ● *v.* make a challenge to; question the truth or rightness of. □ **challenger** *n.*

chamber *n.* a hall used for meetings of a council, parliament etc.; (*old use*) a room, a bedroom; (**chambers**) a set of rooms; a cavity or compartment.

chamberlain *n.* an official managing a sovereign's or noble's household.

chambermaid *n.* a woman cleaner of hotel bedrooms.

chamber music *n.* music written for a small group of players.

chamber pot *n.* a bedroom receptacle for urine.

chameleon (kă-mee-li-ŏn) *n.* a small lizard that changes colour according to its surroundings.

chamfer *v.* bevel the edge of.

chamois *n.* **1** (sham-wah) a small mountain antelope. **2** (sha-mi) a piece of soft leather used for cleaning things.

chamomile var. of **camomile**.

champ *v.* munch noisily, make a chewing action. □ **champ at the bit** show impatience.

champagne *n.* a sparkling white French wine.

champion *n.* a person or thing that defeats all others in a competition; a person who fights or speaks in support of another or of a cause. ● *v.* support as champion. □ **championship** *n.*

chance *n.* the way things happen through no known cause or agency; luck; likelihood; opportunity. ● *adj.* happening by chance. ● *v.* happen; risk.

chancel *n.* the part of a church near the altar.

chancellor *n.* the government minister in charge of the nation's budget; a state or law official of various other kinds; the non-resident head of a university. □ **chancellorship** *n.*

Chancery *n.* a division of the High Court of Justice.

chancy *adj.* (**chancier**) risky, uncertain.

chandelier *n.* a hanging support for several lights.

chandler *n.* a dealer in ropes, canvas, etc. for ships.

change *v.* make or become different; exchange, substitute; put fresh clothes or coverings on; go from one of two (sides, trains, etc.) to another; get or give small money or different currency for. ● *n.* changing; money in small units or returned as balance. □ **changeable** *adj.*

changeling *n.* a child or thing believed to have been substituted secretly for another.

channel *n.* **1** a stretch of water connecting two seas; a passage for water. **2** a medium of communication. **3** a band of broadcasting frequencies. ● v. (**channelled**; Amer. **channeled**) direct through a channel.

chant n. a melody for psalms; a monotonous song; a rhythmic shout. ● v. sing, esp. to a chant; shout rhythmically.

chanter n. the melody-pipe of bagpipes.

chanterelle n. a yellow edible funnel-shaped fungus.

chantry n. a chapel founded for priests to sing Masses for the founder's soul.

chaos n. great disorder. □ **chaotic** adj., **chaotically** adv.

chap n. (colloquial) a man. ● v. (**chapped**) (of the skin) split or crack.

chapati n. (also **chapati**) a thin flat circle of unleavened bread.

chapel n. a place used for Christian worship, other than a cathedral or parish church; a place with a separate altar within a church.

chaperone n. (also **chaperon**) an older woman looking after a young unmarried woman on social occasions. ● v. act as chaperon to.

chaplain n. a clergyman of an institution, private chapel, ship, regiment, etc. □ **chaplaincy** n.

chapter n. **1** a division of a book. **2** the canons of a cathedral.

char n. a charwoman. ● v. (**charred**) make or become black by burning.

charabanc n. an early form of bus with bench seats.

character n. **1** qualities making a person or thing what he, she, or it is; moral strength; a noticeable or eccentric person; a person in a novel, play, etc.; a testimonial; a physical characteristic. **2** a letter or sign used in writing, printing, etc.

characteristic n. a feature forming part of the character of a person or thing. ● adj. typical of or distinguishing a person or thing. □ **characteristically** adv.

characterize v. (also **-ise**) describe the character of; be a characteristic of. □ **characterization** n.

charade (sha-rahd) n. **1** (**charades**) a game involving guessing words from acted clues. **2** an absurd pretence.

charcoal n. a black substance made by burning wood slowly.

charge n. **1** the price asked for goods or services. **2** a quantity of explosive. **3** the electricity contained in a substance. **4** a task, a duty. **5** custody; a person or thing entrusted. **6** an accusation. **7** a rushing attack. ● v. **1** ask as a price or from (a person); record as a debt. **2** load or fill with explosive; give an electric charge to. **3** give as a task or duty, entrust. **4** accuse formally. **5** rush forward in attack. □ **in charge** in command. **take charge** take control.

charge card n. a credit card.

chargé d'affaires n. (pl. **chargés d'affaires**) an ambassador's deputy.

charger n. **1** a cavalry horse. **2** a device for charging a battery.

chariot n. a two-wheeled horse-drawn vehicle used in ancient times in battle and in racing.

charioteer n. a driver of a chariot.

charisma (ka-riz-ma) n. the power to inspire devotion and enthusiasm; great charm.

charismatic adj. having charisma; (of worship) characterized by spontaneity. □ **charismatically** adv.

charitable adj. full of charity; of or belonging to charities. □ **charitably** adv.

charity *n.* loving kindness; unwillingness to think badly of others; an institution or fund for helping the needy, help so given.

charlady *n.* a charwoman.

charlatan *n.* a person falsely claiming to be an expert.

charlotte *n.* a pudding of cooked fruit with a covering of breadcrumbs or sponge.

charm *n.* **1** attractiveness, the power of arousing love or admiration. **2** an act, object, or words believed to have magic power; a small ornament worn on a bracelet etc. ● *v.* **1** give pleasure to. **2** influence by personal charm; influence as if by magic. □ **charmer** *n.*

charming *adj.* delightful.

charnel house *n.* a place containing corpses or bones.

chart *n.* **1** a map for navigators; a table, diagram, or outline map. **2** (**the charts**) a list of recordings that are currently most popular, esp. pop singles. ● *v.* make a chart of.

charter *n.* **1** an official document granting rights. **2** chartering of aircraft etc. ● *v.* **1** grant a charter to. **2** let or hire (an aircraft, ship, or vehicle).

chartered *adj.* (of an accountant, engineer, etc.) qualified according to the rules of an association holding a royal charter.

charter flight *n.* a flight by a chartered aircraft as opposed to a scheduled flight.

chartreuse (shah-trerz) *n.* a fragrant green or yellow liqueur.

charwoman *n.* (*pl.* **-women**) a woman employed to clean a house etc.

chary *adj.* (**charier**) cautious. □ **charily** *adv.*, **chariness** *n.*

chase *v.* go quickly after in order to capture, overtake, or drive away. ● *n.* **1** chasing,

pursuit; hunting. **2** a steeplechase.

chaser *n.* a drink taken after a drink of another kind.

chasm (kaz-ŭm) *n.* a deep cleft.

chassis (sha-see) *n.* (*pl.* **chassis**) the base frame of a vehicle.

chaste *adj.* **1** virgin, celibate; sexually pure. **2** simple in style, not ornate. □ **chastely** *adv.*

chasten *v.* discipline by punishment; subdue the pride of.

chastise *v.* punish, esp. by beating. □ **chastisement** *n.*

chastity *n.* being chaste.

chat *n.* an informal conversation. ● *v.* (**chatted**) have a chat.

chateau *n.* (*pl.* **chateaux**) a French castle or large country house.

chatelaine *n.* the mistress of a large house.

chattel *n.* a movable possession.

chatter *v.* **1** talk quickly and continuously about unimportant matters. **2** (of teeth) rattle together. ● *n.* chattering talk. □ **chatterer** *n.*

chatterbox *n.* a talkative person.

chatty *adj.* (**chattier**) fond of chatting; resembling chat; informal. □ **chattily** *adv.*, **chattiness** *n.*

chauffeur (shoh-fer) *n.* a person employed to drive a car. ● *v.* drive as a chauffeur.

chauvinism (shoh-vin-izm) *n.* exaggerated patriotism. □ **male chauvinism** a prejudiced belief in male superiority over women. □ **chauvinist** *n.*, **chauvinistic** *adj.*

cheap *adj.* low in cost or value; poor in quality. □ **cheaply** *adv.*, **cheapness** *n.*

cheapen *v.* make or become cheap; degrade.

cheapskate *n.* (*colloquial*) a mean stingy person.

cheat *v.* act dishonestly or unfairly to win profit or advantage;

trick, deprive by deceit. ● *n.* a person who cheats; a deception.

check *v.* **1** test or examine, inspect. **2** stop, slow the motion of. **3** (*Amer.*) correspond when compared. ● *n.* **1** an inspection. **2** a pause; a restraint. **3** the exposure of a chess king to capture. **4** (*Amer.*) a bill in a restaurant; a cheque. **5** a pattern of squares or crossing lines. □ **check in** register on arrival. □ **check out** register on departure or dispatch. □ **checked** *adj.*

checker *n.* **1** var. of **chequer**. **2** (**checkers**) (*Amer.*) the game of draughts. **3** a person who checks things.

checkmate *n.* the situation in chess where capture of a king is inevitable; complete defeat, deadlock. ● *v.* put into checkmate; defeat, foil.

checkout *n.* a desk where goods are paid for in a supermarket.

checkpoint *n.* a place or barrier where documents, vehicles, etc. are checked.

cheek *n.* **1** the side of the face below the eye. **2** impudent speech, arrogance. ● *v.* speak cheekily to. □ **cheek by jowl** close together.

cheeky *adj.* (**cheekier**) showing cheerful lack of respect. □ **cheekily** *adv.*

cheep *n.* a weak shrill cry like that of a young bird. ● *v.* make this cry.

cheer *n.* a shout of applause; cheerfulness. ● *v.* utter a cheer, applaud with a cheer; gladden. □ **cheer up** make or become more cheerful.

cheerful *adj.* happy, contented; pleasantly bright. □ **cheerfully** *adv.*, **cheerfulness** *n.*

cheerio *int.* (*colloquial*) goodbye.

cheerless *adj.* gloomy, dreary.

cheery *adj.* (**cheerier**) cheerful. □ **cheerily** *adv.*, **cheeriness** *n.*

cheese *n.* **1** food made from pressed milk curds. **2** thick stiff jam.

cheeseburger *n.* a hamburger with cheese on it.

cheesecake *n.* **1** an open tart filled with flavoured cream cheese or curd cheese. **2** (*slang*) the portrayal of women in a sexually attractive manner.

cheesecloth *n.* a thin loosely-woven cotton fabric.

cheesed off *adj.* (*colloquial*) bored, exasperated.

cheese-paring *adj.* stingy. ● *n.* stinginess.

cheetah *n.* a swift large animal of the cat family with a spotted coat.

chef *n.* a professional cook.

chef-d'oeuvre (shay-**dervr**) *n.* (*pl.* **chefs-d'oeuvre**) a masterpiece.

chemical *adj.* of or made by chemistry. ● *n.* a substance obtained by or used in a chemical process. □ **chemically** *adv.*

chemise (shĕ-**meez**) *n.* a woman's loose-fitting undergarment or dress.

chemist *n.* a person skilled in chemistry; a dealer in medicinal drugs.

chemistry *n.* the study of substances and their reactions; the structure and properties of a substance.

chemotherapy (kee-moh-**the**-ră-pi) *n.* treatment of disease, esp. cancer, by drugs etc.

chenille (shĕ-**neel**) *n.* a fabric with a velvety pile.

cheque *n.* (*Amer.* **check**) a written order to a bank to pay out money from an account; a printed form for this.

cheque card *n.* (also **banker's card**) a card guaranteeing payment of cheques.

chequer *n.* (also **checker**) a pattern of squares, esp. of alternating colours.

chequered adj. **1** marked with a chequer pattern. **2** having frequent changes of fortune.

cherish v. take loving care of; be fond of; cling to (hopes etc.).

cheroot n. a cigar with both ends open.

cherry n. a small soft round fruit with a stone; a tree bearing this or grown for its ornamental flowers; deep red.

cherub n. **1** (pl. **cherubim**) an angelic being. **2** (in art) a chubby infant with wings; an angelic child. □ **cherubic** adj.

chervil n. a cooking herb.

chess n. a game for two players using 32 pieces on a chequered board with 64 squares.

chest n. **1** a large strong box. **2** the upper front surface of the body.

chesterfield n. a sofa with arms and back of the same height.

chestnut n. **1** a tree with a hard brown nut; this nut. **2** reddish brown; a horse of this colour. **3** an old joke or anecdote.

chest of drawers n. a piece of furniture with drawers for clothes etc.

cheval glass (shĕ-val) n. a tall mirror mounted on a tilting frame.

chevron n. a V-shaped symbol.

chew v. work or grind between the teeth; make this movement.

chewing gum n. flavoured gum used for prolonged chewing.

chewy adj. (**chewier**) suitable for chewing. □ **chewiness** n.

chez (shay) prep. at the home of.

chiaroscuro (ki-arŏ-skoor-oh) n. light and shade effects; the use of contrast.

chic (sheek) adj. stylish and elegant. ● n. stylishness, elegance.

chicane (shi-kayn) n. a barrier or obstacle on a motor-racing track.

chicanery n. trickery.

chick n. a newly hatched young bird.

chicken n. a young domestic fowl; its flesh as food. ● adj. (colloquial) cowardly. ● v.

chicken out n. (colloquial) withdraw through cowardice.

chicken feed n. (colloquial) a trifling amount of money.

chickenpox n. an infectious illness with a rash of small red blisters.

chickpea n. a pea with yellow seeds used as a vegetable.

chicory n. a blue-flowered plant grown for its salad leaves and its root which is used as a flavouring with coffee.

chide v. (**chided** or **chid**, **chidden**, **chiding**) (old use) rebuke.

chief n. a leader, a ruler; the person with the highest rank. ● adj. highest in rank; most important.

chiefly adv. mainly.

chieftain n. the chief of a clan or tribe.

chiffon (shif-on) n. a thin almost transparent fabric.

chignon (sheen-yon) n. a coil of hair worn at the back of the head.

chihuahua (chi-wah-wă) n. a very small smooth-haired dog.

chilblain n. a painful swelling caused by exposure to cold.

child n. (pl. **children**) a young human being; a son or daughter. □ **childhood** n.

childbirth n. the process of giving birth to a child.

childish adj. like a child, unsuitable for a grown person.

childless adj. having no children.

childlike adj. simple and innocent.

chill n. an unpleasant coldness; an illness with feverish shivering. ● adj. chilly. ● v. make or become chilly; preserve at a low temperature without freezing.

chiller n. a cold cabinet, a refrigerator.

chilli n. (Amer. **chili**) (pl. **chillies**) a hot-tasting dried pod of red or green pepper.

chilly adj. (**chillier**) rather cold; unfriendly in manner.

chime n. a tuned set of bells; a series of notes from these. ● v. ring as a chime. □ **chime in** put in a remark.

chimera (ky-meer-ă) n. a legendary monster with a lion's head, goat's body, and serpent's tail; a fantastic product of the imagination.

chimney n. (pl. **chimneys**) a structure for carrying off smoke or gases.

chimney breast n. a projecting wall surrounding a chimney.

chimney pot n. a pipe on top of a chimney.

chimp n. (colloquial) a chimpanzee.

chimpanzee n. an African ape.

chin n. the front of the lower jaw.

china n. fine earthenware, porcelain; things made of this.

chinchilla n. a small squirrel-like South American animal; its grey fur.

chine n. **1** an animal's backbone. **2** a ravine in southern England.

Chinese adj. & n. (pl. **Chinese**) (a native, the language) of China.

Chinese leaves n.pl. a vegetable with long pale cabbage-like leaves.

chink n. **1** a narrow opening, a slit. **2** the sound of glasses or coins striking together. ● v. make this sound.

chinless adj. (colloquial) weak or feeble in character.

chinoiserie n. imitation Chinese motifs as decoration.

chintz n. glazed cotton cloth used for furnishings.

chip n. **1** a small piece cut or broken off something hard. **2** a fried oblong strip of potato. **3** a counter used in gambling. ● v. (**chipped**) break or cut the edge or surface of; shape in this way. □ **chip in** (colloquial) interrupt; contribute money.

chipboard n. board made of compressed wood chips.

chipmunk n. a striped squirrel-like animal of North America.

chipolata n. a small sausage.

chippings n.pl. chips of stone etc. for surfacing a path or road.

chiropody (ki-rop-ŏdi) n. treatment of minor ailments of the feet. □ **chiropodist** n.

chiropractic (ky-rŏ-prak-tik) n. treatment of certain physical disorders by manipulation of the joints. □ **chiropractor** n.

chirp n. a short sharp sound made by a small bird or grasshopper. ● v. make this sound.

chirpy adj. (**chirpier**) lively and cheerful.

chisel n. a tool with a sharp bevelled end for shaping wood or stone etc. ● v. (**chiselled**; Amer. **chiseled**) cut with this.

chit n. **1** a young child. **2** a short written note.

chitterlings n.pl. the small intestines of a pig, cooked as food.

chivalry n. courtesy and considerate behaviour, esp. towards weaker people. □ **chivalrous** adj.

chive n. a herb with onion-flavoured leaves.

chivvy v. (colloquial) urge to hurry.

chloride n. a compound of chlorine and another element.

chlorinate v. treat or sterilize with chlorine. □ **chlorination** n.

chlorine n. a chemical element (symbol Cl), a poisonous gas.

chlorofluorocarbon see CFC.

chloroform n. a liquid giving off vapour that causes unconsciousness when inhaled.

chlorophyll n. green colouring matter in plants.

choc n. (colloquial) chocolate.

choc ice n. a bar of ice cream coated with chocolate.

chock n. a block or wedge for preventing something from moving. ● v. wedge with chocks.

chock-a-block adj. & adv. crammed, crowded together.

chocolate n. an edible substance made from cacao seeds; a sweet made or coated with this, a drink made with this; a dark brown colour.

choice n. choosing, the right of choosing; a variety from which to choose; a person or thing chosen. ● adj. of especially good quality.

choir n. an organized band of singers, esp. in church; part of a church where these sit, the chancel.

choirboy n. a boy singer in a church choir.

choke v. stop (a person) breathing by squeezing or blocking the windpipe; be unable to breathe; clog, smother. ● n. a valve controlling the flow of air into a petrol engine.

choker n. a close-fitting necklace.

cholera n. a serious often fatal disease caused by bacteria.

choleric adj. easily angered.

cholesterol n. a fatty animal substance thought to cause hardening of arteries.

chomp v. munch noisily.

choose v. (chose, chosen, choosing) select out of a number of things; decide (on), prefer.

choosy adj. (choosier) (colloquial) careful in choosing, hard to please. □ **choosiness** n.

chop v. (chopped) cut by a blow with an axe or knife; hit with a

short downward movement. ● n. a chopping stroke; a thick slice of meat, usu. including a rib.

chopper n. 1 a chopping tool. 2 (colloquial) a helicopter.

choppy adj. (choppier) full of short broken waves; jerky.

chopstick n. each of a pair of sticks used in China, Japan, etc., to lift food to the mouth.

chop suey n. a Chinese dish of meat chopped and fried with vegetables.

choral adj. for or sung by a chorus.

chorale n. a choral composition using the words of a hymn; a choir.

chord n. 1 a combination of notes sounded together. 2 a straight line joining two points on a curve.

chore n. a routine task.

choreography (ko-ri-og-răfi) n. the composition of stage dances. □ **choreographer** n., **choreographic** adj.

chorister n. a member of a choir.

chortle n. a loud chuckle. ● v. utter a chortle.

chorus n. a group of singers; a group of singing dancers in a musical comedy etc.; something spoken or sung by many together; the refrain of a song. ● v. say as a group.

chose, chosen see choose.

choux pastry (shoo) n. very light pastry enriched with eggs.

chow n. 1 a long-haired dog of a Chinese breed. 2 (slang) food.

chowder n. a thick soup usu. containing clams or fish.

chow mein (mayn) n. a Chinese-style dish of fried noodles and shredded meat etc.

christen v. admit to the Christian Church by baptism; name.

Christendom n. all Christians or Christian countries.

Christian *adj.* of or believing in Christianity; kindly, humane.
● *n.* a believer in Christianity.

Christianity *n.* a religion based on the teachings of Christ.

Christian name *n.* a personal name, a first name.

Christian Science *n.* a religious system by which health and healing are sought by Christian faith, without medical treatment.

Christmas *n.* a festival (25 Dec.) commemorating Christ's birth.

Christmas tree *n.* an evergreen or artificial tree decorated at Christmas.

chromatic *adj.* of colour, in colours. □ **chromatically** *adv.*

chromatic scale *n.* a music scale proceeding by semitones.

chromatography *n.* separation of substances by slow passage through material that absorbs these at different rates so that they appear as layers.

chrome *n.* chromium; a yellow pigment from a compound of this.

chromium *n.* a metallic element (symbol Cr) that does not rust.

chromosome *n.* a threadlike structure carrying genes in animal and plant cells.

chronic *adj.* constantly present or recurring; having a chronic disease or habit; (*colloquial*) bad. □ **chronically** *adv.*

chronicle *n.* a record of events.
● *v.* record in a chronicle. □ **chronicler** *n.*

chronological *adj.* arranged in the order in which things occurred. **chronologically** ● *adv.*

chronology *n.* arrangement of events in order of occurrence.

chronometer *n.* a time-measuring instrument.

chrysalis *n.* (*pl.* **chrysalises** or **chrysalides**) a form of an insect in the stage between larva and adult insect; the case enclosing it.

chrysanthemum *n.* a garden plant flowering in autumn.

chub *n.* a river fish.

chubby *adj.* (**chubbier**) round and plump. □ **chubbiness** *n.*

chuck *v.* (*colloquial*) throw carelessly or casually. ● *n.* **1** part of a lathe holding the drill; part of a drill holding the bit. **2** a cut of beef from neck to ribs.

chuckle *v.* & *n.* (utter) a quiet laugh.

chuff *v.* (of an engine) work with a regular puffing noise.

chuffed *adj.* (*colloquial*) pleased.

chug *v.* (**chugged**) make or move with a dull short repeated sound. ● *n.* this sound.

chukka *n.* (*Amer.* **chukker**) a period of play in a polo game.

chum *n.* (*colloquial*) a close friend. □ **chummy** *adj.*

chump *n.* **1** (*slang*) the head. **2** (*colloquial*) a foolish person.

chump chop *n.* a chop from the thick end of a loin of mutton.

chunk *n.* a thick piece; a substantial amount.

chunky *adj.* (**chunkier**) short and thick; in chunks, containing chunks. □ **chunkiness** *n.*

chunter *v.* (*colloquial*) mutter, grumble.

church *n.* a building for public Christian worship; a religious service in this; (**the Church**) Christians collectively; a particular group of these.

churchwarden *n.* a parish representative, assisting with church business.

churchyard *n.* enclosed land round a church, used for burials.

churlish *adj.* ill-mannered, surly. □ **churlishly** *adv.*, **churlishness** *n.*

churn *n.* a machine in which milk is beaten to make butter; a very large milk can. ● *v.* beat (milk) or make (butter) in a churn; stir or swirl violently. □ **churn out** produce rapidly.

chute n. a sloping channel down which things can be slid or dropped.

chutney n. (pl. **chutneys**) a seasoned mixture of fruit, vinegar, spices, etc.

Ci abbr. curie.

CIA abbr. (in the USA) Central Intelligence Agency.

ciao (chow) int. (colloquial) hello; goodbye.

cicada (si-kah-dǎ) n. a chirping insect resembling a grasshopper.

cicatrice n. a scar.

CID abbr. Criminal Investigation Department.

cider n. a fermented drink made from apples.

cigar n. a roll of tobacco leaf for smoking.

cigarette n. a roll of shredded tobacco in thin paper for smoking.

cinch n. (colloquial) a certainty, an easy task.

cinder n. a piece of partly burnt coal or wood.

cine-camera n. a cinematographic camera.

cinema n. a theatre where films are shown; films as an art form or industry.

cinematography n. the process of making and projecting moving pictures. □ **cinematographic** adj.

cinnamon n. a spice made from the bark of a south-east Asian tree.

cipher n. (also **cypher**) **1** a symbol (0) representing nought or zero; a numeral; a person of no importance. **2** a system of letters or numbers used to represent others for secrecy.

circa prep. about.

circle n. **1** a perfectly round plane figure. **2** a curved tier of seats at a theatre etc. **3** a group with similar interests. ● v. move in a circle; form a circle round.

circlet n. a small circle; a circular band worn as an ornament.

circuit n. a line, route, or distance round a place; the path of an electric current.

circuitous (sir-kew-itǔs) adj. roundabout, indirect. □ **circuitously** adv.

circuitry n. circuits.

circular adj. shaped like or moving round a circle. ● n. a letter or leaflet sent to a circle of people. □ **circularity** n.

circulate v. go or send round.

circulation n. **1** circulating; the movement of blood round the body. **2** the number of copies sold, esp. of a newspaper. □ **circulatory** adj.

circumcise v. cut off the foreskin of. □ **circumcision** n.

circumference n. the boundary of a circle, the distance round this.

circumflex accent n. the accent ˆ.

circumlocution n. a roundabout, verbose, or evasive expression. □ **circumlocutory** adj.

circumnavigate v. sail completely round. □ **circumnavigation** n., **circumnavigator** n.

circumscribe v. draw a line round; restrict.

circumspect adj. cautious and watchful, wary. □ **circumspection** n., **circumspectly** adv.

circumstance n. an occurrence or fact.

circumstantial adj. detailed; consisting of facts that suggest something but do not prove it.

circumvent v. evade (a difficulty etc.). □ **circumvention** n.

circus n. a travelling show with performing animals, acrobats, etc.

cirque n. a bowl-shaped hollow on a mountain.

cirrhosis (si-**roh**-sis) n. disease of the liver.

cirrus n. (pl. **cirri**) a high wispy white cloud.

cistern n. a tank for storing water.

citadel n. a fortress overlooking a city.

cite v. quote or mention as an example etc. □ **citation** n.

citizen n. an inhabitant of a city; a person with full rights in a country. □ **citizenship** n.

citric acid n. acid in the juice of lemons, limes, etc.

citrus n. a tree of a group including lemon, orange, etc.

city n. an important town; a town with special rights given by charter and containing a cathedral.

civet n. (in full **civet cat**) a cat-like animal of central Africa; a musky substance obtained from its glands.

civic adj. of a city or citizenship.

civil adj. **1** of citizens; not of the armed forces or the Church. **2** polite and obliging. □ **civilly** adv.

civil engineering n. the design and construction of roads, bridges, etc.

civilian n. a person not in the armed forces.

civility n. politeness.

civilization n. (also **-isation**) making or becoming civilized; a stage in the evolution of society; civilized conditions.

civilize v. (also **-ise**) cause to improve to a developed stage of society; improve the behaviour of.

Civil List n. an annual allowance for the royal family's household expenses.

civil servant n. an employee of the **civil service**, government departments other than the armed forces.

civil war n. war between citizens of the same country.

civvies n.pl. (slang) ordinary clothes, not uniform.

Cl symb. chlorine.

cl abbr. centilitre(s).

clack n. a short sharp sound; the noise of chatter. ● v. make this sound.

clad adj. clothed.

cladding n. boards or metal plates as a protective covering.

claim v. demand as one's right; assert. ● n. a demand; an assertion; a right or title.

claimant n. a person making a claim.

clairvoyance n. the power of seeing the future. □ **clairvoyant** n. & adj.

clam n. a shellfish with a hinged shell. □ **clam up** (**clammed**) (colloquial) refuse to talk.

clamber v. climb with difficulty.

clammy adj. (**clammier**) unpleasantly moist and sticky.

clamour n. (Amer. **clamor**) a loud confused noise; a loud protest etc. □ **clamorous** adj.

clamp n. a device for holding things tightly; a device for immobilizing an illegally parked car. ● v. grip with a clamp, fix firmly; immobilize (an illegally parked car) with a clamp. □ **clamp down on** become firmer about, put a stop to.

clan n. a group of families with a common ancestor. □ **clannish** adj.

clandestine adj. kept secret, done secretly.

clang n. a loud ringing sound. ● v. make this sound.

clanger n. (slang) a blunder.

clangour n. (Amer. **clangor**) a clanging noise.

clank n. a sound like metal striking metal. ● v. make or cause to make this sound.

clap v. (**clapped**) strike the palms loudly together, esp. in applause; strike or put quickly or vigorously. ● *n.* an act or sound of clapping; a sharp noise of thunder. □ **clapped out** (*slang*) worn out.

clapper *n.* the tongue or striker of a bell.

clapperboard *n.* a device in film-making for making a sharp clap for synchronizing picture and sound at the start of a scene.

claptrap *n.* pretentious talk; nonsense.

claret *n.* a dry red wine.

clarify *v.* **1** make or become clear. **2** remove impurities from (fats) by heating. □ **clarification** *n.*

clarinet *n.* a woodwind instrument with a single reed and finger-holes and keys. □ **clarinettist** *n.*

clarion *adj.* loud, rousing.

clarity *n.* clearness.

clash *n.* a conflict; discordant sounds or colours. ● *v.* make clash; conflict.

clasp *n.* a device for fastening things, with interlocking parts; a grasp, a handshake. ● *v.* fasten, join with a clasp; grasp, embrace closely.

class *n.* a set of people or things with characteristics in common; a standard of quality; a rank of society; a set of students taught together. ● *v.* place in a class.

classic *adj.* **1** of recognized high quality. **2** typical. **3** simple in style. ● *n.* **1** a classic author or work etc. **2** (**classics**) the study of ancient Greek and Roman literature, history, etc. □ **classicism** *n.*, **classicist** *n.*

classical *adj.* classic; of the ancient Greeks and Romans; traditional and standard. □ **classically** *adv.*

classify *v.* **1** arrange systematically, class. **2** designate as

officially secret. □ **classifiable** *adj.*, **classification** *n.*

classless *adj.* without distinctions of social class.

classroom *n.* a room where a class of students is taught.

classy *adj.* (**classier**) (*colloquial*) of high quality, stylish.

clatter *n.* a rattling sound. ● *v.* make this sound.

clause *n.* **1** a single part in a treaty, law, or contract. **2** a distinct part of a sentence, with its own verb.

claustrophobia *n.* abnormal fear of being in an enclosed space. □ **claustrophobic** *adj.*

clavichord *n.* an early small keyboard instrument.

clavicle *n.* the collarbone.

claw *n.* a pointed nail on an animal's or bird's foot; a clawlike device for grappling or holding things. ● *v.* scratch or pull with a claw or hand.

clay *n.* stiff sticky earth, used for making bricks and pottery. □ **clayey** *adj.*

clay pigeon *n.* a breakable disc thrown up as a target for shooting.

clean *adj.* free from dirt or impurities; not soiled or used. ● *v.* make clean. □ **cleaner** *n.*, **cleanly** *adv.*

cleanly (klen-li) *adj.* (**cleanlier**) attentive to cleanness, with clean habits. □ **cleanliness** *n.*

cleanse (klenz) *v.* make clean. □ **cleanser** *n.*

clear *adj.* **1** transparent. **2** free from doubt, difficulties, obstacles, etc. **3** easily seen, heard, or understood. ● *v.* **1** make or become clear. **2** prove innocent. **3** get past or over. **4** make as net profit. □ **clear off** (*colloquial*) go away. **clear out** empty; remove; (*colloquial*) go away. □ **clearly** *adv.*

clearance n. 1 clearing. 2 permission. 3 space allowed for one object to pass another.

clearing n. a space cleared of trees in a forest.

clearing house n. an office at which banks exchange cheques; an agency collecting and distributing information.

clearway n. a road where vehicles must not stop.

cleat n. a projecting piece for fastening ropes to.

cleavage n. a split, a separation; the hollow between full breasts.

cleave v. (**cleaved, clove,** or **cleft; cleaved, cloven,** or **cleft; cleaving**) (*literary*) split.

cleaver n. a butcher's chopper.

clef n. a symbol on a stave in music, showing the pitch of notes.

cleft adj. split. ● n. a split, a cleavage.

cleft palate n. a defect in the roof of the mouth where two sides of the palate failed to join.

clematis n. a climbing plant with showy flowers.

clemency n. mildness, esp. of weather; mercy. □ **clement** adj.

clementine n. a small variety of orange.

clench v. close (the teeth or fingers) tightly.

clerestory n. an upper row of windows in a large church.

clergy n. people ordained for religious duties. □ **clergyman** n.

cleric n. a member of the clergy.

clerical adj. 1 of clerks. 2 of clergy.

clerk n. a person employed to do written work in an office.

clever adj. quick at learning and understanding things; showing skill. □ **cleverly** adv., **cleverness** n.

cliché (klee-shay) n. an overused phrase or idea. □ **clichéd** adj.

click n. a short sharp sound. ● v. 1 make or cause to make a click.

2 (*colloquial*) be a success, be understood.

client n. a person using the services of a professional person; a customer.

clientele (klee-ahn-tel) n. clients.

cliff n. a steep rock face, esp. on a coast.

cliffhanger n. a story or contest full of suspense.

climate n. the regular weather conditions of an area.

climax n. the point of greatest interest or intensity.

climb v. go up or over. ● n. an ascent made by climbing. □ **climber** n.

clime n. a climate; a region.

clinch v. fasten securely; settle conclusively; (of boxers) hold on to each other. ● n. clinching. □ **clincher** n.

cling v. (**clung, clinging**) hold on tightly; stick.

cling film n. thin polythene wrapping.

clinic n. a place or session at which medical treatment is given; a private or specialized hospital.

clinical adj. 1 of or used in treatment of patients. 2 looking bare and hygienic. 3 unemotional and detached. □ **clinically** adv.

clink n. a thin sharp sound. ● v. make this sound.

clinker n. fused coal ash.

clip n. 1 a device for holding things tightly or together. 2 an act of clipping; a piece clipped from something. 3 (*colloquial*) a sharp blow. ● v. (**clipped**) 1 fix or fasten with a clip. 2 cut with shears or scissors. 3 (*colloquial*) hit sharply.

clipper n. 1 a fast sailing ship. 2 (**clippers**) an instrument for clipping things.

clipping n. a piece clipped off; a newspaper cutting.

clique (kleek) *n.* a small exclusive group. □ **cliquey** *adj.*, **cliquish** *adj.*

clitoris *n.* the small erectile part of the female genitals.

Cllr. *abbr.* Councillor.

cloak *n.* a loose sleeveless outer garment. ● *v.* cover, conceal.

cloakroom *n.* a room where outer garments can be left, often containing a lavatory.

clobber (*slang*) *n.* equipment; belongings. ● *v.* hit hard; defeat heavily.

cloche (klosh) *n.* a translucent cover for protecting plants.

clock *n.* an instrument indicating time. □ **clock in** or **on**, **out** or **off** register one's time of arrival or departure. **clock up** achieve.

clockwise *adv. & adj.* moving in the direction of the hands of a clock.

clockwork *n.* a mechanism with wheels and springs.

clod *n.* a lump of earth.

clog *n.* a wooden-soled shoe. ● *v.* (**clogged**) cause an obstruction in; become blocked.

cloister *n.* a covered walk along the side of a church etc.; life in a monastery or convent.

cloistered *adj.* shut away, secluded.

clone *n.* a group of plants or organisms produced asexually from one ancestor. ● *v.* grow in this way.

close (klohss) *adj.* **1** near; near together. **2** dear to each other. **3** dense; concentrated. **4** secretive; stingy. **5** stuffy. ● *adv.* closely; in a near position. ● *v.* (klohz) **1** shut; bring or come to an end. **2** come nearer together. ● *n.* **1** (klohz) a conclusion, an end. **2** (klohss) a street closed at one end; the grounds round a cathedral or abbey. □ **closely** *adv.*, **closeness** *n.*

closet *n.* a cupboard; a storeroom. ● *v.* (**closeted**) shut away in private conference or study.

close-up *n.* a photograph etc. showing a subject at close range.

closure *n.* closing, a closed condition.

clot *n.* **1** a thickened mass of liquid. **2** (*colloquial*) a stupid person. ● *v.* (**clotted**) form clots.

cloth *n.* woven or felted material; a piece of this; a tablecloth; clerical clothes.

clothe *v.* put clothes on, provide with clothes.

clothes *n.pl.* things worn to cover the body.

clothier *n.* a seller of men's clothes.

clothing *n.* clothes for the body.

clotted cream *n.* very thick cream, thickened by scalding.

cloud *n.* a visible mass of watery vapour floating in the sky; a mass of smoke or dust. ● *v.* become covered with clouds or gloom.

cloudburst *n.* a violent storm of rain.

cloudy *adj.* (**cloudier**) covered with clouds; (of liquid) not transparent. □ **cloudiness** *n.*

clout *n.* **1** a blow. **2** (*colloquial*) influence, power of effective action. ● *v.* hit.

clove *n.* **1** a dried bud of a tropical tree, used as spice. **2** one division of a compound bulb such as garlic. *See also* **cleave**.

clove hitch *n.* a knot used to fasten a rope round a pole etc.

cloven hoof *n.* a divided hoof like that of sheep, cows, etc.

clover *n.* a plant with three-lobed leaves. □ **in clover** in luxury.

clown *n.* a person who does comical tricks. ● *v.* perform or behave as a clown.

cloy *v.* sicken by glutting with sweetness or pleasure.

club n. **1** a heavy stick used as a weapon; a stick with a wooden or metal head, used in golf. **2** a playing card of the suit marked with black clover leaves. **3** a group who meet for social or sporting purposes, their premises; an organization offering benefits to subscribers. ● v. (**clubbed**) strike with a club. □ **club together** join in subscribing.

club class n. a class of air travel designed for business travellers

cluck n. the throaty cry of a hen. ● v. utter a cluck.

clue n. a fact or idea giving a guide to the solution of a problem.

clump n. a cluster, a mass. ● v. **1** tread heavily. **2** form into a clump.

clumsy adj. (**clumsier**) large and ungraceful. □ **clumsily** adv., **clumsiness** n.

clung see **cling**.

cluster n. a small close group. ● v. form a cluster.

clutch v. grasp tightly. ● n. **1** a tight grasp. **2** a device for connecting and disconnecting moving parts. **3** a set of eggs for hatching; chickens hatched from these.

clutter n. things lying about untidily. ● v. fill with clutter.

Cm symb. curium.

cm abbr. centimetre.

CND abbr. Campaign for Nuclear Disarmament.

Co symb. cobalt.

Co. abbr. **1** Company. **2** County.

c/o abbr. care of.

co- pref. joint, jointly.

coach n. **1** a long-distance bus; a large horse-drawn carriage. **2** a private tutor; an instructor in sports. ● v. train, teach.

coagulate (koh-ag-yoo-layt) v. change from liquid to semi-solid, clot. □ **coagulant** n., **coagulation** n.

coal n. a hard black mineral burnt as fuel.

coalesce v. combine. □ **coalescence** n.

coalfield n. an area where coal occurs.

coalition n. a union, esp. a temporary union of political parties.

coal tar n. tar produced when gas is made from coal.

coarse adj. composed of large particles; rough in texture; rough or crude in manner, vulgar. □ **coarsely** adv., **coarseness** n.

coarse fish n. any freshwater fish other than salmon and trout.

coarsen v. make or become coarse.

coast n. the seashore and land near it. ● v. **1** sail along a coast. **2** ride a bicycle or drive a motor vehicle without using power. □ **coastal** adj.

coaster n. **1** a ship trading along a coast. **2** a mat for a glass.

coastguard n. an officer of an organization that keeps watch on the coast.

coat n. an outdoor garment with sleeves; the fur or hair covering an animal's body; a covering layer. ● v. cover with a layer.

coating n. a covering layer.

coat of arms n. a design on a shield as the emblem of a family or institution.

coax v. persuade gently; manipulate carefully or slowly.

coaxial (koh-aks-iăl) adj. (of cable) containing two conductors, one surrounding but insulated from the other.

cob n. **1** a sturdy short-legged horse. **2** a hazelnut. **3** the central part of an ear of maize. **4** a small round loaf.

cobalt n. a metallic element (symbol Co); a deep blue pigment made from it.

cobber n. (*Austral. colloquial*) a friend, a mate.

cobble n. a rounded stone formerly used for paving roads. ● v. mend roughly.

cobbler n. a shoe-mender.

cobra n. a poisonous snake of India and Africa.

cobweb n. a network spun by a spider.

cocaine n. a drug used illegally as a stimulant.

coccyx (kok-siks) n. the bone at the base of the spinal column.

cochineal n. red food colouring made from dried insects.

cock n. 1 a male bird. 2 a tap or valve controlling a flow. ● v. tilt or turn upwards; set (a gun) for firing.

cockade n. a rosette worn on a hat as a badge.

cock-a-hoop adj. very pleased, triumphant.

cockatiel n. (also **cockateel**) a small Australian parrot.

cockatoo n. (pl. **cockatoos**) a crested parrot.

cockatrice n. 1 a basilisk. 2 a mythical cock with a serpent's tail.

cockerel n. a young male fowl.

cocker spaniel n. a small silky-coated spaniel.

cock-eyed adj. (*colloquial*) askew; absurd.

cockle n. an edible shellfish.

cockney n. (pl. **cockneys**) a native or dialect of the East End of London.

cockpit n. 1 the compartment for the pilot in a plane, or for the driver in a racing car. 2 a pit for cockfighting.

cockroach n. a beetle-like insect.

cockscomb n. a cock's crest.

cocksure adj. over-confident.

cocktail n. a mixed alcoholic drink. ● **fruit cocktail** mixed chopped fruit.

cocky adj. (**cockier**) conceited and arrogant. □ **cockily** adv.

cocoa n. powder of crushed cacao seeds; a drink made from this.

coconut n. a nut of a tropical palm; its edible lining.

cocoon n. a silky sheath round a chrysalis; a protective wrapping.

COD abbr. cash (or *Amer.* collect) on delivery.

cod n. a large edible sea fish.

coda n. the final part of a musical composition.

coddle v. 1 cherish and protect. 2 cook (eggs) in water just below boiling point.

code n. a set of laws, rules, or signals; a word or phrase used to represent a message for secrecy; a cipher.

codeine n. a substance made from opium, used to relieve pain.

codex n. (pl. **codices** or **codexes**) 1 an ancient manuscript text in book form. 2 a pharmaceutical description of drugs.

codicil n. an appendix to a will.

codify v. arrange (laws etc.) into a code. □ **codification** n.

codpiece n. a flap at the front of men's breeches in 15th- and 16th-century dress.

co-education n. education of boys and girls in the same classes. □ **co-educational** adj.

coefficient n. a multiplier; a mathematical factor.

coelacanth (seel-ă-kanth) n. a large sea fish formerly thought to be extinct.

coeliac disease (seel-iak) n. (*Amer.* **celiac**) a disease causing inability to digest gluten.

coerce v. compel by threats or force. □ **coercion** n., **coercive** adj.

coeval (koh-eev-ăl) adj. of the same age or epoch.

coexist v. exist together, esp. harmoniously. □ **coexistence** n., **coexistent** adj.

coffee n. the beanlike seeds of a tropical shrub, roasted and ground for making a drink; this drink; a pale brown colour.

coffee table n. a small low table.

coffer n. a large strong box for holding money and valuables; (**coffers**) financial resources.

coffer-dam n. an enclosure pumped dry to enable construction work to be done within it.

coffin n. a box in which a corpse is placed for burial or cremation.

cog n. one of a series of projections on the edge of a wheel, engaging with those of another.

cogent adj. convincing. □ **cogently** adv., **cogency** n.

cogitate v. think deeply. □ **cogitation** n.

cognac n. French brandy.

cognate adj. akin, related. ● n. a relative; a cognate word.

cognition n. knowing, perceiving. □ **cognitive** adj.

cognizant adj. aware, having knowledge. □ **cognizance** n.

cognoscente (kon-yŏ-shen-ti) n. (pl. **cognoscenti**) a connoisseur.

cohabit v. live together as man and wife. □ **cohabitation** n.

cohere v. stick together.

coherent adj. cohering; connected logically, not rambling. □ **coherently** adv., **coherence** n.

cohesion n. a tendency to cohere.

cohesive adj. cohering.

cohort n. a tenth part of a Roman legion; people grouped together.

coiffure (kwa-fyoor) n. a hairstyle.

coil v. wind into rings or a spiral. ● n. 1 something coiled; one ring or turn in this. 2 a

contraceptive device inserted into the womb.

coin n. a piece of metal money. ● v. 1 make (coins) by stamping metal; get (money) in large quantities as profit. 2 invent (a word or phrase). □ **coiner** n.

coinage n. 1 coining; coins, a system of these. 2 a coined word or phrase.

coincide v. occupy the same portion of time or space; be in agreement or identical.

coincidence n. coinciding; a remarkable occurrence of similar events at the same time by chance. □ **coincidental** adj., **coincidentally** adv.

coir n. coconut fibre.

coitus n. (also **coition**) sexual intercourse.

coke n. 1 a solid substance left after gas and tar have been extracted from coal, used as fuel. 2 (slang) cocaine.

col n. a depression in a range of mountains.

cola n. (also **kola**) a West African tree with seeds containing caffeine; a drink flavoured with these.

colander n. a bowl-shaped perforated vessel for draining food.

cold adj. 1 at or having a low temperature. 2 not affectionate, not enthusiastic. ● n. 1 low temperature; a cold condition. 2 an illness causing catarrh and sneezing. □ **coldly** adv., **coldness** n.

cold-blooded adj. 1 having a blood temperature varying with that of the surroundings. 2 unfeeling, ruthless.

cold calling n. a sales technique involving contacting people who have not previously shown interest.

cold feet n.pl. (colloquial) loss of nerve or confidence.

cold-shoulder v. treat with deliberate unfriendliness.

cold turkey n. (slang) sudden withdrawal of drugs from an addict.

cold war n. hostility between nations without fighting.

coleslaw n. a salad of shredded raw cabbage in dressing.

coley n. any of several edible fish, e.g. rock salmon.

colic n. severe abdominal pain.

colitis n. inflammation of the colon.

collaborate v. work in partnership. □ **collaboration** n., **collaborator** n., **collaborative** adj.

collage (kol--ahzh) n. an artistic composition in which objects are glued to a backing to form a picture.

collagen n. a protein substance found in bone and tissue.

collapse v. fall down suddenly; lose strength suddenly; fold. ● n. collapsing; breakdown.

collapsible adj. made so as to fold up.

collar n. a band round the neck of a garment; a band holding part of a machine; a cut of bacon from near the head. ● v. (colloquial) seize, take for oneself.

collate v. compare in detail; collect and arrange systematically. □ **collator** n.

collateral adj. parallel; additional but subordinate. ● n. additional security pledged. □ **collaterally** adv.

collation n. 1 collating. 2 a light meal.

colleague n. a fellow worker esp. in a business or profession.

collect v. bring or come together; obtain specimens of, esp. as a hobby; fetch. □ **collectable** or **collectible** adj. & n.

collected adj. calm and controlled.

collection n. collecting; objects or money collected.

collective adj. of or denoting a group taken or working as a unit. □ **collectively** adv.

collective noun n. (Grammar) a noun (singular in form) denoting a group (e.g. army, herd).

collector n. a person who collects things.

colleen n. (Irish) a girl.

college n. an educational establishment for higher or professional education; an organized body of professional people. □ **collegiate** adj.

collide v. come into collision.

collie n. a sheepdog with a pointed muzzle.

colliery n. a coal mine.

collision n. violent striking of one thing against another.

collocate v. place (words) together. □ **collocation** n.

cologne n. eau-de-cologne or other lightly scented liquid.

colloquial adj. suitable for informal speech or writing. □ **colloquially** adv., **colloquialism** n.

collusion n. an agreement made for a deceitful or fraudulent purpose.

colon n. **1** a punctuation mark (:). **2** the lower part of the large intestine. □ **colonic** adj.

colonel (ker-nĕl) n. an army officer next below brigadier.

colonial adj. of a colony or colonies. ● n. an inhabitant of a colony.

colonialism n. a policy of acquiring or maintaining colonies.

colonize v. (also **-ise**) establish a colony in. □ **colonization** n., **colonist** n.

colonnade n. a row of columns.

colony n. a settlement or settlers in a new territory, remaining subject to the parent state; people of one nationality or occupation living in a particular area; birds congregated similarly.

color, colorful etc. Amer. sp. of **colour, colourful** etc.

coloration n. (also **colouration**) colouring.

coloratura n. elaborate ornamentation of a vocal melody; a singer (esp. a soprano) skilled in such singing.

colossal adj. immense. □ **colossally** adv.

colossus n. (pl. **colossi** or **colossuses**) an immense statue.

colostomy n. an operation to form an opening from the colon onto the surface of the abdomen, through which the bowel can empty.

colour (Amer. **color**) n. the sensation produced by rays of light of particular wavelengths; pigment; (usu. **colours**) the flag of a ship or regiment. ● v. put colour on; paint, stain, dye; blush; give a special character or bias to.

colourant n. (also **colorant**) colouring matter.

colour-blind adj. unable to distinguish between certain colours.

coloured adj. (Amer. **colored**) having a colour; wholly or partly of non-white descent.

colourful adj. (Amer. **colorful**) full of colour; with vivid details. □ **colourfully** adv.

colourless adj. (Amer. **colorless**) without colour; lacking vividness.

colt n. a young male horse.

coltsfoot n. a wild plant with yellow flowers.

columbine n. a flower with pointed projections on its petals.

column n. 1 a round pillar. 2 something shaped like this; a vertical division of a page, printed matter in this; a long narrow formation of troops, vehicles, etc.

columnist n. a journalist who regularly writes a column of comments.

coma n. deep unconsciousness.

comatose adj. in a coma; drowsy.

comb n. 1 a toothed strip of stiff material for tidying hair. 2 a fowl's fleshy crest. ● v. tidy with a comb; search thoroughly.

combat n. a battle, a contest. ● v. (**combated**) counter. □ **combative** adj.

combatant n. & adj. (a person or nation) engaged in fighting.

combe var. of **coomb**.

combination n. 1 combining; a set of people or things combined. 2 (**combinations**) an undergarment covering the body and legs.

combination lock n. a lock controlled by a series of positions of dials.

combine v. (kŏm-byn) join into a group, set, or mixture. ● n. (kom-byn) 1 a combination of people or firms acting together. 2 (in full **combine harvester**) a combined reaping and threshing machine.

combining form n. a word or partial word used in combination with another to form a different word, e.g. Anglo-.

combustible adj. capable of catching fire.

combustion n. burning; the process in which substances combine with oxygen and produce heat.

come v. (came, come, coming) move towards the speaker or a place or point; arrive; occur. □ **come about** happen. **come across** meet or find unexpectedly. **come by** obtain. **come into** inherit. **come off** be successful. **come out** become visible; emerge; declare one's feelings openly. **come round** recover from fainting; be con-

verted to the speaker's opinion.
come to regain consciousness.
come up arise for discussion.

comeback n. **1** a return to a former successful position. **2** a retort.

comedian n. a humorous entertainer or actor.

comedienne n. a female comedian.

comedown n. a fall in status.

comedy n. a light amusing drama.

comely adj. (**comelier**) (old use) good-looking. □ **comeliness** n.

comestibles n.pl. (formal) food.

comet n. a heavenly body with a luminous tail of gas and dust.

come-uppance n. deserved punishment.

comfort n. a state of ease and contentment; relief of suffering or grief; a person or thing giving this. ● v. give comfort to. □ **comforter** n.

comfortable adj. providing or having ease and contentment; not close or restricted. □ **comfortably** adv.

comfrey n. a tall plant with bell-like flowers.

comfy adj. (**comfier**) (colloquial) comfortable.

comic adj. causing amusement; of comedy. ● n. **1** a comedian. **2** a children's paper with a series of strip cartoons. □ **comical** adj., **comically** adv.

comic strip n. a sequence of drawings telling a story.

comma n. a punctuation mark (,).

command n. an order, an instruction; the right to control others; mastery; forces or a district under a commander. ● v. give a command to; have authority over.

commandant n. an officer in command of a military establishment.

commandeer v. seize for use.

commander n. a person in command; a naval officer next below captain; a police officer next below commissioner.

commandment n. a God-given rule for living.

commando n. (pl. **commandos**) a member of a military unit specially trained for making raids and assaults.

commemorate v. keep in the memory by a celebration or memorial. □ **commemoration** n., **commemorative** n.

commence v. begin. □ **commencement** n.

commend v. praise; entrust. □ **commendation** n.

commendable adj. worthy of praise. □ **commendably** adv.

commensurable adj. measurable by the same standard. □ **commensurably** adv., **commensurability** n.

commensurate adj. of the same size; proportionate.

comment n. an opinion given; an explanatory note. ● v. make comments or remarks.

commentary n. a series of comments.

commentate v. act as commentator.

commentator n. a person who writes or speaks a commentary.

commerce n. all forms of trade and services (e.g. banking, insurance).

commercial adj. of, engaged in, or financed by commerce. □ **commercially** adv.

commercialize v. (also **-ise**) make commercial; make profitable. □ **commercialization** n.

commiserate v. express pity for; sympathize. □ **commiseration** n.

commissariat n. a military department supplying food.

commission n. **1** committing. **2** the authority to perform a task; a task given; a body of people given such authority; a warrant conferring authority on an officer in the armed forces. **3** a payment to an agent selling goods or services. ● v. **1** give a commission to. **2** place an order for. □ **in commission** ready for service. **out of commission** not in working order.

commissionaire n. a uniformed attendant at the door of a theatre, business premises, etc.

commissioner n. a member of, or a person appointed by, a commission; a government official in charge of a district abroad.

commit v. (**committed**) **1** do, perform. **2** entrust, consign; pledge to a course of action.

commitment n. committing; an obligation or pledge, a state of being involved in this.

committal n. committing to prison etc.

committee n. a group of people appointed to attend to special business or manage the affairs of a club etc.

commode n. **1** a chest of drawers. **2** a chamber pot in a chair or box.

commodious adj. roomy.

commodity n. an article of trade, a product.

commodore n. a naval officer next below rear admiral; a president of a yacht club.

common adj. of or affecting all; occurring often; ordinary; of inferior quality. ● n. an area of unfenced grassland for all to use.

commoner n. one of the common people, not a noble.

common law an unwritten law based on custom and former court decisions.

commonly adv. usually, frequently.

Common Market n. the European Community.

commonplace adj. ordinary; lacking originality.

common room n. a room shared by students or workers for social purposes.

common sense n. normal good sense in practical matters.

commonwealth n. an independent state; a federation of states; (**the Commonwealth**) an international association of independent states that used to be ruled by Britain.

commotion n. fuss and disturbance.

communal adj. shared among a group. □ **communally** adv.

commune v. (kŏ-mewn) communicate mentally or spiritually. ● n. (kom-yoon) **1** a group (not all of one family) sharing accommodation and goods. **2** a district of local government in France etc.

communicable adj. able to be communicated.

communicant n. **1** a person who receives Holy Communion. **2** one who communicates information.

communicate v. make known; transmit; pass news and information to and fro; have or be a means of access. □ **communicator** n.

communication n. communicating; a letter or message; a means of access.

communicative adj. talkative, willing to give information.

communion n. **1** fellowship; social dealings. **2** a branch of the Christian Church; ((**Holy**) **Communion**) a sacrament in which bread and wine are consumed.

communiqué (kŏ-mew-ni-kay) n. an official report; an agreed statement.

communism n. a social system based on common ownership of property, means of production, etc.; a political doctrine or movement seeking a form of this. □ **communist** n.

community n. a body of people living in one district or having common interests or origins.

commutable adj. exchangeable.

commute v. **1** exchange for something else. **2** travel regularly by train or bus to and from one's work. □ **commuter** n.

compact adj. (kŏm-pakt) closely or neatly packed together; concise. ● v. (kŏm-pakt) make compact. ● n. (kom-pakt) **1** a small flat case for face powder. **2** a pact, a contract.

compact disc n. a small disc from which sound etc. is reproduced by laser action.

companion n. a person who accompanies another; a thing that matches or accompanies another. □ **companionship** n.

companionable adj. sociable.

companionway n. a staircase from a ship's deck to cabins etc.

company n. being with another or others; people assembled; guests; associate(s); people working together or united for business purposes, a firm; a subdivision of an infantry battalion.

comparable adj. suitable to be compared, similar. □ **comparability** n., **comparably** adv.

comparative adj. involving comparison; of the grammatical form expressing 'more'. ● n. a comparative form of a word. □ **comparatively** adv.

compare v. estimate the similarity of; liken, declare to be similar; be worthy of comparison.

comparison n. comparing.

compartment n. a partitioned space. □ **compartmental** adj.

compass n. **1** a device showing the direction of the magnetic or true north. **2** range, scope. **3**

(**compasses**) a hinged instrument for drawing circles. ● v. encompass.

compassion n. a feeling of pity. □ **compassionate** adj., **compassionately** adv.

compatible adj. able to exist or be used together; consistent. □ **compatibly** adv., **compatibility** n.

compatriot n. a fellow countryman.

compel v. (**compelled**) force.

compelling adj. rousing strong interest or attention.

compendious adj. giving much information concisely.

compendium n. (pl. **compendia** or **compendiums**) a summary; a collection of information etc.

compensate v. make payment to (a person) in return for loss or damage; counterbalance. □ **compensation** n., **compensatory** adj.

compère n. a person who introduces performers in a variety show. ● v. act as compère to.

compete v. take part in a competition or other contest.

competence n. ability, authority.

competent adj. having ability or authority to do what is required; adequate. □ **competently** adv.

competition n. a friendly contest; competing; those who compete.

competitive adj. involving competition. □ **competitively** adv., **competitiveness** n.

competitor n. one who competes.

compile v. collect and arrange into a list or book; make (a book) in this way. □ **compilation** n., **compiler** n.

complacent adj. self-satisfied. □ **complacently** adv., **complacency** n.

complain v. say one is dis-satisfied; say one is suffering from pain. □ **complainant** n.

complaint n. **1** a statement that one is dissatisfied. **2** an illness.

complaisant adj. willing to please others. □ **complaisance** n.

complement n. that which completes or fills something. ● v. form a complement to. □ **complementary** adj.

complete adj. having all its parts; finished; thorough, in every way. ● v. make complete; fill in (a form etc.). □ **completely** adv., **completeness** n., **completion** n.

complex adj. made up of many parts. ● n. a complex whole; a set of feelings that influence behaviour; a set of buildings. □ **complexity** n.

complexion n. the colour and texture of the skin of the face; the general character of things.

compliant adj. complying, obedient. □ **compliance** n.

complicate v. make complicated. □ **complicated** adj. complex and difficult. **complication** n.

complicity n. involvement in wrongdoing.

compliment n. a polite expression of praise. ● v. pay a compliment to.

complimentary adj. **1** expressing a compliment. **2** free of charge.

compline n. (esp. in the RC Church) the last service of the day.

comply v. act in accordance with a request.

component n. one of the parts of which a thing is composed.

comport v. □ **comport oneself** conduct oneself, behave.

compose v. **1** create in music or literature. **2** arrange in good order. **3** calm. □ **composer** n.

composite adj. made up of parts.

composition n. composing; a thing composed; a compound artificial substance.

compositor n. a typesetter.

compos mentis adj. sane.

compost n. decayed matter used as a soil improver; a mixture of soil or peat for growing seedlings etc.

composure n. calmness.

compote n. fruit in syrup.

compound adj. (kom-pownd) made up of two or more ingredients. ● n. (kom-pownd) **1** a compound substance. **2** a fenced-in enclosure. ● v. (kŏm-pownd) **1** combine; add to. **2** settle by agreement.

comprehend v. **1** understand. **2** include.

comprehensible adj. intelligible. □ **comprehensibly** adv., **comprehensibility** n.

comprehension n. understanding.

comprehensive adj. including much or all. ● n. (in full **comprehensive school**) a school providing secondary education for children of all abilities. □ **comprehensively** adv., **comprehensiveness** n.

compress v. (kŏm-press) squeeze, force into less space. ● n. (kom-press) a pad to stop bleeding or to cool inflammation. □ **compression** n., **compressor** n.

comprise v. include; consist of.

■ **Usage** It is a mistake to use *comprise* to mean 'to compose or make up'.

---

compromise n. a settlement reached by concessions on each side. ● v. **1** make a settlement in this way. **2** expose to suspicion, scandal, or danger.

comptroller (kŏn-troh-ler) n. (in titles) a financial controller.

compulsion n. compelling; being compelled; an irresistible urge. □ **compulsive** adj., **compulsively** adv.

compulsory adj. that must be done, required. □ **compulsorily** adv.

compunction n. regret, scruple.

compute v. calculate; use a computer. □ **computation** n.

computer n. an electronic apparatus for analysing or storing data, making calculations, etc.

computerize v. (also **-ise**) equip with, perform, or operate by computer. □ **computerization** n.

comrade n. a companion, an associate. **comradeship** n.

con v. (**conned**) **1** (colloquial) persuade or swindle after winning confidence. **2** (Amer. **conn**) direct the steering of (a ship). ● n. (slang) a confidence trick. □ **pros and cons** see pro.

concatenation n. combination.

concave adj. curved like the inner surface of a ball.

conceal v. hide, keep secret. □ **concealment** n.

concede v. admit to be true; grant (a privilege etc.); admit defeat in (a contest).

conceit n. too much pride in oneself. □ **conceited** adj.

conceivable adj. able to be imagined or believed true; possible. □ **conceivably** adv.

conceive v. **1** become pregnant. **2** form (an idea etc.) in the mind.

concentrate v. **1** employ all one's thought or effort. **2** bring or come together. **3** make less dilute. ● n. a concentrated substance.

concentration n. concentrating; a concentrated thing.

concentration camp n. a prison camp for political prisoners in Nazi Germany.

concentric adj. having the same centre. □ **concentrically** adv.

concept n. an idea, a general notion.

conception n. **1** conceiving. **2** an idea.

conceptual adj. of concepts.

conceptualize v. (also **-ise**) form a concept of. □ **conceptualization** n.

concern v. be relevant or important to; involve. ● n. a thing that concerns one; anxiety; business.

concerned adj. anxious.

concerning prep. with reference to.

concert n. a musical entertainment.

concerted adj. done in combination.

concertina n. a portable musical instrument with bellows and buttons. ● v. (**concertinaed**, **concertinaing**) fold or collapse like concertina bellows.

concerto (kŏn-**chair**-toh) n. (**concertos** or **concerti**) a musical composition for solo instrument and orchestra.

concession n. conceding; a thing conceded; a special privilege; a right granted.

conch n. a spiral shell.

conciliate v. soothe the hostility of; reconcile. □ **conciliation** n., **conciliatory** adj.

concise adj. brief and comprehensive. □ **concisely** adv., **conciseness** n.

conclave n. a private meeting.

conclude v. end; settle finally; reach an opinion by reasoning.

conclusion n. concluding; an ending; an opinion reached.

conclusive adj. ending doubt; convincing. □ **conclusively** adv.

concoct v. prepare from ingredients; invent. □ **concoction** n.

concomitant adj. accompanying.

concord n. agreement, harmony.

concordance n. agreement; an index of words.

concordant adj. being in concord.

concourse n. **1** a crowd, a gathering. **2** an open area at a railway terminus etc.

concrete n. a mixture of gravel and cement etc. used for building. ● adj. existing in material form; definite. ● v. cover with or embed in concrete; solidify.

concubine n. a secondary wife (in polygamous societies); a woman who lives with a man as his wife.

concur v. (**concurred**) **1** agree in opinion. **2** happen together, coincide. □ **concurrence** n., **concurrent** adj., **concurrently** adv.

concuss v. affect with concussion.

concussion n. an injury to the brain caused by a hard blow.

condemn v. express strong disapproval of; convict; sentence; doom; declare unfit for use. □ **condemnation** n.

condense v. **1** make denser or briefer. **2** change from gas or vapour to liquid. □ **condensation** n.

condenser n. **1** an apparatus for condensing vapour. **2** (*Electricity*) = **capacitor**.

condescend v. consent to do something less dignified or fitting than is usual. ● **condescension** n.

condiment n. a seasoning for food.

condition n. something that must exist if something else is to exist or occur; a state of being;   (**conditions**) circumstances. ● v. bring to the desired condition; have a strong effect on; accustom.

conditional adj. subject to specified conditions. □ **conditionally** adv.

conditioner n. a substance that conditions hair, fabric, etc.

condole v. express sympathy. □ **condolence** n.

condom n. a contraceptive sheath.

condominium n. **1** the joint control of a state's affairs by other states. **2** (*Amer.*) a building containing flats which are individually owned.

condone v. forgive or overlook (a fault etc.).

condor n. a vulture.

conduce v. help to cause or produce. □ **conducive** adj.

conduct v. (kŏn-**dukt**) **1** lead, guide; be the conductor of; manage. **2** transmit (heat or electricity). ● n. (**kon**-dukt) behaviour; a way of conducting business etc.

conduction n. the conducting of heat or electricity. □ **conductive** adj., **conductivity** n.

conductor n. **1** a person who controls an orchestra's or choir's performance. **2** a substance that conducts heat or electricity.

conduit n. a pipe or channel for liquid; a tube protecting wires.

cone n. **1** an object with a circular base, tapering to a point. **2** a dry scaly fruit of a pine or fir.

coney var. of **cony**.

confection n. a prepared dish or delicacy.

confectioner n. a maker or seller of confectionery.

confectionery n. sweets, cakes, and pastries.

confederacy n. a league of states.

confederate adj. joined by treaty or agreement. ● n. a member of a confederacy; an accomplice.

confederation n. a union of states, people, or organizations.

confer v. (**conferred**) grant; hold a discussion. □ **conferment** n.

conference n. a meeting for discussion.

confess v. acknowledge, admit; declare one's sins to a priest.

confession n. acknowledgement of a fact, sin, guilt, etc.; a statement of beliefs.

confessional n. an enclosed stall in a church for hearing confessions.

confessor n. a priest who hears confessions and gives counsel.

confetti n. bits of coloured paper thrown at a bride and bridegroom.

confidant n. a person in whom one confides.

confidante n. a woman in whom one confides.

confide v. tell or talk confidentially; entrust.

confidence n. firm trust; a feeling of certainty, boldness; something told confidentially.

confidence trick n. a swindle worked by gaining a person's trust.

confident adj. feeling confidence. □ **confidently** adv.

confidential adj. to be kept secret; entrusted with secrets. □ **confidentially** adv., **confidentiality** n.

configuration n. a method of arrangement; a shape, an outline.

confine v. keep within limits; keep shut up.

confinement n. 1 confining, being confined. 2 the time of childbirth.

confines n.pl. boundaries.

confirm v. make firmer or definite; corroborate. □ **confirmatory** adj.

confirmation n. confirming; something that confirms.

confiscate v. take or seize by authority. □ **confiscation** n.

conflagration n. a great fire.

conflate v. blend or fuse together. □ **conflation** n.

conflict n. (**kon**-flikt) a fight, a struggle; disagreement. ● v. (kŏn-**flikt**) clash, have a conflict.

confluence n. a place where two rivers unite. □ **confluent** adj.

conform v. be similar; act or be in accordance, keep to rules or custom. □ **conformity** n.

conformable adj. consistent; adaptable. □ **conformably** adv.

conformist n. a person who conforms to rules or custom. □ **conformism** n.

confound v. astonish and perplex; confuse.

confront v. be or come or bring face to face with; face boldly. □ **confrontation** n.

confuse v. throw into disorder; make unclear; bewilder; destroy the composure of. □ **confusion** n.

confute v. prove wrong. □ **confutation** n.

conga n. a dance in which people form a long winding line.

congeal v. coagulate, solidify.

congenial adj. pleasant, agreeable to oneself. □ **congenially** adv.

congenital adj. being so from birth. □ **congenitally** adv.

conger n. a large sea eel.

congest v. make abnormally full; overcrowd. □ **congestion** n.

conglomerate adj. (kŏn-glom-er-ăt) gathered into a mass. ● n. (kŏn-**glom**-er-ăt) a coherent mass; a group formed from a merger of different firms. ● v. (kŏn-**glom**-er-ayt) collect into a coherent mass. □ **conglomeration** n.

congratulate v. express pleasure at the good fortune or happiness or excellence of (a person). □ **congratulation** n., **congratulatory** adj.

congregate | conservation

congregate v. flock together.

congregation n. people assembled at a church service.

congress n. a formal meeting of delegates for discussion; (**Congress**) a law-making assembly, esp. of the USA. □ **congressional** adj.

congruent adj. **1** suitable, consistent. **2** having exactly the same shape and size. □ **congruence** n.

conic adj. of a cone.

conical adj. cone-shaped.

conifer n. a tree bearing cones. □ **coniferous** adj.

conjecture n. & v. (a) guess.

conjugal adj. of marriage.

conjugate v. (Grammar) give the different forms of (a verb).

conjunction n. a word such as 'and' or 'or' that connects others.

conjunctivitis n. inflammation of the membrane connecting the eyeball and eyelid.

conjure v. do sleight-of-hand tricks. □ **conjurer** or **conjuror** n.

conk (slang) n. the nose; the head. ● v. hit. □ **conk out** break down.

conn see **con**.

connect v. join, be joined; associate mentally; (of a train, coach, or flight) arrive so that passengers are in time to catch another. □ **connective** adj.

connection n. **1** connecting; a place where things connect, a connecting part; trains etc. timed to connect with one another; a link, esp. by telephone. **2** (**connections**) influential relatives or associates.

connector n. a thing that connects others.

connive v. **connive at** tacitly consent to. □ **connivance** n.

connoisseur (kon-ŏ-ser) n. a person with expert understanding esp. of artistic subjects.

connote v. imply in addition to its basic meaning. □ **connotation** n.

conquer v. overcome in war or by effort. □ **conqueror** n.

conquest n. conquering; something won by conquering.

conscience n. a person's sense of right and wrong; a feeling of remorse.

conscientious adj. showing careful attention. □ **conscientiously** adv., **conscientiousness** n.

conscientious objector n. a person who refuses to serve in the armed forces for moral reasons.

conscious adj. **1** awake and alert; aware. **2** intentional. □ **consciously** adv., **consciousness** n.

conscript v. (kŏn-skript) summon for compulsory military service. ● n. (kon-skript) a conscripted person. □ **conscription** n.

consecrate v. make sacred; dedicate to the service of God. □ **consecration** n.

consecutive adj. following continuously. □ **consecutively** adv.

consensus n. general agreement.

consent v. say one is willing to do or allow what is asked. ● n. willingness; permission.

consequence n. **1** a result. **2** importance.

consequent adj. resulting.

consequential adj. **1** resulting. **2** important. □ **consequentially** adv.

consequently adv. as a result.

conservancy n. **1** an authority controlling a river etc. **2** official conservation.

conservation n. conserving; preservation of the natural environment.

conservationist n. a person who seeks to preserve the natural environment.

conservative adj. **1** opposed to change; (**Conservative**) of the Conservative Party. **2** (of an estimate) purposely low. ● n. a conservative person. □ **conservatively** adv., **conservatism** n.

Conservative party n. a British political party promoting free enterprise and private ownership.

conservatoire n. a school of music or other arts.

conservatory n. a greenhouse built on to a house.

conserve v. (kŏn-serv) keep from harm, decay, or loss. ● n. (kon-serv) jam. □ **conservator** n.

consider v. **1** think about, esp. in order to decide. **2** allow for. **3** be of the opinion.

considerable adj. great in amount or importance. □ **considerably** adv.

considerate adj. careful not to hurt or inconvenience others. □ **considerately** adv.

consideration n. **1** careful thought; a fact that must be kept in mind; being considerate. **2** a payment given as a reward.

considering prep. taking into account.

consign v. deposit, entrust; send (goods etc.).

consignee n. a person to whom goods are sent.

consignment n. consigning; a batch of goods.

consist v. □ **consist of** be composed of.

consistency n. **1** being consistent. **2** the degree of thickness or solidity.

consistent adj. unchanging; not contradictory. □ **consistently** adv.

consolation n. consoling; something that consoles.

console[1] (kŏn-sohl) v. comfort in time of sorrow. □ **consolable** adj.

console[2] (kon-sohl) n. a bracket supporting a shelf; a frame or panel holding the controls of equipment.

consolidate v. combine; make or become secure and strong. □ **consolidation** n.

consommé (kŏn-som-ay) n. clear soup.

consonant n. a letter other than a vowel; the sound it represents. ● adj. consistent, harmonious. □ **consonance** n.

consort n. (kon-sort) a husband or wife, esp. of a monarch. ● v. (kŏn-sort) keep company.

consortium n. (pl. **consortia** or **consortiums**) a combination of firms acting together.

conspicuous adj. easily seen, attracting attention. □ **conspicuously** adv.

conspiracy n. conspiring; a secret plan.

conspirator n. a person who conspires. □ **conspiratorial** adj., **conspiratorially** adv.

conspire v. **1** plan secretly against others. **2** (of events) seem to combine.

constable n. a police officer of the lowest rank.

constabulary n. the police force.

constancy n. **1** the quality of being unchanging. **2** faithfulness.

constant adj. **1** continuous; occurring repeatedly; unchanging. **2** faithful. ● n. an unvarying quantity. □ **constantly** adv.

constellation n. a group of stars.

consternation n. great surprise and anxiety or dismay.

constipation n. difficulty in emptying the bowels.

constituency n. a body of voters who elect a representative; an area represented in this way.

constituent adj. forming part of a whole. ● n. a constituent part; a member of a constituency.

constitute v. be the parts of.

constitution n. 1 the principles by which a state is organized. 2 the general condition of the body.

constitutional adj. in accordance with a constitution. ● n. a walk taken for exercise.

constrain v. compel, oblige.

constraint n. constraining; a restriction; a strained manner.

constrict v. tighten, make narrower, squeeze. □ **constriction** n., **constrictor** n.

construct v. (kŏn-**strukt**) make by placing parts together. ● n. (**kon**-strukt) something constructed, esp. in the mind. □ **constructor** n.

construction n. 1 constructing; a thing constructed. 2 words put together to form a phrase; an interpretation.

constructive adj. constructing; making useful suggestions. □ **constructively** adv.

construe v. interpret; analyse word for word.

consul n. an official representative of a state in a foreign city. □ **consular** adj.

consulate n. a consul's position or premises.

consult v. seek information or advice from. □ **consultation** n.

consultant n. a specialist consulted for professional advice. □ **consultancy** n.

consultative adj. of or for consultation; advisory.

consume v. 1 use up; eat or drink up. 2 destroy (by fire).

consumer n. a person who buys or uses goods or services.

consummate v. (**kon**-sŭ-mayt) accomplish; complete (esp. a

marriage by sexual intercourse).
● adj. (kon-**sum**-ăt) skilled. □ **consummation** n.

consumption n. 1 consuming. 2 (old use) tuberculosis.

consumptive adj. & n. (a person) suffering from tuberculosis.

cont. abbr. continued.

contact n. touching, meeting, communicating; an electrical connection; a person who may be contacted for information or help. ● v. get in touch with.

contact lens n. a small lens worn directly on the eyeball to correct the vision.

contagion n. spreading of disease by contact. □ **contagious** adj.

contain v. 1 have within itself; include. 2 control, restrain.

container n. a receptacle, esp. of standard design to transport goods.

containment n. prevention of hostile expansion.

contaminate v. pollute. □ **contaminant** n., **contamination** n.

contemplate v. 1 gaze at. 2 consider as a possibility, intend; meditate. □ **contemplation** n.

contemplative adj. meditative; of religious meditation.

contemporaneous adj. existing or occurring at the same time.

contemporary adj. of the same period or age; modern in style. ● n. a person of the same age.

contempt n. despising, being despised; disrespect.

contemptible adj. deserving contempt.

contemptuous adj. showing contempt. □ **contemptuously** adv.

contend v. strive, compete; assert. □ **contender** n.

content¹ (kŏn-**tent**) adj. satisfied with what one has. ● n. being content. ● v. make content. □ **contented** adj., **contentment** n.

content² (kon-tent) *n.* what is contained in something.

contention *n.* contending; an assertion made in argument.

contentious *adj.* quarrelsome; likely to cause contention.

contest *v.* (kŏn-test) compete for or in; dispute. ● *n.* (kon-test) a struggle for victory; a competition. □ **contestant** *n.*

context *n.* what precedes or follows a word or statement and fixes its meaning; circumstances. □ **contextual** *adj.*

contiguous *adj.* adjacent, touching.

continent *n.* one of the main land masses of the earth. ● *adj.* able to control the excretion of one's urine and faeces. □ **continental** *adj.*, **continence** *n.*

contingency *n.* something unforeseen; a thing that may occur.

contingent *adj.* happening by chance; possible but not certain; conditional. ● *n.* a body of troops contributed to a larger group.

continual *adj.* constantly or frequently recurring. □ **continually** *adv.*

continuance *n.* continuing.

continue *v.* not cease; remain in a place or condition; resume. □ **continuation** *n.*

continuo *n.* (*pl.* **continuos**) (*Music*) a (keyboard) accompaniment providing a bass line.

continuous *adj.* without interval, uninterrupted. □ **continuously** *adv.*, **continuity** *n.*

continuum *n.* (*pl.* **continua**) a continuous thing.

contort *v.* force or twist out of normal shape. □ **contortion** *n.*

contortionist *n.* a performer who can twist his or her body dramatically.

contour *n.* an outline; a line on a map showing height above sea level.

contra- *pref.* against.

contraband *n.* smuggled goods.

contraception *n.* the prevention of pregnancy, birth control.

contraceptive *adj.* & *n.* (a drug or device) preventing conception.

contract *n.* (kon-trakt) a formal agreement. ● *v.* (kŏn-trakt) **1** make a contract; arrange (work) to be done by contract. **2** catch (an illness). **3** make or become smaller or shorter. □ **contractor** *n.*, **contractual** *adj.*

contractable *adj.* (of a disease) able to be contracted.

contractile *adj.* able to be shrunk or drawn together.

contractile *adj.* able to contract or to produce contraction.

contraction *n.* contracting; a shortened form of a word or words; a shortening of the uterine muscles during childbirth.

contradict *v.* say that (a statement) is untrue or (a person) is wrong; be contrary to. □ **contradiction** *n.*, **contradictory** *adj.*

contraflow *n.* a flow (esp. of traffic) in a direction opposite to and alongside the usual flow.

contralto *n.* (*pl.* **contraltos**) the lowest female voice.

contraption *n.* a strange device or machine.

contrapuntal *adj.* of or in counterpoint.

contrariwise *adv.* on the other hand; in the opposite way.

contrary (kon-trǎ-ri) *adj.* **1** opposite in nature, tendency, or direction. **2** (kŏn-trair-i) (*colloquial*) perverse, doing the opposite of what is expected. ● *n.* the opposite. ● *adv.* in opposition. □ **on the contrary** as the opposite of what was just stated. □ **contrarily** *adv.*, **contrariness** *n.*

contrast *n.* (kon-trahst) a difference shown by comparison. ● *v.* (kŏn-**trahst**) show a contrast; compare so as to do this.

contravene *v.* break (a rule etc.). □ **contravention** *n.*

contretemps (kon-trĕ-tahn) *n.* (*pl.* **contretemps**) an unfortunate happening.

contribute *v.* give to a common fund or effort; help to bring about. □ **contribution** *n.*, **contributor** *n.*, **contributory** *adj.*

contrite *adj.* penitent, sorry. □ **contritely** *adv.*, **contrition** *n.*

contrivance *n.* contriving; a contrived thing, a device.

contrive *v.* plan, make, or do resourcefully.

control *n.* the power to give orders or restrain something; a means of restraining or regulating; a check. ● *v.* (**controlled**) have control of; regulate; restrain.

controversial *adj.* causing controversy. □ **controversially** *adv.*

controversy *n.* prolonged dispute.

controvert *v.* deny the truth of. □ **controvertible** *adj.*

contumacy *n.* stubborn refusal to obey.

contusion *n.* a bruise.

conundrum *n.* a riddle, a puzzle.

conurbation *n.* a large urban area formed where towns have spread and merged.

convalesce *v.* regain health after illness. □ **convalescence** *n.*, **convalescent** *adj.* & *n.*

convection *n.* the transmission of heat within a liquid or gas by movement of heated particles.

convene *v.* assemble. □ **convener** or **convenor** *n.*

convenience *n.* **1** being convenient; a convenient thing. **2** a lavatory.

convenient *adj.* easy to use or deal with; with easy access. □ **conveniently** *adv.*

convent *n.* a residence of a community of nuns.

convention *n.* **1** an accepted custom. **2** an assembly. **3** a formal agreement. □ **conventional** *adj.*, **conventionally** *adv.*

converge *v.* come to or towards the same point. □ **convergence** *n.*, **convergent** *adj.*

conversant *adj.* □ **conversant with** having knowledge of.

conversation *n.* informal talk between people. □ **conversational** *adj.*, **conversationally** *adv.*

converse[1] *v.* (kŏn-**verss**) hold a conversation.

converse[2] *n.* (kon-verss) *adj.* opposite, contrary. ● *n.* a converse idea or statement. □ **conversely** *adv.*

convert *v.* (kŏn-**vert**) change from one form or use to another; cause to change an attitude or belief. ● *n.* (kon-vert) a person converted, esp. to a religious faith. □ **conversion** *n.*, **converter** or **convertor** *n.*

convertible *adj.* able to be converted. ● *n.* a car with a folding or detachable roof.

convex *adj.* curved like the outer surface of a ball. □ **convexity** *n.*

convey *v.* carry, transport, transmit; communicate as an idea.

conveyance *n.* conveying; a means of transport, a vehicle.

conveyancing *n.* the business of transferring legal ownership of property.

conveyor *n.* a person or thing that conveys; (in full **conveyor belt**) a continuous moving belt conveying objects.

convict *v.* (kŏn-**vikt**) prove or declare guilty. ● *n.* (kon-vikt) a convicted person in prison.

conviction n. **1** convicting or being convicted. **2** a firm opinion.

convince v. make (a person) feel certain that something is true.

convivial adj. sociable and lively.

convocation n. convoking; an assembly.

convoke v. summon to assemble.

convoluted adj. **1** coiled, twisted. **2** complicated.

convolution n. **1** a coil, a twist. **2** complexity.

convolvulus n. a twining plant with trumpet-shaped flowers.

convoy n. ships or vehicles travelling under escort or together.

convulse v. cause violent movement or a fit of laughter in.

convulsion n. a violent involuntary movement of the body; (**convulsions**) uncontrollable laughter. □ **convulsive** adj.

cony n. (also **coney**) rabbit fur.

coo v. make a soft murmuring sound like a dove. ● n. this sound.

cooee int. a cry to attract attention.

cook v. **1** prepare (food) by heating; undergo this process. **2** (colloquial) falsify (accounts etc.). ● n. a person who cooks, esp. as a job. □ **cook up** (colloquial) concoct (a story etc.).

cooker n. a stove for cooking food.

cookery n. the art and practice of cooking.

cookie n. (Amer.) a sweet biscuit.

cool adj. **1** fairly cold. **2** calm, unexcited; not enthusiastic. ● n. **1** coolness. **2** (slang) calmness. ● v. make or become cool. □ **coolly** adv., **coolness** n.

coolant n. fluid for cooling machinery.

cool bag n. (also **cool box**) an insulated container for keeping food cool.

coolie n. (old use) a native labourer in Eastern countries.

coomb (koom) n. (also **combe**) a valley.

coop n. a cage for poultry. ● v. confine, shut in.

co-op n. (colloquial) a cooperative society; a shop run by this.

cooper n. a person who makes or repairs casks and barrels.

cooperate v. work or act together. □ **cooperation** n.

cooperative adj. cooperating; willing to help; (of a firm etc.) run collectively, based on economic cooperation. ● n. a farm or firm etc. run on this basis.

co-opt v. appoint to a committee by invitation of existing members, not election.

coordinate adj. (koh-ord-i-năt) equal in importance. ● n. (koh-ord-i-năt) **1** any of the magnitudes used to give the position of a point. **2** (**coordinates**) matching items of clothing. ● v. (koh-ord-i-nayt) bring into a proper relation; cause to function together efficiently. □ **coordination** n., **coordinator** n.

coot n. a waterbird.

cop (slang) n. a police officer. ● v. (**copped**) catch or arrest. □ **cop it** get into trouble. **cop out** withdraw; go back on a promise.

cope v. □ **cope with** deal successfully with; manage successfully.

copeck n. (also **kopeck, kopek**) a Russian coin, one-hundredth of a rouble.

copier n. a copying machine.

co-pilot n. a second pilot.

coping n. the sloping top row of masonry of a wall.

copious adj. plentiful. □ **copiously** adv.

copper n. **1** a reddish-brown metallic element (symbol Cu); a coin containing this; its colour. **2** (slang) a police officer. ● adj. made of or coloured like copper.

copper-bottomed adj. reliable, esp. financially; genuine.

copperplate n. elaborate clear handwriting.

coppice n. (also **copse**) a group of small trees and undergrowth.

Coptic adj. of the Egyptian branch of the Christian Church.

copula n. (Grammar) the verb be.

copulate v. have sexual intercourse. □ **copulation** n.

copy n. a thing made to look like another; a specimen of a book etc. ● v. make a copy of; imitate.

copyright n. the sole right to publish a work. ● v. secure copyright for.

copywriter n. a person who writes advertising copy.

coquette n. a woman who flirts. □ **coquettish** adj., **coquetry** n.

coracle n. a small wicker boat.

coral n. a hard red, pink, or white substance built by tiny sea creatures; a reddish-pink colour.

cor anglais (kor ahng-lay) n. (pl. **cors anglais**) a woodwind instrument like the oboe but lower in pitch.

corbel n. a stone or wooden support projecting from a wall.

cord n. **1** long thin flexible material made from twisted strands; a piece of this. **2** corduroy.

cordial adj. warm and friendly. ● n. a fruit-flavoured drink. □ **cordially** adv.

cordite n. a smokeless explosive.

cordon n. **1** a line of police, soldiers, etc., enclosing something. **2** a fruit tree pruned to grow as a single stem. ● v. enclose by a cordon.

cordon bleu (kor-don bler) adj. of the highest class in cookery.

corduroy n. cloth with velvety ridges.

core n. the central or most important part; the horny central part of an apple etc., containing seeds. ● v. remove the core from.

co-respondent n. the person with whom the respondent in a divorce suit is said to have committed adultery.

corgi n. a dog of a small Welsh breed with short legs.

coriander n. an aromatic plant with seeds and leaves used for flavouring.

cork n. the light tough bark of a Mediterranean oak; a piece of this used as a float; a bottle stopper. ● v. stop up with a cork.

corkage n. a restaurant's charge for serving wine brought from elsewhere.

corked adj. (of wine) contaminated by a decayed cork.

corkscrew n. a tool for extracting corks from bottles; a spiral thing.

corm n. a bulb-like underground stem from which buds grow.

cormorant n. a large black seabird.

corn n. **1** wheat, oats, or maize; grain. **2** (slang) something corny. **3** a small area of horny hardened skin, esp. on the foot.

corncrake n. a bird with a harsh cry.

cornea n. the transparent outer covering of the eyeball. □ **corneal** adj.

cornelian n. (also **carnelian**) a reddish or white semi-precious stone.

corner n. an angle or area where two lines, sides, or streets meet; a free kick or hit from the corner of the field in football or hockey. ● v. 1 force into a position from which there is no escape. 2 drive round a corner. 3 obtain a monopoly of (a commodity).

cornerstone n. a basis; a vital foundation.

cornet n. a brass instrument like a small trumpet; a cone-shaped wafer holding ice cream.

cornflour n. flour made from maize.

cornflower n. a blue-flowered plant often growing among corn.

cornice n. an ornamental moulding round the top of an indoor wall.

Cornish adj. of Cornwall. ● n. the ancient language of Cornwall.

cornucopia n. a horn-shaped container overflowing with fruit and flowers, a symbol of abundance.

corny adj. (**cornier**) (colloquial) overused; sentimental.

corolla n. the petals of a flower.

corollary n. a proposition that follows logically from another.

corona n. (pl. **coronae**) a ring of light round the sun or moon.

coronary n. one of the arteries supplying blood to the heart; a blockage in this by a blood clot.

coronation n. the ceremony of crowning a monarch or consort.

coroner n. an officer holding inquests.

coronet n. a small crown.

corpora see **corpus**.

corporal n. a non-commissioned officer next below sergeant. ● adj. of the body.

corporal punishment n. whipping or beating.

corporate adj. shared by members of a group; united in a group.

corporation n. a group in business or elected to govern a town.

corporeal adj. having a body, tangible. □ **corporeally** adv.

corps (kor) n. (pl. **corps**) a military unit; an organized body of people.

corpse n. a dead body.

corpulent adj. fat. □ **corpulence** n.

corpus n. (pl. **corpora**) a collection of writings.

corpuscle n. a blood cell.

corral n. (Amer.) an enclosure for cattle. ● v. (**corralled**) put or keep in a corral.

correct adj. true, accurate; in accordance with an approved way of behaving or working. ● v. make correct; mark errors in; reprove; punish. □ **correctly** adv., **correctness** n.

correction n. correcting; an alteration correcting something.

corrective adj. & n. (something) correcting what is bad or harmful.

correlate v. compare, connect, or be connected systematically. □ **correlation** n.

correspond v. 1 be similar, equivalent, or in harmony. 2 write letters to each other.

correspondence n. 1 similarity. 2 writing letters; letters written.

correspondent n. 1 a person who writes letters. 2 a person employed by a newspaper or TV news station to gather news and send reports.

corridor n. a passage in a building or train; a strip of territory giving access to somewhere.

corrie n. a round hollow on a mountainside.

corroborate v. get or give supporting evidence. □ **corroboration** n., **corroborative** adj.

corrode v. destroy (a metal etc.) gradually by chemical action. □ **corrosion** n., **corrosive** adj.

corrugated *adj.* shaped into alternate ridges and grooves. □ **corrugation** *n.*

corrupt *adj.* dishonest, accepting bribes; immoral, wicked; decaying. ● *v.* make corrupt; spoil, taint. □ **corruption** *n.*

corsair *n.* a pirate ship; a pirate.

corset *n.* a close-fitting undergarment worn to shape or support the body.

cortège (kor-*tayzh*) *n.* a funeral procession.

cortex *n.* (*pl.* **cortices**) the outer part of the brain; an outer layer of tissue. □ **cortical** *adj.*

cortisone *n.* a hormone produced by the adrenal glands or synthetically.

corvette *n.* a small fast gunboat.

cos *n.* a long-leaved lettuce. ● *abbr.* cosine.

cosh *n.* a weighted weapon for hitting people. ● *v.* hit with a cosh.

cosine *n.* (*Maths*) the ratio of the side adjacent to an acute angle (in a right-angled triangle) to the hypotenuse.

cosmetic *n.* a substance for beautifying the complexion etc. ● *adj.* improving the appearance.

cosmic *adj.* of the universe.

cosmic rays *n.pl.* (also **cosmic radiation**) radiation from outer space.

cosmogony *n.* (a theory of) the origin of the universe.

cosmology *n.* the science or theory of the universe. □ **cosmological** *adj.*

cosmonaut *n.* a Russian astronaut.

cosmopolitan *adj.* of or from all parts of the world; free from national prejudices. ● *n.* a cosmopolitan person.

cosmos *n.* the universe.

Cossack *n. & adj.* (a member) of a people of southern Russia, famous as horsemen.

cosset *v.* (**cosseted**) pamper.

cost *v.* 1 (**cost**, **costing**) have as its price; involve the sacrifice or loss of. 2 (**costed**, **costing**) estimate the cost of. ● *n.* what a thing costs.

costal *adj.* of the ribs.

co-star *n.* a celebrity performing with another of equal performance.

costermonger *n.* a person selling fruit etc. from a barrow in the street.

costly *adj.* (**costlier**) expensive.

costume *n.* a style of clothes, esp. that of a historical period; garments for a special activity.

cosy *adj.* (*Amer.* **cozy**) (**cosier**) warm and comfortable. ● *n.* a cover to keep a teapot hot. □ **cosily** *adv.*, **cosiness** *n.*

cot *n.* a child's bed with high sides.

cot death *n.* (also **SIDS**) the unexplained death of a sleeping baby.

coterie *n.* a select group.

cotoneaster (kŏ-toh-ni-*ass*-ter) *n.* a shrub with red berries.

cottage *n.* a small simple house in the country.

cottage cheese *n.* soft white cheese made from curds without pressing.

cottage pie *n.* a dish of minced meat topped with mashed potato.

cotton *n.* a soft white substance round the seeds of a tropical plant; this plant; thread or fabric made from cotton. □ **cotton on** (*colloquial*) understand.

cotton wool *n.* fluffy wadding, originally made from raw cotton.

cotyledon (koti-lee-dŏn) the first leaf growing from a seed.

couch *n.* a sofa. ● *v.* express in a specified way.

couchette *n.* a sleeping berth that can be converted from seats in a train.

couch grass n. a weed with long creeping roots.

cougar (koo-ger) n. (Amer.) a puma.

cough v. expel air etc. from the lungs with a sudden sharp sound. ● n. the act or sound of coughing; an illness causing coughing.

could see **can²**.

coulomb (koo-lom) n. a unit of electric charge.

council n. an assembly to govern a town or advise on or organize something.

councillor n. (Amer. **councilor**) a member of a council.

council tax n. a UK local tax based on property value.

counsel n. **1** advice, suggestions. **2** a barrister. ● v. (**counselled**; Amer. **counseled**) advise. □ **counsellor** n.

count v. **1** find the total of; say numbers in order; include or be included in a reckoning. **2** be important. **3** regard as. ● n. **1** counting; a number reached by this. **2** a point being considered. **3** a foreign nobleman. □ **count on** rely on; expect confidently.

countdown n. counting seconds backwards to zero.

countenance n. **1** the face; an expression. **2** approval. ● v. give approval to.

counter n. **1** a flat-topped fitment over which goods are sold or business transacted with customers. **2** a small disc used in board games. ● adv. in the opposite direction. ● adj. opposed. ● v. take opposing action against.

counter- pref. rival; retaliatory; reversed; opposite.

counteract v. reduce or prevent the effects of. □ **counteraction** n.

counter-attack n. & v. (an) attack in reply to an opponent's attack.

counterbalance n. a weight or influence balancing another. ● v. act as a counterbalance to.

counterblast n. a powerful retort.

counterfeit adj. forged, not genuine. ● n. a forgery. ● v. forge.

counterfoil n. a section of a cheque or receipt kept as a record.

countermand v. cancel.

counterpane n. a bedspread.

counterpart n. a person or thing corresponding to another.

counterpoint n. (Music) a technique of combining melodies.

counter-productive adj. having the opposite of the desired effect.

countersign n. a password. ● v. add a confirming signature to.

countersink v. (**countersunk**, **countersinking**) sink (a screwhead) into a shaped cavity so that the surface is level.

counter-tenor n. a male alto.

countess n. a count's or earl's wife or widow; a woman with the rank of count or earl.

countless adj. too many to be counted.

countrified adj. (also **countryfied**) rural or rustic in appearance etc.

country n. **1** a nation's or state's land; the people of this; the state of which one is a member; a region. **2** land consisting of fields etc. with few buildings.

countryman n. (pl. **-men**) a person living or liking to live in the country; a person of one's own country.

countryside n. a rural district.

countrywoman n. (pl. **-women**) a woman living or liking to live in the country; a woman of one's own country.

county n. **1** a major administrative division of a country. **2** families of high social class long established in a county.

coup (koo) *n.* sudden action taken to obtain power etc.

coup de grâce (koo dĕ grahs) *n.* a finishing stroke.

coup d'état (koo day-tah) *n.* (*pl. coups d'état*) the sudden overthrow of a government by force or illegal means.

coupé *n.* (*Amer.* **coupe**) a closed two-door car with a sloping back.

couple *n.* two people or things; a married or engaged pair. ● *v.* fasten or link together; copulate.

couplet *n.* two successive rhyming lines of verse.

coupling *n.* a connecting device.

coupon *n.* a form or ticket entitling the holder to something; an entry form for a football pool.

courage *n.* the ability to control fear when facing danger or pain. □ **courageous** *adj.*, **courageously** *adv.*

courgette *n.* a small vegetable marrow.

courier *n.* a messenger carrying documents; a person employed to guide and assist tourists.

course *n.* 1 onward progress; a direction taken or intended. 2 a series of lessons or treatments. 3 an area on which golf is played or a race takes place. 4 a layer of stone etc. in a building. 5 one part of a meal. ● *v.* move or flow freely. □ **of course** without doubt.

court *n.* 1 a courtyard; an area where tennis, squash, etc., are played. 2 a sovereign's establishment with attendants; a room or building where legal cases are heard or judged. ● *v.* 1 try to win the favour, support, or love of. 2 invite (danger etc.).

courteous *adj.* polite. □ **courteously** *adv.*, **courteousness** *n.*

courtesan *n.* (*literary*) a prostitute with high-class clients.

courtesy *n.* polite behaviour.

courtier *n.* a sovereign's companion or attendant at court.

courtly *adj.* dignified and polite.

court martial *n.* (*pl.* **courts martial**) a court trying offences against military law; a trial by this.

court-martial *v.* (**-martialled**; *Amer.* **-martialed**) try by a court martial.

courtship *n.* courting, wooing.

courtyard *n.* a space enclosed by walls or buildings.

couscous (koos-koos) *n.* a North African dish of crushed wheat, usu. with meat etc. added.

cousin *n.* (also **first cousin**) a child of one's uncle or aunt. □ **second cousin** a child of one's parent's cousin.

couture *n.* the design and making of fashionable clothes.

couturier *n.* a designer of fashionable clothes.

cove *n.* a small bay.

coven (kuv-ĕn) *n.* an assembly of witches.

covenant *n.* a formal agreement; a contract. ● *v.* make a covenant.

cover *v.* 1 place or be or spread over; conceal or protect in this way. 2 travel over (a distance). 3 protect by insurance or a guarantee; be enough to pay for. 4 deal with (a subject etc.); report for (a newspaper etc.). ● *n.* 1 a thing that covers; a wrapper, envelope, or binding of a book; a screen, shelter, or protection. 2 a place laid at a meal. □ **cover up** conceal (a thing or fact). □ **cover-up** *n.*

coverage *n.* the process of covering; an area or risk etc. covered.

covering letter *n.* an explanatory letter enclosed with goods.

coverlet *n.* a cover lying over other bedclothes.

covert *n.* thick undergrowth where animals hide. ● *adj.*

concealed, done secretly. □ **covertly** adv.

covet v. (**coveted**) desire (a thing belonging to another person). □ **covetous** adj.

covey n. (pl. **coveys**) a group of partridges.

cow n. a fully grown female of cattle or certain other large animals (e.g. the elephant or whale). ● v. intimidate.

coward n. a person who lacks courage. □ **cowardly** adj.

cowardice n. lack of courage.

cowboy n. 1 a man in charge of cattle on a ranch. 2 (colloquial) a person with reckless methods in business.

cower v. crouch or shrink in fear.

cowl n. a monk's hood or hooded robe; a hood-shaped covering.

cowling n. a removable metal cover on an engine.

cowrie n. a seashell.

cowslip n. a wild plant with small fragrant yellow flowers.

cox n. a coxswain. ● v. act as cox of (a racing boat).

coxcomb n. a show-off.

coxswain (kok-sŭn) n. a person who steers a boat; a senior petty officer.

coy adj. pretending to be shy or embarrassed. □ **coyly** adv.

coyote (koy-oh-ti) n. a North American wolflike wild dog.

coypu n. a beaver-like aquatic rodent.

cozy Amer. sp. of **cosy**.

CPU abbr. central processing unit.

Cr symb. chromium.

crab n. a ten-legged shellfish.

crab apple n. a sharp sour apple.

crabbed adj. 1 bad-tempered. 2 (of handwriting) hard to read.

crabby adj. (**crabbier**) bad-tempered; sour.

crack n. 1 a sudden sharp noise; a line where a thing is broken but not separated; a sharp blow. 2 (colloquial) a joke. 3 a strong form of cocaine. ● adj. (colloquial) first-rate. ● v. 1 make or cause to make the sound of a crack; break without parting completely; knock sharply; (of the voice) become harsh; give way under strain. 2 find a solution to (a problem). 3 tell (a joke). □ **crack down on** (colloquial) take severe measures against. **crack up** (colloquial) have a physical or mental breakdown; praise.

crack-brained adj. (colloquial) crazy.

crackdown n. (colloquial) severe measures against something.

cracker n. 1 a small explosive firework; a toy paper tube made to give an explosive crack when pulled apart. 2 a thin dry biscuit.

crackers adj. (slang) crazy.

crackle v. make or cause to make a series of light cracking sounds. ● n. these sounds.

crackling n. crisp skin on roast pork.

crackpot (slang) n. an eccentric person. ● adj. crazy; unworkable.

cradle n. 1 a baby's bed usu. on rockers. 2 a place where something originates. 3 a supporting structure. ● v. hold or support gently.

craft n. 1 a skill; an occupation requiring this. 2 cunning, deceit. 3 (pl. **craft**) a ship or boat.

craftsman n. (pl. **-men**) a worker skilled in a craft. □ **craftsmanship** n.

crafty adj. (**craftier**) cunning, using underhand methods. □ **craftily** adv., **craftiness** n.

crag n. a steep or rugged rock.

craggy adj. (**craggier**) rugged.

cram v. (**crammed**) force into too small a space; overfill; study intensively for an examination.

cramp n. **1** a painful involuntary tightening of a muscle. **2** a metal bar with bent ends for holding things together. ● v. keep within too narrow limits.

crampon n. a spiked plate worn on boots for climbing on ice.

cranberry n. a small red sharp-tasting berry.

crane n. **1** a large wading bird. **2** an apparatus for lifting and moving heavy objects. ● v. stretch (one's neck) to see something.

crane-fly n. a long-legged flying insect.

cranium n. (**craniums** or **crania**) the skull.

crank n. an L-shaped part for converting to-and-fro into circular motion. **2** a person with very strange ideas. ● v. turn (an engine etc.) with a crank. □ **cranky** adj.

crankshaft n. a shaft driven by a crank.

cranny n. a crevice.

crap (vulgar slang) n. faeces; nonsense; rubbish. ● v. defecate.

craps n.pl. (Amer.) a gambling game played with a pair of dice.

crash n. a loud noise of breakage; a violent collision; a financial collapse. ● v. **1** make a crash; move with a crash; be or cause to be involved in a crash. **2** (colloquial) gatecrash. ● adj. involving intense effort to achieve something rapidly.

crash helmet n. a padded helmet worn esp. by a motorcyclist to protect the head in a crash.

crashing adj. overwhelming.

crash-land v. land (an aircraft) in an emergency, causing damage.

crass adj. gross; very stupid; insensitive.

crate n. a packing case made of wooden slats; (slang) an old aircraft or car. ● v. pack in crate(s).

crater n. a bowl-shaped cavity.

cravat n. a short scarf; a necktie.

crave v. feel an intense longing (for); ask earnestly for.

craven adj. cowardly.

craving n. an intense longing.

craw n. a bird's crop.

crawfish n. = **crayfish**.

crawl v. **1** move on hands and knees or with the body on the ground; move very slowly. **2** (colloquial) seek favour by servile behaviour. ● n. a crawling movement or pace; an overarm swimming stroke. ● **crawler** n.

crayfish n. (pl. **crayfish**) a freshwater shellfish like a small lobster.

crayon n. a stick of coloured wax etc. for drawing. ● v. draw or colour with crayons.

craze n. a temporary enthusiasm.

crazy adj. (**crazier**) insane; very foolish; (colloquial) madly eager. □ **crazily** adv., **craziness** n.

crazy paving n. paving made of irregular pieces.

creak n. a harsh squeak. ● v. make this sound. □ **creaky** adj.

cream n. the fatty part of milk; its colour, yellowish white; a creamlike substance; the best part. ● adj. yellowish white. ● v. remove the cream from; beat to a creamy consistency. □ **creamy** adj.

cream cheese n. a soft rich cheese.

cream cracker n. a crisp unsweetened biscuit.

creamery n. a factory producing butter and cheese.

cream of tartar n. a compound of potassium used in baking powder.

cream tea n. afternoon tea with scones, jam, and cream.

crease n. a line made by crushing or pressing; a line marking the limit of the bowler's or batsman's position in cricket.

● *v.* make a crease in; develop creases.

create *v.* **1** bring into existence; produce by what one does; give a new rank to. **2** (*colloquial*) make a fuss. □ **creation** *n.*, **creator** *n.*

creative *adj.* having the ability to create things; showing imagination and originality. □ **creatively** *adv.*, **creativity** *n.*

creature *n.* an animal; a person.

crèche (kresh) *n.* a day nursery.

credence *n.* belief.

credentials *n.pl.* documents giving evidence of a person's achievements, qualities, etc.

credible *adj.* believable. □ **credibly** *adv.*, **credibility** *n.*

■ **Usage** *Credible* is sometimes confused with *credulous*, which means 'gullible'.

credit *n.* **1** belief that a thing is true. **2** honour for an achievement. **3** a system of allowing payment to be deferred. **4** the sum at a person's disposal in a bank; an entry in an account for a sum received. **5** an acknowledgement in a book or film. ● *v.* (**credited**) **1** believe. **2** attribute. **3** enter as credit.

creditable *adj.* deserving praise. □ **creditably** *adv.*

credit card *n.* a plastic card containing machine-readable magnetic code enabling the holder to make purchases on credit.

creditor *n.* a person to whom money is owed.

credulous *adj.* too ready to believe things; gullible. □ **credulity** *n.*

creed *n.* a set of beliefs or principles.

creek *n.* a narrow inlet of water, esp. on a coast; (*Amer.*) a tributary.

creep *v.* (**crept**, **creeping**) move with the body close to the ground; move timidly, slowly, or stealthily; develop gradually; (of a plant) grow along the ground or a wall etc.; feel creepy. ● *n.* **1** creeping. **2** (*colloquial*) an unpleasant person. **3** (**the creeps**) (*colloquial*) a nervous sensation.

creepy *adj.* (**creepier**) (*colloquial*) causing an unpleasant shivering sensation.

cremate *v.* burn (a corpse) to ashes. □ **cremation** *n.*

crematorium *n.* (*pl.* **crematoria** or **crematoriums**) a place where corpses are cremated.

crème de la crème (krem dě la krem) *n.* the very best.

crème de menthe (krem dě month) *n.* a peppermint-flavoured liqueur.

crenellated *adj.* having battlements. □ **crenellation** *n.*

Creole *n.* a descendant of European settlers in the West Indies or South America; their dialect; a hybrid language.

creosote *n.* a brown oily liquid distilled from coal tar, used as a preservative for wood.

crêpe (krayp) *n.* **1** a fabric with a wrinkled surface. **2** a pancake.

crept *see* **creep**.

crepuscular *adj.* of or like twilight; active at twilight.

Cres. *abbr.* Crescent.

crescendo (kri-shen-doh) *adv.* gradually becoming louder. ● *n.* (*pl.* **crescendos** or **crescendi**) a gradual increase in loudness.

crescent *n.* a narrow curved shape tapering to a point at each end; a curved street of houses.

cress *n.* a plant with small leaves used in salads.

crest *n.* a tuft or outgrowth on a bird's or animal's head; a plume on a helmet; the top of a slope or hill, a white top of a large wave; a design above a shield on a coat of arms.

crestfallen *adj.* disappointed at failure.

cretaceous adj. chalky.

cretin n. a person who is deformed and mentally undeveloped; (colloquial) a stupid person. □ **cretinous** adj.

crevasse n. a deep open crack esp. in a glacier.

crevice n. a narrow gap in a surface.

crew n. the people working a ship or aircraft; a group working together; a gang. See also **crow**.

crew-cut n. a very short haircut.

crib n. **1** a rack for fodder; a model of the manger scene at Bethlehem; a cot. **2** (colloquial) a translation of a text for students' use. ● v. (**cribbed**) (colloquial) copy unfairly.

cribbage n. a card game.

crick n. a sudden painful stiffness in the neck or back.

cricket n. **1** an outdoor game for two teams of 11 players with ball, bats, and wickets. **2** a brown insect resembling a grasshopper. □ **cricketer** n.

crier n. (also **cryer**) an official making public announcements.

crikey int. (colloquial) an exclamation of astonishment.

crime n. a serious offence, an act that breaks a law; illegal acts.

criminal n. a person guilty of a crime. ● adj. of or involving crime. □ **criminally** adv., **criminality** n.

criminology n. the study of crime. □ **criminologist** n.

crimp v. press into ridges.

crimson adj. & n. deep red.

cringe v. cower; behave obsequiously.

crinkle n. & v. (a) wrinkle.

crinoline n. a light framework formerly worn to make a long skirt stand out.

cripple n. a lame person. ● v. make lame; weaken seriously.

crisis n. (pl. **crises**) a decisive moment; a time of acute difficulty.

crisp adj. **1** brittle; slightly stiff. **2** cold and bracing. **3** brisk and decisive. ● n. a thin slice of potato fried crisp. □ **crisply** adv., **crispness** n., **crispy** adj.

crispbread n. a thin crisp unsweetened biscuit.

criss-cross n. a pattern of crossing lines. ● adj. & adv. in this pattern. ● v. mark, form, or move in this way; intersect.

criterion n. (pl. **criteria**) a standard of judgement.

■ Usage It is a mistake to use the plural form criteria when only one criterion is meant.

critic n. a person who points out faults; one skilled in criticism.

critical adj. **1** looking for faults; expressing criticism. **2** of or at a crisis. □ **critically** adv.

criticism n. the pointing out of faults; an evaluation, esp. of literary or artistic work.

criticize v. (also **-ise**) express criticism of.

critique n. a critical essay.

croak n. a deep hoarse cry or sound like that of a frog. ● v. utter or speak with a croak.

crochet (kroh-shay) n. lacy fabric produced from thread worked with a hooked needle. ● v. (**crocheted**) make by or do such work.

crock n. **1** an earthenware pot; a broken piece of this. **2** (colloquial) a person who is disabled or ill; a worn-out vehicle etc.

crockery n. household china.

crocodile n. a large amphibious tropical reptile.

crocodile tears n.pl. insincere sorrow.

crocus n. a small spring-flowering plant growing from a corm.

croft n. a small rented farm in Scotland.

crofter n. the tenant of a croft.

croissant (krwa-son) n. a rich crescent-shaped roll.

crone n. a withered old woman.

crony n. a close friend or companion.

crook n. **1** a hooked stick; a bent thing. **2** (*colloquial*) a criminal. ● v. bend.

crooked adj. **1** not straight. **2** dishonest. □ **crookedly** adv.

croon v. sing softly. □ **crooner** n.

crop n. **1** a batch of plants grown for their produce; a harvest from this; a group or amount produced at one time. **2** a pouch in a bird's gullet where food is broken up for digestion. **3** the handle of a whip. **4** a very short haircut. ● v. (**cropped**) **1** cut or bite off. **2** produce or gather as harvest. □ **crop up** occur unexpectedly.

cropper n. □ **come a cropper** (*colloquial*) fall heavily; fail badly.

croquet (kroh-kay) n. a game played on a lawn with balls and mallets.

croquette (kroh-ket) n. a fried ball or roll of potato, meat, etc.

crosier n. (also **crozier**) a bishop's hooked staff.

cross n. **1** a mark made by drawing one line across another; something shaped like this; a stake with a transverse bar used in crucifixion; an affliction to be borne with Christian patience. **2** a mixture or compromise between two things; a hybrid animal or plant. **3** a crossing shot in football etc. ● v. **1** go or extend across; draw line(s) across, mark (a cheque) in this way so that it must be paid into an account. **2** cause to interbreed. **3** oppose the wishes of. ● adj. **1** passing from side to side. **2** reciprocal. **3** showing bad temper. □ **at cross purposes** misunderstanding or conflicting. □ **crossly** adv., **crossness** n.

crossbar n. a horizontal bar.

cross-bench n. a bench in Parliament for members not belonging to the government or main opposition.

crossbow n. a mechanical bow fixed across a wooden stock.

cross-breed n. an animal produced by interbreeding. □ **cross-bred** adj.

cross-check v. check again by a different method.

cross-dressing n. wearing the clothes of the opposite sex.

cross-examine v. question (a witness in court) to check a testimony already given. □ **cross-examination** n.

cross-eyed adj. squinting.

crossfire n. gunfire crossing another line of fire.

crossing n. a journey across water; a place where things cross; a place for pedestrians to cross a road.

crosspatch n. (*colloquial*) a bad-tempered person.

cross-ply adj. (of a tyre) having fabric layers with cords lying crosswise.

cross-reference n. a reference to another place in the same book.

crossroads n. a place where roads intersect.

cross-section n. a diagram showing internal structure; a representative sample.

crosswise adv. (also **crossways**) in the form of a cross; intersecting; diagonally.

crossword n. a puzzle in which intersecting words have to be inserted into a diagram.

crotch n. a place where things fork, esp. where legs join the trunk.

crotchet n. a note in music, half a minim.

crotchety adj. peevish, irritable. □ **crotchetiness** n.

crouch v. stoop low with the legs tightly bent. ● n. this position.

croup n. **1** an inflammation of the windpipe in children, causing coughing and breathing difficulty. **2** the rump of a horse etc.

croupier n. a person who rakes in stakes and pays out winnings at a gaming table.

crouton n. a small piece of fried or toasted bread.

crow n. a large black bird; a crowing cry or sound. ● v. (**crowed** or **crew, crowing**) **1** utter a cock's cry. **2** express gleeful triumph.

crowbar n. an iron with a bent end, used as a lever.

crowd n. a large group. ● v. come together in a crowd; fill or occupy fully.

crown n. the top part of a head, hat, or arched thing; a monarch's ceremonial headdress; (**the Crown**) the supreme governing power in a monarchy. ● v. **1** form or cover the top part of; place a crown on; (slang) hit on the head. **2** be a climax to.

Crown prince, Crown princess n. the heir to a throne.

crozier var. of **crosier**.

cruces see **crux**.

crucial adj. very important, decisive. □ **crucially** adv.

crucible n. a pot in which metals are melted.

crucifix n. a model of a cross with a figure of Christ on it.

crucifixion n. crucifying; (**the Crucifixion**) that of Christ.

cruciform adj. cross-shaped.

crucify v. put to death by nailing or binding to a transverse bar; cause extreme pain to.

crude adj. in a natural or raw state; not well finished; lacking good manners; vulgar. □ **crudely** adv., **crudity** n.

crudités n.pl. a starter of mixed raw vegetables, often with a dip.

cruel adj. (**crueller** or **crueler**) feeling pleasure in another's suffering; hard-hearted; causing suffering. □ **cruelly** adv., **cruelty** n.

cruet n. a set of containers for salt, pepper, etc. at the table.

cruise v. **1** sail for pleasure or on patrol. **2** travel at a moderate economical speed. ● n. a cruising voyage.

cruiser n. a fast warship; a motor boat with a cabin.

crumb n. a small fragment of bread etc.

crumble v. break into small fragments. ● n. a pudding of fruit with crumbly topping.

crumbly adj. (**crumblier**) easily crumbled.

crummy adj. (**crummier**) (colloquial) dirty, squalid; inferior.

crumpet n. a flat soft yeast cake eaten toasted.

crumple v. crush or become crushed into creases; collapse.

crunch v. crush noisily with the teeth; make this sound. ● n. **1** the sound of crunching. **2** (colloquial) a decisive event or moment.

crunchy adj. (**crunchier**) able to be crunched; hard and crispy.

crupper n. a strap looped under a horse's tail from the saddle.

crusade n. a medieval Christian military expedition to recover the Holy Land from Muslims; a campaign for a cause. ● v. take part in a crusade. □ **crusader** n.

crush v. press so as to break, injure, or wrinkle; pound into fragments; defeat or subdue completely. ● n. **1** a crowded mass of people. **2** (colloquial) an infatuation.

crust n. a hard outer layer, esp. of bread.

crustacean n. a creature with a hard shell (e.g. a lobster).

crusty adj. (**crustier**) **1** with a crisp crust. **2** having a harsh manner.

crutch n. **1** a support for a lame person. **2** the crotch.

crux n. (pl. **cruces** or **cruxes**) a vital part of a problem; a difficult point.

cry n. a loud wordless sound uttered; an appeal; a rallying call; a spell of weeping. ●v. (**cries, cried, crying**) shed tears; call loudly; appeal for help.

cryer var. of **crier**.

cryogenics n. a branch of physics dealing with very low temperatures. □ **cryogenic** adj.

crypt n. a room below the floor of a church.

cryptic adj. concealing its meaning in a puzzling way.

cryptogram n. something written in cipher.

cryptography n. the study of ciphers. □ **cryptographer** n.

crystal adj. a glasslike mineral; high-quality glass; a symmetrical piece of a solidified substance.

crystalline adj. like or made of crystal; clear.

crystallize v. (also **-ise**) form into crystals; make or become definite in form. □ **crystallization** n.

Cs symb. caesium.

CSE abbr. Certificate of Secondary Education.

CS gas n. a gas causing tears and choking, used to control riots etc.

Cu symb. copper.

cu. abbr. cubic.

cub n. **1** the young of certain animals. **2** (**Cub**, in full **Cub Scout**) a member of the junior branch of the Scout Association.

cubby hole n. a very small room or space.

cube n. **1** a solid object with six equal square sides. **2** the product of a number multiplied by itself twice.

cube root n. a number which produces a given number when cubed.

cubic adj. of three dimensions.

cubicle n. a small division of a large room, screened for privacy.

cubism n. a style of painting in which objects are shown as geometrical shapes. □ **cubist** n.

cuckold n. a man whose wife commits adultery. ●v. make a cuckold of.

cuckoo n. a bird with a call that is like its name.

cucumber n. a long green-skinned fruit eaten as salad.

cud n. food that cattle bring back from the stomach into the mouth and chew again.

cuddle v. hug lovingly; nestle together. ●n. a gentle hug. □ **cuddly** adj.

cudgel n. a short thick stick used as a weapon. ●v. (**cudgelled**; Amer. **cudgeled**) beat with a cudgel.

cue n. **1** a signal to do something. **2** a long rod for striking balls in billiards etc. ●v. (**cued, cueing**) **1** give a signal to (someone). **2** strike with a cue.

cuff n. **1** a band of cloth round the edge of a sleeve. **2** a blow with the open hand. ●v. strike with the open hand.

cuff link n. a device of two linked discs etc. to hold cuff edges together.

cuisine (kwi-zeen) n. a style of cooking.

cul-de-sac n. a street closed at one end.

culinary adj. of, for, or used in cooking.

cull v. gather, select; select and kill (surplus animals).

culminate v. reach its highest point or degree. □ **culmination** n.

culottes n.pl. women's trousers styled to resemble a skirt.

culpable adj. deserving blame. □ **culpably** adv., **culpability** n.

culprit n. a person who has committed a slight offence.

cult n. a system of religious worship; excessive admiration of a person or thing.

cultivate v. **1** prepare and use (land) for crops; produce (crops) by tending them. **2** develop by practice. **3** further one's acquaintance with (a person). □ **cultivation** n., **cultivator** n.

culture n. **1** a developed understanding of literature, art, music, etc.; a type of civilization. **2** artificial rearing of bacteria; bacteria grown for study. ● v. grow in artificial conditions. □ **cultural** adj., **culturally** adv.

culvert n. a drain under a road.

cum prep. with; also used as; a bedroom-cum-study.

cumbersome adj. clumsy to carry or use.

cumin n. (also **cummin**) a plant with aromatic seeds used in cooking.

cummerbund n. a sash for the waist.

cumquat var. of **kumquat**.

cumulative adj. increasing by additions. □ **cumulatively** adv.

cumulus n. (pl. **cumuli**) clouds formed in heaped-up round masses.

cuneiform n. ancient writing done in wedge-shaped strokes cut into stone etc.

cunning adj. skilled at deception, crafty; ingenious. ● n. craftiness, ingenuity. □ **cunningly** adv.

cup n. **1** a drinking vessel usu. with a handle at the side. **2** a prize. **3** wine or fruit juice with added flavourings. ● v. (cupped) form into a cuplike shape. □ **cupful** n.

cupboard n. a recess or piece of furniture with a door, in which things may be stored.

cupidity n. greed for gain.

cupola n. a small dome.

cupreous adj. of or like copper.

cur n. a worthless dog.

curaçao (kewr-ásoh) n. an orange-flavoured liqueur.

curacy n. the position of curate.

curare n. a vegetable poison that induces paralysis.

curate n. a member of the clergy who assists a parish priest.

curate's egg n. something that is good in parts.

curator n. a person in charge of a museum or other collection.

curb n. a means of restraint. ● v. restrain.

curds n.pl. the thick soft substance formed when milk turns sour.

curdle v. form or cause to form curds.

cure v. **1** restore to health; get rid of (a disease or trouble etc.). **2** preserve by salting, drying, etc. ● n. curing; a substance or treatment that cures disease etc.

curette n. a surgical scraping instrument. □ **curettage** n.

curfew n. a signal or time after which people must stay indoors.

curie n. a unit of radioactivity.

curio n. (pl. **curios**) an unusual and interesting object.

curiosity n. a desire to find out and know things; a curio.

curious adj. **1** eager to learn or know something. **2** strange, unusual. □ **curiously** adv., **curiousness** n.

curium n. a radioactive metallic element (symbol Cm).

curl v. curve, esp. in a spiral shape or course. ● n. a curled thing or shape; a coiled lock of hair.

curler *n.* a device for curling hair.

curlew *n.* a wading bird with a long curved bill.

curlicue *n.* a curly ornamental line.

curling *n.* a game like bowls played on ice.

curly *adj.* (**curlier**) full of curls.

curmudgeon *n.* a bad-tempered old person.

currant *n.* a dried grape used in cookery; a small round edible berry, a shrub producing this.

currency *n.* **1** money in use. **2** the state of being widely known.

current *adj.* belonging to the present time; in general use. ● *n.* a body of water or air moving in one direction; a flow of electricity. □ **currently** *adv.*

curriculum *n.* (*pl.* **curricula**) a course of study.

curriculum vitae *n.* a brief account of one's career.

curry *n.* a savoury dish cooked with hot spices. ● *v.* groom (a horse) with a curry-comb. □ **curry favour** win favour by flattery.

curry-comb *n.* a metal device with a serrated edge for grooming horses.

curse *n.* a call for evil to come on a person or thing; a great evil; a violent exclamation of anger. ● *v.* utter a curse (against); afflict. □ **cursed** *adj.*

cursive *adj.* & *n.* (writing) done with joined letters.

cursor *n.* a movable indicator on a VDU screen.

cursory *adj.* hasty and not thorough. □ **cursorily** *adv.*

curt *adj.* noticeably or rudely brief. □ **curtly** *adv.*, **curtness** *n.*

curtail *v.* cut short, reduce. □ **curtailment** *n.*

curtain *n.* a piece of cloth hung as a screen, esp. at a window.

curtsy *n.* (also **curtsey**) a woman's movement of respect

made by bending the knees. ● *v.* make a curtsy.

curvaceous *adj.* (of a woman) having a shapely curved figure.

curvature *n.* curving; a curved form.

curve *n.* a line or surface with no part straight or flat. ● *v.* form (into) a curve.

curvilinear *adj.* contained by or consisting of curved lines.

cushion *n.* a stuffed bag used as a pad, esp. for leaning against; a padded part; a body of air supporting a hovercraft. ● *v.* protect with a pad; lessen the impact of.

cushy *adj.* (**cushier**) (*colloquial*) pleasant and easy.

cusp *n.* a pointed part where curves meet.

cuss *n.* (*colloquial*) a curse; a difficult person. ● *v.* curse.

cussed (**cuss-id**) *adj.* (*colloquial*) stubborn.

custard *n.* a sauce made with milk and eggs or flavoured cornflour.

custodian *n.* a guardian, a keeper.

custody *n.* imprisonment.

custom *n.* **1** the usual way of behaving or acting. **2** regular dealing by customers. **3** (**customs**) duty on imported goods.

customary *adj.* usual. □ **customarily** *adv.*

customer *n.* a person buying goods or services from a shop etc.

cut *v.* (**cut, cutting**) divide, wound, or shape by pressure of a sharp edge; reduce; intersect; divide (a pack of cards); have (a tooth) coming through the gum. ● *n.* a wound or mark made by a sharp edge; a piece cut off; a style of cutting; a reduction; (*colloquial*) a share.

cute *adj.* (*colloquial*) **1** attractive, pretty; quaint. **2** clever, ingenious. □ **cutely** *adv.*, **cuteness** *n.*

cuticle n. the skin at the base of a nail.

cutlass n. a short curved sword.

cutler n. a maker of cutlery.

cutlery n. table knives, forks, and spoons.

cutlet n. a neck-chop; a flat cake of minced meat or nuts and breadcrumbs etc.; a thin piece of veal.

cut-throat adj. merciless. ● n. a murderer.

cutting adj. (of remarks) hurtful. ● n. **1** a passage cut through high ground for a railway etc. **2** a piece of a plant for replanting.

cuttlefish n. a sea creature that ejects black fluid when attacked.

CV abbr. curriculum vitae.

cwt abbr. hundredweight.

cyan n. a greenish-blue colour.

cyanide n. a strong poison.

cyanogen n. an inflammable poisonous gas.

cybernetics n. the science of systems of control and communication in animals and machines.

cyclamen n. a plant with petals that turn back.

cycle n. **1** a recurring series of events. **2** a bicycle or motorcycle. ● v. ride a bicycle. □ **cyclist** n.

cyclic adj. (also **cyclical**) happening in cycles. □ **cyclically** adv.

cyclone n. a violent wind rotating round a central area. □ **cyclonic** adj.

cyclotron n. an apparatus for accelerating charged particles in a spiral path.

cygnet n. a young swan.

cylinder n. an object with straight sides and circular ends. □ **cylindrical** adj., **cylindrically** adv.

cymbal n. a brass plate struck with another or with a stick as a percussion instrument.

cynic n. a person who believes people's motives are usually bad or selfish. □ **cynical** adj., **cynically** adv., **cynicism** n.

cynosure n. a centre of attraction.

cypher var. of **cipher**.

cypress n. an evergreen tree with dark feathery leaves.

cyst n. a sac of fluid or soft matter on or in the body.

cystic adj. of the bladder or gall bladder.

cystic fibrosis n. a hereditary disease usu. resulting in respiratory infections.

cystitis n. inflammation of the bladder.

cytology n. the study of biological cells. □ **cytological** adj.

czar var. of **tsar**.

Dd

D noun (as a Roman numeral) 500. ● symb. deuterium.

d. abbr. (until 1971) penny, pence.

dab n. **1** a small amount of something applied; a light blow or tap. **2** a small flatfish. ● v. (**dabbed**) strike or press lightly or feebly.

dabble v. **1** splash about gently or playfully. **2** work at something in a casual or superficial way.

dab hand n. (colloquial) an expert.

da capo adv. (Music) repeat from the beginning.

dace n. (pl. **dace**) a small freshwater fish.

dacha n. a Russian country cottage.

dachshund n. a small dog with a long body and short legs.

dad n. (colloquial) father.

daddy n. (colloquial) father.

daddy-long-legs n. a crane-fly.

dado n. (pl. **dados**) the lower part of a wall decorated differently from the upper part.

daffodil n. a yellow flower with a trumpet-shaped central part.

daft adj. silly, crazy.

dagger n. a short pointed two-edged weapon used for stabbing.

dago n. (pl. **dagos** or **dagoes**) (offensive slang) a person from southern Europe (esp. Spain or Italy).

daguerrotype n. an early kind of photograph.

dahlia n. a garden plant with bright flowers.

Dáil (doil) n. (in full **Dáil Éireann**) the lower House of Parliament in the Republic of Ireland.

daily adj. happening or appearing on every day or every weekday. ● adv. once a day. ● n. **1** a daily newspaper. **2** (colloquial) a domestic cleaner.

dainty adj. (**daintier**) **1** small and pretty. **2** choosy, particular. □ **daintily** adv. **daintiness** n.

daiquiri (da-ki-ri) n. a cocktail of rum and lime juice.

dairy n. a place where milk and its products are processed or sold.

dais (day-iss) n. a low platform, esp. at the end of a hall.

daisy n. a flower with many petal-like rays.

daisy wheel n. a printing device with radiating spokes.

dal var. of **dhal**.

dale n. a valley.

dally v. idle, dawdle; flirt. □ **dalliance** n.

Dalmatian n. a dog of a large white breed with dark spots.

dam n. **1** a barrier built across a river to hold back water. **2** the mother of an animal, esp. a mammal. ● v. (**dammed**) hold back with a dam; obstruct (a flow).

damage n. **1** something done that reduces the value or usefulness of the thing affected or spoils its appearance. **2** (**damages**) money as compensation for injury. ● v. cause damage to.

damask n. a fabric woven with a pattern visible on either side.

dame n. **1** (**Dame**) the title of a woman with an order of knighthood. **2** (Amer. slang) a woman.

damn v. condemn to hell; condemn as a failure; swear at. ● int. (colloquial) an exclamation of annoyance. ● adj. & adv. (also **damned**) (colloquial) annoying(ly), extreme(ly).

damnable adj. hateful, annoying. □ **damnably** adv.

damnation n. eternal punishment in hell. ● int. (colloquial) an exclamation of annoyance.

damp n. moisture. ● adj. slightly wet. ● v. **1** make damp. **2** discourage; stop the vibration of. □ **dampness** n.

dampen v. make or become damp.

damper n. a plate controlling the draught in a flue; a depressing influence; a pad that damps the vibration of a piano string.

damsel n. (old use) a young woman.

damson n. a small purple plum.

dance v. move with rhythmical steps and gestures, usu. to music; move in a quick or lively way. ● n. a piece of dancing, music for this; a social gathering for dancing. ● **dance attendance on** follow about and help dutifully. □ **dancer** n.

d. and c. abbr. dilatation and curettage; a minor operation to clean the womb.

dandelion n. a wild plant with bright yellow flowers.

dandified adj. like a dandy.

dandle v. dance or nurse (a child) in one's arms.

dandruff n. flakes of dead skin from the scalp.

dandy n. a man who pays excessive attention to his appearance. ●adj. (colloquial) splendid. □ **dandyism** n.

Dane n. a native or inhabitant of Denmark.

danger n. likelihood of harm or death; something causing this.

dangerous adj. causing danger, not safe. □ **dangerously** adv.

dangle v. hang or swing loosely; hold out temptingly. □ **dangler** n., **dangly** adj.

Danish adj. & n. (the language) of Denmark.

dank adj. damp and cold. □ **dankly** adv., **dankness** n.

daphne n. a flowering shrub.

dapper adj. neat and precise, esp. in dress or movement.

dapple v. mark with patches of colour or shade.

dapple grey (of a horse etc.) grey or white with darker spots.

dare v. be bold enough (to do something); challenge (to do something risky). ●n. this challenge.

daredevil n. a recklessly daring person. ●adj. recklessly daring.

daring adj. bold. ●n. boldness. □ **daringly** adv.

dark adj. 1 with little or no light; closer to black than to white; having dark hair or skin. 2 gloomy; secret; mysterious. ●n. absence of light; a time of darkness, night. □ **darkly** adv., **darkness** n.

darken v. make or become dark.

dark horse n. a successful competitor of whom little is known.

darkroom n. a darkened room for processing photographs.

darling n. a loved or lovable person or thing; a favourite. ●adj. beloved, lovable; favourite.

darn v. mend (a hole) by weaving thread across. ●n. a darned area in material. ●adj. (also **darned** (colloquial)) damn.

dart n. 1 a small pointed missile, esp. for throwing at the target in the game of **darts**. 2 a darting movement. 3 a tapering tuck. ●v. run suddenly; send out (a glance etc.) rapidly.

dartboard n. a target in the game of darts.

dash v. 1 run rapidly, rush; knock or throw forcefully against something. 2 destroy (hopes). ●n. 1 a rapid run, a rush. 2 a small amount of liquid or flavouring added. 3 a dashboard. 4 a lively spirit or appearance. 5 a punctuation mark (—) marking a pause or break in the sense or representing omitted letters.

dashboard n. the instrument panel of a motor vehicle.

dashing adj. spirited, showy.

dastardly adj. contemptible.

DAT abbr. digital audio tape.

data n. facts on which a decision is to be based; facts to be processed by computer.

■ **Usage** Strictly speaking, data is a plural noun (singular datum), and it is wrong to say a data or every data or to make the plural form datas. However, in some contexts, especially in the computing world, it is treated as a singular noun, as in Much useful data has been collected.

data bank n. a large store of computerized data.

database n. an organized store of computerized data.

date¹ n. 1 the day, month, or year of a thing's occurrence; a period to which a thing belongs. 2 (colloquial) an appointment to meet a person socially; (esp. Amer.) the person to be met. 3 a small brown edible fruit. ● v. **1** mark with a date; assign a date to; originate from a particular date; become out of date. **2** (*colloquial*) make a social appointment (with). □ **to date** until now.

date rape n. the rape of a woman by a person with whom she is on a date.

dative n. the grammatical case expressing the indirect object.

datum n. (*pl.* **data**) an item of data.

■ **Usage** See note at **data**.

daub v. smear roughly. ● n. a crudely painted picture; a smear.

daughter n. a female in relation to her parents.

daughter-in-law n. (*pl.* **daughters-in-law**) a son's wife.

daunt v. dismay or discourage.

dauntless adj. brave, not daunted.

dauphin n. the title of the eldest son of former kings of France.

Davenport n. **1** a small desk with a sloping top. **2** (*Amer.*) a large sofa.

davit n. a small crane on a ship.

dawdle v. walk slowly and idly, take one's time. □ **dawdler** n.

dawn n. the first light of day; a beginning. ● v. begin to grow light; become apparent. □ **dawning** n.

day n. the time while the sun is above the horizon; a period of 24 hours; hours given to work during a day; a specified day; a time, a period.

daybreak n. the first light of day.

daydream n. pleasant idle thoughts. ● v. have daydreams.

daze v. cause to feel stunned or bewildered. ● n. a dazed state.

dazzle v. blind temporarily with bright light; impress with splendour. □ **dazzlement** n.

dB abbr. decibel(s).

DC abbr. direct current.

DD abbr. Doctor of Divinity.

DDT abbr. a chlorinated hydrocarbon used as an insecticide.

de- pref. implying removal or reversal.

deacon n. a member of the clergy ranking below priest; a lay person attending to church business in Nonconformist churches.

dead adj. **1** no longer alive. **2** no longer used. **3** without brightness or resonance or warmth; dull. **4** exact. ● adv. completely, exactly.

dead beat adj. tired out.

deaden v. deprive of or lose vitality, loudness, feeling, etc.

dead end n. a road closed at one end.

dead heat n. a race in which two or more competitors finish exactly even.

deadline n. a time limit.

dead letter n. a law or rule no longer observed.

deadlock n. a state when no progress can be made. ● v. reach this.

deadly adj. (**deadlier**) **1** causing death or serious damage. **2** deathlike; very dreary. ● adv. **1** as if dead. **2** extremely. □ **deadliness** n.

deadly nightshade n. a plant with poisonous black berries.

deadpan adj. expressionless.

deaf adj. wholly or partly unable to hear; refusing to listen. □ **deafness** n.

deafen v. make unable to hear by a very loud noise.

deal v. (**dealt, dealing**) **1** distribute; hand out (cards) to players in a card game; give, inflict. **2** do business; trade. ● n. **1** a player's turn to deal. **2** a business transaction. **3** (*colloquial*) a large amount. **4** fir or pine timber. □ **deal with** take action about; be about or concerned with.

dealer n. a person who deals; a trader.

dean n. **1** a clergyman who is head of a cathedral chapter. **2** a university official.

deanery n. a dean's position or residence.

dear adj. **1** much loved, cherished. **2** expensive. ● n. a dear person. ● int. an exclamation of surprise or distress. □ **dearly** adv., **dearness** n.

dearth n. a scarcity, a lack.

death n. the process of dying, the end of life; the state of being dead; ending; destruction.

death duty n. a tax levied on property after the owner's death.

deathly adj. (**deathlier**) like death.

death trap n. a very dangerous place.

death-watch beetle n. a beetle whose larvae bore into wood and make a ticking sound.

deb n. (colloquial) a debutante.

debacle (day-bah-kl) n. an utter failure or defeat; a sudden collapse.

debar v. (**debarred**) exclude.

debase v. lower in quality or value. □ **debasement** n.

debatable adj. questionable.

debate n. a formal discussion. ● v. hold a debate about, consider.

debauchery n. over-indulgence in harmful or immoral pleasures. □ **debauched** adj.

debenture n. a certificate acknowledging a debt on which fixed interest is paid.

debilitate v. weaken. □ **debilitation** n.

debility n. weakness of health.

debit n. an entry in an account for a sum owing. ● v. (**debited**) enter as a debit, charge.

debonair adj. having a carefree self-confident manner.

debouch v. come out from a narrow into an open area.

debrief v. question to obtain facts about a completed mission.

debris (deb-ree) n. scattered broken pieces or rubbish.

debt (det) n. something owed. □ **in debt** owing something.

debtor n. a person who owes money.

debug v. (**debugged**) remove bugs from.

debunk v. (colloquial) show up as exaggerated or false.

debut (day-bew) n. a first public appearance.

debutante n. a young woman making her first appearance in society.

Dec. abbr. December.

deca- comb. form ten.

decade n. a ten-year period.

decadent adj. in a state of moral deterioration. □ **decadence** n.

decaffeinated adj. with caffeine removed or reduced.

decagon n. a geometric figure with ten sides.

Decalogue n. the Ten Commandments.

decamp v. go away suddenly or secretly.

decant v. pour (liquid) into another container, leaving sediment behind.

decanter n. a bottle into which wine may be decanted before serving.

decapitate v. behead. □ **decapitation** n.

decarbonize v. (also **-ise**) remove carbon deposit from (an engine). □ **decarbonization** n.

decathlon n. an athletic contest involving ten events.

decay v. rot; lose quality or strength. ● n. decaying, rot.

decease n. death.

deceased adj. dead.

deceit n. deceiving, deception.

deceitful adj. intending to deceive. □ **deceitfully** adv.

deceive v. **1** cause to believe something that is not true. **2** be sexually unfaithful to. □ **deceiver** n.

decelerate v. reduce the speed (of). □ **deceleration** n.

December n. the twelfth month.

decennial adj. happening every tenth year; lasting ten years. □ **decennially** adv.

decent adj. conforming to accepted standards of what is proper; respectable; (colloquial) kind, obliging. □ **decently** adv., **decency** n.

decentralize v. (also **-ise**) transfer from central to local control. □ **decentralization** n.

deception n. deceiving; a trick.

deceptive adj. deceiving; misleading. □ **deceptively** adv.

deci- comb. form one-tenth.

decibel n. a unit for measuring the relative loudness of sound.

decide v. make up one's mind; settle a contest or argument.

decided adj. having firm opinions; clear, definite. □ **decidedly** adv.

deciduous adj. (of a tree) shedding its leaves annually.

decimal adj. reckoned in tens or tenths. ● n. a decimal fraction.

decimal currency n. that with each unit 10 or 100 times the value of the one next below it.

decimal fraction n. a fraction based on powers of ten, shown as figures after a dot.

decimalize v. (also **-ise**) convert into a decimal. □ **decimalization** n.

decimal point n. the dot used in a decimal fraction.

decimate v. destroy one-tenth of; destroy a large proportion of. □ **decimation** n.

decipher v. make out the meaning of (code, bad handwriting).

decision n. deciding, a judgement reached; the ability to decide.

decisive adj. conclusive; showing decision and firmness. □ **decisively** adv., **decisiveness** n.

deck n. a floor or storey of a ship or bus. ● v. decorate, dress up.

deckchair n. a folding canvas chair.

declaim v. speak or say impressively. □ **declamation** n., **declamatory** adj.

declare v. announce openly or formally; state firmly. □ **declaration** n., **declaratory** adj.

declassify v. cease to classify as officially secret. □ **declassification** n.

declension n. (Grammar) a class of nouns and adjectives having the same inflectional forms.

decline v. **1** refuse. **2** slope downwards; decrease, lose strength or vigour. ● n. a gradual decrease or loss of strength.

declivity n. a downward slope.

declutch v. disengage the clutch of a motor.

decoct v. make a decoction of.

decoction n. boiling to extract essence; the essence itself.

decode v. put (a coded message) into plain language; make (an electronic signal) intelligible. □ **decoder** n.

décolleté adj. having a low neckline.

decompose v. (cause to) rot or decay. □ **decomposition** n.

decompress v. release from compression; reduce air pressure in. □ **decompression** n.

decongestant n. a medicinal substance that relieves congestion.

decontaminate v. rid of contamination. □ **decontamination** n.

decor n. the style of decoration used in a room.

decorate v. **1** make attractive by adding objects or details; paint or paper the walls of. **2** confer a medal or award on. □ **decoration** n.

decorative adj. ornamental. □ **decoratively** adv.

decorator n. a person who paints and papers rooms etc. professionally.

decorous adj. polite and well-behaved, decent. □ **decorously** adv.

decorum n. correctness and dignity of behaviour.

decoy n. a person or animal used to lure others into danger. ● v. lure by a decoy.

decrease v. make or become smaller or fewer. ● n. decreasing; the amount of this.

decree n. an order given by a government or other authority. ● v. order by decree.

decrepit adj. made weak by age or use; dilapidated. □ **decrepitude** n.

decriminalize v. (also **-ise**) cease to treat (an action) as criminal.

decry v. disparage.

dedicate v. devote to a person, use, or cause. □ **dedication** n.

deduce v. arrive at (a conclusion) by reasoning; infer. □ **deducible** adj.

deduct v. subtract.

deduction n. **1** deducting; something deducted. **2** deducing; a conclusion deduced.

deductive adj. based on reasoning.

deed n. **1** something done, an act. **2** a legal document.

deem v. consider to be.

deep adj. **1** going or situated far down or in; intense; low-pitched. **2** absorbed. **3** profound. □ **deeply** adv., **deepness** n.

deepen v. make or become deeper.

deer n. (pl. **deer**) a hoofed animal, the male of which usu. has antlers.

deerstalker n. a cloth cap with a peak in front and at the back.

deface v. spoil or damage the surface of. □ **defacement** n.

de facto adj. & adv. (existing) in fact, whether by right or not.

defame v. attack the good reputation of. □ **defamation** n., **defamatory** adj.

default v. fail to fulfil one's obligations or to appear. ● n. this failure. □ **defaulter** n.

defeat v. win victory over; cause to fail. ● n. defeating; being defeated.

defeatist n. a person who pessimistically expects or accepts defeat. □ **defeatism** n.

defecate v. discharge faeces from the body. □ **defecation** n.

defect n. (dee-fekt) a deficiency, an imperfection. ● v. (di-fekt) desert one's country or cause. □ **defection** n., **defector** n.

defective adj. having defect(s); incomplete. □ **defectively** adv., **defectiveness** n.

defence n. (Amer. **defense**) defending; protection; arguments against an accusation. □ **defenceless** adj., **defencelessness** n.

defend v. protect from attack; uphold by argument; represent (the defendant). □ **defender** n.

defendant n. a person accused or sued in a lawsuit.

defensible adj. able to be defended. □ **defensibility** n., **defensibly** adv.

defensive adj. intended for defence; in an attitude of defence.

defer v. (**deferred**) **1** postpone. **2** yield to a person's wishes or authority. □ **deferment** n., **deferral** n.

deference n. polite respect. □ **deferential** adj., **deferentially** adv.

defiance n. defying; open disobedience. □ **defiant** adj., **defiantly** adv.

deficiency n. lack, shortage; a thing or amount lacking.

deficient adj. not having enough; insufficient, lacking.

deficit n. an amount by which a total falls short of what is required.

defile v. make dirty, pollute. ● n. a narrow pass or gorge.

define v. state or explain precisely; mark the boundary of.

definite adj. unmistakable, clearly defined; certain. □ **definitely** adv.

definite article see **article**.

definition n. a statement of precise meaning; making or being distinct, clearness of outline.

definitive adj. finally fixing or settling something; most authoritative. □ **definitively** adv.

deflate v. (cause to) collapse through release of air. □ **deflation** n.

deflect v. turn aside. □ **deflection** or **deflexion** n., **deflector** n.

deflower v. (literary) deprive of virginity.

defoliate v. remove the leaves of. □ **defoliant** n., **defoliation** n.

deforest v. clear of trees. □ **deforestation** n.

deform v. spoil the shape of. □ **deformation** n.

deformity n. abnormality of shape, esp. of a part of the body.

defraud v. deprive by fraud.

defray v. provide money to pay (costs). □ **defrayal** n.

defrost v. remove ice from; thaw.

deft adj. skilful, handling things neatly. □ **deftly** adv.

defunct adj. dead; no longer existing or functioning.

defuse v. remove the fuse from (an explosive); reduce the dangerous tension in (a situation).

defy v. resist; refuse to obey; challenge to do something.

degenerate v. (di-**jen**-er-ayt) become worse. ● adj. (di-**jen**-er-ǎt) having degenerated. □ **degeneracy** n., **degeneration** n.

degrade v. **1** reduce to a lower rank; humiliate. **2** decompose. □ **degradation** n.

degree n. **1** a stage in a series or of intensity. **2** an academic award for proficiency. **3** a unit of measurement for angles or temperature.

dehumanize v. (also **-ise**) remove human qualities from; make impersonal. □ **dehumanization** n.

dehydrate v. dry; lose moisture. □ **dehydration** n., **dehydrator** n.

de-ice v. remove ice from. □ **de-icer** n.

deify v. treat as a god. □ **deification** n.

deign v. condescend.

deity n. a god, a goddess.

déjà vu (day-zhah voo) n. a feeling of having experienced a present situation before.

dejected adj. in low spirits.

dejection n. lowness of spirits.

de jure (dee joo-ri) adj. & adv. rightful, by right.

delay v. make or be late; postpone. ● n. delaying.

delectable adj. delightful. □ **delectably** adv.

delectation n. enjoyment.

delegate n. (**del**-i-gǎt) a representative. ● v. (**del**-i-gayt) entrust (a task or power) to an agent.

delegation *n.* delegating; a group of representatives.

delete *v.* strike out (a word etc.). □ **deletion** *n.*

deleterious *adj.* harmful.

delft *n.* glazed earthenware, typically decorated in blue.

deliberate *adj.* (di-lib-ĕr-ăt) intentional; slow and careful. ● *v.* (di-lib-ĕr-ayt) think over or discuss carefully. □ **deliberately** *adv.*, **deliberation** *n.*

delicacy *n.* 1 being delicate. 2 a choice food.

delicate *adj.* 1 fine, exquisite. 2 not robust or strong. 3 requiring or using tact. □ **delicately** *adv.*

delicatessen *n.* a shop selling speciality groceries, cheeses, cooked meats, etc.

delicious *adj.* delightful, esp. to taste or smell. □ **deliciously** *adv.*

delight *n.* a great pleasure; something giving this. ● *v.* please greatly; feel delight. □ **delightful** *adj.*, **delightfully** *adv.*

delimit *v.* determine the limits or boundaries of. □ **delimitation** *n.*

delineate *v.* outline. □ **delineation** *n.*, **delineator** *n.*

delinquent *adj.* & *n.* (a person) guilty of persistent lawbreaking. □ **delinquency** *n.*

deliquesce *v.* become liquid, melt.

delirium *n.* a disordered state of mind, esp. during fever; wild excitement. □ **delirious** *adj.*, **deliriously** *adv.*

deliver *v.* 1 take to an addressee or purchaser; hand over; utter (a speech etc.); aim (a blow or attack). 2 rescue, set free. 3 assist in the birth (of). □ **deliverer** *n.*, **delivery** *n.*

deliverance *n.* rescue, freeing.

dell *n.* a small wooded hollow.

delphinium *n.* a tall garden plant with usu. blue flowers.

delta *n.* 1 the fourth letter of the Greek alphabet (Δ, δ). 2 a triangular patch of deposited earth at the mouth of a river, formed by its diverging outlets.

delude *v.* deceive.

deluge *n.* a flood; a heavy fall of rain; something coming in an overwhelming rush. ● *v.* flood; come down on like a deluge.

delusion *n.* a false belief or impression. □ **delusory** *adj.*

delusive *adj.* deceptive, raising false hopes.

de luxe *adj.* of superior quality; luxurious.

delve *v.* search deeply.

demagogue *n.* a person who wins support by appealing to popular feelings and prejudices. □ **demagogic** *adj.*, **demagogy** *n.*

demand *n.* a firm or official request; customers' desire for goods or services; a claim. ● *v.* make a demand for; need.

demanding *adj.* making many demands; requiring great skill or effort.

demarcation *n.* the marking of a boundary or limits, esp. of work for different trades.

demean *v.* lower the dignity of.

demeanour *n.* (*Amer.* **demeanor**) the way a person behaves.

demented *adj.* driven mad, crazy.

dementia *n.* a mental disorder.

demerara *n.* brown raw cane sugar.

demesne *n.* a domain; a landed estate.

demi- *pref.* half.

demilitarize *v.* (also **-ise**) remove military forces from. □ **demilitarization** *n.*

demise *n.* death; termination.

demisemiquaver *n.* a note equal to half a semiquaver.

demist *v.* clear mist from (a windscreen etc.). □ **demister** *n.*

demo n. (pl. **demos**) (colloquial) a demonstration.

demob (colloquial) v. (**de-mobbed**) demobilize. ●n. demobilization.

demobilize v. (also **-ise**) release from military service. □ **demobilization** n.

democracy n. government by all the people, usu. through elected representatives; a country governed in this way.

democrat n. a person favouring democracy.

democratic adj. of or according to democracy. □ **democratically** adv.

demography n. the statistical study of human populations. □ **demographic** adj.

demolish v. pull or knock down; destroy. □ **demolition** n.

demon n. a devil, an evil spirit; a cruel or forceful person. □ **demonic** adj., **demoniac** adj., **demoniacal** adj.

demonstrable adj. able to be demonstrated. □ **demonstrability** n., **demonstrably** adv.

demonstrate v. 1 show evidence of, prove; show the working of. 2 take part in a public protest. □ **demonstration** n., **demonstrator** n.

demonstrative adj. showing, proving; showing one's feelings. □ **demonstratively** adv.

demoralize v. (also **-ise**) dishearten. □ **demoralization** n.

demote v. reduce to a lower rank or category. □ **demotion** n.

demur v. (**demurred**) raise objections. ●n. an objection raised.

demure adj. quiet and serious or pretending to be so. □ **demurely** adv., **demureness** n.

den n. a wild animal's lair; a person's small private room.

denary adj. of ten; decimal.

denationalize v. (also **-ise**) privatize. □ **denationalization** n.

denature v. 1 change the properties of. 2 make (alcohol) unfit for drinking.

dendrochronology n. the dating of timber by study of the annual growth rings.

deniable adj. able to be denied.

denial n. denying; a statement that a thing is not true.

denier (den-yer) n. a unit of weight for measuring the fineness of yarn.

denigrate v. blacken the reputation of. □ **denigration** n.

denim n. a strong twilled fabric; (**denims**) trousers made of this.

denizen n. a person or plant living in a specified place.

denominate v. name; describe as.

denomination n. 1 a name, a title. 2 a specified Church or sect. 3 a class of units of measurement or money. □ **denominational** adj.

denominator n. a number below the line in a vulgar fraction.

denote v. be the sign, symbol, or name of; indicate. □ **denotation** n.

denouement (day-noo-mahn) n. the final outcome of a play or story.

denounce v. speak against; inform against.

dense adj. thick; closely massed; stupid. □ **densely** adv., **denseness** n.

density n. denseness; the relation of weight to volume.

dent n. a hollow left by a blow or pressure. ●v. make or become dented.

dental adj. of or for teeth; of dentistry.

dental floss n. thread for cleaning between the teeth.

dentate adj. toothed, notched.

dentifrice n. a substance for cleaning teeth.

dentine n. the hard tissue forming the teeth.

dentist n. a person qualified to treat decay and malformations of teeth.

dentistry n. a dentist's work.

dentition n. the arrangement of the teeth.

denture n. a set of artificial teeth.

denude v. strip of covering or property. □ **denudation** n.

denunciation n. denouncing; a public condemnation.

deny v. say that (a thing) is untrue or does not exist; disown; prevent from having.

deodorant n. a substance that removes or conceals unwanted odours. ● adj. deodorizing.

deodorize v. (also **-ise**) destroy the odour of. □ **deodorization** n.

deoxyribonucleic acid see DNA.

depart v. go away, leave.

department n. a section of an organization. □ **departmental** adj.

department store n. a large shop selling many kinds of goods.

departure n. departing; setting out on a new course of action.

depend v. □ **depend on** be determined by; be unable to do without; trust confidently.

dependable adj. reliable.

dependant n. one who depends on another for support.

dependence n. depending.

dependency n. a dependent state.

dependent adj. depending; controlled by another.

depict v. represent in a picture or in words. □ **depiction** n.

depilatory adj. & n. (a substance) removing hair.

deplete v. reduce by using quantities of. □ **depletion** n.

deplorable adj. regrettable; very bad. □ **deplorably** adv.

deplore v. find or call deplorable.

deploy v. spread out, organize for effective use. □ **deployment** n.

depopulate v. reduce the population of. □ **depopulation** n.

deport v. remove (a person) from a country. □ **deportation** n.

deportment n. behaviour, bearing.

depose v. remove from power.

deposit v. (**deposited**) put down; leave as a layer of matter; entrust for safe keeping; pay as a deposit. ● n. something deposited; a sum entrusted or left as a guarantee. □ **depositor** n.

depositary n. a person to whom something is entrusted.

deposition n. **1** deposing; depositing. **2** a sworn statement.

depository n. **1** a storehouse. **2** a depositary.

depot (de-poh) n. a storage area, esp. for vehicles; (Amer.) a bus or railway station.

deprave v. make morally bad, corrupt.

depravity n. moral corruption, wickedness.

deprecate v. **1** express disapproval of. **2** disclaim politely. □ **deprecation** n., **deprecatory** adj.

■ **Usage** Deprecate is often confused with depreciate.

depreciate v. diminish in value. □ **depreciation** n., **depreciatory** adj.

■ **Usage** Depreciate is often confused with deprecate.

depredation n. plundering, destruction.

depress v. **1** press down. **2** reduce (trade etc.). **3** make sad. □ **depressant** adj. & n.

depression n. **1** pressing down; a sunken place. **2** a state of sadness. **3** a long period of inactivity in trading. **4** an area of low atmospheric pressure. □ **depressive** adj.

deprive v. prevent from using or enjoying something. □ **deprivation** n.

depth n. deepness, a measure of this; the deepest or most central part. □ **in depth** thoroughly. **out of one's depth** in water too deep to stand in.

depth charge n. a bomb that will explode under water.

deputation n. a body of people sent to represent others.

depute v. appoint to act as one's representative.

deputize v. (also **-ise**) act as deputy.

deputy n. a person appointed to act as a substitute or representative.

derail v. cause (a train) to leave the rails. □ **derailment** n.

derange v. disrupt; make insane. □ **derangement** n.

derelict adj. left to fall into ruin.

dereliction n. abandonment; neglect (of duty).

derestrict v. remove restrictions from.

deride v. scoff at.

de rigueur (dĕ rig-er) adj. required by custom or etiquette.

derision n. scorn, ridicule.

derisive adj. scornful, showing derision. □ **derisively** adv.

derisory adj. showing derision; deserving derision.

derivative adj. derived from another source; not original. ● n. a derived thing.

derive v. obtain from a source; have its origin. □ **derivation** n.

dermatitis n. inflammation of the skin.

dermatology n. the study of the skin and its diseases. □ **dermatologist** n.

derogatory adj. disparaging.

derrick n. a crane with a pivoted arm; a framework over an oil well etc.

derris n. an insecticide made from the root of a tropical plant.

derv n. fuel for diesel engines.

dervish n. a member of a Muslim religious order known for their whirling dance.

DES abbr. Department of Education and Science.

desalinate v. remove salt from (esp. sea water). □ **desalination** n.

descant n. a treble accompaniment to a main melody.

descend v. go or come down; stoop to unworthy behaviour. □ **be descended from** have as one's ancestor(s).

descendant n. a person descended from another.

descent n. descending; a downward route or slope; lineage.

describe v. **1** give a description of. **2** mark the outline of.

description n. **1** a statement of what a person or thing is like. **2** a kind of thing.

descriptive adj. describing.

descry v. catch sight of, discern.

desecrate v. treat (a sacred thing) irreverently. □ **desecration** n., **desecrator** n.

desegregate v. abolish segregation in or of. □ **desegregation** n.

deselect v. reject (an already selected candidate). □ **deselection** n.

desert[1] (dez-ert) n. a barren uninhabited often sandy area.

desert[2] (di-zert) v. abandon; leave one's service in the armed forces without permission. □ **deserter** n., **desertion** n.

deserts (di-zerts) *n.pl.* what one deserves.

deserve *v.* be worthy of or entitled to. □ **deservedly** *adv.*

déshabillé (day-za-bee-ay) *n.* (also **dishabille**) the state of being partly dressed.

desiccate *v.* dry out moisture from. □ **desiccation** *n.*

desideratum *n.* (*pl.* **desiderata**) something required.

design *n.* a drawing that shows how a thing is to be made; a general form or arrangement; lines or shapes forming a decoration; a mental plan. ● *v.* prepare a design for; plan, intend. □ **designedly** *adv.*, **designer** *n.*

designate *adj.* (dez-ig-nǎt) appointed but not yet installed. ● *v.* (dez-ig-nayt) name as; specify; appoint to a position. □ **designation** *n.*

designing *adj.* scheming, crafty.

desirable *adj.* **1** arousing desire, attractive. **2** advisable. □ **desirability** *n.*

desire *n.* a feeling of wanting something strongly; a thing desired. ● *v.* feel a desire for.

desirous *adj.* desiring.

desist *v.* cease, stop.

desk *n.* a piece of furniture for reading or writing at; a counter; a section of a newspaper office etc.

desktop *n.* **1** the working surface of a desk. **2** (in full **desktop computer**) a computer small enough for use on a desk.

desktop publishing *n.* producing documents, booklets, etc. with a desktop computer and high-quality printer.

desolate *adj.* lonely; deserted, uninhabited. □ **desolation** *n.*

desolated *adj.* feeling very distressed.

despair *n.* complete lack of hope. ● *v.* feel despair.

despatch var. of **dispatch**.

desperado *n.* (*pl.* **desperadoes**; *Amer.* **desperados**) a reckless criminal.

desperate *adj.* reckless through despair. □ **desperately** *adv.*, **desperation** *n.*

despicable *adj.* contemptible. □ **despicably** *adv.*

despise *v.* regard as worthless.

despite *prep.* in spite of.

despoil *v.* (*literary*) plunder. □ **despoilment** *n.*, **despoliation** *n.*

despondent *adj.* dejected. □ **despondently** *adv.*, **despondency** *n.*

despot *n.* a dictator. □ **despotic** *adj.*, **despotically** *adv.*, **despotism** *n.*

dessert *n.* the sweet course of a meal.

dessertspoon *n.* a medium-sized spoon for eating puddings etc. □ **dessertspoonful** *n.*

destabilize *v.* (also **-ise**) make unstable; seriously weaken (a government).

destination *n.* the place to which a person or thing is going.

destine *v.* settle the future of; set apart for a purpose.

destiny *n.* fate; one's future destined by fate.

destitute *adj.* extremely poor; without means to live. □ **destitution** *n.*

destroy *v.* pull or break down; ruin; kill (an animal). □ **destruction** *n.*, **destructive** *adj.*

destroyer *n.* a person or thing that destroys; a fast warship.

destruct *v.* destroy (esp. a rocket) deliberately.

destructible *adj.* able to be destroyed.

desuetude *n.* disuse.

desultory *adj.* going from one subject to another, not systematic. □ **desultorily** *adv.*

detach *v.* release or separate. □ **detachable** *adj.*

detached adj. **1** not joined to another. **2** free from bias or emotion.

detachment n. **1** detaching; being detached. **2** a military group.

detail n. **1** a small fact or item; such items collectively. **2** a small military detachment. ● v. **1** relate in detail. **2** assign to a special duty.

detain v. keep in confinement; cause delay to. □ **detainment** n.

detainee n. a person detained in custody.

detect v. discover the presence of. □ **detection** n., **detector** n.

detective n. a person whose job is to investigate crimes.

détente (day-tahnt) n. an easing of tension between nations.

detention n. detaining; imprisonment.

deter v. (**deterred**) discourage from action. □ **determent** n.

detergent n. a cleansing substance, esp. other than soap.

deteriorate v. become worse. □ **deterioration** n.

determinant n. a decisive factor.

determination n. firmness of purpose; the process of deciding.

determine v. decide; calculate precisely; resolve firmly.

determined adj. full of determination.

determinism n. a theory that actions are determined by external forces.

deterrent n. something that deters. □ **deterrence** n.

detest v. dislike intensely. □ **detestable** adj., **detestation** n.

dethrone v. remove from a throne. □ **dethronement** n.

detonate v. explode. □ **detonation** n., **detonator** n.

detour n. a deviation from a direct or intended course.

detoxify v. remove harmful substances from.

detract v. □ **detract from** reduce the credit that is due to; lessen. □ **detraction** n.

detractor n. a person who criticizes something.

detriment n. harm. □ **detrimental** adj., **detrimentally** adv.

detritus n. debris; loose stones.

de trop (dĕ troh) adj. not wanted.

deuce n. **1** a score of 40 all in tennis. **2** (colloquial) (in exclamations) the Devil.

deuterium n. a heavy form of hydrogen.

Deutschmark (doich-mark) n. the unit of money in Germany.

devalue v. reduce the value of. □ **devaluation** n.

devastate v. cause great destruction to. □ **devastation** n.

devastating adj. overwhelming; causing destruction.

develop v. (**developed**) **1** make or become larger or more mature or organized; bring or come into existence. **2** make usable or profitable, build on (land). **3** treat (a film) so as to make a picture visible. □ **developer** n., **development** n.

deviant adj. & n. (a person or thing) deviating from normal behaviour.

deviate v. turn aside from a course of action, truth, etc. □ **deviation** n.

device n. a thing made or used for a purpose; a scheme.

devil n. an evil spirit, (**the Devil**) the supreme spirit of evil; a cruel or annoying person; a person of mischievous energy or cleverness; (colloquial) a difficult person or problem. □ **devilish** adj.

devilled adj. (Amer. **deviled**) cooked with hot spices.

devilment n. mischief.

devilry n. wickedness; devilment.

devil's advocate n. a person who tests a proposition by arguing against it.

devious adj. indirect; underhand. □ **deviously** adv., **deviousness** n.

devise v. plan; invent. □ **devisor** n.

devoid adj. □ **devoid of** lacking, free from.

devolution n. devolving; delegation of power from central to local administration.

devolve v. pass or be passed to a deputy or successor.

devote v. give or use for a particular purpose.

devoted adj. showing devotion.

devotee n. an enthusiast.

devotion n. great love or loyalty; zeal; worship; (**devotions**) prayers.

devotional adj. used in worship.

devour v. eat hungrily or greedily; consume; take in avidly. □ **devourer** n.

devout adj. earnestly religious; earnest, sincere. □ **devoutly** adv.

dew n. drops of condensed moisture forming on cool surfaces at night.

dewclaw n. a small claw on the inner side of a dog's leg.

dewlap n. a fold of loose skin at the throat of cattle etc.

dexterity n. skill.

dextrose n. a form of glucose.

dexterous adj. (also **dextrous**) skilful. □ **dexterously** adv.

dhal n. (also **dal**) a small pea-like pulse; an Indian dish made with this.

di- pref. two; double.

diabetes n. a disease in which sugar and starch are not properly absorbed by the body. □ **diabetic** adj. & n.

diabolic adj. of the Devil.

diabolical adj. very cruel, wicked, or cunning. □ **diabolically** adv.

diabolism n. worship of the Devil.

diachronic adj. of the historical development of a subject.

diaconate n. the office of deacon; a body of deacons. □ **diaconal** adj.

diacritic n. a sign (e.g. an accent) on a letter of a word.

diadem n. a crown.

diaeresis n. a mark over a vowel sounded separately.

diagnose v. make a diagnosis of.

diagnosis n. (pl. **diagnoses**) the identification of a disease or condition after observing its signs. □ **diagnostic** adj., **diagnostician** n.

diagonal adj. & n. (a line) crossing from corner to corner. □ **diagonally** adv.

diagram n. a drawing that shows the parts or operation of something. □ **diagrammatic** adj., **diagrammatically** adv.

dial n. the face of a clock or watch; a similar plate or disc with a movable pointer; a movable disc manipulated to connect one telephone with another. □ v. (**dialled**; Amer. **dialed**) select or operate by using a dial or numbered buttons.

dialect n. a local form of a language. □ **dialectal** adj.

dialectic n. an investigation of truths in philosophy etc. by systematic reasoning. □ **dialectical** adj.

dialogue n. (Amer. **dialog**) a conversation or discussion.

dialysis n. purification of blood by causing it to flow through a suitable membrane.

diamanté (dee-ă-**mon**-tay) adj. decorated with artificial jewels.

diameter n. a straight line from side to side through the centre of a circle or sphere; its length.

diametrical adj. 1 of or along a diameter. 2 (of opposites) complete; absolutely opposed. □ **diametrically** *adv.*

diamond *n.* **1** a very hard brilliant precious stone. **2** a four-sided figure with equal sides and with angles that are not right angles; a playing card marked with such shapes.

diamond wedding *n.* a 60th anniversary.

diaper *n.* (*Amer.*) a baby's nappy.

diaphanous *adj.* almost transparent. □ **diaphanously** *adv.*

diaphragm (dy-å-fram) *n.* **1** the muscular partition between the chest and abdomen. **2** a contraceptive cap fitting over the cervix.

diarrhoea (dyå-ree-å) (*Amer.* **diarrhea**) *n.* a condition with frequent fluid faeces.

diary *n.* a daily record of events; a book for noting these. □ **diarist** *n.*

diatribe *n.* a violent verbal attack.

dibber *n.* (also **dibble**) a tool to make holes in the ground for young plants.

dice *n.* (*pl.* **dice**) a small cube marked on each side with 1-6 spots, used in games of chance. ● *v.* cut into small cubes. □ **dice with death** take great risks.

dicey *adj.* (**dicier**) (*slang*) risky; unreliable.

dichotomy (dy-kot-ŏmi) *n.* a division into two parts or kinds.

dicky *adj.* (**dickier**) (*slang*) shaky, unsound. ● *n.* a false shirt-front.

dicky bow *n.* a bow tie.

dicta *see* **dictum**.

dictate *v.* say (words) aloud to be written or recorded; state or order authoritatively; give orders officiously. ● *n.pl.* (**dictates**) commands. □ **dictation** *n.*

dictator *n.* a ruler with unrestricted authority; a domineering person. □ **dictatorship** *n.*

dictatorial *adj.* of or like a dictator. □ **dictatorially** *adv.*

diction *n.* a manner of uttering or pronouncing words.

dictionary *n.* a book that lists and explains the words of a language or the topics of a subject.

dictum *n.* (*pl.* **dicta**) a formal saying.

did *see* **do**.

didactic *adj.* meant or meaning to instruct. □ **didactically** *adv.*

diddle *v.* cheat, swindle.

die *v.* (**died, dying**) cease to be alive; cease to exist or function; fade away. ● *n.* a device that stamps a design or that cuts or moulds material into shape. □ **be dying for** or **to** feel an intense longing for or to.

diehard *n.* a very conservative or stubborn person.

diesel *n.* a diesel engine; a vehicle driven by this.

diesel engine *n.* an oil-burning engine in which ignition is produced by the heat of compressed air.

diet *n.* **1** usual food; a restricted selection of food. **2** a congress, a parliamentary assembly in certain countries. ● *v.* restrict what one eats. □ **dietary** *adj.*, **dieter** *n.*

dietetic *adj.* of diet and nutrition. ● *n.* (**dietetics**) the study of diet and nutrition.

dietitian *n.* an expert in dietetics.

differ *v.* be unlike; disagree.

difference *n.* being different; the amount of this; the remainder after subtraction; a disagreement.

different *adj.* not the same; separate; unusual. □ **differently** *adv.*

■ **Usage** It is safer to use *different from*, but *different to* is common in informal use.

differential *adj.* of, showing, or depending on a difference. ● *n.* **1** an agreed difference in wage-rates. **2** an arrangement of gears allowing a vehicle's wheels to revolve at different speeds when cornering.

differentiate *v.* be a difference between; distinguish between; develop differences. □ **differentiation** *n.*

difficult *adj.* needing much effort or skill to do, deal with, or understand; troublesome. □ **difficulty** *n.*

diffident *adj.* lacking self-confidence. □ **diffidently** *adv.*, **diffidence** *n.*

diffract *v.* break up a beam of light into a series of coloured or dark-and-light bands. □ **diffraction** *n.*, **diffractive** *adj.*

diffuse *adj.* (di-fewss) not concentrated. ● *v.* (di-fewz) spread widely or thinly. □ **diffusely** *adv.*, **diffuser** *n.*, **diffusion** *n.*, **diffusive** *adj.* **diffusible** *adj.*

dig *v.* (**dug**, **digging**) **1** break up and move soil; make (a way or hole) in this way; excavate. **2** find by investigation. **3** poke. ● *n.* **1** excavation. **2** a poke. **3** a cutting remark. **4** (**digs**) (colloquial) lodgings.

digest *v.* (dy-**jest**) break down (food) in the body; absorb into the mind. ● *n.* (**dy**-jest) a methodical summary. □ **digester** *n.*

digestible *adj.* able to be digested. □ **digestibility** *n.*

digestion *n.* the process or power of digesting food.

digestive *adj.* of or aiding digestion.

digestive biscuit *n.* a wholemeal biscuit.

digger *n.* a person who digs; a mechanical excavator.

digit *n.* **1** any numeral from 0 to 9. **2** a finger or toe.

digital *adj.* of or using digits; (of a clock) showing the time by a row of figures; (of a recording) converting sound into electrical pulses. □ **digitally** *adv.*

digitalis *n.* a heart stimulant prepared from foxglove leaves.

dignified *adj.* showing dignity.

dignify *v.* give dignity to.

dignitary *n.* a person holding high rank or position.

dignity *n.* a calm and serious manner; high rank or position.

digress *v.* depart from the main subject temporarily. □ **digression** *n.*, **digressive** *adj.*

dike var. of **dyke**.

diktat *n.* a firm statement or order.

dilapidated *adj.* in disrepair.

dilapidation *n.* a dilapidated state.

dilate *v.* make or become wider. □ **dilation**, **dilatation** *n.*, **dilator** *n.*

dilatory (**dil**-ă-ter-i) *adj.* delaying, not prompt.

dilemma *n.* a situation in which a difficult choice has to be made.

■ *Usage Dilemma* is sometimes also used to mean 'a difficult situation or predicament', but this is considered incorrect by some people.

dilettante (dili-**tan**-ti) *n.* (*pl.* **dilettanti** or **dilettantes**) a person who dabbles in a subject for pleasure.

diligent *adj.* working or done with care and effort. □ **diligently** *adv.*, **diligence** *n.*

dill *n.* a herb with feathery leaves and spicy seeds.

dilly-dally *v.* (colloquial) dawdle; waste time by indecision.

dilute *v.* reduce the strength of (fluid) by adding water etc.; reduce the forcefulness of. □ **dilution** *n.*

dim *adj.* (**dimmer**) **1** lit faintly; indistinct. **2** (colloquial) stupid. ● *v.* (**dimmed**) make or become dim. □ **dimly** *adv.*, **dimness** *n.*

dime n. a 10-cent coin of the USA.

dimension n. a measurable extent; scope. □ **dimensional** adj.

diminish v. make or become less.

diminuendo n. (pl. **diminuendos** or **diminuendi**) (Music) a gradual decrease in loudness.

diminution n. a decrease.

diminutive adj. tiny. ● n. an affectionate form of a name.

dimple n. a small dent, esp. in the skin. ● v. show dimples; produce dimples in.

din n. a loud annoying noise. ● v. (**dinned**) force (information) into a person by constant repetition.

dinar (dee-nar) n. a unit of money in some Balkan and Middle Eastern countries.

dine v. eat dinner. □ **diner** n.

ding-dong n. the sound of bells. ● adj. & adv. (colloquial) with vigorous action between contestants.

dinghy n. a small open boat or inflatable rubber boat.

dingle n. a deep dell.

dingo n. (pl. **dingoes**) an Australian wild dog.

dingy adj. (**dingier**) dirty-looking. □ **dingily** adv., **dinginess** n.

dining room n. a room in which meals are eaten.

dinky adj. (**dinkier**) (colloquial) attractively small and neat.

dinner n. the chief meal of the day; a formal evening meal.

dinner jacket n. a man's usu. black jacket for evening wear.

dinosaur n. an extinct prehistoric reptile, often of enormous size.

dint n. a dent. □ **by dint of** by means of.

diocese n. a district under the care of a bishop. □ **diocesan** adj.

diode n. a semiconductor allowing the flow of current in one direction only and having two terminals.

dioptre (dy-op-ter) n. (Amer. **diopter**) a unit of refractive power of a lens.

dioxide n. an oxide with two atoms of oxygen to one of a metal or other element.

dip v. (**dipped**) plunge briefly into liquid; lower, go downwards. ● n. dipping; a short bathe; a liquid or mixture into which something is dipped; a downward slope. □ **dip into** read briefly from (a book).

Dip. Ed. abbr. Diploma in Education.

diphtheria n. an infectious disease with inflammation of the throat.

diphthong n. a compound vowel sound (as ou in loud).

diploma n. a certificate awarded on completion of a course of study.

diplomacy n. handling of international relations; tact.

diplomat n. **1** an official representing a country abroad. **2** a tactful person. □ **diplomatic** adj., **diplomatically** adv.

dipper n. **1** a diving bird. **2** a ladle.

dipsomania n. an uncontrollable craving for alcohol. □ **dipsomaniac** n.

diptych (dip-tik) n. a pair of pictures on two panels hinged together.

dire adj. dreadful; ominous; extreme and urgent.

direct adj. straight, not roundabout; with nothing or no one between; straightforward, frank. ● adv. by a direct route. ● v. tell how to do something or reach a place; address (a letter etc.); guide; control; command. □ **directness** n.

direct debit n. an instruction allowing an organization to take regular payments from one's bank account.

direction n. **1** directing. **2** the line along which a thing moves or faces. **3** an instruction. □ **directional** adj.

directive n. a general instruction issued by an authority.

directly adv. **1** in a direct line or manner. **2** very soon. ● conj. as soon as.

direct object n. (Grammar) the primary object of a transitive verb, the person or thing directly affected.

director n. a supervisor; a member of a board directing a business; one who supervises acting and filming. □ **directorship** n.

directorate n. the office of director; a board of directors.

directory n. a list of telephone subscribers, members, etc.

direct speech n. words actually spoken, not reported.

dirge n. a mournful song.

dirham n. the unit of money in Morocco and the United Arab Emirates.

dirigible n. an airship.

dirndl n. a full gathered skirt.

dirt n. unclean matter; soil; foul words, scandal.

dirty adj. (**dirtier**) soiled, not clean; producing pollution; dishonourable; obscene. ● v. make or become dirty. □ **dirtily** adv., **dirtiness** n.

disability n. something that disables someone; a physical incapacity.

disable v. deprive of some ability, make unfit.

disabled adj. having a physical disability. □ **disablement** n.

disabuse v. disillusion.

disadvantage n. an unfavourable condition. □ **disadvantaged** adj., **disadvantageous** adj.

disaffected adj. discontented, no longer feeling loyalty. □ **disaffection** n.

disagree v. have a different opinion; fail to agree; quarrel. □ **disagreement** n.

disagreeable adj. unpleasant; bad-tempered. □ **disagreeably** adv.

disallow v. refuse to sanction.

disappear v. pass from sight or existence. □ **disappearance** n.

disappoint v. fail to do what was desired or expected. □ **disappointment** n.

disapprobation n. disapproval.

disapprove v. consider bad or immoral. □ **disapproval** n.

disarm v. **1** deprive of weapons; reduce armed forces. **2** make less hostile; win over.

disarmament n. a reduction of a country's forces or weapons.

disarrange v. put into disorder. □ **disarrangement** n.

disarray n. disorder, confusion.

disassociate v. = dissociate.

disaster n. a sudden great misfortune or failure. □ **disastrous** adj., **disastrously** adv.

disavow v. disclaim. □ **disavowal** n.

disband v. separate, disperse.

disbar v. (**disbarred**) deprive (a barrister) of the right to practise law.

disbelieve v. refuse or be unable to believe. □ **disbelief** n.

disburse v. pay out (money). □ **disbursement** n.

disc n. a thin circular plate; a record bearing recorded sound; a layer of cartilage between the vertebrae; (Computing) = disk.

discard v. (dis-kard) reject as useless or unwanted.

discern v. perceive with the mind or senses. □ **discernment** n., **discernible** adj., **discernibly** adv.

discerning *adj.* perceptive, showing sensitive understanding.

discharge *v.* (**discharged**) **1** send or flow out; release; dismiss. **2** pay (a debt), perform (a duty etc.). ● *n.* discharging; a substance discharged.

disciple *n.* a person accepting the teachings of another; one of the original followers of Christ.

disciplinarian *n.* a person who enforces strict discipline.

disciplinary *adj.* of or for discipline.

discipline *n.* **1** orderly or controlled behaviour; training or control producing this. **2** a branch of learning. ● *v.* train to be orderly; punish.

disc jockey *n.* a person who introduces and plays pop records on the radio or at a disco.

disclaim *v.* disown.

disclaimer *n.* a statement disclaiming something.

disclose *v.* reveal. □ **disclosure** *n.*

disco *n.* (*pl.* **discos**) a place where recorded pop music is played for dancing; equipment for playing this.

discolour *v.* (*Amer.* **discolor**) change in colour; stain. □ **discoloration** *n.*

discomfit *v.* (**discomfited**) disconcert. □ **discomfiture** *n.*

discomfort *n.* being uncomfortable; something causing this.

discommode *v.* inconvenience.

disconcert *v.* upset the self-confidence of, fluster.

disconnect *v.* break the connection of; cut off the power supply of. □ **disconnection** *n.*

disconsolate *adj.* unhappy, disappointed. □ **disconsolately** *adv.*

discontent *n.* dissatisfaction. □ **discontented** *adj.*

discontinue *v.* put an end to; cease. □ **discontinuance** *n.*

discontinuous *adj.* not continuous. □ **discontinuity** *n.*

discord *n.* **1** disagreement, quarrelling. **2** harsh noise. □ **discordance** *n.*, **discordant** *adj.*

discotheque *n.* a disco.

discount *n.* (**dis-kownt**) an amount of money taken off the full price. ● *v.* (**dis-kownt**) disregard partly or wholly.

discourage *v.* dishearten; dissuade (from). □ **discouragement** *n.*

discourse *n.* (**dis-korss**) a conversation, a lecture; a treatise. ● *v.* (**dis-korss**) utter or write a discourse.

discourteous *adj.* lacking courtesy. □ **discourteously** *adv.*, **discourtesy** *n.*

discover *v.* obtain sight or knowledge of. □ **discovery** *n.*

discredit *v.* (**discredited**) damage the reputation of; cause to be disbelieved. ● *n.* (something causing) damage to a reputation.

discreditable *adj.* bringing discredit.

discreet *adj.* prudent; not giving away secrets; unobtrusive. □ **discreetly** *adv.*

discrepancy *n.* a failure to tally; a difference.

discrete *adj.* separate, not continuous. □ **discretely** *adv.*

discretion *n.* **1** being discreet. **2** freedom to decide something.

discretionary *adj.* done or used at a person's discretion.

discriminate *v.* make a distinction (between). □ **discriminate against** treat unfairly. □ **discrimination** *n.*, **discriminatory** *adj.*

discriminating *adj.* having good judgement.

discursive *adj.* rambling, not keeping to the main subject.

discus *n.* a heavy disc thrown in an athletic contest.

discuss v. examine by argument, talk or write about. □ **discussion** n.

disdain v. & n. scorn. □ **disdainful** adj., **disdainfully** adv.

disease n. an unhealthy condition; a specific illness. □ **diseased** adj.

disembark v. put or go ashore. □ **disembarkation** n.

disembodied adj. (of a voice) apparently not produced by anyone.

disembowel v. (**disembowelled**; Amer. **disemboweled**) take out the bowels of. □ **disembowelment** n.

disenchant v. free from enchantment, disillusion. □ **disenchantment** n.

disenfranchise v. (also **disfranchise**) deprive of the right to vote. □ **disenfranchisement** n.

disengage v. separate; detach. □ **disengagement** n.

disentangle v. free from tangles or confusion; separate. □ **disentanglement** n.

disestablish v. end the established state of; deprive (the Church) of its official connection with the state.

disfavour n. (Amer. **disfavor**) dislike, disapproval.

disfigure v. spoil the appearance of. □ **disfigurement** n.

disfranchise var. of **disenfranchise**.

disgorge v. eject, pour forth. □ **disgorgement** n.

disgrace n. (something causing) loss of respect. ● v. bring disgrace upon. □ **disgraceful** adj., **disgracefully** adv.

disgruntled adj. discontented, resentful. □ **disgruntlement** n.

disguise v. conceal the identity of. ● n. disguising, a disguised condition; something that disguises.

disgust n. a strong dislike. ● v. cause disgust in. □ **disgusting** adj.

dish n. a shallow bowl, esp. for food; food prepared for the table. □ **dish out** (colloquial) distribute. **dish up** serve out food.

dishabille var. of **déshabillé**.

disharmony n. lack of harmony.

dishearten v. cause to lose hope or confidence.

dished adj. concave.

dishevelled adj. (Amer. **disheveled**) ruffled and untidy. □ **dishevelment** n.

dishonest adj. not honest. □ **dishonestly** adv., **dishonesty** n.

dishonour v. & n. (Amer. **dishonor**) disgrace.

dishonourable adj. (Amer. **dishonorable**) not honourable, shameful. □ **dishonourably** adv.

dishwasher n. a machine for washing dishes.

disillusion v. rid of pleasant but mistaken beliefs. □ **disillusionment** n.

disincentive n. something that discourages an action or effort.

disinclination n. unwillingness.

disincline v. cause to feel reluctant or unwilling.

disinfect v. cleanse by destroying harmful bacteria. □ **disinfection** n.

disinfectant n. a substance used for disinfecting things.

disinformation n. deliberately misleading information.

disingenuous adj. insincere.

disinherit v. reject from being one's heir. □ **disinheritance** n.

disintegrate v. break into small pieces. □ **disintegration** n.

disinter v. (**disinterred**) dig up, unearth. □ **disinterment** n.

disinterested adj. 1 unbiased, impartial. 2 uninterested. □ **disinterestedly** adv.

━━ **Usage** The use of *disinterested* to mean 'uninterested' is common in informal use but is widely considered incorrect. The use of the noun *disinterest* to mean 'lack of interest' is also objected to, but it is rarely used in any other sense and the alternative *uninterest* is rare.

disjointed *adj.* lacking orderly connection.

disjunctive *adj.* involving separation; expressing an alternative.

disk *n.* a flat circular device on which computer data can be stored.

diskette *n.* (*Computing*) a small floppy disk.

dislike *n.* a feeling of not liking something. ● *v.* feel dislike for.

dislocate *v.* displace from its position; disrupt. □ **dislocation** *n.*

dislodge *v.* move or force from an established position.

disloyal *adj.* not loyal. □ **disloyally** *adv.*, **disloyalty** *n.*

dismal *adj.* gloomy; (*colloquial*) feeble. □ **dismally** *adv.*

dismantle *v.* take to pieces.

dismay *n.* a feeling of anxiety or despair. ● *v.* cause dismay to.

dismember *v.* remove the limbs of; split into pieces. □ **dismemberment** *n.*

dismiss *v.* send away from one's presence or employment; reject. □ **dismissal** *n.*, **dismissive** *adj.*

dismount *v.* get off a thing on which one is riding.

disobedient *adj.* not obedient. □ **disobediently** *adv.*, **disobedience** *n.*

disobey *v.* disregard orders.

disoblige *v.* fail to help or oblige. □ **disobliging** *adj.*

disorder *n.* **1** lack of order or of discipline. **2** an ailment. ● *v.* throw into disorder; upset. □ **disorderly** *adj.*, **disorderliness** *n.*

disorganize *v.* (also **-ise**) upset the orderly arrangement of. □ **disorganization** *n.*

disorientate *v.* (also **disorient**) cause (a person) to lose his or her sense of direction. □ **disorientation** *n.*

disown *v.* refuse to acknowledge; reject all connection with.

disparage *v.* speak slightingly of. □ **disparagement** *n.*

disparate *adj.* different in kind. □ **disparately** *adv.*

disparity *n.* inequality, difference.

dispassionate *adj.* not emotional; impartial. □ **dispassionately** *adv.*

dispatch (also **despatch**) *v.* **1** send off to a destination or for a purpose. **2** complete (a task) quickly. **3** kill. ● *n.* **1** dispatching. **2** promptness. **3** an official message. **4** a news report.

dispatch box *n.* a container for carrying official documents.

dispatch rider *n.* a messenger who travels by motorcycle.

dispel *v.* (**dispelled**) drive away; disperse.

dispensable *adj.* not essential.

dispensary *n.* a place where medicines are dispensed.

dispensation *n.* dispensing; distributing; exemption.

dispense *v.* deal out; prepare and give out (medicine etc.). □ **dispense with** do without; make unnecessary. □ **dispenser** *n.*

disperse *v.* go or send in different directions, scatter. □ **dispersal** *n.*, **dispersion** *n.*

dispirited *adj.* dejected. □ **dispiriting** *adj.*

displace *v.* shift; take the place of; oust. □ **displacement** *n.*

display *v.* show, arrange conspicuously. ● *n.* displaying; thing(s) displayed.

displease *v.* irritate; annoy.

displeasure *n.* disapproval.

disport v. (also **disport oneself**) frolic; amuse oneself.

disposable adj. **1** at one's disposal. **2** designed to be thrown away after use. □ **disposability** n.

disposal n. disposing. □ **at one's disposal** available for one's use.

dispose v. **1** place, arrange. **2** make willing or ready to do something. □ **dispose of** get rid of; finish off. **be well disposed** be friendly or favourable.

disposition n. **1** arrangement. **2** a person's character. **3** a tendency.

dispossess v. deprive of the possession of. □ **dispossession** n.

disproportionate adj. relatively too large or too small. □ **disproportionately** adv.

disprove v. show to be wrong.

disputable adj. questionable. □ **disputably** adv.

disputant n. a person engaged in a dispute.

disputation n. an argument, a debate.

disputatious adj. fond of arguing.

dispute v. argue, debate; quarrel; question the validity of. ● n. a debate; a quarrel.

disqualify v. make ineligible or unsuitable. □ **disqualification** n.

disquiet n. uneasiness, anxiety. ● v. cause disquiet to.

disquisition n. a long elaborate account.

disregard v. pay no attention to. ● n. lack of attention.

disrepair n. bad condition caused by lack of repair.

disreputable adj. not respectable. □ **disreputably** adv.

disrepute n. discredit.

disrespect n. lack of respect. □ **disrespectful** adj., **disrespectfully** adv.

disrobe v. undress.

disrupt v. cause to break up; interrupt the flow or continuity of. □ **disruption** n., **disruptive** adj.

dissatisfaction n. lack of satisfaction or of contentment.

dissatisfied adj. not satisfied.

dissect v. cut apart so as to examine the internal structure. □ **dissection** n., **dissector** n.

dissemble v. conceal (feelings). □ **dissemblance** n.

disseminate v. spread widely. □ **dissemination** n.

dissension n. disagreement that gives rise to strife.

dissent v. have a different opinion. ● n. difference in opinion. □ **dissenter** n.

dissertation n. a lengthy essay; a speech.

disservice n. an unhelpful or harmful action.

dissident adj. disagreeing. ● n. a person who disagrees, esp. with an established government. □ **dissidence** n.

dissimilar adj. unlike. □ **dissimilarity** n., **dissimilitude** n.

dissimulate v. dissemble. □ **dissimulation** n.

dissipate v. dispel; fritter away. □ **dissipation** n.

dissipated adj. living a dissolute life.

dissociate v. regard as separate; declare to be unconnected. □ **dissociation** n.

dissolute adj. lacking moral restraint or self-discipline.

dissolution n. the dissolving of an assembly or partnership.

dissolve v. make or become liquid or dispersed in liquid; disappear gradually; disperse (an assembly); end (a partnership, esp. marriage).

dissonant adj. harsh-toned; discordant. □ **dissonance** n., **dissonantly** adv.

dissuade v. persuade against a course of action. □ **dissuasion** n.

distaff n. a cleft stick holding wool etc. in spinning. □ **the distaff side** the mother's side of the family.

distance n. the length of space between two points; a distant part; remoteness. ● v. separate.

distant adj. at a specified or considerable distance away; aloof. □ **distantly** adv.

distaste n. dislike, disapproval.

distasteful adj. arousing distaste. □ **distastefully** adv.

distemper n. 1 a disease of dogs. 2 a kind of paint for use on walls. ● v. paint with distemper.

distend v. swell from pressure within. □ **distension** n.

distil v. (Amer. **distill**) (**distilled**) treat or make by distillation; undergo distillation.

distillation n. the process of vaporizing and condensing a liquid so as to purify it or to extract elements; something distilled.

distiller n. one who makes alcoholic liquor by distillation.

distillery n. a place where alcohol is distilled.

distinct adj. 1 clearly perceptible. 2 different in kind. □ **distinctly** adv.

distinction n. 1 distinguishing; difference; something that differentiates. 2 a mark of honour; excellence.

distinctive adj. distinguishing, characteristic. □ **distinctively** adv.

distinguish v. 1 be or see a difference between; discern. 2 make notable. □ **distinguishable** adj.

distinguished adj. having distinction; famous for great achievements.

distort v. pull out of shape; misrepresent. □ **distortion** n.

distract v. draw away the attention of.

distracted adj. distraught.

distraction n. 1 distracting; something that distracts; an entertainment. 2 a distraught state.

distrain v. levy a distraint (on a person or goods).

distraint n. seizure of a debtor's possessions as payment for the debt.

distraught adj. nearly crazy with grief or worry.

distress n. 1 suffering, unhappiness. 2 = **distraint**. ● v. 1 cause distress to. 2 make (wood or furniture) look old and worn. □ **in distress** in danger and needing help.

distribute v. divide and share out; scatter, place at different points. □ **distribution** n.

distributor n. a person or thing that distributes; a device in an engine for passing electric current to the spark plugs.

district n. an area of a country, county, or city) with a particular feature or regarded as an administrative unit.

distrust n. lack of trust; suspicion. ● v. feel distrust in. □ **distrustful** adj., **distrustfully** adv.

disturb v. break the quiet, rest, or calm of; cause to move from a settled position. □ **disturbance** n.

disturbed adj. mentally or emotionally unstable or abnormal.

disuse n. a state of not being used.

disused adj. no longer used.

ditch n. a long narrow trench for drainage. ● v. 1 make or repair ditches. 2 (colloquial) abandon.

dither v. hesitate indecisively.

ditto n. (in lists) the same again.

ditty n. a short simple song.

diuretic n. a drug that causes more urine to be excreted.

diurnal adj. of or in the day.

diva n. a famous female singer.

divan n. a couch without a back or arms; a bed resembling this.

dive v. plunge head first into water; plunge or move quickly downwards; go under water; rush headlong. ● n. **1** diving; a sharp downward movement or fall. **2** (colloquial) a disreputable nightclub etc.

diver n. one who dives; a person who works underwater.

diverge v. separate and go in different directions; depart from a path etc. □ **divergence** n., **divergent** adj.

diverse adj. of differing kinds.

diversify v. introduce variety into; vary. □ **diversification** n.

diversion n. **1** diverting; something that diverts attention; an entertainment. **2** a route round a closed road.

diversity n. variety.

divert v. **1** turn from a course or route. **2** entertain, amuse.

divest v. □ **divest** of strip of.

divide v. **1** separate into parts or from something else. **2** cause to disagree. **3** find how many times one number contains another; be able to be divided. ● n. a dividing line.

dividend n. a share of profits payable; a benefit from an action.

divider n. **1** a thing that divides. **2** (**dividers**) measuring compasses.

divination n. divining.

divine adj. **1** of, from, or like God or a god. **2** (colloquial) excellent, beautiful. ● v. discover by intuition or magic. □ **divinely** adv., **diviner** n.

divining rod n. a dowser's stick.

divinity n. being divine; a god.

divisible adj. able to be divided. □ **divisibility** n.

division n. dividing; a dividing line, a partition; one of the parts into which a thing is divided. □ **divisional** adj.

divisive adj. tending to cause disagreement.

divisor n. a number by which another is to be divided.

divorce n. the legal termination of a marriage; separation. ● v. end the marriage of (a person) by divorce; separate.

divorcee n. a divorced person.

divot n. a tuft of grass cut out by a golf club.

divulge v. reveal (information).

Diwali n. a Hindu festival at which lamps are lit, held between September and November.

DIY abbr. do-it-yourself.

dizzy adj. (**dizzier**) giddy, feeling confused; causing giddiness. ● **dizzily** adv., **dizziness** n.

DJ abbr. **1** disc jockey. **2** dinner jacket.

djellaba (jel-ăbă) n. an Arab cloak.

D.Litt. abbr. Doctor of Letters.

DM abbr. Deutschmark.

D.Mus. abbr. Doctor of Music.

DNA abbr. deoxyribonucleic acid, a substance storing genetic information.

D-notice n. an official order not to publish specific items for security reasons.

do v. (**does, did, done, doing**) **1** perform, complete; deal with; act, proceed. **2** fare. **3** be suitable; suffice. ● v.aux. used to form the present or past tense, for emphasis, or to avoid repeating a verb just used. ● n. (pl. **dos** or **do's**) an entertainment, a party. **do away with** abolish, get rid of. **do for** (colloquial) ruin, destroy. **do in** (colloquial) ruin, kill; tire out. **do out** clean, redecorate. **do up** fasten, wrap; repair, redecorate. **do without** manage without.

Dobermann pinscher n. a dog of a large smooth-coated breed.

docile *adj.* submissive, easily managed. □ **docilely** *adv.*, **docility** *n.*

dock *n.* 1 an enclosed body of water where ships are loaded, unloaded, or repaired. 2 an enclosure for the prisoner in a criminal court. 3 a weed with broad leaves. ● *v.* 1 (of a ship) come into dock. 2 connect (spacecraft) in space; be joined in this way. 3 cut short; reduce, take away part of.

docker *n.* a labourer who loads and unloads ships in a dockyard.

docket *n.* a document listing goods delivered; a voucher. ● *v.* (**docketed**) label with a docket.

dockyard *n.* the area and buildings round a shipping dock.

doctor *n.* 1 a person qualified to give medical treatment. 2 a person holding a doctorate. ● *v.* 1 treat medically. 2 castrate or spay. 3 patch up; tamper with, falsify.

doctorate *n.* the highest degree at a university. □ **doctoral** *adj.*

doctrinaire *adj.* applying theories or principles rigidly.

doctrine *n.* a principle or the beliefs of a religious, political, or other group. □ **doctrinal** *adj.*

docudrama *n.* a television drama based on real events.

document *n.* a piece of paper giving information or evidence. ● *v.* provide or prove with documents. □ **documentation** *n.*

documentary *adj.* 1 consisting of documents. 2 giving a factual report. ● *n.* a documentary film.

dodder *v.* totter because of age or frailty. □ **dodderer** *n.*, **doddery** *adj.*

dodecagon *n.* a geometric figure with twelve sides.

dodge *v.* move quickly to one side so as to avoid (a thing);

evade. ● *n.* a dodging movement; (*colloquial*) a clever trick. □ **dodger** *n.*

dodgem *n.* one of the small cars in an enclosure at a funfair, driven so as to bump or dodge others.

dodo *n.* (*pl.* **dodos**) a large extinct bird.

DoE *abbr.* Department of the Environment.

doe *n.* the female of the deer, hare, or rabbit.

does *see* **do**.

doff *v.* take off (one's hat).

dog *n.* 1 a four-legged carnivorous wild or domesticated animal; the male of this or of the fox or wolf. 2 (**the dogs**) (*colloquial*) greyhound racing. ● *v.* (**dogged**) follow persistently.

dog cart *n.* a two-wheeled cart with back-to-back seats.

dog collar *n.* (*colloquial*) a clerical collar fastening at the back of the neck.

dog-eared *adj.* with page-corners crumpled through use.

dogfish *n.* a small shark.

dogged (dog-id) *adj.* determined. □ **doggedly** *adv.*

doggerel *n.* bad verse.

doggo *adv.* □ **lie doggo** (*slang*) remain motionless or making no sign.

doggy *adj.* of or like a dog. ● *n.* (also **doggie**) (*colloquial*) a little dog.

doggy bag *n.* a bag for taking home leftovers from a restaurant etc.

doghouse *n.* (*Amer.*) a dog's kennel. □ **in the doghouse** (*slang*) in disgrace or disfavour.

dogma *n.* doctrines put forward by authority to be accepted without question.

dogmatic *adj.* of or like dogmas; stating things in an authoritative way. □ **dogmatically** *adv.*

do-gooder n. a well-meaning but unrealistic promoter of social work or reform.

dog rose n. a wild hedge-rose.

dogsbody n. (colloquial) a drudge.

dogwood n. a shrub with dark red or golden branches and whitish flowers.

doh n. (Music) the first note of a major scale, or the note C.

doily n. (also **doyley**) a small ornamental mat.

Dolby n. (trade mark) a system for reducing unwanted sounds in a tape-recording.

doldrums n.pl. low spirits; a feeling of depression.

dole n. (colloquial) unemployment benefit. □ **dole out** distribute.

doleful adj. mournful. □ **dolefully** adv., **dolefulness** n.

doll n. a small model of a human figure, esp. as a child's toy.

dollar n. the unit of money in the USA and various other countries.

dollop n. (colloquial) a mass of a soft substance.

dolly n. **1** a child's name for a doll. **2** a movable platform for a cine-camera.

dolman sleeve n. a tapering sleeve cut in one piece with the body of a garment.

dolmen n. a megalithic structure of a large flat stone laid on two upright ones.

dolomite n. a type of limestone rock. □ **dolomitic** adj.

dolour n. (Amer. **dolor**) (literary) sorrow. □ **dolorous** adj.

dolphin n. a sea animal like a large porpoise, with a beaklike snout.

dolt n. a stupid person. □ **doltish** adj.

domain n. an area under a person's control; a field of activity.

dome n. a rounded roof with a circular base; something shaped like this. □ **domed** adj.

domestic adj. of home or household; of one's own country; domesticated. ● n. a servant in a household. □ **domestically** adv.

domesticate v. train (an animal) to live with humans; accustom to household work and home life. □ **domestication** n.

domesticity n. domestic life.

domicile n. a place of residence. □ **domiciliary** adj.

dominant adj. dominating. □ **dominance** n.

dominate v. have a commanding influence over; be the most influential or conspicuous person or thing; tower over. □ **domination** n.

domineer v. behave forcefully, making others obey.

dominion n. authority to rule, control; a ruler's territory.

domino n. (pl. **dominoes**) a small oblong piece marked with 0-6 pips, used in the game of **dominoes**.

don v. (**donned**) put on. ● n. **1** a head, fellow, or tutor of a college. **2** (**Don**) a Spanish title put before a man's Christian name. □ **donnish** adj.

donate v. give as a donation.

donation n. a gift (esp. of money) to a fund or institution.

done see **do**. adj. (colloquial) socially acceptable.

doner kebab (do-ner, doh-ner) n. spiced lamb cooked on a spit and served in slices.

donkey n. (pl. **donkeys**) an animal of the horse family, with long ears.

donkey jacket n. a thick weatherproof jacket.

donkey's years n.pl. (colloquial) a very long time.

donkey work n. drudgery.

Donna n. the title of an Italian, Spanish, or Portuguese lady.

donor *n.* one who gives or donates something.

donut Amer. sp. of **doughnut**.

doodle *v.* scribble idly. ● *n.* the drawing or marks made.

doom *n.* a grim fate; death or ruin. ● *v.* destine to a grim fate.

doomsday *n.* the day of the Last Judgement.

door *n.* a hinged, sliding, or revolving barrier closing an opening; a doorway.

doorway *n.* an opening filled by a door.

dope (*colloquial*) *n.* **1** a drug; a narcotic. **2** information. **3** a stupid person. ● *v.* drug.

dopey *adj.* (also **dopy**) (**dopier**) half asleep; stupid.

dormant *adj.* sleeping; temporarily inactive. □ **dormancy** *n.*

dormer *n.* an upright window under a small gable on a sloping roof.

dormitory *n.* a room with several beds in a school, hostel, etc.

dormitory town *n.* a town from which most residents travel to work elsewhere.

Dormobile *n.* (*trade mark*) a motor caravan.

dormouse *n.* (*pl.* **dormice**) a mouselike animal that hibernates.

dorsal *adj.* of or on the back.

dory *n.* an edible sea fish.

DOS *abbr.* a computer operating system.

dosage *n.* the size of a dose.

dose *n.* an amount of medicine to be taken at one time; an amount of radiation received. ● *v.* give a dose of medicine to.

doss *v.* (*slang*) sleep in a doss-house or on a makeshift bed etc. □ **dosser** *n.*

doss-house *n.* a cheap hostel.

dossier *n.* a set of documents about a person or event.

DoT *abbr.* Department of Transport.

dot *n.* a small round mark. ● *v.* (**dotted**) mark with dots; scatter here and there. □ **on the dot** exactly on time.

dotage *n.* senility.

dote *v.* □ **dote on** feel great fondness for. □ **doting** *adj.*

dotty *adj.* (**dottier**) (*colloquial*) feeble-minded; eccentric; silly. □ **dottily** *adv.*, **dottiness** *n.*

double *adj.* consisting of two things or parts; twice as much or as many; designed for two people or things. ● *adv.* twice as much; in twos. ● *n.* **1** a double quantity or thing; (**doubles**) a game with two players on each side. **2** a person or thing very like another. ● *v.* **1** make or become twice as much or as many; fold in two; act two parts; have two uses. **2** turn back sharply. □ **at the double** running, hurrying. □ **doubly** *adv.*

double bass *n.* the largest and lowest-pitched instrument of the violin family.

double-breasted *adj.* (of a coat) with fronts overlapping.

double chin *n.* a chin with a roll of fat below.

double cream *n.* thick cream.

double-cross *v.* cheat, deceive.

double-dealing *n.* deceit, esp. in business.

double-decker *n.* a bus with two decks.

double Dutch *n.* (*colloquial*) incomprehensible talk.

double entendre (doobl ahn-tahndr) *n.* a phrase with two meanings, one of which is usu. indecent.

double figures *n.pl.* numbers from 10 to 99.

double glazing *n.* two sheets of glass in a window, designed to reduce heat loss.

double negative *n.* (*Grammar*) a negative statement (incorrectly) containing two negative elements (see note below).

■ **Usage** Double negatives like *He didn't do nothing* and *I'm never going nowhere like that* are mistakes in standard English because one negative element is redundant. However, two negatives are perfectly acceptable in, for instance, *a not ungenerous sum* (meaning 'quite a generous sum').

doublet n. **1** each of a pair of similar things. **2** (*old use*) a man's close-fitting jacket.

double take n. a delayed reaction just after one's first reaction.

double-talk n. talk with deliberately ambiguous meaning.

double whammy n. a situation in which two unwanted things happen together.

doubloon n. a former Spanish gold coin.

doubt n. a feeling of uncertainty or disbelief; being undecided. ● v. feel doubt about, hesitate to believe. □ **doubter** n.

doubtful adj. feeling or causing doubt; unlikely. □ **doubtfully** adv.

doubtless adj. certainly.

douche (doosh) n. a jet of water applied to the body; a device for applying this. ● v. use a douche (on).

dough (doh) n. **1** a thick mixture of flour etc. and liquid, for baking. **2** (*slang*) money. □ **doughy** adj.

doughnut (doh-nut) n. (*Amer.* **donut**) a small cake of fried sweetened dough.

doughty (dow-ti) adj. (**doughtier**) valiant, stout-hearted.

dour (door) adj. stern, gloomy-looking. □ **dourly** adv., **dourness** n.

douse (dowss) v. extinguish (a light); throw water on, put into water.

dove n. a bird with a thick body and short legs; a person favouring negotiation rather than violence.

dovecote n. (also **dovecot**) a shelter for domesticated pigeons.

dovetail n. a wedge-shaped joint interlocking two pieces of wood. ● v. combine neatly.

dowager n. a woman holding a title or property from her dead husband.

dowdy adj. (**dowdier**) dull, not stylish; dressed in dowdy clothes. □ **dowdily** adv., **dowdiness** n.

dowel n. a headless wooden or metal pin holding pieces of wood or stone together.

dowelling n. (*Amer.* **doweling**) rods for cutting into dowels.

down adv. **1** to, in, or at a lower place or state etc.; to a smaller size. **2** from an earlier to a later time. **3** recorded in writing. **4** to the source or place where a thing is. **5** as (partial) payment at the time of purchase. ● prep. downwards along or through or into; at a lower part of. ● adj. **1** directed downwards. **2** travelling away from a central place. ● v. (*colloquial*) knock or bring or put down; swallow. ● n. **1** very fine soft furry feathers or short hairs. **2** an area of open undulating land; (esp. **downs**) chalk uplands. □ **have a down on** (*colloquial*) show hostility towards. **down under** in the antipodes, esp. Australia.

down-and-out adj. destitute. ● n. a destitute person.

downbeat adj. dismal; gloomy. ● n. (*Music*) an accented beat.

downcast adj. dejected; (of eyes) looking downwards.

downfall n. fall from prosperity or power; something causing this.

downgrade v. reduce to a lower grade.

downhearted adj. in low spirits.

downhill adj. & adv. going or sloping downwards.

download v. transfer (data) onto one's computer from another using a direct link.

downmarket adj. & adv. of or towards the cheaper end of the market.

downpour n. a great fall of rain.

downright adj. frank, straightforward; thorough. ● adv. thoroughly.

downside n. a negative aspect.

Down's syndrome n. a congenital disorder characterized by a broad face and learning difficulties.

downstairs adv. & adj. to or on a lower floor.

downstream adj. & adv. in the direction in which a stream flows.

down-to-earth adj. sensible and practical.

downtown adj. & n. (Amer.) (of) the lower or more central part of a city.

downtrodden adj. oppressed.

downward adj. moving or leading down. ● adv. (also **downwards**) towards what is lower, less important, or later.

downy adj. (**downier**) of, like, or covered with soft down.

dowry n. property or money brought by a bride to her husband.

dowse (dowz) v. search for underground water or minerals by using a stick which dips when these are present. □ **dowser** n.

doxology n. a formula of praise to God.

doyen n. a man who is the senior member of his staff or profession.

doyenne n. a woman who is the senior member of her staff or profession.

doyley var. of **doily**.

doze v. sleep lightly. ● n. a short light sleep.

dozen n. a set of twelve; (**dozens**) very many.

dozy adj. (**dozier**) drowsy; stupid.

D.Phil abbr. Doctor of Philosophy.

DPP abbr. Director of Public Prosecutions.

Dr abbr. Doctor.

drab adj. (**drabber**) dull, uninteresting.

drachm (dram) n. one-eighth of an ounce or of a fluid ounce.

drachma n. (pl. **drachmas** or **drachmae**) the unit of money in Greece.

draconian adj. (of laws) harsh.

draft n. 1 a preliminary written version. 2 a written order to a bank to pay money. 3 (Amer.) military conscription. 4 Amer. sp. of **draught**. ● v. 1 prepare a draft of. 2 (Amer.) conscript for military service.

drag v. (**dragged**) pull along; trail on the ground; bring or proceed with effort; search (water) with nets or hooks. ● n. 1 something that slows progress. 2 (slang) a draw on a cigarette. 3 (slang) women's clothes worn by men.

dragnet n. a net for dragging water.

dragon n. a mythical reptile able to breathe out fire; a fierce person.

dragonfly n. a long-bodied insect with gauzy wings.

dragoon n. a cavalryman or (formerly) mounted infantryman. ● v. force into action.

drag race n. an acceleration race between cars over a short distance.

drain v. 1 draw off (liquid) by channels or pipes etc.; flow away. 2 deprive gradually of strength or resources. 3 drink all of. ●n. 1 a channel or pipe carrying away water or sewage. 2 something that drains one's strength etc.

drainage n. draining; a system of drains; what is drained off.

drake n. a male duck.

dram n. a drachm; a small drink of spirits.

drama n. a play or plays for acting on the stage or broadcasting; a dramatic quality or series of events.

dramatic adj. 1 of drama. 2 exciting, striking, impressive. □ **dramatically** adv.

dramatist n. a writer of plays.

dramatize v. (also **-ise**) make into a drama. □ **dramatization** n.

drank see **drink**.

drape v. cover or arrange loosely. ●n.pl. (**drapes**) (Amer.) curtains.

drastic adj. having a strong or violent effect. □ **drastically** adv.

drat int. (colloquial) an expression of annoyance.

dratted adj. (colloquial) cursed; damn.

draught n. (Amer. **draft**) 1 a current of air. 2 pulling. 3 the depth of water needed to float a ship. 4 an amount of liquid swallowed at one time. 5 (**draughts**) a game played with 24 round pieces on a chessboard.

draught beer n. beer drawn from a cask.

draughtsman n. (pl. **-men**) a person who draws plans or sketches.

draughty adj. (Amer. **drafty**) (**draughtier**) letting in sharp currents of air. □ **draughtily** adv., **draughtiness** n.

draw v. (**drew, drawn, drawing**) 1 pull; attract; take in (breath etc.); take from or out. 2 obtain by a lottery. 3 finish a contest with scores equal. 4 require (a specified depth) in which to float. 5 produce (a picture or diagram) by making marks. 6 promote or allow a draught of air (in). 7 make one's way, come. 8 infuse. ●n. 1 an act of drawing. 2 something that draws custom or attention. 3 the drawing of lots. 4 a drawn game. □ **draw in** (of days) become shorter. **draw out** prolong; (of days) become longer. **draw the line at** refuse to do or tolerate. **draw up** come to a halt; compose (a contract etc.).

drawback n. a disadvantage.

drawbridge n. a bridge over a moat, hinged for raising.

drawer n. 1 a horizontal sliding compartment. 2 a person who draws. 3 a person who writes a cheque. 4 (**drawers**) knickers, underpants.

drawing n. a picture made with a pencil or pen.

drawing pin n. a pin for fastening paper to a surface.

drawing room n. a formal sitting room.

drawl v. speak lazily or with drawn-out vowel sounds. ●n. a drawling manner of speaking.

drawn see **draw**. adj. looking strained from tiredness or worry.

drawstring n. a string that can be pulled to tighten an opening.

dray n. a low cart for heavy loads.

dread n. great fear. ●v. fear greatly. ●adj. dreaded.

dreadful adj. very bad. □ **dreadfully** adv.

dream n. a series of pictures or events in a sleeping person's mind; a fantasy. ●v. (**dreamed** or **dreamt**, **dreaming**) have a dream; have an ambition; think of as a possibility. □ **dream up**

imagine; invent. □ **dreamer** n., **dreamless** adj.

dreamy adj. (**dreamier**) **1** daydreaming; vague, impractical. **2** (colloquial) delightful. □ **dreamily** adv., **dreaminess** n.

dreary adj. (**drearier**) dull, boring; gloomy. □ **drearily** adv., **dreariness** n.

dredge v. **1** remove (silt) from (a river or channel). **2** sprinkle with flour or sugar.

dredger n. **1** a boat that dredges. **2** a container with a perforated lid for sprinkling flour or sugar.

dregs n.pl. a sediment at the bottom of liquid; the worst and useless part.

drench v. wet all through.

dress n. **1** outer clothing. **2** a woman's or girl's garment with a bodice and skirt. ● v. **1** put clothes on; clothe oneself. **2** arrange, decorate, trim. **3** put a dressing on.

dressage (dress-ahzh) n. the management of a horse to show its obedience and deportment.

dress circle n. the first gallery in a theatre.

dresser n. **1** a person who helps actors with their costumes. **2** a sideboard with shelves above for dishes etc.

dressing n. **1** a sauce for food. **2** a bandage or ointment etc. for a wound. **3** fertilizer etc. spread over land.

dressing down n. a scolding.

dressing gown n. a loose robe worn when one is not fully dressed.

dressing table n. a table with a mirror, for use while dressing.

dressmaker n. a person, esp. a woman, who makes women's clothes. □ **dressmaking** n.

dress rehearsal n. a final rehearsal of a play etc., in costume.

dress shirt n. a shirt for wearing with evening dress.

dressy adj. (**dressier**) wearing stylish clothes; elegant, elaborate.

drew see **draw**.

drey n. (pl. **dreys**) a squirrel's nest.

dribble v. **1** have saliva flowing from the mouth; flow or let flow in drops. **2** (in football etc.) move the ball forward with slight touches. ● n. the act or flow of dribbling.

dried adj. (of food) preserved by removal of moisture.

drier n. (also **dryer**) a device for drying things.

drift v. be carried by a current of water or air; go casually or aimlessly. ● n. **1** a drifting movement. **2** a mass of snow piled up by the wind. **3** the general meaning of a speech etc.

drifter n. an aimless person.

driftwood n. wood floating on the sea or washed ashore.

drill n. **1** a tool or machine for boring holes or sinking wells. **2** training; routine procedure. **3** a strong twilled cotton fabric. ● v. **1** use a drill, make (a hole) with a drill. **2** train, be trained.

drily adv. (also **dryly**) in a dry way.

drink v. (**drank, drunk, drinking**) swallow (liquid); take alcoholic drink, esp. in excess; pledge good wishes (to) by drinking. ● n. a liquid for drinking; alcoholic liquor. □ **drink in** watch or listen to eagerly. □ **drinker** n.

drink-driver n. a person who drives having drunk more than the legal limit of alcohol.

drip v. (**dripped**) fall or let fall in drops. ● n. liquid falling in drops; the sound of this; (also **drip-feed**) an apparatus for administering a liquid at a very

drip-dry | drum majorette

slow rate into the body, esp. intravenously.

drip-dry v. hang up to dry when dripping wet without wringing or ironing. ● adj. (of clothing) able to be drip-dried.

dripping n. fat melted from roast meat.

drive v. (**drove, driven, driving**) 1 send or urge onwards; propel. 2 operate (a vehicle) and direct its course; travel or convey in a private vehicle. 3 cause, compel. 4 make (a bargain). ● n. 1 a journey in a private vehicle. 2 transmission of power to machinery. 3 energy; organized effort. 4 a track for a car, leading to a house, hotel, etc. □ **drive at** intend to convey as a meaning.

drive-in adj. (of a bank etc.) able to be used without getting out of one's car.

drivel n. silly talk, nonsense.

driver n. 1 a person who drives. 2 a golf club for driving from a tee.

drizzle n. & v. rain in very fine drops.

droll adj. amusing in an odd way. □ **drolly** adv., **drollery** n.

dromedary n. a camel with one hump, bred for riding.

drone n. a male bee; a deep humming sound. ● v. make this sound; speak monotonously.

drool v. 1 slaver, dribble. 2 show gushing appreciation.

droop v. bend or hang down limply. ● n. a drooping attitude. □ **droopy** adj.

drop n. 1 a small rounded mass of liquid; (**drops**) medicine measured by drops; a very small quantity. 2 a fall; a steep descent, the distance of this. ● v. (**dropped**) 1 fall; shed, let fall; make or become lower. 2 utter casually. 3 omit; reject, give up. □ **drop in** pay a casual visit.

drop off fall asleep. **drop out** cease to participate.

droplet n. small drop of liquid.

drop-out n. a person who drops out from a course of study or from conventional society.

dropper n. a device for releasing liquid in drops.

droppings n.pl. animal dung.

dropsy n. an illness in which fluid collects in the body. □ **dropsical** adj.

dross n. scum on molten metal; impurities, rubbish.

drought n. a long spell of dry weather.

drove see **drive**. n. a moving herd, flock, or crowd.

drover n. a person who drives cattle.

drown v. kill or be killed by suffocating in water or other liquid; flood, drench; deaden (grief etc.) with drink; overpower (sound) with greater loudness.

drowse v. be lightly asleep. □ **drowsy** adj., **drowsily** adv., **drowsiness** n.

drub v. (**drubbed**) thrash; defeat thoroughly.

drudge n. a person who does laborious or menial work. ● v. do such work. □ **drudgery** n.

drug n. a substance used in medicine or as a stimulant or narcotic. ● v. (**drugged**) add or give a drug to.

drugstore n. (Amer.) a chemist's shop also selling various goods.

Druid n. an ancient Celtic priest. □ **Druidic, Druidical** adj.

drum n. a percussion instrument, a round frame with skin etc. stretched across; a cylindrical object. ● v. (**drummed**) beat or tap continuously. □ **drum up** obtain by vigorous effort.

drum majorette n. a female member of a parading group.

drummer n. a person who plays drums.

drumstick n. **1** a stick for beating a drum. **2** the lower part of a cooked fowl's leg.

drunk see **drink**. adj. excited or stupefied by alcoholic drink. ● n. a drunken person.

drunkard n. a person who is often drunk.

drunken adj. intoxicated, often in this condition. □ **drunkenly** adv., **drunkenness** n.

dry adj. (**drier**) **1** without water, moisture, or rainfall; thirsty. **2** uninteresting; expressed with pretended seriousness. **3** (of a country etc.) not allowing the sale of alcohol. **4** (of wine etc.) not sweet. ● v. (**dries, dried, drying**) make or become dry; preserve (food) by removing its moisture. □ **dry up** dry washed dishes; (colloquial) cease talking. □ **dryness** n.

dryad n. a wood nymph.

dry-clean v. clean with solvents without using water.

dryly var. of **drily**.

dry rot n. decay of wood that is not ventilated.

dry run n. (colloquial) a rehearsal.

drystone wall n. a wall built without mortar.

DSS abbr. Department of Social Security.

DTI abbr. Department of Trade and Industry.

DTP abbr. desktop publishing.

dual adj. composed of two parts; double. □ **duality** n.

dual carriageway n. a road with a dividing strip between traffic travelling in opposite directions.

dub v. (**dubbed**) **1** replace the soundtrack of a film. **2** give a nickname to.

dubbin n. (also **dubbing**) a thick grease for softening and waterproofing leather.

dubiety n. (literary) a feeling of doubt.

dubious adj. doubtful. □ **dubiously** adv.

ducal adj. of a duke.

ducat n. a former gold coin of various European countries.

duchess n. a woman with the rank of duke; a duke's wife or widow.

duchy n. the territory of a duke.

duck n. **1** a swimming bird of various kinds; the female of this. **2** a batsman's score of 0. **3** a quick dip or lowering of the head. ● v. **1** push (a person) or dip one's head under water. **2** bob down, esp. to avoid being seen or hit. **3** dodge (a task etc.).

duckboards n.pl. boards forming a narrow path.

duckling n. a young duck.

duckweed n. a plant growing on the surface of ponds.

duct n. a channel or tube conveying liquid or air. □ **ductless** adj.

ductile adj. (of metal) able to be drawn into fine strands; easily moulded. □ **ductility** n.

dud (slang) n. something that is counterfeit or fails to work; an ineffectual person or thing. ● adj. useless; counterfeit.

dude n. (esp. Amer. slang) a fellow; a dandy.

dudgeon n. indignation.

due adj. **1** owed; payable immediately. **2** merited. **3** scheduled to do something or to arrive. ● adv. directly or exactly; due north. ● n. a person's right, what is owed to him or her; (**dues**) fees. □ **due to** because of, caused by.

■ **Usage** Many people believe that due to, meaning 'because of', should only be used after the verb to be, as in The mistake was due to ignorance, and not as in All trains may be delayed due to a signal failure. Instead, ow-

ing to a signal failure could be used.

duel n. a fight or contest between two persons or sides. ● v. (**duelled**; *Amer.* **dueled**) fight a duel. □ **duellist** n.

duenna n. an elderly female chaperone, esp. in Spain.

duet n. a musical composition for two performers.

duff adj. (*slang*) dud.

duffel coat n. a heavy woollen coat with a hood.

duffer n. (*slang*) an inefficient or stupid person.

dug *see* **dig.** n. an udder, a teat.

dugong (dew-gong) n. an Asian sea mammal.

dugout n. **1** an underground shelter. **2** a canoe made from a hollowed tree trunk.

duke n. a nobleman of the highest hereditary rank; a ruler of certain small states. □ **dukedom** n.

dulcet adj. sounding sweet.

dulcimer n. a musical instrument with strings struck by two hammers.

dull adj. not bright; stupid; boring; not sharp; not resonant. ● v. make or become dull. □ **dully** adv., **dullness** n.

dullard n. a stupid person.

duly adv. in due time; in the proper way.

dumb adj. **1** unable to speak; silent. **2** (*colloquial*) stupid. □ **dumbly** adv., **dumbness** n.

dumb-bell n. a short bar with weighted ends, lifted to exercise muscles.

dumbfound v. astonish.

dumdum bullet n. a soft-nosed bullet that expands on impact.

dummy n. **1** a sham article; a model of the human figure, used to display clothes. **2** a rubber teat for a baby to suck. ● adj. sham.

dummy run n. a trial attempt; a rehearsal.

dump v. **1** deposit as rubbish; put down carelessly. **2** sell (goods) abroad at a lower price. ● n. a rubbish heap; a temporary store; (*colloquial*) a dull place.

dumpling n. a ball of dough cooked in stew or with fruit inside.

dun adj. & n. greyish brown.

dunce n. a person slow at learning.

dune n. a mound of drifted sand.

dung n. animal excrement.

dungarees n.pl. overalls of coarse cotton cloth.

dungeon n. a strong underground cell for prisoners.

dunk v. dip into liquid.

duo n. (*pl.* **duos**) a pair of performers; a duet.

duodecimal adj. reckoned in twelves or twelfths.

duodenum n. the part of the intestine next to the stomach. □ **duodenal** adj.

dupe v. deceive, trick. ● n. a duped person.

duple adj. having two parts; (in music) having two beats to the bar.

duplex adj. having two elements.

duplicate n. (**dew-pli-kăt**) an exact copy. ● adj. (**dew-pli-kăt**) exactly like another. ● v. (**dew-pli-kayt**) make or be a duplicate; do twice. □ **duplication** n., **duplicator** n.

duplicity n. deceitfulness.

durable adj. likely to last. ● n.pl. (**durables**) durable goods. □ **durably** adv., **durability** n.

duration n. the time during which a thing continues.

duress n. the use of force or threats.

during prep. throughout; at a point in the duration of.

dusk n. a darker stage of twilight.

dusky adj. (**duskier**) shadowy; dark-coloured. □ **duskiness** n.

dust n. fine particles of earth or other matter. ● v. **1** sprinkle with dust or powder. **2** clear of dust by wiping, clean a room etc. in this way.

dustbin n. a bin for household rubbish.

dust bowl n. an area denuded of vegetation and reduced to desert.

dust jacket n. (also **dust cover**) a paper cover used to protect a book.

duster n. a cloth for dusting things.

dustman n. (pl. **-men**) a person employed to empty dustbins.

dustpan n. a container into which dust is brushed from a floor.

dusty adj. (**dustier**) like dust; covered with dust. □ **dustiness** n.

Dutch adj. & n. (the language) of the Netherlands. □ **go Dutch** share expenses on an outing. □ **Dutchman** n., **Dutchwoman** n.

Dutch courage n. false courage obtained by drinking alcohol.

dutiable adj. on which customs or other duties must be paid.

dutiful adj. doing one's duty, showing due obedience. □ **dutifully** adv.

duty n. **1** a moral or legal obligation. **2** a tax on goods or imports. □ **on duty** at work.

duvet (doo-vay) n. a thick soft bed quilt.

dwarf n. (pl. **dwarfs** or **dwarves**) a person or thing much below the usual size; (in fairy tales) a small being with magic powers. ● adj. very small. ● v. stunt; make (a thing) seem small.

dwell v. (**dwelt, dwelling**) live as an inhabitant. □ **dwell on** write, speak, or think lengthily about. □ **dweller** n.

dwelling n. a house etc. to live in.

dwindle v. become less or smaller.

Dy symb. dysprosium.

dye v. (**dyed, dyeing**) colour, esp. by dipping in liquid. ● n. a substance used for dyeing things; a colour given by dyeing. □ **dyer** n.

dying see **die**.

dyke n. (also **dike**) a wall or embankment to prevent flooding; a drainage ditch.

dynamic adj. of force producing motion; energetic, forceful. □ **dynamically** adv.

dynamics n. the branch of mechanics dealing with the motion of bodies under the action of forces.

dynamism n. energizing power.

dynamite n. a powerful explosive made of nitroglycerine. ● v. fit with a charge of dynamite, blow up with dynamite.

dynamo n. (pl. **dynamos**) a small generator producing electric current.

dynasty n. a line of hereditary rulers. □ **dynastic** adj.

dysentery n. a disease causing severe diarrhoea.

dysfunction n. a malfunction.

dyslexia n. a condition causing difficulty in reading and spelling. □ **dyslexic** adj. & n.

dyspepsia n. indigestion. □ **dyspeptic** adj. & n.

dysprosium n. a metallic element (symbol Dy).

dystrophy n. wasting of a part of the body. See also **muscular dystrophy**.

Ee

E abbr. (also **E**) east; eastern.

each adj. & pron. every one of two or more. □ **each way** (of a bet) backing a horse etc. to win or to be placed.

eager adj. full of desire, enthusiastic. □ **eagerly** adv., **eagerness** n.

eagle n. a large bird of prey.

ear n. **1** the organ of hearing; the external part of this; the ability to distinguish sounds accurately. **2** the seed-bearing part of corn.

eardrum n. a membrane inside the ear, vibrating when sound waves strike it.

earl n. a British nobleman ranking between marquess and viscount. □ **earldom** n.

early adj. (**earlier**) ● adj. & adv. before the usual or expected time; not far on in development or in a series.

earmark n. a distinguishing mark. ● v. put such a mark on; set aside for a particular purpose.

earn v. get or deserve for work or merit; (of money) gain as interest.

earnest adj. showing serious feeling or intention. □ **in earnest** seriously. □ **earnestly** adv., **earnestness** n.

earphone n. a device applied to the ear to aid hearing or receive radio or telephone communications.

earring n. a piece of jewellery worn on the ear.

earshot n. the distance over which something can be heard.

earth n. **1** the planet we live on; its surface, dry land; soil. **2** a fox's den. **3** a connection of an electrical circuit to ground. ● v. connect an electrical circuit to earth. □ **run to earth** find after a long search.

earthen adj. made of earth or of baked clay.

earthenware n. pottery made of coarse baked clay.

earthly adj. of this earth, of man's life on it.

earthquake n. a violent movement of part of the earth's crust.

earthwork n. a bank built of earth.

earthworm n. a worm living in the soil.

earthy adj. (**earthier**) **1** like earth or soil. **2** (of humour etc.) gross, coarse.

earwig n. a small insect with pincers at the end of its body.

ease n. freedom from pain, worry, or effort. ● v. **1** relieve from pain etc.; make or become less tight or severe. **2** move gently or gradually.

easel n. a frame to support a painting, blackboard, etc.

easement n. a right of way over another's property.

east n. the point on the horizon where the sun rises; the direction in which this lies; an eastern part. ● adj. in the east; (of wind) from the east. ● adv. towards the east.

Easter n. a festival commemorating Christ's resurrection.

Easter egg n. a chocolate egg given at Easter.

easterly adj. towards or blowing from the east.

eastern adj. of or in the east.

easternmost adj. furthest east.

eastward adj. towards the east. □ **eastwards** adv.

easy adj. (**easier**) done or got without great effort; free from pain, trouble, or anxiety. ● adv. in an easy way. □ **easily** adv., **easiness** n.

easy chair n. a large comfortable chair.

easygoing adj. relaxed in manner, not strict.

eat v. (**ate, eaten, eating**) 1 chew and swallow (food); have a meal. 2 destroy gradually by corrosion etc. □ **eater** n.

eatable adj. fit to be eaten. ● n.pl. (**eatables**) food.

eau de Cologne (oh-dĕ-kŏ-lohn) n. a delicate perfume.

eau-de-vie (oh-dĕ-vee) n. spirits, esp. brandy.

eaves n.pl. the overhanging edge of a roof.

eavesdrop v. (**eavesdropped**) listen secretly to a private conversation. □ **eavesdropper** n.

ebb n. 1 the outward movement of the tide, away from the land. 2 a decline. ● v. 1 flow away. 2 decline.

ebonite n. vulcanite.

ebony n. the hard black wood of a tropical tree. ● adj. black as ebony.

ebullient adj. full of high spirits. □ **ebulliently** adv., **ebullience** n.

EC abbr. European Community; European Commission.

eccentric adj. 1 unconventional. 2 not concentric; (of an orbit or wheel) not circular. ● n. an eccentric person. □ **eccentrically** adv., **eccentricity** n.

ecclesiastical adj. of the Church or clergy.

ECG abbr. electrocardiogram.

echelon (esh-ĕlon) n. a wedge-shaped formation of troops etc.; a level of rank or authority.

echo n. (pl. **echoes**) a repetition of sound by reflection of sound waves; a close imitation. ● v. (**echoed, echoing**) repeat by an echo; imitate.

éclair n. a finger-shaped pastry cake with cream filling.

eclampsia n. a condition involving convulsions, affecting women in pregnancy.

eclectic adj. choosing or accepting from various sources.

eclipse n. the blocking of light from one heavenly body by another; a loss of brilliance or power etc. ● v. cause an eclipse of; outshine.

ecliptic n. the sun's apparent path.

eclogue n. a short pastoral poem.

eco- comb. form relating to ecology.

ecology n. (the study of) relationships of living things to their environment; the protection of the natural environment. □ **ecological** adj., **ecologically** adv., **ecologist** n.

economic adj. of economics; enough to give a good return for money or effort outlaid.

economical adj. thrifty, avoiding waste. □ **economically** adv.

economics n. the science of the production and use of goods or services; (as pl.) the financial aspects of something.

economist n. an expert in economics.

economize v. (also **-ise**) use or spend less.

economy n. being economical; a community's system of using its resources to produce wealth; the state of a country's prosperity.

ecosystem n. a system of interacting organisms and their environemt.

ecru n. a light fawn colour.

ecstasy n. 1 intense delight. 2 (usu. **Ecstasy**) (slang) a hallucinogenic drug. □ **ecstatic** adj., **ecstatically** adv.

ECT abbr. electroconvulsive therapy.

ectopic pregnancy n. an unsuccessful pregnancy in which

the egg develops outside the womb.

ecu n. (also **Ecu**) a European currency unit.

ecumenical adj. of the whole Christian Church; seeking world-wide Christian unity.

eczema n. a skin disease causing scaly itching patches.

eddy n. a swirling patch of water or air etc. ● v. swirl in eddies.

edelweiss (ay-dĕl-vys) n. an alpine plant with woolly white bracts.

edema Amer. sp. of **oedema**.

edge n. **1** a rim, the narrow surface of a thin or flat object. **2** a sharpened side of a blade; sharpness. **3** the outer limit of an area. ● v. **1** border. **2** move gradually.

edgeways adv. (also **edgewise**) with the edge forwards or outwards.

edging n. something placed round an edge to define or decorate it.

edgy adj. (**edgier**) tense and irritable. □ **edgily** adv., **edginess** n.

edible adj. suitable for eating. □ **edibility** n.

edict (ee-dikt) n. an order proclaimed by authority.

edifice n. a large building.

edify v. be an uplifting influence on the mind of. □ **edification** n.

edit v. (**edited**) prepare for publication; prepare (a film) by arranging sections in sequence.

edition n. the form in which something is published; a number of objects issued at one time.

editor n. a person responsible for the contents of a newspaper etc. or a section of this; a person who edits.

editorial adj. of an editor. ● n. a newspaper article giving the editor's comments.

educate v. train the mind and abilities of; provide with training for. □ **education** n., **educational** adj.

Edwardian adj. & n. (a person) of the reign of Edward VII (1901–10).

EEC abbr. European Economic Community.

EEG abbr. electroencephalogram.

eel n. a snakelike fish.

eerie adj. (**eerier**) mysterious and frightening. □ **eerily** adv., **eeriness** n.

efface v. rub out, obliterate; make inconspicuous. □ **effacement** n.

effect n. **1** a change produced by an action or cause; an impression. **2** a state of being operative. **3** (**effects**) property. ● v. bring about.

■ **Usage** As a verb, *effect* should not be confused with *affect*. *He effected an entrance* means 'He got in (somehow)', but *This won't affect me* means 'My life won't be changed by this'.

---

effective adj. **1** producing an effect; striking. **2** operative. □ **effectively** adv., **effectiveness** n.

effectual adj. answering its purpose. □ **effectually** adv.

effeminate adj. (of a man) feminine in appearance or manner. □ **effeminacy** n., **effeminately** adv.

effervesce v. give off bubbles, fizz. □ **effervescence** n., **effervescent** adj.

effete adj. having lost vitality; feeble. □ **effeteness** n.

efficacious adj. producing the desired result. □ **efficaciously** adv., **efficacy** n.

efficient adj. producing results with little waste of effort. □ **efficiently** adv., **efficiency** n.

effigy n. a model of a person.

effloresce v. flower. □ **efflorescence** n.

effluent n. outflow, sewage.

effluvium n. (pl. **effluvia**) an unpleasant or harmful smell or outflow.

effort n. use of energy; an attempt. □ **effortless** adj.

effrontery n. bold insolence.

effusion n. an outpouring.

effusive adj. expressing emotion in an unrestrained way. □ **effusively** adv., **effusiveness** n.

EFL abbr. English as a foreign language.

EFTA abbr. European Free Trade Association.

e.g. abbr. (Latin exempli gratia) for example.

egalitarian adj. & n. (a person) holding the principle of equal rights for all. □ **egalitarianism** n.

egg n. a hard-shelled oval body produced by the female of birds, esp. that of the domestic hen; an ovum. □ **egg on** (colloquial) urge on.

egghead n. (colloquial) an intellectual person.

eggshell n. the shell of an egg. ● adj. **1** (of china) fragile. **2** (of paint) with a slightly glossy finish.

eggplant n. (esp. Amer.) an aubergine.

ego n. self; self-esteem.

egocentric adj. self-centred.

egoism n. self-centredness.

egoist n. a self-centred person. □ **egoistic** adj.

egotism n. the practice of talking too much about oneself, conceit.

egotist n. a conceited person. □ **egotistic** adj., **egotistical** adj.

egregious (i-gree-jŭs) adj. outstandingly bad, shocking.

egress n. departure; a way out.

egret n. a kind of heron.

Egyptian adj. & n. (a native) of Egypt.

Egyptology n. the study of Egyptian antiquities. □ **Egyptologist** n.

eider n. a large northern duck.

eiderdown n. a quilt stuffed with soft material.

eight adj. & n. one more than seven (8, VIII). □ **eighth** adj. & n.

eighteen adj. & n. one more than seventeen (18, XVIII). □ **eighteenth** adj. & n.

eighty adj. & n. ten times eight (80, LXXX). □ **eightieth** adj. & n.

einsteinium n. a radioactive metallic element (symbol Es).

eisteddfod n. a Welsh festival of music, poetry, and dance.

either adj. & pron. one or other of two; each of two. ● adv. & conj. as the first alternative; likewise.

ejaculate v. **1** utter suddenly. **2** eject (semen). □ **ejaculation** n.

eject v. send out forcefully. □ **ejection** n., **ejector** n.

eke v. **eke out** supplement; make (a living) laboriously.

elaborate adj. (i-lab-er-ăt) with many parts or details. ● v. (i-lab-er-ayt) add detail to. □ **elaborately** adv., **elaboration** n.

élan (ay-lan) n. vivacity, vigour.

eland (ee-lănd) n. a large African antelope.

elapse v. (of time) pass away.

elastic adj. going back to its original length or shape after being stretched or squeezed; adaptable. ● n. cord or material made elastic by interweaving strands of rubber etc. □ **elasticity** n.

elate v. cause to feel very pleased or proud. □ **elated** adj., **elation** n.

elbow n. the joint between the forearm and upper arm; the part of a sleeve covering this; a sharp bend. ● v. thrust with one's elbow.

elbow grease n. vigorous polishing, hard work.

elbow room n. enough space to move or work in.

elder adj. older. ● n. **1** an older person; an official in certain Churches. **2** a tree with small dark berries.

elderberry n. the fruit of the elder tree.

elderly adj. old.

eldest adj. first-born, oldest.

eldorado n. (pl. **eldorados**) a place of prosperity, abundance, etc.

elect v. choose by vote; choose as a course. ● adj. chosen.

election n. electing; the process of electing representatives.

electioneer v. busy oneself in an election campaign.

elective adj. chosen by election; entitled to elect; optional.

elector n. a person entitled to vote in an election. □ **electoral** adj.

electorate n. a body of electors.

electric adj. of, producing, or worked by electricity.

electrical adj. of electricity. □ **electrically** adv.

electrician n. a person whose job is to deal with electrical equipment.

electricity n. a form of energy occurring in certain particles; a supply of electric current.

electrics n.pl. electrical fittings.

electrify v. charge with electricity; convert to the use of electric power. □ **electrification** n.

electrocardiogram n. a record of the electric current generated by heartbeats.

electroconvulsive therapy n. treatment of mental illness by electric shocks producing convulsions.

electrocute v. kill by electricity. □ **electrocution** n.

electrode n. a solid conductor through which electricity enters or leaves a vacuum tube etc.

electroencephalogram n. a record of the electrical activity of the brain.

electrolysis n. the decomposition of a substance by the application of an electric current; the destruction of hair-roots etc. by this process.

electrolyte n. a solution that conducts an electric current.

electromagnet n. a magnet consisting of a metal core magnetized by a current-carrying coil round it.

electromagnetic adj. having both electrical and magnetic properties. □ **electromagnetism** n., **electromagnetically** adv.

electromotive adj. producing an electric current.

electron n. a particle with a negative electric charge.

electronic adj. produced or worked by a flow of electrons; of electronics. □ **electronically** adv.

electronics n. the use of electronic devices; (as pl.) electronic circuits.

electronic tagging n. the attaching of electronic markers to people or goods enabling them to be traced.

electron microscope n. a very powerful one using a focused beam of electrons instead of light.

electroplate v. coat with a thin layer of silver etc. by electrolysis.

elegant adj. tasteful and dignified. □ **elegantly** adv., **elegance** n.

elegy n. a sorrowful or serious poem. □ **elegiac** adj.

element n. **1** a component part. **2** a substance that cannot be broken down into other substances. **3** a suitable or satisfying environment. **4** a trace. **5** a wire that gives out heat in an electrical appliance. **6 (the elements)** atmospheric forces. **7 (elements)** basic principles. □ **elemental** adj.

elementary adj. dealing with the simplest facts of a subject.

elephant n. a very large animal with a trunk and ivory tusks.

elephantine adj. of or like elephants; very large; clumsy.

elevate v. raise to a higher position or level.

elevation n. **1** elevating; altitude; a hill. **2** a drawing showing one side of a structure.

elevator n. a device that hoists something; a lift.

eleven adj. & n. one more than ten (11, XI). □ **eleventh** adj. & n.

elevenses n. (colloquial) a mid-morning snack.

elf n. (pl. **elves**) an imaginary small being with magic powers. □ **elfin** adj.

elicit v. draw out.

eligible adj. qualified to be chosen or allowed something. □ **eligibility** n.

eliminate v. get rid of; exclude. □ **elimination** n., **eliminator** n.

elision n. omission of part of a word in pronunciation.

elite (ay-leet) n. a group regarded as superior and favoured.

elitism (ay-leet-izm) n. favouring of or dominance by a selected group. □ **elitist** n. & adj.

elixir n. a fragrant liquid used as a medicine or flavouring.

Elizabethan adj. & n. (a person) of Elizabeth I's reign (1558–1603).

elk n. a large deer.

ellipse n. a regular oval.

ellipsis n. (pl. **ellipses**) omission of words.

elliptical adj. **1** shaped like an ellipse. **2** having omissions. □ **elliptically** adv.

elm n. a tree with rough serrated leaves; its wood.

elocution n. a style of speaking, the art of speaking expressively. □ **elocutionary** adj.

elongate v. lengthen.

elope v. run away secretly with a lover. □ **elopement** n.

eloquence n. fluent and effective use of language. □ **eloquent** adj., **eloquently** adv.

else adv. besides; otherwise.

elsewhere adv. somewhere else.

elucidate v. throw light on, explain. □ **elucidation** n.

elude v. escape skilfully from; avoid; escape the memory or understanding of. □ **elusion** n.

elusive adj. eluding, escaping.

elver n. a young eel.

emaciated adj. thin from illness or starvation. □ **emaciation** n.

e-mail n. (also **email**) electronic mail, messages sent from one computer user to another and displayed on-screen.

emanate v. issue, originate from a source. □ **emanation** n.

emancipate v. liberate, free from restraint. □ **emancipation** n.

emasculate v. deprive of force, weaken. □ **emasculation** n.

embalm v. preserve (a corpse) by using spices or chemicals. □ **embalmment** n.

embankment n. a bank or stone structure to keep a river from spreading or to carry a railway.

embargo n. (pl. **embargoes**) an order forbidding commerce or other activity.

embark v. board a ship. □ **embark on** begin (an undertaking). □ **embarkation** n.

embarrass v. cause to feel awkward or ashamed. □ **embarrassment** n.

embassy n. an ambassador and staff; their headquarters.

embattled adj. **1** prepared for battle. **2** involved in a conflict or difficult undertaking.

embed v. (also **imbed**) (**embedded**) fix firmly in a surrounding mass.

embellish v. ornament; improve (a story) with invented details. □ **embellishment** n.

embers n.pl. small pieces of live coal or wood in a dying fire.

embezzle v. take (company funds etc.) fraudulently for one's own use. □ **embezzlement** n., **embezzler** n.

embitter v. rouse bitter feelings in. □ **embitterment** n.

emblem n. a symbol, a design used as a badge etc.

emblematic adj. serving as an emblem. □ **emblematically** adv.

embody v. **1** express (principles or ideas) in visible form. **2** incorporate. □ **embodiment** n.

embolden v. make bold, encourage.

embolism n. obstruction of a blood vessel by a clot or air bubble.

emboss v. decorate by a raised design; mould in relief.

embrace v. **1** hold closely and lovingly, hold each other in this way. **2** accept, adopt; include. ● n. an act of embracing, a hug.

embrocation n. liquid for rubbing on the body to relieve aches.

embroider v. ornament with needlework; embellish (a story). □ **embroidery** n.

embroil v. involve in an argument or quarrel etc.

embryo n. (pl. **embryos**) an animal developing in a womb or egg; something in an early stage of development. □ **embryonic** adj.

embryology n. the study of embryos.

emend v. alter to remove errors. □ **emendation** n., **emendatory** adj.

emerald n. a bright green precious stone; its colour.

emerge v. come up or out into view; become known. □ **emergence** n., **emergent** adj.

emergency n. a serious situation needing prompt attention.

emeritus adj. retired and retaining a title as an honour.

emery n. a coarse abrasive.

emery board n. a strip of cardboard coated with emery, used for filing the nails.

emetic n. a medicine used to cause vomiting.

emf abbr. electromotive force.

emigrate v. leave one country and go to settle in another. □ **emigration** n., **emigrant** n.

émigré (em-i-gray) n. an emigrant, a political exile.

eminence n. **1** the state of being eminent; (**Eminence**) a title of a cardinal. **2** a piece of rising ground.

eminent adj. famous, distinguished. □ **eminently** adv.

emir (em-eer) n. a Muslim ruler.

emirate n. the territory of an emir.

emissary n. a person sent to conduct negotiations.

emit v. (**emitted**) send out (light, heat, fumes, etc.); utter. □ **emission** n., **emitter** n.

emollient adj. softening, soothing. ● n. an emollient substance.

emolument n. a fee; a salary.

emotion n. intense mental feeling.

emotional adj. of emotion(s); showing great emotion. □ **emotionally** adv.

■ **Usage** See note at **emotive**.

emotive *adj.* rousing emotion.

■ **Usage** Although the senses of *emotive* and *emotional* overlap, *emotive* is more common in the sense 'arousing emotion', as in *an emotive issue*, and only *emotional* can mean 'especially liable to emotion', as in *a highly emotional person*.

empathize *v.* (also **-ise**) show empathy; treat with empathy.

empathy *n.* the ability to identify oneself mentally with, and so understand, a person or thing.

emperor *n.* a male ruler of an empire.

emphasis *n.* (*pl.* **emphases**) special importance; vigour of expression etc.; stress on a sound or word.

emphasize *v.* (also **-ise**) lay emphasis on.

emphatic *adj.* using or showing emphasis. □ **emphatically** *adv.*

emphysema *n.* enlargement of the air sacs in the lungs, causing breathlessness.

empire *n.* a group of countries ruled by a supreme authority; a large organization controlled by one person or group.

empirical *adj.* based on observation or experiment, not on theory. □ **empirically** *adv.*, **empiricism** *n.*, **empiricist** *n.*

emplacement *n.* a place or platform for a gun or battery of guns.

employ *v.* give work to; use the services of; make use of. □ **employment** *n.*, **employer** *n.*

employee *n.* a person employed by another in return for wages.

emporium *n.* (*pl.* **emporia** or **emporiums**) a centre of commerce; a large shop.

empower *v.* authorize, enable.

empress *n.* a female ruler of an empire; the wife of an emperor.

empty *adj.* (**emptier**) **1** containing nothing; without occupants. **2** meaningless. ● *v.* make or become empty. □ **emptiness** *n.*

emu *n.* a large Australian flightless bird resembling an ostrich.

emulate *v.* try to do as well as. □ **emulation** *n.*, **emulator** *n.*

emulsify *v.* convert or be converted into emulsion. □ **emulsification** *n.*, **emulsifier** *n.*

emulsion *n.* **1** a creamy liquid. **2** a light-sensitive coating on photographic film.

enable *v.* give the means or authority to do something.

enact *v.* **1** make into a law. **2** play (a part or scene). □ **enactment** *n.*

enamel *n.* **1** a glasslike coating for metal or pottery. **2** glossy paint. **3** the hard outer covering of teeth. ● *v.* (**enamelled**; *Amer.* **enameled**) coat with enamel.

enamoured *adj.* (*Amer.* **enamored**) fond.

en bloc (ahn blok) *adv.* as a whole, all at the same time.

encamp *v.* settle in a camp.

encampment *n.* a camp.

encapsulate *v.* **1** enclose (as) in a capsule. **2** summarize. □ **encapsulation** *n.*

encase *v.* enclose in a case.

encephalitis *n.* inflammation of the brain.

enchant *v.* bewitch. □ **enchanter** *n.*, **enchantment** *n.*, **enchantress** *n.*

encircle *v.* surround. □ **encirclement** *n.*

enclave *n.* a small territory wholly within the boundaries of another.

enclose *v.* **1** shut in on all sides, seclude. **2** include with other contents.

enclosure *n.* enclosing; an enclosed area; something enclosed.

encode v. put into code; put (data) into computerized form. □ **encoder** n.

encomium n. (pl. **encomiums** or **encomia**) a formal expression of praise.

encompass v. encircle; include.

encore n. a (call for) repetition of a performance. ● int. this call.

encounter v. meet by chance; be faced with. ● n. a chance meeting; a battle.

encourage v. give hope, confidence, or stimulus to; urge. □ **encouragement** n.

encroach v. intrude on someone's territory or rights. □ **encroachment** n.

encrust v. cover with a crust of hard material. □ **encrustation** n.

encumber v. be a burden to, hamper. □ **encumbrance** n.

encyclical n. a pope's letter for circulation to churches.

encyclopedia n. (also **encyclopaedia**) a book of information on many subjects. □ **encyclopedic** adj.

end n. **1** a limit; the furthest point or part; the final part. **2** destruction, death. **3** purpose. ● v. bring or come to an end. □ **make ends meet** live within one's income.

endanger v. cause danger to.

endear v. cause to be loved.

endearment n. words expressing love.

endeavour v. & n. (Amer. **endeavor**) (make) an earnest attempt.

endemic adj. commonly found in a specified area or people.

ending n. the final part.

endive n. a curly-leaved plant used in salads; (Amer.) chicory.

endless adj. without end, continual. □ **endlessly** adv.

endocrine gland n. a gland secreting hormones into the blood.

endorse v. **1** sign the back of (a cheque); note an offence on (a driving licence etc.). **2** declare approval of. □ **endorsement** n.

endow v. provide with a permanent income. □ **endowment** n.

endurance n. the power of enduring.

endure v. experience and survive (pain or hardship); tolerate; last. □ **endurable** adj.

enema n. liquid injected into the rectum, esp. to empty the bowels.

enemy n. one who is hostile to and seeks to harm another.

energetic adj. full of energy; done with energy. □ **energetically** adv.

energize v. (also **-ise**) give energy to; cause electricity to flow into.

energy n. capacity for vigorous activity; the ability of matter or radiation to do work; oil etc. as fuel.

enervate v. cause to lose vitality. □ **enervation** n.

enfant terrible (ahn-fahn te-reebl) n. (pl. **enfants terribles**) a person whose behaviour is embarrassing or irresponsible.

enfeeble v. make feeble. □ **enfeeblement** n.

enfold v. wrap up; clasp.

enforce v. compel obedience to. □ **enforceable** adj., **enforcement** n.

enfranchise v. give the right to vote. □ **enfranchisement** n.

engage v. **1** employ (a person); reserve; occupy the attention of. **2** begin a battle with; interlock.

engaged adj. **1** having promised to marry a specified person. **2** occupied; in use.

engagement n. **1** engaging something. **2** a promise to marry a specified person. **3** an appointment. **4** a battle.

engaging adj. attractive.

engender v. give rise to.

engine n. a machine using fuel and supplying power; a railway locomotive.

engineer n. a person skilled in engineering; one in charge of machines and engines. ●v. contrive; bring about.

engineering n. the application of science for the design and building of machines and structures.

English n. the language of Britain, the USA, and several other countries; **(the English)** the people of England. ●adj. of England or its language. □ **Englishman** n., **Englishwoman** n.

engrave v. cut a (design) into a hard surface; ornament in this way. □ **engraver** n.

engraving n. a print made from an engraved metal plate.

engross v. occupy fully by absorbing the attention. □ **engrossment** n.

engulf v. swamp.

enhance v. increase the quality or power etc. of. □ **enhancement** n.

enigma n. a mysterious person or thing. □ **enigmatic** adj., **enigmatically** adv.

enjoy v. get pleasure from; have as an advantage or benefit. □ **enjoyable** adj., **enjoyment** n.

enlarge v. make or become larger. □ **enlarge upon** say more about. □ **enlargement** n., **enlarger** n.

enlighten v. inform; free from ignorance. □ **enlightenment** n.

enlist v. enrol for military service; get the support of. □ **enlistment** n.

enliven v. make more lively. □ **enlivenment** n.

en masse (ahn mass) adv. all together; in a mass.

enmesh v. entangle.

enmity n. hostility; hatred.

ennoble v. make noble. □ **ennoblement** n.

ennui (ahn-wee) n. boredom.

enormity n. **1** great wickedness. **2** great size.

■ **Usage** Many people believe it is wrong to use **enormity** to mean 'great size'.

enormous adj. very large.

enough adj., adv., & n. as much or as many as necessary.

enquire v. ask. □ **enquiry** n.

enrage v. make furious.

enrapture v. delight intensely.

enrich v. make richer. □ **enrichment** n.

enrol v. (Amer. **enroll**) (**enrolled**) admit as or become a member. □ **enrolment** n.

en route (ahn root) adv. on the way.

ensconce v. establish securely or comfortably.

ensemble (ahn-sahmbl) n. a thing viewed as a whole; a set of performers; an outfit.

enshrine v. set in a shrine. □ **enshrinement** n.

ensign n. a military or naval flag.

enslave v. make (a person) a slave. □ **enslavement** n.

ensnare v. snare; trap.

ensue v. happen afterwards or as a result.

en suite (ahn sweet) adv. & adj. forming a single unit.

ensure v. make safe or certain.

entail v. make necessary.

entangle v. tangle; entwine and trap. □ **entanglement** n.

entente (on-tont) n. (also **entente cordiale**) friendly understanding between countries.

enter v. **1** go or come in or into. **2** put on a list or into a record etc.; register as a competitor.

enteritis n. inflammation of the intestines.

enterprise n. a bold undertaking; an initiative; a business activity.

enterprising adj. full of initiative.

entertain v. **1** amuse, occupy pleasantly. **2** receive with hospitality. **3** consider favourably. □ **entertainer** n., **entertainment** n.

enthral v. (Amer. **enthrall**) (**enthralled**) hold spellbound. □ **enthralment** n.

enthrone v. place on a throne. □ **enthronement** n.

enthuse v. fill with or show enthusiasm.

enthusiasm n. eager liking or interest. □ **enthusiastic** adj., **enthusiastically** adv.

enthusiast n. a person who is full of enthusiasm for something.

entice v. attract by offering something pleasant. □ **enticement** n.

entire adj. complete. □ **entirely** adv.

entirety n. □ **in its entirety** as a whole.

entitle v. **1** give (a person) a right or claim. **2** give (a book etc.) the title of. □ **entitlement** n.

entity n. a separate thing.

entomology n. the study of insects. □ **entomological** adj., **entomologist** n.

entourage (on-toor-ahzh) n. people accompanying an important person.

entr'acte (on-trakt) n. an interval between acts of a play.

entrails n.pl. intestines.

entrance[1] (en-trâns) n. entering; a door or passage by which one enters; a right of admission, the fee for this.

entrance[2] (en-trahns) v. fill with intense delight.

entreat v. request earnestly or emotionally. □ **entreaty** n.

entrée (on-tray) n. **1** a dish served between the fish and meat courses of a meal. **2** the right of admission.

entrench v. establish firmly. □ **entrenchment** n.

entrepreneur n. a person who organizes a commercial undertaking, esp. involving risk. □ **entrepreneurial** adj.

entropy n. a measure of the amount of a system's thermal energy not available for conversion into mechanical work.

entrust v. give as a responsibility, place in a person's care.

entry n. entering; an entrance; an item entered in a list etc. or for a competition.

entwine v. twine round.

E-number n. E plus a number, the EC designation for permitted food additives.

enumerate v. mention (items) one by one. □ **enumeration** n.

enunciate v. pronounce; state clearly. □ **enunciation** n.

envelop v. (**enveloped**) wrap, cover on all sides. □ **envelopment** n.

envelope n. a folded gummed cover for a letter.

enviable adj. desirable enough to arouse envy. □ **enviably** adv.

envious adj. full of envy. □ **enviously** adv.

environment n. surroundings; the natural world. □ **environmental** adj., **environmentally** adv.

environmentalist adj. & n. (a person) seeking to protect the natural environment.

environs n.pl. the surrounding districts, esp. of a town.

envisage v. imagine; foresee.

envoy n. a messenger, esp. to a foreign government.

envy n. discontent aroused by another's possessions or success; the object of this. ● v. feel envy of.

enzyme n. a protein formed in living cells (or produced synthetically) and assisting chemical processes.

eon var. of **aeon.**

epaulette n. an ornamental shoulder-piece.

ephemera n.pl. things of only short-lived usefulness.

ephemeral adj. lasting only a short time. □ **ephemerally** adv.

epic n. a long poem, story, or film about heroic deeds or history. ● adj. of or like an epic.

epicene adj. of or for both sexes.

epicentre n. (Amer. **epicenter**) the point where an earthquake reaches the earth's surface.

epicure n. a person who enjoys delicate food and drink. □ **epicurean** adj. & n., **epicureanism** n.

epidemic n. an outbreak of a disease etc. spreading through a community. □ **epidemiology** n.

epidermis n. the outer layer of the skin.

epidural n. a spinal anaesthetic affecting the lower part of the body.

epiglottis n. a cartilage that covers the larynx in swallowing.

epigram n. a short witty saying. □ **epigrammatic** adj.

epilepsy n. a disorder of the nervous system, causing fits. □ **epileptic** adj. & n.

epilogue n. a short concluding section.

episcopal adj. of or governed by bishops.

episcopalian adj. & n. (a member) of an episcopal church.

episiotomy n. a cut made at the opening of the vagina during childbirth.

episode n. an event forming one part of a sequence; one part of a serial. □ **episodic** adj., **episodically** adv.

epistle n. a letter. □ **epistolary** adj.

epitaph n. words inscribed on a tomb or describing a dead person.

epithet n. descriptive word(s).

epitome (i-pit-ōmi) n. a perfect model or example.

epitomize v. (also **-ise**) be a perfect example of. □ **epitomization** n.

epoch (ee-pok) n. a particular period.

eponymous adj. after whom something is named.

equable adj. free from extremes; even-tempered. □ **equably** adv.

equal adj. the same in size, amount, value, etc.; having the same rights or status. ● n. a person or thing equal to another. ● v. (**equalled**; Amer. **equaled**) be the same in size etc. as; do something equal to. □ **equally** adv., **equality** n.

■ **Usage** It is a mistake to say equally as, as in She was equally as guilty. The correct version is She was equally guilty or possibly, for example, She was as guilty as he was.

equalize v. (also **-ise**) make or become equal; equal an opponent's score. □ **equalization** n.

equalizer n. (also **-iser**) an equalizing goal etc.

equal opportunity n. the opportunity to compete equally for jobs regardless of race, sex, etc.

equanimity n. calmness of mind or temper.

equate v. consider to be equal or equivalent.

equation n. a mathematical statement that two expressions are equal.

equator n. an imaginary line round the earth at an equal distance from the North and South Poles. □ **equatorial** adj.

equerry n. an officer attending the British royal family.

equestrian adj. of horse-riding; on horseback.

equidistant adj. at an equal distance.

equilateral adj. having all sides equal.

equilibrium n. (pl. **equilibria** or **equilibriums**) a balanced state.

equine (ek-wyn) adj. of or like a horse.

equinox n. the time of year when night and day are of equal length. □ **equinoctial** adj.

equip v. (**equipped**) supply with what is needed.

equipage n. requisites, an outfit; a carriage, horses, and attendants.

equipment n. equipping; tools or an outfit etc. needed for a job or expedition.

equipoise n. equilibrium.

equitable adj. fair and just. □ **equitably** adv.

equitation n. horse-riding.

equity n. **1** fairness, impartiality. **2** (**equities**) stocks and shares not bearing fixed interest.

equivalent adj. equal in amount, value, meaning, etc. ● n. an equivalent thing. □ **equivalence** n.

equivocal adj. ambiguous; questionable, suspect. □ **equivocally** adv.

equivocate v. use words ambiguously. □ **equivocation** n.

ER abbr. Elizabeth Regina.

Er symb. erbium.

era n. a period of history.

eradicate v. wipe out. □ **eradication** n., **eradicable** adj.

erase v. rub out. □ **eraser** n., **erasure** n.

erbium n. a soft metallic element (symbol Er).

ere prep. & conj. (poetic) before.

erect adj. upright; rigid from sexual excitement. ● v. set up, build. □ **erectile** adj.

erection n. erecting; becoming erect; a thing erected, a building.

erg n. a unit of work or energy.

ergo adv. therefore.

ergonomics n. the study of work and its environment in order to improve efficiency. □ **ergonomic** adj., **ergonomically** adv.

ERM abbr. Exchange Rate Mechanism.

ermine n. a stoat; its white winter fur.

erode v. wear away gradually. □ **erosion** n., **erosive** adj.

erogenous adj. of or arousing sexual excitement.

erotic adj. of or arousing sexual desire. □ **erotically** adv., **eroticism** n.

err v. (**erred**) make a mistake; be incorrect; sin.

errand n. a short journey to take or fetch something; its purpose.

errant adj. misbehaving.

erratic adj. irregular, uneven. □ **erratically** adv.

erratum n. an error in printing or writing.

erroneous adj. incorrect. □ **erroneously** adv.

error n. a mistake; being wrong; the amount of inaccuracy.

ersatz adj. made in imitation.

erstwhile adj. former.

eructation n. belching.

erudite adj. learned. □ **erudition** n.

erupt v. break out or through; (of a volcano) eject lava. □ **eruption** n.

erythrocyte n. a red blood cell.

Es symb. einsteinium.

escalate v. increase in intensity or extent. □ **escalation** n.

escalator n. a moving staircase.

escalope n. a slice of boneless meat, esp. veal.

escapade n. a piece of reckless or mischievous conduct.

escape v. get free; get out of its container; avoid; be forgotten or unnoticed by. ● n. an act or means of escaping.

escapee n. one who escapes.

escapement n. a mechanism regulating a clock movement.

escapism n. escape from the realities of life. □ **escapist** n. & adj.

escapologist n. a person who entertains by escaping from confinement. □ **escapology** n.

escarpment n. a steep slope at the edge of a plateau etc.

eschew v. (literary) abstain from, avoid.

escort n. (ess-kort) a group of people or vehicles accompanying another as a protection or honour; a person accompanying a person of the opposite sex socially. ● v. (i-skort) act as escort to.

escritoire n. a writing desk with drawers.

escudo n. (pl. escudos) a unit of money in Portugal.

escutcheon n. 1 a shield bearing a coat of arms. 2 the protective plate around a keyhole or door handle.

Eskimo n. (pl. Eskimos or Eskimo) a member of a people living near the Arctic coast of America and eastern Siberia; their language.

■ **Usage** The Eskimos of North America prefer the term Inuit.

esophagus Amer. sp. of **oesophagus**.

esoteric adj. intended only for people with special knowledge or interest.

ESP abbr. extrasensory perception.

espadrille n. a canvas shoe with a sole of plaited fibre.

espalier n. a trellis; a shrub or tree trained on this.

esparto n. a coarse grass used in making paper.

especial adj. special, outstanding. □ **especially** adv.

Esperanto n. an artificial international language.

espionage n. spying.

esplanade n. a promenade.

espouse v. 1 support (a cause). 2 marry. □ **espousal** n.

espresso n. (also **expresso**) (pl. espressos) strong coffee made by forcing steam through powdered coffee beans.

esprit de corps (es-pree dè **kor**) n. loyalty uniting a group.

espy v. catch sight of.

Esq. abbr. Esquire, a courtesy title placed after a man's surname.

essay n. (ess-ay) a short literary composition in prose. ● v. (e-say) attempt.

essence n. the basic nature of something; an indispensable quality or element; a concentrated extract.

essential adj. unable to be dispensed with; fundamental. ● n. an essential thing. □ **essentially** adv.

establish v. set up; settle; cause to be accepted; prove.

established adj. officially recognized as the national Church.

establishment n. 1 establishing. 2 a staff of employees; a firm or institution; (**the Establishment**) people in positions of power.

estate n. landed property; a residential or industrial district planned as a unit; property left at one's death.

estate car n. a car with a door at the back and extended luggage space.

esteem v. think highly of. ● n. a favourable opinion, respect.

esthete, esthetic Amer. sp. of **aesthete, aesthetic**.

estimable adj. worthy of esteem.

estimate n. (ess-ti-măt) judgement of a thing's approximate value, amount, cost, etc. ●v. (ess-ti-mayt) form an estimate of. □ **estimation** n.

estrange v. cause to be no longer friendly or loving. □ **estrangement** n.

estrogen Amer. sp. of **oestrogen**.

estuary n. the mouth of a large river, affected by tides. □ **estuarine** adj.

ETA abbr. estimated time of arrival.

et al. abbr. and others.

etc. abbr. et cetera, and other similar things.

etch v. **1** produce (a picture) by engraving a metal plate with acid. **2** impress deeply (on the mind). □ **etcher** n., **etching** n.

eternal adj. existing always; unchanging. □ **eternally** adv.

eternity n. infinite time; the endless period of life after death; a long time.

eternity ring n. a jewelled finger ring symbolizing eternal love.

ethanol n. alcohol.

ether n. **1** the upper air. **2** a liquid used as an anaesthetic and solvent.

ethereal adj. **1** light and delicate. **2** heavenly. □ **ethereally** adv.

ethic n. a moral principle; (**ethics**) moral philosophy.

ethical adj. of ethics; morally correct, honourable. □ **ethically** adv.

ethnic adj. of a group sharing a common origin, culture, or language. □ **ethnically** adv., **ethnicity** n.

ethnic cleansing n. the mass expulsion or killing of people from opposing ethnic groups.

ethnology n. the study of human races and their characteristics. □ **ethnological** adj., **ethnologist** n.

ethos (ee-thoss) n. the characteristic spirit and beliefs of a community.

ethylene n. a hydrocarbon occurring in natural gas, used in manufacturing polythene.

etiolate (ee-ti-ŏ-layt) v. make pale through lack of light. □ **etiolation** n.

etiology Amer. sp. of **aetiology**.

etiquette n. rules of correct behaviour.

étude n. a short musical composition.

etymology n. an account of a word's origin and development. □ **etymological** adj., **etymologically** adv., **etymologist** n.

EU abbr. European Union.

Eu symbol. europium.

eucalyptus n. (also **eucalypt**) a tree, native to Australia, with leaves that yield a strong-smelling oil.

Eucharist n. the Christian sacrament commemorating the Last Supper, in which bread and wine are consumed; this bread and wine. □ **Eucharistic** adj.

eugenics n. the science of improving the human race by breeding.

eulogy n. a piece of praise. □ **eulogistic** adj., **eulogize** v.

eunuch n. a castrated man.

euphemism n. a mild expression substituted for an improper or blunt one. □ **euphemistic** adj., **euphemistically** adv.

euphonium n. a tenor tuba.

euphony n. pleasantness of sounds, esp. in words.

euphoria n. a feeling of happiness. □ **euphoric** adj., **euphorically** adv.

Eurasian adj. of Europe and Asia; of mixed European and Asian parentage. ●n. a Eurasian person.

eureka (yoor-eek-ă) int. I have found it! (announcing a discovery etc.).

Euro- *comb. form* European.

European *adj.* of Europe or its people. ● *n.* a European person.

eurhythmics *n.* (*Amer.* **eurythmics**) harmony of bodily movement developed with music and dance.

europium *n.* a soft metallic element (symbol Eu).

eurythmics *Amer.* sp. of **eurhythmics**.

Eustachian tube (yoo-stay-shän) *n.* the passage between the ear and the throat.

euthanasia *n.* bringing about an easy death, esp. to end suffering.

evacuate *v.* send away from a dangerous place; empty. □ **evacuation** *n.*

evacuee *n.* an evacuated person.

evade *v.* avoid by cleverness or trickery.

evaluate *v.* find out or state the value of; assess. □ **evaluation** *n.*

evanesce *v.* fade from sight.

evanescent *adj.* quickly fading. □ **evanescence** *n.*

evangelical *adj.* of or preaching the gospel. | **evangelicalism** *n.*

evangelist *n.* any of the authors of the four Gospels; a person who preaches the gospel. □ **evangelism** *n.*, **evangelistic** *adj.*

evaporate *v.* **1** turn into vapour. **2** cease to exist. □ **evaporation** *n.*

evasion *n.* evading; an evasive answer or excuse.

evasive *adj.* evading; not frank. □ **evasively** *adv.*, **evasiveness** *n.*

eve *n.* an evening, day, or time just before a special event.

even *adj.* **1** level, smooth; uniform. **2** calm. **3** equal. **4** exactly divisible by two. ● *v.* make or become even. ● *adv.* used for emphasis or in comparing things; *even faster*. □ **evenly** *adv.*, **evenness** *n.*

evening *n.* the latter part of the day, before nightfall.

evensong *n.* an evening service in the Church of England.

event *n.* something that happens, esp. something important; an item in a sports programme.

eventful *adj.* full of incidents.

eventual *adj.* coming at last, ultimate. □ **eventually** *adv.*

eventuality *n.* a possible event.

ever *adv.* always; at any time.

evergreen *adj.* having green leaves throughout the year. ● *n.* an evergreen tree or shrub.

everlasting *adj.* lasting for ever or for a very long time.

evermore *adv.* for ever, always.

every *adj.* each one without exception; each in a series; all possible.

everybody *pron.* every person.

everyday *adj.* worn or used on ordinary days; ordinary.

everyone *pron.* everybody.

everything *pron.* all things; all that is important.

everywhere *adv.* in every place.

evict *v.* expel (a tenant) by legal process. □ **eviction** *n.*, **evictor** *n.*

evidence *n.* anything that gives reason for believing something; statements made in a law court to support a case. ● *v.* be evidence of. □ **be in evidence** be conspicuous. □ **evidential** *adj.*

evident *adj.* obvious to the eye or mind. □ **evidently** *adv.*

evil *adj.* morally bad; harmful; very unpleasant. ● *n.* an evil thing, a sin, harm. □ **evilly** *adv.*

evince *v.* show, indicate.

eviscerate *v.* disembowel. □ **evisceration** *n.*

evoke *v.* bring to one's mind. □ **evocation** *n.*, **evocative** *adj.*

evolution *n.* the process of developing into a different form; the origination of living things by such development. □ **evolutionary** *adj.*

evolve v. develop or work out gradually. □ **evolvement** n.

ewe n. a female sheep.

ewer n. a pitcher, water jug.

ex prep. **1** (of goods) as sold from (a factory etc.). **2** without, excluding. ● n. (colloquial) a former husband, wife, or partner.

ex- pref. former.

exacerbate (ig-zass-er-bayt) v. make worse; irritate. □ **exacerbation** n.

exact adj. accurate; giving all details. ● v. insist on and obtain. □ **exactness** n.

exacting adj. making great demands, requiring great effort.

exactitude n. exactness.

exactly adv. in an exact manner; quite so, as you say.

exaggerate v. make seem greater than it really is. □ **exaggeration** n., **exaggerator** n.

exalt v. raise in rank; praise highly; make joyful. □ **exaltation** n.

exam n. (colloquial) an examination.

examination n. examining; a formal test of knowledge or ability.

examine v. look at closely; question formally. □ **examiner** n.

examinee n. a person being tested in an examination.

example n. a fact illustrating a general rule; a typical specimen; a person or thing worthy of imitation. □ **make an example of** punish as a warning to others.

exasperate v. annoy greatly. □ **exasperation** n.

excavate v. make (a hole) by digging, dig out; reveal by digging. □ **excavation** n., **excavator** n.

exceed v. be greater than; go beyond the limit of.

exceedingly adv. very.

excel v. (**excelled**) be or do better than; be very good at something.

Excellency n. the title of an ambassador, governor, etc.

excellent adj. extremely good. □ **excellently** adv., **excellence** n.

except prep. not including. ● v. exclude from a statement etc.

excepting prep. except.

exception n. excepting; something that does not follow the general rule. □ **take exception to** object to.

exceptionable adj. open to objection.

exceptional adj. very unusual; outstandingly good. □ **exceptionally** adv.

excerpt n. an extract from a book, film, etc.

excess n. the exceeding of due limits; the amount by which one quantity etc. exceeds another. ● adj. exceeding a limit.

excessive adj. too much. □ **excessively** adv.

exchange v. give or receive in place of another thing. ● n. exchanging; the price at which one currency is exchanged for another; a place where merchants, brokers, or dealers assemble to do business; a centre where telephone lines are connected. □ **exchangeable** adj.

exchequer n. a national treasury; a supply of money.

excise n. (ek-syz) duty or tax on certain goods and licences. ● v. (ek-syz) cut out or away. □ **excision** n.

excitable adj. easily excited. □ **excitably** adv., **excitability** n.

excitation n. exciting, arousing; stimulation.

excite v. rouse the emotions of, make eager; cause (a feeling or reaction). □ **excitement** n.

exclaim v. cry out or utter suddenly.

exclamation n. exclaiming; words exclaimed. □ **exclamatory** adj.

exclamation mark n. a punctuation mark (!) placed after an exclamation.

exclude v. keep out from a place, group, privilege, etc.; omit, ignore as irrelevant; make impossible. □ **exclusion** n.

exclusive adj. excluding others; catering only for the wealthy; not obtainable elsewhere. □ **exclusive** of not including. □ **exclusively** adv., **exclusiveness** n.

excommunicate v. officially exclude from a Church or its sacraments. □ **excommunication** n.

excoriate v. strip skin from; criticize severely. □ **excoriation** n.

excrement n. faeces.

excrescence n. an outgrowth on an animal or plant.

excreta n.pl. matter (esp. faeces) excreted from the body.

excrete v. expel (waste matter) from the body or tissues. □ **excretion** n., **excretory** adj.

excruciating adj. intensely painful.

excursion n. a short trip or outing, returning to the starting point.

excusable adj. able to be excused. □ **excusably** adv.

excuse v. (eks-skewz) pardon (a person); overlook (a fault); justify; exempt. ● n. (eks-skewss) a reason put forward to justify a fault etc.

ex-directory adj. deliberately not listed in a telephone directory.

execrable adj. abominable. □ **execrably** adv.

execrate v. express loathing for; utter curses. □ **execration** n.

execute v. **1** carry out (an order); produce (a work of art). **2** put (a condemned person) to death. □ **execution** n.

executioner n. an official who executes condemned people.

executive n. a person or group with managerial powers, or with authority to put government decisions into effect. ● adj. having such power or authority.

executor n. a person appointed to carry out the terms of a will.

exemplar n. a model, a type; an instance.

exemplary adj. outstandingly good; serving as a warning to others.

exemplify v. serve as an example of. □ **exemplification** n.

exempt adj. free from a customary obligation, payment, etc. ● v. make exempt. □ **exemption** n.

exercise n. **1** use of one's powers or rights. **2** activity, esp. designed to train the body or mind. ● v. **1** use (powers etc.) **2** (cause to) take exercise.

exercise book n. a book for writing in.

exert v. bring into use. □ **exert oneself** make an effort.

exertion n. exerting; a great effort.

exeunt (eks-iunt) v. (as a stage direction) they (actors) leave the stage.

exfoliate v. come off in scales or layers; remove dead layers of skin. □ **exfoliation** n.

ex gratia (eks gray-shǎ) adj. & adv. done or given as a favour, without legal obligation.

exhale v. breathe out; give off in vapour. □ **exhalation** n.

exhaust v. use up completely; tire out. ● n. waste gases from an engine etc; a device through which they are expelled. □ **exhaustible** adj.

exhaustion n. exhausting; being tired out.

exhaustive adj. thorough; comprehensive. □ **exhaustively** adv.

exhibit v. display, present for the public to see. ● n. a thing exhibited. □ **exhibitor** n.

exhibition n. **1** exhibiting; a public display. **2** a grant to a student from college funds etc.

exhibitionism n. a tendency to behave in a way designed to attract attention. □ **exhibitionist** n.

exhilarate v. make joyful or lively. □ **exhilaration** n.

exhort v. urge or advise earnestly. □ **exhortation** n., **exhortative** adj.

exhume v. dig up (a buried corpse). □ **exhumation** n.

exigency n. (also **exigence**) an urgent need; an emergency.

exigent adj. urgent; requiring much, exacting.

exiguous adj. very small, scanty.

exile n. banishment or long absence from one's country or home, esp. as a punishment; an exiled person. ● v. send into exile.

exist v. have being; be real; maintain life. □ **existence** n., **existent** adj.

existentialism n. a philosophical theory emphasizing that individuals are free to choose their actions. □ **existentialist** n.

exit n. departure from a stage or place; a way out. ● v. (as a stage direction) he or she (an actor) leaves the stage.

exodus n. a departure of many people.

ex officio (eks ŏ-fish-ioh) adv. & adj. because of one's official position.

exonerate v. declare or show to be blameless. □ **exoneration** n.

exorbitant adj. (of a price or demand) much too great. □ **exorbitantly** adv., **exorbitance** n.

exorcize v. (also **-ise**) drive out (an evil spirit) by prayer; free (a person or place) of an evil spirit. □ **exorcism** n., **exorcist** n.

exotic adj. brought from abroad; colourful, unusual. □ **exotically** adv.

exotica n. strange or rare objects.

expand v. **1** make or become larger; spread out; give a fuller account of. **2** become genial. □ **expandable** adj., **expander** n., **expansion** n.

expanse n. a wide area or extent.

expansive adj. **1** able to expand. **2** genial, communicative. □ **expansiveness** n.

expatiate (eks-pay-shi-ayt) v. speak or write at length about a subject.

expatriate adj. living abroad. ● n. an expatriate person.

expect v. believe that (a person or thing) will come or (a thing) will happen; be confident of receiving; think, suppose.

expectant adj. **1** filled with expectation. **2** pregnant. □ **expectantly** adv., **expectancy** n.

expectation n. expecting; something expected; a probability.

expectorant n. a medicine for causing a person to expectorate.

expectorate v. cough and spit phlegm; spit. □ **expectoration** n.

expedient adj. advantageous rather than right or just. □ **expediency** n.

expedite v. help or hurry the progress of.

expedition n. a journey for a purpose; people and equipment for this. □ **expeditionary** adj.

expeditious adj. speedy and efficient. □ **expeditiously** adv.

expel v. (**expelled**) send or drive out; compel to leave.

expend v. spend; use up.

expendable adj. able to be expended; not worth saving.

expenditure n. the expending of money etc.; an amount expended.

expense n. cost; a cause of spending money; (**expenses**) the amount spent doing a job; reimbursement of this.

expensive adj. involving great expenditure; costing or charging more than average. □ **expensively** adv., **expensiveness** n.

experience n. observation of facts or events, practice in doing something; knowledge or skill gained by this. ● v. feel or have an experience of.

experienced adj. having had much experience.

experiment n. a test to find out or prove something; trial of something new. ● v. conduct an experiment. □ **experimentation** n.

experimental adj. of or used in experiments; still being tested. □ **experimentally** adv.

expert n. a person with great knowledge or skill in a particular thing. ● adj. having great knowledge or skill. □ **expertly** adv.

expertise n. expert knowledge or skill.

expiate v. make amends for. □ **expiation** n., **expiatory** adj.

expire v. 1 die; cease to be valid. 2 breathe out (air). □ **expiration** n.

expiry n. termination of validity.

explain v. make clear, show the meaning of; account for. □ **explanation** n., **explanatory** adj.

expletive n. a violent exclamation, an oath.

explicable adj. able to be explained. □ **explicability** n.

explicit adj. stated plainly. □ **explicitly** adv., **explicitness** n.

explode v. (cause to) expand and break with a loud noise; show sudden violent emotion; increase suddenly. □ **explosion** n.

exploit n. (**eks**-ployt) a notable deed. ● v. (eks-**ployt**) make good use of; use selfishly. □ **exploitable** adj., **exploitation** n., **exploiter** n.

explore v. travel into (a country etc.) in order to learn about it; examine. □ **exploration** n., **exploratory** adj., **explorer** n.

explosive adj. & n. (a substance) able or liable to explode.

exponent n. 1 one who favours a specified theory etc. 2 a raised figure beside a number indicating the number of times the number is to be multiplied by itself.

exponential adj. of a mathematical exponent; (of an increase) more and more rapid.

export v. (eks-**port**) send (goods etc.) to another country for sale. ● n. (**eks**-port) exporting; a thing exported. □ **exportation** n., **exporter** n.

expose v. leave uncovered or unprotected; subject to a risk etc.; allow light to reach (film etc.); reveal. □ **exposure** n.

exposé n. (eks-**poh**-zay) a statement of facts; a disclosure.

exposition n. 1 expounding; an explanation. 2 a large exhibition.

expostulate v. protest, remonstrate. □ **expostulation** n., **expostulatory** adj.

expound v. explain in detail.

express adj. 1 definitely stated. 2 travelling rapidly, designed for high speed. ● adv. at high speed. ● n. a fast train or bus making few or no stops. ● v. 1 make (feelings or qualities) known; put into words; represent by symbols. **2** press or squeeze out. □ **expressible** *adj.*

expression *n.* expressing; a word or phrase; a look or manner that expresses feeling.

expressionism *n.* a style of art seeking to express feelings rather than represent objects realistically. □ **expressionist** *n.*

expressive *adj.* expressing something; full of expression. □ **expressively** *adv.*

expressly *adv.* explicitly; for a particular purpose.

expresso var. of espresso.

expressway *n.* an urban motorway.

expropriate *v.* seize (property); dispossess. □ **expropriation** *n.*

expulsion *n.* expelling; being expelled. □ **expulsive** *adj.*

expunge *v.* wipe out.

expurgate *v.* remove (objectionable matter) from (a book etc.). □ **expurgation** *n.*, **expurgator** *n.*, **expurgatory** *adj.*

exquisite *adj.* having exceptional beauty; acute, keenly felt. □ **exquisitely** *adv.*

extant *adj.* still existing.

extemporize *v.* (also **-ise**) speak, perform, or produce without preparation. □ **extemporization** *n.*

extend *v.* **1** make longer; stretch; reach; enlarge. **2** offer.

extendable *adj.* (also **extendible, extensible**) able to be extended.

extension *n.* **1** extending; extent. **2** an additional part or period; a subsidiary telephone, its number.

extensive *adj.* large in area or scope. □ **extensively** *adv.*

extensor *n.* a muscle that extends a part of the body.

extent *n.* the space over which a thing extends; scope; a large area.

extenuate *v.* make (an offence) seem less great by providing a partial excuse. □ **extenuation** *n.*

exterior *adj.* on or coming from the outside. ● *n.* an exterior surface or appearance.

exterminate *v.* destroy all members or examples of. □ **extermination** *n.*, **exterminator** *n.*

external *adj.* of or on the outside. □ **externally** *adv.*

externalize *v.* (also **-ise**) give or attribute external existence to.

extinct *adj.* no longer burning or active or existing in living form.

extinction *n.* extinguishing; making or becoming extinct.

extinguish *v.* put out (a light or flame); end the existence of.

extinguisher *n.* a device for discharging liquid chemicals or foam to extinguish a fire.

extirpate *v.* root out, destroy. □ **extirpation** *n.*

extol *v.* (**extolled**) praise enthusiastically.

extort *v.* obtain by force or threats. □ **extortion** *n.*, **extortioner** *n.*

extortionate *adj.* excessively high in price, exorbitant. □ **extortionately** *adv.*

extra *adj.* additional, more than is usual or expected. ● *adv.* more than usually; in addition. ● *n.* an extra thing; a person employed as one of a crowd in a film.

extra- *pref.* outside, beyond.

extract *v.* (eks-**trakt**) take out or obtain by force or effort; obtain by suction or pressure or chemical treatment. ● *n.* (eks-**trakt**) a substance extracted from another; a passage extracted from a book, film, etc. □ **extractor** *n.*

extraction *n.* **1** extracting. **2** lineage.

extra-curricular *adj.* not part of the normal curriculum.

extradite *v.* hand over (an accused person) for trial in the

extramarital *adj.* of sexual relationships outside marriage.

extramural *adj.* for students who are not members of a university.

extraneous *adj.* of external origin; not relevant. □ **extraneously** *adv.*

extraordinary *adj.* remarkable; beyond what is usual. □ **extraordinarily** *adv.*

extrapolate *v.* estimate on the basis of available data. □ **extrapolation** *n.*

extrasensory *adj.* achieved by some means other than the known senses.

extraterrestrial *adj.* of or from outside the earth or its atmosphere.

extravagant *adj.* spending or using excessively; going beyond what is reasonable. □ **extravagantly** *adv.*, **extravagance** *n.*

extravaganza *n.* a lavish spectacular display or entertainment.

extreme *adj.* **1** at the end(s); outermost. **2** very great or intense; going to great lengths in actions or views. ● *n.* **1** either end of a thing. **2** an extreme degree, act, or condition. □ **extremely** *adv.*

extremist *n.* a person holding extreme views. □ **extremism** *n.*

extremity *n.* an extreme point; an extreme degree of need or danger; **(the extremities)** the hands and feet.

extricable *adj.* able to be extricated.

extricate *v.* free from an entanglement or difficulty. □ **extrication** *n.*

extrinsic *adj.* not intrinsic; extraneous. □ **extrinsically** *adv.*

extrovert *n.* a lively sociable person. □ **extroversion** *n.*

extrude *v.* thrust or squeeze out. □ **extrusion** *n.*, **extrusive** *adj.*

exuberant *adj.* **1** full of high spirits. **2** growing profusely. □ **exuberantly** *adv.*, **exuberance** *n.*

exude *v.* ooze; give off like sweat or a smell. □ **exudation** *n.*

exult *v.* rejoice greatly. □ **exultant** *adj.* exulting. **exultation** *n.*

eye *n.* the organ of sight; the iris of this; the region round it; the power of seeing; something like an eye. ● *v.* **(eyed, eyeing)** look at, watch.

eyeball *n.* the whole of the eye within the eyelids.

eyebrow *n.* the fringe of hair on the ridge above the eye socket.

eyelash *n.* one of the hairs fringing the eyelids.

eyelet *n.* a small hole; a ring strengthening this.

eyelid *n.* either of the two folds of skin that can be moved together to cover the eye.

eyeliner *n.* a cosmetic applied in a line around the eye.

eye-opener *n.* (*colloquial*) something that brings enlightenment or great surprise.

eyepiece *n.* the lens to which the eye is applied in a telescope or microscope etc.

eye-shade *n.* a device to protect the eyes from strong light.

eyeshadow *n.* a cosmetic applied to the skin round the eyes.

eyesight *n.* the ability to see; the range of vision.

eyesore *n.* an ugly thing, esp. a building.

eye-tooth *n.* a canine tooth in the upper jaw, below the eye.

eyewash *n.* (*slang*) talk or behaviour intended to give a misleadingly good impression.

eyewitness *n.* a person who actually saw something happen.

eyrie (eer-i) n. an eagle's nest, built high up; a house etc. perched high up.

....

Ff

....

F abbr. Fahrenheit. ● symb. fluorine.

f abbr. **1** female; feminine. **2** (Music) forte.

FA abbr. Football Association.

fable n. a story not based on fact, often with a moral. □ **fabled** adj.

fabric n. **1** cloth or knitted material. **2** the walls etc. of a building.

fabricate v. **1** construct, manufacture. **2** invent (a story etc.). □ **fabrication** n., **fabricator** n.

fabulous adj. incredibly great; marvellous. □ **fabulously** adv.

façade (fă-sahd) n. the front of a building; an outward appearance.

face n. **1** the front of the head; an expression shown by its features; a grimace. **2** an outward aspect; the front or main side. **3** the dial of a clock. **4** the surface of a coal-seam. ● v. **1** have or turn the face towards. **2** meet firmly. **3** put a facing on.

facecloth n. a small towelling cloth for washing the face and body.

faceless adj. without identity; purposely not identifiable.

facelift n. an operation for tightening the skin of the face; an alteration that improves the appearance.

facet n. one of many sides of a cut stone or jewel; one aspect.

facetious adj. intended or intending to be amusing. □ **facetiously** adv., **facetiousness** n.

facia (fay-shă) n. (also **fascia**) **1** the instrument panel of a vehicle. **2** a nameplate over a shop front.

facial adj. of the face. ● n. a beauty treatment for the face.

facile (fa-syl) adj. easily achieved but of little value; glib.

facilitate v. make easy or easier. □ **facilitation** n.

facility n. absence of difficulty; a means for doing something.

facing n. an outer covering; a layer of material at the edge of a garment for strengthening, contrast, neatening, etc.

facsimile (fak-sim-ili) n. a reproduction of a document etc.

fact n. a thing known to have happened or to be true.

faction n. a small united group within a larger one. □ **factious** adj.

factitious adj. artificial; not genuine.

factor n. **1** a circumstance that contributes towards a result. **2** a number by which a given number can be divided exactly.

factory n. a building in which goods are manufactured.

factotum n. a general servant or assistant.

facts of life n.pl. information about sex and reproduction.

factual adj. based on or containing facts. □ **factually** adv.

faculty n. **1** any of the powers of the body or mind. **2** a department teaching a specified subject in a university or college.

fad n. a craze, a whim.

faddy adj. (**faddier**) having petty likes and dislikes, esp. about food.

fade v. (cause to) lose colour, freshness, or vigour; disappear gradually.

faeces (fee-seez) n.pl. (Amer. **feces**) waste matter discharged from the bowels. □ **faecal** adj.

faff n. (colloquial) a bother; something that is too tricky. □ **faff about** fuss; dither.

fag (colloquial) v. (fagged) toil; make tired. ● n. 1 tiring work, drudgery. 2 a cigarette.

faggot n. (Amer. fagot) 1 a tied bundle of sticks or twigs. 2 a ball of chopped seasoned liver etc., baked or fried.

fah n. (Music) the fourth note of a major scale, or the note F.

Fahrenheit adj. of a temperature scale with the freezing point of water at 32° and boiling point at 212°.

faience (fy-ahns) n. painted glazed earthenware.

fail v. be unsuccessful; become weak, cease functioning; neglect or be unable; disappoint; become bankrupt; declare to be unsuccessful. ● n. a failure.

failing n. a weakness or fault. ● prep. if (a thing) does not happen.

failure n. failing, lack of success; a person or thing that fails.

fain adv. (old use) willingly.

faint adj. 1 indistinct; not intense; weak, feeble. 2 about to faint. ● v. collapse unconscious. ● n. the act or state of fainting. □ **faintly** adv., **faintness** n.

faint-hearted adj. timid.

fair n. 1 a funfair. 2 a gathering for a sale of goods, often with entertainments; an exhibition of commercial goods. ● adj. 1 light in colour, having light-coloured hair. 2 (of weather) fine, (of wind) favourable. 3 just, unbiased. 4 of moderate quality or amount. ● adv. fairly.

fairground n. an open space where a fair is held.

fairing n. a streamlining structure added to a ship, vehicle, etc.

fairly adv. 1 in a fair way. 2 moderately.

fairway n. 1 a navigable channel. 2 part of a golf course between tee and green.

fairy n. an imaginary small being with magical powers.

fairy godmother n. a benefactress providing a sudden unexpected gift.

fairyland n. the world of fairies; a very beautiful place.

fairy lights n.pl. strings of small coloured lights used as decorations.

fairy story n. (also **fairy tale**) a tale about fairies or magic; a falsehood.

fait accompli (fayt ă-kom-plee) n. something already done and not reversible.

faith n. reliance, trust; belief in religious doctrine; loyalty, sincerity.

faith healing n. a cure etc. dependent on faith. □ **faith healer** n.

faithful adj. loyal, trustworthy; true, accurate. □ **faithfully** adv., **faithfulness** n.

faithless adj. disloyal.

fake n. a person or thing that is not genuine. ● adj. faked. ● v. make an imitation of; pretend. □ **faker** n.

fakir (fay-keer) n. a Muslim or Hindu religious mendicant or ascetic.

falcon n. a small long-winged hawk.

falconry n. the breeding and training of hawks. □ **falconer** n.

fall v. (**fell, fallen, falling**) 1 come or go down freely; decrease. 2 pass into a specified state; occur. 3 (of the face) show dismay. 4 lose one's position or office; be captured or conquered; die in battle. ● n. 1 falling; the amount of this. 2 (Amer.) autumn. 3 (falls) a waterfall. □ **fall back** have recourse to. **fall for** (colloquial) fall in love with; be deceived by. **fall out** quarrel;

happen. **fall short** be inadequate. **fall through** (of a plan) fail to be achieved.

fallacy n. a false belief, false reasoning. □ **fallacious** adj.

fallible adj. liable to make mistakes. □ **fallibility** n.

Fallopian tube n. either of the two tubes from the ovary to the womb.

fallout n. airborne radioactive debris.

fallow adj. (of land) left unplanted for a time. ● n. such land.

fallow deer n. a reddish-brown deer with white spots.

false adj. incorrect; deceitful; unfaithful; not genuine, sham. □ **falsely** adv., **falseness** n.

falsehood n. a lie; telling lies.

falsetto n. (pl. **falsettos**) a voice above one's natural range.

falsify v. alter fraudulently; misrepresent. □ **falsification** n.

falsity n. falseness; falsehood.

falter v. go or function unsteadily; become weaker; speak hesitantly.

fame n. the condition of being known to many people; good reputation.

familial adj. of a family.

familiar adj. **1** well known; well acquainted. **2** too informal. □ **familiarly** adv., **familiarity** n.

familiarize v. (also **-ise**) make familiar. □ **familiarization** n.

family n. parents and their children; a person's children; a set of relatives; a group of related plants, animals, or things.

famine n. extreme scarcity (esp. of food) in a region.

famished adj. extremely hungry.

famous adj. **1** known to very many people. **2** (colloquial) excellent. □ **famously** adv.

fan n. **1** a hand-held or mechanical device to create a current of air. **2** an enthusiastic admirer or supporter. ● v. (**fanned**) fan

cool with a fan. **2** spread from a central point.

fanatic n. a person with excessive enthusiasm for something. □ **fanatical** adj., **fanatically** adv., **fanaticism** n.

fan belt n. a belt driving a fan that cools a car engine.

fancier n. a person with special knowledge and love of something.

fanciful adj. imaginative; imaginary. □ **fancifully** adv.

fancy n. **1** imagination; something imagined; an unfounded idea. **2** a desire; a liking. ● adj. (**fancier**) ornamental, elaborate. ● v. **1** imagine; suppose. **2** (colloquial) find attractive.

fancy dress n. a costume representing an animal, historical character, etc., worn for a party.

fandango n. (pl. **fandangos** or **fandangoes**) a lively Spanish dance.

fanfare n. a short ceremonious sounding of trumpets.

fang n. a long sharp tooth; a snake's tooth that injects venom.

fanlight n. a small window above a door or larger window.

fantasia n. an imaginative musical or other composition.

fantasize v. (also **-ise**) daydream.

fantastic adj. **1** absurdly fanciful. **2** (colloquial) excellent. □ **fantastically** adv.

fantasy n. **1** imagination; something imagined. **2** a fanciful design.

far adv. at, to, or by a great distance. ● adj. distant, remote.

farad n. a unit of electrical capacitance.

farce n. a light comedy; an absurd and useless situation; a pretence. □ **farcical** adj., **farcically** adv.

fare n. **1** the price charged for a passenger to travel; a passenger paying this. **2** food provided.

● v. get on or be treated (well, badly, etc.).

Far East n. China, Japan, and other countries of east Asia.

farewell int. goodbye. ● n. leave-taking; saying goodbye.

far-fetched adj. unconvincing, very unlikely.

farinaceous adj. starchy.

farm n. a unit of land used for raising crops or livestock. ● v. grow crops, raise livestock; use (land) for this. □ **farmer** n.

farmhouse n. a farmer's house.

farmstead n. a farm and its buildings.

farmyard n. an enclosed area round farm buildings.

farrago (fă-rah-goh) n. (pl. **farragos** or **farragoes**) a hotch-potch.

farrier n. a smith who shoes horses.

farrow v. give birth to young pigs. ● n. farrowing; a litter of pigs.

fart v. (vulgar) send out wind from the anus.

farther, farthest vars. of **further, furthest**.

farthingale n. a hooped petti-coat.

fascia var. of **facia**.

fascinate v. 1 attract and hold the interest of; charm greatly. 2 make (a victim) motionless with fear. □ **fascination** n.

Fascism (fash-izm) n. (also **fascism**) a system of extreme right-wing dictatorship. □ **Fascist** n.

fashion n. a manner or way of doing something; a style popular at a given time. ● v. shape, make.

fashionable adj. in or using a currently popular style; used by stylish people. □ **fashionably** adv.

fast adj. 1 moving or done quickly; allowing quick movement; showing a time ahead of

the correct one. 2 firmly fixed. ● adv. 1 quickly. 2 firmly, tightly. ● v. go without food. ● n. fasting; a period appointed for fasting.

fastback n. a car with a long sloping back.

fasten v. fix firmly, tie or join together; become fastened.

fastener n. (also **fastening**) a device used for fastening something.

fastidious adj. 1 choosing only what is good; fussy. 2 easily disgusted. □ **fastidiously** adv., **fastidiousness** n.

fastness n. a stronghold, a fortress.

fat n. a white or yellow substance found in animal bodies and certain seeds. ● adj. (**fatter**) 1 excessively plump; containing much fat; fattened; thick. 2 fertile; richly rewarded. □ **fatness** n., **fatty** adj.

fatal adj. causing or ending in death or disaster; fateful. □ **fatally** adv.

fatalist n. a person who submits to what happens, regarding it as inevitable. □ **fatalism** n., **fatalistic** adj.

fatality n. a death caused by accident or in war etc.

fate n. a power thought to control all events; a person's destiny.

fated adj. destined by fate; doomed.

fateful adj. bringing future usu. unpleasant events. □ **fatefully** adv.

father n. a male parent or ancestor; a founder, an originator; a title of certain priests. ● v. beget; originate. □ **fatherhood** n., **fatherly** adj.

father-in-law n. (pl. **fathers-in-law**) the father of one's wife or husband.

fatherland n. one's native country.

fatherless adj. without a living or known father.

fathom n. a measure (1.82 m) of the depth of water. ● v. understand. □ **fathomable** adj.

fathomless adj. **1** too deep to measure. **2** incomprehensible.

fatigue n. **1** tiredness. **2** weakness in metal etc., caused by stress. **3** a soldier's non-military task. ● v. cause fatigue to.

fatstock n. livestock fattened for slaughter as food.

fatten v. make or become fat.

fatuous adj. foolish, silly. □ **fatuously** adv., **fatuousness** n.

fatwa n. a ruling made by an Islamic leader.

faucet n. (esp. Amer.) a tap.

fault n. **1** a defect, an imperfection; an offence. **2** responsibility for something wrong. **3** a break in layers of rock. ● v. find fault in; make imperfect. □ **at fault** responsible for a mistake etc. □ **faultless** adj., **faulty** adj.

faun n. a Roman god of the countryside with a goat's legs and horns.

fauna n. (pl. **faunas** or **faunae**) the animals of an area or period.

faux pas (foh pah) n. (pl. **faux pas**) an embarrassing blunder.

favour (Amer. **favor**) n. **1** liking, approval. **2** a kindly or helpful act beyond what is due. **3** favouritism. ● v. **1** regard or treat with favour; oblige. **2** resemble (one parent etc.).

favourable adj. (Amer. **favorable**) giving or showing approval; pleasing, satisfactory; advantageous. □ **favourably** adv.

favourite adj. (Amer. **favorite**) liked above others. ● n. a favoured person or thing; a competitor expected to win.

favouritism n. (Amer. **favoritism**) unfair favouring of one at the expense of others.

fawn n. **1** a deer in its first year. **2** light yellowish brown. ● adj. fawn-coloured. ● v. try to win favour by obsequiousness; (of a dog) show extreme affection.

fax n. transmission of exact copies of documents by electronic scanning; a copy produced in this way; a machine for sending and receiving faxes. ● v. transmit (a document) by this process.

fay n. (literary) a fairy.

faze v. fluster; daunt.

FBI abbr. (in the USA) Federal Bureau of Investigation.

FC abbr. Football Club.

Fe symb. iron.

fealty n. loyalty, allegiance.

fear n. an unpleasant sensation caused by nearness of danger or pain. ● v. feel fear of; be afraid.

fearful adj. **1** terrible. **2** feeling fear. **3** (colloquial) extremely unpleasant. □ **fearfully** adv.

fearless adj. feeling no fear. □ **fearlessly** adv., **fearlessness** n.

fearsome adj. frightening, alarming.

feasible adj. able to be done, possible. □ **feasibly** adv., **feasibility** n.

■ **Usage** Feasible should not be used to mean 'likely'. Possible or probable should be used instead.

feast n. a large elaborate meal; a joyful festival; a treat. ● v. eat heartily; give a feast to.

feat n. a remarkable achievement.

feather n. each of the structures with a central shaft and fringe of fine strands, growing from a bird's skin. ● v. **1** cover or fit with feathers. **2** turn (an oar) to pass through the air edgeways. □ **feather one's nest** enrich oneself. □ **feathery** adj.

feather-bed v. (-bedded) make things financially easy for.

featherweight n. a very light-weight thing or person; a weight between bantamweight and lightweight.

feature n. **1** a distinctive part of the face. **2** a noticeable quality. **3** a prominent article in a newspaper etc. **4** a full-length cinema film. ● v. give prominence to; be a feature of or in.

Feb. abbr. February.

febrile (fee-bryl) adj. of fever, feverish.

February n. the second month.

feces Amer. sp. of **faeces**.

feckless adj. incompetent and irresponsible. □ **fecklessness** n.

fecund adj. fertile. □ **fecundity** n.

fed see **feed**. adj. ● **fed up** (colloquial) discontented or bored.

federal adj. of a system in which states unite under a central authority but are independent in internal affairs. □ **federalism** n., **federalist** n., **federally** adv.

federate v. (fed-er-ayt) unite on a federal basis or for a common purpose. ● adj. (fed-er-ăt) united in this way. □ **federative** adj.

federation n. federating; a federated society or group of states.

fee n. a sum payable for professional services, or for a privilege.

feeble adj. weak; ineffective. □ **feebly** adv., **feebleness** n.

feed v. (**fed**, **feeding**) give food to; give as food; (of animals) take food; nourish; supply. ● n. a meal; food for animals.

feedback n. return of part of a system's output to its source; return of information about a product, a piece of work, etc. to the producer.

feeder n. **1** a person or thing that feeds; a baby's feeding bottle; a feeding apparatus in a machine. **2** a road, railway line, etc. linking outlying areas to a central system.

feel v. (**felt**, **feeling**) **1** explore or perceive by touch. **2** be conscious of (being); give a sensation; have a vague conviction or impression; have as an opinion. ● n. the sense of touch; an act of feeling; a sensation produced by a thing touched. □ **feel like** be in the mood for.

feeler n. **1** a long slender organ of touch in certain animals. **2** a tentative suggestion.

feeling n. **1** the power to feel things. **2** mental or physical awareness; an opinion or belief not based on reasoning; (**feelings**) emotional susceptibilities. **3** sympathy.

feet see **foot**.

feign (fayn) v. pretend.

feint (faynt) n. a sham attack made to divert attention. ● v. make a feint. ● adj. (of ruled lines) faint.

feisty adj. (**feistier**) (colloquial) aggressive, excitable.

feldspar n. (also **felspar**) a white or red mineral containing silicates.

felicitate v. congratulate. □ **felicitation** n.

felicitous adj. well-chosen, apt. □ **felicitously** adv., **felicitousness** n.

felicity n. happiness; a pleasing manner or style.

feline adj. of cats, catlike. ● n. an animal of the cat family.

fell n. a stretch of moor or hilly land, especially in north England. ● v. strike or cut down. See also **fall**.

fellow n. **1** an associate, a comrade. **2** a thing like another. **3** a member of a learned society or governing body of a college. **4** (colloquial) a man or boy.

fellowship n. **1** friendly association with others. **2** a society;

membership of this. **3** the position of a college fellow.

felon n. a person who has committed a serious violent crime. □ **felony** n.

felspar var. of **feldspar**.

felt n. cloth made by matting and pressing fibres. ● v. make or become matted; cover with felt. *See also* **feel**.

felt-tip pen n. (also **felt-tipped pen**) a pen with a writing point made of fibre.

female adj. **1** of the sex that can bear offspring or produce eggs; (of plants) fruit-bearing. **2** (of a connection etc.) hollow. ● n. a female animal or plant.

feminine adj. of, like, or traditionally considered suitable for women; having the grammatical form of the female gender. □ **femininity** n.

feminist n. a supporter of women's claims to be given rights equal to those of men. □ **feminism** n.

femme fatale (fam fã-tahl) n. (pl. **femmes fatales**) a dangerously seductive woman.

femur n. (pl. **femurs** or **femora**) the thigh bone. □ **femoral** adj.

fen n. a low-lying marshy or flooded tract of land.

fence n. **1** a barrier round the boundary of a field or garden etc. **2** (slang) a person who deals in stolen goods. ● v. **1** surround with a fence. **2** engage in the sport of fencing. □ **fencer** n.

fencing n. **1** fences, their material. **2** the sport of fighting with foils.

fend v. □ **fend for** look after (esp. oneself). **fend off** ward off.

fender n. **1** a low frame bordering a fireplace. **2** a pad hung over a moored vessel's side to protect against bumping. **3** (*Amer.*) the mudguard or bumper of a motor vehicle.

fennel n. an aniseed-flavoured herb; its bulbous stem, used as a vegetable.

fenugreek n. a plant with fragrant seeds used for flavouring.

feral adj. wild.

ferment v. (fer-ment) **1** undergo fermentation; cause fermentation in. **2** seethe with excitement. ● n. (fer-ment) **1** fermentation. **2** excitement; agitation.

fermentation n. a chemical change caused by an organic substance, producing effervescence and heat.

fermium n. a radioactive metallic element (symbol Fm).

fern n. a flowerless plant with feathery green leaves. □ **ferny** adj.

ferocious adj. fierce, savage. □ **ferociously** adv., **ferocity** n.

ferrel var. of **ferrule**.

ferret n. a small animal of the weasel family. ● v. (**ferreted**) search, rummage. □ **ferret out** discover by searching. □ **ferrety** adj.

ferric adj. (also **ferrous**) of or containing iron.

Ferris wheel n. a giant revolving vertical wheel with passenger cars for funfair rides.

ferroconcrete n. reinforced concrete.

ferrule n. (also **ferrel**) a metal cap strengthening the end of a stick or tube.

ferry v. convey in a boat across water; transport. ● n. a boat used for ferrying; the place where it operates; the service it provides.

fertile adj. **1** able to produce vegetation, fruit, or young; capable of developing into a new plant or animal. **2** (of the mind) inventive. □ **fertility** n.

fertilize v. (also **-ise**) make fertile; introduce pollen or sperm into. □ **fertilization** n.

fertilizer *n.* (also **fertiliser**) material added to soil to make it more fertile.

fervent *adj.* showing intense feeling. □ **fervently** *adv.*, **fervency** *n.*

fervid *adj.* fervent. □ **fervidly** *adv.*

fervour *n.* (*Amer.* **fervor**) intensity of feeling.

fester *v.* **1** make or become septic. **2** cause continuing resentment.

festival *n.* **1** a day or period of celebration. **2** a series of performances of music or drama etc.

festive *adj.* of or suitable for a festival; gaily decorated. □ **festively** *adv.*, **festiveness** *n.*

festivity *n.* a festive occasion; celebration.

festoon *n.* a hanging chain of flowers or ribbons etc. ● *v.* decorate with hanging ornaments.

feta *n.* (also **fetta**) a white salty Greek cheese.

fetch *v.* **1** go for and bring back; cause to come out. **2** be sold for (a price).

fetching *adj.* attractive.

fête (fayt) *n.* a festival; an outdoor entertainment or sale, esp. in aid of charity. ● *v.* entertain in celebration of an achievement.

fetid *adj.* (also **foetid**) stinking.

fetish *n.* an object worshipped as having magical powers; something given excessive respect.

fetlock *n.* a horse's leg above and behind the hoof.

fetta var. of **feta**.

fetter *n.* a shackle for the ankles; a restraint. ● *v.* put into fetters; restrain.

fettle *n.* condition, trim.

fetus Amer. sp. of **foetus**.

feud *n.* lasting hostility. ● *v.* conduct a feud.

feudal *adj.* of or like the feudal system. □ **feudalism** *n.*, **feudalistic** *adj.*

feudal system *n.* a medieval system of holding land by giving one's services to the owner.

fever *n.* an abnormally high body temperature; a disease causing it; nervous excitement. □ **fevered** *adj.*, **feverish** *adj.*, **feverishly** *adv.*

few *adj.* & *n.* not many. □ **a few** some. **a good few** a fairly large number.

fey *adj.* strange, other-worldly; clairvoyant. □ **feyness** *n.*

fez *n.* (*pl.* **fezzes**) a high flat-topped red cap worn by some Muslim men.

ff *abbr.* (*Music*) fortissimo.

ff. *abbr.* the following pages.

fiancé, fiancée *n.* a man (*fiancé*) or woman (*fiancée*) one is engaged to marry.

fiasco *n.* (*pl.* **fiascos**) a ludicrous failure.

fiat (fy-at) *n.* an order; an authorization.

fib *n.* a trivial lie. ● *v.* tell a fib. □ **fibber** *n.*

fibre *n.* (*Amer.* **fiber**) **1** a threadlike strand; a substance formed of fibres; fibrous material in food, roughage. **2** strength of character. □ **fibrous** *adj.*

fibreglass *n.* (*Amer.* **fiberglass**) material made of or containing glass fibres.

fibre optics *n.* transmission of information by light along thin flexible glass fibres.

fibril *n.* a small fibre.

fibroid *adj.* consisting of fibrous tissue. ● *n.* a benign fibroid tumour.

fibrositis *n.* rheumatic pain in tissue other than bones and joints.

fibula *n.* (*pl.* **fibulae** or **fibulas**) the bone on the outer side of the shin.

fiche (feesh) n. (pl. **fiche** or **fiches**) a microfiche.

fickle adj. often changing, not loyal. □ **fickleness** n.

fiction n. an invented story; a class of literature consisting of books containing such stories. □ **fictional** adj.

fictitious adj. imaginary, not true.

fiddle n. **1** a violin. **2** (colloquial) a swindle. ● v. **1** fidget with something. **2** (colloquial) cheat, falsify. □ **fiddler** n.

fiddlesticks int. nonsense.

fiddly adj. (**fiddlier**) awkward to do or use.

fidelity n. **1** faithfulness, loyalty. **2** accuracy.

fidget v. (**fidgeted**) make small restless movements; make or be uneasy. ● n. a person who fidgets. □ **fidgety** adj.

fiduciary adj. held or given etc. in trust. ● n. a trustee.

fief n. land held under the feudal system; a domain.

field n. **1** a piece of open ground, esp. for pasture or cultivation; a sports ground. **2** an area rich in a natural product. **3** a sphere of action or interest. **4** all competitors in a race or contest. ● v. **1** be a fielder, stop and return (a ball). **2** put (a team) into a contest.

field day n. a day of much activity.

fielder n. a person who fields a ball; a member of the side not batting.

field events n. athletic contests other than races.

field glasses n.pl. binoculars.

field marshal n. an army officer of the highest rank.

fieldwork n. practical work done outside libraries and laboratories by surveyors, social workers, etc. □ **fieldworker** n.

fiend (feend) n. **1** an evil spirit; a wicked, mischievous, or annoying person. **2** (colloquial) a devotee or addict; a fitness fiend.

fiendish adj. cruel; extremely difficult. □ **fiendishly** adv.

fierce adj. violent in manner or action; eager, intense. □ **fiercely** adv., **fierceness** n.

fiery adj. (**fierier**) **1** consisting of or like fire. **2** intense, spirited. □ **fierily** adv., **fieriness** n.

fiesta n. a festival in Spanish-speaking countries.

fife n. a small shrill flute.

fifteen adj. & n. one more than fourteen (15, XV). □ **fifteenth** adj. & n.

fifth adj. & n. the next after fourth. □ **fifthly** adv.

fifty adj. & n. five times ten (50, L). □ **fiftieth** adj. & n.

fifty-fifty adj. & adv. half-and-half, equally.

fig n. a tree with broad leaves and soft pear-shaped fruit; this fruit.

fig. abbr. figure.

fight v. (**fought**, **fighting**) struggle against, esp. in physical combat or war; contend; strive to obtain or accomplish something or to overcome. ● n. fighting; a battle, a contest, a struggle; a boxing match.

fighter n. one who fights; an aircraft designed for attacking others.

figment n. something that does not exist except in the imagination.

figurative adj. metaphorical. □ **figuratively** adv.

figure n. **1** a written symbol of a number; a value, an amount of money; (**figures**) arithmetic. **2** bodily shape; a representation of a person or animal. **3** a diagram; a geometric shape. ● v. **1** appear or be mentioned; form part of a plan etc. **2** work out by arithmetic or logic.

figured adj. with a woven pattern.

figurehead *n.* a carved image at the prow of a ship; a leader with only nominal power.

figure of speech *n.* an expression used for effect and not literally.

figurine *n.* a statuette.

filament *n.* a strand; a fine wire giving off light in an electric lamp.

filbert *n.* the nut of a cultivated hazel.

filch *v.* pilfer, steal.

file *n.* **1** a cover or box etc. for keeping documents; its contents. **2** a set of data in a computer. **3** a line of people or things one behind another. **4** a tool with a rough surface for smoothing things. ● *v.* **1** place (a document) in a file; place on record. **2** march in a long line. **3** shape or smooth (a surface) with a file.

filial *adj.* of or due from a son or daughter. □ **filially** *adv.*

filibuster *v.* delay the passage of a bill by making long speeches. ● *n.* a person who does this; this action.

filigree *n.* ornamental work of fine gold or silver wire.

filings *n.pl.* particles filed off.

Filipino *n. & adj.* (*pl.* **Filipinos**) (a native) of the Philippine Islands.

fill *v.* **1** make or become full; block. **2** occupy; appoint to (a vacant post). ● *n.* enough to fill a thing; enough to satisfy a person's appetite or desire. □ **fill in** complete; act as substitute. **fill out** enlarge; become enlarged or plump. **fill up** fill completely.

filler *n.* a thing or material used to fill a gap or increase bulk.

fillet *n.* a piece of boneless meat or fish. ● *v.* (**filleted**) remove bones from.

filling *n.* a substance used to fill a cavity etc.

filling station *n.* a place selling petrol to motorists.

fillip *n.* **1** a stimulus or incentive. **2** a quick blow with the finger.

filly *n.* a young female horse.

film *n.* **1** a thin layer. **2** a sheet or rolled strip of light-sensitive material for taking photographs; a motion picture. ● *v.* **1** make a film of. **2** cover or become covered with a thin layer.

filmstrip *n.* a series of transparencies in a strip for projection.

filmy *adj.* (**filmier**) thin and almost transparent.

filo *n.* pastry in very thin sheets.

Filofax *n.* (*trade mark*) a portable loose-leaf personal filing system.

filter *n.* a device or substance for holding back impurities in liquid or gas passing through it; a screen for absorbing or modifying light or electrical or sound waves; an arrangement for filtering traffic. ● *v.* pass through a filter, remove impurities in this way; pass gradually in or out; (of traffic) be allowed to pass while other traffic is held up.

filth *n.* disgusting dirt; obscenity. □ **filthy** *adj.*, **filthily** *adv.*, **filthiness** *n.*

filtrate *n.* a filtered liquid. ● *v.* filter. □ **filtration** *n.*

fin *n.* a thin projection from a fish's body, used for propelling and steering itself; a similar projection to improve the stability of aircraft etc.

final *adj.* at the end, coming last; conclusive. ● *n.* the last contest in a series; the last edition of a day's newspaper; (**finals**) final examinations. □ **finally** *adv.*

finale (fi-**nah**-li) *n.* the final section of a performance or musical composition.

finalist *n.* a competitor in a final.

finality n. the quality or fact of being final.

finalize v. (also **-ise**) bring to an end; put in final form. □ **finalization** n.

finance n. management of money; money resources. ● v. provide money for. □ **financial** adj., **financially** adv.

financier n. a person engaged in financing businesses.

finch n. a small bird.

find v. (**found**, **finding**) **1** discover; obtain. **2** (of a jury etc.) decide and declare. ● n. a discovery; a thing found. □ **find out** get information about; detect, discover. □ **finder** n.

fine n. a sum of money to be paid as a penalty. ● v. punish by a fine. ● adj. **1** of high quality or merit. **2** bright, free from rain. **3** slender, in small particles; delicate, subtle. **4** excellent. ● adv. finely. □ **finely** adv., **fineness** n.

finery n. showy clothes etc.

fines herbes (feenz **airb**) n.pl. mixed cooking herbs.

finesse n. delicate manipulation; tact.

finger n. each of the five parts extending from each hand; any of these other than the thumb; a finger-like object or part; a measure (about 20 mm) of alcohol in a glass. ● v. touch or feel with the fingers.

fingerboard n. a flat strip on a stringed instrument against which the strings are pressed with the fingers to produce different notes.

fingerprint n. an impression of the ridges on the pad of a finger.

finger-stall n. a sheath to cover an injured finger.

fingertip n. the tip of a finger.

finial n. an ornament at the apex of a gable, pinnacle, etc.

finicky adj. (also **finicking**) fiddly; over-particular.

finish v. bring or come to an end, complete; reach the end of a task or race etc; consume all of; put final touches to. ● n. the last stage; the point where a race etc. ends; a completed state. □ **finisher** n.

finite adj. limited.

Finn n. a native of Finland.

Finnish n. & adj. (the language) of Finland.

fiord var. of **fjord**.

fir n. an evergreen cone-bearing tree.

fire n. **1** combustion; destructive burning; a heating device with a flame or glow. **2** the firing of guns. **3** angry or excited feeling. ● v. **1** send a bullet or shell from (a gun); launch (a missile). **2** dismiss from a job. **3** set fire to; catch fire. **4** bake (pottery etc.). **5** excite.

firearm n. a gun, pistol, etc.

firebrand n. a person who causes trouble.

firebreak n. an open space as an obstacle to the spread of fire.

fire brigade n. an organized body of people employed to extinguish fires.

firecracker n. (Amer.) an explosive firework.

firedamp n. an explosive mixture of methane and air in mines.

firedog n. an iron support for logs in a fireplace.

fire engine n. a vehicle with equipment for putting out fires.

fire escape n. a special staircase or apparatus for escape from a burning building.

firefly n. a phosphorescent beetle.

fireman n. (pl. **-men**) (also **firefighter**) a member of a fire brigade.

fireplace n. a recess with a chimney for a domestic fire.

fireside n. the area round a fireplace; this as the centre of a home.

firework n. a device containing chemicals that burn or explode spectacularly.

firing squad n. a group ordered to shoot a condemned person.

firkin n. a small barrel.

firm adj. not yielding when pressed or pushed; steady, not shaking; securely established; resolute. ● adv. firmly. ● v. make or become firm. □ a business company.

firmament n. the sky with its clouds and stars.

first adj. coming before all others in time, order, or importance. ● n. the first thing or occurrence; the first day of a month. ● adv. before all others or another; for the first time. □ at first at the beginning.

first aid n. basic treatment given for an injury etc. before a doctor arrives.

first-class adj. & adv. of the best quality; in the best category of accommodation.

first cousin see cousin.

first-hand adv. & adj. directly from the original source.

firstly adv. first.

first name n. a personal name.

first-rate adj. excellent.

firth n. (also **frith**) an estuary or narrow inlet of the sea in Scotland.

fiscal adj. of public revenue.

fish n. (pl. **fish** or **fishes**) a cold-blooded vertebrate living wholly in water; its flesh as food. ● v. 1 try to catch fish (from). 2 make a search for by reaching into something.

fishery n. an area of sea where fishing is done; the business of fishing.

fishmeal n. ground dried fish used as a fertilizer or animal feed.

fishmonger n. a shopkeeper who sells fish.

fishnet adj. (of fabric) of a coarse open mesh.

fishy adj. (**fishier**) 1 like fish. 2 causing disbelief or suspicion.

fissile adj. tending to split; capable of undergoing nuclear fission.

fission n. splitting (esp. of an atomic nucleus, with release of energy).

fissure n. a cleft.

fist n. a tightly closed hand.

fisticuffs n.pl. fighting with fists.

fistula n. a pipe-like ulcer; a pipe-like passage in the body.

fit adj. (**fitter**) 1 suitable; right and proper. 2 in good health. ● v. (**fitted**) 1 be or adjust to be the right shape and size for. 2 put into place. 3 make or be suitable or competent. ● n. 1 the way a thing fits. 2 a sudden attack of illness or its symptoms, or of convulsions or loss of consciousness; a short period of a feeling or activity. □ **fitness** n.

fitful adj. occurring in short periods not steadily. □ **fitfully** adv.

fitment n. a piece of fixed furniture.

fitter n. 1 a person who supervises the fitting of clothes. 2 a mechanic.

fitting adj. right and proper. ● n. 1 the process of having a garment fitted. 2 (**fittings**) fixtures and fitments.

five adj. & n. one more than four (5, V).

fiver n. (colloquial) a five-pound note.

fix v. 1 make firm, stable, or permanent. 2 direct (the eyes or attention) steadily. 3 establish, specify. 4 repair. 5 (colloquial) deal with or arrange. ● n. 1 an awkward situation. 2 a position determined by taking bearings. □ **fix up** organize; provide for.

fixated adj. having an obsession.

fixation n. **1** fixing. **2** an obsession.

fixative n. a substance for keeping things in position, or preventing fading or evaporation.

fixedly adv. intently.

fixity n. a fixed state, stability, permanence.

fixture n. a thing fixed in position; a firmly established person or thing.

fizz v. hiss or splutter, esp. when gas escapes in bubbles from a liquid. ●n. this sound; a fizzing drink. □ **fizziness** n., **fizzy** adj.

fizzle v. hiss or splutter feebly. □ **fizzle out** end feebly.

fjord (fi-ord) n. (also **fiord**) a narrow inlet of sea between cliffs, esp. in Norway.

fl. abbr. **1** floruit. **2** fluid.

flab n. (colloquial) flabbiness, fat.

flabbergast v. (colloquial) astound.

flabby adj. (**flabbier**) fat and limp, not firm. □ **flabbiness** n.

flaccid adj. hanging loose or wrinkled, not firm. □ **flaccidly** adv., **flaccidity** n.

flag n. a piece of cloth attached by one edge to a staff or rope as a signal or symbol; a similarly shaped device. ●v. (**flagged**) **1** mark or signal (as) with a flag. **2** droop; lose vigour.

flag day n. a day on which small emblems are sold for a charity.

flagellate v. whip, flog. □ **flagellant** n., **flagellation** n.

flageolet n. a small wind instrument.

flagged adj. paved with flagstones.

flagon n. a large bottle for wine or cider; a vessel with a handle, lip, and lid for serving wine.

flagrant (flay-grănt) adj. (of an offence or offender) very bad and obvious. □ **flagrantly** adv.

flagship n. an admiral's ship; a principal vessel, shop, product, etc.

flagstone n. a large paving stone.

flail n. an implement formerly used for threshing grain. ●v. thrash or swing about wildly.

flair n. natural ability.

flak n. **1** anti-aircraft shells. **2** adverse criticism.

flake n. a small thin piece. ●v. come off in flakes. □ **flake out** (colloquial) faint, fall asleep from exhaustion. □ **flaky** adj., **flakiness** n.

flambé adj. (of food) served covered with flaming alcohol.

flamboyant adj. showy in appearance or manner. □ **flamboyantly** adv., **flamboyance** n.

flame n. a bright tongue-shaped portion of gas burning visibly. ●v. burn with flames; become bright red. □ **old flame** (colloquial) a former sweetheart.

flamenco n. (pl. **flamencos**) a Spanish style of singing and dancing.

flamingo n. (pl. **flamingos** or **flamingoes**) a wading bird with long legs and pink feathers.

flammable adj. able to be set on fire. □ **flammability** n.

■ **Usage** Flammable is often used because inflammable could be taken to mean 'not flammable'. The negative of flammable is non-flammable.

flan n. an open pastry or sponge case with filling.

flange n. a projecting rim. □ **flanged** adj.

flank n. a side, esp. of the body between ribs and hip. ●v. place or be at the side of.

flannel n. **1** woollen fabric. **2** a facecloth. **3** (flannels) trousers of flannel. **4** (slang) nonsense, flattery. ●v. (**flannelled**; Amer. **flanneled**) (slang) flatter.

flannelette n. a heavy brushed cotton fabric.

flap v. (**flapped**) **1** sway or move up and down with a sharp sound. **2** (*colloquial*) show agitation. ● n. **1** the act or sound of flapping. **2** a hanging or hinged piece. **3** (*colloquial*) agitation.

flapjack n. a biscuit made with oats.

flare v. **1** blaze suddenly; burst into activity or anger. **2** widen outwards. ● n. **1** a sudden blaze; device producing flame as a signal or illumination. **2** a flared shape.

flash v. give out a sudden bright light; show suddenly or ostentatiously; come suddenly into sight or mind; move rapidly; cause to shine briefly. ● n. a sudden burst of flame or light; a sudden show of wit or feeling; a very brief time; a brief news item; a device producing a brief bright light in photography. ● adj. (*colloquial*) flashy.

flashback n. a change of scene in a story or film to an earlier period.

flasher n. (*slang*) a man who indecently exposes himself.

flash flood n. a sudden destructive flood.

flashing n. a strip of metal covering a joint in a roof etc.

flashlight n. an electric torch.

flashpoint n. the temperature at which a vapour ignites.

flashy adj. (**flashier**) showy, gaudy. □ **flashily** adv., **flashiness** n.

flask n. a narrow-necked bottle; a vacuum flask.

flat adj. (**flatter**) **1** horizontal, level; lying at full length. **2** absolute. **3** monotonous; dejected. **4** having lost effervescence or power to generate electric current. **5** below the correct pitch in music. ● adv. **1** in a flat manner. **2** (*colloquial*) exactly.

● n. **1** a flat surface, level ground. **2** a set of rooms on one floor, used as a residence. **3** (*Music*) (a sign indicating) a note lowered by a semitone. □ **flat out** at top speed; with maximum effort.

flatfish n. a sea fish with a flattened body and both eyes on one side.

flatlet n. a small flat.

flatmate n. a person with whom one shares a flat.

flatten v. make or become flat.

flatter v. compliment insincerely; exaggerate the good looks of. □ **flatterer** n., **flattery** n.

flatulent adj. causing or suffering from formation of gas in the digestive tract. □ **flatulence** n.

flaunt v. display proudly or ostentatiously; show off.

■ **Usage** Flaunt is often confused with *flout*, which means 'to disobey contemptuously'.

flautist n. a flute-player.

flavour (*Amer.* **flavor**) n. a distinctive taste; a special characteristic. ● v. give flavour to. □ **flavourless** adj.

flavouring n. (*Amer.* **flavoring**) a substance used to give flavour to food.

flaw n. an imperfection. ● v. spoil with a flaw. □ **flawless** adj.

flax n. a blue-flowered plant; a textile fibre from its stem.

flaxen adj. made of flax; pale yellow like dressed flax.

flay v. strip off the skin or hide of; criticize severely.

flea n. a small jumping bloodsucking insect.

flea market n. a market for second-hand goods.

fleck n. a very small mark; a speck. ● v. mark with flecks.

fled see **flee**.

fledged *adj.* (of a young bird) with fully grown wing-feathers, able to fly; (of a person) fully trained.

fledgeling *n.* a bird just fledged.

flee *v.* (**fled, fleeing**) run or hurry away (from).

fleece *n.* a sheep's woolly hair. ● *v.* rob by trickery. □ **fleecy** *adj.*

fleet *n.* a navy; ships sailing together; vehicles or aircraft under one command or ownership. ● *adj.* moving swiftly, nimble. □ **fleetly** *adv.*, **fleetness** *n.*

fleeting *adj.* passing quickly, brief.

flesh *n.* **1** the soft substance of animal bodies; the body as opposed to the mind or soul. **2** the pulpy part of fruits and vegetables. □ **flesh and blood** human nature; one's relatives.

fleshy *adj.* (**fleshier**) of or like flesh; having much flesh, plump, pulpy.

fleur-de-lis (fler dĕ lee) *n.* (also **fleur-de-lys**) (*pl.* **fleurs-de-lis**) a heraldic design of a lily with three petals.

flew *see* **fly**.

flex *n.* a flexible insulated wire for carrying electric current. ● *v.* bend; move (a muscle) so that it bends a joint. □ **flexion** *n.*

flexible *adj.* able to bend easily; adaptable, able to be changed. □ **flexibly** *adv.*, **flexibility** *n.*

flexitime *n.* a system of flexible working hours.

flibbertigibbet *n.* a gossiping or frivolous person.

flick *n.* a quick light blow or stroke. ● *v.* move, strike, or remove with a flick.

flicker *v.* burn or shine unsteadily; occur briefly; quiver. ● *n.* a flickering light or movement; a brief occurrence.

flick knife *n.* a knife with a blade that springs out.

flier var. of **flyer**.

flight *n.* **1** flying; the movement or path of a thing through the air; a journey by air; a group of birds or aircraft. **2** a series of stairs. **3** feathers etc. on a dart or arrow. **4** fleeing.

flight deck *n.* **1** the cockpit of a large aircraft. **2** the deck of an aircraft carrier.

flightless *adj.* unable to fly.

flight recorder *n.* an electronic device in an aircraft recording details of its flight.

flighty *adj.* (**flightier**) frivolous. □ **flightily** *adv.*, **flightiness** *n.*

flimsy *adj.* (**flimsier**) light and thin; fragile; unconvincing. □ **flimsily** *adv.*, **flimsiness** *n.*

flinch *v.* draw back in fear, wince; shrink from, avoid.

fling *v.* (**flung, flinging**) move or throw violently or hurriedly. ● *n.* a spell of indulgence in pleasure.

flint *n.* very hard stone; a piece of a hard alloy producing sparks when struck.

flintlock *n.* an old type of gun discharged by a spark from a flint.

flip *v.* (**flipped**) flick; toss with a sharp movement. ● *n.* an action of flipping; (*colloquial*) a short flight, a quick tour. ● *adj.* (*colloquial*) glib, flippant.

flippant *adj.* not showing proper seriousness. □ **flippantly** *adv.*, **flippancy** *n.*

flipper *n.* a sea animal's limb used in swimming; a large flat rubber attachment to the foot for underwater swimming.

flirt *v.* behave in a frivolously amorous way; toy (with an idea etc.). ● *n.* a person who flirts. □ **flirtation** *n.*

flirtatious *adj.* flirting; fond of flirting. □ **flirtatiously** *adv.*

flit *v.* (**flitted**) fly or move lightly and quickly; disappear stealthily. ● *n.* an act of flitting.

flitch n. a side of bacon.

flitter v. flit about.

float v. **1** rest or drift on the surface of liquid. **2** have or allow (currency) to have a variable rate of exchange. **3** start (a company or scheme). ● n. **1** a thing designed to float on liquid. **2** money for minor expenditure or giving change.

floatation var. of **flotation**.

flocculent adj. like tufts of wool.

flock n. **1** a number of animals or birds together; a large number of people, a congregation. **2** a tuft of wool or cotton; wool or cotton waste as stuffing. ● v. gather or go in a group.

floe n. a sheet of floating ice.

flog v. (**flogged**) **1** beat severely. **2** (slang) sell. □ **flogging** n.

flood n. an overflow of water on a place usually dry; a great outpouring; the inflow of the tide. ● v. cover or fill with a flood; overflow; come in great quantities.

floodlight n. a lamp producing a broad bright beam. ● v. (**floodlit, floodlighting**) illuminate with this.

floor n. **1** the lower surface of a room; a storey. **2** the right to speak in an assembly. ● v. **1** provide with a floor. **2** knock down; baffle.

flooring n. material for a floor.

floor show n. a cabaret.

floozie n. (also **floozy**) (colloquial) a disreputable woman.

flop v. (**flopped**) **1** hang or fall heavily and loosely. **2** (slang) be a failure. ● n. **1** a flopping movement or sound. **2** (slang) failure.

floppy adj. (**floppier**) tending to flop, not firm.

floppy disk n. a magnetic disk for storing machine-readable data.

flora n. the plants of an area or period.

floral adj. of flowers.

floret n. each of the small flowers of a composite flower.

floribunda n. a rose etc. with dense clusters of small flowers.

florid adj. **1** ornate. **2** ruddy. □ **floridity** n.

florin n. a former British coin worth two shillings; a Dutch guilder.

florist n. a person who sells flowers.

floruit n. & v. (the period at which a person) was alive and working.

floss n. **1** a mass of silky fibres. **2** dental floss. □ **flossy** adj.

flotation n. (also **floatation**) floating; the launching of a commercial venture.

flotilla n. a small fleet; a fleet of small ships.

flotsam n. floating wreckage. □ **flotsam and jetsam** odds and ends.

flounce v. go in an impatient annoyed manner. ● n. **1** a flouncing movement. **2** a deep frill. □ **flounced** adj.

flounder v. move clumsily, as in mud; become confused when trying to do something. ● n. a small flatfish.

flour n. fine powder made from grain, used in cooking. ● v. cover with flour. □ **floury** adj.

flourish v. **1** grow vigorously; prosper, be successful. **2** wave dramatically. ● n. a dramatic gesture; an ornamental curve; a fanfare.

flout v. disobey (a law etc.) contemptuously.

■ Usage Flout is often confused with flaunt, which means 'to display proudly or show off'.

flow v. glide along as a stream; proceed evenly, hang loosely; gush out. ● n. a flowing movement or mass; an amount flowing; the inflow of the tide.

flow chart *n.* a diagram showing the sequence of events in a process.

flower *n.* **1** the part of a plant where fruit or seed develops; a plant grown for this. **2** the best part. ● *v.* produce flowers.

flowered *adj.* ornamented with a design of flowers.

flowerpot *n.* a pot in which plants are grown.

flowery *adj.* **1** full of flowers. **2** full of ornamental phrases.

flown *see* fly.

flu *n.* (*colloquial*) influenza.

fluctuate *v.* vary irregularly. □ **fluctuation** *n.*

flue *n.* a smoke-duct in a chimney; a channel for conveying heat.

fluent *adj.* speaking or spoken smoothly and readily. □ **fluently** *adv.*, **fluency** *n.*

fluff *n.* a soft mass of fibres or down. ● *v.* **1** shake into a soft mass. **2** (*colloquial*) bungle, make a mistake. □ **fluffy** *adj.*, **fluffiness** *n.*

fluid *adj.* consisting of particles that move freely among themselves; not stable. ● *n.* a fluid substance. □ **fluidity** *n.*

fluid ounce *n.* one-twentieth (in the USA, one-sixteenth) of a pint (about 28 ml, in the USA 35 ml).

fluke *n.* **1** a lucky accident. **2** the barbed arm of an anchor etc.; a lobe of a whale's tail. **3** a flatfish. **4** a flat parasitic worm.

flummery *n.* **1** a sweet milk pudding. **2** empty talk.

flummox *v.* (*colloquial*) baffle, bewilder.

flung *see* fling.

flunk *v.* (*colloquial*) fail.

flunkey *n.* (also **flunky**) (*pl.* **flunkeys** or **flunkies**) a liveried servant; a person who does menial work.

fluoresce *v.* be or become fluorescent.

fluorescent *adj.* taking in radiations and sending them out as light. □ **fluorescence** *n.*

fluoridate *v.* add fluoride to (a water supply). □ **fluoridation** *n.*

fluoride *n.* a compound of fluorine with metal.

fluorine *n.* a chemical element (symbol F), a pungent corrosive gas.

fluorspar *n.* calcium fluoride as a mineral.

flurry *n.* a short rush of wind, rain, or snow; a commotion; nervous agitation. ● *v.* fluster.

flush *v.* **1** become red in the face. **2** cleanse or dispose of with a flow of water. **3** drive out from cover. ● *n.* **1** a blush. **2** a rush of emotion. **3** a rush of water. ● *adj.* **1** level, in the same plane. **2** (*colloquial*) well supplied with money.

fluster *v.* make nervous or confused. ● *n.* a flustered state.

flute *n.* **1** a wind instrument consisting of a pipe with holes along it and a mouth-hole at the side. **2** an ornamental groove.

flutter *v.* move wings hurriedly; wave or flap quickly; (of the heart) beat irregularly. ● *n.* a fluttering movement or beat; nervous excitement; a stir.

fluvial *adj.* of or found in rivers.

flux *n.* **1** a flow. **2** a continuous succession of changes. **3** a substance mixed with metal etc. to assist fusion.

fly *v.* (**flew, flown, flying**) **1** move through the air on wings or in an aircraft; control the flight of. **2** display (a flag). **3** go quickly; flee. ● *n.* **1** flying. **2** a two-winged insect. **3** (*flies*) a fastening down the front of trousers. □ **with flying colours** with great credit or success.

flyblown *adj.* tainted by flies' eggs.

flycatcher *n.* a bird that catches flying insects.

flyer n. (also **flier**) **1** a thing that flies; an airman or airwoman. **2** a fast animal or vehicle. **3** a small handbill.

flying adj. able to fly.

flying buttress n. a buttress based on a separate structure, usu. forming an arch.

flying fish n. a tropical fish with winglike fins for gliding through the air.

flying fox n. a large fruit-eating bat.

flying saucer n. an unidentified object reported as seen in the sky.

flying squad n. a group of police etc. organized for rapid movement.

flyleaf n. (pl. **flyleaves**) a blank leaf at the beginning or end of a book.

flyover n. a bridge carrying one road or railway over another.

fly-post v. display (posters etc.) in unauthorized places.

flysheet n. an outer cover for a tent.

fly-tip v. (**fly-tipped**) dump (waste) illegally.

flyweight n. a weight below bantamweight, in amateur boxing between 48 and 51 kg.

flywheel n. a heavy wheel revolving on a shaft to regulate machinery.

FM abbr. frequency modulation.

Fm symb. fermium.

foal n. the young of a horse or related animal. ● v. give birth to a foal.

foam n. **1** a collection of small bubbles. **2** spongy rubber or plastic. ● v. form foam. □ **foamy** adj.

fob n. an ornament hanging from a watch chain; a tab on a key-ring. □ **fob off** (**fobbed**) palm off; get (a person) to accept something inferior.

focal adj. of or at a focus.

fo'c's'le var. of **forecastle**.

focus n. (pl. **focuses** or **foci**) **1** the point where rays meet. **2** the distance at which an object is most clearly seen; an adjustment on a lens to produce a sharp image. **3** the centre of activity or interest. ● v. (**focused** or **focussed**) **1** adjust the focus of; bring into focus. **2** concentrate.

fodder n. food for animals.

foe n. an enemy.

foetid var. of **fetid**.

foetus (fee-tŭs) n. (Amer. **fetus**) (pl. **foetuses**) a developed embryo in a womb or egg. □ **foetal** adj.

fog n. thick mist. ● v. (**fogged**) cover or become covered with fog or condensed vapour. □ **foggy** adj., **fogginess** n.

fogey n. (also **fogy**) (pl. **fogeys** or **fogies**) an old-fashioned person.

foghorn n. a deep-sounding instrument for warning ships in fog.

foible n. a harmless peculiarity in a person's character.

foil n. **1** metal in a paper-thin sheet. **2** a person or thing emphasizing another's qualities by contrast. **3** a long thin sword with a button on the point. ● v. thwart, frustrate.

foist v. cause a person to accept (an inferior or unwelcome thing).

fold v. **1** bend so that one part lies on another. **2** clasp; envelop. **3** cease to function. **4** enclose (sheep) in a fold. ● n. **1** a folded part; a line or hollow made by folding. **2** an enclosure for sheep.

folder n. a folding cover for loose papers; a leaflet.

foliage n. leaves.

foliate adj. having leaves; leaf-like. □ **foliation** n.

folio n. (pl. **folios**) a folded sheet of paper making two leaves of a book; the page-number of a book.

folk n. (pl. **folk** or **folks**) people; one's relatives. ● adj. (of music, song, etc.) in the traditional style of a country.

folklore n. the traditional beliefs and tales of a community.

folksy adj. (**folksier**) 1 informal and friendly. 2 of folk culture.

follicle n. a very small cavity containing a hair-root. □ **follicular** adj.

follow v. 1 go or come after; go along (a road etc.). 2 accept the ideas of; take an interest in the progress of; grasp the meaning of. 3 be a natural consequence of. **follow suit** follow a person's example. **follow up** pursue; supplement. □ **follower** n.

following n. a body of believers or supporters. ● adj. about to be mentioned, next. ● prep. as a sequel to.

folly n. 1 foolishness, a foolish act. 2 an ornamental building.

foment v. stir up (trouble). □ **fomentation** n.

fond adj. 1 affectionate; doting. 2 (of hope) unlikely to be fulfilled. □ **fondly** adv., **fondness** n.

fondant n. a soft sugary sweet.

fondle v. handle lovingly.

fondue n. a dish of flavoured melted cheese.

font n. 1 a basin in a church, holding water for baptism. 2 (also **fount**) one size and style of printing type.

fontanelle n. (Amer. **fontanel**) a soft spot where the bones of an infant's skull have not yet grown together.

food n. a substance (esp. solid) that can be taken into the body of an animal or plant to maintain its life.

foodie n. a gourmet.

food processor n. a machine for chopping and mixing food.

foodstuff n. a substance used as food.

fool n. 1 a foolish person. 2 a creamy fruit-flavoured pudding. ● v. joke, tease; play about idly; trick.

foolery n. foolish acts.

foolhardy adj. taking foolish risks.

foolish adj. lacking good sense or judgement; ridiculous. □ **foolishly** adv., **foolishness** n.

foolproof adj. simple and easy to use; unable to go wrong.

foolscap n. a large size of paper.

foot n. (pl. **feet**) 1 the part of the leg below the ankle; a lower part or end. 2 a measure of length, = 12 inches (30.48 cm). 3 a unit of rhythm in verse. ● v. 1 walk. 2 be the one to pay (a bill).

footage n. 1 a length measured in feet. 2 an amount of film.

foot-and-mouth disease n. a contagious virus disease of cattle.

football n. a large round or elliptical inflated ball; a game played with this. □ **footballer** n.

football pools n.pl. a form of gambling on the results of football matches.

footfall n. the sound of footsteps.

foothills n.pl. low hills near the bottom of a mountain or range.

foothold n. a place just wide enough for one's foot; a small but secure position gained.

footing n. 1 a foothold; balance. 2 status; conditions.

footlights n.pl. a row of lights along the front of a stage floor.

footling adj. (colloquial) trivial.

footloose adj. independent, without responsibilities.

footman n. (pl. **-men**) a manservant, usu. in livery.

footnote n. a note printed at the bottom of a page.

footpath n. a path for pedestrians, a pavement.

footplate *n.* the platform for the crew of a locomotive.

footprint *n.* an impression left by a foot or shoe.

footsie *n.* (*colloquial*) flirtatious touching of another's feet with one's own.

footsore *adj.* with feet sore from walking.

footstep *n.* a step; the sound of this.

footstool *n.* a stool for resting the feet on while sitting.

footwear *n.* shoes, socks, etc.

footwork *n.* a manner of moving or using the feet in sports etc.

fop *n.* an affectedly fashionable man; a dandy. ● **foppery** *n.*, **foppish** *adj.*

for *prep.* **1** in place of. **2** as the price or penalty of. **3** in defence or favour of. **4** with a view to. **5** in the direction of. **6** intended to be received or used by. **7** because of. **8** during. ● *conj.* because.

forage *v.* go searching; rummage. ● *n.* **1** foraging. **2** food for horses and cattle.

foray *n.* a sudden attack, a raid. ● *v.* make a foray.

forbade *see* **forbid**.

forbear *v.* (**forbore, forborne, forbearing**) refrain (from).

forbearance *n.* patience, tolerance.

forbearing *adj.* patient, tolerant.

forbid *v.* (**forbade, forbidden, forbidding**) order not to; refuse to allow.

forbidding *adj.* having an uninviting appearance, stern.

force *n.* **1** strength; intense effort. **2** an influence tending to cause movement. **3** a body of troops or police; an organized or available group. **4** compulsion; effectiveness. ● *v.* use force upon, esp. in order to get or do something; break open by force; strain to the utmost, overstrain; impose; produce by effort.

forceful *adj.* powerful and vigorous. □ **forcefully** *adv.*, **forcefulness** *n.*

forcemeat *n.* finely chopped seasoned meat used as stuffing.

forceps *n.* (*pl.* **forceps**) pincers used in surgery etc.

forcible *adj.* done by force. □ **forcibly** *adv.*

ford *n.* a shallow place where a stream may be crossed by wading or driving through. ● *v.* cross in this way.

fore *adj.* & *adv.* in, at, or towards the front. ● *n.* the fore part. □ **to the fore** in front; conspicuous.

forearm *n.* (**for-arm**) the arm from the elbow downwards. ● *v.* (for-**arm**) arm or prepare in advance against possible danger.

forebears *n.pl.* ancestors.

foreboding *n.* a feeling that trouble is coming.

forecast *v.* (**forecast, forecasting**) tell in advance (what is likely to happen). ● *n.* a statement that does this. □ **forecaster** *n.*

forecastle (fohk-**sŭl**) *n.* (also **fo'c's'le**) the forward part of certain ships.

foreclose *v.* take possession of property when a loan secured on it is not repaid. □ **foreclosure** *n.*

forecourt *n.* an enclosed space in front of a building.

forefathers *n.pl.* ancestors.

forefinger *n.* the finger next to the thumb.

forefoot *n.* (*pl.* **forefeet**) an animal's front foot.

forefront *n.* the very front.

foregather *v.* (also **forgather**) assemble.

forego *var.* of **forgo**.

foregoing *adj.* preceding.

foregone conclusion *n.* a predictable result.

foreground n. the part of a scene etc. that is nearest to the observer.

forehand n. a stroke played with the palm of the hand turned forwards. ● adj. of or made with this stroke. □ **forehanded** adj.

forehead n. the part of the face above the eyes.

foreign adj. of, from, or dealing with a country that is not one's own; not belonging naturally.

foreigner n. a person born in or coming from another country.

foreknowledge n. knowledge of a thing before it occurs.

foreleg n. an animal's front leg.

forelock n. a lock of hair just above the forehead.

foreman n. (pl. -men) a worker supervising others; the president and spokesman of a jury.

foremost adj. most advanced in position or rank; most important. ● adv. in the foremost position.

forename n. a first name.

forenoon n. (literary) the morning.

forensic adj. of or used in law courts.

forensic medicine n. medical knowledge used in police investigations etc.

foreplay n. stimulation preceding sexual intercourse.

forerunner n. a person or thing that comes in advance of another which it foreshadows.

foresee v. (foresaw, foreseen, foreseeing) be aware of or realize beforehand. ● **foreseeable** adj.

foreshadow v. be an advance sign of (a future event etc.).

foreshore n. the part of the shore between high and low water marks.

foreshorten v. show or portray with apparent shortening giving an effect of distance.

foresight n. the ability to foresee and prepare for future needs.

foreskin n. the fold of skin covering the end of the penis.

forest n. trees and undergrowth covering a large area.

forestall v. prevent or foil by taking action first.

forester n. an officer in charge of a forest or of growing timber.

forestry n. the science of planting and caring for forests.

foretaste n. an experience in advance of what is to come.

foretell v. (foretold, foretelling) forecast.

forethought n. careful thought and planning for the future.

forever adv. continually, without end.

forewarn v. warn beforehand.

foreword n. introductory remarks at the beginning of a book.

forfeit n. something that has to be paid or given up as a penalty. ● v. give or lose as a forfeit. ● adj. forfeited. □ **forfeiture** n.

forgather var. of **foregather**.

forgave see **forgive**.

forge n. a blacksmith's workshop; a furnace where metal is heated. ● v. **1** shape (metal) by heating and hammering. **2** make a fraudulent copy of. **3** advance by effort. □ **forger** n.

forgery n. forging; something forged.

forget v. (forgot, forgotten, forgetting) cease to remember or think about.

forgetful adj. tending to forget. □ **forgetfully** adv., **forgetfulness** n.

forget-me-not n. a plant with small blue flowers.

forgive v. (forgave, forgiven, forgiving) cease to feel angry or bitter towards or about. ● **forgivable** adj., **forgiveness** n.

forgo v. (also **forego**) (**forwent, forgone, forgoing**) give up; go without.

fork n. a pronged instrument or tool; a thing or part divided like this; each of its divisions. ● v. **1** lift or dig with a fork. **2** separate into two branches; follow one of these branches.

fork-lift truck n. a truck with a forked device for lifting and carrying loads.

forlorn adj. left alone and unhappy. □ **forlornly** adv.

forlorn hope n. the only faint hope left; a desperate enterprise.

form n. **1** shape, appearance; the way in which a thing exists. **2** a school class. **3** etiquette. **4** a document with blank spaces for details. **5** a bench. ● v. shape, produce; bring into existence, constitute; take shape; develop.

formal adj. **1** conforming to accepted rules or customs. **2** of form. **3** regular in design. □ **formally** adv.

formaldehyde n. a colourless gas used in solution as a preservative and disinfectant.

formalin n. a solution of formaldehyde in water, used as a preservative for biological specimens.

formalism n. strict adherence to outward form (as opposed to content).

formality n. being formal; a formal act, esp. one required by rules.

formalize v. (also **-ise**) make formal or official. □ **formalization** n.

format n. the shape and size of a book etc.; a style of arrangement. ● v. (**formatted**) arrange in a format.

formation n. forming; a thing formed; a particular arrangement.

formative adj. forming; of formation.

former adj. of an earlier period; mentioned first of two.

formerly adv. in former times.

Formica n. (trade mark) a hard heat-resistant plastic laminate.

formic acid n. a colourless acid in fluid emitted by ants.

formidable adj. inspiring fear or awe; difficult to do. □ **formidably** adv.

formula n. (pl. **formulae** or **formulas**) **1** symbols showing chemical constituents or a mathematical statement. **2** a fixed series of words for use on social or ceremonial occasions. **3** a list of ingredients. **4** a classification of a racing car. □ **formulaic** adj.

formulate v. express systematically. □ **formulation** n.

fornicate v. (old use) have sexual intercourse while unmarried. □ **fornication** n., **fornicator** n.

forsake v. (**forsook, forsaken, forsaking**) withdraw one's help or companionship from; abandon.

forsooth adv. (old use) indeed.

forswear v. (**forswore, forsworn, forswearing**) renounce on oath.

forsythia n. a shrub with yellow flowers.

fort n. a fortified place or building.

forte (for-tay) n. a person's strong point. ● adv. (Music) loudly.

forth adv. out; onwards. □ **back and forth** to and fro.

forthcoming adj. **1** about to occur or appear. **2** communicative.

forthright adj. frank, outspoken.

forthwith adv. immediately.

fortification n. fortifying; a defensive wall or building etc.

fortify v. **1** strengthen against attack. **2** increase the vigour, food value, or alcohol content of.

fortissimo *adv.* (*Music*) very loudly.

fortitude *n.* courage in bearing pain or trouble.

fortnight *n.* a period of two weeks.

fortnightly *adj. & adv.* (happening or appearing) once a fortnight.

Fortran *n.* a computer programming language used esp. for scientific work.

fortress *n.* a fortified building or town.

fortuitous *adj.* happening by chance. □ **fortuitously** *adv.*

fortunate *adj.* lucky. □ **fortunately** *adv.*

fortune *n.* 1 chance as a power in people's affairs; destiny. 2 prosperity; success; much wealth.

fortune-teller *n.* a person who claims to foretell future events in people's lives.

forty *adj. & n.* four times ten (40, XL). □ **fortieth** *adj. & n.*

forty winks *n.pl.* (*colloquial*) a nap.

forum *n.* a place or meeting where a public discussion is held.

forward *adj.* 1 directed towards the front or its line of motion. 2 having made more than normal progress. 3 presumptuous. ● *n.* an attacking player in football or hockey. ● *adv.* forwards; towards the future; in advance, ahead. ● *v.* 1 send on (a letter, goods) to a final destination. 2 advance (interests). □ **forwardness** *n.*

forwards *adv.* towards the front; with forward motion; so as to make progress; with the front foremost.

fosse *n.* a long fortification ditch.

fossil *n.* the hardened remains or traces of a prehistoric animal or plant. ● *adj.* 1 of or like a fossil. 2 (of fuel) extracted from the ground.

fossilize *v.* (also **-ise**) turn or be turned into a fossil. □ **fossilization** *n.*

foster *v.* 1 promote the growth of. 2 bring up (a child that is not one's own).

foster child *n.* a child brought up by parents other than its own.

foster home *n.* a home in which a foster child is reared.

foster parent *n.* a person who fosters a child.

fought *see* **fight**.

foul *adj.* 1 causing disgust. 2 against the rules of a game. ● *adv.* unfairly. ● *n.* an action that breaks rules. ● *v.* 1 make or become foul. 2 entangle or collide with; obstruct. 3 commit a foul against. □ **foully** *adv.*, **foulness** *n.*

found[1] 1 establish (an institution etc.); base. 2 melt or mould (metal or glass); make (an object) in this way. *See also* **find[1]**.

found[2] *see* **find**.

foundation *n.* 1 founding. 2 an institution or fund founded. 3 a base, a first layer. 4 an underlying principle.

founder *v.* stumble or fall; (of a ship) sink; fail completely. ● *n.* a person who has founded an institution etc.

foundling *n.* a deserted child of unknown parents.

foundry *n.* a workshop where metal or glass founding is done.

fount[1] *n.* 1 (*literary*) a fountain; a source. 2 var. of **font** (*sense* 2).

fountain *n.* a spring or jet of water; a structure provided for this; a source.

fountainhead *n.* a source.

fountain pen *n.* a pen that can be filled with a supply of ink.

four *adj. & n.* one more than three (4, IV).

fourfold *adj. & adv.* four times as much or as many.

four-poster n. a bed with four posts that support a canopy.

foursome n. a party of four people.

fourteen adj. & n. one more than thirteen (14, XIV). □ **fourteenth** adj. & n.

fourth adj. next after the third. ● n. a fourth thing, class, etc.; a quarter. □ **fourthly** adv.

four-wheel drive n. motive power acting on all four wheels of a vehicle.

fowl n. a bird kept to supply eggs and flesh for food.

fox n. **1** a wild animal of the dog family with a bushy tail; its fur. **2** a cunning person. ● v. deceive or puzzle by acting cunningly.

foxglove n. a tall plant with flowers like glove-fingers.

foxhole n. a small trench as a military shelter.

foxhound n. a hound bred to hunt foxes.

foxtrot n. a dance with slow and quick steps; music for this.

foyer (foi-yay) n. an entrance hall of a theatre, cinema, or hotel.

Fr symb. francium.

fracas (fra-kah) n. (pl. **fracas**) a noisy quarrel or disturbance.

fraction n. a number that is not a whole number; a small part or amount. □ **fractional** adj., **fractionally** adv.

fractious adj. irritable, peevish. □ **fractiously** adv., **fractiousness** n.

fracture n. a break, esp. of a bone. ● v. break.

fragile adj. easily broken or damaged; not strong. □ **fragility** n.

fragment n. (frag-mĕnt) a piece broken off something; an isolated part. ● v. (frag-ment) break into fragments. □ **fragmentation** n.

fragmentary adj. consisting of fragments.

fragrant adj. having a pleasant smell. □ **fragrance** n.

frail adj. not strong; physically weak. □ **frailty** n.

frame n. **1** a rigid structure supporting other parts. **2** an open case or border enclosing a picture or pane of glass etc. **3** a single exposure on a cinema film. ● v. **1** put or form a frame round. **2** construct. **3** (colloquial) arrange false evidence against. □ **frame of mind** a temporary state of mind.

framework n. a supporting frame.

franc n. a unit of money in France, Belgium, Switzerland, etc.

franchise n. **1** the right to vote in public elections. **2** authorization to sell a company's goods or services in a certain area. ● v. grant a franchise to.

francium n. a radioactive metallic element (symbol Fr).

Franco- comb. form French.

frangipani n. a fragrant tropical shrub.

frank adj. showing one's thoughts and feelings unmistakably. ● v. mark (a letter etc.) to show that postage has been paid. □ **frankly** adv., **frankness** n.

frankfurter n. a smoked sausage.

frankincense n. a sweet-smelling gum burnt as incense.

frantic adj. wildly excited or agitated. □ **frantically** adv.

fraternal adj. of a brother or brothers. □ **fraternally** adv.

fraternity n. brotherhood.

fraternize v. (also **-ise**) associate with others in a friendly way. □ **fraternization** n.

fratricide n. the killing of one's own brother or sister; a person who does this. □ **fratricidal** adj.

Frau (frow) n. the title of a German married woman.

fraud n. criminal deception; a dishonest trick; a person carrying this out. □ **fraudulence** n., **fraudulent** adj., **fraudulently** adv.

fraught adj. causing or suffering anxiety. □ **fraught with** filled with, involving.

Fräulein (froi-lyn) n. the title of a German unmarried woman.

fray v. make or become worn, esp. producing loose threads in woven material; strain (nerves or temper). ● n. a fight, a conflict.

frazzle n. (colloquial) an exhausted state.

freak n. an abnormal person or thing. □ **freak out** (colloquial) (cause to) hallucinate or become wildly excited. □ **freakish** adj., **freaky** adj.

freckle n. a light brown spot on the skin. ● v. spot or become spotted with freckles. □ **freckled** adj.

free adj. (**freer**) **1** not in the power of another, not a slave; having freedom. **2** not fixed. **3** without; not subject to. **4** without charge. **5** not occupied, not in use. **6** lavish. ● v. **1** make free. **2** rid of. **3** clear, disentangle. □ **free from** not containing.

freebie n. (colloquial) something provided free.

freebooter n. a pirate.

freedom n. **1** being free; independence. **2** frankness. **3** unrestricted use. **4** honorary citizenship.

free fall n. unrestricted falling under the force of gravity, esp. the part of a parachute descent before the parachute opens.

Freefone n. (also **Freephone**) a system whereby certain telephone calls may be made free of charge.

freehand adj. (of drawing) done by hand without ruler or compasses etc.

free hand n. the right to take what action one chooses.

freehold n. the holding of land or a house etc. in absolute ownership. ● adj. owned in this way. □ **freeholder** n.

free house n. a public house not controlled by one brewery.

freelance adj. & n. (a person) selling services to various employers.

freeloader n. (slang) a sponger.

Freemason n. a member of a fraternity for mutual help, with elaborate secret rituals. □ **Freemasonry** n.

Freephone var. of **Freefone**.

Freepost n. a system in which postage is paid by the addressee.

free-range adj. (of hens) allowed to range freely in search of food; (of eggs) from such hens.

freesia n. a fragrant flower.

freestyle n. a swimming race allowing any stroke.

freeway n. (Amer.) a motorway.

freewheel v. ride a bicycle without pedalling.

freeze v. (**froze**, **frozen**, **freezing**) **1** change from liquid to solid by extreme cold; be so cold that water turns to ice; preserve by refrigeration; chill or be chilled by extreme cold or fear. **2** make (assets) unable to be realized; hold (prices or wages) at a fixed level. **3** stop, stand very still. ● n. **1** a period of freezing weather. **2** the freezing of prices etc.

freeze-dry v. freeze and dry by evaporation of ice in a vacuum.

freezer n. a refrigerated container for preserving and storing food.

freight n. cargo; the transport of goods. ● v. load with freight; transport as freight.

freighter n. a ship or aircraft carrying mainly freight.

freightliner n. a train carrying goods in containers.

French adj. & n. (the language) of France. □ **Frenchman** n., **Frenchwoman** n.

French bread n. white bread in a long crisp loaf.

French dressing n. a salad dressing of oil and vinegar.

French fries n.pl. potato chips.

French horn n. a brass wind instrument with a coiled tube.

French leave n. absence without permission.

French polish v. polish (wood) with shellac polish.

French window n. a window reaching to the ground, used also as a door.

frenetic adj. in a state of frenzy. □ **frenetically** adv.

frenzy n. wild excitement or agitation; mental derangement. □ **frenzied** adj.

frequency n. frequent occurrence; the rate of repetition; (Physics) the number of cycles of a carrier wave per second; a band or group of these.

frequent adj. (free-kwent) happening or appearing often. ● v. (fri-kwent) go frequently to, be often in (a place). □ **frequently** adv.

fresco n. (pl. **frescos** or **frescoes**) a picture painted on a wall or ceiling before the plaster is dry.

fresh adj. **1** new, not stale or faded; not preserved by tinning or freezing etc. **2** not salty. **3** refreshing, vigorous. □ **freshly** adv., **freshness** n.

freshen v. make or become fresh. □ **freshener** n.

fresher n. (also **freshman**) a first-year university student.

freshwater adj. of fresh water, not of the sea.

fret v. (**fretted**) worry; vex; show anxiety. ● n. each of the ridges on the fingerboard of a guitar etc.

fretful adj. constantly worrying or crying. □ **fretfully** adv.

fretsaw n. a narrow saw used for fretwork.

fretwork n. woodwork cut in decorative patterns.

Fri. abbr. Friday.

friable adj. easily crumbled. □ **friability** n.

friar n. a member of certain religious orders of men.

friary n. a monastery of friars.

fricassee n. a dish of pieces of meat served in a thick sauce. ● v. make a fricassee of.

friction n. **1** rubbing; resistance of one surface to another that moves over it. **2** conflict of people who disagree. □ **frictional** adj.

Friday n. the day following Thursday.

fridge n. (colloquial) a refrigerator.

fried see **fry**.

friend n. a person (other than a relative or lover) with whom one is on terms of mutual affection; a helper, a sympathizer. □ **friendship** n.

friendly adj. (**friendlier**) like a friend; favourable. □ **friendliness** n.

frieze n. a band of decoration round a wall.

frigate n. a small fast naval ship.

fright n. **1** sudden great fear. **2** a ridiculous-looking person or thing.

frighten v. cause fright to; feel fright; drive or compel by fright.

frightened adj. afraid.

frightful adj. causing horror; ugly; (colloquial) extremely great or bad. □ **frightfully** adv., **frightfulness** n.

frigid adj. intensely cold; very cold in manner; unresponsive sexually. ◻ **frigidly** adv., **frigidity** n.

frill n. a gathered or pleated strip of trimming attached at one edge; an unnecessary extra. ◻ **frilled** adj., **frilly** adj.

fringe n. **1** an ornamental edging of hanging threads or cords; front hair cut short to hang over the forehead. **2** the edge of an area or group etc. ● v. edge.

fringe benefit n. one provided in addition to wages.

frippery n. showy unnecessary finery or ornament.

frisbee n. (trade mark) a plastic disc for skimming through the air as an outdoor game.

frisk v. **1** leap or skip playfully. **2** feel over or search (a person) for concealed weapons etc. ● n. a playful leap or skip.

frisky adj. (**friskier**) lively, playful. ◻ **friskily** adv., **friskiness** n.

frisson (free-son) n. a thrill.

frith var. of **firth**.

fritillary n. **1** a plant with speckled bell-shaped flowers. **2** a butterfly.

fritter v. waste little by little on trivial things. ● n. a fried batter-coated slice of fruit or meat etc.

frivolous adj. lacking a serious purpose, pleasure-loving. ◻ **frivolously** adv., **frivolity** n.

frizz v. form (hair) into a mass of small curls. ◻ **frizzy** adj., **frizziness** n.

frizzle v. fry until crisp.

frock n. a woman's or girl's dress.

frock coat n. a man's long-skirted coat not cut away in front.

frog n. a small amphibian with long web-footed hind legs. ◻ **frog in one's throat** (colloquial) hoarseness.

frogman n. (pl. **-men**) a swimmer with a rubber suit and oxygen supply for working under water.

frogmarch v. hustle (a person) forcibly, holding the arms.

frogspawn n. the eggs of a frog, surrounded by transparent jelly.

frolic v. (**frolicked**, **frolicking**) play about in a lively way. ● n. such play.

from prep. **1** having as the starting point, source, or cause. **2** as separated, distinguished, or unlike. ◻ **from time to time** at intervals of time.

fromage frais (from-azh fray) n. a smooth low-fat soft cheese.

frond n. a long leaf or leaflike part of a fern, palm tree, etc.

front n. **1** the side or part normally nearer or towards the spectator or line of motion. **2** an outward appearance; a cover for secret activities. **3** a promenade of a seaside resort. **4** a boundary between warm and cold airmasses. **5** a battle line. ● adj. of or at the front. ● v. **1** face, have the front towards. **2** (colloquial) serve as a cover for secret activities. ◻ **in front** at the front.

frontage n. the front of a building; land bordering this.

frontal adj. of or on the front.

frontbencher n. an MP entitled to sit on the front benches in Parliament, reserved for ministers and the Shadow Cabinet.

frontier n. a boundary between countries.

frontispiece n. an illustration opposite the title-page of a book.

front runner n. the contestant most likely to win.

frost n. a freezing weather condition; white frozen dew or vapour. ● v. **1** injure with frost. **2** cover with frost or frosting. **3** make (glass) opaque by roughening its surface.

frostbite n. injury to body tissues due to freezing. ◻ **frostbitten** adj.

frosting n. (Amer.) sugar icing.

frosty adj. (**frostier**) **1** cold with frost; covered with frost. **2** unfriendly. □ **frostily** adv., **frostiness** n.

froth n. & v. foam. □ **frothy** adj.

frown v. wrinkle one's brow in thought or disapproval. ● n. a frowning movement or look. □ **frown on** disapprove of.

frowsty adj. (**frowstier**) fusty, stuffy.

frowzy adj. (also **frowsy**) (**frowzier**) fusty; dingy.

froze, frozen see freeze.

fructose n. a sugar found in honey and fruit.

frugal adj. careful and economical; scanty; costing little. □ **frugally** adv., **frugality** n.

fruit n. **1** the seed-containing part of a plant; this used as food. **2** (**fruits**) the product of labour. ● v. produce or allow to produce fruit.

fruiterer n. a shopkeeper selling fruit.

fruitful adj. producing much fruit or good results. □ **fruitfully** adv., **fruitfulness** n.

fruition (froo-ish-ŏn) n. the fulfilment of hopes; the results of work.

fruitless adj. producing little or no result. □ **fruitlessly** adv., **fruitlessness** n.

fruit machine n. a coin-operated gambling machine.

fruity adj. (**fruitier**) like fruit in smell or taste. □ **fruitiness** n.

frump n. a dowdy woman. □ **frumpish** adj., **frumpy** adj.

frustrate v. prevent from achieving something or from being achieved. □ **frustration** n.

fry v. (**fries, fried, frying**) cook or be cooked in very hot fat. ● n. (pl. **fry**) young fish. ● **small fry** people of little importance. □ **fryer** n.

ft abbr. foot or feet (as a measure).

FT-SE abbr. Financial Times-Stock Exchange 100 share index.

fuchsia (few-shā) n. an ornamental plant with drooping flowers.

fuck (vulgar slang) v. have sexual intercourse (with). □ **fuck off** go away.

fucking (vulgar slang) adj. & adv. damned.

fuddle v. confuse or stupefy, esp. with alcoholic drink.

fuddy-duddy adj. & n. (colloquial) (a person who is) old-fashioned.

fudge n. a soft sweet made of milk, sugar, and butter. ● v. put together in a makeshift or dishonest way; fake.

fuel n. material burnt as a source of energy; something that increases anger etc. ● v. (**fuelled**; Amer. **fueled**) supply with fuel.

fug n. a stuffy atmosphere in a room etc. □ **fuggy** adj., **fugginess** n.

fugitive n. a person who is fleeing or escaping. ● adj. fleeing.

fugue (fewg) n. a musical composition using repeated themes in increasingly complex patterns.

fulcrum n. (pl. **fulcra** or **fulcrums**) the point of support on which a lever pivots.

fulfil v. (Amer. **fulfill**) (**fulfilled**) accomplish, carry out (a task); satisfy, do what is required by (a contract etc.). □ **fulfil oneself** develop and use one's abilities fully. □ **fulfilment** n.

full adj. holding or having as much as is possible; copious; complete; plump; made with material hanging in folds; (of tone) deep and mellow. ● adv. completely; exactly. □ **fully** adv., **fullness** n.

full-blooded adj. vigorous, hearty.

full-blown adj. fully developed.

full moon n. the moon with the whole disc illuminated.

full-scale adj. of actual size, not reduced.

full stop n. a dot used as a punctuation mark at the end of a sentence or abbreviation; a complete stop.

fulmar n. an Arctic seabird.

fulminate v. protest loudly and bitterly. □ **fulmination** n.

fulsome adj. excessive; insincere.

■ Usage Fulsome is sometimes wrongly used to mean 'generous', as in fulsome praise, or 'generous with praise', as in a fulsome tribute.

fumble v. touch or handle awkwardly; grope about.

fume n. pungent smoke or vapour. ● v. 1 emit fumes; subject to fumes. 2 seethe with anger.

fumigate v. disinfect with fumes. □ **fumigation** n., **fumigator** n.

fun n. light-hearted amusement. □ **make fun of** cause people to laugh at.

function n. 1 the special activity or purpose of a person or thing. 2 an important ceremony. 3 (in mathematics) a quantity whose value depends on varying values of others. ● v. perform a function; be in action.

functional adj. 1 of function(s). 2 practical, not decorative. 3 able to function. □ **functionally** adv.

functionary n. an official.

fund n. 1 a sum of money for a special purpose; (**funds**) money resources. 2 a stock, a supply. ● v. provide with money.

fundamental adj. basic; essential. ● n. a fundamental fact or principle. □ **fundamentally** adv.

fundamentalist n. a person who upholds a strict or literal interpretation of traditional religious beliefs. □ **fundamentalism** n.

fundholder n. a medical practice controlling its own budget.

funeral n. a ceremony of burial or cremation; a procession to this.

funerary adj. of or used for a burial or funeral.

funereal adj. suitable for a funeral; dismal, dark.

funfair n. a fair consisting of amusements and sideshows.

fungicide n. a substance that kills fungus. □ **fungicidal** adj.

fungus n. (pl. **fungi**) a plant without green colouring matter (e.g. a mushroom or mould). □ **fungal** adj., **fungous** adj.

funicular adj. (of a railway) operating by cable up and down a mountainside.

funk n. (slang) dance music with a heavy rhythmical beat. □ **funky** adj.

funnel n. 1 a tube with a wide top for pouring liquid into small openings. 2 a chimney on a steam engine or ship. ● v. (**funnelled**; Amer. **funneled**) move through a narrowing space.

funny adj. (**funnier**) 1 causing amusement. 2 puzzling, odd. □ **funnily** adv.

funny bone n. the part of the elbow where a very sensitive nerve passes.

fur n. 1 the short fine hair of certain animals; a skin with this used for clothing. 2 a coating, an incrustation. ● v. (**furred**) cover or become covered with fur.

furbish v. clean up; renovate.

furious adj. 1 full of anger. 2 violent, intense. □ **furiously** adv.

furl v. roll up and fasten.

furlong n. an eighth of a mile.

furlough (fer-loh) n. leave of absence.

furnace n. an enclosed fireplace for intense heating or smelting.

furnish v. 1 equip with furniture. 2 provide, supply.

furnishings n.pl. furniture and fitments etc.

furniture n. movable articles (e.g. chairs, beds) for use in a room.

furore (few-**ror**-i) n. (Amer. **furor**) an uproar of enthusiastic admiration or fury.

furrier n. a person who deals in furs or fur clothes.

furrow n. a long cut in the ground; a groove. ● v. make furrows in.

furry adj. (**furrier**) like fur; covered with fur. □ **furriness** n.

further adv. & adj. (also **farther**) 1 at or to a greater distance; more distant. 2 to a greater extent. 3 additional(ly). ● v. help the progress of. □ **furtherance** n.

further education n. education provided for persons above school age but usu. below degree level.

furthermore adv. moreover.

furthest (also **farthest**) adj. most distant. ● adv. at or to the greatest distance.

furtive adj. sly, stealthy. □ **furtively** adv., **furtiveness** n.

fury n. wild anger, rage; violence.

furze n. gorse.

fuse v. 1 blend (metals etc.); become blended; unite. 2 fit with a fuse; stop functioning through melting of a fuse. ● n. 1 a strip of wire placed in an electric circuit to melt and interrupt the current when the circuit is overloaded. 2 (also **fuze**) a length of easily burnt material for igniting a bomb or explosive.

fuselage n. the body of an aeroplane.

fusible adj. able to be fused. □ **fusibility** n.

fusilier n. a soldier of certain regiments.

fusillade n. a continuous discharge of firearms; an outburst of criticism etc.

fusion n. fusing; the union of atomic nuclei, with release of energy.

fuss n. unnecessary excitement or activity; a vigorous protest. ● v. make a fuss; agitate.

fussy adj. (**fussier**) 1 often fussing; fastidious. 2 with much unnecessary detail or decoration. □ **fussily** adv., **fussiness** n.

fustian n. 1 thick twilled cotton cloth. 2 pompous language.

fusty adj. (**fustier**) 1 smelling stale and stuffy. 2 old-fashioned. □ **fustiness** n.

futile adj. producing no result. □ **futilely** adv., **futility** n.

futon (**foo**-ton) n. a Japanese quilted mattress laid on the floor for use as a bed; this with a wooden frame convertible into a sofa.

future adj. belonging to the time after the present. ● n. 1 future time, events, or condition. 2 a prospect of success etc. □ **in future** from now on.

futuristic adj. looking suitable for the distant future, not traditional. □ **futuristically** adv.

fuze var. of **fuse** n. (sense 2).

fuzz n. 1 fluff, a fluffy or frizzy thing. 2 (slang) the police; a police officer.

fuzzy adj. (**fuzzier**) 1 like or covered with fuzz. 2 blurred, indistinct. □ **fuzzily** adv., **fuzziness** n.

Gg

G *abbr.* giga-; gauss.

g *abbr.* gram(s); gravity.

Ga *symb.* gallium.

gabardine *n.* (also **gaberdine**) a strong twilled fabric; a raincoat.

gabble *v.* talk quickly and indistinctly.

gable *n.* a triangular part of a wall, between sloping roofs. □ **gabled** *adj.*

gad *v.* (**gadded**) □ **gad about** go about idly in search of pleasure.

gadabout *n.* an idle pleasure-seeker.

gadfly *n.* a fly that bites cattle.

gadget *n.* a small mechanical device or tool. □ **gadgetry** *n.*

gadolinium *n.* a metallic element (symbol Gd).

Gaelic (gay-lik) *n.* the Celtic language of the Scots or Irish.

gaff *n.* a hooked stick for landing large fish.

gaffe *n.* a blunder.

gaffer *n.* (*colloquial*) an elderly man; a boss, a foreman.

gag *n.* **1** something put over a person's mouth to silence them; a surgical device to hold the mouth open. **2** a joke. ●*v.* (**gagged**) **1** put a gag on; deprive of freedom of speech. **2** retch.

gaga *adj.* senile; crazy.

gage Amer. sp. of **gauge**.

gaggle *n.* a flock (of geese); a disorderly group.

gaiety *n.* cheerfulness, bright appearance; merrymaking.

gaily *adv.* cheerfully; thoughtlessly.

gain *v.* **1** obtain, secure; acquire gradually; profit. **2** (of a clock) become fast. **3** reach. ●*n.* an increase in wealth or value.

gain on get nearer (a person pursued).

gainful *adj.* profitable. □ **gainfully** *adv.*

gainsay *v.* (**gainsaid, gainsaying**) deny, contradict.

gait *n.* a manner of walking or running.

gaiter *n.* a covering for the lower leg.

gala *n.* a fête; a sports gathering.

galaxy *n.* a system of stars, esp. (**the Galaxy**) the one containing the sun and the earth; the Milky Way. □ **galactic** *adj.*

gale *n.* a very strong wind; a noisy outburst.

gall *n.* **1** bile; bitterness of feeling; impudence. **2** a sore made by rubbing; an abnormal growth on a plant. ●*v.* rub and make sore; annoy.

gallant *adj.* brave; chivalrous; fine. □ **gallantly** *adv.*, **gallantry** *n.*

gall bladder *n.* an organ attached to the liver, storing bile.

galleon *n.* a large Spanish sailing ship of the 15th-17th centuries.

galleria *n.* a group of small shops under one roof.

gallery *n.* a building for showing works of art; a long room or passage, used for a special purpose; a balcony in a hall or theatre.

galley *n.* (*pl.* **galleys**) **1** an ancient ship, usu. rowed by slaves; a kitchen on a boat or aircraft. **2** (also **galley proof**) a printer's proof before division into pages.

Gallic *adj.* of ancient Gaul; French.

galling (gawl-ing) *adj.* annoying.

gallium *n.* a metallic element (symbol Ga).

gallivant *v.* go about looking for fun.

gallon *n.* a measure for liquids, = 8 pints (4.546 litres, or 3.785 litres in the USA).

gallop n. a horse's fastest pace; a ride at this pace. ● v. (**galloped**) go at a gallop; go fast.

gallows n. a framework with a noose for hanging criminals.

gallstone n. a small hard mass forming in the gall bladder.

Gallup poll n. = **opinion poll**.

galore adv. in plenty.

galosh n. a rubber overshoe.

galvanize v. (also **-ise**) **1** stimulate into activity. **2** coat with zinc.

galvanometer n. an instrument measuring electric current.

gambit n. an opening move; (Chess) an opening involving the sacrifice of a pawn.

gamble v. play games of chance for money; risk in hope of gain. ● n. gambling; a risky undertaking. □ **gambler** n.

gambol v. (**gambolled**; Amer. **gamboled**) jump about playfully.

game n. **1** a form of play or sport; a section of this as a scoring unit. **2** a scheme. **3** wild animals hunted for sport or food; their flesh as food. ● adj. brave; willing. □ **gamely** adv.

gamekeeper n. a person employed to protect and breed game.

gamelan n. a SE Asian percussion orchestra; a xylophone used in this.

gamesmanship n. the art of winning games by upsetting an opponent's confidence.

gamete n. a sexual cell.

gamine n. a girl with mischievous charm.

gamma n. the third letter of the Greek alphabet (Γ,γ); a third-class mark.

gammon n. cured or smoked ham.

gammy adj. lame, crippled.

gamut n. the whole range or scope.

gamy adj. smelling or tasting of game kept till it is high.

gander n. **1** a male goose. **2** (slang) a look.

gang n. a group of people. □ **gang up (on)** combine in a group (against a person).

gangling adj. tall and awkward.

ganglion n. (pl. **ganglia** or **ganglions**) a group of nerve cells; a cyst on a tendon.

gangplank n. a plank for walking to or from a boat.

gang rape n. the rape of one person by several men.

gangrene n. decay of body tissue. □ **gangrenous** adj.

gangster n. a member of a gang of violent criminals.

gangway n. a passage, esp. between rows of seats; a movable bridge from a ship to land.

gannet n. a large seabird.

gantry n. an overhead framework supporting railway signals, road signs, a crane etc.

gaol etc. var. of **jail** etc.

gap n. **1** a space, an opening; an interval. **2** a deficiency; a wide difference.

gape v. open one's mouth wide; be wide open.

garage n. a building for storing a vehicle; an establishment selling petrol or repairing and selling vehicles. ● v. put or keep in a garage.

garb n. clothing. ● v. clothe.

garbage n. rubbish.

garbled adj. (of a message or story) distorted or confused.

garden n. a piece of cultivated ground by a house; (**gardens**) ornamental public grounds. ● v. tend a garden. □ **gardener** n.

gardenia n. a shrub with fragrant white or yellow flowers.

gargantuan adj. gigantic.

gargle v. wash the throat with liquid held there by breathing out through it. ● n. liquid for this.

gargoyle n. a waterspout in the form of a grotesque carved face on a building.

garish adj. gaudy; too bright. □ **garishly** adv.

garland n. a wreath of flowers as a decoration. ● v. decorate with garlands.

garlic n. an onion-like plant. □ **garlicky** adj.

garment n. a piece of clothing.

garner v. store up, collect.

garnet n. a red semi-precious stone.

garnish v. decorate (food). ● n. something used for garnishing.

garret n. an attic.

garrison n. troops stationed in a town or fort; the building they occupy. ● v. occupy in this way.

garrotte n. (also **garotte**; *Amer.* **garrote**) a wire or a metal collar used to strangle a victim. ● v. strangle with this.

garrulous adj. talkative. □ **garrulously** adv., **garrulousness** n.

garter n. a band worn round the leg to keep up a stocking; **(the Garter)** the highest order of English knighthood.

gas n. (pl. **gases**) an airlike substance (not a solid or liquid); such a substance used as fuel; (*Amer.*) petrol. ● v. (**gassed**) **1** kill or overcome by poisonous gas. **2** (*colloquial*) talk lengthily.

gas chamber n. a room filled with poisonous gas to kill people.

gas mask n. a device worn over the face as protection against poisonous gas.

gaseous adj. of or like a gas.

gash n. a long deep cut. ● v. make a gash in.

gasify v. change into gas.

gasket n. a piece of rubber etc. sealing a joint between metal surfaces.

gasoline n. (*Amer.*) petrol.

gasp v. draw in breath sharply; speak breathlessly. ● n. a breath drawn in sharply.

gastric adj. of the stomach.

gastroenteritis n. inflammation of the stomach and intestines.

gastropod n. a mollusc, such as a snail, that moves by means of a ventral organ.

gate n. **1** a movable barrier in a wall or fence; an entrance. **2** the number of spectators paying to attend a sporting event; the amount of money taken.

-gate comb. form describing scandals relating to the Watergate scandal; *Irangate*.

gateau (gat-oh) n. (pl. **gateaux** or **gateaus**) a large rich cream cake.

gatecrash v. go to (a private party) uninvited. □ **gatecrasher** n.

gateway n. an opening or structure framing a gate; an entrance.

gather v. **1** bring or come together; collect; obtain gradually; draw together in folds. **2** understand; conclude.

gathering n. people assembled.

GATT abbr. General Agreement on Tariffs and Trade, an international trade treaty.

gauche (gohsh) adj. socially awkward. □ **gaucherie** n.

gaucho (gow-choh) n. (pl. **gauchos**) a South American cowboy.

gaudy adj. (**gaudier**) showy or bright, but tasteless. □ **gaudily** adv., **gaudiness** n.

gauge (gayj) n. (*Amer.* **gage**) a standard measure of contents or thickness; a measuring device; the distance between pairs of rails or wheels. ● v. measure; estimate.

gaunt adj. lean and haggard; grim, desolate.

gauntlet n. a glove with a long wide cuff. □ **run the gauntlet** be exposed to continuous criticism or risk.

gauss (gowss) n. (pl. **gauss**) a unit of magnetic flux density.

gauze n. thin transparent fabric; fine wire mesh. □ **gauzy** adj.

gave see **give**.

gavel n. a mallet used by an auctioneer or chairman to call for attention.

gawky adj. (**gawkier**) awkward and ungainly.

gay adj. **1** happy and full of fun; brightly coloured. **2** homosexual. ● n. a homosexual person. □ **gayness** n.

gaze v. look long and steadily. ● n. a long steady look.

gazebo (gă-zee-boh) n. (pl. **gazebos**) a turret or summer house with a wide view.

gazelle n. a small antelope.

gazette n. the title of certain newspapers or of journals containing public notices.

gazetteer n. an index of places, rivers, mountains, etc.

gazpacho n. a Spanish cold vegetable soup.

gazump v. raise the price of a property after accepting an offer from (a buyer).

GB abbr. Great Britain.

Gb abbr. gigabyte.

GBH abbr. grievous bodily harm.

GCE abbr. General Certificate of Education.

GCSE abbr. General Certificate of Secondary Education.

Gd abbr. gadolinium.

GDP abbr. gross domestic product.

Ge symb. germanium.

gear n. **1** equipment; apparatus. **2** a set of toothed wheels working together in machinery. ● v. **1** provide with gear(s). **2** adapt (to a purpose); prepare. □ **in gear** with the gear mechanism engaged.

gearbox n. a case enclosing a gear mechanism.

gecko n. (pl. **geckos**) a tropical lizard.

geese see **goose**.

Geiger counter (gy-ger) n. a device for detecting radioactivity.

geisha (gay-shă) n. a Japanese hostess trained to entertain men.

gel n. a jelly-like substance.

gelatin n. (also **gelatine**) a clear substance made by boiling bones. □ **gelatinous** adj.

geld v. castrate.

gelding n. a castrated horse.

gelignite n. an explosive containing nitroglycerine.

gem n. a precious stone; something of great beauty or excellence.

gender n. one's sex; a grammatical classification corresponding roughly to sex.

gene n. each of the factors controlling heredity, carried by a chromosome.

genealogy n. a list of ancestors; the study of family pedigrees. □ **genealogical** adj., **genealogist** n.

genera see **genus**.

general adj. **1** of or involving all or most parts, things, or people; not detailed or specific. **2** (in titles) chief. ● n. an army officer next below field marshal. □ **generally** adv.

general election n. an election of parliamentary representatives from the whole country.

generality n. being general; a statement without details.

generalize v. (also **-ise**) draw a general conclusion; speak in general terms. □ **generalization** n.

general practitioner n. a community doctor treating cases of all kinds.

generate v. bring into existence, produce.

generation n. **1** generating. **2** a single stage in descent or pedigree; all people born at about the same time; a period of about 30 years.

generator n. a machine converting mechanical energy into electricity.

generic adj. of a whole genus or group. □ **generically** adv.

generous adj. giving or given freely. □ **generously** adv., **generosity** n.

genesis n. a beginning or origin.

genetic adj. of genes or genetics. ● n. (**genetics**) the science of heredity. □ **genetically** adv., **geneticist** n.

genetic engineering n. manipulation of DNA to change hereditary features.

genetic fingerprinting n. identifying individuals by their DNA patterns.

genial adj. kindly and cheerful; pleasant. □ **genially** adv., **geniality** n.

genie n. (pl. **genii**) a magical spirit or goblin.

genital adj. of animal reproduction or sex organs. ● n.pl. (**genitals**) (also **genitalia**) the external sex organs.

genitive n. the grammatical case expressing possession or source.

genius n. (pl. **geniuses**) exceptionally great natural ability; a person with this.

genocide n. deliberate extermination of a race of people.

genre (zhahnr) n. a style of art or literature.

genteel adj. affectedly polite and refined. □ **genteelly** adv.

gentian (gen-shăn) n. an alpine plant with deep blue flowers.

Gentile n. a non-Jewish person.

gentility n. good manners and elegance.

gentle adj. mild, moderate, not rough or severe. □ **gently** adv., **gentleness** n.

gentleman n. (pl. **-men**) a well-mannered man; a man of good

social position. □ **gentlemanly** adj.

gentrify v. alter (an area) to conform to middle-class tastes.

gentry n.pl. people next below nobility.

genuflect v. bend the knee and lower the body, esp. in worship.

genuine adj. really what it is said to be. □ **genuinely** adv., **genuineness** n.

genus n. (pl. **genera**) a group of similar animals or plants, usu. containing several species; a kind.

geocentric adj. having the earth as a centre; as viewed from the earth's centre.

geode n. a cavity lined with crystals; a rock containing this.

geodesy n. the study of the earth's shape and size.

geography n. the study of the earth's physical features, climate, etc.; the features and arrangement of a place. □ **geographer** n., **geographical** adj., **geographically** adv.

geology n. the study or features of the earth's crust. □ **geological** adj., **geologically** adv., **geologist** n.

geometry n. the branch of mathematics dealing with lines, angles, surfaces, and solids. □ **geometric** adj., **geometrical** adj., **geometrically** adv.

georgette n. thin dress material.

Georgian adj. of the time of the Georges, kings of England, esp. 1714–1830.

geranium n. a cultivated flowering plant.

gerbil n. (also **jerbil**) a rodent with long hind legs.

geriatric adj. of old people. ● n. an old person; (**geriatrics**) the branch of medicine dealing with the diseases and care of old people.

germ n. **1** a micro-organism causing disease. **2** a portion of an organism capable of developing into a new organism; a basis from which a thing may develop.

German *adj. & n.* (a native, the language) of Germany.

German measles *n.* (also **rubella**) a disease like mild measles.

germane *adj.* relevant.

Germanic *adj.* having German characteristics; of the Scandinavians, Anglo-Saxons, or Germans.

germanium *n.* a semi-metallic element (symbol Ge).

germinate *v.* begin or cause to grow. □ **germination** *n.*

gerontology *n.* the study of ageing.

gerrymander *v.* arrange boundaries of (a constituency) to gain unfair electoral advantage.

gerund *n.* (*Grammar*) (in English) a verbal noun ending in *-ing*.

Gestapo *n.* the German secret police of the Nazi regime.

gestation *n.* the period in the womb between conception and birth.

gesticulate *v.* make expressive movements with the hands and arms. □ **gesticulation** *n.*

gesture *n.* an expressive movement or action. ● *v.* make a gesture.

get *v.* (**got**, **getting**) **1** come into possession of; receive. **2** fetch; capture, catch. **3** prepare (a meal). **4** bring or come into a certain state. **5** (*colloquial*) understand. **6** (*colloquial*) annoy. □ **get by**; manage to survive. **get off** be acquitted. **get on** make progress; be on harmonious terms; advance in age. **get out of** evade, avoid. **get over** recover from. **get round** persuade; evade, avoid. **get up** stand up; get out of bed; prepare, organize.

getaway *n.* an escape after a crime.

get-together *n.* a social gathering.

get-up *n.* an outfit.

geyser (gee-zer) *n.* **1** a spring spouting hot water or steam. **2** a water heater.

ghastly *adj.* (**ghastlier**) **1** causing horror; pale and ill-looking. **2** (*colloquial*) very bad. □ **ghastliness** *n.*

ghee *n.* Indian clarified butter.

gherkin *n.* a small pickled cucumber.

ghetto *n.* (*pl.* **ghettos**) a slum area occupied by a particular group.

ghetto blaster *n.* a large portable stereo radio etc.

ghost *n.* a dead person's spirit. ● *v.* write as a ghost writer. □ **ghostly** *adj.*, **ghostliness** *n.*

ghost writer *n.* a person who writes a book etc. for another to pass off as his or her own.

ghoul (gool) *n.* **1** a person who enjoys gruesome things. **2** (in Muslim stories) a spirit that robs and devours corpses. □ **ghoulish** *adj.*

giant *n.* (in fairy tales) a being of superhuman size; an abnormally large person, animal, or thing; a person of outstanding ability. ● *adj.* very large.

gibber *v.* make meaningless sounds in shock or terror.

gibberish *n.* unintelligible talk; nonsense.

gibbet *n.* a gallows.

gibbon *n.* a long-armed ape.

gibe *v.* (also **jibe**) jeer. ● *n.* a jeering remark.

giblets *n.pl.* edible organs from a bird.

giddy *adj.* (**giddier**) having or causing the feeling that everything is spinning; excitable, flighty. □ **giddily** *adv.*, **giddiness** *n.*

gift *n.* something given or received without payment; a natural ability; an easy task.

gifted *adj.* having great natural ability.

gig *n.* **1** a light two-wheeled horse-drawn carriage. **2** (*colloquial*) an engagement to perform.

giga- *comb. form* one thousand million, 10⁹.

gigantic *adj.* very large.

giggle *v.* laugh quietly. ● *n.* this laughter.

gigolo (jig-ŏ-loh) *n.* (*pl.* **gigolos**) a man paid by a woman to be her escort or lover.

gild *v.* cover with a thin layer of gold or gold paint. ● *n.* var. of **guild**.

gill (jil) *n.* one-quarter of a pint.

gills (gilz) *n.* **1** the organ with which a fish breathes. **2** the vertical plates on the underside of a mushroom.

gilt *adj.* gilded, gold-coloured. ● *n.* **1** a substance used in gilding. **2** a gilt-edged investment.

gilt-edged *adj.* (of an investment etc.) very safe.

gimbals *n.pl.* a contrivance of rings to keep instruments horizontal in ship.

gimcrack (jim-krak) *adj.* cheap and flimsy.

gimlet *n.* a small tool with a screw-like tip for boring holes.

gimmick *n.* a trick or device to attract attention.

gin *n.* an alcoholic spirit flavoured with juniper berries.

ginger *n.* **1** the hot-tasting root of a tropical plant. **2** reddish yellow. ● *adj.* ginger-coloured.

ginger ale *n.* (also **ginger beer**) a ginger-flavoured fizzy drink.

gingerbread *n.* ginger-flavoured cake.

ginger group *n.* a group urging a more active policy.

gingerly *adj. & adv.* cautious(ly).

gingham *n.* cotton fabric with a checked or striped pattern.

gingivitis (jin-ji-vy-tiss) *n.* inflammation of the gums.

ginseng *n.* a medicinal plant with a fragrant root.

gipsy var. of **gypsy**.

giraffe *n.* a long-necked African animal.

gird *v.* encircle or attach with a belt or band.

girder *n.* a metal beam supporting a structure.

girdle *n.* a cord worn round the waist; an elastic corset; a ring of bones in the body. ● *v.* surround.

girl *n.* a female child; a young woman. □ **girlhood** *n.*

girlfriend *n.* a female friend, esp. a man's usual companion.

giro (jy-roh) *n.* (*pl.* **giros**) a banking system in which payment can be made by transferring credit from one account to another; a cheque or payment made by this.

girth *n.* the distance round something; a band under a horse's belly, holding a saddle in place.

gismo var. of **gizmo**.

gist (jist) *n.* the essential points or general sense of a speech etc.

gîte (zheet) *n.* (in France) a holiday cottage.

give *v.* (**gave**, **given**, **giving**) **1** cause to receive or have. **2** utter; make or perform (an action or effort). **3** yield as a product or result. **4** be flexible. ● *n.* springiness, elasticity. □ **give away** give as a gift; reveal (a secret etc.) unintentionally. **give in** acknowledge defeat. **give off** emit. **give out** announce; become exhausted or used up; emit. **give over** (*colloquial*) cease. **give up** cease; part with, hand over; abandon hope or an attempt. **give way** yield;

allow other traffic to go first; collapse. □ **giver** n.

give-away n. **1** a free gift. **2** (colloquial) an unintentional disclosure.

given see **give**. adj. **1** specified. **2** having a tendency; given to swearing.

given name n. a first name (given in addition to the family name).

gizmo n. (also **gismo**) (pl. **gizmos**) a gadget.

gizzard n. a bird's second stomach, in which food is ground.

glacé (gla-say) adj. iced with sugar; preserved in sugar.

glacial adj. icy; of or from glaciers. □ **glacially** adv.

glaciated adj. covered with or affected by a glacier. □ **glaciation** n.

glacier n. a mass or river of ice moving very slowly.

glad adj. pleased, joyful. □ **gladly** adv., **gladness** n.

gladden v. make glad.

glade n. an open space in a forest.

gladiator n. a man trained to fight at public shows in ancient Rome. □ **gladiatorial** adj.

gladiolus n. (pl. **gladioli**) a garden plant with spikes of flowers.

glamour n. (Amer. **glamor**) alluring beauty; attractive exciting qualities. □ **glamorize** v., **glamorous** adj., **glamorously** adv.

glance v. look briefly; strike and glide off. ● n. a brief look.

gland n. an organ that secretes substances to be used or expelled by the body. □ **glandular** adj.

glare v. shine with a harsh dazzling light; stare angrily or fiercely. ● n. a glaring light or stare.

glaring adj. conspicuous.

glasnost (glaz-nost) n. a policy of more openness in news reporting etc.

glass n. **1** a hard brittle transparent substance; things made of this; a mirror; a glass drinking vessel; a barometer. **2** (**glasses**) spectacles; binoculars. □ **glassy** adj.

glasshouse n. **1** a greenhouse. **2** (slang) a military prison.

glaucoma n. a condition causing weakening of sight.

glaze v. **1** fit or cover with glass. **2** coat with a glossy surface. **3** become glassy. ● n. a shiny surface or coating.

glazier n. a person whose job is to fit glass in windows.

gleam n. **1** a beam or ray of soft light. **2** a brief show of a quality. ● v. send out gleams.

glean v. pick up (grain left by harvesters); gather scraps of. □ **gleaner** n., **gleanings** n.pl.

glebe n. a portion of land allocated to, and providing revenue for, a clergyman.

glee n. lively or triumphant joy. □ **gleeful** adj., **gleefully** adv.

glen n. a narrow valley.

glib adj. (**glibber**) ready with words but insincere or superficial.

glide v. move smoothly; fly in a glider or aircraft without engine power. ● n. a gliding movement.

glider n. an aeroplane with no engine.

glimmer n. a faint gleam. ● v. gleam faintly.

glimpse n. a brief view. ● v. catch a glimpse of.

glint n. a brief flash of light. ● v. send out a glint.

glisten v. shine like something wet.

glitch n. (colloquial) a hitch, a problem.

glitter v. & n. sparkle.

glitterati n.pl. rich famous people.

glitz n. showy glamour.

gloaming n. evening twilight.

gloat v. be full of greedy or malicious delight.

global adj. worldwide; of a whole group of items. □ **globally** adv.

global warming n. an increase in the temperature of the earth's atmosphere.

globe n. a ball-shaped object, esp. with a map of the earth on it; the world.

globe artichoke n. an edible artichoke flower.

globe-trotting n. & adj. travelling widely as a tourist.

globular adj. shaped like a globe.

globule n. a small round drop.

globulin n. a protein found in animal and plant tissue.

glockenspiel n. a musical instrument of metal bars or tubes struck by hammers.

gloom n. **1** semi-darkness. **2** a feeling of sadness and depression. □ **gloomy** adj., **gloomily** adv.

glorify v. **1** praise highly; worship. **2** make (something) seem grander than it is. □ **glorification** n.

glorious adj. having or bringing glory; splendid. □ **gloriously** adv.

glory n. fame, honour, and praise; something deserving this; magnificence. ● v. rejoice; pride oneself.

gloss n. a shine on a smooth surface. ● v. make glossy. □ **gloss over** cover up (a mistake etc.).

glossary n. a list of technical or special words with definitions.

glossy adj. (**glossier**) shiny. □ **glossily** adv., **glossiness** n.

glottis n. the opening at the upper end of the windpipe between the vocal cords. □ **glottal** adj.

glove n. a covering for the hand with separate divisions for fingers and thumb.

glow v. send out light and heat without flame; have a warm or flushed look, colour, or feeling. ● n. a glowing state.

glower v. scowl.

glow-worm n. a beetle that can give out a greenish light.

glucose n. a form of sugar found in fruit juice.

glue n. a sticky substance used for joining things. ● v. (**glued**, **gluing**) fasten with glue; attach closely. □ **gluey** adj.

glue-sniffing n. inhaling glue fumes for their narcotic effects.

glum adj. (**glummer**) sad and gloomy. □ **glumly** adv., **glumness** n.

glut v. (**glutted**) supply with more than is needed; satisfy fully with food. ● n. an excessive supply.

gluten n. a sticky protein found in cereals.

glutinous adj. glue-like, sticky. □ **glutinously** adv.

glutton n. a greedy person; one who is eager for something. □ **gluttonous** adj., **gluttony** n.

glycerine n. (Amer. **glycerin**) a thick sweet liquid used in medicines etc.

glycerol n. = glycerine.

gm abbr. gram(s).

GMT abbr. Greenwich Mean Time.

gnarled adj. knobbly; twisted and misshapen.

gnash v. (of teeth) strike together, grind.

gnat n. a small biting fly.

gnaw v. bite persistently (at something hard).

gnome n. a dwarf in fairy tales.

gnomic (noh-mik) adj. (of language) characterized by brief maxims.

gnomon (noh-mon) n. the rod of a sundial.

gnostic (noss-tik) *adj.* of knowledge; having mystical knowledge.

GNP *abbr.* gross national product.

gnu (noo) *n.* an ox-like antelope.

GNVQ *abbr.* General National Vocational Qualification.

go *v.* (**goes, went, gone, going**) **1** move; depart; extend. **2** be functioning; make (a specified movement or sound); (of time) pass. **3** belong in a specified place. **4** become; proceed. **5** be sold; be spent or used up. **6** collapse, fail, die. ● *n.* (*pl.* **goes**) **1** a turn or try. **2** energy. **3** a Japanese board game. □ **go back on** fail to keep (a promise). **go for** like, choose; (*slang*) attack. **go off** explode. **go out** be extinguished. **go round** be enough for everyone. **go slow** work at a deliberately slow pace as a protest. **go under** succumb; fail. **go with** match, harmonize with. **on the go** in constant motion; active. **to go** (*Amer.*) (of food) to be taken away for eating.

goad *n.* a pointed stick for driving cattle; a stimulus to activity. ● *v.* stimulate by annoying.

go-ahead *n.* a signal to proceed. ● *adj.* enterprising.

goal *n.* **1** a structure or area into which players send the ball to score a point in certain games; a point scored. **2** an objective.

goalie *n.* (*colloquial*) a goalkeeper.

goalkeeper *n.* a player whose job is to keep the ball out of the goal.

goalpost *n.* either of the posts marking the limit of a goal.

goat *n.* a small horned animal.

go-between *n.* a person who acts as a messenger or negotiator.

gobble *v.* **1** eat quickly and greedily. **2** make a throaty sound like a turkeycock.

gobbledegook *n.* (*colloquial*) language that is unintelligible and often pompous.

goblet *n.* a drinking glass with a stem and a foot.

goblin *n.* a mischievous ugly elf.

go-cart var. of **go-kart**.

god *n.* a superhuman being worshipped as having power over nature and human affairs; a person or thing greatly admired or adored; (**God**) the creator and ruler of the universe in Christian, Jewish, and Muslim teaching.

godchild *n.* (*pl.* **-children**) a child in relation to its godparent(s).

god-daughter *n.* a female godchild.

goddess *n.* a female god.

godetia *n.* an annual plant with showy flowers.

godfather *n.* a male godparent.

God-fearing *adj.* sincerely religious.

godforsaken *adj.* wretched, dismal.

godhead *n.* divine nature; a deity.

godmother *n.* a female godparent.

godparent *n.* a person who promises at a child's baptism to see that it is brought up as a Christian.

godsend *n.* a piece of unexpected good fortune.

godson *n.* a male godchild.

go-getter *n.* a pushy enterprising person.

goggle *v.* stare with wide-open eyes.

goggles *n.pl.* spectacles for protecting the eyes.

goitre (goy-ter) *n.* (*Amer.* **goiter**) an enlarged thyroid gland.

go-kart *n.* (also **go-cart**) a miniature racing car.

gold *n.* a chemical element (symbol Au); a yellow metal of high value; coins or articles made of this; its colour; a gold

medal (awarded as first prize).
● *adj.* made of or coloured like gold.

golden *adj.* **1** gold. **2** precious, excellent.

golden handshake *n.* a generous cash payment given on redundancy or early retirement.

golden jubilee *n.* the 50th anniversary of a sovereign's reign.

golden wedding *n.* the 50th anniversary of a wedding.

goldfield *n.* an area where gold is found.

goldfinch *n.* a songbird with a band of yellow across each wing.

goldfish *n.* (*pl.* **goldfish** or **goldfishes**) a small reddish carp kept in a bowl or pond.

gold leaf *n.* gold beaten into a very thin sheet.

gold rush *n.* a rush to a newly discovered goldfield.

goldsmith *n.* a person who makes gold articles.

golf *n.* a game in which a ball is struck with clubs into a series of holes. □ **golfer** *n.*

golf course *n.* (also **golf links**) an area of land on which golf is played.

golliwog *n.* a black-faced soft doll with fuzzy hair.

gonad *n.* an animal organ producing gametes.

gondola *n.* a boat with high pointed ends, used on canals in Venice.

gondolier *n.* a man who propels a gondola with a pole.

gone *see* **go**.

gong *n.* a metal plate that resounds when struck.

gonorrhoea *n.* (*Amer.* **gonorrhea**) a venereal disease with a discharge from the genitals.

goo *n.* (*colloquial*) a sticky wet substance.

good *adj.* (**better**, **best**) **1** having the right or desirable qualities; proper, expedient; morally correct, kindly; well-behaved. **2** enjoyable, beneficial. **3** efficient; thorough. **4** considerable, full. ● *n.* **1** that which is morally right. **2** profit, benefit. **3** (**goods**) movable property; articles of trade; things to be carried by road or rail. □ **as good as** practically, almost.

goodbye *int.* & *n.* an expression used when parting.

good-for-nothing *adj.* worthless. ● *n.* a worthless person.

Good Friday *n.* the Friday before Easter, commemorating the Crucifixion.

goodness *n.* the quality of being good; the good part of something.

goodwill *n.* friendly feeling; the established popularity of a business, treated as a saleable asset.

goody *n.* (also **goodie**) (*colloquial*) something good or attractive, esp. to eat; a good person, a hero.

goody-goody *n.* a smugly virtuous person.

gooey *adj.* (**gooier**) (*colloquial*) wet and sticky.

goose *n.* (*pl.* **geese**) a web-footed bird larger than a duck; the female of this.

gooseberry *n.* a thorny shrub; its edible (usu. green) berry.

goose-flesh *n.* (also **goose pimples**) bristling bumpy skin caused by cold or fright.

goose-step *n.* a way of marching without bending the knees.

gopher *n.* an American burrowing rodent.

gore *n.* **1** clotted blood from a wound. **2** a triangular or tapering section of a skirt or sail. ● *v.* pierce with a horn or tusk.

gorge *n.* a narrow steep-sided valley. ● *v.* eat greedily; fill full, choke up.

gorgeous *adj.* **1** (*colloquial*) very pleasant, beautiful. **2** richly coloured; magnificent. □ **gorgeously** adv.

gorgon n. a terrifying woman.

Gorgonzola n. a strong blue-veined cheese.

gorilla n. a large powerful ape.

gorse n. a wild evergreen thorny shrub with yellow flowers.

gory adj. (**gorier**) covered with blood; involving bloodshed.

goshawk n. a larg hawk with short wings.

gosling n. a young goose.

go-slow n. working slowly as a protest.

gospel n. **1** the teachings of Christ; (**Gospel**) any of the first four books of the New Testament. **2** something regarded as definitely true.

gossamer n. a fine piece of cobweb; flimsy delicate material.

gossip n. casual talk about other people's affairs; a person fond of such talk. ● v. (**gossiped**) engage in gossip.

got see **get**.

Gothic adj. **1** of an architectural style of the 12th–16th centuries, with pointed arches. **2** (of a novel etc.) in a horrific style popular in the 18th–19th centuries.

gotten (Amer.) = **got**.

gouache n. painting with opaque pigments in water thickened with a gluey substance.

gouge n. a chisel with a concave blade. ● v. cut out with a gouge; scoop or force out.

goulash n. a stew of meat and vegetables, seasoned with paprika.

gourd n. a fleshy fruit of a climbing plant; a container made from its dried rind.

gourmand n. a food lover; a glutton.

gourmet (goor-may) n. a connoisseur of good food and drink.

gout n. a disease causing inflammation of the joints.

govern v. rule with authority; keep under control; influence; direct. □ **governable** adj., **governor** n.

governance n. governing, control.

governess n. a woman employed to teach children in a private household.

government n. governing; a group or organization governing a country. □ **governmental** adj.

gown n. a loose flowing garment; a woman's long dress; an official robe.

GP abbr. general practitioner.

grab v. (**grabbed**) grasp suddenly; take greedily. ● n. a sudden clutch or attempt to seize something; a mechanical device for gripping things.

grace n. **1** attractiveness and elegance of manner or movement. **2** favour; mercy. **3** a short prayer of thanks for a meal. ● v. confer honour or dignity on; be an ornament to.

graceful adj. having or showing grace. □ **gracefully** adv.

graceless adj. inelegant.

gracious adj. kind and pleasant towards inferiors. □ **graciously** adv., **graciousness** n.

gradation n. a stage in a process of gradual change; this process.

grade n. **1** a level of rank or quality; a mark indicating standard of work. **2** (Amer.) a class in school. ● v. arrange in grades; assign a grade to.

gradient n. a slope; the angle of a slope.

gradual adj. taking place by degrees, not sudden. □ **gradually** adv.

graduate n. (grad-yoo-ăt) a person who has a university degree. ● v. (grad-yoo-ayt) **1** obtain a university degree. **2** divide into graded sections; mark into regular divisions. □ **graduation** n.

graffiti n.pl. words or drawings scribbled or sprayed on a wall.

graft n. **1** a plant shoot fixed into a cut in another plant to form a new growth; living tissue transplanted surgically. **2** (slang) hard work. ● v. **1** put a graft in or on; join inseparably. **2** (slang) work hard.

grain n. **1** small seed(s) of a food plant such as wheat or rice; these plants; a small hard particle. **2** a unit of weight (about 65 mg). **3** the texture or pattern made by fibres or particles. □ **grainy** adj.

gram n. (also **gramme**) one-thousandth of a kilogram.

grammar n. (the rules governing) the use of words in their correct forms and relationships.

■ **Usage** Grammar is not spelt with an e.

grammatical adj. conforming to the rules of grammar. □ **grammatically** adv.

grampus n. a dolphin-like sea animal.

gran n. (colloquial) grandmother.

granary n. a storehouse for grain.

grand adj. great; splendid; imposing; (colloquial) very good. ● n. **1** a grand piano. **2** (slang) a thousand dollars or pounds. □ **grandly** adv., **grandness** n.

grandchild n. (pl. **-children**) a child of one's son or daughter.

granddad n. (colloquial) grandfather.

granddaughter n. a female grandchild.

grandeur n. splendour, grandness.

grandfather n. a male grandparent.

grandfather clock n. a clock in a tall wooden case.

grandiloquent adj. using pompous language. □ **grandiloquence** n.

grandiose adj. imposing; planned on a large scale.

grandma n. (colloquial) grandmother.

grandmother n. a female grandparent.

grandpa n. (colloquial) grandfather.

grandparent n. a parent of one's father or mother.

grand piano n. a large piano with horizontal strings.

grand slam n. the winning of all of a group of matches.

grandson n. a male grandchild.

grandstand n. the principal stand for spectators at a sports ground.

grange n. a country house with farm buildings.

granite n. a hard grey stone.

granny n. (also **grannie**) (colloquial) grandmother.

granny flat n. self-contained accommodation in one's house for a relative.

grant v. **1** give or allow as a privilege. **2** admit to be true. ● n. a sum of money given from public funds to a student, for house repairs, etc. □ **take for granted** assume to be true or sure to happen or continue.

granular adj. like grains.

granulate v. **1** form into grains. **2** roughen the surface of.

granule n. a small grain.

grape n. a green or purple berry growing in clusters, used for making wine.

grapefruit n. a large round yellow citrus fruit.

grapevine n. **1** a vine bearing grapes. **2** the way news spreads unofficially.

graph n. a diagram showing the relationship between quantities.

graphic adj. **1** of drawing, painting, or engraving. **2** giving a vivid description. **3** (**graphics**) diagrams used in calculation and design; drawings; computer images. □ **graphically** adv.

graphical adj. using diagrams or graphs.

graphic equalizer n. a device controlling individual frequency bands of a stereo system.

graphite n. a form of carbon.

graphology n. the study of handwriting. □ **graphologist** n.

grapnel n. a small anchor with several hooks; a hooked device for dragging a river bed.

grapple v. **1** seize, hold firmly. **2** struggle.

grappling iron n. a grapnel.

grasp v. **1** seize and hold. **2** understand. ●n. **1** a firm hold or grip. **2** an understanding.

grasping adj. greedy, avaricious.

grass n. **1** a plant with green blades; a species of this (e.g. a cereal plant); ground covered with grass; (slang) marijuana. **2** (slang) an informer. ●v. **1** cover with grass. **2** act as an informer. □ **grassy** adj.

grasshopper n. a jumping insect that makes a chirping noise.

grassland n. a wide grass-covered area with few trees.

grass roots n.pl. the fundamental level or source; ordinary people, rank-and-file members.

grass widow n. a wife whose husband is absent for some time.

grate n. a metal framework keeping fuel in a fireplace; a hearth. ●v. **1** shred finely by rubbing against a jagged surface; make a harsh noise by rubbing. **2** have an irritating effect.

grateful adj. feeling that one values a kindness or benefit received. □ **gratefully** adv.

grater n. a device for grating food.

gratify v. give pleasure to; satisfy (wishes). □ **gratification** n.

grating n. a screen of spaced bars placed across an opening.

gratis adj. & adv. free of charge.

gratitude n. being grateful.

gratuitous adj. free, without payment; uncalled for. □ **gratuitously** adv.

gratuity n. money given for services rendered, a tip.

grave n. a hole dug to bury a corpse. ●adj. serious, causing great anxiety; solemn. □ **gravely** adv.

grave accent (grahv) n. the accent `.

gravel n. small stones, used for paths etc.

gravelly adj. **1** like gravel. **2** rough-sounding.

graven adj. **1** carved. **2** firmly fixed (in the memory).

gravestone n. a stone placed over a grave.

graveyard n. a burial ground.

gravitate v. move or be attracted (towards).

gravitation n. gravitating; the force of gravity. □ **gravitational** adj.

gravity n. **1** seriousness; solemnity. **2** the force that attracts bodies towards the centre of the earth.

gravy n. juice from cooked meat; sauce made from this.

gravy train n. a source of easy money.

gray Amer. sp. of **grey**.

graze v. **1** feed on growing grass; pasture animals in (a field). **2** injure by scraping the skin; touch or scrape lightly in passing. ●n. a grazed place on the skin.

grease n. a fatty or oily substance, a lubricant. ●v. put grease on. □ **greasy** adj.

greasepaint n. make-up used by actors.

great adj. much above average in size, amount, or intensity; of remarkable ability or character, important; (colloquial) very good. □ **greatness** n.

great- comb. form (of a family relationship) one generation removed in ancestry or descent.

greatly adv. very much.

grebe n. a diving bird.

Grecian adj. Greek.

Grecian nose n. a straight nose.

greed n. excessive desire for food or wealth. □ **greedy** adj., **greedily** adv.

Greek adj. & n. (an inhabitant, the language) of Greece.

green adj. **1** of the colour of growing grass. **2** concerned with protecting the environment. **3** unripe; inexperienced; easily deceived. ● n. a green colour or thing; a piece of grassy public land; (**greens**) green vegetables. □ **greenness** n.

green belt n. an area of open land round a town.

green card n. an international insurance document for motorists.

greenery n. green foliage or plants.

greenfinch n. a green and yellow finch.

green fingers n. skill in growing plants.

greenfly n. (pl. **greenfly**) a small green insect that sucks juices from plants.

greengage n. a round plum with a greenish skin.

greengrocer n. a shopkeeper selling vegetables and fruit.

greenhorn n. an inexperienced person.

greenhouse n. a glass building for rearing plants.

greenhouse effect n. the trapping of the sun's radiation by pollution in the atmosphere.

greenhouse gas n. a gas contributing to the greenhouse effect.

greenish adj. rather green.

green light n. a signal or permission to proceed.

Green Paper n. a government report of proposals being considered.

green room n. a room in a theatre used by actors when off stage.

greenstick fracture n. a bent and partially broken bone.

greet v. address politely on meeting; react to; present itself to (sight or hearing). □ **greeting** n.

gregarious adj. living in flocks or communities; fond of company. □ **gregariousness** n.

gremlin n. (colloquial) a mischievous spirit blamed for faults or problems.

grenade n. a small bomb thrown by hand or fired from a rifle.

grenadine n. a syrup made from pomegranates etc.

grew see **grow**.

grey adj. (Amer. **gray**) of the colour between black and white. ● n. a grey colour or thing. □ **greyish** adj., **greyness** n.

greyhound n. a slender smooth-haired dog noted for its swiftness.

grey matter n. the brain.

grid n. a grating; a system of numbered squares for map references; a network of lines, power cables, etc.; a gridiron.

gridiron n. a framework of metal bars for cooking on; a field for American football, marked with parallel lines.

grief n. deep sorrow. □ **come to grief** meet with disaster; fail.

grievance n. a cause for complaint.

grieve v. cause or feel grief.

grievous adj. causing grief; serious. □ **grievously** adv.

griffin n. (also **gryphon**) a mythological creature with an eagle's head and wings and a lion's body.

griffon n. **1** a small terrier-like dog. **2** a vulture. **3** a griffin.

grill n. **1** a metal grid, a grating, a grille; a device on a cooker for radiating heat downwards. **2** grilled food. ● v. **1** cook under a grill or on a gridiron. **2** question closely and severely.

grille n. (also **grill**) a grating; a grid protecting a vehicle's radiator.

grim adj. (**grimmer**) stern, severe; without cheerfulness, unattractive. □ **grimly** adv., **grimness** n.

grimace n. a contortion of the face in pain or amusement. ● v. make a grimace.

grime n. ingrained dirt. ● v. blacken with grime. □ **grimily** adv., **griminess** n., **grimy** adj.

grin v. (**grinned**) smile broadly. ● n. a broad smile.

grind v. (**ground**, **grinding**) **1** crush into grains or powder; crush or oppress by cruelty. **2** sharpen or smooth by friction; rub harshly together. ● n. a grinding process; hard monotonous work.

grindstone n. a revolving disc for sharpening or grinding things.

grip v. (**gripped**) take firm hold of; hold the attention of. ● n. a firm grasp or hold; a method of holding; understanding; a travelling bag.

gripe v. (colloquial) grumble. ● n. **1** colic pain. **2** (colloquial) a grievance.

grisly adj. (**grislier**) causing fear, horror, or disgust.

grist n. grain to be ground. □ **grist to the mill** a source of advantage.

gristle n. tough tissue of animal bodies, esp. in meat. □ **gristly** adj.

grit n. **1** particles of stone or sand. **2** (colloquial) courage and endurance. ● v. (**gritted**) **1** make a grating sound. **2** clench (the teeth). **3** spread grit on.□ **gritty** adj., **grittiness** n.

grizzle v. whimper, whine; complain.

grizzled adj. grey-haired.

grizzly bear n. a large brown bear of North America.

groan v. make a long deep sound in pain or disapproval; make a deep creaking sound. ● n. this sound.

grocer n. a shopkeeper selling food and household goods.

grocery n. a grocer's shop or goods.

grog n. a drink of spirits mixed with water.

groggy adj. (**groggier**) weak and unsteady, esp. after illness. □ **groggily** adv.

groin n. **1** the groove where each thigh joins the trunk; the curved edge where two vaults meet. **2** Amer. sp. of **groyne**.

grommet n. **1** an insulating washer. **2** a tube placed through the eardrum.

groom n. **1** a person employed to look after horses. **2** a bridegroom. ● v. **1** clean and brush (an animal); make neat and tidy. **2** prepare (a person) for a career or position.

groove n. a long narrow channel.

grope v. feel about as one does in the dark.

gross adj. **1** thick, large-bodied. **2** vulgar; outrageous. **3** total, without deductions. ● n. (pl. **gross**) twelve dozen. ● v. produce or earn as total profit. □ **grossly** adv.

grotesque adj. very odd or ugly; ● n. a comically distorted figure;

a design using fantastic forms. □ **grotesquely** adv., **grotesqueness** n.

grotto n. (pl. **grottoes** or **grottos**) a picturesque cave.

grouch v. (colloquial) grumble. ● n. a grumbler; a complaint.

ground see grind. n. **1** the solid surface of the earth; an area, position, or distance on this; (**grounds**) the land belonging to a large house. **2** (**grounds**) the reason for a belief or action. **3** (**grounds**) coffee dregs. ● v. **1** prevent (an aircraft or pilot) from flying. **2** base. **3** give basic training to.

ground glass n. glass made opaque by grinding.

grounding n. basic training.

groundless adj. without basis or good reason.

groundnut n. a peanut.

ground rent n. rent paid for land leased for building.

groundsheet n. a waterproof sheet for spreading on the ground.

groundsman n. (pl. **-men**) a person employed to look after a sports ground.

groundswell n. **1** slow heavy waves. **2** an increasingly forceful public opinion.

groundwork n. preliminary or basic work.

group n. a number of people or things near, categorized, or working together. ● v. form or gather into group(s).

grouse n. **1** a game bird. **2** a complaint. ● v. grumble.

grout n. thin fluid mortar. ● v. fill with grout.

grove n. a group of trees.

grovel v. (**grovelled**; Amer. **groveled**) crawl face downwards; humble oneself.

grow v. (**grew**, **grown**, **growing**) increase in size or amount; exist as a living plant; become; allow to grow; produce by cultivation.

□ **grow up** become adult or mature. □ **grower** n.

growl v. make a low threatening sound as a dog does. ● n. this sound.

grown see grow. adj. adult, fully developed.

grown-up adj. adult. ● n. an adult.

growth n. the process of growing; something that grows or has grown; a tumour.

groyne n. (Amer. **groin**) a solid structure projecting towards the sea to prevent erosion.

grub n. **1** the worm-like larva of certain insects. **2** (slang) food. ● v. (**grubbed**) dig the surface of soil; dig up by the roots; rummage.

grubby adj. (**grubbier**) dirty. □ **grubbiness** n.

grudge n. a feeling of resentment or ill will. ● v. begrudge, resent.

gruel n. thin oatmeal porridge.

gruelling adj. (Amer. **grueling**) very tiring.

gruesome adj. horrifying, disgusting.

gruff adj. (of the voice) low and hoarse; surly. □ **gruffly** adv., **gruffness** n.

grumble v. **1** complain in a bad-tempered way. **2** rumble. ● n. **1** a complaint. **2** a rumble. □ **grumbler** n.

grumpy adj. (**grumpier**) bad-tempered. □ **grumpily** adv., **grumpiness** n.

grunt n. a gruff snorting sound made by a pig. ● v. make this sound.

gryphon n. = griffin.

G-string n. a narrow strip of cloth covering the genitals, attached to a string round the waist.

G-suit n. a pressurized suit worn by astronauts and some pilots.

guano (gwah-noh) n. the dung of seabirds, used as manure; artificial fish manure.

guarantee n. a formal promise to do something or that a thing is of a specified quality; something offered as security; a guarantor. ●v. give or be a guarantee of or to.

guarantor n. the giver of a guarantee.

guard v. watch over and protect or supervise; restrain; take precautions. ●n. a state of watchfulness for danger; a defensive attitude in boxing, cricket, etc.; a person guarding something; a railway official in charge of a train; a protecting part or device.

guarded adj. cautious, discreet.

guardian n. one who guards or protects; a person undertaking legal responsibility for an orphan. □ **guardianship** n.

guardsman n. (pl. -men) a soldier acting as guard.

guava (gwah-vă) n. the orange-coloured fruit of a tropical American tree.

gudgeon n. 1 a small freshwater fish. 2 a pivot; a socket for a rudder; a metal pin.

guerrilla n. (also **guerilla**) a member of a small fighting force, taking independent irregular action.

guess v. form an opinion without definite knowledge; think likely. ●n. an opinion formed by guessing.

guesstimate n. (also **guestimate**) an estimate based on guesswork and reasoning.

guesswork n. guessing.

guest n. a person entertained at another's house, or staying at a hotel; a visiting performer.

guest house n. a private house offering paid accommodation.

guestimate var. of **guesstimate**.

guffaw n. a coarse noisy laugh. ●v. laugh in this way.

guidance n. guiding; advice.

guide n. 1 a person who shows others the way; one employed to point out sights to travellers. 2 a book of information. 3 something directing actions or movement. ●v. act as a guide to.

guidebook n. a book of information about a place.

guild n. (also **gild**) a society for mutual aid or with a common purpose; an association of craftsmen or merchants.

guilder n. a unit of money of the Netherlands.

guile n. treacherous cunning, craftiness. □ **guileless** adj.

guillemot (gil-I-mot) n. a seabird nesting on cliffs.

guillotine n. 1 a machine for beheading criminals; a machine for cutting paper or metal. 2 the fixing of times for voting in Parliament, to prevent a lengthy debate. ●v. use a guillotine on.

guilt n. the fact of having committed an offence; a feeling that one is to blame. □ **guiltless** adj.

guilty adj. (**guiltier**) having done wrong; feeling or showing guilt. □ **guiltily** adv.

guinea n. a former British coin worth 21 shillings (£1.05); this amount.

guinea pig n. 1 a small domesticated rodent. 2 a person or thing used as a subject for an experiment.

guise n. a false outward manner or appearance; a pretence.

guitar n. a stringed musical instrument. □ **guitarist** n.

gulf n. a large area of sea partly surrounded by land; a deep hollow; a wide difference in opinion.

gull n. a seabird with long wings.

gullet n. the passage by which food goes from mouth to stomach.

gullible adj. easily deceived. □ **gullibility** n.

gully n. a narrow channel cut by water or carrying rainwater from a building.

gulp v. swallow (food etc.) hastily or greedily; make a gulping movement. ● n. the act of gulping; a large mouthful of liquid gulped.

gum n. **1** the firm flesh in which teeth are rooted. **2** a sticky substance exuded by certain trees; adhesive; chewing gum. ● v. (**gummed**) smear or stick together with gum. □ **gummy** adj.

gum tree n. a tree that exudes gum, esp. a eucalyptus.

gumboots n.pl. rubber boots, wellingtons.

gumdrop n. a hard gelatin sweet.

gumption n. (colloquial) common sense.

gun n. a weapon that fires shells or bullets from a metal tube; a device forcing out a substance through a tube. ● v. (**gunned**) **1** shoot with a gun. **2** search for determinedly to rebuke; he's gunning for you.

gunboat diplomacy n. diplomacy backed by the threat of force.

gunfire n. the firing of guns.

gunge n. a sticky mass.

gunman n. (pl. -**men**) a person armed with a gun.

gunnel var. of **gunwale**.

gunner n. an artillery soldier; a naval officer in charge of a battery of guns.

gunnery n. the construction and operating of large guns.

gunny n. coarse material for making sacks; a sack.

gunpowder n. an explosive of saltpetre, sulphur, and charcoal.

gunrunning n. the smuggling of firearms. □ **gunrunner** n.

gunshot n. a shot fired from a gun.

gunsmith n. a maker and repairer of small firearms.

gunwale (gun-ǎl) n. (also **gunnel**) the upper edge of a boat's side.

guppy n. a small brightly-coloured tropical fish.

gurdwara (gerd-wah-rǎ) n. a Sikh temple.

gurgle n. a low bubbling sound. ● v. make this sound.

Gurkha n. a Hindu of Nepal; a Nepalese soldier serving in the British army.

guru n. (pl. **gurus**) a Hindu spiritual teacher; a revered teacher.

gush v. flow suddenly or in great quantities; talk effusively. ● n. a sudden or great outflow; effusiveness.

gusset n. a piece of cloth inserted to strengthen or enlarge a garment. □ **gusseted** adj.

gust n. a sudden rush of wind, rain, or smoke. ● v. blow in gusts. □ **gusty** adj.

gusto n. zest, enthusiasm.

gut n. **1** the intestine; thread made from animal intestines. **2** (**guts**) the abdominal organs. **3** (**guts**) courage and determination. ● v. (**gutted**) remove guts from (fish); remove or destroy the internal fittings or parts of.

gutsy adj. (**gutsier**) (colloquial) **1** courageous. **2** greedy.

gutta-percha n. a rubbery substance made from the juice of various Malaysian trees.

gutter n. **1** a trough round a roof, or a channel at a roadside, for carrying away rainwater. **2** a slum environment. ● v. (of a candle) burn unsteadily.

guttersnipe n. a street urchin.

guttural adj. throaty, harsh-sounding. □ **gutturally** adv.

guv n. (slang) governor.

guy n. **1** (colloquial) a man. **2** an effigy of Guy Fawkes burnt on 5 Nov. **3** a rope or chain to keep a thing steady or secured. ● v. ridicule by imitation.

guzzle v. eat or drink greedily.

gybe v. (Amer. **jibe**) (of a sail or boom) swing across; (of a boat) change course in this way. ● n. this movement.

gym n. a gymnasium; gymnastics.

gymkhana n. a horse-riding competition.

gymnasium n. (pl. **gymnasia** or **gymnasiums**) a room equipped for physical training and gymnastics.

gymnast n. an expert in gymnastics.

gymnastics n. & n.pl. exercises to develop the muscles or demonstrate agility. □ **gymnastic** adj.

gynaecology (gy-ni-**kol**-ŏji) n. (Amer. **gynecology**) the study of the physiological functions and diseases of women. □ **gynaecological** adj., **gynaecologist** n.

gypsum n. a chalk-like substance.

gypsy n. (also **gipsy**) a member of a travelling people.

gyrate v. move in circles or spirals, revolve. □ **gyration** n.

gyratory adj. gyrating, following a circular or spiral path.

gyrocompass n. a navigation compass using a gyroscope.

gyroscope n. a rotating device used to keep navigation instruments steady. □ **gyroscopic** adj.

....

Hh

....

H abbr. (of pencil lead) hard. ● symb. hydrogen.

Ha symb. hahnium.

ha int. an exclamation of triumph. ● abbr. hectare(s).

habeas corpus n. an order requiring a person to be brought to court.

haberdashery n. sewing goods, ribbons, etc.

habit n. 1 a regular way of behaving. 2 a monk's or nun's long dress; a woman's riding-dress.

habitable adj. suitable for living in.

habitat n. an animal's or plant's natural environment.

habitation n. a place to live in.

habitual adj. done regularly or constantly; usual. □ **habitually** adv.

habituate v. accustom.

hacienda n. a ranch or large estate in South America.

hack n. 1 a person doing routine work, esp. as a writer. 2 a horse for ordinary riding. ● v. 1 cut, chop, or hit roughly. 2 gain unauthorized access to computer files. 3 ride on horseback at an ordinary pace.

hacker n. (colloquial) a computer enthusiast, esp. one gaining unauthorized access to files.

hacking adj. (of a cough) dry and frequent.

hackles n.pl. the hairs on the back of an animal's neck, raised in anger.

hackneyed adj. (of sayings) over-used and lacking impact.

hacksaw n. a saw for metal.

had see **have**.

haddock n. (pl. **haddock**) an edible sea fish like a small cod.

hadji var. of **hajji**.

haem- pref. (Amer. **hem-**) of the blood.

haematology (heemă-) n. the study of blood. □ **haematologist** n.

haemoglobin (heem-ŏ-) n. the red oxygen-carrying substance in blood.

haemophilia (heemŏ-) n. failure of the blood to clot causing excessive bleeding. □ **haemophil-iac** *n.*

haemorrhage (hemŏ-) *n.* heavy bleeding. ● *v.* bleed heavily.

haemorrhoids (hemŏ-) *n.pl.* varicose veins at or near the anus.

hafnium *n.* a metallic element (symbol Hf).

haft *n.* the handle of a knife or dagger.

hag *n.* an ugly old woman.

haggard *adj.* looking exhausted and distraught.

haggis *n.* a Scottish dish made from offal boiled in a sheep's stomach.

haggle *v.* argue about the price or terms of a deal.

ha-ha *n.* a ditch with a wall in it.

hahnium *n.* a radioactive element (symbol Ha).

haiku (hy-koo) *n.* a Japanese three-line poem of 17 syllables.

hail *n.* a shower of frozen rain; a shower of blows, questions, etc. ● *v.* **1** pour down as or like hail. **2** greet; call to; summon.

hailstone *n.* a pellet of frozen rain.

hair *n.* the fine thread-like strands growing from the skin.

haircut *n.* shortening of hair by cutting it; a style of this.

hairdo (hy-koo) *n.* (*pl.* **hairdos**) an arrangement of the hair.

hairdresser *n.* a person who cuts and arranges hair. □ **hairdressing** *n.*

hairgrip *n.* a springy hairpin.

hairline *n.* **1** the edge of the hair on the forehead etc. **2** a very narrow crack or line.

hairpin *n.* a U-shaped pin for keeping hair in place.

hairpin bend *n.* a sharp U-shaped bend.

hair-raising *adj.* terrifying.

hair-trigger *n.* a trigger operated by the slightest pressure.

hairy *adj.* (**hairier**) **1** covered with hair. **2** (*slang*) hair-raising,

unpleasant, difficult. □ **hairiness** *n.*

Haitian (hay-shǎn) *adj.* & *n.* (a native) of Haiti.

hajji *n.* (also **hadji**) a Muslim who has been to Mecca on pilgrimage.

haka *n.* a Maori war dance with chanting.

hake *n.* (*pl.* **hake**) an edible sea fish of the cod family.

halal *n.* (also **hallal**) meat from an animal killed according to Muslim law.

halcyon *adj.* (of a period) happy and peaceful.

hale *adj.* strong and healthy.

half *n.* (*pl.* **halves**) **1** each of two equal parts; this amount. **2** a half-back, a half-pint, etc. ● *adj.* amounting to a half. ● *adv.* to the extent of a half, partly. □ **half a dozen** six. **half and half** half one thing and half another.

half-back *n.* a player between forwards and full-back(s).

half board *n.* bed, breakfast, and evening meal.

half-brother *n.* a brother with whom one has only one parent in common.

half-caste *n.* a person of mixed race.

half-hearted *adj.* not very enthusiastic.

half-life *n.* the time after which radioactivity etc. is half its original level.

half-mast *adj.* (of a flag) lowered in mourning.

half nelson *n.* a wrestling hold.

halfpenny (hayp-ni) *n.* (*pl.* **halfpennies** for single coins, **halfpence** for a sum of money) a coin worth half a penny.

half-sister *n.* a sister with whom one has only one parent in common.

half-term *n.* a short holiday halfway through a school term.

half-timbered *adj.* built with a timber frame and brick or plaster filling.

half-time *n.* the interval between two halves of a game.

half-tone *n.* a black and white illustration with grey shades shown by dots.

half-volley *n.* the return of the ball as soon as it bounces.

halfway *adj.* & *adv.* at a point equidistant between two others.

half-wit *n.* a stupid person. □ **half-witted** *adj.*

halibut *n.* (*pl.* **halibut**) a large edible flatfish.

halitosis *n.* breath that smells unpleasant.

hall *n.* a large room or building for meetings, concerts, etc.; a large country house; the space inside the front entrance of a house.

hallal var. of **halal**.

hallelujah var. of **alleluia**.

halliard var. of **halyard**.

hallmark *n.* an official mark on precious metals to indicate their standard; a distinguishing characteristic. □ **hallmarked** *adj.*

hallo *int.* & *n.* var. of **hello**.

Hallowe'en *n.* 31 Oct., eve of All Saints' Day.

hallucinate *v.* experience hallucinations.

hallucination *n.* an illusion of seeing or hearing something. □ **hallucinatory** *adj.*

hallucinogenic *adj.* causing hallucinations.

halm var. of **haulm**.

halo *n.* (*pl.* **haloes**) a circle of light esp. round the head of a sacred figure.

halogen *n.* any of a group of various non-metallic elements.

halon *n.* a gaseous compound of halogens used to extinguish fires.

halt *v.* stop. ● *n.* a temporary stop; a minor stopping place on a railway.

halter *n.* a strap round the head of a horse for leading or holding it.

halting *adj.* slow and hesitant.

halve *v.* divide equally between two; reduce by half.

halyard *n.* (also **halliard**) a rope for raising or lowering a sail or flag.

ham *n.* **1** (meat from) a pig's thigh, salted or smoked. **2** (*colloquial*) a poor actor or performer; an amateur radio operator. ● *v.* (**hammed**) (*colloquial*) overact.

hamburger *n.* a flat round cake of minced beef.

ham-fisted *adj.* clumsy.

hamlet *n.* a small village.

hammer *n.* a tool with a head for hitting nails etc.; a metal ball attached to a wire, thrown in an athletic contest. ● *v.* hit or beat with a hammer; strike loudly.

hammock *n.* a hanging bed of canvas or netting.

hamper *n.* a basketwork packing case; a selection of food packed as a gift. ● *v.* prevent free movement or activity of, hinder.

hamster *n.* a small rodent with cheek-pouches.

hamstring *n.* a tendon at the back of a knee or hock. ● *v.* (**hamstrung**, **hamstringing**) cripple by cutting the hamstring(s); cripple the activity of.

hand *n.* **1** the part of the arm below the wrist; control, influence, or help; a manual worker; a style of handwriting. **2** a pointer on a dial etc.; a side or direction; a round of a card game, a player's cards. ● *v.* give or pass. □ **at hand** close by. **hands down** easily. **on hand** available. **out of hand** out of control. **to hand** within reach.

handbag n. a bag to hold a purse and personal articles.

handball n. **1** a game with a ball thrown by hand. **2** the intentional touching of the ball in football (a foul).

handbill n. a printed notice circulated by hand.

handbook n. a small book giving useful facts.

handcuff n. a metal ring linked to another, for securing a prisoner's wrists. ● v. put handcuffs on.

handful n. **1** a quantity that fills the hand; a few. **2** (colloquial) a difficult person or task.

handicap n. a disadvantage imposed on a superior competitor to equalize chances; a race etc. in which handicaps are imposed; something that makes progress difficult; a physical or mental disability. ● v. (**handicapped**) impose or be a handicap on.

handkerchief n. (pl. **-chiefs** or **-chieves**) a small square of cloth for wiping the nose etc.

handle n. a part by which a thing is held, carried, or controlled. ● v. touch or move with the hands; deal with; manage.

handlebar n. a steering bar of a bicycle etc.

handler n. a person in charge of a trained dog etc.

handout n. something distributed free of charge.

handrail n. a rail beside stairs etc.

handshake n. the act of shaking hands as a greeting etc.

handsome adj. **1** good-looking. **2** generous; (of a price etc.) very large.

handstand n. balancing on one's hands with feet in the air.

handwriting n. writing by hand with pen or pencil; a style of this.

handy adj. (**handier**) **1** convenient. **2** clever with one's hands. □ **handily** adv., **handiness** n.

handyman n. (pl. **-men**) a person who does odd jobs.

hang v. (**hung, hanging**; in sense 2 **hanged, hanging**) **1** support or be supported from above. **2** kill or be killed by suspension from a rope round the neck. **3** droop; be present oppressively, remain. ● n. the way a thing hangs. □ **get the hang of** (colloquial) get the knack of, understand. **hang about** loiter. **hang back** hesitate. **hang on** hold tightly; depend on; (slang) wait.

hangar n. a building for aircraft.

hangdog adj. shamefaced.

hanger n. a loop or hook by which a thing is hung; a shaped piece of wood etc. to hang a garment on.

hang-glider n. the frame used in hang-gliding.

hang-gliding n. the sport of gliding in an airborne frame holding one person.

hangings n.pl. draperies hung on walls.

hangman n. (pl. **-men**) a person whose job is to hang people condemned to death.

hangnail n. torn skin at the base of a fingernail.

hangover n. unpleasant after-effects from drinking too much alcohol.

hank n. a coil or length of thread.

hanker v. crave, feel a longing.

hanky n. (colloquial) a handkerchief.

Hanukkah n. a Jewish festival of lights, beginning in December.

haphazard adj. chosen or chosen at random. □ **haphazardly** adv.

hapless adj. unlucky.

happen v. occur; chance. □ **happen to** be the fate or experience of.

happy adj. (**happier**) contented, pleased; fortunate. □ **happily** adv., **happiness** n.

happy-go-lucky adj. cheerfully casual.

harangue v. lecture earnestly and at length.

harass v. worry or annoy continually; make repeated attacks on. □ **harassment** n.

harbour (Amer. **harbor**) n. a place of shelter for ships. ● v. shelter; keep in one's mind.

hard adj. **1** firm, not easily cut. **2** difficult; not easy to bear; harsh; strenuous. **3** (of drugs) strong and addictive; (of currency) not likely to drop suddenly in value; (of drinks) strongly alcoholic; (of water) containing minerals that prevent soap from lathering freely. ● adv. intensively; so as to be hard. □ **hardness** n.

hardbitten adj. tough and tenacious.

hard-boiled adj. **1** (of eggs) boiled until the yolk and white are set. **2** (of people) callous.

hardboard n. stiff board made of compressed wood pulp.

hard copy n. material produced in printed form from a computer.

harden v. make or become hard or hardy.

hard-headed adj. shrewd and practical.

hard-hearted adj. unfeeling.

hardly adv. only with difficulty; scarcely.

hard of hearing adj. slightly deaf.

hard sell n. aggressive salesmanship.

hardship n. harsh circumstances.

hard shoulder n. an extra strip of road beside a motorway, for use in an emergency.

hard up adj. short of money.

hardware n. tools and household implements sold by a shop; weapons; machinery used in a computer system.

hardwood n. the hard heavy wood of deciduous trees.

hardy adj. (**hardier**) capable of enduring cold or harsh conditions. □ **hardiness** n.

hare n. a field animal like a large rabbit. ● v. run rapidly.

hare-brained adj. wild and foolish, rash.

harelip n. a deformed lip with a vertical slit.

harem (har-eem) n. the women of a Muslim household; their quarters.

haricot bean (ha-ri-koh) n. the edible white dried seed of a kind of bean.

hark v. listen. □ **hark back** return to an earlier subject.

harlequin adj. in varied colours.

harm n. damage, injury. ● v. cause harm to. □ **harmful** adj., **harmless** adj.

harmonic adj. full of harmony.

harmonica n. a mouth organ.

harmonious adj. forming a pleasing or consistent whole; free from ill feeling; sweet-sounding. □ **harmoniously** adv.

harmonium n. a musical instrument like a small organ.

harmonize v. (also **-ise**) make or be harmonious; add notes to form chords. □ **harmonization** n.

harmony n. being harmonious; the combination of musical notes to form chords; melodious sound.

harness n. straps and fittings by which a horse is controlled; similar fastenings. ● v. put harness on, attach by this; control and use.

harp n. a musical instrument with strings in a triangular frame. □ **harp on** talk repeatedly about. □ **harpist** n.

harpoon n. a spear-like missile with a rope attached. ● v. spear with a harpoon.

harpsichord n. a piano-like instrument.

harpy n. a grasping unscrupulous woman.

harridan n. a bad-tempered old woman.

harrier n. 1 a hound used for hunting hares. 2 a falcon. 3 a group of cross-country runners.

harrow n. a heavy frame with metal spikes or discs for breaking up soil. ● v. 1 draw a harrow over (soil). 2 distress greatly.

harry v. harass.

harsh adj. rough and disagreeable; severe, cruel. □ **harshly** adv., **harshness** n.

hart n. an adult male deer.

hartebeest n. a large African antelope.

harvest n. the gathering of crop(s); the season for this; a season's yield of a natural product. ● v. gather a crop; reap. □ **harvester** n.

has see **have**.

has-been n. (colloquial) a person or thing that has lost a former importance or popularity.

hash n. 1 a dish of chopped recooked meat. 2 a jumble.

hashish n. hemp dried for chewing or smoking as a narcotic.

hasp n. a clasp fitting over a U-shaped staple, secured by a pin or padlock.

hassle n. & v. (colloquial) a quarrel or struggle; trouble, inconvenience.

hassock n. a thick firm cushion for kneeling on.

haste n. hurry.

hasten v. hurry.

hasty adj. (**hastier**) hurried; acting or done too quickly. □ **hastily** adv., **hastiness** n.

hat n. a covering for the head.

hatch n. an opening in a door, floor, ship's deck, etc.; its cover. ● v. 1 emerge or produce (young) from an egg. 2 devise (a plot). 3 mark with close parallel lines.

hatchback n. a car with a back door that opens upwards.

hatchery n. a place for hatching eggs, especially for fish.

hatchet n. a small axe. □ **bury the hatchet** stop quarrelling.

hatchway n. an opening in a ship's deck for loading cargo.

hate n. hatred. ● v. feel hatred towards; dislike greatly.

hateful adj. arousing hatred.

hatred n. extreme dislike or enmity.

hat-trick n. three successes in a row, esp. in sports.

haughty adj. (**haughtier**) proud and looking down on others. □ **haughtily** adv., **haughtiness** n.

haul v. pull or drag forcibly; transport by truck etc. ● n. the process of hauling; an amount gained by effort, booty.

haulage n. transport of goods.

haulier n. a person or firm transporting goods by road.

haulm n. (also **halm**) a stalk or stem.

haunch n. the fleshy part of the buttock and thigh; a leg and loin of meat.

haunt v. linger in the mind of; (esp. of a ghost) repeatedly visit (a person or place). ● n. a place often visited by a particular person.

haute couture (oht koo-**tewr**) n. high fashion.

haute cuisine (oht kwi-**zeen**) n. high-class cookery.

have v. (**has, had, having**) 1 possess; contain. 2 experience, undergo; give birth to. **3** cause to be or do or be done; allow; be compelled (to do). **4** (*colloquial*) cheat, deceive. ● *v.aux.* used with the past participle to form past tenses; *he has gone.* □ **have it out** settle a problem by frank discussion. **have up** bring (a person) to trial. **haves and have-nots** people with and without wealth or privilege.

haven *n.* a refuge.

haversack *n.* a strong bag carried on the back or shoulder.

havoc *n.* great destruction or disorder.

haw *n.* a hawthorn berry.

hawk *n.* **1** a bird of prey. **2** a person who favours an aggressive policy.

hawk-eyed *adj.* having very keen sight.

hawser *n.* a heavy rope or cable for mooring or towing a ship.

hawthorn *n.* a thorny tree with small red berries.

hay *n.* grass cut and dried for fodder.

hay fever *n.* an allergy caused by pollen and dust.

haymaking *n.* cutting grass and spreading it to dry for fodder.

haystack *n.* a pile of hay stacked in a large block shape for storing.

haywire *adj.* badly disorganized.

hazard *n.* a risk, a danger; an obstacle. ● *v.* risk; venture. □ **hazardous** *adj.*

haze *n.* thin mist.

hazel *n.* a tree with small edible nuts; light brown. □ **hazelnut** *n.*

hazy *adj.* (**hazier**) misty; indistinct; vague. □ **hazily** *adv.*, **haziness** *n.*

HB *abbr.* (of a pencil lead) hard black.

H-bomb *n.* a hydrogen bomb.

He *symb.* helium.

he *pron.* the male previously mentioned. ● *n.* a male.

head *n.* **1** the part of the body containing the eyes, nose, mouth, and brain; the intellect. **2** an individual person or animal. **3** something like the head in form or position, the top or leading part or position; foam on beer etc.; the chief person; a headteacher. **4** (**heads**) the side of a coin showing a head, turned upwards after being tossed. **5** a body of water or steam confined for exerting pressure. ● *v.* **1** be at the head or top of. **2** strike (a ball) with one's head. **3** go in a particular direction. □ **head off** force to turn by getting in front.

headache *n.* a continuous pain in the head; a worrying problem.

headdress *n.* an ornamental covering worn on the head.

header *n.* a dive with the head first; a heading of the ball in football.

headgear *n.* a hat or headdress.

headhunt *v.* seek to recruit (senior staff) from another firm.

heading *n.* word(s) at the top of written matter as a title.

headlamp *n.* a headlight.

headland *n.* a promontory.

headlight *n.* a powerful light on the front of a vehicle etc.; its beam.

headline *n.* a heading in a newspaper; (**headlines**) a summary of broadcast news.

headlong *adj.* & *adv.* falling or plunging with the head first; in a hasty and rash way.

headmaster *n.* a man who is a headteacher.

headmistress *n.* a woman who is a headteacher.

head-on *adj.* & *adv.* with the head or front foremost.

headphones *n.pl.* a set of earphones for listening to audio equipment.

headquarters *n.pl.* a place from which an organization is controlled.

headstone *n.* a stone set up at the head of a grave.

headstrong *adj.* self-willed and obstinate.

head teacher *n.* the principal teacher in a school, responsible for organizing it.

headway *n.* progress.

headwind *n.* a wind blowing from directly in front.

heady *adj.* (**headier**) **1** likely to cause intoxication. **2** exciting.

heal *v.* make or become healthy after injury; cure. □ **healer** *n.*

health *n.* the state of being well and free from illness; the condition of the body.

health centre *n.* a doctors' surgery with several doctors, a nurse, a pharmacy, etc.

health farm *n.* an establishment offering controlled regimes of diet, exercise, massage, etc. to improve health.

health foods *n.pl.* natural unprocessed foods.

health visitor *n.* a nurse trained to visit people at home.

healthy *adj.* (**healthier**) having, showing, or producing good health; functioning well. □ **healthily** *adv.*, **healthiness** *n.*

heap *n.* **1** a number of things or particles lying one on top of another. **2** (**heaps**) (*colloquial*) plenty. ● *v.* pile or become piled in a heap; load with large quantities.

hear *v.* (**heard, hearing**) perceive (sounds) with the ear; pay attention to; receive information. ● **hear! hear!** I agree. □ **hearer** *n.*

hearing *n.* ability to hear; an opportunity of being heard; a case being judged by a law court or tribunal.

hearing aid *n.* a small sound-amplifier worn by a partially deaf person to improve the hearing.

hearsay *n.* rumour or gossip.

hearse *n.* a vehicle carrying the coffin at a funeral.

heart *n.* **1** the muscular organ that keeps blood circulating. **2** the centre of a person's emotions or inner thoughts; courage; enthusiasm; a central part. **3** a figure representing a heart; a playing card of the suit marked with these. ● **break a person's heart** cause him or her overwhelming grief. **by heart** memorized thoroughly.

heartache *n.* mental pain, sorrow.

heart attack *n.* (also **heart failure**) sudden failure of the heart to function normally.

heartbeat *n.* the pulsation of the heart.

heartbreak *n.* overwhelming grief.

heartbroken *adj.* very upset.

heartburn *n.* a burning sensation in the lower part of the chest from indigestion.

hearten *v.* encourage.

heartfelt *adj.* felt deeply, sincere.

hearth *n.* the floor of a fireplace; the fireside.

heartless *adj.* not feeling pity or sympathy. □ **heartlessly** *adv.*

heart-rending *adj.* very distressing.

heart-searching *n.* examination of one's own feelings and motives.

heart-throb *n.* (*colloquial*) a person for whom one has romantic feelings.

heart-to-heart *adj.* frank and personal.

heart-warming *adj.* emotionally moving and encouraging.

heartwood *n.* the dense, hardest, inner part of a tree trunk.

hearty *adj.* (**heartier**) **1** vigorous; enthusiastic. **2** (of meals) large. □ **heartily** *adv.*, **heartiness** *n.*

heat *n.* **1** a form of energy produced by movement of molecules; hotness; passion, anger. **2** a preliminary contest. ● *v.* make or become hot. □ **on heat** (of female mammals) ready to mate.

heated *adj.* (of a person or discussion) angry. □ **heatedly** *adv.*

heater *n.* a device supplying heat.

heath *n.* flat uncultivated land with low shrubs.

heathen *n.* a person who does not believe in an established religion.

heather *n.* an evergreen plant with purple, pink, or white flowers.

heatstroke *n.* an illness caused by overexposure to sun.

heatwave *n.* a long period of hot weather.

heave *v.* **1** lift or haul with great effort; (*colloquial*) throw. **2** utter (a sigh); rise and fall like waves; pant, retch. ● *n.* an act of heaving. □ **heave in sight** (**hove**, **heaving**) come into view. **heave to** (**hove**, **heaving**) bring a ship to a standstill with its head to the wind.

heaven *n.* **1** the abode of God; a place or state of bliss. **2** (**the heavens**) the sky.

heavenly *adj.* of heaven; divine; of or in the heavens; (*colloquial*) very pleasing.

heavenly bodies *n.pl.* the sun, moon, stars, etc.

heavy *adj.* (**heavier**) **1** having great weight, force, or intensity. **2** dense; stodgy. **3** serious. □ **heavily** *adv.*, **heaviness** *n.*

heavy-hearted *adj.* sad.

heavy industry *n.* industry producing metal or heavy machines etc.

heavy metal *n.* a type of loud rock music.

heavyweight *adj.* having great weight or influence. ● *n.* a heavyweight person.

Hebrew *n. & adj.* (a member) of a Semitic people in ancient Palestine; (of) their language or a modern form of this. □ **Hebraic** *adj.*

heckle *v.* interrupt (a public speaker) with aggressive questions and abuse. □ **heckler** *n.*

hectare *n.* a unit of area, 10,000 sq. metres (2.471 acres).

hectic *adj.* with feverish activity. □ **hectically** *adv.*

hectogram *n.* 100 grams.

hector *v.* intimidate by bullying.

hedge *n.* a barrier of bushes or shrubs. ● *v.* **1** surround with a hedge; make or trim hedges. **2** avoid giving a direct answer or commitment.

hedgehog *n.* a small animal covered in stiff spines.

hedgerow *n.* bushes etc. forming a hedge.

hedonist *n.* a person who believes pleasure is the chief good. □ **hedonism** *n.*, **hedonistic** *adj.*

heed *v.* pay attention to. ● *n.* careful attention. □ **heedful** *adj.*, **heedless** *adj.*, **heedlessly** *adv.*

heel *n.* **1** the back part of the human foot; part of a stocking or shoe covering this. **2** (*slang*) a scoundrel. ● *v.* **1** make or repair the heel(s) of. **2** (of a boat) tilt to one side. □ **down at heel** shabby. **take to one's heels** run away.

hefty *adj.* (**heftier**) large and heavy. □ **heftily** *adv.*, **heftiness** *n.*

hegemony (hi-gem-ŏni) *n.* leadership, esp. by one country.

Hegira (hej-iră) (also **Hejira**) *n.* Muhammad's flight from Mecca (AD 622), from which the Muslim era is reckoned.

heifer (hef-er) *n.* a young cow.

height *n.* measurement from base to top or head to foot; the distance above ground or sea level; the highest degree of something.

heighten *v.* make or become higher or more intense.

heinous (hay-nŭss) *adj.* very wicked.

heir (air) *n.* a person entitled to inherit property or a rank etc.

heiress (air-ess) *n.* a female heir.

heirloom (air-loom) *n.* a possession handed down in a family for several generations.

Hejira var. of **Hegira**.

held see **hold**.

helical *adj.* like a helix.

helicopter *n.* an aircraft with blades that revolve horizontally.

heliport *n.* a helicopter station.

helium *n.* a light colourless gas (symbol He) that does not burn.

helix *n.* (*pl.* **helices**) a spiral.

hell *n.* a place of punishment for the wicked after death; a place or state of misery. □ **hell for leather** very fast.

hell-bent *adj.* recklessly determined.

hello *int.* & *n.* (also **hallo, hullo**) (*pl.* **hellos**) an exclamation used in greeting or to call attention.

helm *n.* the tiller or wheel by which a ship's rudder is controlled.

helmet *n.* a protective head-covering.

helmsman *n.* (*pl.* **-men**) a person controlling a ship's helm.

help *v.* do part of another's work; be useful (to); make easier; serve with food. ● *n.* an act of helping; a person or thing that helps. □ **helper** *n.*

helpful *adj.* giving help, useful. □ **helpfully** *adv.*, **helpfulness** *n.*

helping *n.* a portion of food served.

helpless *adj.* unable to manage without help; powerless. □ **helplessly** *adv.*, **helplessness** *n.*

helpline *n.* a telephone service providing help with problems.

helter-skelter *adv.* in disorderly haste. ● *n.* a spiral slide at a funfair.

hem *n.* an edge of cloth turned under and sewn down. ● *v.* (**hemmed**) sew a hem on. □ **hem in** surround and restrict.

hem-, hematology, etc. Amer. sp. of **haem-, haematology**, etc.

hemisphere *n.* half a sphere; half of the earth. □ **hemispherical** *adj.*

hemlock *n.* a poisonous plant.

hemp *n.* a plant with coarse fibres used in making rope and cloth; a narcotic drug made from this.

hen *n.* a female bird, esp. of the domestic fowl.

hence *adv.* **1** from this time; for this reason. **2** (*old use*) from here.

henceforth *adv.* (also **henceforward**) from this time on, in future.

henchman *n.* (*pl.* **-men**) a trusty supporter.

henna *n.* a reddish dye used esp. on the hair; the tropical plant from which it is obtained. □ **hennaed** *adj.*

hen-party *n.* (*colloquial*) a party for women only.

henpecked *adj.* (of a man) nagged by his wife.

henry *n.* a unit of electric inductance.

hepatic *adj.* of the liver.

hepatitis *n.* inflammation of the liver.

heptagon *n.* a geometric figure with seven sides. □ **heptagonal** *adj.*

heptathlon *n.* an athletic contest involving seven events.

her pron. the objective case of she. ● adj. belonging to her.

herald n. a person or thing heralding something. ● v. proclaim the approach of.

heraldry n. the study of coats of arms. □ **heraldic** adj.

herb n. a plant used as a flavouring or in medicine.

herbaceous adj. soft-stemmed.

herbaceous border n. a border containing esp. perennial plants.

herbal adj. of herbs. ● n. a book about herbs.

herbalist n. a dealer in medicinal herbs.

herbicide n. a substance used to destroy plants.

herbivore n. a herbivorous animal.

herbivorous adj. feeding on plants.

herculean adj. needing or showing great strength or effort.

herd n. a group of animals feeding or staying together; a mob. ● v. gather, stay, or drive as a group; tend (a herd). □ **herdsman** n. (pl. -men).

here adv. in, at, or to this place; at this point. ● n. this place.

hereabouts adv. near here.

hereafter adv. from now on. ● n. the future; the next world.

hereby adv. by this means; as a result of this.

hereditary adj. inherited; holding a position by inheritance.

heredity n. inheritance of characteristics from parents.

herein adv. in this place or book etc.

heresy n. an opinion contrary to accepted beliefs.

heretic n. a person who holds a heresy. □ **heretical** adj., **heretically** adv.

hereto adv. to this.

herewith adv. with this.

heritage n. things inherited; a nation's historic buildings etc.

hermaphrodite n. a creature with male and female sexual organs.

hermetic adj. with an airtight closure. □ **hermetically** adv.

hermit n. a person living in solitude.

hermitage n. a hermit's dwelling.

hernia n. a protrusion of part of an organ through the wall of the cavity (esp. the abdomen) containing it.

hero n. (pl. **heroes**) a man admired for his brave deeds; the chief male character in a story.

heroic adj. very brave. ● n.pl. (**heroics**) over-dramatic behaviour. □ **heroically** adv.

heroin n. a powerful addictive drug prepared from morphine.

heroine n. a woman admired for her brave deeds; the chief female character in a story.

heroism n. heroic conduct.

heron n. a long-legged wading bird.

herpes (her-peez) n. a virus disease causing blisters.

Herr n. (pl. **Herren**) the title of a German man, corresponding to Mr.

herring n. an edible North Atlantic fish.

herringbone n. a zigzag pattern or arrangement.

hers poss.pron. belonging to her.

herself pron. the emphatic and reflexive form of she and her.

hertz n. (pl. **hertz**) a unit of frequency of electromagnetic waves.

hesitant adj. hesitating. □ **hesitantly** adv., **hesitancy** n.

hesitate v. pause doubtfully; be reluctant, scruple. □ **hesitation** n.

hessian n. a strong coarse cloth of hemp or jute.

heterodox adj. not orthodox.

heterogeneous *adj.* made up of people or things of various sorts. □ **heterogeneity** *n.*

heterosexual *adj.* & *n.* (a person) sexually attracted to people of the opposite sex. □ **heterosexuality** *n.*

hew *v.* (**hewed, hewn** or **hewed, hewing**) chop or cut with an axe etc.; cut into shape.

hex *n.* a magic spell; a curse.

hexadecimal *adj.* (*Computing*) of a number system using 16 rather than 10 as a base.

hexagon *n.* a geometric figure with six sides. □ **hexagonal** *adj.*

hexagram *n.* a six-pointed star formed by two intersecting triangles.

hey *int.* an exclamation of surprise or inquiry, or calling attention.

heyday *n.* the time of greatest success.

Hf *symb.* hafnium.

Hg *symb.* mercury.

hg *abbr.* hectogram.

HGV *abbr.* heavy goods vehicle.

HH *abbr.* (of pencil lead) double-hard.

hi *int.* an exclamation calling attention or greeting.

hiatus (hy-ay-tus) *n.* (*pl.* **hiatuses**) a break or gap in a sequence.

hibernate *v.* spend the winter in a sleep-like state. □ **hibernation** *n.*

Hibernian *adj.* & *n.* (a native) of Ireland.

hibiscus *n.* a shrub with large flowers.

hiccup *n.* (also **hiccough**) a sudden stopping of breath with a 'hic' sound; a brief hitch. ● *v.* (**hiccuped, hiccuping**) make this sound.

hide *v.* (**hid, hidden, hiding**) put or keep out of sight; keep secret; conceal oneself. ● *n.* **1** a hiding place used when birdwatching etc. **2** an animal's skin.

hidebound *adj.* rigidly conventional.

hideous *adj.* very ugly. □ **hideously** *adv.*, **hideousness** *n.*

hideout *n.* a hiding place.

hiding *n.* (*colloquial*) a thrashing.

hierarchy *n.* a system with grades ranking one above another. □ **hierarchical** *adj.*

hieroglyph *n.* a pictorial symbol used in ancient writing. □ **hieroglyphic** *adj.*, **hieroglyphics** *n.pl.*

hi-fi *adj.* high fidelity. ● *n.* a set of hi-fi equipment.

higgledy-piggledy *adj.* & *adv.* in complete confusion.

high *adj.* **1** extending far upwards or a specified distance upwards; far above ground or sea level. **2** ranking above others; greater than normal; (of sound or a voice) not deep or low. **3** (of meat) slightly decomposed. **4** (*slang*) intoxicated, under the influence of drugs. ● *n.* a high level; an area of high pressure. ● *adv.* in, at, or to a high level.

highbrow *adj.* very intellectual, cultured. ● *n.* a highbrow person.

higher education *n.* education at university etc.

highfalutin *adj.* (also **highfaluting**) pompous, pretentious.

high-handed *adj.* using authority arrogantly.

highlands *n.pl.* a mountainous region. □ **highland** *adj.*, **highlander** *n.*

highlight *n.* a bright area in a picture; a light streak in the hair; the best feature. ● *v.* emphasize.

highlighter *n.* a coloured marker pen.

highly *adv.* in a high degree, extremely; very favourably.

highly-strung adj. nervous, easily upset.

high-rise adj. with many storeys.

high road n. a main road.

high school n. a secondary school.

high seas n. the sea outside a country's territorial waters.

high season n. the busiest season.

high-spirited adj. lively.

high street n. the principal shopping street.

high tea n. an early evening meal with tea and cooked food.

high-tech adj. (also **hi-tech**) involving advanced technology and electronics.

high tide n. the tide at its highest level.

high water n. = **high tide**.

highway n. a public road; a main route.

highwayman n. (pl. **-men**) a person (usu. on horseback) who robbed travellers in former times.

hijack v. seize control illegally of (a vehicle or aircraft in transit). ● n. hijacking. □ **hijacker** n.

hike n. a long walk. ● v. go for a hike.

hilarious adj. very funny; boisterous and merry. □ **hilariously** adv., **hilarity** n.

hill n. a raised part of the earth's surface, lower than a mountain; a slope in a road etc.

hillock n. a small hill.

hilt n. the handle of a sword or dagger. □ **to the hilt** completely.

him pron. the objective case of he.

Himalayan adj. of the Himalaya Mountains.

himself pron. the emphatic and reflexive form of he and him.

hind n. a female deer. ● adj. situated at the back.

hinder v. delay the progress of.

Hindi n. a group of languages of northern India.

hindmost adj. furthest behind.

hindrance n. something that hinders; hindering; being hindered.

hindsight n. wisdom about an event after it has occurred.

Hindu n. a person whose religion is Hinduism. ● adj. of Hindus.

Hinduism n. the principal religion and philosophy of India.

Hindustani n. the language of much of northern India and Pakistan.

hinge n. a movable joint such as that on a door or lid. ● v. attach or be attached by hinge(s). □ **hinge on** depend on.

hint n. a slight indication; an indirect suggestion; a piece of practical information. ● v. make a hint.

hinterland n. the district behind a coast etc. or served by a port or other centre.

hip n. **1** the projection of the pelvis on each side of the body. **2** the fruit of the rose.

hip hop n. a youth culture combining rap, graffiti, and break-dancing.

hippie n. (also **hippie**) a young person rejecting convention, supporting peace and free love.

hippopotamus n. (pl. **hippopotamuses** or **hippopotami**) a large African river animal with a thick skin.

hire v. purchase the temporary use of. ● n. hiring. □ **hire out** grant temporary use of for payment.

hireling n. (derog.) a hired helper.

hire purchase n. a system of purchase by payment in instalments.

hirsute (herss-yoot) adj. hairy, shaggy.

his adj. & poss.pron. belonging to him.

Hispanic *adj. & n.* (a native) of Spain or a Spanish-speaking country.

hiss *n.* a sound like 's'. ● *v.* make this sound; utter with a hiss; express disapproval in this way.

histamine *n.* a substance present in the body associated with allergic reactions.

histology *n.* the study of organic tissues.

historian *n.* an expert on history.

historic *adj.* famous in history.

historical *adj.* of or concerned with history. ● **historically** *adv.*

history *n.* past events; a methodical record of these; the study of past events; a story. □ **make history** do something memorable.

histrionic *adj.* of acting; theatrical in manner. ● *n.pl.* (**histrionics**) theatrical behaviour.

hit *v.* (**hit, hitting**) **1** strike with a blow or missile, come forcefully against. **2** affect badly. **3** reach; find. ● *n.* a blow, a stroke; a shot that hits its target; a success. □ **hit it off** get on well together. □ **hitter** *n.*

hitch *v.* **1** move (a thing) with a slight jerk; fasten with a loop or hook. **2** hitch-hike, obtain (a lift) in this way. ● *n.* **1** a slight jerk. **2** a noose or knot of various kinds. **3** a problem or snag.

hitch-hike *v.* travel by seeking free lifts in passing vehicles. □ **hitch-hiker** *n.*

hit list *n.* (*slang*) a list of prospective victims.

hi-tech *adj.* = **high-tech.**

hither *adv.* to or towards this place.

hitherto *adv.* until this time.

HIV *abbr.* human immunodeficiency virus (causing Aids).

hive *n.* **1** a structure in which bees live. **2** a skin eruption, esp.

nettle-rash. □ **hive off** separate from a larger group.

HMG *abbr.* Her (or His) Majesty's Government.

HMS *abbr.* Her (or His) Majesty's Ship.

HMSO *abbr.* Her (or His) Majesty's Stationery Office.

HNC *abbr.* Higher National Certificate.

HND *abbr.* Higher National Diploma.

Ho *symb.* holmium.

hoard *v.* save and store away. ● *n.* things hoarded.

hoarding *n.* a fence of boards, often bearing advertisements.

hoar frost *n.* white frost.

hoarse *adj.* (of a voice) rough and dry-sounding. □ **hoarsely** *adv.*, **hoarseness** *n.*

hoary *adj.* (**hoarier**) grey with age; (of a joke etc.) old.

hoax *v.* deceive jokingly. ● *n.* a joking deception. □ **hoaxer** *n.*

hob *n.* a cooking surface with hotplates.

hobble *v.* **1** walk lamely. **2** fasten the legs of (a horse) to limit its movement. ● *n.* **1** a hobbling walk. **2** a rope etc. used to hobble a horse.

hobby *n.* something done for pleasure in one's spare time.

hobby horse *n.* a stick with a horse's head, as a toy; a favourite topic.

hobgoblin *n.* a mischievous imp.

hobnail *n.* a heavy-headed nail for boot-soles.

hobnob *v.* (**hobnobbed**) spend time together in a friendly way.

Hobson's choice no alternative to the thing offered.

hock *n.* **1** the middle joint of an animal's hind leg. **2** German white wine.

hockey *n.* **1** a field game played with curved sticks and a small hard ball. **2** ice hockey.

hocus-pocus *n.* trickery.

hod *n.* **1** a trough on a pole for carrying mortar or bricks. **2** a tall container for coal.

hodgepodge var. of **hotchpotch**.

hoe *n.* a tool for loosening soil or scraping up weeds. ●*v.* (**hoed, hoeing**) dig or scrape with a hoe.

hog *n.* **1** a castrated male pig reared for meat. **2** a greedy person. ●*v.* (**hogged**) take greedily; hoard selfishly.

hogmanay *n.* (*Scot.*) New Year's Eve.

hoick *v.* lift or bring out with a jerk.

hoi polloi *n.* ordinary people.

hoist *v.* raise or haul up. ●*n.* an apparatus for hoisting things.

hoity-toity *adj.* haughty.

hokum *n.* (*slang*) sentimental or unreal material in a story; nonsense.

hold *v.* (**held, holding**) **1** keep in one's arms or hands etc. or in one's possession or control; keep in a position or condition; contain; bear the weight of. **2** remain unbroken under strain; continue, remain valid. **3** cause to take place. **4** believe. ●*n.* **1** an act, manner, or means of holding; a means of exerting influence. **2** a storage cavity below a ship's deck. □ **hold out** offer; last; continue to make a demand. **hold up** hinder; stop and rob. **hold with** approve of. □ **holder** *n.*

holdall *n.* a large soft travel bag.

holding *n.* something held or owned; land held by an owner or tenant.

hold-up *n.* a delay; a robbery.

hole *n.* **1** a hollow place; a burrow; an aperture. **2** a wretched place; an awkward situation. ●*v.* make hole(s) in.

holey *adj.* full of holes.

holiday *n.* a period of recreation. ●*v.* spend a holiday.

holiness *n.* being holy.

holism *n.* (also **wholism**) treating the whole person rather than just the symptoms. ● **holistic** *adj.*

hollandaise sauce *n.* a yellow sauce containing butter, egg yolks, and vinegar.

hollow *adj.* **1** empty within, not solid; sunken; echoing as if in something hollow. **2** worthless. ●*n.* a cavity; a sunken place; a valley. ●*v.* make hollow. □ **hollowly** *adv.*, **hollowness** *n.*

holly *n.* an evergreen shrub with prickly leaves and red berries.

hollyhock *n.* a plant with large flowers on a tall stem.

holmium *n.* a metallic element (symbol Ho).

holocaust *n.* large-scale destruction, esp. by fire.

hologram *n.* a three-dimensional photographic image.

holograph *adj.* & *n.* (a document) written wholly in the handwriting of the author.

holography *n.* the study of holograms.

holster *n.* a leather case holding a pistol or revolver.

holy *adj.* (**holier**) belonging or devoted to God and revered; consecrated.

homage *n.* things said or done as a mark of respect or loyalty.

home *n.* the place where one lives; a dwelling house; an institution where those needing care may live. ●*adj.* of one's home or country; played on one's own ground. ●*adv.* at or to one's home; to the point aimed at. ●*v.* make its way home or to a target.

homeland *n.* one's native land.

homeless *adj.* lacking a home. □ **homelessness** *n.*

homely *adj.* (**homelier**) simple and informal; (*Amer.*) plain, not beautiful. □ **homeliness** *n.*

homeopathy Amer. sp. of **homoeopathy**.

homesick adj. longing for home.

home truth n. an unpleasant truth about oneself.

homeward adj. & adv. going towards home. □ **homewards** adv.

homework n. work set for a pupil to do away from school.

homicide n. killing of one person by another. □ **homicidal** adj.

homily n. a moralizing lecture.

hominid adj. & n. (a member) of the family of existing and fossil man.

homoeopathy (hohm-i-op-ăthi) n. (Amer. **homeopathy**) treatment of a disease by very small doses of a substance that would produce the same symptoms in a healthy person. □ **homoeopath** n., **homoeopathic** adj.

homogeneous adj. of the same kind, uniform. □ **homogeneously** adv., **homogeneity** n.

■ **Usage** Homogeneous is often confused with homogenous, but that is a term in biology meaning 'similar owing to common descent'.

homogenize v. (also -**ise**) treat (milk) so that cream does not separate and rise to the top.

homograph n. (also **homonym**) a word spelt the same as another.

homophobia n. hatred or fear of homosexuals.

homophone n. a word with the same sound as another.

Homo sapiens (hoh-moh sap-i-enz) n. modern humans.

homosexual adj. sexually attracted to people of the same sex. ● n. a homosexual person. □ **homosexuality** n.

hone v. sharpen on a whetstone.

honest adj. truthful, trustworthy; fairly earned. □ **honesty** n.

honestly adv. in an honest way; really.

honey n. (pl. **honeys**) **1** a sweet substance made by bees from nectar. **2** (esp. Amer.) darling.

honey bee n. the common hive-bee.

honeycomb n. a bees' wax structure holding their honey and eggs; a pattern of six-sided sections.

honeydew n. **1** a sticky substance on plants, secreted by aphids. **2** a variety of melon.

honeyed adj. flattering, pleasant.

honeymoon n. **1** a holiday for a newly married couple. **2** an initial period of goodwill. ● v. spend a honeymoon.

honeysuckle n. a climbing shrub with fragrant pink and yellow flowers.

honk n. the sound of an old-fashioned car horn; the cry of a wild goose. ● v. make this noise.

honor Amer. sp. of **honour**.

honorarium n. (pl. **honorariums** or **honoraria**) a voluntary payment make where no fee is legally required.

honorary adj. given as an honour; unpaid.

honour n. (Amer. **honor**) great respect or public regard; a mark of this, a privilege; good personal character or reputation. ● v. feel honour for; confer an honour on; pay (a cheque) or fulfil (a promise etc.).

honourable adj. (Amer. **honorable**) deserving, possessing, or showing honour. □ **honourably** adv.

honours degree n. a degree of a higher standard than a pass.

hood n. **1** a covering for the head and neck. **2** a hood-like thing or cover; a folding roof over a car; (*Amer.*) a car bonnet. **3** (*Amer.*) a gangster or gunman.

hoodlum *n.* a young thug or hooligan.

hoodoo *n.* (*Amer.*) bad luck; something causing this.

hoodwink *v.* deceive.

hoof *n.* (*pl.* **hoofs** or **hooves**) the horny part of a horse's foot.

hook *n.* **1** a curved device for catching hold or hanging things on. **2** a short blow made with the elbow bent. ● *v.* **1** grasp, catch, or fasten with a hook. **2** scoop or propel with a curving movement. □ **off the hook** freed from a difficulty; (of a telephone receiver) off its rest.

hookah *n.* an oriental tobacco pipe with a long tube passing through water.

hooked *adj.* **1** hook-shaped. **2** addicted.

hook-up *n.* an interconnection of broadcasting equipment.

hookworm *n.* a parasitic worm with hooklike mouthparts.

hooligan *n.* a young ruffian. □ **hooliganism** *n.*

hoop *n.* a circular band of metal or wood; a metal croquet arch.

hoopla *n.* a game in which rings are thrown to encircle a prize.

hoopoe *n.* a bird with a crest and striped plumage.

hooray *int.* & *n.* = **hurrah**.

hoot *n.* **1** an owl's cry; the sound of a hooter; a cry of laughter or disapproval. **2** something funny. ● *v.* make a hoot.

hooter *n.* a siren or steam whistle; a car horn.

Hoover *n.* (*trade mark*) a vacuum cleaner. ● *v.* (**hoover**) clean with a vacuum cleaner.

hop *v.* (**hopped**) **1** jump on one foot or (of an animal) from both or all feet. **2** (*colloquial*) make a quick short trip. ● *n.* **1** a hopping movement; an informal dance; a short flight. **2** a plant

cultivated for its cones, used to give a bitter flavour to beer.

hope *n.* a feeling of expectation and desire; something giving cause for this; what one hopes for. ● *v.* feel hope. □ **hopeful** *adj.*

hopefully *adv.* **1** in a hopeful way. **2** it is to be hoped.

■ **Usage** The use of *hopefully* to mean 'it is to be hoped' is common, but it is considered incorrect by some people.

hopeless *adj.* **1** without hope. **2** inadequate, incompetent. □ **hopelessly** *adv.*, **hopelessness** *n.*

hopper *n.* **1** one who hops. **2** a container with an opening at the base for discharging its contents.

hopscotch *n.* a game involving hopping over marked squares.

horde *n.* a large group or crowd.

horizon *n.* **1** the line at which earth and sky appear to meet. **2** the limit of knowledge or interests.

horizontal *adj.* parallel to the horizon, going straight across. □ **horizontally** *adv.*

hormone *n.* a substance produced by the body or a plant to stimulate an organ or growth. □ **hormonal** *adj.*

horn *n.* **1** a hard pointed growth on the heads of certain animals; the substance of this; a similar projection. **2** a wind instrument with a trumpet-shaped end; a device for sounding a warning signal.

hornblende *n.* a dark mineral constituent of granite etc.

hornet *n.* a large wasp.

hornpipe *n.* a lively solo dance traditionally performed by sailors.

horn-rimmed *adj.* with frames of material like horn or tortoiseshell.

horny adj. (**hornier**) **1** of or like horn; hardened and calloused. **2** (slang) sexually excited.

horology n. the art of making clocks. □ **horologist** n.

horoscope n. a forecast of events based on the positions of stars.

horrendous adj. horrifying. □ **horrendously** adv.

horrible adj. causing horror; (colloquial) unpleasant. □ **horribly** adv.

horrid adj. horrible.

horrific adj. horrifying. □ **horrifically** adv.

horrify v. arouse horror in.

horror n. loathing and fear; intense dislike or dismay; a person or thing causing horror.

hors d'oeuvre (or **dervr**) n. food served as an appetizer.

horse n. **1** a four-legged animal with a mane and tail. **2** a padded structure for vaulting over in a gym. □ **horse around** fool about.

horseback n. □ **on horseback** riding on a horse.

horsebox n. a vehicle for transporting horses.

horse chestnut n. a brown shiny nut; the tree bearing this.

horsefly n. a large biting fly.

horseman n. (pl. -men) a rider on horseback. □ **horsemanship** n.

horseplay n. boisterous play.

horsepower n. a unit for measuring the power of an engine.

horseradish n. a plant with a hot-tasting root used to make sauce.

horse sense n. common sense.

horseshoe n. a U-shaped strip of metal nailed to a horse's hoof; something shaped like this.

horsewoman n. (pl. -women) a woman rider on horseback.

horsy adj. of or like a horse; interested in horses.

horticulture n. the art of garden cultivation. □ **horticultural** adj., **horticulturist** n.

hose n. **1** (also **hosepipe**) a flexible tube for conveying water. **2** stockings and socks. ● v. water or spray with a hosepipe.

hosiery n. stockings, socks, etc.

hospice n. a hospital or home for the terminally ill.

hospitable adj. giving hospitality, welcoming. □ **hospitably** adv.

hospital n. an institution for treatment of sick or injured people.

hospitality n. friendly and generous entertainment of guests.

hospitalize v. (also **-ise**) send or admit to a hospital. □ **hospitalization** n.

host n. **1** a large number of people or things. **2** a person entertaining guests. **3** an organism on which another lives as a parasite. ● v. act as host to.

hostage n. a person held as security that the holder's demands will be satisfied.

hostel n. a lodging house for students, nurses, homeless people, etc.

hostess n. a woman entertaining guests.

hostile adj. of an enemy; unfriendly.

hostility n. being hostile, enmity; (**hostilities**) acts of warfare.

hot adj. (**hotter**) **1** at or having a high temperature. **2** producing a burning sensation to the taste. **3** eager, angry; excited, excitable. □ **hot up** (**hotted**) (colloquial) make or become hot or exciting. **in hot water** in trouble or disgrace.

hot air n. excited or boastful talk.

hotbed n. a place encouraging vice, intrigue, etc.

hotchpotch n. (also **hodgepodge**) a jumble.

hot dog *n.* a hot sausage in a bread roll.

hotel *n.* a building where meals and rooms are provided for travellers.

hotelier *n.* a hotel-keeper.

hotfoot *adv.* in eager haste.

hothead *n.* an impetuous person. □ **hot-headed** *adj.*

hothouse *n.* a heated greenhouse.

hotline *n.* a direct telephone line for speedy communication.

hotplate *n.* a heated surface on a cooker or hob.

hommus var. of **hummus**.

hound *n.* a dog used in hunting. ● *v.* pursue, harass.

hour *n.* one twenty-fourth part of a day and night; a point in time; (**hours**) a period for daily work.

hourglass *n.* a glass containing sand that takes one hour to trickle from the upper to the lower section.

houri (**hoor**-i) *n.* a beautiful young woman of the Muslim paradise.

hourly *adj.* done or occurring once an hour; continual. ● *adv.* every hour.

house *n.* (howss) **1** a building for people to live in, or for a specific purpose; a household. **2** a legislative assembly; a business firm; a theatre audience or performance; a family or dynasty. ● *v.* (howz) provide accommodation or storage space for; encase.

house arrest *n.* detention in one's own home.

houseboat *n.* a boat fitted up as a dwelling.

housebound *adj.* unable to leave one's house, esp. through illness.

housebreaker *n.* a burglar. □ **housebreaking** *n.*

housecoat *n.* a woman's dressing gown.

household *n.* the occupants of a house living as a group.

householder *n.* a person owning or renting a house or flat.

household word *n.* a familiar saying or name.

housekeeper *n.* a person employed to look after a household.

housekeeping *n.* management of household affairs; money to be used for this.

housemaster *n.* a male teacher in charge of a school boarding house.

housemistress *n.* a female teacher in charge of a school boarding house.

house-proud *adj.* giving great attention to the appearance of one's home.

house-trained *adj.* (of animals) trained to be clean in the house.

house-warming *n.* a party to celebrate occupation of a new home.

housewife *n.* (*pl.* **-wives**) a woman managing a household.

housework *n.* cleaning and cooking etc. in a house.

housing *n.* **1** accommodation. **2** a rigid case enclosing machinery.

hove *see* **heave**.

hovel *n.* a small miserable dwelling.

hover *v.* **1** (of a bird etc.) remain in one place in the air. **2** linger, wait close at hand.

hovercraft *n.* a vehicle supported by air thrust downwards from its engines.

how *adv.* **1** by what means, in what way; to what extent or amount etc. **2** in what condition.

howdah *n.* a seat with a canopy on an elephant's back.

however *adv.* in whatever way, to whatever extent; nevertheless.

howitzer *n.* a short gun firing shells at high elevation.

howl *n.* a long loud wailing cry or sound. ● *v.* make or utter with a howl; weep loudly.

howler *n.* **1** (*colloquial*) a stupid mistake. **2** a person or animal that howls.

hoyden *n.* a boisterous girl.

h.p. *abbr.* hire purchase; horse-power.

HQ *abbr.* headquarters.

HRH *abbr.* His or Her Royal Highness.

HRT *abbr.* hormone replacement therapy.

hub *n.* the the central part of a wheel; the centre of activity.

hubbub *n.* a confused noise of voices.

hubcap *n.* the cover for the hub of a car wheel.

hubris (hew-bris) *n.* arrogant pride.

huckleberry *n.* a low shrub with dark blue fruit, common in North America.

huddle *v.* crowd into a small place. ● *n.* a close mass.

hue *n.* a colour, a tint. □ **hue and cry** an outcry.

huff *n.* a fit of annoyance. ● *v.* blow. □ **huffy** *adj.* **huffily** *adv.*

hug *v.* (**hugged**) squeeze tightly in one's arms; keep close to. ● *n.* a hugging movement.

huge *adj.* extremely large. □ **hugely** *adv.*

hula *n.* a Polynesian women's dance.

hula hoop *n.* a large hoop for spinning round the body.

hulk *n.* the body of an old ship; a large clumsy-looking person or thing.

hulking *adj.* (*colloquial*) large and clumsy.

hull *n.* **1** the framework of a ship. **2** the pod of a pea or bean; the cluster of leaves on a strawberry. ● *v.* remove the hulls of (beans, strawberries, etc.).

hullabaloo *n.* an uproar.

hullo var. of **hello**.

hum *v.* (**hummed**) **1** sing with closed lips; make a similar sound. **2** be in a state of activity. ● *n.* a humming sound.

human *adj.* of mankind; of people. ● *n.* a human being. □ **humanly** *adv.*

humane *adj.* kind-hearted, merciful. □ **humanely** *adv.*

humanism *n.* a system of thought concerned with human affairs and ethics (not theology); promotion of human welfare. □ **humanist** *n.*, **humanistic** *adj.*

humanitarian *adj.* promoting human welfare and reduction of suffering. □ **humanitarianism** *n.*

humanity *n.* **1** human nature or qualities; kindness; the human race. **2** (**humanities**) arts subjects.

humanize *v.* (also **-ise**) make human; make humane.

humble *adj.* having a modest opinion of one's own importance; of low rank; not large or expensive. ● *v.* lower the rank or self-importance of. □ **humbly** *adv.*

humbug *n.* **1** misleading behaviour or talk. **2** a hard usu. peppermint-flavoured sweet.

humdrum *adj.* dull, commonplace.

humerus *n.* (*pl.* **humeri**) the bone in the upper arm. □ **humeral** *adj.*

humid *adj.* (of air) damp. □ **humidity** *n.*

humidify *v.* keep air moist in a room etc. □ **humidifier** *n.*

humiliate *v.* cause to feel disgraced. □ **humiliation** *n.*

humility *n.* a humble condition or attitude of mind.

hummock *n.* a hump in the ground.

hummus *n.* (also **hoommos**) a paste or dip of ground chickpeas, sesame oil, lemon juice, and garlic.

humour *n.* (*Amer.* **humor**) **1** the quality of being amusing; the ability to perceive and enjoy this. **2** a state of mind. ● *v.* keep (a person) contented by doing as he or she wishes. □ **humorous** *adj.*, **humorously** *adv.*

hump *n.* a rounded projecting part; a curved deformity of the spine. ● *v.* **1** form into a hump. **2** hoist and carry.

humpback *n.* a hunchback; a whale with a hump on its back.

humpback bridge *n.* a small steeply arched bridge.

humus *n.* rich dark organic material in soil, formed by decay of dead leaves and plants.

hunch *v.* bend into a hump. ● *n.* **1** a hump. **2** an intuitive feeling.

hunchback *n.* a person with a humped back.

hundred *n.* ten times ten (100, C). □ **hundredth** *adj.* & *n.*

hundredfold *adj.* & *adv.* 100 times as much or as many.

hundredweight *n.* a measure of weight, 112 lb (50.802 kg), or in America 100 lb (45.359 kg); a metric unit of weight equal to 50 kg.

hung see **hang**. *adj.* (of a council, parliament, etc.) with no party having a clear majority.

hung-over *adj.* (*colloquial*) having a hangover.

Hungarian *adj.* & *n.* a (native, the language) of Hungary.

hunger *n.* **1** pain or discomfort felt when one has not eaten for some time. **2** a strong desire. ● *v.* feel hunger.

hunger strike *n.* refusal of food as a form of protest.

hungry *adj.* (**hungrier**) feeling hunger. □ **hungrily** *adv.*

hunk *n.* a large or clumsy piece.

hunt *v.* **1** pursue (wild animals) for food or sport; pursue with hostility; seek; search. **2** (of an engine) run unevenly. ● *n.* hunting; a hunting group.

hunter *n.* one who hunts; a horse used for hunting.

hurdle *n.* a portable fencing panel; a frame to be jumped over in a race; an obstacle, a difficulty. □ **hurdler** *n.*

hurl *v.* throw violently. ● *n.* a violent throw.

hurly-burly *n.* rough bustle.

hurrah *int.* & *n.* (also **hurray**) an exclamation of joy or approval.

hurricane *n.* a violent storm-wind.

hurricane lamp *n.* a lamp with the flame protected from the wind.

hurried *adj.* done with great haste. □ **hurriedly** *adv.*

hurry *v.* act or move with eagerness or too quickly; cause to do this. ● *n.* hurrying.

hurt *v.* cause pain, harm, or injury (to); feel pain. ● *n.* injury, harm. □ **hurtful** *adj.*

hurtle *v.* move or hurl rapidly.

husband *n.* a married man in relation to his wife. ● *v.* use economically, try to save.

husbandry *n.* farming; management of resources.

hush *v.* make or become silent. ● *n.* silence.

husk *n.* the dry outer covering of certain seeds and fruits. ● *v.* remove the husk from.

husky *adj.* (**huskier**) **1** dry; hoarse. **2** burly. ● *n.* an Arctic sledge-dog. □ **huskily** *adv.*, **huskiness** *n.*

hustle *v.* push roughly; hurry. ● *n.* hustling.

hustings *n.* election campaigning.

hut *n.* a small simple or roughly made house or shelter.

hutch *n.* a box-like cage for rabbits.

hyacinth *n.* a plant with fragrant bell-shaped flowers.

hyaena var. of **hyena**.

hybrid *n.* the offspring of two different species or varieties; something made by combining different elements. ● *adj.* produced in this way. □ **hybridism** *n.*

hybridize *v.* (also **-ise**) cross-breed; produce hybrids; interbreed. □ **hybridization** *n.*

hydrangea *n.* a shrub with pink, blue, or white flowers in clusters.

hydrant *n.* a pipe from a water main in a street to which a hose can be attached.

hydrate *n.* a chemical compound of water with another substance.

hydraulic *adj.* operated by pressure of fluid conveyed in pipes; hardening under water. ● *n.* (**hydraulics**) the science of hydraulic operations. □ **hydraulically** *adv.*

hydride *n.* a compound of hydrogen with an element.

hydrocarbon *n.* a compound of hydrogen and carbon.

hydrochloric acid *n.* a corrosive acid containing hydrogen and chlorine.

hydrodynamic *adj.* of the forces exerted by liquids in motion.

hydroelectric *adj.* using water-power to produce electricity.

hydrofoil *n.* a boat with a structure that raises its hull out of the water when in motion.

hydrogen *n.* an odourless gas, the lightest element (symbol H).

hydrogen bomb *n.* a powerful bomb releasing energy by fusion of hydrogen nuclei.

hydrolysis *n.* decomposition by chemical reaction with water. □ **hydrolytic** *adj.*

hydrometer *n.* a device measuring the density of liquids.

hydrophobia *n.* abnormal fear of water; rabies.

hydroponics *n.* growing plants in water impregnated with chemicals.

hydrostatic *adj.* of the pressure and other characteristics of liquid at rest.

hydrotherapy *n.* use of water to treat diseases etc.

hydrous *adj.* containing water.

hyena *n.* (also **hyaena**) a wolf-like animal with a howl that sounds like laughter.

hygiene *n.* cleanliness as a means of preventing disease. □ **hygienic** *adj.*, **hygienically** *adv.*, **hygienist** *n.*

hymen *n.* the membrane partly closing the opening of the vagina of a virgin girl or woman.

hymn *n.* a Christian song of praise.

hype *n.* (*slang*) intensive promotion of a product.

hyper- *pref.* excessively.

hyperactive *adj.* abnormally active. □ **hyperactivity** *n.*

hyperbola *n.* (*pl.* **hyperbolas** or **hyperbolae**) the curve produced by a cut made through a cone at an angle with the base greater than that of the side of the cone.

hyperbole *n.* (hy-per-bŏli) an exaggerated statement.

hyperglycaemia (hy-per-gly-see-miă) *n.* (*Amer.* **hyperglycemia**) excess glucose in the blood.

hypermarket *n.* a very large self-service store.

hypersonic *adj.* of speeds more than five times that of sound.

hypertension *n.* **1** abnormally high blood pressure. **2** extreme tension.

hypertext *n.* the provision of several texts on a computer screen.

hyperventilation *n.* abnormally rapid breathing.

hyphen *n.* a sign (-) used to join words together or divide a word into parts.

hyphenate *v.* join or divide with a hyphen. □ **hyphenation** *n.*

hypnosis *n.* a sleep-like condition produced in a person who then obeys suggestions.

hypnotic *adj.* of or producing hypnosis. □ **hypnotically** *adv.*

hypnotism *n.* hypnosis.

hypnotize *v.* (also **-ise**) produce hypnosis in; fascinate, dominate the mind or will of. □ **hypnotist** *n.*

hypo- *pref.* under; below normal.

hypochondria *n.* the state of constantly imagining that one is ill. □ **hypochondriac** *n.*

hypocrisy *n.* falsely pretending to be virtuous; insincerity.

hypocrite *n.* a person guilty of hypocrisy. □ **hypocritical** *adj.*, **hypocritically** *adv.*

hypodermic *adj.* injected beneath the skin; used for such injections. ● *n.* a hypodermic syringe.

hypotenuse *n.* the longest side of a right-angled triangle.

hypothermia *n.* the condition of having an abnormally low body temperature.

hypothesis *n.* (*pl.* **hypotheses**) a supposition put forward as a basis for reasoning or investigation.

hypothetical *adj.* supposed but not necessarily true. □ **hypothetically** *adv.*

hysterectomy *n.* the surgical removal of the womb.

hysteria *n.* wild uncontrollable emotion. □ **hysterical** *adj.*, **hysterically** *adv.*

hysterics *n.pl.* a hysterical outburst; (*colloquial*) uncontrollable laughter.

Hz *abbr.* hertz.

I *pron.* the person speaking or writing and referring to himself or herself.

iambic *adj.* & *n.* (of verse) using iambuses, metrical feet of one long and one short syllable.

Iberian *adj.* of the peninsula comprising Spain and Portugal.

ibex *n.* (*pl.* **ibex** or **ibexes**) a mountain goat.

ibid. *abbr.* in the same book, passage, etc.

ibis *n.* a wading bird.

ice *n.* frozen water; an ice cream. ● *v.* **1** become frozen; make very cold. **2** decorate with icing.

iceberg *n.* a mass of ice floating in the sea.

icebox *n.* the freezing compartment in a fridge; (*Amer.*) a fridge.

ice cream *n.* a sweet creamy frozen food.

ice hockey *n.* a game like hockey played on ice.

Icelandic *adj.* & *n.* (the language) of Iceland. □ **Icelander** *n.*

ichthyology (ik-thi-ol-ŏji) *n.* the study of fishes. □ **ichthyologist** *n.*

icicle *n.* a piece of ice hanging downwards.

icing *n.* a mixture of powdered sugar etc. used to decorate cakes.

icon *n.* (also **ikon**) a sacred painting or mosaic; an image or statue; (*Computing*) a graphic symbol on a computer screen.

iconoclast *n.* a person who attacks cherished beliefs. □ **iconoclasm** *n.*, **iconoclastic** *adj.*

icy adj. (**icier**) **1** very cold; covered with ice. **2** very unfriendly. □ **icily** adv., **iciness** n.

ID abbr. identification.

idea n. a plan etc. formed in the mind; an opinion; a mental impression; a vague belief.

ideal adj. satisfying one's idea of what is perfect. ● n. a person or thing regarded as perfect or as a standard to aim at. □ **ideally** adv.

idealist n. a person with high ideals. □ **idealism** n., **idealistic** adj.

idealize v. (also **-ise**) regard or represent as perfect.

identical adj. the same; exactly alike. □ **identically** adv.

identify v. recognize as being a specified person or thing; associate (oneself) closely in feeling or interest (with a person, idea, etc.). □ **identifiable** adj., **identification** n.

identikit n. a set of pictures of features that can be put together to form a likeness of a person.

identity n. who or what a person or thing is; sameness.

ideogram n. a symbol or picture representing an idea, e.g. Chinese characters or road signs.

ideology n. ideas that form the basis of a political or economic theory. □ **ideological** adj.

idiocy n. the state of being an idiot; extreme foolishness.

idiom n. a phrase or usage peculiar to a language.

idiomatic adj. full of idioms. □ **idiomatically** adv.

idiosyncrasy n. a person's own characteristic way of behaving. □ **idiosyncratic** adj.

idiot n. a very stupid person. □ **idiotic** adj., **idiotically** adv.

idle adj. not employed or in use; lazy; aimless. ● v. be idle, pass (time) aimlessly; (of an engine) run slowly in neutral gear. □ **idly** adv., **idleness** n., **idler** n.

idol n. an image worshipped as a god; an idolized person or thing.

idolatry n. worship of idols. □ **idolater** n., **idolatrous** adj.

idolize v. (also **-ise**) love or admire excessively.

idyll (**id-il**) n. a peaceful or romantic scene; a description of this, usu. in verse. □ **idyllic** adj., **idyllically** adv.

i.e. abbr. that is.

if conj. on condition that; supposing that; whether. ● n. a condition or supposition.

igloo n. a dome-shaped Eskimo snow house.

igneous adj. (of rock) formed by volcanic action.

ignite v. set fire to; catch fire.

ignition n. igniting; a mechanism producing a spark to ignite the fuel in an engine.

ignoble adj. not noble in character, aims, or purpose. □ **ignobly** adv.

ignominy n. disgrace, humiliation. □ **ignominious** adj., **ignominiously** adv.

ignoramus n. (pl. **ignoramuses**) an ignorant person.

ignorant adj. lacking knowledge; behaving rudely through not knowing good manners. □ **ignorantly** adv., **ignorance** n.

ignore v. take no notice of.

iguana n. a tropical tree-climbing lizard.

ikebana n. the art of Japanese flower arranging.

ikon var. of **icon**.

ileum n. part of the small intestine.

ilk n. □ **of that ilk** of that kind.

ill adj. unwell; bad; harmful; hostile, unkind. ● adv. badly. ● n. evil, harm, injury. □ **ill at ease** uncomfortable, embarrassed.

ill-advised adj. unwise.

illegal adj. against the law. □ **illegally** adv., **illegality** n.

illegible *adj.* not readable. □ **illegibly** *adv.*, **illegibility** *n.*

illegitimate *adj.* born of parents not married to each other; contrary to a law or rule. □ **illegitimately** *adv.*, **illegitimacy** *n.*

ill-gotten *adj.* gained by evil or unlawful means.

illicit *adj.* unlawful, not allowed. □ **illicitly** *adv.*

illiterate *adj.* unable to read and write; uneducated. □ **illiteracy** *n.*

ill-mannered *adj.* having bad manners.

illness *n.* the state of being ill; a particular form of ill health.

illogical *adj.* not logical. □ **illogically** *adv.*, **illogicality** *n.*

ill-treat *v.* treat badly or cruelly.

illuminate *v.* light up; throw light on (a subject); decorate with lights. □ **illumination** *n.*

illumine *v.* light up; enlighten.

illusion *n.* a false belief; something wrongly supposed to exist.

illusionist *n.* a conjuror.

illusory *adj.* (also **illusive**) based on illusion; deceptive.

illustrate *v.* supply (a book etc.) with drawings or pictures; make clear by example(s) or picture(s); serve as an example of. □ **illustration** *n.*, **illustrative** *adj.*, **illustrator** *n.*

illustrious *adj.* distinguished.

ill will *n.* hostility, unkind feeling.

image *n.* an optical appearance of a thing produced in a mirror or through a lens; a likeness; a mental picture; a reputation.

imaginable *adj.* able to be imagined.

imaginary *adj.* existing only in the imagination, not real.

imagination *n.* imagining; the ability to imagine or to plan creatively. □ **imaginative** *adj.*, **imaginatively** *adv.*

imagine *v.* form a mental image of; think, suppose; guess.

imago (i-may-goh) *n.* (*pl.* **imagines** or **imagos**) an insect in its fully developed adult stage.

imam *n.* a Muslim spiritual leader.

imbalance *n.* lack of balance.

imbecile *n.* a stupid person.

imbed var. of **embed**.

imbibe *v.* drink; absorb (ideas).

imbroglio (im-broh-lyoh) *n.* (*pl.* **imbroglios**) a confused situation.

imbue *v.* fill with feelings, qualities, or emotions.

IMF *abbr.* International Monetary Fund.

imitable *adj.* able to be imitated.

imitate *v.* try to act or be like; copy. □ **imitation** *n.*, **imitator** *n.*

imitative *adj.* imitating.

immaculate *adj.* spotlessly clean and tidy; free from blemish or fault. □ **immaculately** *adv.*

immanent *adj.* inherent; present in everything. □ **immanence** *n.*

immaterial *adj.* **1** having no physical substance. **2** of no importance.

immature *adj.* not mature. □ **immaturity** *n.*

immeasurable *adj.* not measurable, immense. □ **immeasurably** *adv.*

immediate *adj.* **1** with no delay. **2** nearest, with nothing between. □ **immediately** *adv.* & *conj.*, **immediacy** *n.*

immemorial *adj.* existing from before what can be remembered.

immense *adj.* extremely great. □ **immensely** *adv.*, **immensity** *n.*

immerse *v.* put completely into liquid; involve deeply.

immersion *n.* immersing.

immersion heater *n.* an electric heater placed in the liquid to be heated.

immigrate *v.* come into a foreign country as a permanent resident. □ **immigrant** adj. & n., **immigration** n.

imminent adj. about to occur. □ **imminently** adv., **imminence** n.

immiscible adj. not able to be mixed (with another substance).

immobile adj. unable to be moved; not moving. □ **immobility** n.

immobilize v. (also **-ise**) make or keep immobile. □ **immobilization** n.

immoderate adj. excessive. □ **immoderately** adv.

immolate v. kill as a sacrifice.

immoral adj. morally wrong. □ **immorally** adv., **immorality** n.

immortal adj. living for ever, not mortal; famous for all time. □ **immortality** n.

immortalize v. (also **-ise**) make immortal.

immovable adj. unable to be moved; unyielding. □ **immovably** adv.

immune adj. having immunity; protected.

immunity n. the ability to resist infection; special exemption.

immunize v. (also **-ise**) make immune to infection, esp. by vaccination. □ **immunization** n.

immunodeficiency n. a reduction in normal resistance to infection.

immunology n. the study of resistance to infection. □ **immunological** adv., **immunologist** n.

immure v. imprison, shut in.

immutable adj. unchangeable. □ **immutably** adv., **immutability** n.

imp n. a small devil; a mischievous child.

impact n. (im-pakt) a collision, the force of this; a strong effect. ● v. (im-pakt) press or wedge firmly. □ **impaction** n.

impair v. damage, weaken. □ **impairment** n.

impala n. (pl. **impala** or **impalas**) a small African antelope.

impale v. fix or pierce with a pointed object. □ **impalement** n.

impalpable adj. not easily grasped by the mind; unable to be felt by touch. □ **impalpably** adv.

impart v. give; make (information etc.) known.

impartial adj. not favouring one more than another. □ **impartially** adv., **impartiality** n.

impassable adj. impossible to travel on or over.

impasse (am-pahss) n. a deadlock.

impassioned adj. passionate.

impassive adj. not feeling or showing emotion. □ **impassively** adv.

impatient adj. feeling or showing lack of patience; intolerant. □ **impatiently** adv., **impatience** n.

impeach v. accuse of a serious crime against the state and bring for trial. □ **impeachment** n.

impeccable adj. faultless. □ **impeccably** adv., **impeccability** n.

impecunious adj. having little or no money.

impedance n. resistance of an electric circuit to the flow of current.

impede v. hinder.

impediment n. a hindrance or obstruction; a defect in speech e.g. a lisp or stammer.

impel v. (**impelled**) urge; drive forward.

impending adj. imminent.

impenetrable adj. unable to be penetrated; incomprehensible. □ **impenetrably** adv., **impenetrability** n.

imperative *adj.* expressing a command; essential. ● *n.* a command; an essential thing.

imperceptible *adj.* too slight to be noticed. □ **imperceptibly** *adv.*

imperfect *adj.* 1 not perfect. 2 (*Grammar*) (of a tense) implying action going on but not completed. □ **imperfectly** *adv.*, **imperfection** *n.*

imperial *adj.* 1 of an empire; majestic. 2 (of measures) belonging to the British official non-metric system. □ **imperially** *adv.*

imperialism *n.* the policy of having or extending an empire. □ **imperialist** *n.*, **imperialistic** *adj.*

imperil *v.* (**imperilled**; *Amer.* **imperiled**) endanger.

imperious *adj.* commanding, bossy. □ **imperiously** *adv.*, **imperiousness** *n.*

impermeable *adj.* not able to be penetrated by liquid.

impersonal *adj.* not showing or influenced by personal feeling. □ **impersonally** *adv.*, **impersonality** *n.*

impersonate *v.* pretend to be (another person). □ **impersonation** *n.*, **impersonator** *n.*

impertinent *adj.* not showing proper respect. □ **impertinently** *adv.*, **impertinence** *n.*

imperturbable *adj.* not excitable, calm.

impervious *adj.* □ **impervious to** not able to be penetrated or influenced by. □ **imperviousness** *n.*

impetigo (im-pi-ty-goh) *n.* a contagious skin disease.

impetuous *adj.* acting or done on impulse. □ **impetuously** *adv.*, **impetuosity** *n.*

impetus *n.* a moving or driving force.

impiety *n.* lack of reverence.

impinge *v.* make an impact; encroach.

impious *adj.* not reverent, wicked. □ **impiously** *adv.*

implacable *adj.* unable to be persuaded or placated; relentless. □ **implacably** *adv.*, **implacability** *n.*

implant *v.* (im-plahnt) fix into; insert (tissue) in a living thing. ● *n.* (im-plahnt) implanted tissue. □ **implantation** *n.*

implausible *adj.* not plausible.

implement *n.* a tool. ● *v.* put into effect. □ **implementation** *n.*

implicate *v.* show or cause to be involved in a crime etc.

implication *n.* implicating; implying; something implied.

implicit *adj.* implied, not explicit; absolute, unquestioning. □ **implicitly** *adv.*

implode *v.* (cause to) burst inwards. □ **implosion** *n.*

implore *v.* request earnestly.

imply *v.* suggest without stating directly; mean.

impolite *adj.* bad-mannered, rude.

impolitic *adj.* unwise; inexpedient.

imponderable *adj.* not able to be estimated. ● *n.* an imponderable thing.

import *v.* (im-port) 1 bring in from abroad or from an outside source. 2 imply, indicate. ● *n.* (im-port) 1 importing; something imported. 2 meaning; importance. □ **importation** *n.*, **importer** *n.*

important *adj.* having a great effect; having great authority or influence. □ **importance** *n.*

importunate *adj.* making persistent requests. □ **importunity** *n.*

importune *v.* make insistent requests to; solicit.

impose | impulsive

impose v. levy (a tax); inflict; force acceptance of. □ **impose on** take unfair advantage of.

imposing adj. impressive.

imposition n. the act of imposing something; something imposed; a burden imposed unfairly.

impossible adj. not possible; unendurable. □ **impossibly** adv., **impossibility** n.

impostor n. a person who fraudulently pretends to be someone else.

imposture n. a fraudulent deception.

impotent adj. powerless; (of a male) unable to copulate successfully. □ **impotence** n.

impound v. take (property) into legal custody; confiscate.

impoverish v. cause to become poor; exhaust the strength or fertility of. □ **impoverishment** n.

impracticable adj. not able to be put into practice.

impractical adj. not practical; unwise.

imprecation n. a spoken curse.

imprecise adj. not precise.

impregnable adj. safe against attack. □ **impregnability** n.

impregnate v. introduce sperm or pollen into and fertilize; penetrate all parts of. □ **impregnation** n.

impresario n. (pl. **impresarios**) an organizer of public entertainment.

impress v. **1** cause to form a strong (usu. favourable) opinion; fix in the mind. **2** press a mark into.

impression n. **1** an effect produced on the mind; an uncertain idea. **2** an imitation done for entertainment. **3** an impressed mark; a reprint.

impressionable adj. easily influenced.

impressionism n. a style of art giving a general impression without detail. □ **impressionist** n., **impressionistic** adj.

impressive adj. making a strong favourable impression. □ **impressively** adv., **impressiveness** n.

imprint n. (im-print) a mark made by pressing on a surface; a publisher's name etc. on a title-page. ● v. (im-print) impress or stamp a mark etc. on.

imprison v. put into prison; keep in confinement. □ **imprisonment** n.

improbable adj. not likely to be true or to happen. □ **improbably** adv., **improbability** n.

improbity n. dishonesty.

impromptu adj. & adv. without preparation or rehearsal.

improper adj. unsuitable, wrong; not conforming to social conventions. □ **improperly** adv., **impropriety** n.

improve v. make or become better. □ **improvement** n.

improvident adj. not providing for future needs. □ **improvidently** adv., **improvidence** n.

improvise v. provide from whatever materials are at hand; act or speak without a script or rehearsal. □ **improvisation** n.

imprudent adj. unwise, rash. □ **imprudently** adv., **imprudence** n.

impudent adj. cheeky, impertinent. □ **impudently** adv., **impudence** n.

impugn (im-pewn) v. express doubts about the truth or honesty of.

impulse n. a sudden urge to do something; a push; impetus; a stimulating force in a nerve.

impulsion n. impelling; impulse; impetus.

impulsive adj. acting or done on impulse, without planning or

prior thought. □ **impulsively**
adv., **impulsiveness** *n.*

impunity *n.* freedom from punishment or injury.

impure *adj.* not pure.

impurity *n.* being impure; a substance that makes another impure.

impute *v.* attribute (a fault etc.); blame. □ **imputation** *n.*

In *symb.* indium.

in *prep.* having as a position or state within (limits of space, time, surroundings, etc.); having as a state or manner; into, or towards. ● *adv.* **1** in or to a position bounded by limits; inside. **2** in fashion, season, or office. ● *adj.* **1** internal; living etc. inside. **2** fashionable. □ **in for** about to experience; competing in. **ins and outs** details. **in so far as** to the extent that.

in. *abbr.* inch(es).

inability *n.* being unable.

in absentia *adv.* in his, her, or their absence.

inaccessible *adj.* not accessible; unfriendly.

inaccurate *adj.* not accurate. □ **inaccurately** *adv.*, **inaccuracy** *n.*

inaction *n.* lack of action.

inactive *adj.* not active. □ **inactivity** *n.*

inadequate *adj.* not adequate; not sufficiently able. □ **inadequately** *adv.*, **inadequacy** *n.*

inadmissible *adj.* not allowable.

inadvertent *adj.* unintentional. □ **inadvertently** *adv.*

inalienable *adj.* not able to be given or taken away.

inane *adj.* silly, lacking sense. □ **inanely** *adv.*, **inanity** *n.*

inanimate *adj.* lacking animal life; showing no sign of being alive.

inappropriate *adj.* unsuitable.

inarticulate *adj.* not expressed in words; unable to speak distinctly; unable to express ideas clearly.

inasmuch *adv.* □ **inasmuch as** seeing that, because.

inattentive *adj.* not paying attention.

inaudible *adj.* unable to be heard.

inaugurate *v.* admit to office ceremonially; begin (an undertaking), open (a building etc.) formally; be the beginning of. □ **inaugural** *adj.*, **inauguration** *n.*, **inaugurator** *n.*

inboard *adj. & adv.* (of an engine) inside a boat.

inborn *adj.* existing in a person or animal from birth, natural.

inbred *adj.* produced by inbreeding; inborn.

inbreeding *n.* breeding from closely related individuals.

inbuilt *adj.* built-in.

Inc. *abbr.* Incorporated.

incalculable *adj.* unable to be calculated. □ **incalculably** *adv.*

incandescent *adj.* glowing with heat. □ **incandescence** *n.*

incantation *n.* words or sounds uttered as a magic spell.

incapable *adj.* not capable; helpless. □ **incapability** *n.*

incapacitate *v.* disable; make ineligible.

incapacity *n.* inability, lack of sufficient strength or power.

incarcerate *v.* imprison. □ **incarceration** *n.*

incarnate *adj.* embodied, esp. in human form.

incarnation *n.* embodiment, esp. in human form; **(the Incarnation)** that of God as Christ.

incautious *adj.* rash.

incendiary *adj.* designed to cause fire. ● *n.* an incendiary bomb; an arsonist.

incense[1] (in-sens) *n.* a substance burnt to produce fragrant smoke, esp. in religious ceremonies; this smoke.

incense² (in-sens) v. make angry.

incentive n. something that encourages action or effort.

inception n. the beginning of something.

incessant adj. not ceasing. □ **incessantly** adv.

incest n. sexual intercourse between very closely related people. □ **incestuous** adj.

inch n. a measure of length (= 2.54 cm). ● v. move gradually.

inchoate (in-koh-ăt) adj. just begun; not fully developed.

incidence n. 1 the rate at which a thing occurs. 2 the falling of a ray, line, etc. on a surface.

incident n. an event, esp. one causing trouble.

incidental adj. 1 occurring in connection with something else. 2 minor, not essential.

incidentally adv. in an incidental way; by the way.

incidental music n. background music composed for a film.

incinerate v. burn to ashes. □ **incineration** n., **incinerator** n.

incipient adj. in its early stages; beginning.

incise v. make a cut in; engrave. □ **incision** n.

incisive adj. clear and decisive. □ **incisively** adv., **incisiveness** n.

incisor n. a sharp-edged front tooth.

incite v. urge on to action; stir up. □ **incitement** n.

incivility n. rudeness.

inclement adj. (of weather) unpleasant.

inclination n. 1 a slope; bending. 2 a tendency; a liking or preference.

incline v. (in-klyn) 1 slope; bend. 2 have or cause a certain tendency, influence. ● n. (in-klyn) a slope.

include v. have or treat as part of a whole; put into a specified category. □ **inclusion** n.

inclusive adj. & adv. including what is mentioned; including everything. □ **inclusively** adv., **inclusiveness** n.

incognito adj. & adv. with one's identity kept secret.

incoherent adj. rambling in speech or reasoning. □ **incoherently** adv., **incoherence** n.

incombustible adj. not able to be burnt.

income n. money received as wages, interest, etc.

incoming adj. coming in.

incommunicado adj. not allowed or not wishing to communicate with others.

incomparable adj. beyond comparison, without an equal.

incompatible adj. conflicting, inconsistent. □ **incompatibility** n.

incompetent adj. lacking skill. □ **incompetence** n.

incomplete adj. not complete.

incomprehensible adj. not able to be understood. □ **incomprehension** n.

inconceivable adj. unable to be imagined; most unlikely.

inconclusive adj. not fully convincing; not decisive. □ **inconclusively** adv.

incongruous adj. unsuitable, out of place. □ **incongruously** adv., **incongruity** n.

inconsequential adj. unimportant; not following logically. □ **inconsequentially** adv.

inconsiderable adj. of small size or value.

inconsiderate adj. thoughtless.

inconsistent adj. not consistent. □ **inconsistency** n.

inconsolable adj. not able to be consoled. □ **inconsolably** adv.

inconstant adj. fickle; changeable; irregular. □ **inconstancy** n.

incontestable *adj.* indisputable.

incontinent *adj.* unable to control one's excretion of urine and/or faeces. □ **incontinence** *n.*

incontrovertible *adj.* indisputable, undeniable. □ **incontrovertibly** ● *adv.*

inconvenience *n.* lack of convenience; something causing this. ● *v.* cause inconvenience to.

inconvenient *adj.* not convenient, slightly troublesome. □ **inconveniently** *adv.*

incorporate *v.* **1** include as a part. **2** form into a corporation. □ **incorporation** *n.*

incorrect *adj.* wrong; not correct; improper.

incorrigible *adj.* not able to be reformed or improved. □ **incorrigibly** *adv.*

incorruptible *adj.* not corruptible morally; not liable to decay.

increase *v.* (in-**kreess**) make or become greater. ● *n.* (**in**-kreess) increasing; the amount by which a thing increases.

increasingly *adv.* more and more.

incredible *adj.* unbelievable; very surprising. □ **incredibly** *adv.*

incredulous *adj.* unbelieving, showing disbelief. □ **incredulously** *adv.*, **incredulity** *n.*

increment *n.* an increase, an added amount. □ **incremental** *adj.*

incriminate *v.* indicate as involved in wrongdoing. □ **incrimination** *n.*, **incriminatory** *adj.*

incrustation *n.* encrusting; a crust or deposit formed on a surface.

incubate *v.* hatch (eggs) by warmth; cause (bacteria etc.) to develop. □ **incubation** *n.*

incubator *n.* an apparatus for incubating eggs or bacteria; an enclosed heated compartment in which a premature baby can be kept.

inculcate *v.* implant (ideas or habits) by constant urging.

incumbent *adj.* forming an obligation or duty. ● *n.* the holder of an office, esp. a rector or a vicar.

incur *v.* (**incurred**) bring (something unpleasant) on oneself.

incurable *adj.* unable to be cured.

incursion *n.* a brief invasion, a raid.

indebted *adj.* owing money or gratitude.

indecent *adj.* offending against standards of decency; unseemly. □ **indecently** *adv.*, **indecency** *n.*

indecent assault *n.* sexual assault not involving rape.

indecent exposure *n.* exposing one's genitals in public.

indecipherable *adj.* unable to be read or deciphered.

indecision *n.* inability to decide, hesitation.

indecorous *adj.* improper; not in good taste.

indeed *adv.* in truth, really.

indefatigable *adj.* untiring.

indefensible *adj.* unable to be defended; not justifiable.

indefinable *adj.* unable to be defined or described clearly.

indefinite *adj.* not clearly stated or fixed; vague.

indefinite article *see* **article**.

indefinitely *adv.* in an indefinite way; for an unlimited period.

indelible *adj.* (of a mark) unable to be removed or washed away; (of ink) making such a mark. □ **indelibly** *adv.*

indelicate *adj.* slightly indecent; tactless. □ **indelicately** *adv.*, **indelicacy** *n.*

indemnify *v.* protect or insure (a person) against penalties that he

indemnity | indiscreet

or she might incur; compensate. □ **indemnification** n.

indemnity n. protection against penalties incurred by one's actions; compensation for injury.

indent v. **1** start inwards from a margin. **2** place an official order (for goods etc.). □ **indentation** n.

indenture n. a written contract, esp. of apprenticeship. ● v. bind by this.

independent adj. not dependent on or not controlled by another person or thing. □ **independently** adv., **independence** n.

indescribable adj. unable to be described; too great or bad etc. □ **indescribably** adv.

indestructible adj. unable to be destroyed.

indeterminable adj. impossible to discover or decide.

indeterminate adj. not fixed in extent or character; vague. □ **indeterminacy** n.

index n. (pl. **indexes** or **indices**) **1** a list (usu. alphabetical) of names, subjects, etc., with references. **2** a figure indicating the current level of prices etc. compared with a previous level. **3** (Maths) the exponent of a number. ● v. **1** make an index to; enter in an index. **2** adjust (wages etc.) according to a price index.

indexation n. the practice of making wages and benefits index-linked.

index finger n. the finger next to the thumb.

index-linked adj. (of wages and benefits) increased in line with the cost-of-living index.

Indian adj. of India or Indians. ● n. **1** a native of India. **2** any of the original inhabitants of the American continent or their descendants.

Indian ink n. ink made with a black pigment.

Indian summer n. dry sunny weather in autumn.

India rubber n. a rubber for rubbing out pencil or ink marks.

indicate v. point out; be a sign of; state briefly. □ **indication** n., **indicative** adj.

indicator n. a thing that indicates; a pointer; a flashing light on a vehicle showing when it is going to turn.

indices see **index**.

indict (in-dyt) v. make a formal accusation against. □ **indictable** adj., **indictment** n.

indifferent adj. **1** showing no interest or sympathy. **2** neither good nor bad; not very good. □ **indifferently** adv., **indifference** n.

indigenous adj. native.

indigent adj. needy. □ **indigence** n.

indigestible adj. difficult or impossible to digest.

indigestion n. discomfort caused by difficulty in digesting food.

indignant adj. feeling or showing indignation. □ **indignantly** adv.

indignation n. anger aroused by something unjust or wicked.

indignity n. unworthy treatment, humiliation.

indigo n. a deep blue dye or colour.

indirect adj. not direct. □ **indirectly** n.

indirect object n. (Grammar) a person or thing indirectly affected by the verb.

indirect speech n. = **reported speech**.

indirect taxes n.pl. taxes paid on goods and services, not on income or capital.

indiscernible adj. unable to be discerned.

indiscreet adj. revealing secrets; not cautious. □ **indiscreetly** adv., **indiscretion** n.

indiscriminate *adj.* not discriminating, not making a careful choice. □ **indiscriminately** *adv.*

indispensable *adj.* essential.

indisposed *adj.* slightly ill; unwilling. □ **indisposition** *n.*

indisputable *adj.* undeniable. □ **indisputably** *adv.*

indissoluble *adj.* firm and lasting, not able to be destroyed.

indistinct *adj.* unclear; obscure. □ **indistinctly** *adv.*

indistinguishable *adj.* not distinguishable.

indium *n.* a metallic element (symbol In).

individual *adj.* single; separate; characteristic of one particular person or thing. ● *n.* one person, animal, or plant considered separately; a person. □ **individually** *adv.*, **individuality** *n.*

individualist *n.* a person who is very independent in thought or action.

indivisible *adj.* not able to be divided.

Indo- *comb. form* Indian (and).

indoctrinate *v.* fill (a person's mind) with particular ideas or doctrines. □ **indoctrination** *n.*

indolent *adj.* lazy. □ **indolently** *adv.*, **indolence** *n.*

indomitable *adj.* unyielding, untiringly persistent. □ **indomitably** *adv.*

indoor *adj.* situated, used, or done inside a building. ● *adv.* (**indoors**) inside a building.

indubitable *adj.* that cannot reasonably be doubted. □ **indubitably** *adv.*

induce *v.* **1** persuade; cause. **2** bring on (labour) artificially.

inducement *n.* inducing; an attraction, an incentive.

induct *v.* install (a clergyman) ceremonially into a benefice.

inductance *n.* the property of producing an electric current by induction.

induction *n.* **1** inducting; inducing. **2** reasoning (from observed examples) that a general law exists. **3** the production of an electric or magnetic state by proximity of an electrified or magnetic object. **4** drawing of a fuel mixture into the cylinder(s) of an engine. **5** formal introduction to a new job. □ **inductive** *adj.*

indulge *v.* allow (a person) to have what he or she wishes; gratify. □ **indulgence** *n.*

indulgent *adj.* indulging a person's wishes too freely; kind, lenient. □ **indulgently** *adv.*

industrial *adj.* of, for, or full of industries. □ **industrially** *adv.*

industrial action *n.* a strike or similar protest.

industrial estate *n.* an area of land developed for business and industry.

industrialism *n.* a system in which manufacturing industries are predominant.

industrialist *n.* an owner or manager of an industrial business.

industrialized *adj.* (also **-ised**) full of industries.

industrial relations *n.pl.* relations between management and workers.

industrious *adj.* hard-working. □ **industriously** *adv.*

industry *n.* **1** manufacture or production of goods; business activity. **2** being industrious.

inebriated *adj.* drunk.

inedible *adj.* not edible, not suitable for eating.

ineducable *adj.* incapable of being educated.

ineffable *adj.* too great to be described.

ineffective *adj.* not effective; inefficient. □ **ineffectively** *adv.*

ineffectual *adj.* having no effect. □ **ineffectually** *adv.*

inefficient *adj.* not efficient; wasteful. □ **inefficiently** *adv.* **inefficiency** *n.*

inelegant *adj.* not elegant; unrefined. □ **inelegantly** *adv.*

ineligible *adj.* not eligible or qualified.

ineluctable *adj.* inescapable, unavoidable.

inept *adj.* unsuitable, absurd; unskilful. □ **ineptly** *adv.*, **ineptitude** *n.*, **ineptness** *n.*

inequable *adj.* unfair; not the same, not uniform.

inequality *n.* lack of equality.

inequitable *adj.* unfair, unjust. □ **inequitably** *adv.*

inert *adj.* without power to move; without active properties; not moving or taking action. □ **inertly** *adv.*, **inertness** *n.*

inertia *n.* **1** being inert, slowness to act. **2** the property by which matter continues in its existing state of rest or line of motion unless acted upon by a force.

inertia reel *n.* a reel holding one end of a safety belt, allowing free movement unless it is pulled suddenly.

inescapable *adj.* unavoidable. □ **inescapably** *adv.*

inessential *adj.* not essential. ● *n.* an inessential thing.

inestimable *adj.* too great or intense to be estimated. □ **inestimably** *adv.*

inevitable *adj.* not able to be prevented, sure to happen or appear. □ **inevitably** *adv.*, **inevitability** *n.*

inexact *adj.* not exact. □ **inexactly** *adv.*, **inexactitude** *n.*

inexcusable *adj.* unable to be excused or justified.

inexhaustible *adj.* available in unlimited quantity.

inexorable *adj.* relentless; unable to be persuaded. □ **inexorably** *adv.*

inexpensive *adj.* not expensive, good value.

inexperience *n.* lack of experience. □ **inexperienced** *adj.*

inexpert *adj.* not expert, unskilful. □ **inexpertly** *adv.*

inexplicable *adj.* unable to be explained. □ **inexplicably** *adv.*, **inexplicability** *n.*

inexpressible *adj.* unable to be expressed in words.

in extremis at the point of death; in an emergency.

inextricable *adj.* inescapable; unable to be disentangled. □ **inextricably** *adv.*

infallible *adj.* incapable of being wrong; never failing. □ **infallibly** *adv.*, **infallibility** *n.*

infamous (in-fǎ-mǔs) *adj.* having a bad reputation. □ **infamously** *adv.*, **infamy** *n.*

infancy *n.* early childhood, babyhood; an early stage of development.

infant *n.* a child during the earliest stage of its life.

infanticide *n.* the killing or killer of an infant soon after its birth.

infantile *adj.* of infants or infancy; very childish.

infantry *n.* troops who fight on foot.

infatuated *adj.* filled with intense unreasoning love. □ **infatuation** *n.*

infect *v.* affect or contaminate with a disease or its germs; affect with one's feeling.

infection *n.* infecting, being infected; a disease spread in this way.

infectious *adj.* (of disease) able to spread by air or water; infecting others. □ **infectiousness** *n.*

infer *v.* (**inferred**) work out from facts or reasoning; deduce. □ **inference** *n.*

■ **Usage** It is a mistake to use *infer* to mean 'imply', as in *Are you inferring that I'm a liar?*

inferior *adj.* low or lower in rank, importance, quality, or ability. ● *n.* a person inferior to another. ▢ **inferiority** *n.*

infernal *adj.* **1** of hell. **2** (*colloquial*) detestable, tiresome. ▢ **infernally** *adv.*

inferno *n.* (*pl.* **infernos**) a raging fire; an intensely hot place; hell.

infertile *adj.* not fertile; unable to have offspring.

infest *v.* be numerous or troublesome in (a place). ▢ **infestation** *n.*

infidel *n.* a person who does not believe in a religion.

infidelity *n.* unfaithfulness, esp. adultery.

infighting *n.* conflict within an organization.

infill *n.* (also **infilling**) material used to fill a gap; filling gaps in a row of buildings.

infiltrate *v.* enter gradually and without being noticed. ▢ **infiltration** *n.*, **infiltrator** *n.*

infinite *adj.* having no limit; very great, very many. ▢ **infinitely** *adv.*

infinitesimal *adj.* very small. ▢ **infinitesimally** *adv.*

infinitive *n.* the form of a verb not indicating tense, number, or person (e.g. *to go*).

infinity *n.* an infinite number, extent, or time.

infirm *adj.* weak from age or illness. ▢ **infirmity** *n.*

infirmary *n.* a hospital.

inflame *v.* **1** arouse strong feeling in. **2** cause inflammation in.

inflammable *adj.* easily set on fire. ▢ **inflammability** *n.*

■ **Usage** Because *inflammable* could be taken to mean 'not easily set on fire', *flammable* is often used instead. The negative of *inflammable* is *non-inflammable*.

inflammation *n.* redness and heat in a part of the body.

inflammatory *adj.* arousing strong feeling or anger.

inflatable *adj.* able to be inflated. ● *n.* an inflatable object.

inflate *v.* **1** fill with air or gas so as to swell. **2** increase artificially.

inflation *n.* inflating; a general increase in prices and fall in the purchasing power of money. ▢ **inflationary** *adj.*

inflect *v.* change the pitch of (a voice) in speaking; change the ending or form of (a word) grammatically. ▢ **inflection** *n.*

inflexible *adj.* not flexible; unable to be altered; not adaptable. ▢ **inflexibly** *adv.*, **inflexibility** *n.*

inflict *v.* cause (a blow, penalty, etc.) to be suffered. ▢ **infliction** *n.*

inflorescence *n.* flowering; the arrangement of flowers on a stem.

influence *n.* power to produce an effect, esp. on character, beliefs, or actions; a person or thing with this power. ● *v.* exert influence on.

influential *adj.* having great influence. ▢ **influentially** *adv.*

influenza *n.* a virus disease causing fever, muscular pain, and catarrh.

influx *n.* an inward flow.

inform *v.* give information to; reveal secret or criminal activities to police etc. ▢ **informer** *n.*

informal *adj.* not formal, without formality or ceremony. ▢ **informally** *adv.*, **informality** *n.*

informant *n.* a giver of information.

information *n.* facts told or discovered.

information technology *n.* (also **information science**) the study and use of computers, microelectronics, etc. for storing and transferring information.

informative *adj.* giving information. □ **informatively** *adv.*

informed *adj.* having a good knowledge of something.

infra dig *adj.* (*colloquial*) beneath one's dignity.

infra-red *adj.* of or using radiation with a wavelength longer than that of visible light rays.

infrastructure *n.* the basic structural parts of something; roads, sewers, etc. regarded as a country's basic facilities.

infrequent *adj.* not frequent. □ **infrequently** *adv.*

infringe *v.* break (a rule or agreement); encroach. □ **infringement** *n.*

infuriate *v.* make very angry.

infuse *v.* **1** fill (with a quality). **2** soak (tea or herbs) to bring out flavour.

infusion *n.* infusing; a liquid made by this.

ingenious *adj.* clever at inventing things; cleverly contrived. □ **ingeniously** *adv.*, **ingenuity** *n.*

■ **Usage** *Ingenious* is sometimes confused with *ingenuous*.

ingenuous *adj.* without artfulness, unsophisticated. □ **ingenuously** *adv.*

■ **Usage** *Ingenuous* is sometimes confused with *ingenious*.

ingest *v.* take in as food.

inglenook *n.* a space for sitting within a very large fireplace.

inglorious *adj.* **1** shameful. **2** not famous.

ingot *n.* a brick-shaped lump of cast metal.

ingrained *adj.* deeply embedded in a surface or in a person's character.

ingratiate *v.* bring (oneself) into a person's favour, esp. to gain advantage.

ingratitude *n.* lack of gratitude.

ingredient *n.* any of the parts in a mixture.

ingress *n.* going in; the right of entry.

ingrowing *adj.* growing abnormally into the flesh.

inhabit *v.* live in as one's home. □ **inhabitable** *adj.*, **inhabitant** *n.*

inhalant *n.* a medicinal substance to be inhaled.

inhale *v.* breathe in; draw tobacco smoke into the lungs.

inhaler *n.* a device producing a medicinal vapour to be inhaled.

inherent *adj.* existing in a thing as a natural or permanent quality. □ **inherently** *adv.*

inherit *v.* receive from a predecessor, esp. from someone who has died; receive (a characteristic) from one's parents. □ **inheritance** *n.*

inhibit *v.* restrain, prevent; cause inhibitions in. □ **inhibitor** *n.*

inhibition *n.* inhibiting; resistance to an impulse or feeling.

inhospitable *adj.* not hospitable; (of a place) with a harsh climate or landscape.

in-house *adj.* & *adv.* within an organization.

inhuman *adj.* (also **inhumane**) brutal, extremely cruel. □ **inhumanly** *adv.*, **inhumanity** *n.*

inimical *adj.* hostile. □ **inimically** *adv.*

inimitable *adj.* impossible to imitate. □ **inimitably** *adv.*

iniquity *n.* great injustice; wickedness. □ **iniquitous** *adj.*

initial *n.* the first letter of a word or name. ● *v.* (**initialled**) *Amer.* **initialed**) mark or sign with initials. ● *adj.* of the beginning. □ **initially** *adv.*

initiate *v.* (in-ish-i-ayt) **1** cause to begin. **2** admit into membership. **3** give instruction to. ● *n.* (in-ish-i-ăt) an instructed person. □ **initiation** *n.*, **initiator** *n.*, **initiatory** *adj.*

initiative n. **1** the first step in a process. **2** readiness to initiate things.

inject v. force (liquid) into the body with a syringe. □ **injection** n.

injudicious adj. unwise. □ **injudiciously** adv.

injunction n. a court order.

injure v. cause injury to.

injurious adj. causing injury.

injury n. damage, harm; a form of this; a wrong or unjust act.

injustice n. lack of justice; an unjust action.

ink n. coloured liquid used in writing, printing, etc. ● v. apply ink to. □ **inky** adj.

inkling n. a slight suspicion.

inlaid see **inlay**.

inland adj. & adv. in or towards the interior of a country.

in-laws n.pl. (colloquial) one's relatives by marriage.

inlay v. (in-lay) (**inlaid, inlaying**) set (one thing in another) so that the surfaces are flush. ● n. (in-lay) inlaid material or design.

inlet n. **1** a strip of water extending inland. **2** a way in (e.g. for water into a tank).

in loco parentis adv. acting in the place of a parent.

inmate n. a person living in a prison or other institution.

in memoriam in memory of (a dead person).

inmost adj. furthest inward.

inn n. a pub, esp. one in the country offering accommodation.

innards n.pl. (colloquial) the stomach and bowels; the inner parts.

innate adj. inborn; natural. □ **innately** adv.

inner adj. nearer to the centre or inside; interior, internal.

inner city n. the central densely populated area of a city.

innermost adj. furthest inward.

innings n. in cricket, a batsman's or side's turn at batting.

innocent adj. not guilty, free of evil; foolishly trustful. □ **innocently** adv., **innocence** n.

innocuous adj. harmless. □ **innocuously** adv.

innovate v. introduce something new. □ **innovation** n., **innovative** adj., **innovator** n.

innuendo n. (pl. **innuendoes** or **innuendos**) a suggestive indirect remark.

innumerable adj. too many to be counted.

innumerate adj. without knowledge of basic maths. □ **innumeracy** n.

inoculate v. protect against disease with vaccines or serums. □ **inoculation** n.

inoperable adj. unable to be cured by surgical operation.

inoperative adj. not functioning.

inopportune adj. happening at an unsuitable time.

inordinate adj. excessive. □ **inordinately** adv.

inorganic adj. of mineral origin, not organic. □ **inorganically** adv.

in-patient n. a patient staying in a hospital during treatment.

input n. what is put in. ● v. (**input** or **inputted, inputting**) put in; supply (data etc.) to a computer.

inquest n. a judicial investigation, esp. of a sudden death.

inquire v. make an inquiry.

inquiry n. an investigation.

inquisition n. detailed or relentless questioning. □ **inquisitor** n., **inquisitorial** adj.

inquisitive adj. curious; prying. □ **inquisitively** adv., **inquisitiveness** n.

inroad n. a hostile attack. □ **make inroads on** or **into** use up a lot of (resources).

insalubrious adj. unhealthy.

insane *adj.* mad; extremely foolish. □ **insanely** *adv.*, **insanity** *n.*

insanitary *adj.* not clean, not hygienic.

insatiable *adj.* unable to be satisfied. □ **insatiably** *adv.*, **insatiability** *n.*

inscribe *v.* write or engrave.

inscription *n.* words inscribed.

inscrutable *adj.* baffling, impossible to interpret. □ **inscrutably** *adv.*, **inscrutability** *n.*

insect *n.* a small creature with six legs, no backbone, and a segmented body.

insecticide *n.* a substance for killing insects.

insectivorous *adj.* insect-eating.

insecure *adj.* unsafe, not secure; lacking confidence.

inseminate *v.* insert semen into. □ **insemination** *n.*

insensible *adj.* **1** unconscious; unaware. **2** imperceptible. □ **insensibly** *adv.*

insensitive *adj.* not sensitive. □ **insensitively** *adv.*, **insensitivity** *n.*

inseparable *adj.* unable to be separated or kept apart. □ **inseparably** *adv.*, **inseparability** *n.*

insert *v.* (in-sert) put into or between or among. ● *n.* (in-sert) something inserted. □ **insertion** *n.*

in-service *adj.* (of training) for people working in the profession concerned.

inset *v.* (in-set) (**inset, insetting**) place in; decorate with an inset. ● *n.* (in-set) something set into a larger thing.

inshore *adj.* & *adv.* near or nearer to the shore.

inside *n.* the inner side, surface, or part. ● *adj.* of or from the inside. ● *adv.* on, in, or to the inside. ● *prep.* on or to the inside of; within. **inside out**

with the inner side outwards; thoroughly.

insider dealing *n.* the illegal practice of using confidential information to gain advantage in buying stocks and shares.

insidious *adj.* spreading or developing without being noticed with a harmful effect. □ **insidiously** *adv.*, **insidiousness** *n.*

insight *n.* perception and understanding of a thing's nature.

insignia *n.pl.* symbols of authority or office; an identifying badge.

insignificant *adj.* unimportant. □ **insignificantly** *adv.*, **insignificance** *n.*

insinuate *v.* insert gradually or craftily; hint artfully. □ **insinuation** *n.*, **insinuator** *n.*

insipid *adj.* lacking flavour, interest, or liveliness.

insist *v.* demand or state emphatically.

insistent *adj.* insisting; forcing itself on one's attention. □ **insistently** *adv.*, **insistence** *n.*

in situ (in sit-yoo) *adv.* in its original place.

insolent *adj.* disrespectful, arrogant. □ **insolently** *adv.*, **insolence** *n.*

insoluble *adj.* unable to be dissolved; unable to be solved.

insolvent *adj.* unable to pay one's debts. □ **insolvency** *n.*

insomnia *n.* inability to sleep. □ **insomniac** *n.*

insouciant *adj.* carefree, unconcerned. □ **insouciance** *n.*

inspect *v.* examine critically or officially. □ **inspection** *n.*

inspector *n.* **1** a person who inspects. **2** a police officer above sergeant.

inspiration *n.* inspiring; an inspiring influence; a sudden brilliant idea. □ **inspirational** *adj.*

inspire *v.* stimulate to activity; instil (a feeling or idea) into; animate.

inst. *abbr.* instant, of the current month.

instability *n.* lack of stability.

install *v.* place (a person) into office ceremonially; set in position and ready for use; establish.

installation *n.* the process of installing; an apparatus etc. installed.

instalment *n.* (*Amer.* **installment**) one of the regular payments made to clear a debt paid over a period of time; one part of a serial.

instance *n.* an example; a particular case. ● *v.* mention as an instance.

instant *adj.* immediate; (of food) quickly and easily prepared. ● *n.* an exact moment; a very short time. □ **instantly** *adv.*

instantaneous *adj.* occurring or done instantly. □ **instantaneously** *adv.*

instead *adv.* as an alternative.

instep *n.* the middle part of the foot; part of a shoe etc. covering this.

instigate *v.* incite; initiate. □ **instigation** *n.*, **instigator** *n.*

instil *v.* (*Amer.* **instill**) (**instilled**) introduce (ideas etc.) into a person's mind gradually. □ **instillation** *n.*

instinct *n.* an inborn impulse; a natural tendency or ability. □ **instinctive** *adj.*, **instinctively** *adv.*

institute *n.* an organization for promotion of a specified activity; its premises. ● *v.* set up; establish.

institution *n.* 1 the process of instituting. 2 an institute. 3 a home for people with special needs. 4 an established rule or custom. □ **institutional** *adj.*

institutionalize *v.* (also **-ise**) accustom to living in an institution. □ **institutionalized** *adj.*

instruct *v.* 1 teach (a person) a subject or skill. 2 give instructions to. □ **instructor** *n.*

instruction *n.* 1 the process of teaching. 2 a statement telling a person what to do. □ **instructional** *adj.*

instructive *adj.* giving instruction, enlightening. □ **instructively** *adv.*

instrument *n.* 1 a tool for delicate work. 2 a measuring device of an engine or vehicle. 3 a device for producing musical sounds.

instrumental *adj.* 1 serving as a means. 2 performed on musical instruments.

instrumentalist *n.* a player of a musical instrument.

insubordinate *adj.* disobedient, rebellious. □ **insubordination** *n.*

insubstantial *adj.* lacking reality or solidity.

insufferable *adj.* unbearable. □ **insufferably** *adv.*

insufficient *adj.* not enough; inadequate.

insular *adj.* 1 of an island. 2 narrow-minded. □ **insularity** *n.*

insulate *v.* 1 cover with a substance that prevents the passage of electricity, sound, or heat. 2 isolate from influences. □ **insulation** *n.*, **insulator** *n.*

insulin *n.* a hormone controlling the body's absorption of sugar.

insult *v.* (in-sult) speak or act so as to offend (a person). ● *n.* (insult) an insulting remark or action. □ **insulting** *adj.*

insuperable *adj.* unable to be overcome. □ **insuperably** *adv.*

insupportable *adj.* unbearable.

insurance *n.* a contract to provide compensation for loss, damage, or death; a sum payable as a premium or in compensation; a safeguard against loss or failure.

insure *v.* 1 protect by insurance. 2 (*Amer.*) ensure. □ **insurer** *n.*

insurgent *adj.* rebellious, rising in revolt. ● *n.* a rebel. □ **insurgency** *n.*

insurmountable *adj.* unable to be overcome.

insurrection *n.* a rebellion. □ **insurrectionist** *n.*

intact *adj.* undamaged, complete.

intake *n.* taking things in; a place or amount of this.

intangible *adj.* unable to be touched or grasped mentally.

integer *n.* a whole number, not a fraction.

integral *adj.* forming or necessary to form a whole.

integrate *v.* combine (parts) into a whole; bring or come into full membership of a community. □ **integration** *n.*

integrity *n.* honesty; wholeness, soundness.

intellect *n.* the mind's power of reasoning and acquiring knowledge.

intellectual *adj.* of or using the intellect; having a strong intellect. ● *n.* an intellectual person. □ **intellectually** *adv.*

intelligence *n.* **1** mental ability to learn and understand things. **2** information, esp. that of military value; people collecting this.

intelligent *adj.* having mental ability. □ **intelligently** *adv.*

intelligentsia *n.* intellectual people.

intelligible *adj.* able to be understood. □ **intelligibly** *adv.*, **intelligibility** *n.*

intend *v.* have in mind as what one wishes to achieve.

intense *adj.* strong in quality or degree; feeling strong emotion. □ **intensely** *adv.*, **intensity** *n.*

■ **Usage** *Intense* is sometimes confused with *intensive*, and wrongly used to describe a course of study etc.

intensifier *n.* (*Grammar*) a word used to give emphasis, e.g. *really* in *I'm really hot.*

intensify *v.* make or become more intense. □ **intensification** *n.*

intensive *adj.* **1** using a lot of effort; concentrated. **2** intended to increase production in relation to cost. □ **intensively** *adv.*, **intensiveness** *n.*

intensive care *n.* medical treatment with constant attention for a seriously ill patient.

intent *n.* intention. ● *adj.* with concentrated attention. □ **intent on** determined to. □ **intently** *adv.*, **intentness** *n.*

intention *n.* what one intends to do.

intentional *adj.* done on purpose. □ **intentionally** *adv.*

inter *v.* (**interred**) bury (a dead body).

inter- *pref.* between, among.

interact *v.* have an effect upon each other. □ **interaction** *n.*, **interactive** *adj.*

inter alia *adv.* among other things.

interbreed *v.* (**interbred**, **interbreeding**) breed with each other, cross-breed.

intercede *v.* intervene on someone's behalf.

intercept *v.* stop or catch between a starting point and destination. □ **interception** *n.*, **interceptor** *n.*

intercession *n.* interceding.

interchange *v.* (inter-chaynj) **1** cause to change places. **2** alternate. ● *n.* (**inter**-chaynj) **1** a process of interchanging. **2** a road junction designed so that streams of traffic do not cross on the same level. □ **interchangeable** *adj.*

intercom *n.* a communication system operating by radio or telephone.

interconnect v. connect with each other. □ **interconnection** n.

intercontinental adj. between continents.

intercourse n. **1** dealings between people or countries. **2** sexual intercourse, copulation.

interdenominational adj. involving more than one religious denomination.

interdependent adj. dependent on each other.

interdict n. a formal prohibition.

interdisciplinary adj. involving different subjects or disciplines.

interest n. **1** a feeling of curiosity or concern; the object of this. **2** a legal share. **3** money paid for use of money borrowed. **4** an advantage or benefit; *it is in my interest to go.* ● v. arouse the interest of.

interested adj. feeling interest; having an interest, not impartial.

interesting adj. arousing interest.

interface n. **1** a place where interaction occurs. **2** (*Computing*) a program or apparatus connecting two machines or enabling a user to use a program.

interfere v. meddle, intervene; be an obstruction. □ **interfere with** molest sexually.

interference n. interfering; disturbance of radio signals.

interferon n. a protein preventing the development of a virus.

intergalactic adj. between galaxies.

interim n. an intervening period. ● adj. of or in such a period, temporary.

interior adj. inner. ● n. the inner part; the inside.

interject v. put in (a remark) when someone is speaking.

interjection n. the process of interjecting; a remark interjected; an exclamation.

interlace v. weave or lace together.

interlink v. link together.

interlock v. fit into each other. ● n. a fine machine-knitted fabric.

interloper n. an intruder.

interlude n. an interval; something happening or performed in this.

intermarry v. marry members of the same or another group. □ **intermarriage** n.

intermediary n. a mediator, a messenger. ● adj. acting as an intermediary; intermediate.

intermediate adj. coming between two things in time, place, or order.

interment n. burial.

intermezzo (inter-**mets**-oh) n. (pl. **intermezzos** or **intermezzi**) a short piece of music.

interminable adj. very long and boring. □ **interminably** adv.

intermission n. an interval, a pause.

intermittent adj. occurring at intervals. □ **intermittently** adv.

intern v. compel (esp. an enemy alien) to live in a special area. ● n. (also **interne**) (esp. *Amer.*) a resident junior doctor at a hospital.

internal adj. of or in the inside; of a country's domestic affairs; applying within an organization. □ **internally** adv.

internal-combustion engine n. an engine producing power from fuel exploded within a cylinder.

internalize v. (also **-ise**) learn, absorb into the mind.

international adj. between countries. ● n. a sports contest between players of different countries; one of these players. □ **internationally** adv.

interne var. of **intern** n.

internecine (inter-**nee**-syn) adj. mutually destructive.

internee n. an interned person.

internment *n.* interning, being interned.

interplay *n.* interaction.

interpolate *v.* interject; insert. □ **interpolation** *n.*

interpose *v.* insert; intervene.

interpret *v.* explain the meaning of; understand in a particular way; act as interpreter. □ **interpretation** *n.*, **interpretative** *adj.*, **interpretive** *adj.*

interpreter *n.* a person who orally translates speech between people speaking different languages.

interracial *adj.* involving different races.

interregnum *n.* a period between the rule of two successive rulers.

interrelated *adj.* related to each other.

interrogate *v.* question closely. □ **interrogation** *n.*, **interrogator** *n.*

interrogative *adj.* forming or in the form of a question. □ **interrogatively** *adv.*

interrupt *v.* break the continuity of; break the flow of (speech etc.) by a remark. □ **interruption** *n.*

intersect *v.* divide or cross by passing or lying across. □ **intersection** *n.*

intersperse *v.* insert here and there.

interstate *adj.* between states, esp. of the USA.

interval *n.* a time or pause between two events or parts of an action; a space between two things; a difference in musical pitch.

intervene *v.* **1** occur between events. **2** become involved in a situation to change its course or resolve it. □ **intervention** *n.*

interview *n.* a formal meeting with a person to assess his or her merits or obtain information. ● *v.* hold an interview with. □ **interviewee** *n.*, **interviewer** *n.*

interweave *v.* (**interwove**, **interwoven**, **interweaving**) weave together; blend.

intestate *adj.* not having made a valid will. □ **intestacy** *n.*

intestine *n.* a long tubular section of the alimentary canal between the stomach and anus. □ **intestinal** *adj.*

intimate[1] (in-ti-măt) *adj.* closely acquainted or familiar; having a sexual relationship (esp. outside marriage); private and personal. ● *n.* an intimate friend. □ **intimately** *adv.*, **intimacy** *n.*

intimate[2] (in-ti-mayt) *v.* make known, esp. by hinting. □ **intimation** *n.*

intimidate *v.* influence by frightening. □ **intimidation** *n.*

into *prep.* **1** to the inside of, to a point within. **2** to a particular state or occupation. **3** dividing (a number) mathematically. **4** (*colloquial*) interested and involved in.

intolerable *adj.* unbearable. □ **intolerably** *adv.*

intonation *n.* intoning; the pitch of the voice in speaking.

intone *v.* chant, esp. on one note.

intoxicant *adj.* causing intoxication. ● *n.* an intoxicating substance.

intoxicate *v.* make drunk; make greatly excited. □ **intoxication** *n.*

intra- *pref.* within.

intractable *adj.* hard to deal with or control. □ **intractability** *n.*

intramural *adj.* **1** within the walls of an institution etc. **2** part of ordinary university work.

intransigent *adj.* stubborn. □ **intransigently** *adv.*, **intransigence** *n.*

intransitive *adj.* (*Grammar*) (of a verb) not followed by a direct object.

intrauterine *adj.* within the uterus.

intravenous *adj.* into a vein. □ **intravenously** *adv.*

in-tray *n.* a tray holding documents that need attention.

intrepid *adj.* fearless, brave. □ **intrepidly** *adv.*

intricate *adj.* very complicated. □ **intricately** *adv.*, **intricacy** *n.*

intrigue *v.* **1** plot secretly. **2** rouse the interest of. ● *n.* a plot, plotting; a secret love affair. □ **intriguing** *adj.*

intrinsic *adj.* existing in a thing as a natural or permanent quality; essential. □ **intrinsically** *adv.*

introduce *v.* **1** make (a person) known to another; present to an audience. **2** bring into use. **3** insert.

introduction *n.* introducing; an introductory part.

introductory *adj.* introducing a person or thing; preliminary.

introspection *n.* examination of one's own thoughts and feelings. □ **introspective** *adj.*

introvert *n.* an introspective and shy person. □ **introverted** *adj.*

intrude *v.* come or join in without being invited or wanted; thrust in. □ **intrusion** *n.*, **intrusive** *adj.*

intruder *n.* a person who intrudes; a burglar.

intuition *n.* the power of knowing without learning or reasoning. □ **intuitive** *adj.*, **intuitively** *adv.*

Inuit (in-yoo-it) *n.* (*pl.* **Inuit** or **Inuits**) a North American Eskimo.

inundate *v.* flood; overwhelm.

inure *v.* **1** accustom, esp. to something unpleasant. **2** (in law) take effect.

invade *v.* enter (territory) with hostile intent; crowd into; penetrate harmfully. □ **invader** *n.*

invalid[1] (in-vă-lid) *n.* a person suffering from ill health.

invalid[2] (in-val-id) *adj.* not valid.

invalidate *v.* make no longer valid.

invaluable *adj.* having a value too great to be measured.

invariable *adj.* not variable, always the same. □ **invariably** *adv.*

invasion *n.* a hostile or harmful intrusion.

invective *n.* abusive language.

invent *v.* make or design (something new); make up (a lie, a story). □ **inventor** *n.*

inventive *adj.* able to invent things. □ **inventiveness** *n.*

inventory *n.* a detailed list of goods or furniture.

inverse *adj.* inverted. □ **inversely** *adv.*

invert *v.* turn upside down; reverse the position, order, or relationship of. □ **inversion** *n.*

invertebrate *n.* & *adj.* (an animal) having no backbone.

inverted commas *n.pl.* quotation marks.

invest *v.* **1** use (money, time, etc.) to earn interest or bring profit. **2** confer rank or power upon. **3** endow with a quality. □ **investment** *n.*, **investor** *n.*

investigate *v.* study carefully; inquire into. □ **investigation** *n.*, **investigative** *adj.*, **investigator** *n.*

investiture *n.* investing a person with honours or rank.

inveterate *adj.* habitual; firmly established.

invidious *adj.* liable to cause resentment. □ **invidiously** *adv.*

invigilate *v.* supervise examination candidates. □ **invigilator** *n.*

invigorate *v.* fill with vigour, give strength or courage to.

invincible *adj.* unconquerable.

inviolable *adj.* not able to be violated.

inviolate *adj.* not violated; safe.

invisible *adj.* not able to be seen. □ **invisibly** *adv.*, **invisibility** *n.*

invite *v.* ask (a person) politely to come or to do something; ask for; attract. □ **invitation** *n.*

inviting *adj.* pleasant and tempting. □ **invitingly** *adv.*

in vitro *adj.* & *adv.* in a test tube or other laboratory environment; *in vitro fertilization.*

invocation *n.* invoking.

invoice *n.* a bill for goods or services. ● *v.* send an invoice to.

invoke *v.* call for the help or protection of; summon (a spirit).

involuntary *adj.* done without intention. □ **involuntarily** *adv.*

involve *v.* have as a consequence; include or affect; implicate. □ **involvement** *n.*

involved *adj.* complicated; concerned in something.

invulnerable *adj.* not vulnerable.

inward *adj.* situated on or going towards the inside; in the mind or spirit. ● *adv.* inwards. □ **inwardly** *adv.*, **inwards** *adv.*

iodine *n.* a chemical (symbol I) used in solution as an antiseptic.

iodize *v.* (also **-ise**) treat or impregnate with iodine.

ion *n.* an electrically charged atom that has lost or gained an electron.

ionize *v.* (also **-ise**) convert or be converted into ions.

ionosphere *n.* the ionized region of the atmosphere. □ **ionospheric** *adj.*

iota *n.* **1** a Greek letter (I, ι). **2** a very small amount.

IOU *n.* a signed paper given as a receipt for money borrowed.

ipso facto *adv.* by that very fact.

IQ *abbr.* intelligence quotient; a number showing how a person's intelligence compares with the average.

Ir *symb.* iridium.

irascible *adj.* hot-tempered. □ **irascibly** *adv.*

irate *adj.* angry. □ **irately** *adv.*

ire *n.* anger.

iridescent *adj.* coloured like a rainbow; shimmering. □ **iridescence** *n.*

iridium *n.* a metallic element (symbol Ir).

iris *n.* **1** the coloured part of the eyeball, round the pupil. **2** a plant with showy flowers and sword-shaped leaves.

Irish *adj.* & *n.* (the language) of Ireland. □ **Irishman** *n.*, **Irishwoman** *n.*

irk *v.* annoy, be tiresome to.

irksome *adj.* tiresome.

iron *n.* **1** a metallic element (symbol Fe); a tool made of this. **2** an implement with a flat base heated for smoothing cloth. **3** (**irons**) fetters. ● *adj.* made of iron; as strong as iron. ● *v.* smooth (clothes etc.) with an iron.

ironic *adj.* (also **ironical**) using or characterized by irony. □ **ironically** *adv.*

ironmonger *n.* a shopkeeper selling tools and household implements.

iron rations *n.pl.* a small emergency food supply.

ironstone *n.* **1** a hard iron ore. **2** a kind of hard white pottery.

irony *n.* the expression of meaning by use of words normally conveying the opposite; the apparent perversity of fate or circumstances.

irradiate *v.* throw light or other radiation on; treat (food) by radiation. □ **irradiation** *n.*

irrecoverable *adj.* unable to be recovered. □ **irrecoverably** *adv.*

irredeemable *adj.* unable to be redeemed; hopeless.

irrefutable adj. unable to be refuted. □ **irrefutably** adv.

irregular adj. not regular; contrary to rules or custom. □ **irregularly** adv., **irregularity** n.

irrelevant adj. not relevant. □ **irrelevantly** adv., **irrelevance** n.

irreparable adj. unable to be repaired. □ **irreparably** adv.

irreplaceable adj. unable to be replaced.

irrepressible adj. unable to be repressed. □ **irrepressibly** adv.

irreproachable adj. blameless, faultless. □ **irreproachably** adv.

irresistible adj. too strong or delightful to be resisted. □ **irresistibly** adv., **irresistibility** n.

irresolute adj. unable to make up one's mind. □ **irresolutely** adv.

irrespective adj. **irrespective of** not taking (a thing) into account.

irresponsible adj. not showing a proper sense of responsibility. □ **irresponsibly** adv., **irresponsibility** n.

irretrievable adj. unable to be retrieved or restored.

irreverent adj. not reverent; not respectful. □ **irreverently** adv., **irreverence** n.

irreversible adj. not reversible, unable to be altered or revoked. □ **irreversibly** adv.

irrevocable adj. unalterable. □ **irrevocably** adv.

irrigate v. supply (land) with water by streams, pipes, etc. □ **irrigation** n.

irritable adj. easily annoyed, bad-tempered. □ **irritably** adv., **irritability** n.

irritant adj. & n. (something) causing irritation.

irritate v. **1** annoy. **2** cause to itch. □ **irritation** n.

irrupt v. make a violent entry.

ISBN abbr. international standard book number.

isinglass n. gelatin obtained from fish.

Islam n. the Muslim religion; the Muslim world. □ **Islamic** adj.

island n. a piece of land surrounded by water.

islander n. an inhabitant of an island.

isle n. an island.

islet n. a small island.

ism n. (colloquial) a doctrine or practice.

isobar n. a line on a map, connecting places with the same atmospheric pressure. □ **isobaric** adj.

isolate v. place apart or alone; separate from others or from a compound. □ **isolation** n.

isolationism n. the policy of holding aloof from other countries or groups. □ **isolationist** n.

isomer n. one of two or more substances whose molecules have the same atoms arranged differently.

isosceles (I-sos-i-leez) adj. (of a triangle) having two sides equal.

isotherm n. a line on a map, connecting places with the same temperature.

isotope n. one of two or more forms of a chemical element differing in their atomic weight. □ **isotopic** adj.

issue n. **1** an outflow; issuing; a quantity issued. **2** one edition (e.g. of a magazine). **3** an important topic. **4** offspring. ● v. **1** flow out. **2** supply for use; send out. **3** publish.

isthmus (iss-muss) n. (pl. **isthmuses**) a narrow strip of land connecting two larger masses of land.

it pron. **1** the thing mentioned or being discussed. **2** used as the subject of an impersonal verb; it is raining.

Italian adj. & n. (a native, the language) of Italy.

italic *adj.* (of type) sloping like *this.* ● *n.pl.* (**italics**) italic type.

italicize *v.* (also **-ise**) print in italics.

itch *n.* a tickling sensation in the skin, causing a desire to scratch; a restless desire. ● *v.* have or feel an itch. □ **itchy** *adj.*

item *n.* a single thing in a list or collection; a single piece of news.

itemize *v.* (also **-ise**) list, state the individual items of. □ **itemization** *n.*

iterate *v.* repeat. □ **iterative** *adj.*

itinerant *adj.* travelling.

itinerary *n.* a route, a list of places to be visited on a journey.

its *poss.pron.* of it.

it's it is, it has.

■ **Usage** Because it has an apostrophe, *it's* is easily confused with *its*. Both are correctly used in *Where's the dog? — It's in its kennel, and it's eaten its food* (= *It is in its kennel, and it has eaten its food.*)

---

itself *pron.* the emphatic and reflexive form of *it.*

ITV *abbr.* Independent Television.

IUD *abbr.* intrauterine device; a contraceptive coil placed inside the womb.

IVF *abbr.* in vitro fertilization.

ivory *n.* a hard creamy-white substance forming the tusks of an elephant etc.; an object made of this; its colour. ● *adj.* creamy white.

ivory tower *n.* a place providing seclusion from the harsh realities of life.

ivy *n.*

Jj

J *abbr.* joule(s).

jab *v.* (**jabbed**) poke roughly. ● *n.* a rough poke; (*colloquial*) an injection.

jabber *v.* talk rapidly, often unintelligibly.

jacaranda *n.* a tropical tree with scented wood.

jack *n.* **1** a portable device for raising heavy weights off the ground. **2** a playing card next below queen. **3** a ship's small flag showing nationality. **4** an electrical connection with a single plug. **5** a small ball aimed at in bowls. ● *v.* □ **jack up** raise with a jack.

jackal *n.* a dog-like wild animal.

jackass *n.* **1** a male ass. **2** a stupid person.

jackboot *n.* **1** a military boot reaching above the knee. **2** bullying behaviour.

jackdaw *n.* a bird of the crow family.

jacket *n.* a short coat; an outer covering.

jackknife *n.* (*pl.* **-knives**) a large folding knife. ● *v.* (of an articulated vehicle) fold against itself in an accident.

jackpot *n.* a large prize of money that has accumulated until won. □ **hit the jackpot** (*colloquial*) have a sudden success.

Jacobean *adj.* of the reign of James I of England (1603–25).

Jacobite *n.* a supporter of James II of England or of the exiled Stuarts.

jacuzzi *n.* (*trade mark*) a large bath with underwater jets of water.

jade *n.* a hard green, blue, or white stone; its green colour.

jaded adj. tired and bored.

jagged adj. having sharp projections.

jaguar n. a large flesh-eating animal of the cat family.

jail n. (also **gaol**) prison. ● v. put into jail.

jailbird n. (also **gaolbird**) a person who has been in prison; a habitual criminal.

jailer n. (also **gaoler**) a person in charge of a jail or its prisoners.

Jain n. a member of an Indian religion. □ **Jainism** n.

jalopy n. an old battered car.

jam n. **1** a thick sweet substance made by boiling fruit with sugar. **2** a squeeze, a crush; a crowded mass making movement difficult. **3** (colloquial) a difficult situation. ● v. (**jammed**) **1** squeeze or wedge into a space; become wedged; crowd (an area). **2** make (a broadcast) unintelligible by causing interference.

jamb n. the side post of a door or window.

jamboree n. a large party; a rally.

Jan. abbr. January.

jangle n. a harsh metallic sound. ● v. make or cause to make this sound; upset by discord.

janitor n. the caretaker of a building.

January n. the first month.

japan n. a hard usu. black varnish. ● v. (**japanned**) coat with this.

Japanese adj. & n. (a native, the language) of Japan.

jar n. a cylindrical glass or earthenware container; (colloquial) a glass of beer. ● v. (**jarred**) jolt; have a harsh or disagreeable effect (upon).

jardinière (zhar-din-**yair**) n. a large ornamental pot for growing plants.

jargon n. words or expressions developed for use within a particular group of people.

jasmine n. a shrub with white or yellow flowers.

jasper n. a kind of quartz.

jaundice n. a condition in which the skin becomes abnormally yellow.

jaundiced adj. **1** affected by jaundice. **2** filled with resentment.

jaunt n. a short pleasure trip.

jaunty adj. (**jauntier**) cheerful, self-confident. □ **jauntily** adv., **jauntiness** n.

javelin n. a light spear thrown in sport (formerly as a weapon).

jaw n. **1** the bone(s) forming the framework of the mouth; (colloquial) lengthy talk. **2** (**jaws**) the gripping-parts of a tool; the mouth of a valley, channel, etc. ● v. (colloquial) talk lengthily.

jay n. a bird of the crow family.

jaywalking n. crossing a road carelessly. □ **jaywalker** n.

jazz n. a type of music with strong rhythm and much syncopation.

JCB n. (trade mark) a mechanical excavator.

jealous adj. resentful towards a rival; taking watchful care. □ **jealously** adv., **jealousy** n.

jeans n.pl. denim trousers.

jeep n. (trade mark) a small sturdy motor vehicle with four-wheel drive.

jeer v. laugh or shout rudely or scornfully (at). ● n. a jeering shout.

jehad var. of **jihad**.

Jehovah n. the name of God in the Old Testament.

jejune adj. scanty, poor; unsatisfying to the mind; (of land) barren.

jell v. (colloquial) set as a jelly; take definite form.

jelly n. a soft solid food made of liquid set with gelatin; a substance of similar consistency; jam made of strained fruit juice.

jellyfish *n.* a sea animal with a jelly-like body.

jemmy *n.* a burglar's short crowbar.

jenny *n.* a female donkey.

jeopardize (jep-er-dyz) *v.* (also **-ise**) endanger.

jeopardy (jep-er-di) *n.* danger.

jerbil var. of **gerbil**.

jerboa *n.* a rat-like desert animal with long hind legs.

jerk *n.* a sudden sharp movement or pull. ● *v.* move, pull, or stop with a jerk. □ **jerky** *adj.,* **jerkily** *adv.,* **jerkiness** *n.*

jerkin *n.* a sleeveless jacket.

jeroboam *n.* a wine bottle of 4 to 12 times the ordinary size.

jerry-built *adj.* poorly built.

jerrycan *n.* a five-gallon can for petrol or water.

jersey *n.* (*pl.* **jerseys**) a knitted woollen pullover with sleeves; machine-knitted fabric.

jest *n.* a joke. ● *v.* make jokes.

jester *n.* a clown at a medieval court.

Jesuit *n.* a member of the Society of Jesus, a Roman Catholic religious order.

jet¹ *n.* a hard black mineral; glossy black.

jet² *n.* a stream of water, gas, or flame from a small opening; a burner on a gas cooker; an engine or aircraft using jet propulsion. ● *v.* (**jetted**) travel by jet.

jet lag *n.* delayed tiredness etc. after a long flight.

jet-propelled *adj.* propelled by jet engines.

jet propulsion *n.* forward movement provided by engines sending out jets of gas at the back.

jetsam *n.* goods jettisoned by a ship and washed ashore.

jettison *v.* throw overboard; eject; discard.

jetty *n.* a breakwater; a landing-stage.

Jew *n.* a person of Hebrew descent or whose religion is Judaism. □ **Jewish** *adj.*

jewel *n.* a precious stone cut or set as an ornament; a highly valued person or thing. □ **jewelled** *adj.*

jeweller *n.* (*Amer.* **jeweler**) a person who makes or deals in jewels or jewellery.

jewellery *n.* (also **jewelry**) jewels or similar ornaments to be worn.

Jewry *n.* the Jewish people.

jib *n.* a triangular sail stretching forward from a mast; a projecting arm of a crane. ● *v.* (**jibbed**) refuse to proceed. □ **jib at** object to.

jibe¹ var. of **gibe**. **2** *Amer.* sp. of **gybe**.

jiffy *n.* (*colloquial*) a moment.

Jiffy bag *n.* (*trade mark*) a padded envelope.

jig *n.* **1** a lively dance. **2** a device that holds work and guides tools working on it; a template. ● *v.* (**jigged**) move quickly up and down.

jiggery-pokery *n.* (*colloquial*) trickery.

jiggle *v.* rock or jerk lightly.

jigsaw *n.* **1** a machine fretsaw. **2** (also **jigsaw puzzle**) a picture cut into pieces to be shuffled and reassembled for amusement.

jihad *n.* (also **jehad**) (in Islam) a holy war.

jilt *v.* abandon (a lover).

jingle *v.* (cause to) make a ringing or clinking sound. ● *n.* this sound; a simple rhyme.

jingoism *n.* excessive patriotism and contempt for other countries. □ **jingoistic** *adj.*

jinx *n.* (*colloquial*) an influence causing bad luck.

jitters *n.pl.* (*colloquial*) nervousness. □ **jittery** *adj.*

jiu-jitsu var. of **ju-jitsu**.

jive n. fast lively jazz music; a dance done to this. ● v. dance to this music.

Jnr. abbr. Junior.

job n. a piece of work; a paid position of employment; (colloquial) a difficult task.

jobber n. a principal or wholesaler dealing on the Stock Exchange.

jobbing adj. doing single pieces of work for payment.

jobcentre n. (in the UK) a government office displaying information about available jobs.

jobless adj. out of work.

job lot n. miscellaneous articles sold together.

jockey n. (pl. **jockeys**) a person who rides in horse races. ● v. manoeuvre to gain advantage.

jockstrap n. a protective support for the male genitals, worn while taking part in sports.

jocose adj. joking.

jocular adj. joking. □ **jocularly** adv., **jocularity** n.

jocund adj. merry, cheerful.

jodhpurs n.pl. riding-breeches fitting closely from knee to ankle.

jog v. (**jogged**) nudge; stimulate; run at a slow regular pace. ● n. a slow run; a nudge. □ **jogger** n.

joggle v. shake slightly. ● n. a slight shake.

jogtrot n. a slow regular trot.

joie de vivre (zhwah dè veevr) n. great enjoyment of life.

join v. **1** unite; connect. **2** come into the company of; take one's place in; become a member of. ● n. a place where things join. □ **join up** enlist in the forces.

joiner n. a maker of wooden doors, windows, etc. □ **joinery** n.

joint adj. shared by two or more people. ● n. **1** a join; a structure where parts or bones fit together; a large piece of meat. **2** (slang) a marijuana cigarette.

● v. connect by a joint or joints; divide into joints. □ **out of joint** dislocated; in disorder. □ **jointly** adv.

jointure n. an estate settled on a widow for her lifetime.

joist n. one of the beams supporting a floor or ceiling.

jojoba (hŏ-hoh-bă) n. a plant with seeds containing oil used in cosmetics.

joke n. something said or done to cause laughter; a ridiculous person or thing. ● v. make jokes.

joker n. a person who jokes; an extra playing card with no fixed value.

jollification n. merrymaking.

jollity n. being jolly; merrymaking.

jolly adj. (**jollier**) cheerful, merry; very pleasant. ● adv. (colloquial) very. □ **jolly along** keep (a person) in good humour.

jolt v. shake or dislodge with a jerk; move jerkily; shock. ● n. a jolting movement; a shock.

jonquil n. a narcissus with fragrant flowers.

josh v. (Amer. slang) hoax, tease.

joss stick n. a thin stick that burns with a smell of incense.

jostle v. push roughly.

jot n. a very small amount. ● v. (**jotted**) write down briefly.

jotter n. a notepad.

joule n. a unit of energy (symbol J).

journal n. a daily record of events; a newspaper or periodical.

journalese n. a style of language used in inferior journalism.

journalist n. a person employed to write for a newspaper or magazine. □ **journalism** n.

journey n. (pl. **journeys**) an act of going from one place to another. ● v. make a journey.

journeyman n. (pl. **-men**) a qualified craftsman who works for another.

joust | jumper

joust *v.* fight on horseback with lances.

jovial *adj.* full of cheerful good humour. □ **jovially** *adv.*, **joviality** *n.*

jowl *n.* the jaw, the cheek; an animal's dewlap.

joy *n.* great pleasure; something causing delight.

joyful *adj.* full of joy. □ **joyfully** *adv.*, **joyfulness** *n.*

joyous *adj.* joyful. □ **joyously** *adv.*

joyride *n.* (*colloquial*) a car ride taken for pleasure in a stolen car. □ **joyriding** *n.*

joystick *n.* an aircraft's control lever; a device for moving a cursor on a VDU screen.

JP *abbr.* Justice of the Peace.

Jr. *abbr.* Junior.

jubilant *adj.* joyful, rejoicing. □ **jubilantly** *adv.*, **jubilation** *n.*

jubilee *n.* a special anniversary.

Judaic *adj.* Jewish.

Judaism *n.* the religion of the Jewish people.

judder *v.* shake noisily or violently. ● *n.* this movement.

judge *n.* a public officer appointed to hear and try cases in law courts; a person who decides who has won a contest; a person able to give an authoritative opinion. ● *v.* try (a case) in a law court; act as judge of.

judgement *n.* (also **judgment**) judging; a judge's decision. □ **judgemental** *adj.*

judicial *adj.* of the administration of justice; of a judge or judgement. □ **judicially** *adv.*

judiciary *n.* the whole body of judges in a country.

judicious *adj.* judging wisely, showing good sense. □ **judiciously** *adv.*

judo *n.* a Japanese system of unarmed combat.

jug *n.* a container with a handle and a shaped lip, for holding and pouring liquids. □ **jugful** *n.*

juggernaut *n.* a very large transport vehicle; an overwhelmingly powerful object or institution.

juggle *v.* toss and catch several objects skilfully for entertainment; manipulate skilfully. □ **juggler** *n.*

jugular vein *n.* either of the two large veins in the neck.

juice *n.* the fluid content of fruits, vegetables, or meat; fluid secreted by an organ of the body. □ **juicy** *adj.*

ju-jitsu *n.* (also **jiu-jitsu, jujutsu**) a Japanese system of unarmed combat.

jukebox *n.* a coin-operated record player.

Jul. *abbr.* July.

julep *n.* a drink of spirits and water flavoured esp. with mint.

julienne *n.* vegetables cut into thin strips.

July *n.* the seventh month.

jumble *v.* mix in a confused way. ● *n.* jumbled articles; items for a jumble sale.

jumble sale *n.* a sale of second-hand articles to raise money for charity.

jumbo *n.* (*pl.* **jumbos**) something that is very large of its kind; (also **jumbo jet**) a very large jet aircraft.

jump *v.* move up off the ground etc. by movement of the legs; make a sudden upward movement; pass over by jumping; leave (rails or track) accidentally; pounce on. ● *n.* **1** a jumping movement. **2** a sudden rise or change; a gap in a series. **3** an obstacle to be jumped. □ **jump the gun** act before the permitted time. **jump the queue** obtain something without waiting one's turn.

jumper *n.* **1** a knitted garment for the upper part of the body; (*Amer.*) a pinafore dress. **2** one who jumps; a short wire used to make or break an electrical circuit.

jump lead *n.* a cable for carrying electric current from one battery through another.

jumpsuit *n.* a one-piece garment for the whole body.

jumpy *adj.* (**jumpier**) nervous.

Jun. *abbr.* June.

junction *n.* a join; a place where roads or railway lines meet.

junction box *n.* a box containing a junction of electric cables.

juncture *n.* a point of time; a convergence of events.

June *n.* the sixth month.

jungle *n.* **1** a tropical forest with tangled vegetation. **2** a scene of ruthless struggle.

junior *adj.* younger in age; lower in rank or authority; for younger children. ● *n.* a junior person.

juniper *n.* an evergreen shrub with dark berries.

junk *n.* **1** useless or discarded articles, rubbish. **2** a flat-bottomed ship with sails, used in the China seas.

junket *n.* a sweet custard-like food made of milk and rennet.

junk food *n.* food with low nutritional value.

junkie *n.* (*slang*) a drug addict.

junk mail *n.* unrequested advertising matter sent by post.

junk shop *n.* a shop selling cheap second-hand goods.

junta *n.* a group who combine to rule a country, esp. after a revolution.

jurisdiction *n.* the authority to administer justice or exercise power.

jurisprudence *n.* the science or philosophy of law.

jurist *n.* an expert in law.

juror *n.* a member of a jury.

jury *n.* a group of people sworn to give a verdict on a case in a court of law.

jury-rigged *adj.* with makeshift rigging.

just *adj.* fair to all concerned; right in amount etc.; deserved. ● *adv.* exactly; by only a short amount etc.; only a moment ago; merely; positively. □ **just now** a very short time ago. □ **justly** *adv.*

justice *n.* just treatment, fairness; legal proceedings; a judge.

Justice of the Peace *n.* a non-professional magistrate.

justifiable *adj.* able to be justified. □ **justifiably** *adv.*

justify *v.* **1** show to be right or reasonable; be sufficient reason for. **2** adjust (a line of type) to fill a space neatly. □ **justification** *n.*

jut *v.* (**jutted**) □ **jut out** protrude, project over.

jute *n.* fibre from the bark of certain tropical plants.

juvenile *adj.* youthful, childish; for young people. ● *n.* a young person.

juvenile delinquent *n.* a young offender, too young to be held legally responsible for his or her actions.

juxtapose *v.* put (things) side by side. □ **juxtaposition** *n.*

Kk

K *abbr.* kelvin(s); King; one thousand; (*Computing*) a unit of 1024 bits or bytes. ● *symb.* potassium.

kaftan *n.* (also **caftan**) a long tunic worn by men in the Near East; a long loose dress.

kaiser *n.* (*hist.*) an emperor of Germany or Austria.

Kalashnikov *n.* a Russian rifle or sub-machine gun.

kale *n.* a green vegetable with curly leaves.

kaleidoscope *n.* a toy tube containing coloured fragments reflected to produce changing patterns. □ **kaleidoscopic** *adj.*

kamikaze (kami-**kah**-zi) *n.* (in the Second World War) a Japanese explosive-laden aircraft deliberately crashed on its target.

kangaroo *n.* an Australian animal with a pouch to carry its young and strong hind legs for jumping.

kangaroo court *n.* a court formed illegally by a group to settle disputes among themselves.

kaolin *n.* fine white clay used in porcelain and medicine.

kapok *n.* a fluffy fibre used as padding.

kaput *adj.* (*slang*) broken; ruined.

karaoke (kari-**oh**-ki) *n.* entertainment in which customers in a bar sing along to a backing track.

karate (kă-**rah**-ti) *n.* a Japanese system of unarmed combat.

karma *n.* (in Buddhism and Hinduism) a person's actions as affecting his or her next reincarnation.

kasbah *n.* (also **casbah**) the citadel of a North African city.

kayak (**ky**-ak) *n.* a small covered canoe.

kc/s *abbr.* kilocycle(s) per second.

kebab *n.* small pieces of meat cooked on a skewer.

kedge *n.* a small anchor.

kedgeree *n.* a cooked dish of fish, rice, hard-boiled eggs, etc.

keel *n.* a timber or steel structure along the base of a ship. □ **keel over** overturn; become tilted.

keen *adj.* **1** sharp; penetrating; piercingly cold. **2** intense; very eager. □ **keen on** (*colloquial*)

liking greatly. □ **keenly** *adv.,* **keenness** *n.*

keep *v.* (**kept, keeping**) **1** retain possession of; have charge of. **2** remain or cause to remain in a specified state or position; detain. **3** put aside for a future time. **4** provide with food and other necessities; own and look after (animals); manage (a shop etc.). **5** continue doing something; remain in good condition. ● *n.* **1** a person's food and other necessities. **2** a strongly fortified structure in a castle. □ **keep up** progress at the same pace as others; continue; maintain.

keeper *n.* a person who keeps or looks after something.

keeping *n.* custody, charge. □ **in keeping with** suited to.

keepsake *n.* something kept in memory of the giver.

keg *n.* a small barrel.

keg beer *n.* beer from a pressurized metal container.

kelp *n.* a large brown seaweed used as manure.

kelvin *n.* a degree of the **Kelvin scale** of temperature with zero at absolute zero (−273.15°C).

kendo *n.* a Japanese sport of fencing with bamboo swords.

kennel *n.* a shelter for a dog; (**kennels**) a boarding place for dogs.

kept *see* **keep.**

keratin *n.* a protein forming the basis of horn, nails, and hair.

kerb *n.* a stone edging to a pavement.

kerchief *n.* a headscarf or neckerchief.

kerfuffle *n.* (*colloquial*) a fuss, a commotion.

kermes *n.* an insect used in making a red dye.

kernel *n.* a seed within a husk, nut, or fruit stone; the central or important part.

kerosene *n.* (also **kerosine**) paraffin oil.

kestrel *n.* a small falcon.

ketch *n.* a two-masted sailing boat.

ketchup *n.* (*Amer.* **catsup**) a thick tomato sauce.

kettle *n.* a container with a spout and handle, for boiling water.

kettledrum *n.* a drum with parchment stretched over a large metal bowl.

key *n.* **1** a piece of metal shaped for moving the bolt of a lock, winding a clock, etc.; something giving access, control, or insight. **2** a system of related notes in music. **3** a lever for a finger to press on a piano, typewriter, etc. □ **keyed up** stimulated, nervously tense.

keyboard *n.* a set of keys on a piano, typewriter, or computer. ● *v.* enter (data) using a keyboard. □ **keyboarder** *n.*

keyhole *n.* a hole for a key in a lock.

keyhole surgery *n.* surgery carried out through a very small incision.

keynote *n.* **1** the note on which a key in music is based. **2** the prevailing idea of a speech, conference etc.

keypad *n.* a small device with buttons for operating electronic equipment, a telephone, etc.

keyring *n.* a ring on which keys are threaded.

keystone *n.* the central stone of an arch, locking others into position.

keystroke *n.* a single depression of a key on a keyboard.

keyword *n.* the key to a cipher etc.

kg *abbr.* kilogram(s).

KGB *abbr.* the secret police of the former USSR.

khaki *adj.* & *n.* dull brownish yellow, the colour of some military uniforms.

khan *n.* the title of rulers and officials in central Asia.

kHz *abbr.* kilohertz.

kibbutz *n.* (*pl.* **kibbutzim**) a communal settlement in Israel.

kick *v.* **1** strike or propel with the foot. **2** (of a gun) recoil when fired. ● *n.* **1** an act of kicking; a blow with the foot. **2** (*colloquial*) a thrill. □ **kick out** (*colloquial*) drive out forcibly; dismiss. **kick up** (*colloquial*) create (a fuss or noise).

kick-off *n.* the start of a football game.

kick-start *n.* a lever pressed with the foot to start a motorcycle.

kid *n.* a young goat; (*slang*) a child. ● *v.* (**kidded**) (*colloquial*) deceive, tease.

kidnap *v.* (**kidnapped**; *Amer.* **kidnaped**) carry off (a person) illegally to obtain a ransom. □ **kidnapper** *n.*

kidney *n.* (*pl.* **kidneys**) either of a pair of organs that remove waste products from the blood and secrete urine.

kidney bean *n.* a red-skinned dried bean.

kill *v.* **1** cause the death of; put an end to. **2** spend (time) unprofitably when waiting. ● *n.* a killing; an animal killed by a hunter. □ **killer** *n.*

killjoy *n.* a person who spoils others' enjoyment.

kiln *n.* an oven for hardening or drying pottery or hops, or burning lime.

kilo *n.* (*pl.* **kilos**) a kilogram.

kilo- *comb. form* one thousand.

kilobyte *n.* (*Computing*) 1,024 bytes.

kilocalorie *n.* the amount of heat needed to raise the temperature of 1Kg of water by 1°C.

kilocycle *n.* (*old use*) a kilohertz.

kilogram *n.* a unit of weight or mass in the metric system (2.205 lb).

kilohertz n. a unit of frequency of electromagnetic waves, = 1,000 cycles per second.

kilojoule n. 1,000 joules.

kilometre n. 1,000 metres (0.62 mile).

kilovolt n. 1,000 volts.

kilowatt n. 1,000 watts.

kilt n. a knee-length pleated skirt of tartan, as traditionally worn by Highland men.

kimono n. (pl. **kimonos**) a loose Japanese robe worn with a sash.

kin n. a person's relatives.

kind n. a class of similar things. ●adj. gentle and considerate towards others. □ **in kind** (of payment) in goods etc., not money. □ **kind-hearted** adj., **kindness** n.

kindergarten n. a school for very young children.

kindle v. set on fire; arouse, stimulate.

kindling n. small pieces of wood for lighting fires.

kindly adj. (**kindlier**) kind. ●adv. in a kind way; please. □ **kindliness** n.

kindred n. a person's relatives. ●adj. related; of a similar kind.

kinetic adj. of movement.

king n. **1** a male ruler of a country by right of birth; a man or thing regarded as supreme. **2** a chess piece; a playing card next above queen.

kingdom n. a country ruled by a king or queen; a division of the natural world.

kingfisher n. a bird with bright blue feathers that dives to catch fish.

kingpin n. an indispensable person or thing.

king-size adj. (also **king-sized**) extra large.

kink n. a short twist in thread or wire etc.; a mental peculiarity. ●v. form or cause to form a kink or kinks. □ **kinky** adj.

kinsfolk n.pl. a person's relatives. □ **kinsman** n., **kinswoman** n.

kiosk n. a booth where newspapers or refreshments are sold, or one containing a public telephone.

kip n. (colloquial) a sleep.

kipper n. a smoked herring.

kirk n. (Scot.) a church.

kirsch (keersh) n. a colourless liqueur made from cherries.

kismet n. destiny, fate.

kiss n. & v. touch or caress with the lips.

kissogram n. a novelty greetings message delivered with a kiss.

kit n. an outfit of clothing; a set of tools; a set of parts to be assembled. ●v. (**kitted**) equip with kit.

kitbag n. a bag for holding kit.

kitchen n. a room where meals are prepared.

kitchenette n. a small kitchen.

kitchen garden n. a garden for vegetables, fruit, and herbs.

kite n. **1** a large bird of the hawk family. **2** a light framework on a string, for flying in the wind as a toy.

Kitemark n. an official mark on goods approved by British Standards.

kith and kin n. relatives.

kitsch (kich) n. pretentiousness and lack of good taste.

kitten n. a young of cat, rabbit, or ferret.

kittiwake n. a small seagull.

kitty n. a communal fund.

kiwi n. a New Zealand bird that does not fly.

kJ abbr. kilojoule(s).

Klaxon n. (trade mark) an electric horn.

Kleenex n. (trade mark) a paper handkerchief.

kleptomania n. a compulsive desire to steal. □ **kleptomaniac**

km *abbr.* kilometre(s).

knack *n.* the ability to do something skilfully.

knacker *n.* a person who buys and slaughters old horses, cattle, etc. ● *v.* (*slang*) exhaust; wear out; damage.

knapsack *n.* a bag worn strapped on the back.

knave *n.* (*old use*) a rogue; a jack in playing cards.

knead *v.* press and stretch (dough) with the hands; massage.

knee *n.* the joint between the thigh and the lower leg; part of a garment covering this. ● *v.* (**kneed, kneeing**) hit with the knee.

kneecap *n.* the small bone over the front of the knee. ● *v.* (**kneecapped**) shoot in the knee as a punishment.

knee-jerk *adj.* (of a reaction) automatic and predictable.

kneel *v.* (**knelt** or **kneeled, kneeling**) rest on the knees.

knees-up *n.* (*colloquial*) a lively party.

knell *n.* the sound of a bell tolled after a death.

knelt *see* **kneel**.

knew *see* **know**.

knickerbockers *n.pl.* loose breeches gathered at the knee.

knickers *n.pl.* underpants.

knick-knack *n.* a small ornament.

knife *n.* (*pl.* **knives**) a cutting or spreading instrument with a blade and handle. ● *v.* stab with a knife.

knight *n.* **1** a man given a rank below baronet, with the title 'Sir'. **2** a chess piece shaped like a horse's head. ● *v.* confer a knighthood on.

knighthood *n.* the rank of knight.

knit *v.* (**knitted** or **knit, knitting**) make a garment from yarn formed into interlocking loops;

grow together so as to unite. □ **knitter** *n.*, **knitting** *n.*

knob *n.* a rounded projecting part, esp. as a handle; a small lump. □ **knobbly** *adj.*

knock *v.* **1** hit with an audible sharp blow; strike a door to gain admittance; drive in by hitting. **2** (*slang*) criticize. ● *n.* the act or sound of knocking; a sharp blow. □ **knock about** treat roughly; wander casually. **knock off** (*colloquial*) cease work; complete quickly; (*slang*) steal. **knock out** make unconscious; eliminate; disable. **knock up** rouse by knocking at a door; make or arrange hastily.

knock-down *adj.* (of a price) very low.

knocker *n.* one who knocks; a hinged device for knocking on a door.

knock-kneed *adj.* having knees that bend inwards.

knock-on effect *n.* a secondary, indirect effect.

knockout *n.* **1** knocking a person or team out. **2** (*colloquial*) an outstanding person or thing.

knock-up *n.* a practice game at tennis etc.

knoll *n.* a small hill.

knot *n.* **1** an intertwining of one or more pieces of thread or rope etc. as a fastening; a tangle; a cluster. **2** a hard round spot in timber formed where a branch joined the trunk. **3** a unit of speed used by ships and aircraft, = one nautical mile per hour. ● *v.* (**knotted**) tie or fasten with a knot; entangle.

knotty *adj.* (**knottier**) **1** full of knots. **2** puzzling; difficult.

know *v.* (**knew, known, knowing**) have in one's mind or memory; feel certain; recognize, be familiar with; understand. □ **in the know** (*colloquial*) having inside information.

know-how n. practical knowledge or skill.

knowing adj. aware; cunning. □ **knowingly** adv.

knowledge n. knowing about things; all a person knows; all that is known, information.

knowledgeable adj. intelligent; well-informed.

knuckle n. a finger joint; an animal's leg joint as meat. □ **knuckle under** yield, submit.

knuckleduster n. a metal device worn over the knuckles to increase the effect of a blow.

koala n. an Australian bearlike tree-climbing animal with thick grey fur.

kohl n. black powder used as eye make-up.

kohlrabi n. (pl. **kohlrabies**) a cabbage with an edible turnip-like stem.

kola var. of **cola**.

koodoo var. of **kudu**.

kookaburra n. an Australian giant kingfisher with a harsh cry.

kopeck, kopek vars. of **copeck**.

Koran n. the sacred book of Islam.

kosher adj. **1** conforming to Jewish dietary laws. **2** (colloquial) genuine, legitimate.

kowtow v. behave with exaggerated respect.

k.p.h. abbr. kilometres per hour.

Kr symb. krypton.

kremlin n. a citadel in a Russian town; (**the Kremlin**) the Russian government.

krill n. tiny plankton crustaceans that are food for whales etc.

krona n. the unit of money in Sweden (pl. **kronor**) and Iceland (pl. **krónur**).

krone n. (pl. **kroner**) the unit of money in Denmark and Norway.

krugerrand n. a South African gold coin.

krypton n. a chemical element (symbol Kr), a colourless, odourless gas.

Ku symb. kurchatovium.

kudos n. (colloquial) honour and glory.

kudu n. (also **koodoo**) an African antelope.

kummel (kuu-měl) n. a liqueur flavoured with caraway seeds.

kumquat n. (also **cumquat**) a tiny variety of orange.

kung fu n. a Chinese form of unarmed combat.

kurchatovium n. the Russian name (symbol Ku) for the element rutherfordium.

Kurd n. a member of a people of SW Asia. □ **Kurdish** adj.

kV abbr. kilovolt(s).

kW abbr. kilowatt(s).

LI

L abbr. large; learner driver. ● n. (Roman numeral) 50.

l abbr. litre(s).

lab n. (colloquial) a laboratory.

label n. a note fixed on an object to show its nature, destination, etc. ● v. (**labelled**; Amer. **labeled**) fix a label to; describe as.

labia n.pl. the lips of the female genitals.

labial adj. of the lips.

labor etc. Amer. sp. of **labour** etc.

laboratory n. a room or building equipped for scientific work.

laborious adj. needing or showing much effort. □ **laboriously** adv.

labour n. (Amer. **labor**) **1** work, exertion; workers. **2** contractions of the womb at childbirth. ● v. work hard; emphasize lengthily.

laboured adj. (Amer. **labored**) showing signs of great effort; not spontaneous.

labourer n. (Amer. **laborer**) a person employed to do unskilled work.

Labour Party n. a British political party formed to represent the interests of ordinary working people.

Labrador n. a dog of the retriever breed with a black or golden coat.

laburnum n. a tree with hanging clusters of yellow flowers.

labyrinth n. a maze. □ **labyrinthine** adj.

lace n. **1** ornamental openwork fabric or trimming. **2** a cord threaded through holes or hooks to pull opposite edges together. ● v. **1** fasten with a lace or laces; intertwine. **2** add a spirit to (a drink).

lacerate v. tear (flesh); wound (feelings). □ **laceration** n.

lachrymose adj. tearful.

lack n. the state or fact of not having something. ● v. be without.

lackadaisical adj. lacking vigour or determination, unenthusiastic.

lackey n. (pl. **lackeys**) a footman, a servant; a servile follower.

lacking adj. undesirably absent; without.

lacklustre adj. (Amer. **lackluster**) lacking brightness, enthusiasm, or conviction.

laconic adj. using few words; terse. □ **laconically** adv.

lacquer n. a hard glossy varnish. ● v. coat with lacquer.

lacrosse n. a game similar to hockey played with nets on sticks.

lactation n. suckling; secretion of milk.

lactic acid n. acid found in sour milk and produced in the muscles during exercise.

lactose n. a sugar present in milk.

lacuna n. (pl. **lacunas** or **lacunae**) a gap.

lacy adj. (**lacier**) of or like lace.

lad n. a boy.

ladder n. a set of crossbars between uprights, used for climbing up; a vertical ladder-like flaw where stitches become undone in a stocking. ● v. cause or develop a ladder (in).

laden adj. loaded.

ladle n. a deep long-handled spoon for transferring liquids. ● v. transfer with a ladle.

lady n. a woman; a well-mannered woman; (**Lady**) the title of wives, widows, or daughters of certain noblemen.

ladybird n. a small flying beetle, usu. red with black spots.

lady-in-waiting n. a lady attending a queen or princess.

ladylike adj. polite and appropriate to a lady.

ladyship n. the title used to or about a woman with the rank of Lady.

lag¹ v. (**lagged**) **1** go too slow, not keep up. **2** encase in material that prevents loss of heat. ● n. a delay.

lager n. a light-coloured beer.

lager lout n. (colloquial) a drunken rowdy youth.

laggard n. a person who lags behind.

lagging n. material used to lag a boiler etc.

lagoon n. a salt-water lake beside the sea.

lah n. (Music) the sixth note of a major scale, or the note A.

laid see **lay²**.

laid-back adj. easygoing, relaxed.

lain see **lie²**.

lair n. a place where a wild animal rests.

laird n. (*Scot.*) a landowner.

laissez-faire (less-ay-**fair**) n. a policy of non-interference, esp. in politics or economics.

laity n. lay people.

lake n. **1** a large body of water surrounded by land. **2** a reddish pigment.

lam v. (**lammed**) (*slang*) hit hard.

lama n. a Buddhist priest in Tibet and Mongolia.

lamb n. **1** a young sheep; its flesh as food. **2** a gentle or endearing person. ● v. give birth to a lamb.

lambaste v. thrash; reprimand severely.

lame adj. unable to walk normally; weak, unconvincing. ● v. make lame. □ **lamely** adv., **lameness** n.

lamé (lah-may) n. a fabric interwoven with gold or silver thread.

lament n. an expression of grief; a song or poem expressing grief. ● v. feel or express grief or regret. □ **lamentation** n.

lamentable adj. regrettable, deplorable. □ **lamentably** adv.

laminate n. laminated material.

laminated adj. made of layers joined one upon another.

lamp n. a device for giving light.

lamplight n. light from a lamp.

lampoon n. a piece of writing that attacks a person with ridicule. ● v. ridicule in a lampoon.

lamp-post n. the tall post of a street lamp.

lamprey n. (pl. **lampreys**) a small eel-like water animal.

lampshade n. a shade on a lamp screening its light.

lance n. a long spear. ● v. prick or cut open with a lancet.

lance corporal n. an army rank below corporal.

lancet n. a surgeon's pointed two-edged knife; a tall narrow pointed arch or window.

land n. **1** the part of the earth's surface not covered by water; ground, soil. **2** a country or state. ● v. **1** set or go ashore; come or bring to the ground. **2** bring to or reach a place or situation. **3** deal (a person) a blow. **4** obtain (a prize, appointment, etc.).

landau (lan-daw) n. a horse-drawn carriage.

landed adj. owning land; consisting of land.

landfall n. the approach to land after a journey by sea or air.

landfill n. waste material used in landscaping or reclaiming ground; the use of this.

landing n. **1** coming or bringing ashore or to ground; a place for this. **2** a level area at the top of a flight of stairs.

landing stage n. a platform for landing from a boat.

landlady n. a woman who lets land or a house or room to a tenant; a woman who runs a public house.

landlocked adj. surrounded by land.

landlord n. a person who lets land or a house or room to a tenant; one who runs a public house.

landlubber n. a person not accustomed to sailing.

landmark n. a conspicuous feature of a landscape; an event marking a stage in a thing's development.

landscape n. the scenery of a land area; a picture of this. ● v. lay out (an area) attractively with natural-looking features.

landslide n. a landslip; an overwhelming majority of votes.

landslip n. a sliding down of a mass of land on a slope.

landward *adj. & adv.* towards the land.

lane *n.* a narrow road, track, or passage; a strip of road for a single line of traffic; a track to which ships or aircraft etc. must keep.

language *n.* words and their use; a system of this used by a nation or group.

languid *adj.* lacking vigour or vitality.

languish *v.* lose or lack vitality; live under miserable conditions.

languor *n.* the state of being languid. □ **languorous** *adj.*

lank *adj.* tall and lean; (of hair) straight and limp. □ **lanky** *adj.*, **lankiness** *n.*

lanolin *n.* a fat extracted from sheep's wool, used in ointments.

lantern *n.* a transparent case for holding and shielding a light.

lanthanum *n.* a metallic element (symbol La).

lanyard *n.* a short rope for securing things on a ship; a cord for hanging a whistle etc. round the neck or shoulder.

lap *n.* **1** a flat area over the thighs of a seated person. **2** a single circuit of a racecourse; a section of a journey. ● *v.* (**lapped**) **1** take up (liquid) by movements of the tongue; flow (against) with ripples. **2** wrap round; be one or more laps ahead of (a competitor).

laparoscope *n.* a fibre optic instrument inserted through the abdomen to view the internal organs. □ **laparoscopy** *n.*

lapdog *n.* a small pampered dog.

lapel *n.* a flap folded back at the front of a coat etc.

lapidary *adj.* of stones.

lapis lazuli *n.* a blue semi-precious stone.

Lapp *n.* a native or the language of Lapland.

lapse *v.* fail to maintain one's position or standard; become void or no longer valid. ● *n.* a slight error; lapsing; the passage of time.

laptop *n.* a portable computer.

lapwing *n.* a plover.

larch *n.* a deciduous tree of the pine family.

lard *n.* a white greasy substance prepared from pig-fat. ● *v.* put strips of fat bacon in or on (meat) before cooking.

larder *n.* a storeroom for food.

large *adj.* of great size or extent. □ **at large** free to roam about; in general. □ **largeness** *n.*

largely *adv.* to a great extent.

largess *n.* (also **largesse**) money or gifts generously given.

lariat *n.* a lasso.

lark *n.* **1** a small brown bird, a skylark. **2** a light-hearted amusing incident. □ **lark about** play light-heartedly.

larkspur *n.* a plant with spur-shaped blue or pink flowers.

larva *n.* (*pl.* **larvae**) an insect in the first stage of its life after coming out of the egg. □ **larval** *adj.*

laryngitis *n.* inflammation of the larynx.

larynx *n.* the part of the throat containing the vocal cords.

lasagne (lä-zan-yǎ) *n.* pasta formed into sheets; a dish of pasta layered with sauces of cheese, meat, tomato, etc.

lascivious *adj.* lustful. □ **lasciviously** *adv.*, **lasciviousness** *n.*

laser *n.* a device emitting an intense narrow beam of light.

lash *v.* move in a whip-like movement; beat with a whip; strike violently; fasten with a cord etc. ● *n.* the flexible part of a whip; a stroke with this; an eyelash. □ **lash out** attack with blows or words; spend lavishly.

lashings *n.pl.* (*colloquial*) a lot.

lass *n.* (also **lassie**) (*Scot. & N. Engl.*) a girl, a young woman.

Lassa fever *n.* a serious virus disease of tropical Africa.

lassitude *n.* tiredness, listlessness.

lasso *n.* (*pl.* **lassos** or **lassoes**) a rope with a noose for catching cattle. ● *v.* (**lassoed, lassoing**) catch with a lasso.

last *adj. & adv.* coming after all others; most recent(ly). ● *n.* **1** the last person or thing. **2** a foot-shaped block used in making and repairing shoes. ● *v.* continue, endure; suffice for a period of time. □ **at last, at long last** after much delay. □ **lasting** *adj.*

lastly *adv.* finally.

last post *n.* a military bugle call sounded at sunset or military funerals.

last straw *n.* an addition to existing difficulties, making them unbearable.

last word *n.* the final statement in a dispute; the latest fashion.

latch *n.* a bar lifted from its catch by a lever, used to fasten a gate etc; a spring-lock that catches when a door is closed. ● *v.* fasten with a latch.

latchkey *n.* a key of an outer door.

late *adj. & adv.* **1** after the proper or usual time; far on in a day or night or period. **2** recent; no longer living or holding a position. □ **of late** lately. □ **lateness** *n.*

lately *adv.* recently.

latent *adj.* existing but not active or developed or visible. □ **latency** *n.*

lateral *adj.* of, at, to, or from the side(s). □ **laterally** *adv.*

latex *n.* a milky fluid from certain plants, esp. the rubber tree; a similar synthetic substance.

lath *n.* (*pl.* **laths**) a narrow thin strip of wood, in a trellis or partition wall.

lathe *n.* a machine for holding and turning pieces of wood or metal while they are worked.

lather *n.* froth from soap and water; frothy sweat. ● *v.* cover with or form lather.

Latin *n.* the language of the ancient Romans. ● *adj.* of or in Latin; speaking a language based on Latin.

Latin America the parts of Central and South America where Spanish or Portuguese is the main language.

latitude *n.* **1** the distance of a place from the equator, measured in degrees; a region. **2** freedom from restrictions. □ **latitudinal** *adj.*

latrine *n.* a lavatory in a camp or barracks.

latter *adj.* mentioned after another; nearer to the end; recent.

latter-day *adj.* modern, recent.

latterly *adv.* recently; nowadays.

lattice *n.* a framework of crossed strips.

laudable *adj.* praiseworthy.

laudanum *n.* opium prepared for use as a sedative.

laudatory *adj.* praising.

laugh *v.* make sounds and facial movements expressing amusement or scorn. ● *n.* the act or manner of laughing; (*colloquial*) an amusing incident.

laughable *adj.* ridiculous.

laughing stock *n.* a person or thing that is ridiculed.

laughter *n.* the act or sound of laughing.

launch *v.* put or go into action; send (a ship) into the water or (a rocket or missile) into the air. ● *n.* the process of launching something; a large motor boat.

launder *v.* **1** wash and iron (clothes etc.). **2** transfer (funds) to conceal their origin.

launderette *n.* an establishment fitted with washing machines to be used for a fee.

laundry n. a place where clothes etc. are laundered; clothes etc. for washing.

laurel n. an evergreen shrub with smooth glossy leaves; (**laurels**) victories or honours gained.

lava n. flowing or hardened molten rock from a volcano.

lavatory n. a fixture into which urine and faeces are discharged; a room equipped with this.

lavender n. a shrub with fragrant purple flowers; light purple.

laver n. edible seaweed.

lavish adj. giving or producing something in large quantities; plentiful. ● v. bestow lavishly. □ **lavishly** adv., **lavishness** n.

law n. a rule established by authority; a set of such rules; their influence or operation; a statement of what always happens in certain circumstances.

law-abiding adj. obeying the law.

law court n. a room or building in which legal trials are held.

lawful adj. permitted or recognized by law. □ **lawfully** adv., **lawfulness** n.

lawless adj. disregarding the law; uncontrolled. □ **lawlessness** n.

lawn n. 1 an area of closely cut grass. 2 fine woven cotton fabric.

lawnmower n. a machine for cutting grass.

lawn tennis n. tennis played with a soft ball on outdoor grass or a hard court.

lawrencium n. a radioactive metallic element (symbol Lw).

lawsuit n. the process of bringing a problem before a court of law for settlement.

lawyer n. a person qualified in legal matters.

lax adj. slack, not strict or severe. □ **laxity** n.

laxative adj. & n. (a medicine) stimulating the bowels to empty.

lay¹ adj. not ordained into the clergy; non-professional.

lay² v. (**laid**, **laying**) 1 place on a surface; arrange ready for use. 2 cause to be in a certain condition. 3 (of a hen) produce (eggs). ● n. the way a thing lies. □ **lay bare** expose, reveal. **lay into** (slang) thrash; scold harshly. **lay off** discharge (workers) temporarily; (colloquial) stop. **lay on** provide. **lay out** arrange; prepare (a body) for burial; spend (money) for a purpose; knock unconscious. **lay up** store; cause (a person) to be ill. **lay waste** devastate (an area). See also **lie²**.

■ **Usage** It is incorrect in standard English to use *lay* to mean 'lie', as in *She was laying on the floor.*

layabout n. a lazy person.

lay-by n. an area beside the road where vehicles may stop.

layer n. 1 one thickness of a substance covering a surface. 2 an attached shoot fastened down to take root. 3 a hen that lays eggs. ● v. 1 arrange or cut in layers. 2 propagate (a plant) by layers.

layette n. an outfit for a newborn baby.

lay figure n. an artist's jointed model of the human body.

layman n. (pl. **-men**) a non-professional person.

layout n. an arrangement of parts etc. according to a plan.

laze v. spend time idly.

lazy adj. (**lazier**) unwilling to work, doing little work; showing lack of energy. □ **lazily** adv., **laziness** n.

lb abbr. pound(s) weight.

LCD abbr. liquid crystal display; lowest common denominator.

LEA *abbr.* Local Education Authority.

lea *n.* (*poetic*) a piece of meadow or arable land.

leach *v.* percolate (liquid) through soil etc.; remove (soluble matter) or be removed in this way.

lead[1] (leed) *v.* (**led, leading**) **1** guide; influence into an action, opinion, or state. **2** be a route or means of access. **3** pass (one's life). **4** be or go first; be ahead. ● *n.* **1** guidance; a clue. **2** the leading place; the amount by which one competitor is in front. **3** a wire conveying electric current; a strap or cord for leading an animal. **4** a chief role in a play etc. □ **lead up to** serve as introduction to or preparation for; direct conversation towards.

lead[2] (led) *n.* a chemical element (symbol Pb), a heavy grey metal; graphite in a pencil; a lump of lead used for sounding depths.

leaden *adj.* made of lead; heavy, slow-moving; (of the sky) dark grey.

leader *n.* one that leads; a newspaper article giving editorial opinions. □ **leadership** *n.*

leading question *n.* one worded to prompt the desired answer.

leaf *n.* (*pl.* **leaves**) **1** a flat (usu. green) organ growing from the stem or root of a plant. **2** a single thickness of paper, a page; a very thin sheet of metal. **3** a hinged flap or extra section of a table. □ **leaf through** turn over the leaves of (a book).

leaflet *n.* **1** a printed sheet of paper giving information. **2** a small leaf of a plant. ● *v.* distribute leaflets.

leaf mould *n.* soil or compost consisting of decayed leaves.

leafy *adj.* (**leafier**) having many leaves.

league *n.* a union of people or countries; an association of sports clubs which compete against each other; a class of contestants. □ **in league with** conspiring with.

leak *n.* a hole through which liquid or gas escapes accidentally; liquid etc. passing through this; a similar escape of an electric charge; a disclosure of secret information. ● *v.* escape or let out from a container; disclose. □ **leakage** *n.*, **leaky** *adj.*

lean *v.* (**leaned** or **leant, leaning**) put or be in a sloping position; rest against. ● *adj.* without much flesh; (of meat) with little or no fat; scanty. ● *n.* the lean part of meat. □ **lean on** depend on for help; (*colloquial*) influence by intimidating. □ **leanness** *n.*

leaning *n.* an inclination, a preference.

lean-to *n.* a shed etc. against the side of a building.

leap *v.* (**leaped** or **leapt, leaping**) jump vigorously. ● *n.* a vigorous jump.

leapfrog *n.* a game in which each player vaults over another who is bending down. ● *v.* (**leapfrogged**) perform this vault (over); overtake alternately.

leap year *n.* a year with an extra day (29 Feb.).

learn *v.* (**learned** or **learnt, learning**) gain knowledge of or skill in; become aware of. □ **learner** *n.*

learned (ler-nid) *adj.* having or showing great learning.

learning *n.* knowledge obtained by study.

lease *n.* a contract allowing the use of land or a building for a specified time. ● *v.* allow, obtain, or hold by lease. □ **leasehold** *n.*, **leaseholder** *n.*

leash *n.* a dog's lead.

least *adj.* smallest in amount or degree; lowest in importance. ● *n.* the least amount etc. ● *adv.* in the least degree.

leather *n.* material made by treating animal skins; a piece of soft leather for polishing with. ● *v.* thrash.

leatherette *n.* imitation leather.

leatherjacket *n.* a crane-fly grub with tough skin.

leathery *adj.* leather-like; tough.

leave *v.* (**left, leaving**) **1** go away (from); go away finally or permanently. **2** let remain; deposit; entrust with; abandon. ● *n.* permission; permission to be absent from duty; the period for which this lasts. □ **leave out** not insert or include.

leaven *n.* a raising agent; an enlivening influence. ● *v.* add leaven to; enliven.

lecher *n.* a lecherous man.

lechery *n.* unrestrained indulgence in sexual lust. □ **lecherous** *adj.*

lecithin *n.* a compound found in plants and animals used as a food emulsifier and stabilizer.

lectern *n.* a stand with a sloping top from which a bible etc. is read.

lecture *n.* a speech giving information about a subject; a lengthy reproof or warning. ● *v.* give a lecture or lectures; reprove at length. □ **lecturer** *n.*

LED *abbr.* light-emitting diode.

led *see* **lead¹**.

ledge *n.* a narrow horizontal projection or shelf.

ledger *n.* a book used as an account book.

lee *n.* the sheltered side; shelter in this.

leech *n.* a small blood-sucking worm.

leek *n.* a plant related to the onion, with a cylindrical white bulb.

leer *v.* look slyly, maliciously, or lustfully. ● *n.* a leering look.

lees *n.pl.* sediment in wine.

leeward *adj. & n.* (on) the side away from the wind.

leeway *n.* a degree of freedom of action.

left¹ *see* **leave**.

left² *adj. & adv.* of, on, or towards the side of the body which is on the west when one is facing north. ● *n.* **1** the left side or region; the left hand or foot. **2** people supporting a more extreme form of socialism than others in their group.

left-handed *adj.* using the left hand.

leftovers *n.pl.* things remaining when the rest is finished.

leg *n.* **1** each of the limbs on which a person, animal, etc. stands or moves; part of a garment covering a person's leg; a support of piece of furniture. **2** one section of a journey or contest.

legacy *n.* something left to someone in a will, or handed down by a predecessor.

legal *adj.* of or based on law; authorized or required by law. □ **legalistic** *adj.*, **legally** *adv.*, **legality** *n.*

legal aid *n.* help from public funds towards the cost of legal action.

legalize *v.* (also **-ise**) make legal. □ **legalization** *n.*

legate *n.* an envoy.

legatee *n.* the recipient of a legacy.

legation *n.* a diplomatic minister and staff; their headquarters.

legato *adv.* (*Music*) smoothly and evenly.

legend *n.* a story handed down from the past; such stories collectively; an inscription on a coin or medal; an explanation of the symbols on a map.

legendary *adj.* of or described in legend; famous.

legerdemain (lej-er-dĕ-mayn) *n.* sleight of hand.

leggings *n.pl.* a close-fitting stretchy garment covering the legs.

legible *adj.* clear enough to be deciphered, readable. □ **legibly** *adv.*, **legibility** *n.*

legion *n.* a division of the ancient Roman army; an organized group; a multitude.

legionnaire *n.* a member of a legion.

legionnaires' disease *n.* a form of bacterial pneumonia.

legislate *v.* make laws. □ **legislator** *n.*

legislation *n.* legislating; law(s) made.

legislative *adj.* making laws.

legislature *n.* a country's legislative assembly.

legitimate *adj.* **1** in accordance with a law or rule; justifiable. **2** born of parents married to each other. □ **legitimately** *adv.*, **legitimacy** *n.*

legitimize *v.* (also **-ise**) make legitimate. □ **legitimization** *n.*

legless *adj.* **1** without legs. **2** (*slang*) drunk.

legume *n.* a plant of the family bearing seeds in pods; a pod of this. □ **leguminous** *adj.*

leisure *n.* time free from work. □ **at one's leisure** when one has time.

leisured *adj.* having plenty of leisure.

leisurely *adj.* & *adv.* without hurry.

leitmotif (lyt-moh-teef) *n.* (also **leitmotiv**) a theme associated with a particular person or idea throughout an artistic work.

lemming *n.* a mouse-like Arctic rodent (said to rush headlong into the sea and drown in its migration).

lemon *n.* **1** a yellow oval fruit with acid juice; the tree bearing it; a pale yellow colour. **2** (*colloquial*) an unsatisfactory person or thing.

lemonade *n.* a lemon-flavoured fizzy drink.

lemon sole *n.* a kind of plaice.

lemur *n.* a nocturnal monkey-like animal of Madagascar.

lend *v.* (**lent**, **lending**) give or allow to use temporarily; provide (money) temporarily in return for payment of interest; contribute as a help or effect. □ **lend itself to** be suitable for. □ **lender** *n.*

length *n.* the measurement or extent from end to end; a great extent; a piece (of cloth etc.). □ **at length** after or taking a long time.

lengthen *v.* make or become longer.

lengthways *adv.* & *adj.* (also **lengthwise**) in the direction of a thing's length.

lengthy *adj.* (**lengthier**) very long; long and boring. □ **lengthily** *adv.*

lenient *adj.* merciful, not severe. □ **leniently** *adv.*, **lenience** *n.*

lens *n.* a piece of glass or similar substance shaped for use in an optical instrument; the transparent part of the eye, behind the pupil.

Lent *n.* the Christian period of fasting and repentance before Easter.

lent *see* **lend**.

lentil *n.* a kind of bean.

leonine *adj.* of or like a lion.

leopard (lep-erd) *n.* a large flesh-eating animal of the cat family, with a dark-spotted yellowish or a black coat.

leotard *n.* a close-fitting stretchy garment worn by dancers, gymnasts, etc.

leper *n.* a person with leprosy.

leprechaun n. (in Irish folklore) an elf resembling a little old man.

leprosy n. an infectious disease affecting the skin and nerves and causing deformities. □ **leprous** adj.

lesbian n. a homosexual woman. ● adj. of lesbians. □ **lesbianism** n.

lese-majesty (leez) n. an insult to a ruler; treason.

lesion n. a harmful change in the tissue of an organ of the body.

less adj. not so much of; smaller in amount or degree. ● adv. to a smaller extent. ● n. a smaller amount. ● prep. minus.

■ **Usage** The use of less to mean 'fewer', as in There are less people than yesterday, is incorrect in standard English.

lessee n. a person holding property by lease.

lessen v. make or become less.

lesser adj. not so great as the other.

lesson n. an amount of teaching given at one time; something to be learnt by a pupil; an experience by which one can learn; a passage from the Bible read aloud.

lessor n. a person who lets property on lease.

lest conj. for fear that.

let v. (**let**, **letting**) allow or cause to; allow the use of (rooms or land) in return for payment. ● v.aux. used in requests, commands, assumptions, or challenges; let's try. ● n. (in tennis etc.) an obstruction of the ball. □ **let alone** refrain from interfering with or doing; not to mention. **let down** let out air from (a tyre etc.); fail to support, disappoint. **let off** fire or explode (a weapon etc.); excuse from. **let on** (slang) reveal a secret. **let**

out allow to go out; make looser. **let up** (colloquial) relax.

lethal adj. causing death.

lethargy n. extreme lack of energy or vitality. □ **lethargic** adj., **lethargically** adv.

letter n. **1** a symbol representing a speech sound. **2** a written message sent by post; (**letters**) literature. ● v. inscribe letters (on).

letter box n. a slit in a door, with a movable flap, through which letters are delivered; a postbox.

letterhead n. a printed heading on stationery.

lettuce n. a plant with broad crisp leaves used as salad.

leucocyte n. a white blood cell.

leukaemia (lew-kee-miǎ) n. a disease in which leucocytes multiply uncontrollably.

levee n. (Amer.) an embankment against floods; a quay.

level adj. **1** horizontal. **2** without projections or hollows. **3** on a level with. **4** steady, uniform. ● n. **1** a horizontal line or plane; a device for testing this. **2** a measured height or value etc.; a relative position. **3** a level surface or area. ● v. (**levelled**; Amer. **leveled**) **1** make or become level; knock down or flatten (a building). **2** aim (a gun etc.). □ **on the level** honest(ly).

level crossing n. a place where a road and railway cross at the same level.

level-headed adj. sensible.

level pegging n. equality in score.

lever n. a bar pivoted on a fixed point to lift something; a pivoted handle used to operate machinery; a means of power or influence. ● v. use a lever; lift by this.

leverage n. the action or power of a lever; power, influence.

leveret n. a young hare.

leviathan n. something of enormous size and power.

levitate v. rise or cause to rise and float in the air. □ **levitation** n.

levity n. a humorous attitude.

levy v. impose or collect (a payment etc.) by authority or force. ● n. levying; a payment levied.

lewd adj. treating sexual matters vulgarly; lascivious. □ **lewdly** adv., **lewdness** n.

lexical adj. of words.

lexicography n. the process of compiling dictionaries. □ **lexicographer** n.

lexicon n. a dictionary; a vocabulary.

Li symb. lithium.

liability n. being liable; (colloquial) a disadvantage; (**liabilities**) debts.

liable adj. held responsible by law; legally obliged to pay a tax or penalty etc.; likely to do or suffer something.

liaise v. (colloquial) act as a link or go-between.

liaison n. communication and cooperation; an illicit sexual relationship.

liana n. a climbing plant of tropical forests.

liar n. a person who tells lies.

libation n. a drink-offering to a god.

libel n. a published false statement that damages a person's reputation; the act of publishing it. ● v. (**libelled**; Amer. **libeled**) publish a libel against. □ **libellous** adj.

liberal adj. generous; tolerant, open-minded; (in politics) advocating liberal policies. □ **liberally** adv.

liberalize v. (also **-ise**) make less strict. □ **liberalization** n.

liberate v. set free. □ **liberation** n., **liberator** n.

libertarian n. a person favouring absolute liberty of thought and action.

libertine n. a person who lives an irresponsible immoral life.

liberty n. freedom. □ **take liberties** behave with undue freedom or familiarity.

librarian n. a person in charge of or assisting in a library.

library n. a collection of books (or records, films, etc.) for consulting or borrowing; a room or building containing these.

libretto n. (pl. **librettos** or **libretti**) the words of an opera.

lice see **louse**.

licence n. (Amer. **license**) **1** an official permit to own or do something; permission. **2** disregard of rules etc.

license v. grant a licence to or for. ● n. Amer. sp. of **licence**.

licensee n. a holder of a licence.

licentiate n. a holder of a certificate of competence in a profession.

licentious adj. sexually immoral. □ **licentiousness** n.

lichee var. of **lychee**.

lichen (ly-kĕn) n. a low-growing dry plant that grows on rocks etc.

lich-gate n. (also **lych-gate**) a roofed gateway to a churchyard.

lick v. **1** pass the tongue over; (of waves or flame) touch lightly. **2** (colloquial) defeat. ● n. **1** an act of licking; a slight application (of paint etc.). **2** (colloquial) a fast pace.

licorice var. of **liquorice**.

lid n. a hinged or removable cover for a box, pot, etc.; an eyelid.

lie[1] n. a statement the speaker knows to be untrue. ● v. (**lied**, **lying**) tell a lie.

lie[2] v. (**lay**, **lain**, **lying**) **1** have or put one's body in a flat or resting position; be at rest on something. **2** be in a specified state; be situated. ● *n.* the way a thing lies. □ **lie in** lie in bed late in the morning. **lie low** conceal oneself or one's intentions.

■ **Usage** It is incorrect in standard English to use *lie* to mean 'lay', as in *Lie her on the bed.*

liege *n.* (*hist.*) one's feudal superior; one's king.

lien (lyen) *n.* (*Law*) the right to hold another person's property until a debt on it is paid.

lieu (lew) *n.* □ **in lieu** instead.

lieutenant (lef-ten-ănt) *n.* an army officer next below major; a naval officer next below lieutenant commander; a rank just below a specified officer; a chief assistant.

life *n.* (*pl.* **lives**) **1** animals' and plants' ability to function and grow; being alive; a period of this; living things. **2** liveliness. **3** a manner of living; a biography.

lifebelt *n.* a belt of buoyant material to keep a person afloat.

lifeboat *n.* a boat for rescuing people at sea; a ship's boat for emergency use.

lifebuoy *n.* a buoyant device to keep a person afloat.

life cycle *n.* a series of forms into which a living thing changes.

lifeguard *n.* an expert swimmer employed to rescue bathers in danger.

life jacket *n.* a buoyant or inflatable jacket for keeping a person afloat in water.

lifeless *adj.* without life; dead; unconscious; lacking liveliness.

lifelike *adj.* exactly like a real person or thing.

lifeline *n.* a rope used in rescue; a vital means of communication.

lifelong *adj.* for all one's life.

life sciences *n.pl.* biology and related subjects.

life-size *adj.* (also **life-sized**) of the same size as a real person or thing.

life-support *adj.* (of equipment etc.) enabling the body to function in a hostile environment or in cases of physical failure.

lifetime *n.* the duration of a person's life.

lift *v.* **1** raise; take up; rise. **2** remove (restrictions). **3** steal, plagiarize. ● *n.* **1** lifting; an apparatus for transporting people or goods from one level to another. **2** a free ride in a motor vehicle. **3** a feeling of elation.

lift-off *n.* vertical take-off of a spacecraft etc.

ligament *n.* a tough flexible tissue holding bones together.

ligature *n.* a thing that ties something, esp. in surgical operations. ● *v.* tie with a ligature.

light¹ *n.* **1** a kind of radiation that stimulates sight; brightness, a light part of a picture etc.; a source of light, an electric lamp; a flame or spark. **2** enlightenment. **3** an aspect; the way something appears to the mind. ● *adj.* full of light, not in darkness; pale. ● *v.* (**lit** or **lighted, lighting**) **1** set burning; begin to burn. **2** provide with light; brighten. □ **bring to light** reveal. **come to light** be revealed. **light on** find accidentally. **light up** put lights on at dusk; brighten; begin to smoke a cigarette etc.

light² *adj.* **1** having little weight; not heavy, easy to lift, carry, or do; of less than average weight, force, or intensity. **2** cheerful; not profound or serious. **3** (of food) easy to digest. ● *adv.* lightly; with little load. □ **make light of** treat as unimportant. □ **lightly** *adv.*, **lightness** *n.*

lighten¹ *v.* shed light on; make or become brighter.

lighten² *v.* make or become less heavy.

lighter *n.* **1** a device for lighting cigarettes and cigars. **2** a flat-bottomed boat for unloading ships.

light-fingered *adj.* apt to steal.

light-headed *adj.* feeling slightly faint; delirious.

light-hearted *adj.* cheerful, not serious.

lighthouse *n.* a tower with a beacon light to warn or guide ships.

light industry *n.* that producing small or light articles.

lighting *n.* a means of providing light; the light itself.

lightning *n.* a flash of bright light produced from cloud by natural electricity. ● *adj.* very quick.

light pen *n.* a light-emitting device for reading bar codes; a pen-shaped device for drawing on a computer screen.

lights *n.pl.* the lungs of certain animals, used as animal food.

lightship *n.* a moored ship with a light, serving as a lighthouse.

lightweight *adj.* not having great weight or influence. ● *n.* a lightweight person; a boxing weight between featherweight and welterweight.

light year *n.* the distance light travels in one year, about 6 million million miles.

lignite *n.* a brown coal of a woody texture.

lignum vitae *n.* a hardwood tree.

like¹ *adj.* having the qualities or appearance of; characteristic of. ● *prep.* in the manner of; to the same degree as. ● *conj.* (*colloquial*) as, as if (see note below). ● *adv.* (*colloquial*) likely. ● *n.* a person or thing like another.

■ **Usage** It is incorrect in standard English to use *like* as a conjunction, as in *Tell it like it is* or *He's spending money like it was going out of fashion.*

like² *v.* find pleasant or satisfactory; wish for. ● *n.pl.* (**likes**) things one likes or prefers.

likeable *adj.* (also **likable**) pleasant, easy to like.

likelihood *n.* a probability.

likely *adj.* (**likelier**) such as may reasonably be expected to occur or be true; seeming to be suitable or have a chance of success. ● *adv.* probably. □ **likeliness** *n.*

liken *v.* point out the likeness of (one thing to another).

likeness *n.* being like; a copy, a portrait.

likewise *adv.* also; in the same way.

liking *n.* what one likes; one's taste for something.

lilac *n.* a shrub with fragrant purple or white flowers; pale purple. ● *adj.* pale purple.

lilt *n.* a light pleasant rhythm; a tune with this. □ **lilting** *adj.*

lily *n.* a plant growing from a bulb, with large flowers.

limb *n.* an arm, leg, or wing; a large branch of a tree.

limber *v.* ● **limber up** exercise in preparation for athletic activity.

limbo¹ *n.* an intermediate inactive or neglected state.

limbo² *n.* (*pl.* **limbos**) a West Indian dance in which the dancer bends back to pass under a bar.

lime¹ *n.* **1** a white substance used in making cement etc. **2** a round yellowish-green fruit like a lemon; its colour. **3** a tree with heart-shaped leaves.

limelight *n.* the glare of publicity.

limerick *n.* a humorous poem with five lines.

limestone *n.* rock from which lime is obtained.

limit *n.* a point beyond which something does not continue; the greatest amount allowed.

● *v.* set or serve as a limit; keep within limits. □ **limitation** *n.*

limousine *n.* a large luxurious car.

limp *v.* walk or proceed lamely. ● *n.* a limping walk. ● *adj.* not stiff or firm; wilting. □ **limply** *adv.*, **limpness** *n.*

limpet *n.* a small shellfish that sticks tightly to rocks.

limpid *adj.* (of liquids) clear.

linchpin *n.* a pin passed through the end of an axle to secure a wheel; a vital person or thing.

linctus *n.* a soothing cough mixture.

linden *n.* a lime tree.

line *n.* **1** a long narrow mark; an outline; a boundary; a row of people or things; a row of words; a brief letter; (**lines**) the words of an actor's part. **2** a series, several generations of a family. **3** a direction, a course; a railway track; a type of activity, business, or goods; a service of ships, buses, or aircraft. **4** a piece of cord for a particular purpose; an electrical or telephone cable, connection by this. **5** each of a set of military fieldworks. ● *v.* **1** mark with lines; arrange in line(s). **2** cover the inside surface of. □ **line one's pockets** make money, esp. in underhand ways.

lineage *n.* a line of ancestors or descendants.

lineal *adj.* of or in a line.

linear *adj.* of a line; of length; arranged in a line.

linen *n.* cloth made of flax; household articles (e.g. sheets, tablecloths) formerly made of this.

liner *n.* a passenger ship or aircraft; a removable lining.

linesman *n.* (*pl.* **-men**) an umpire's assistant at the boundary line; a workman who maintains railway, electrical, or telephone lines.

ling *n.* **1** heather. **2** a sea fish of north Europe.

linger *v.* delay departure; dawdle.

lingerie (lahn-*zher*-ee) *n.* women's underwear.

lingua franca *n.* a common language used among people whose native languages are different.

lingual *adj.* of the tongue; of speech or languages.

linguist *n.* a person who is skilled in languages or linguistics.

linguistic *adj.* of language. ● *n.* (**linguistics**) the study of language. □ **linguistically** *adv.*

liniment *n.* embrocation.

lining *n.* a layer of material or substance covering an inner surface.

link *n.* each ring of a chain; a person or thing connecting others. ● *v.* connect; intertwine. □ **linkage** *n.*

linnet *n.* a kind of finch.

lino *n.* linoleum.

linocut *n.* a design cut in relief on a block of linoleum; a print made from this.

linoleum *n.* a smooth covering for floors.

linseed *n.* the seed of flax.

lint *n.* a soft fabric for dressing wounds; fluff.

lintel *n.* a horizontal timber or stone over a doorway etc.

lion *n.* a large flesh-eating animal of the cat family.

lionize *v.* (also **-ise**) treat as a celebrity.

lip *n.* either of the fleshy edges of the mouth-opening; the edge of a container or opening; a slight projection shaped for pouring from.

lip-read *v.* understand what is said from movements of a speaker's lips.

lipsalve *n.* ointment for the lips.

lipstick *n.* a cosmetic for colouring the lips.

liquefy *v.* make or become liquid. □ **liquefaction** *n.*

liqueur (lik-yoor) *n.* a strong sweet alcoholic spirit with fragrant flavouring.

liquid *n.* a flowing substance like water or oil. ● *adj.* **1** in the form of liquid. **2** (of assets) easy to convert into cash. □ **liquidity** *n.*

liquidate *v.* **1** pay (a debt). **2** close down (a business) and divide its assets between creditors. **3** get rid of, esp. by killing. □ **liquidation** *n.*, **liquidator** *n.*

liquidize *v.* (also **-ise**) reduce to a liquid.

liquidizer *n.* (also **-iser**) a machine for making purées etc.

liquor *n.* alcoholic drink; juice from cooked food.

liquorice *n.* (esp. *Amer.* **licorice**) a black substance used in medicine and as a sweet, made from a plant root.

lira *n.* (*pl.* **lire**) a unit of money in Italy and Turkey.

lisp *n.* a speech defect in which *s* and *z* are pronounced like *th*. ● *v.* speak or utter with a lisp.

lissom *adj.* lithe, agile.

list *n.* a written or printed series of names, items, figures, etc. ● *v.* **1** make a list of; enter in a list. **2** (of a ship) lean over to one side. ● *n.* a listing position.

listen *v.* make an effort to hear; pay attention; be persuaded by advice or a request. □ **listen in** overhear a conversation; listen to a broadcast. □ **listener** *n.*

listeria *n.* a type of bacteria causing food poisoning.

listless *adj.* without energy or enthusiasm. □ **listlessly** *adv.*, **listlessness** *n.*

lit *see* **light**[1].

litany *n.* a set form of prayer; a long monotonous recital.

litchi var. of **lychee**.

liter *Amer.* sp. of **litre**.

literacy *n.* being literate.

literal *adj.* taking the basic meaning of a word, not a metaphorical or exaggerated one. □ **literally** *adv.*, **literalness** *n.*

literary *adj.* of literature.

literate *adj.* able to read and write.

literati *n.pl.* learned people.

literature *n.* great novels, poetry, and plays.

lithe *adj.* supple, agile.

lithium *n.* a light metallic element (symbol Li).

litho *n.* the lithographic process.

lithograph *n.* a picture printed by lithography.

lithography *n.* printing from a plate treated so that ink sticks only to the design. □ **lithographic** *adj.*

litigant *adj.* & *n.* (a person) involved in or initiating a lawsuit.

litigate *v.* carry on a lawsuit; contest in law.

litigious *adj.* fond of litigation.

litmus *n.* a substance turned red by acids and blue by alkalis.

litotes *n.* an ironic understatement.

litre *n.* (*Amer.* **liter**) a metric unit of capacity (1.76 pints) for measuring liquids.

litter *n.* **1** rubbish left lying about. **2** young animals born at one birth; material used as bedding for animals or to absorb their excrement. ● *v.* **1** scatter as litter; make untidy by litter. **2** give birth to (a litter).

little *adj.* small in size, amount, or intensity etc. ● *n.* a small amount, time, or distance. ● *adv.* to a small extent; not at all.

littoral *adj.* of or by the shore.

liturgy *n.* a set form of public worship. □ **liturgical** *adj.*

live[1] (lyv) *adj.* **1** alive. **2** burning; unexploded; charged with electricity. **3** (of broadcasts) transmitted while actually happening.

live² (liv) v. have life, remain alive; have one's home; conduct (one's life) in a certain way; enjoy life fully. □ **live down** live until (scandal etc.) is forgotten. **live on** keep oneself alive on.

livelihood n. a means of earning or providing enough food etc. to sustain life.

lively adj. (**livelier**) full of energy or action. □ **liveliness** n.

liven v. make or become lively.

liver n. a large organ in the abdomen, secreting bile; an animal's liver as food.

liveried adj. wearing livery.

livery n. a distinctive uniform; a colour scheme in which a company's vehicles are painted.

livestock n. farm animals.

livid adj. **1** (colloquial) furiously angry. **2** bluish grey.

living adj. having life, not dead; currently in use. ● n. being alive; a manner of life; a livelihood.

living room n. a room for general daytime use.

lizard n. a reptile with four legs and a long tail.

llama n. a South American animal related to the camel.

load n. **1** a thing or quantity carried; a burden of responsibility or worry. **2** the amount of electric current supplied. **3** (**loads**) (colloquial) plenty. ● v. put a load in or on; receive a load; fill heavily; put ammunition into (a gun) or film into (a camera); put (data) into (a computer).

loaf n. (pl. **loaves**) a mass of bread shaped in one piece; (slang) the head. ● v. spend time idly, stand or saunter about.

loam n. rich soil.

loan n. lending; something lent, esp. money. ● v. grant a loan of.

loan shark n. a person lending money at very high rates of interest.

loath adj. unwilling.

loathe v. feel hatred and disgust for. □ **loathing** n., **loathsome** adj.

lob v. (**lobbed**) send or strike (a ball) slowly in a high arc. ● n. a lobbed ball.

lobar adj. of a lobe, esp. of the lung.

lobby n. **1** a porch, entrance hall, or ante-room. **2** a body of people seeking to influence legislation. ● v. seek to persuade (an MP etc.) to support one's cause.

lobbyist n. a person who lobbies an MP etc.

lobe n. a rounded part or projection; the lower soft part of the ear.

lobelia n. a garden plant.

lobotomy n. an incision into the frontal lobe of the brain.

lobster n. a shellfish with large claws; its flesh as food.

local adj. of or affecting a particular place or small area. ● n. **1** an inhabitant of a particular district. **2** (colloquial) one's nearest public house. □ **locally** adv.

locale n. the scene of an event.

local government n. the administration of a district by representatives elected locally.

locality n. the site or neighbourhood of something.

localize v. (also **-ise**) confine within an area; decentralize. □ **localization** n.

locate v. discover the position of; situate in a particular place.

■ **Usage** In standard English, it is incorrect to use locate to mean merely 'find' as in I can't locate my key.

location n. locating; a place where a thing is situated. □ **on location** (of filming) in a suitable environment, not in a film studio.

loch n. (Scot.) a lake, an arm of the sea.

loci see **locus**.

lock n. **1** a device (opened by a key) for fastening a door or lid etc. **2** a gated section of a canal where the water level can be changed. **3** a secure hold; interlocking. **4** a portion of hair that hangs together; (**locks**) hair. ● v. fasten with a lock; shut into a locked place; make or become rigidly fixed. □ **lockable** adj.

locker n. a lockable cupboard where things can be stowed securely.

locket n. a small ornamental case worn on a chain round the neck.

lockjaw n. tetanus.

lockout n. expulsion of employees during a dispute.

locksmith n. a maker and mender of locks.

lock-up n. lockable premises; a place where prisoners can be kept temporarily.

locomotion n. the ability to move from place to place.

locomotive n. a self-propelled engine for drawing trains. ● adj. of or effecting locomotion.

locum n. a temporary stand-in for a doctor, clergyman, etc.

locus n. (pl. **loci**) a thing's exact place; a line or curve etc. formed by certain points or by the movement of a point or line.

locust n. a grasshopper that devours vegetation.

lode n. a vein of metal ore.

lodestar n. a star (esp. the pole star) used as a guide in navigation.

lodestone n. an oxide of iron used as a magnet.

lodge n. **1** a cabin for use by hunters, skiers, etc.; a gatekeeper's house; a porter's room at the entrance to a building. **2** the members or meeting place of a branch of certain societies. **3** a beaver's or otter's lair. ● v. **1** provide with sleeping quarters or temporary accommodation; live as a lodger. **2** deposit; be or become embedded.

lodger n. a person paying for accommodation in another's house.

lodging n. a place where one lodges; (**lodgings**) a room or rooms rented for living in.

loft n. a space under a roof; a gallery in a church. ● v. hit, throw, or kick (a ball) in a high arc.

lofty adj. (**loftier**) very tall; noble; haughty. □ **loftily** adv.

log n. **1** a piece cut from a trunk or branch of a tree. **2** a device for gauging a ship's speed; a logbook, an entry in this. **3** a logarithm. ● v. (**logged**) enter (facts) in a logbook. □ **log on** or **off**, **log in** or **out** open or close one's on-line access to a computer system.

loganberry n. a large dark red fruit resembling a blackberry.

logarithm n. one of a series of numbers set out in tables, used to simplify calculations.

logbook n. a book for recording details of a journey.

loggerheads n.pl. □ **at loggerheads** disagreeing or quarrelling.

logic n. a science or method of reasoning; correct reasoning.

logical adj. of or according to logic; reasonable; reasoning correctly. □ **logically** adv., **logicality** n.

logician n. a person skilled in logic.

logistics n.pl. the organization of supplies and services. □ **logistical** adj.

logo n. (pl. **logos**) a design used as an emblem.

loin *n.* the side and back of the body between the ribs and hip bone.

loincloth *n.* a cloth worn round the body at the hips.

loiter *v.* linger, stand about idly. □ **loiterer** *n.*

loll *v.* stand, sit, or rest lazily; hang loosely.

lollipop *n.* a large usu. flat boiled sweet on a small stick.

lollipop lady, lollipop man *n.* an official using a circular sign on a stick to stop traffic for children to cross a road.

lollop *v.* (**lolloped**) (*colloquial*) move in clumsy bounds; flop.

lolly *n.* **1** (*colloquial*) a lollipop, an ice lolly. **2** (*slang*) money.

lone *adj.* solitary.

lonely *adj.* (**lonelier**) **1** solitary; sad because one lacks friends. **2** not much frequented. □ **loneliness** *n.*

loner *n.* a person who prefers not to associate with others.

lonesome *adj.* lonely.

long *adj.* of great or specified length. ● *adv.* for a long time; throughout a specified time. ● *v.* feel a longing. □ **as** or **so long as** provided that.

long-distance *adj.* travelling or operated between distant places.

longevity (lon-jev-iti) *n.* long life.

long face *n.* a dismal expression.

longhand *n.* ordinary writing, not shorthand or typing etc.

longhorn *n.* one of a breed of cattle with long horns.

longing *n.* an intense wish.

longitude *n.* the distance east or west (measured in degrees on a map) from the Greenwich meridian.

longitudinal *adj.* of longitude; of length, lengthwise. □ **longitudinally** *adv.*

long johns *n.pl.* (*colloquial*) close-fitting underpants with long legs.

long-life *adj.* (of milk etc.) treated to prolong its shelf-life.

long-lived *adj.* living or lasting for a long time.

long-range *adj.* having a relatively long range; relating to a long period of future time.

longshoreman *n.* (*pl.* **-men**) (*Amer.*) a docker.

long shot *n.* a wild guess or venture.

long-sighted *adj.* able to see clearly only what is at a distance.

long-standing *adj.* having existed for a long time.

long-suffering *adj.* bearing provocation patiently.

long-term *adj.* of or for a long period.

long ton see **ton**.

long wave *n.* a radio wave of frequency less than 300 kHz.

longways *adv.* (also **longwise**) lengthways.

long-winded *adj.* talking or writing at tedious length.

loo *n.* (*colloquial*) a lavatory.

loofah *n.* the dried pod of a gourd, used as a rough sponge.

look *v.* **1** use or direct one's eyes in order to see, search, or examine; face. **2** seem. ● *n.* an act of looking; an inspection, a search; the appearance of something. □ **look after** take care of; attend to. **look down on** despise. **look forward to** await eagerly. **look into** investigate. **look out** be vigilant. **look up** search for information about; improve in prospects; go to visit. **look up to** admire and respect.

looker-on *n.* (*pl.* **lookers-on**) a mere spectator.

lookout *n.* **1** a watch; a guard; an observation post. **2** a prospect; a person's own concern.

loom *v.* appear, esp. close at hand or threateningly. ● *n.* an apparatus for weaving cloth.

loop | lounge suit

loop n. a curve that is U-shaped or that crosses itself; something shaped like this. ● v. form into loop(s); fasten or join with loop(s); enclose in a loop. □ **loop the loop** fly in a vertical circle.

loophole n. a means of evading a rule or contract.

loose adj. not tight; not fastened, held, or fixed; not packed together. ● adv. loosely. ● v. release; untie, loosen. □ **at a loose end** without a definite occupation. □ **loosely** adv., **looseness** n.

loose box n. a stall for a horse.

loose-leaf adj. with each page removable.

loosen v. make or become loose or looser.

loot n. goods taken from an enemy or by theft. ● v. take loot (from); take as loot.

lop v. (**lopped**) cut off, esp. branches or twigs.

lope v. run with a long bounding stride. ● n. this stride.

lop-eared adj. with drooping ears.

lopsided adj. with one side lower, smaller, or heavier than the other.

loquacious adj. talkative. □ **loquacity** n.

lord n. a master, a ruler; a nobleman; the title of certain peers or high officials; (**Lord**) God or Christ.

lordship n. the title used of a man with the rank of Lord.

lore n. a body of traditions and knowledge.

lorgnette (lorn-yet) n. eyeglasses or opera-glasses held on a long handle.

lorry n. a large motor vehicle for transporting heavy loads.

lose v. (**lost, losing**) cease to have or maintain; become unable to find; fail to get; get rid of; be defeated in a contest etc.; suffer loss (of); cause the loss of. □ **loser** n.

loss n. losing; a person, thing, or amount lost; a disadvantage caused by losing something. □ **be at a loss** not know what to do or say.

loss-leader n. an article sold at a loss to attract customers.

lost see **lose**. adj. strayed or separated from its owner.

lot n. **1** a large number or amount; (also (colloquial) **lots**) much, a good deal. **2** each of a set of objects drawn at random to decide something; a person's share or destiny; a piece of land; an item being sold at auction. □ **the lot** the total quantity.

■ Usage A lot of, as in a lot of people, is fairly informal, though acceptable in serious writing, but lots of people is not acceptable.

lotion n. a medicinal or cosmetic liquid applied to the skin.

lottery n. a system of raising money by selling numbered tickets and giving prizes to holders of numbers drawn at random; something where the outcome is governed by luck.

lotus n. (pl. **lotuses**) a tropical water lily; a mythical fruit.

loud adj. producing a lot of noise, easily heard; gaudy. ● adv. loudly. □ **loudly** adv., **loudness** n.

loud hailer n. an electronically operated megaphone.

loudspeaker n. an apparatus that converts electrical impulses into audible sound.

lough (lok) n. (Irish) = **loch**.

lounge v. loll; sit or stand about idly. ● n. a sitting room; a waiting room at an airport etc.

lounge suit n. a man's ordinary suit for day wear.

lour v. (also **lower**) frown, scowl; (of clouds) look dark and threatening.

louse n. (pl. **lice**) a small parasitic insect; (pl. **louses**) a contemptible person.

lousy adj. (**lousier**) **1** (slang) very bad. **2** infested with lice.

lout n. a clumsy ill-mannered person. □ **loutish** adj.

louvre n. (also **louver**) each of a set of overlapping slats arranged to let in air but exclude light or rain. □ **louvred** adj.

lovable adj. easy to love.

lovage n. a herb used for flavouring.

love n. **1** a warm liking or affection; sexual passion; a loved person. **2** (in games) no score, nil. ● v. feel love for; like greatly. □ **in love** feeling (esp. sexual) love for another person. **make love** have sexual intercourse.

love affair n. a romantic or sexual relationship between people who are in love.

lovebird n. a small parakeet that shows great affection for its mate.

lovelorn adj. pining with love.

lovely adj. (**lovelier**) beautiful, attractive; delightful. □ **loveliness** n.

lover n. a person in love with another or having a love affair; one who likes or enjoys something.

loving adj. feeling or showing love. □ **lovingly** adv.

low adj. **1** not high, not extending or lying far up; ranking below others; less than normal in amount or intensity; not loud or shrill. **2** lacking vigour, depressed. **3** ignoble, vulgar. ● n. **1** a low level; an area of low pressure. **2** a deep sound made by cattle. ● v. (of cattle) make a deep mooing sound. ● adv. in, at, or to a low level.

lowbrow adj. not intellectual or cultured.

low-down adj. dishonourable. ● n. (slang) relevant information.

lower v. let or haul down; make or become lower.

lower case n. letters that are not capitals.

low-key adj. restrained, not intense or emotional.

lowlands n.pl. low-lying land. □ **lowland** adj., **lowlander** n.

lowly adj. (**lowlier**) of humble rank or condition.

low-rise adj. (of a building) having few storeys.

low season n. the season that is least busy.

low-tech adj. using technology based on cheap simple components and local resources.

loyal adj. firm in one's allegiance. □ **loyally** adv., **loyalty** n.

loyalist n. a person who is loyal, esp. while others revolt.

lozenge n. a small tablet to be dissolved in the mouth; a four-sided diamond-shaped figure.

LP abbr. a long-playing record.

Lr symb. lawrencium.

LSD n. a powerful hallucinogenic drug.

Ltd. abbr. Limited.

Lu symb. lutetium.

lubricant n. a lubricating substance.

lubricate v. oil or grease (machinery etc.). □ **lubrication** n.

lubricious adj. slippery; lewd.

lucerne n. alfalfa, a clover-like fodder plant.

lucid adj. clearly expressed; sane. □ **lucidly** adv., **lucidity** n.

luck n. good or bad fortune; chance thought of as a force bringing this.

luckless adj. unlucky.

lucky adj. (**luckier**) having, bringing, or resulting from good luck. □ **luckily** adv.

lucky dip n. a tub containing articles which people may choose at random.

lucrative adj. profitable, producing much money.

lucre (loo-ker) n. (derog.) money.

Luddite n. a person opposing the introduction of new technology or working methods.

ludicrous adj. ridiculous. □ **ludicrously** adv.

ludo n. a simple game played with counters on a special board.

lug v. (**lugged**) drag or carry with great effort. ● n. an ear-like projection; (colloquial) an ear.

luge n. a light toboggan, ridden sitting upright.

luggage n. suitcases and bags holding a traveller's possessions.

lugubrious adj. dismal, mournful. □ **lugubriously** adv.

lugworm n. a large marine worm used as bait.

lukewarm adj. only slightly warm; not enthusiastic.

lull v. soothe or send to sleep; calm; become quiet. ● n. a period of quiet or inactivity.

lullaby n. a soothing song for sending a child to sleep.

lumbago n. rheumatic pain in muscles of the lower back.

lumbar adj. of the lower back.

lumber n. useless or unwanted articles, esp. furniture; (Amer.) timber sawn into planks. ● v. encumber; move heavily and clumsily.

lumberjack n. (Amer.) a person who cuts or transports lumber.

luminary n. a natural light-giving body, esp. the sun or moon; an eminent person.

luminescent adj. emitting light without heat. □ **luminescence** n.

luminous adj. emitting light, glowing in the dark. □ **luminosity** n.

lump n. a hard or compact mass; a swelling. ● v. put or consider together.

lumpectomy n. the surgical removal of a lump from the breast.

lumpy adj. (**lumpier**) full of lumps; covered in lumps. □ **lumpiness** n.

lunacy n. insanity; great folly.

lunar adj. of the moon.

lunar month n. the period between new moons (29½ days), four weeks.

lunate adj. crescent-shaped.

lunatic n. an insane person; a very foolish or reckless person.

lunation n. a lunar month.

lunch n. a midday meal. ● v. eat lunch.

luncheon n. lunch.

luncheon meat n. tinned cured meat ready for serving.

luncheon voucher n. a voucher given to an employee, exchangeable for food.

lung n. either of the pair of breathing-organs in the chest.

lunge n. **1** a sudden forward movement of the body; a thrust. **2** a long rope attached to a horse while it is being trained. ● v. make a lunge forward.

lupin n. a garden plant with tall spikes of flowers.

lupine adj. like a wolf.

lupus n. a skin disease producing ulcers.

lurch v. & n. (make) an unsteady swaying movement, stagger. □ **leave in the lurch** leave (a person) in difficulties.

lurcher n. a dog, a cross between a sheepdog or retriever and a greyhound.

lure v. entice. ● n. an enticement; a bait to attract wild animals.

Lurex n. (trade mark) a yarn or fabric containing shiny metallic thread.

lurid adj. in glaring colours; vivid and sensational or shocking. □ **luridly** adv.

lurk v. wait furtively or out of sight; be latent or lingering.

luscious adj. delicious; voluptuously attractive. □ **lusciously** adv., **lusciousness** n.

lush adj. (of grass etc.) growing thickly and strongly; luxurious. □ **lushly** adv., **lushness** n.

lust n. intense sexual desire; any intense desire. ● v. feel lust. □ **lustful** adj., **lustfully** adv.

lustre n. (Amer. **luster**) soft brightness of a surface; brilliance, glory; a metallic glaze on pottery. □ **lustrous** adj.

lusty adj. (**lustier**) strong and vigorous. □ **lustily** adv.

lute n. a guitar-like instrument with a pear-shaped body. □ **lutenist** n.

lutetium n. a metallic element (symbol Lu).

lux n. a unit of illumination.

luxuriant adj. growing profusely. □ **luxuriantly** adv., **luxuriance** n.

luxuriate v. feel great enjoyment in something.

luxurious adj. supplied with luxuries, very comfortable. □ **luxuriously** adv., **luxuriousness** n.

luxury n. choice and costly surroundings, food, etc.; self-indulgence; something that is enjoyable but not essential.

LV abbr. luncheon voucher.

Lw symb. lawrencium.

lx abbr. lux.

lych-gate var. of lich-gate.

lychee n. (also **lichee**, **litchi**) a sweet white fruit with a brown spiny skin.

Lycra n. (trade mark) an elastic fabric.

lye n. water made alkaline with wood ashes.

lying see lie[1], lie[2].

lymph n. a colourless fluid containing white blood cells. □ **lymphatic** adj.

lymphatic system n. the network of vessels carrying lymph, protecting against infection.

lymphoma n. (pl. **lymphomas** or **lymphomata**) a tumour of the lymph glands.

lynch v. execute or punish violently by a mob, without trial.

lynx n. (pl. **lynx** or **lynxes**) a wild animal of the cat family with keen sight.

lyre n. an ancient musical instrument with strings in a U-shaped frame.

lyre-bird n. an Australian bird with a lyre-shaped tail.

lyric adj. of poetry that expresses the poet's thoughts and feelings. ● n. a lyric poem; (**lyrics**) the words of a song.

lyrical adj. resembling or using language suitable for lyric poetry; (colloquial) expressing oneself enthusiastically. □ **lyrically** adv.

lyricist n. a person who writes lyrics.

Mm

M abbr. motorway; mega-; Monsieur. ● n. (as a Roman numeral) 1,000.

m abbr. **1** metre(s); mile(s); million(s). **2** male; masculine.

MA abbr. Master of Arts.

ma'am n. madam.

mac n. (also **mack**) (colloquial) a mackintosh.

macabre adj. gruesome.

macadam n. layers of broken stone used in road-making.

macadamize v. (also **-ise**) surface with macadam.

macaque n. a monkey with prominent cheek-pouches.

macaroni *n.* tube-shaped pasta.

macaroon *n.* a small almond biscuit.

macaw *n.* an American parrot.

mace *n.* **1** a ceremonial staff. **2** a spice made from the outer covering of nutmeg.

macerate *v.* soften by soaking.

Mach (mahk) *n.* (in full **Mach number**) the ratio of the speed of a moving body to the speed of sound.

machete (mă-she-ti) *n.* a broad heavy knife.

machiavellian *adj.* elaborately cunning or deceitful.

machinations *n.pl.* clever scheming.

machine *n.* an apparatus for applying mechanical power; something operated by this; the controlling system of an organization. ● *v.* produce or work on with a machine.

machine code *n.* a computer language that controls the computer directly, interpreting instructions passing between the software and the machine.

machine-gun *n.* a gun that can fire continuously.

machine-readable *adj.* in a form that a computer can process.

machinery *n.* machines; a mechanism; a system.

machine tool *n.* a power-driven engineering machine such as a lathe.

machinist *n.* a person who works machinery.

machismo *n.* manly courage; a show of this.

macho *adj.* aggressively masculine.

mack var. of **mac**.

mackerel *n.* (*pl.* **mackerel** or **mackerels**) an edible sea fish.

mackintosh *n.* (also **macintosh**) a raincoat; a waterproof material of rubber and cloth.

macramé (mă-krah-mi) *n.* the art of knotting cord in patterns; items made in this way.

macro- *comb. form* large-scale; large; long.

macrobiotic *adj.* of a dietary system comprising wholefoods grown in close harmony with nature.

macrocosm *n.* the universe; any great whole.

mad *adj.* (**madder**) not sane; extremely foolish; wildly enthusiastic; frenzied; (*colloquial*) very annoyed. □ **madly** *adv.*, **madness** *n.*

madam *n.* a polite form of address to a woman.

Madame (mă-dahm) *n.* (*pl.* **Mesdames**) a title or form of address for a French-speaking woman, corresponding to Mrs or madam.

madcap *adj.* wildly impulsive.

mad cow disease *n.* (*colloquial*) = **BSE**.

madden *v.* make mad or angry.

madder *n.* a red dye obtained from the root of a plant.

made see **make**.

Madeira *n.* **1** a fortified wine from Madeira. **2** rich plain cake.

Mademoiselle (ma-dĕ-mwă-zel) *n.* (*pl.* **Mesdemoiselles**) a title or form of address for an unmarried French-speaking woman, corresponding to Miss or madam.

madonna *n.* a picture or statue of the Virgin Mary.

madrigal *n.* a part-song for unaccompanied voices.

maelstrom (mayl-strŏm) *n.* a great whirlpool.

maestro (my-stroh) *n.* (*pl.* **maestri** or **maestros**) a great musical conductor or composer; a master of any art.

magazine *n.* **1** an illustrated periodical. **2** a store for arms or explosives; a chamber holding cartridges in a gun, slides in a projector, etc.

magenta *adj.* & *n.* purplish red.

maggot *n.* a larva, esp. of the bluebottle.

magic *n.* the supposed art of controlling things by supernatural power. ● *adj.* using or used in magic. □ **magical** *adj.*, **magically** *adv.*

magician *n.* a person skilled in magic; a conjuror.

magisterial *adj.* of a magistrate; authoritative, imperious.

magistrate *n.* an official or citizen with authority to hold preliminary hearings and judge minor cases. □ **magistracy** *n.*

magma *n.* molten rock under the earth's crust.

magnanimous *adj.* noble and generous, not petty. □ **magnanimously** *adv.*, **magnanimity** *n.*

magnate *n.* a wealthy influential business person.

magnesia *n.* a compound of magnesium used in medicine.

magnesium *n.* a white metallic element (symbol Mg) that burns with an intensely bright flame.

magnet *n.* a piece of iron or steel that can attract iron and point north when suspended; a powerful attraction.

magnetic *adj.* having the properties of a magnet; produced or acting by magnetism. □ **magnetically** *adv.*

magnetic tape *n.* a strip of plastic coated with magnetic particles, used in recording, computers, etc.

magnetism *n.* the properties and effects of magnetic substances; great charm and attraction.

magnetize *v.* (also **-ise**) make magnetic; attract.

magneto (mag-nee-toh) *n.* (*pl.* **magnetos**) a small electric generator using magnets.

magnification *n.* magnifying.

magnificent *adj.* splendid; excellent. □ **magnificently** *adv.*, **magnificence** *n.*

magnify *v.* make (an object) appear larger by use of a lens; exaggerate. □ **magnifier** *n.*

magnitude *n.* largeness; size; importance.

magnolia *n.* a tree with large white or pink flowers.

magnum *n.* a bottle holding two quarts (2.27 litres) of wine or spirits.

magpie *n.* a black and white bird of the crow family.

Magyar *adj.* & *n.* (a member, the language) of a people now predominant in Hungary.

maharaja *n.* (also **maharajah**) the former title an Indian prince.

maharanee *n.* (also **maharani**) a maharaja's wife or widow.

maharishi *n.* a Hindu man of great wisdom.

mahatma *n.* (in India etc.) a title of a man regarded with reverence.

mah-jong *n.* (also **mah-jongg**) a Chinese game played with 136 or 144 pieces (tiles).

mahogany *n.* a very hard reddish-brown wood; its colour.

mahout *n.* an elephant-driver.

maid *n.* a woman servant.

maiden *n.* (*old use*) a young unmarried woman, a virgin. ● *adj.* unmarried; first. □ **maidenhood** *n.*, **maidenly** *adj.*

maidenhair *n.* a fern with very thin stalks and delicate foliage.

maiden name *n.* a woman's family name before she married.

maiden over *n.* an over in cricket with no runs scored.

maidservant *n.* a female servant.

mail *n.* **1** post, letters; messages transmitted by computer from one user to another. **2** body-armour made of metal rings or chains. ● v. send by post or electronic mail.

mail order n. purchase of goods by post.

mailbox n. (Amer.) a letter box.

mailshot n. advertising material sent to potential customers.

maim v. injure so that a part of the body is useless.

main adj. principal, most important, greatest in size or extent. ● n. a main pipe or channel conveying water, gas, or (usu. **mains**) electricity. □ **mainly** adv.

main clause n. (Grammar) a clause that can stand as a complete sentence.

mainframe n. a large computer.

mainland n. a country or continent without its adjacent islands.

mainline v. (slang) take drugs intravenously.

mainmast n. the principal mast.

mainsail n. the lowest sail or the sail set on the after part of the mainmast.

mainspring n. the chief spring of a watch or clock; the chief motivating force.

mainstay n. the cable securing a mainmast; the chief support.

mainstream n. the dominant trend of opinion or style etc.

maintain v. 1 cause to continue, keep in existence; keep in repair; bear the expenses of. 2 assert.

maintenance n. the process of maintaining something; provision of means to support life, an allowance of money for this.

maiolica n. (also **majolica**) white pottery decorated with metallic colours.

maisonette n. part of a house (usu. not all on one floor) used as a separate dwelling.

maître d'hôtel n. a hotel manager; a head waiter.

maize n. a tall cereal plant bearing grain on large cobs; its grain.

majestic adj. stately and dignified, imposing. □ **majestically** adv.

majesty n. impressive stateliness; sovereign power; the title of a king or queen.

majolica var. of **maiolica**.

major adj. greater; very important. ● n. an army officer next below lieutenant colonel. ● v. (Amer.) specialize (in a subject) at college.

majorette n. = drum majorette.

major general n. an army officer next below lieutenant general.

majority n. the greatest part of a group or class; the number by which votes for one party etc. exceed those for the next or for all combined; the age when a person legally becomes an adult.

■ **Usage** *Majority* should strictly be used of a number of people or things, as in *the majority of people*, and not of a quantity of something, as in *the majority of the work*.

make v. (made, making) 1 form, prepare, produce; cause to exist, be, or become. 2 succeed in arriving at or achieving; gain, acquire. 3 reckon to be. 4 compel. 5 perform (an action etc.). ● n. a brand of goods. □ **make do** manage with something not satisfactory. **make for** try to reach; tend to bring about. **make good** become successful; repair or pay compensation for. **make off** go away hastily. **make off with** carry away, steal. **make out** write out (a list etc.); manage to see or understand; assert to be, pretend. **make over** transfer the ownership of; refashion (a garment etc.). **make up** form, constitute, invent (a story); compensate (for a loss etc.); become reconciled after a quarrel; complete (an amount); apply cosmetics (to). **make up one's mind** decide.

make-believe n. pretence.

make-up n. **1** cosmetics applied to the skin. **2** the way a thing is made; a person's character.

maker n. one who makes something.

makeshift adj. & n. (something) used as an improvised substitute.

makeweight n. something added to make up for a deficiency.

mal- comb. form bad, badly; faulty.

malachite (mal-ă-kyt) n. a green mineral.

maladjusted adj. not well adapted to one's circumstances.

maladminister v. manage (business or public affairs) badly or improperly.

maladroit adj. bungling, clumsy.

malady n. illness, disease.

malaise n. a feeling of illness or uneasiness.

malapropism n. a comical confusion of words.

malaria n. a disease causing recurring fever. □ **malarial** adj.

Malay adj. & n. (a member, the language) of a people of Malaysia and Indonesia.

malcontent n. a discontented person.

male adj. of the sex that can fertilize egg cells produced by a female; (of a plant) producing pollen, not seeds; (of a machine part) for insertion into a corresponding hollow part. ● n. a male person, animal, or plant.

malediction n. a curse. □ **maledictory** adj.

malefactor (mal-i-fak-ter) n. a wrongdoer.

malevolent adj. wishing harm to others. □ **malevolently** adv., **malevolence** n.

malfeasance n. misconduct.

malformation n. faulty formation. □ **malformed** adj.

malfunction n. faulty functioning. ● v. function faultily.

malice n. a desire to harm others. □ **malicious** adj., **maliciously** adv.

malign (mă-lyn) adj. harmful; showing malice. ● v. say unpleasant and untrue things about.

malignant adj. **1** showing great ill-will. **2** (of a tumour) growing harmfully and uncontrollably. □ **malignantly** adv., **malignancy** n.

malinger v. pretend illness to avoid work. □ **malingerer** n.

mall n. a sheltered walk or promenade; a shopping precinct.

mallard n. a wild duck, the male of which has a glossy green head.

malleable adj. able to be hammered or pressed into shape; easy to influence. □ **malleability** n.

mallet n. a hammer, usu. of wood; an instrument for striking the ball in croquet or polo.

malmsey n. a strong sweet wine.

malnutrition n. weakness resulting from lack of nutrition.

malodorous adj. stinking.

malpractice n. wrongdoing; improper professional behaviour.

malt n. barley or other grain prepared for brewing or distilling; whisky made with this.

maltreat v. ill-treat. □ **maltreatment** n.

mamba n. a poisonous South African tree-snake.

mammal n. a member of the class of animals that suckle their young. □ **mammalian** adj.

mammary adj. of the breasts.

mammography n. an X-ray of the breasts.

mammoth n. a large extinct elephant with curved tusks. ● adj. huge.

man n. (pl. **men**) an adult male person; a human being; mankind; an individual person; a male servant or employee; an ordinary soldier etc., not an officer; one of the small objects used in board games. ● v. (**manned**) supply with people to do something. □ **man to man** with frankness.

manacle n. a handcuff.

manage v. have control of; be manager of; operate (a tool etc.) effectively; contrive; deal with (a person) tactfully. □ **manageable** adj.

management n. managing; the people who manage a business.

manager n. a person in charge of a business etc. □ **managerial** adj.

manageress n. a woman in charge of a business etc.

mañana adv. tomorrow; at some indefinite time in the future.

manatee n. a large tropical aquatic mammal.

mandarin n. **1** a senior influential official. **2** a variety of small orange. **3** (**Mandarin**) a northern variety of the Chinese language.

mandate n. & v. (give) authority to perform certain tasks.

mandatory adj. compulsory.

mandible n. a jaw or jaw-like part.

mandolin n. a guitar-like musical instrument with a pear-shaped body.

mandrake n. a poisonous plant with a large yellow fruit.

mandrel n. a shaft holding work in a lathe.

mandrill n. a large baboon.

mane n. long hair on the neck of a horse or lion.

maneuver Amer. sp. of **manoeuvre**.

manful adj. brave, resolute. □ **manfully** adv.

manganese n. a hard brittle grey metallic element (symbol Mn) or its black oxide.

mange n. a skin disease affecting hairy animals.

mangel-wurzel n. (also **mangold**) a large beet used as cattle food.

manger n. an open trough for horses or cattle to feed from.

mangetout n. a variety of pea, eaten with the pod.

mangle n. a clothes wringer. ● v. damage by cutting or crushing roughly, mutilate.

mango n. (pl. **mangoes** or **mangos**) a tropical fruit with orange juicy flesh.

mangold n. a mangel-wurzel.

mangrove n. a tropical tree or shrub growing in shore-mud and swamps.

mangy adj. (**mangier**) having mange; shabby.

manhandle v. **1** move by human effort alone. **2** treat roughly.

manhole n. an opening through which a person can enter a drain etc. to inspect it.

manhood n. the state of being a man; manly qualities.

man-hour n. one hour's work by one person.

manhunt n. an organized search for a person, esp. a criminal.

mania n. violent madness; extreme enthusiasm for something.

maniac n. a person behaving wildly; a person affected with mania.

maniacal adj. of or like a mania or maniac.

manic adj. of or affected by mania.

manicure n. cosmetic care of the hands and fingernails. ● v. apply such treatment to. □ **manicurist** n.

manifest adj. clear and unmistakable. ● v. show clearly, give signs of. ● n. a list of cargo or passengers carried by a ship or aircraft. □ **manifestation** n.

manifesto n. (pl. **manifestos**) a public declaration of policy.

manifold adj. of many kinds. ● n. (in a machine) a pipe or chamber with several openings.

manikin n. a little man, a dwarf.

manila n. brown paper used for envelopes and wrapping paper.

manipulate v. handle or manage in a skilful or cunning way. □ **manipulation** n., **manipulator** n.

mankind n. human beings in general.

manly adj. (**manlier**) brave, strong; considered suitable for a man. □ **manliness** n.

man-made adj. synthetic, not natural.

mannequin n. a female model; a window dummy.

manner n. **1** the way a thing is done or happens; a person's way of behaving towards others. **2** kind or sort. **3** (**manners**) polite social behaviour.

mannered adj. having manners of a certain kind; stilted.

mannerism n. a distinctive personal habit or way of doing something.

manoeuvre n. (Amer. **maneuver**) a planned movement of a vehicle, troops, etc.; a skilful or crafty move or plan. ● v. perform manoeuvres; move or guide skilfully or craftily. □ **manoeuvrable** adj., **manoeuvrability** n.

manor n. a large country house, usu. with lands. □ **manorial** adj.

manpower n. the number of people available for work or service.

manqué (**mahn**-kay) adj. that might have been but is not.

mansard roof n. one with the lower part steeper than the upper part on all four sides.

manse n. a church minister's house, esp. in Scotland.

manservant n. (pl. **menservants**) a male servant.

mansion n. a large stately house.

manslaughter n. the act of killing a person unlawfully but not intentionally, or by negligence.

mantelpiece n. the shelf above a fireplace.

mantilla n. a Spanish lace veil worn over the hair and shoulders.

mantis n. (pl. **mantis** or **mantises**) a grasshopper-like insect.

mantle n. a loose cloak; a covering.

mantra n. a phrase repeated to aid concentration during meditation.

manual adj. of the hands; done or operated by the hand(s). ● n. a handbook. □ **manually** adv.

manufacture v. make or produce (goods) on a large scale by machinery; invent. ● n. the process of manufacturing. □ **manufacturer** n.

manure n. a substance, esp. dung, used as a fertilizer. ● v. apply manure to.

manuscript n. a book or document written by hand or typed, not printed.

Manx adj. & n. (the language) of the Isle of Man.

many adj. numerous. ● n. many people or things.

Maori (mow-ri) n. & adj. (pl. **Maori** or **Maoris**) (a member, the language) of the aboriginal people of New Zealand.

map n. a representation of the earth's surface or a part of it. ● v. (**mapped**) make a map of. □ **map out** plan in detail.

maple n. a tree with broad leaves.

mar v. (**marred**) damage, spoil.

Mar. abbr. March.

maracas n.pl. club-like gourds containing beads etc., shaken as a musical instrument.

maraschino n. a liqueur made from small black cherries.

marathon n. a long-distance running race; a long test of endurance.

marauding adj. going about in search of plunder. □ **marauder** n.

marble n. crystalline limestone that can be polished; a piece of sculpture in this; a small ball of glass or clay used in children's games. ●v. give a veined or mottled appearance to.

marcasite n. crystals of iron pyrites, used in jewellery.

march v. walk in a regular rhythm or an organized column; walk purposefully; cause to march or walk; progress steadily. ●n. **1** the act of marching; the distance covered by marching; music suitable for marching to; progress. **2** (**March**) the third month. □ **marcher** n.

marches n.pl. border regions.

marchioness n. the wife or widow of a marquess; a woman with the rank of marquess.

mare n. the female of the horse or a related animal.

margarine n. a substance made from animal or vegetable fat and used like butter.

marge n. (colloquial) margarine.

margin n. the edge or border of a surface; a blank space round the edges of a page; an amount over the essential minimum.

marginal adj. of or in a margin; near a limit; only very slight. □ **marginally** adv.

marginalize v. (also **-ise**) make or treat as insignificant.

marguerite n. a large daisy.

marigold n. a plant with golden daisy-like flowers.

marijuana (ma-ri-wah-nă) n. (also **marihuana**) dried hemp, smoked as a hallucinogenic drug.

marimba n. a xylophone of Africa and Central America.

marina n. a harbour for yachts and pleasure boats.

marinade n. a flavoured liquid in which savoury food is soaked before cooking. ●v. soak in a marinade.

marinate v. marinade.

marine adj. of the sea; of shipping. ●n. a soldier trained to serve on land or sea; a country's shipping.

mariner n. a sailor, a seaman.

marionette n. a puppet worked by strings.

marital adj. of marriage.

maritime adj. living or found near the sea; of seafaring.

marjoram n. a fragrant herb.

mark n. **1** a thing that visibly breaks the uniformity of a surface; a distinguishing feature; something indicating the presence of a quality or feeling; a symbol. **2** a point given for merit; a target; a line or object serving to indicate a position; a numbered design of a piece of equipment etc. **3** a unit of money in Germany, the Deutschmark. ●v. **1** make a mark on; characterize. **2** assign marks of merit to. **3** notice, watch carefully; keep close to and ready to hamper (an opponent in football etc.). □ **mark time** move the feet as if marching but without advancing.

marked adj. clearly noticeable. □ **markedly** adv.

marker n. a person or object that marks something; a broad felt-tipped pen.

market n. a gathering or place for the sale of provisions, livestock, etc.; demand (for a product). ● v. sell in a market; offer for sale. □ **on the market** offered for sale. □ **marketable** adj.

market garden n. a small farm producing vegetables.

marking n. mark(s); the colouring of an animal's skin, feathers, or fur.

marksman n. (pl. -men) a person who is a skilled shot. □ **marksmanship** n.

marl n. soil composed of clay and lime, used as a fertilizer.

marmalade n. a jam made from citrus fruit, esp. oranges.

marmoset n. a small bushy-tailed monkey of tropical America.

marmot n. a small burrowing animal of the squirrel family.

maroon n. 1 a brownish-red colour. 2 an explosive device used as a warning signal. ● adj. brownish-red. ● v. put and leave (a person) ashore in a desolate place; leave stranded.

marquee n. a large tent used for a party or exhibition etc.

marquess n. a nobleman ranking between duke and earl.

marquetry n. inlaid work in wood, ivory, etc.

marquis n. a rank in some European nobilities; a marquess.

marram n. shore grass that binds sand.

marriage n. the legal union of a man and woman; the act or ceremony of marrying.

marriageable adj. suitable or old enough for marriage.

marrow n. a soft fatty substance in the cavities of bones; a gourd used as a vegetable.

marrowfat n. a large variety of pea.

marry v. unite or give or take in marriage; unite (things).

marsh n. low-lying watery ground. □ **marshy** adj.

marshal n. a high-ranking officer; an official controlling an event or ceremony. ● v. (**marshalled**; Amer. **marshaled**) arrange in proper order; assemble; usher.

marshmallow n. a soft sweet made from sugar, egg white, and gelatin.

marsupial n. a mammal that usu. carries its young in a pouch.

mart n. a market.

marten n. a weasel-like animal with thick soft fur.

martial adj. of war, warlike.

martial law n. military government suspending ordinary law.

martin n. a bird of the swallow family.

martinet n. a person who demands strict obedience.

martyr n. a person who undergoes death or suffering for his or her beliefs. ● v. kill or torment as a martyr. □ **martyrdom** n.

marvel n. a wonderful thing. ● v. (**marvelled**; Amer. **marveled**) feel wonder.

marvellous adj. (Amer. **marvelous**) wonderful. □ **marvellously** adv.

Marxism n. the socialist theories of Karl Marx. □ **Marxist** adj. & n.

marzipan n. an edible paste made from ground almonds.

mascara n. a cosmetic for darkening the eyelashes.

mascot n. an object believed to bring good luck to its owner.

masculine adj. of, like, or traditionally considered suitable for men; having the grammatical form of the male gender. □ **masculinity** n.

mash *n.* a soft mixture of grain or bran; mashed potatoes. ● *v.* beat into a soft mass.

mask *n.* a covering worn over the face as a disguise or protection. ● *v.* cover with a mask; disguise, screen, conceal.

masochism *n.* pleasure in suffering pain. □ **masochist** *n.*, **masochistic** *adj.*

mason *n.* a person who builds or works with stone.

masonry *n.* stonework.

masque *n.* a musical drama with mime.

masquerade *n.* a false show or pretence. ● *v.* pretend to be what one is not.

mass *n.* **1** a coherent unit of matter; a large quantity, heap, or expanse; the quantity of matter a body contains. **2** (**the masses**) ordinary people. **3** (usu. **Mass**) a celebration (esp. in the RC Church) of the Eucharist; a form of liturgy used in this. ● *v.* gather or assemble into a mass.

massacre *n.* a great slaughter. ● *v.* slaughter in large numbers.

massage *n.* rubbing and kneading of the body to reduce pain or stiffness. ● *v.* treat in this way; adjust statistics to give an improved result.

masseur *n.* a man who practises massage professionally.

masseuse *n.* a woman who practises massage professionally.

massif *n.* a compact group of mountain heights.

massive *adj.* large and heavy or solid; huge. □ **massively** *adv.*

mass-produce *v.* manufacture in large quantities by a standardized process.

mast *n.* **1** a tall pole, esp. supporting a ship's sails. **2** the fruit of the beech, oak, chestnut, etc., used as food for pigs.

mastectomy *n.* surgical removal of a breast.

master *n.* **1** a man who has control of people or things; a male teacher. **2** a person with great skill, a great artist. **3** something from which a series of copies is made. **4** (**Master**) the title of a boy not old enough to be called *Mr.* ● *adj.* superior; principal; controlling others. ● *v.* bring under control; acquire knowledge or skill in.

masterclass *n.* a class given by a famous musician, artist, etc.

masterful *adj.* domineering. □ **masterfully** *adv.*

master key *n.* a key that opens several different locks.

masterly *adj.* very skilful.

mastermind *n.* a person of outstanding mental ability; the person directing an enterprise. ● *v.* plan and direct.

Master of Arts, Master of Science *n.* a university degree, above a first degree but below a Ph.D.

masterpiece *n.* an outstanding piece of work.

master stroke *n.* a very skilful act of policy.

mastery *n.* control, supremacy; thorough knowledge or skill.

mastic *n.* gum or resin from certain trees; a type of cement.

masticate *v.* chew.

mastiff *n.* a large strong dog.

mastitis *n.* inflammation of the breast or udder.

mastodon *n.* an extinct animal resembling an elephant.

mastoid *n.* a projecting piece of a bone behind the ear.

masturbate *v.* stimulate the genitals with the hand. □ **masturbation** *n.*

mat *n.* **1** a piece of material placed on a floor or other surface as an ornament or to protect it. **2** var. of **matt**.

(matted) make or become tangled into a thick mass.

matador n. a bullfighter.

match n. **1** a short stick tipped with material that catches fire when rubbed on a rough surface. **2** a contest in a game or sport; a person or thing exactly like or corresponding or equal to another; a marriage. ● v. set against each other in a contest; equal in ability or achievement; be alike; find a match for.

matchmaking n. scheming to arrange marriages. □ **matchmaker** n.

matchstick n. the stick of a match.

matchwood n. wood that splinters easily; wood broken into splinters.

mate n. **1** a companion or fellow worker; each of a pair of mated animals; a merchant ship's officer. **2** checkmate. ● v. put or come together as a pair; come or bring (animals) together to breed.

material n. that from which something is or can be made; cloth, fabric. ● adj. **1** of matter; of the physical (not spiritual) world. **2** significant. □ **materially** adv.

materialism n. belief that only the material world exists; excessive concern with material possessions. □ **materialist** n., **materialistic** adj.

materialize v. (also **-ise**) appear, become visible; become a fact, happen.

maternal adj. of a mother; motherly; related through one's mother. □ **maternally** adv.

maternity n. motherhood. ● adj. of or for women in pregnancy and childbirth.

math n. (Amer. colloquial) mathematics.

mathematician n. a person skilled in mathematics.

mathematics n. (as sing.) the science of numbers, quantities, and measurements; (as pl.) the use of this. □ **mathematical** adj., **mathematically** adv.

maths n. (colloquial) mathematics.

matinée n. an afternoon performance.

matinée coat n. a baby's jacket.

matins n. (also **mattins**) morning prayer.

matriarch n. the female head of a family or tribe. □ **matriarchal** adj.

matriarchy n. a social organization in which a female is head of the family.

matrices see **matrix**.

matricide n. killing one's mother. □ **matricidal** adj.

matriculate v. enrol at a college or university. □ **matriculation** n.

matrimony n. marriage. □ **matrimonial** adj.

matrix n. (pl. **matrices** or **matrixes**) a mould in which a thing is cast or shaped; a rectangular array of mathematical quantities.

matron n. a married woman; a woman in charge of domestic affairs or nursing in a school etc.

matronly adj. like or characteristic of a staid or dignified married woman.

matt adj. (also **mat**) dull, not shiny.

matter n. **1** that which occupies space in the visible world; a specified substance or material, specified things; a situation or business being considered. **2** pus. ● v. be of importance.

mattins var. of **matins**.

mattress n. a fabric case filled with padding or springy material, used on or as a bed.

maturation n. maturing.

mature *adj.* fully grown or developed; (of a life assurance policy etc.) due for payment. ● *v.* make or become mature. □ **maturely** *adv.*, **maturity** *n.*

matzo *n.* (*pl.* **matzos**) a wafer of unleavened bread.

maudlin *adj.* sentimental in a silly or tearful way.

maul *v.* treat roughly, injure by rough handling.

maunder *v.* talk in a rambling way.

mausoleum *n.* a magnificent tomb.

mauve (mohv) *adj.* & *n.* pale purple.

maverick *n.* an unorthodox or undisciplined person.

mawkish *adj.* sentimental in a sickly way. □ **mawkishly** *adv.*, **mawkishness** *n.*

maxim *n.* a general truth or rule of conduct.

maximize *v.* (also **-ise**) increase to a maximum. □ **maximization** *n.*

maximum *adj.* & *n.* (*pl.* **maxima**) the greatest (amount) possible. □ **maximal** *adj.*, **maximally** *adv.*

may[1] *v.aux.* used to express a wish, possibility, or permission.

■ **Usage** Both *can* and *may* are used for asking permission, as in *Can I move?* and *May I move?*, but *may* is better in formal English because *Can I move?* also means 'Am I physically able to move?'

may[2] *n.* **1** hawthorn blossom. **2** (**May**) the fifth month.

maya *n.* (in Hinduism) illusion, magic.

maybe *adv.* perhaps.

mayday *n.* an international radio signal of distress.

May Day *n.* 1 May, esp. as a festival.

mayfly *n.* an insect with long hairlike tails, living in spring.

mayhem *n.* violent action, havoc.

mayonnaise *n.* a cold creamy sauce made with eggs and oil.

mayor *n.* the head of the municipal corporation of a city or borough. □ **mayoral** *adj.*, **mayoralty** *n.*

mayoress *n.* a female mayor; a mayor's wife, or other woman with her ceremonial duties.

maypole *n.* a tall pole for dancing round on May Day.

maze *n.* a complex and baffling network of paths, lines, etc.

MB *abbr.* Bachelor of Medicine; (*Computing*; also **Mb**) megabyte.

MBE *abbr.* Member of the Order of the British Empire.

MC *abbr.* Master of Ceremonies; Member of Congress.

MD *abbr.* Doctor of Medicine; Managing Director.

Md *symb.* mendelevium.

ME *abbr.* myalgic encephalomyelitis, a condition characterized by prolonged fatigue.

me[1] *pron.* the objective case of *I.*

■ **Usage** Some people consider it correct to use only *It is I*, but this is very formal or old-fashioned in most situations, and *It is me* is normally quite acceptable. On the other hand, it is not standard English to say *Me and him went* rather than *He and I went.*

me[2] *n.* (*Music*) the third note of a major scale, or the note E.

mea culpa *int.* an acknowledgement of error or guilt.

mead *n.* an alcoholic drink made from honey and water.

meadow *n.* a field of grass.

meagre *adj.* (*Amer.* **meager**) scanty in amount.

meal *n.* **1** an occasion when food is eaten; the food itself. **2** coarsely ground grain.

mealie n. (in South Africa) maize.

mealy adj. of or like meal.

mealy-mouthed adj. trying excessively to avoid offending people.

mean¹ adj. miserly; selfish; unkind; poor in quality or appearance; low in rank; (Amer.) vicious. □ **meanly** adv., **meanness** n.

mean² adj. & n. (something) midway between two extremes; an average.

mean³ v. (**meant**, **meaning**) intend; convey or express; be likely to result in; be of specified importance.

meander v. follow a winding course; wander in a leisurely way. ● n. a winding course.

meaning n. what is meant. ● adj. expressive. □ **meaningful** adj., **meaningless** adj.

means n. (as sing. or pl.) that by which a result is brought about; (as pl.) resources. **by all means** certainly. **by no means** not at all.

means test n. an official inquiry to establish need before giving financial help from public funds.

meant see **mean³**.

meantime adv. meanwhile.

meanwhile adv. in the intervening period; at the same time.

measles n. an infectious disease producing red spots on the body.

measly adj. (**measlier**) (slang) meagre.

measurable adj. able to be measured. □ **measurably** adv.

measure n. 1 a size or quantity found by measuring; a unit, standard, device, or system used in measuring. 2 an action taken for a purpose, a law. 3 rhythm. ● v. find the size etc. of by comparison with a known standard; be of a certain size; mark or deal (a measured amount). □ **measure up to** reach the standard required by.

measured adj. rhythmical; carefully considered.

measurement n. measuring; a size etc. found by measuring.

meat n. animal flesh as food.

meaty adj. (**meatier**) 1 like meat; full of meat. 2 full of subject-matter. □ **meatiness** n.

mechanic n. a skilled workman who uses or repairs machines.

mechanical adj. of or worked by machinery; done without conscious thought. □ **mechanically** adv.

mechanics n. the study of motion and force; the science of machinery; (as pl.) the way a thing works.

mechanism n. the way a machine works, its parts; a process.

mechanize v. (also -**ise**) equip with machinery; use machines for. □ **mechanization** n.

medal n. a coin-like piece of metal commemorating an event or awarded for an achievement.

medallion n. a large medal; a circular ornamental design.

medallist n. (Amer. **medalist**) the winner of a medal.

meddle v. interfere in people's affairs; tinker. □ **meddler** n.

meddlesome adj. often meddling.

media see **medium**. n.pl. (**the media**) newspapers and broadcasting as conveying information to the public.

■ **Usage** It is a mistake to use media with a singular verb, as in The media is biased.

mediaeval adj. var. of **medieval**.

medial adj. situated in the middle. □ **medially** adv.

median adj. in or passing through the middle. ● n. a median point or line.

mediate v. act as peacemaker between opposing sides; bring about (a settlement) in this way. □ **mediation** n., **mediator** n.

medic n. (colloquial) a doctor.

medical adj. of the science of medicine. ● n. (colloquial) a medical examination. □ **medically** adv.

medicament n. any medicine, ointment, etc.

medicate v. treat with a medicinal substance. □ **medication** n.

medicinal adj. having healing properties. □ **medicinally** adv.

medicine n. the science of the prevention and cure of disease; a substance used to treat disease.

medicine man n. a witch-doctor.

medieval adj. (also **mediaeval**) of the Middle Ages.

mediocre adj. of medium quality; second-rate. □ **mediocrity** n.

meditate v. think deeply. □ **meditation** n., **meditative** adj., **meditatively** adv.

medium n. (pl. **media** or **mediums**) a middle size, quality, etc.; the substance or surroundings in which a thing exists; an agency or means. See also **media**. ● adj. intermediate; average; moderate.

medium wave n. a radio wave between 300 kHz and 3 MHz.

medlar n. a fruit like a small brown apple, eaten when it begins to decay.

medley n. (pl. **medleys**) an assortment; excerpts of music from various sources.

medulla n. spinal or bone marrow; the hindmost segment of the brain. □ **medullary** adj.

meek adj. quiet and obedient, not protesting. □ **meekly** adv., **meekness** n.

meerkat n. a South African mongoose.

meerschaum (meer-shăm) n. a tobacco pipe with a white clay bowl.

meet v. (**met**, **meeting**) come into contact (with); be present at the arrival of; make the acquaintance of; experience; satisfy (needs etc.). ● n. an assembly for a hunt etc. ● adj. (old use) suitable, proper.

meeting n. coming together; an assembly for discussion.

mega adj. (slang) excellent; enormous. ● adv. extremely.

mega- comb. form large; one million (as in megavolts, megawatts); (colloquial) extremely; very big.

megabyte n. (Computing) 1,048,576 (i.e. 2^{20}) bytes.

megahertz n. one million cycles per second, as a unit of frequency of electromagnetic waves.

megalith n. a large stone, esp. as a prehistoric monument. □ **megalithic** adj.

megalomania n. excessive self-esteem, esp. as a form of insanity. □ **megalomaniac** adj. & n.

megaphone n. a funnel-shaped device for amplifying the voice.

megaton n. a unit of explosive power equal to one million tons of TNT.

melamine n. a resilient plastic used esp. for laminated coatings.

melancholy n. mental depression, sadness; gloom. ● adj. sad, gloomy.

Melba toast n. thin crisp toast.

mêlée (mel-ay) n. a confused fight; a muddle.

melanin n. a dark pigment in the skin, hair, etc.

melanoma n. a malignant skin tumor.

meld v. merge, blend.

mellifluous adj. sweet-sounding.

mellow adj. (of fruit) ripe and sweet; (of sound or colour) soft and rich; (of people) having become kindly with age. ● v. make or become mellow.

melodeon n. (also **melodion**) a harmonium; a small button accordion.

melodic adj. of melody. □ **melodically** adv.

melodious adj. full of melody. □ **melodiously** adv.

melodrama n. a sensational drama. □ **melodramatic** adj., **melodramatically** adv.

melody n. sweet music; the main part in a piece of harmonized music.

melon n. a large sweet fruit.

melt v. make into or become liquid, esp. by heat; soften through pity or love; fade away.

meltdown n. the melting of an overheated reactor core.

member n. a person or thing belonging to a particular group or society. □ **membership** n.

membrane n. a thin flexible skin-like tissue. □ **membranous** adj.

memento n. (pl. **mementoes** or **mementos**) a souvenir.

memo n. (pl. **memos**)(colloquial) a memorandum.

memoir (mem-wahr) n. a written account of events etc. that one remembers.

memorable adj. worth remembering, easy to remember. □ **memorably** adv., **memorability** n.

memorandum n. (pl. **memoranda** or **memorandums**) a note written as a reminder; a written message from one colleague to another.

memorial n. an object or custom etc. established in memory of an event or person(s). ● adj. serving as a memorial.

memorize v. (also **-ise**) learn (a thing) so as to know it from memory.

memory n. the ability to remember things; a thing remembered; the storage capacity of a computer, RAM.

men see **man**.

menace n. a threat; an annoying or troublesome person or thing. ● v. threaten. □ **menacingly** adv.

ménage (may-nahzh) n. a household.

menagerie n. a collection of wild or strange animals for exhibition.

mend v. repair; make or become better. ● n. a repaired place.

mendacious adj. untruthful. □ **mendaciously** adv., **mendacity** n.

mendelevium n. a radioactive metallic element (symbol Md).

mendicant adj. & n. (a person) living by begging.

menfolk n. men in general; the men of one's family.

menhir (men-heer) n. a tall upright stone set up in prehistoric times.

menial adj. lowly, degrading. ● n. a person who does menial tasks. □ **menially** adv.

meningitis n. inflammation of the membranes covering the brain and spinal cord.

meniscus n. the curved surface of a liquid; a lens convex on one side and concave on the other.

menopause n. the time of life when a woman finally ceases to menstruate. □ **menopausal** adj.

menorah n. a seven-armed candelabrum used in Jewish worship.

menstrual adj. of or in menstruation.

menstruate v. experience a monthly discharge of blood from the womb. □ **menstruation** n.

mensuration n. measuring; mathematical rules for this.

mental adj. **1** of, in, or performed by the mind. **2** (colloquial) mad. □ **mentally** adv.

mentality *n.* a characteristic attitude of mind.

menthol *n.* a peppermint-flavoured substance, used medicinally.

mentholated *adj.* impregnated with menthol.

mention *v.* speak or write about briefly; refer to by name. ● *n.* mentioning, being mentioned; a reference.

mentor *n.* a trusted adviser.

menu *n.* (*pl.* **menus**) a list of dishes to be served; a list of options displayed on a computer screen.

meow (also **miaow**) = **miaow**.

MEP *abbr.* Member of the European Parliament.

mercantile *adj.* trading, of trade or merchants.

mercenary *adj.* working merely for money or reward; grasping. ● *n.* a professional soldier hired by a foreign country.

mercerized *adj.* (also **-ised**) (of cotton) treated with a slightly glossy substance that adds strength.

merchandise *n.* goods bought and sold or for sale. ● *v.* trade; promote sales of (goods).

merchant *n.* a wholesale trader; (*Amer. & Scot.*) a retail trader.

merchantable *adj.* saleable.

merchant bank *n.* one dealing in commercial loans and the financing of businesses.

merchantman *n.* (*pl.* **-men**) a merchant ship.

merchant navy *n.* shipping employed in commerce.

merchant ship *n.* a ship carrying merchandise.

merciful *adj.* showing mercy; giving relief from pain and suffering.

mercifully *adv.* in a merciful way; (*colloquial*) thank goodness.

merciless *adj.* showing no mercy. □ **mercilessly** *adv.*

mercurial *adj.* **1** of or caused by mercury. **2** lively in temperament; liable to sudden changes of mood.

mercury *n.* a heavy silvery usu. liquid metallic element (symbol Hg). □ **mercuric** *adj.*

mercy *n.* kindness shown to an offender or enemy etc. who is in one's power; a merciful act. □ **at the mercy of** wholly in the power of or subject to.

mere¹ *adj.* no more or no better than what is specified. □ **merely** *adv.*

mere² *n.* (*poetic*) a lake.

merest *adj.* very small or insignificant.

meretricious *adj.* showily attractive but cheap or insincere.

merge *v.* combine into a whole; blend gradually.

merger *n.* the combining of commercial companies etc. into one.

meridian *n.* any of the great semicircles on the globe, passing through the North and South Poles.

meringue (mĕ-rang) *n.* a baked mixture of sugar and egg white; a small cake of this.

merino *n.* (*pl.* **merinos**) a breed of sheep with fine soft wool; soft woollen fabric.

merit *n.* a feature or quality that deserves praise; worthiness. ● *v.* (**merited**) deserve.

meritocracy *n.* government or control by people selected for merit.

meritorious *adj.* deserving praise.

merlin *n.* a small falcon.

mermaid *n.* an imaginary sea creature, a woman with a fish's tail instead of legs.

merry *adj.* (**merrier**) **1** cheerful and lively. **2** slightly drunk. □ **merrily** *adv.*, **merriment** *n.*

merry-go-round *n.* a round-about at a funfair.

merrymaking n. revelry.

mescaline n. (also **mescalin**) a hallucinogenic drug.

mesh n. the space between threads in a net, sieve, etc.; a network fabric or structure. ● v. (of a toothed wheel) engage with another.

mesmerize v. (also **-ise**) hypnotize, dominate the attention or will of.

mesolithic adj. of the period between palaeolithic and neolithic.

meson n. an unstable elementary particle.

mess n. **1** a dirty or untidy condition; an untidy collection of things; something spilt; a difficult or confused situation, trouble. **2** (in the armed forces) a group who eat together, their dining room. ● v. **1** (usu. **mess up**) make untidy or dirty; muddle, bungle. **2** (in the armed forces) eat with a group. □ **mess about** potter; fool about.

message n. a spoken or written communication; moral or social teaching.

messenger n. the bearer of a message.

Messiah n. the deliverer expected by Jews; Christ as this. □ **Messianic** adj.

Messrs see **Mr**.

messy adj. (**messier**) untidy or dirty; slovenly; complicated and difficult. □ **messily** adv., **messiness** n.

met see **meet**.

metabolism n. the process by which food is digested and energy supplied. □ **metabolic** adj., **metabolically** adv.

metabolize v. (also **-ise**) process (food) in metabolism.

metacarpus n. (pl. **metacarpi**) the set of bones between the wrist and fingers.

metal n. **1** any of a class of mineral substances such as gold,

silver, iron, etc., or an alloy of these. **2** road metal. ● v. (**metalled**; Amer. **metaled**) make or mend (a road) with road metal.

metallic adj. of or like metal.

metallography n. the study of the internal structure of metals.

metallurgy n. the study of the properties of metals; the science of extracting and working metals.

metamorphic adj. (of rock) changed in form or structure by heat, pressure, etc; of metamorphosis.

metamorphose v. change by metamorphosis.

metamorphosis n. (pl. **metamorphoses**) a change of form or character.

metaphor n. the application of a word or phrase to something that it does not apply to literally (e.g. the evening of one's life, food for thought). □ **metaphorical** adj., **metaphorically** adv.

metaphysics n. the branch of philosophy dealing with the nature of existence and knowledge. □ **metaphysical** adj.

metatarsus n. (pl. **metatarsi**) the set of bones between the ankle and the toes.

mete out v. give what is due.

meteor n. a small mass of matter from outer space.

meteoric adj. of meteors; swift and brilliant. ● **meteorically** adv.

meteorite n. a meteor fallen to earth.

meteorology n. the study of atmospheric conditions in order to forecast weather. □ **meteorological** adj., **meteorologist** n.

meter n. **1** a device measuring and indicating the quantity supplied, distance travelled, time elapsed, etc. **2** Amer. sp. of **metre**. ● v. measure by a meter.

methadone *n.* a narcotic pain-killing drug.

methanal *n.* = **formaldehyde**.

methane *n.* a colourless inflammable gas.

methanol *n.* a colourless inflammable liquid hydrocarbon, used as a solvent.

method *n.* a procedure or way of doing something; orderliness.

methodical *adj.* orderly, systematic. □ **methodically** *adv.*

Methodist *n.* & *adj.* (a member) of a Protestant religious denomination based on the teachings of John and Charles Wesley.

methodology *n.* a particular set of methods used; the science of method.

meths *n.* (*colloquial*) methylated spirit.

methyl *n.* a chemical unit present in methane and many organic compounds.

methylated spirit a form of alcohol used as a solvent and for heating.

meticulous *adj.* very careful and exact. □ **meticulously** *adv.*, **meticulousness** *n.*

métier *n.* one's profession; what one does best.

metre *n.* (*Amer.* **meter**) a metric unit of length (about 39.4 inches); rhythm in poetry.

metric *adj.* of or using the metric system; of poetic metre.

metrical *adj.* of or composed in rhythmic metre; not prose.

metricate *v.* convert to a metric system. □ **metrication** *n.*

metric system *n.* a decimal system of weights and measures, using the metre, litre, and gram as units.

metric ton *see* **ton**.

metro *n.* an underground railway.

metronome *n.* a device used to indicate tempo while practising music.

metropolis *n.* the chief city of country or region.

metropolitan *adj.* of a metropolis.

mettle *n.* courage, strength of character.

mettlesome *adj.* spirited, brave.

mew *n.* a cat's characteristic cry. ● *v.* make this sound.

mews *n.* a set of stables converted into houses.

mezzanine *n.* an extra storey set between two others.

mezzo *adv.* (in music) half; moderately. ● *n.* a mezzo-soprano, a voice between soprano and contralto.

mezzotint *n.* a method of engraving.

Mg *symb.* magnesium.

mg *abbr.* milligram(s).

MHz *abbr.* megahertz.

mi. *abbr.* (*Amer.*) mile(s).

miaow (also **meow**) = **mew**.

miasma *n.* unpleasant or unwholesome air.

mica (my-kǎ) *n.* a mineral substance used as an electrical insulator.

mice *see* **mouse**.

micro- *comb. form* extremely small; one-millionth part of (as in *microgram*).

microbe *n.* a micro-organism. □ **microbial** *adj.*

microbiology *n.* the study of micro-organisms. □ **microbiologist** *n.*

microchip *n.* a small piece of a semiconductor holding a complex electronic circuit.

microclimate *n.* the climatic conditions of a small area, e.g. of part of a garden.

microcomputer *n.* a computer in which the central processor is contained on microchips.

microcosm *n.* something regarded as resembling something else but on a very small scale.

microfiche (my-kroh-feesh) n. (pl. **-fiche** or **-fiches**) a small sheet of microfilm.

microfilm n. a length of film bearing miniature photographs of documents. ● v. photograph on this.

microlight n. a motorized hang-glider.

micromesh n. fine-meshed material, esp. nylon.

micrometer n. an instrument measuring small lengths or angles.

micron n. one-millionth of a metre.

micro-organism n. an organism invisible to the naked eye.

microphone n. an instrument for picking up sound waves for recording or broadcasting.

microprocessor n. an integrated circuit containing the functions of a computer's central processing unit.

microscope n. an instrument with lenses that magnify very small things, making them visible.

microscopic adj. of a microscope; very small; too small to be seen without a microscope. □ **microscopically** adv., **microscopy** n.

microsurgery n. surgery using a microscope.

microwave n. an electromagnetic wave of length between about 50 cm and 1 mm; an oven using such waves to heat food quickly.

mid adj. middle.

midday n. noon.

midden n. a heap of dung or rubbish.

middle adj. occurring at an equal distance from extremes or outer limits. ● n. the middle point, position, area, etc.

middle age n. the part of life between youth and old age. □ **middle-aged** adj.

Middle Ages n.pl. 5th c.–1453, or c. 1000–1453.

middle class n. the class of society between upper and working classes. □ **middle-class** adj.

Middle East n. the area from Egypt to Iran inclusive.

middleman n. (pl. **-men**) a trader handling a commodity between the producer and consumer.

middleweight n. a weight above welterweight, in amateur boxing between 71 and 75 kg.

middling adj. moderately good.

midfield n. the part of a football pitch away from the goals.

midge n. a small biting insect.

midget n. a very small person or thing.

MIDI n. (also **midi**) an interface allowing electronic musical instruments and computers to be connected.

midi system n. a compact hi-fi system of stacking components.

Midlands n.pl. the inland counties of central England. □ **midland** adj.

midnight n. 12 o'clock at night.

midriff n. the front part of the body just above the waist.

midshipman n. (pl. **-men**) a naval rank just below sub-lieutenant.

midst n. the middle.

midway adv. halfway.

midwife n. (pl. **midwives**) a person trained to assist at childbirth.

mien (meen) n. a person's manner or bearing.

might¹ n. great strength or power.

might² v.aux. used to request permission or (like *may*) to express possibility.

mighty adj. (**mightier**) very strong or powerful; very great. □ **mightily** adv.

migraine n. a severe form of headache.

migrant *adj.* migrating. ● *n.* a migrant person or animal.

migrate *v.* leave one place and settle in another; (of animals) go from one place to another each season. □ **migration** *n.*, **migratory** *n.*

mike *n.* (*colloquial*) a microphone.

mil *n.* one-thousandth of an inch.

milage var. of **mileage**.

milch *adj.* giving milk.

mild *adj.* moderate in intensity, not harsh or drastic; gentle; not strongly flavoured. □ **mildly** *adv.*, **mildness** *n.*

mildew *n.* tiny fungi forming a coating on things exposed to damp. □ **mildewed** *adj.*

mile *n.* a measure of length, 1760 yds (about 1.609 km).

mileage *n.* (also **milage**) a distance in miles.

milestone *n.* a stone showing the distance to a certain place; a significant event or stage reached.

milieu (meel-yer) *n.* (*pl.* **milieus** or **milieux**) environment, surroundings.

militant *adj.* & *n.* (a person) prepared to take aggressive action. □ **militancy** *n.*

militarism *n.* reliance on military attitudes. □ **militaristic** *n.*

military *adj.* of soldiers or the army or all armed forces.

militate *v.* have force or effect.

■ **Usage** *Militate* is often confused with *mitigate*, which means 'to make less intense or severe'.

militia (mil-ish-ă) *n.* a military force, esp. of trained civilians available in an emergency.

milk *n.* a white fluid secreted by female mammals as food for their young; cow's milk; a milk-like liquid. ● *v.* draw milk from; exploit.

milkman *n.* (*pl.* **-men**) a person who delivers milk to customers.

milksop *n.* a weak or timid man.

milk teeth *n.* the first (temporary) teeth in young mammals.

milky *adj.* (**milkier**) of or like milk; containing much milk.

mill *n.* machinery for grinding or processing specified material; a building containing this. ● *v.* process in a mill; produce grooves in (metal); move in a confused mass. □ **miller** *n.*

millennium *n.* (*pl.* **millenniums** or **millennia**) **1** a period of 1,000 years. **2** a period of great happiness for everyone.

millepede var. of **millipede**.

millet *n.* a cereal plant; its seeds.

milli- *comb. form* one-thousandth part of (as in **milligram**, **millilitre**, **millimetre**).

millibar *n.* one-thousandth of a bar, a unit of meteorological pressure.

milliner *n.* a person who makes or sells women's hats. □ **millinery** *n.*

million *n.* one thousand thousand (1,000,000). □ **millionth** *adj.* & *n.*

millionaire *n.* a person who has over a million pounds, dollars, etc.

millipede *n.* (also **millepede**) a small crawling creature with many legs.

millstone *n.* a heavy circular stone for grinding corn; a great burden.

milometer *n.* an instrument measuring the distance in miles travelled by a vehicle.

milt *n.* sperm discharged by a male fish over eggs laid by the female.

mime *n.* acting with gestures without words. ● *v.* act with mime.

mimic *v.* (**mimicked, mimicking**) imitate, esp. playfully or for entertainment. ● *n.* a person who is clever at mimicking others. □ **mimicry** n.

mimosa n. a tropical shrub with small ball-shaped flowers.

mina var. of **mynah**.

minaret n. a tall slender tower on or beside a mosque.

mince v. 1 cut into small pieces in a mincer. 2 walk or speak with affected refinement. ● n. minced meat.

mincemeat n. a mixture of dried fruit, sugar, etc. used in pies.

mince pie n. a pie containing mincemeat.

mincer n. a machine with revolving blades for cutting food into very small pieces.

mind n. the ability to be aware of things and to think and reason; a person's attention, remembrance, intention, or opinion; sanity. ● v. 1 have charge of. 2 object to. 3 remember and be careful (about).

minded adj. having inclinations or interests of a certain kind.

minder n. a person whose job is to have charge of a person or thing; (slang) a bodyguard.

mindful adj. taking thought or care (of something).

mindless adj. without intelligence. □ **mindlessness** n.

mine¹ adj. & poss.pron. belonging to me.

mine² n. 1 an excavation for extracting metal or coal etc; an abundant source. 2 an explosive device laid in or on the ground or in water. ● v. 1 dig for minerals, extract in this way. 2 lay explosive mines under or in.

minefield n. an area where explosive mines have been laid; a situation full of difficulties.

miner n. a person who works in a mine.

mineral n. 1 an inorganic natural substance. 2 a fizzy soft drink. ● adj. of or containing minerals.

mineralogy n. the study of minerals. □ **mineralogist** n.

mineral water n. water naturally containing dissolved mineral salts or gases.

minestrone n. (mini-stroh-ni) soup containing vegetables and pasta.

minesweeper n. a ship for clearing away mines laid in the sea.

mingle v. blend together; mix socially.

mini- comb. form miniature, small.

miniature adj. very small, on a small scale. ● n. a small-scale portrait, copy, or model.

miniaturize v. (also **-ise**) make miniature, produce in a very small version.

minibus n. a small bus for about twelve people.

minicab n. a cab like a taxi that can be booked but not hailed in the street.

minicomputer n. a computer that is bigger than a microcomputer, but smaller than a mainframe.

minim n. a note in music, lasting half as long as a semibreve; onesixtieth of a fluid drachm.

minima see **minimum**.

minimal adj. very small, the least possible. □ **minimally** adv.

minimalism n. the use of simple basic design forms; including only the minimum. □ **minimalist** adj. & n.

minimize v. (also **-ise**) reduce to a minimum; represent as small or unimportant.

minimum adj. & n. (pl. **minima**) the smallest (amount) possible.

minion n. (derog.) an assistant.

miniseries n. (pl. **miniseries**) a short series of television programmes on a common theme.

minister *n.* the head of a government department; a clergyman; a senior diplomatic representative. □ **minister to** attend to the needs of. □ **ministerial** *adj.*

ministration *n.* help, service; ministering.

ministry *n.* a government department headed by a minister; a period of government under one leader; the work of a clergyman.

mink *n.* a small stoat-like animal; its valuable fur; a coat made of this.

minnow *n.* a small fish of the carp family.

minor *adj.* lesser; not very important. ● *n.* a person not yet legally of adult age.

minority *n.* the smallest part of a group or class; a small group differing from others; the state of being below the legal adult age.

minster *n.* a large, important church.

minstrel *n.* a medieval singer and musician.

mint¹ *n.* a place authorized to make a country's coins. ● *v.* make (coins). □ **in mint condition** as new.

mint² *n.* a fragrant herb; peppermint, a sweet flavoured with this. □ **minty** *adj.*

minuet (min-yoo-et) *n.* a slow stately dance.

minus *prep.* reduced by subtraction of; below zero; (*colloquial*) without. ● *adj.* less than zero, negative; less than the amount indicated. ● *n.* the sign (–).

minuscule *adj.* very small.

minute¹ (min-it) *n.* one-sixtieth of an hour or degree; a moment of time. **2** (**minutes**) an official summary of the proceedings of a meeting. ● *v.* record in the minutes.

minute² (my-newt) *adj.* extremely small; very precise. □ **minutely** *adv.*, **minuteness** *n.*

minutiae (min-yoo-shi-ee) *n.pl.* very small details.

minx *n.* a mischievous girl.

miracle *n.* a wonderful event attributed to a supernatural being; a remarkable event or thing. □ **miraculous** *adj.*, **miraculously** *adv.*

mirage *n.* an optical illusion caused by atmospheric conditions.

MIRAS *abbr.* mortgage interest relief at source.

mire *n.* swampy ground, a bog; mud or sticky dirt.

mirror *n.* glass coated so that reflections can be seen in it. ● *v.* reflect in a mirror.

mirth *n.* merriment, laughter.

mis- *pref.* badly, wrongly.

misadventure *n.* a piece of bad luck.

misanthrope *n.* (also **misanthropist**) a person who dislikes people in general. □ **misanthropy** *n.*, **misanthropic** *adj.*

misapprehend *v.* misunderstand. □ **misapprehension** *n.*

misappropriate *v.* take dishonestly. □ **misappropriation** *n.*

misbehave *v.* behave badly.

miscalculate *v.* calculate incorrectly. □ **miscalculation** *n.*

miscarriage *n.* an abortion occurring naturally.

miscarry *v.* have a miscarriage; go wrong, be unsuccessful.

miscegenation (mis-i-jin-ay-shŏn) *n.* interbreeding of races.

miscellaneous *adj.* assorted.

miscellany *n.* a collection of assorted items.

mischance *n.* misfortune.

mischief *n.* annoying but not malicious behaviour; playful malice; harm, damage.

mischievous *adj.* full of mischief. □ **mischievously** *adv.*, **mischievousness** *n.*

misconception *n.* a wrong interpretation.

misconduct *n.* bad behaviour; mismanagement.

misconstrue *v.* misinterpret. □ **misconstruction** *n.*

miscreant *n.* a wrongdoer.

misdeed *n.* a wrongful act.

misdemeanour *n.* (*Amer.* **misdemeanor**) a misdeed, a wrongdoing.

miser *n.* a person who hoards money and spends as little as possible. □ **miserly** *adj.*, **miserliness** *n.*

miserable *adj.* full of misery; wretchedly poor in quality or surroundings. □ **miserably** *adv.*

misery *n.* great unhappiness or discomfort; (*colloquial*) a discontented or disagreeable person.

misfire *v.* (of a gun or engine) fail to fire correctly; (of a plan etc.) go wrong.

misfit *n.* a person not well suited to his or her environment.

misfortune *n.* bad luck, an unfortunate event.

misgiving *n.* a slight feeling of doubt, fear, or mistrust.

misguided *adj.* mistaken in one's opinions or actions.

mishap *n.* an unlucky accident.

misinform *v.* give wrong information to. □ **misinformation** *n.*

misinterpret *v.* interpret incorrectly. □ **misinterpretation** *n.*

misjudge *v.* form a wrong opinion or estimate of.

mislay *v.* (**mislaid, mislaying**) lose temporarily.

mislead *v.* (**misled, misleading**) cause to form a wrong impression.

mismanage *v.* manage badly or wrongly. □ **mismanagement** *n.*

misnomer *n.* a wrongly applied name or description.

misogynist *n.* a person who hates women. □ **misogyny** *n.*

misplace *v.* put in a wrong place; place (confidence etc.) unwisely; mislay.

misprint *n.* an error in printing.

misquote *v.* quote incorrectly. □ **misquotation** *n.*

misread *v.* (**misread, misreading**) read or interpret incorrectly.

misrepresent *v.* represent in a false way. □ **misrepresentation** *n.*

misrule *n.* bad government.

Miss *n.* (*pl.* **Misses**) the title of a girl or unmarried woman.

miss *v.* fail to hit, catch, see, hear, understand, etc.; notice or regret the absence or loss of. ●*n.* a failure to hit or attain what is aimed at.

missel thrush var. of **mistle thrush**.

misshapen *adj.* badly shaped.

missile *n.* an object thrown or fired at a target.

missing *adj.* not present; not in its place, lost.

mission *n.* **1** a task that a person or group is sent to perform; this group. **2** a missionaries' headquarters.

missionary *n.* a person sent to spread religious faith.

misspell *v.* (**misspelt** or **misspelled, misspelling**) spell incorrectly.

mist *n.* water vapour near the ground or clouding a window etc; something resembling this. ●*v.* cover or become covered with mist.

mistake *n.* an incorrect idea or opinion; something done incorrectly. ●*v.* (**mistook, mistaken, mistaking**) misunderstand; choose or identify wrongly.

mistle thrush n. (also **missel thrush**) a large thrush.

mistletoe n. a plant with white berries, growing on trees.

mistral n. a cold north or north-west wind in southern France.

mistress n. a woman who has control of people or things; a female teacher; a man's female lover.

mistrial n. a trial invalidated by an error in procedure etc.

mistrust v. feel no trust in. ● n. lack of trust. □ **mistrustful** adj.

misty adj. (**mistier**) full of mist; indistinct. □ **mistily** adv., **mistiness** n.

misunderstand v. (**misunderstood**, **misunderstanding**) fail to understand correctly; misinterpret. □ **misunderstanding** n.

misuse v. (mis-yooz) use wrongly; treat badly. ● n. (mis-yooss) wrong use.

mite n. a very small spider-like animal; a small creature, esp. a child; a small amount.

mitigate v. make less intense or severe. □ **mitigation** n.

■ Usage Mitigate is often confused with militate, which means 'to have force or effect'.

mitre n. (Amer. **miter**) the pointed headdress of bishops and abbots; a join with tapered ends that form a right angle. ● v. (**mitred**, **mitring**) join in this way.

mitt n. a mitten.

mitten n. a glove with no partitions between the fingers.

mix v. **1** combine (different things); blend; prepare by doing this. **2** be compatible; be sociable. ● n. a mixture. □ **mix up** mix thoroughly; confuse. □ **mixer** n.

mixed adj. composed of various elements; of or for both sexes.

mixed-up adj. (colloquial) muddled; not well adjusted emotionally.

mixture n. something made by mixing; the process of mixing things.

mizzen-mast n. the mast that is next aft of the mainmast.

mizzen-sail n. the lowest sail, set lengthways, on a mizzen-mast.

ml abbr. millilitre(s).

M.Litt. abbr. Master of Letters.

mm abbr. millimetre(s).

Mn symb. manganese.

mnemonic (nim-ON-ik) adj. & n. (a verse etc.) aiding the memory.

Mo symb. molybdenum.

moan n. a low mournful sound; a grumble. ● v. make or utter with a moan. □ **moaner** n.

moat n. a deep wide water-filled ditch round a castle etc.

mob n. a large disorderly crowd; (slang) a gang. ● v. (**mobbed**) crowd round in great numbers.

mobile adj. able to move or be moved easily. ● n. an artistic hanging structure whose parts move in currents of air. □ **mobility** n.

mobilize v. (also **-ise**) assemble (troops etc.) for active service. □ **mobilization** n.

moccasin n. a soft flat-soled leather shoe.

mocha n. a kind of coffee.

mock v. make fun of by imitating; jeer. ● adj. imitation.

mockery n. mocking, ridicule; an absurd or unsatisfactory imitation.

mock-up n. a model for testing or study.

MOD abbr. Ministry of Defence.

mod cons n.pl. (colloquial) modern conveniences.

mode n. the way a thing is done; the current fashion.

model n. **1** a three-dimensional reproduction, usu. on a smaller scale; a pattern. **2** an exemplary person or thing. **3** a person employed to pose for an artist or display clothes by wearing them. ●*adj.* exemplary. ●*v.* (**modelled**; *Amer.* **modeled**) **1** make a model of; shape. **2** work as an artist's or fashion model, display (clothes) in this way.

modem *n.* a device for transmitting computer data via a telephone line.

moderate *adj.* (mod-er-åt) medium; not extreme or excessive. ●*n.* (**mod-er-åt**) a holder of moderate views. ●*v.* (mod-er-ayt) make or become moderate. □ **moderately** *adv.*

moderation *n.* moderating. □ **in moderation** in moderate amounts.

moderator *n.* an arbitrator; a Presbyterian minister presiding over a church assembly.

modern *adj.* of present or recent times; in current style. □ **modernity** *n.*

modernist *n.* a person who favours modern ideas or methods. □ **modernism** *n.*

modernize *v.* (also **-ise**) make modern; adapt to modern ways or needs. □ **modernization** *n.*, **modernizer** *n.*

modest *adj.* not vain or boastful; moderate in size etc., not showy; showing regard for conventional decencies. □ **modestly**, *adv.*, **modesty** *n.*

modicum *n.* a small amount.

modify *v.* make less severe; make partial changes in. □ **modification** *n.*

modish (moh-dish) *adj.* fashionable.

modulate *v.* regulate, moderate; vary in tone or pitch. □ **modulation** *n.*

module *n.* a standardized part or independent unit; a unit of training or education. □ **modular** *adj.*

modus operandi *n.* a method of working.

mogul *n.* (*colloquial*) an important or influential person.

mohair *n.* the fine silky hair of the angora goat; a yarn made from this.

Mohammedan var. of **Muhammadan**.

moiety (moy-åti) *n.* half.

moist *adj.* slightly wet.

moisten *v.* make or become moist.

moisture *n.* water or other liquid diffused through a substance or as vapour or condensed on a surface.

moisturize *v.* (also **-ise**) make (skin) less dry. □ **moisturizer** *n.*

molar *n.* a back tooth with a broad top.

molasses *n.* syrup from raw sugar; (*Amer.*) treacle.

mold etc. Amer. sp. of **mould** etc.

mole *n.* **1** a small burrowing animal with dark fur; a spy established within an organization. **2** a small dark spot on human skin.

molecule *n.* a very small unit (usu. a group of atoms) of a substance. □ **molecular** *adj.*

molehill *n.* a mound of earth thrown up by a mole.

molest *v.* attack or interfere with, esp. sexually. □ **molestation** *n.*

mollify *v.* soothe the anger of.

mollusc *n.* an animal with a soft body and often a hard shell.

mollycoddle *v.* pamper.

Molotov cocktail *n.* an improvised incendiary bomb, a bottle filled with inflammable liquid.

molt Amer. sp. of **moult**.

molten *adj.* liquefied by heat.

molto *adv.* (*Music*) very.

molybdenum *n.* a hard metallic element (symbol Mo) used in steel.

moment *n.* a point or brief portion of time; importance.

momentary *adj.* lasting only a moment. □ **momentarily** *adv.*

momentous *adj.* of great importance.

momentum *n.* impetus gained by movement.

Mon. *abbr.* Monday.

monarch *n.* a ruler with the title of king, queen, emperor, or empress. □ **monarchic** *adj.*, **monarchical** *adj.*

monarchist *n.* a supporter of monarchy. □ **monarchism** *n.*

monarchy *n.* a form of government with a monarch as the supreme ruler; a country governed in this way.

monastery *n.* the residence of a community of monks.

monastic *adj.* of monks or monasteries. □ **monasticism** *n.*

Monday *n.* the day of the week following Sunday.

monetarist *n.* a person who advocates control of the money supply to curb inflation. □ **monetarism** *n.*

monetary *adj.* of money or currency.

money *n.* current coins; coins and banknotes; (*pl.* **moneys**) any form of currency; wealth.

moneyed *adj.* wealthy.

money-spinner *n.* a profitable thing.

Mongol *n.* & *adj.* (a native) of Mongolia.

mongoose *n.* (*pl.* **mongooses**) a stoat-like tropical animal that can attack and kill snakes.

mongrel *n.* an animal (esp. a dog) of mixed breed. ● *adj.* of mixed origin or character.

monitor *n.* a device used to observe or test the operation of something; a school pupil with special duties. ● *v.* keep watch over; record and test or control.

monk *n.* a member of a male religious community.

monkey *n.* (*pl.* **monkeys**) an animal of a group closely related to man; a mischievous person. ● *v.* (**monkeyed, monkeying**) tamper mischievously.

monkey-nut *n.* a peanut.

monkey-puzzle *n.* an evergreen tree with needle-like leaves and intertwining branches.

monkey wrench *n.* a wrench with an adjustable jaw.

monkfish *n.* a large edible sea fish.

mono *adj.* monophonic.

mono- *comb. form* one, alone, single.

monochrome *adj.* done in only one colour or in black and white.

monocle *n.* an eyeglass for one eye only.

monocular *adj.* with or for one eye.

monogamy *n.* the system of being married to only one person at a time. □ **monogamous** *adj.*

monogram *n.* letters (esp. a person's initials) combined in a design. □ **monogrammed** *adj.*

monograph *n.* a scholarly treatise on a single subject.

monolith *n.* a large single upright block of stone; a massive organization etc. □ **monolithic** *adj.*

monologue *n.* a long speech.

monomania *n.* an obsession with one idea or interest. □ **monomaniac** *n.*

monophonic *adj.* using only one transmission channel for reproduction of sound.

monoplane *n.* an aeroplane with only one set of wings.

monopolize *v.* (also **-ise**) have a monopoly of; not allow others to share in. □ **monopolization** *n.*

monopoly *n.* sole possession or control of something, esp. of trade in a specified commodity.

monorail n. a railway in which the track is a single rail.

monosodium glutamate n. a substance added to food to enhance its flavour.

monosyllable n. a word of one syllable. □ **monosyllabic** adj.

monotheism n. the doctrine that there is only one God. □ **monotheist** n., **monotheistic** adj.

monotone n. a level unchanging tone of voice.

monotonous adj. dull because lacking in variety or variation. □ **monotonously** adv., **monotony** n.

monoxide n. an oxide with one atom of oxygen.

Monsieur (mŏ-syer) n. (pl. **Messieurs**) a title or form of address for a French-speaking man, corresponding to Mr or sir.

monsoon n. a seasonal wind in South Asia; the rainy season accompanying this.

monster n. something that is huge or very abnormal in form; a huge ugly or frightening creature; a cruel or wicked person.

monstrosity n. a monstrous thing.

monstrous adj. like a monster, huge; outrageous, absurd.

montage n. the making of a composite picture from pieces of others; this picture; joining of disconnected shots in a cinema film.

month n. each of the twelve portions into which the year is divided; a period of 28 days.

monthly adj. & adv. (produced or occurring) once a month. ● n. a monthly periodical.

monument n. an object commemorating a person or event etc.; a structure of historical importance.

monumental adj. of or serving as a monument; massive; extremely great.

moo n. a cow's low deep cry. ● v. make this sound.

mooch v. (colloquial) walk slowly and aimlessly.

mood n. a temporary state of mind or spirits; a fit of bad temper or depression.

moody adj. (**moodier**) gloomy, sullen; liable to become like this. □ **moodily** adv., **moodiness** n.

mooli n. a large white radish.

moon n. the earth's satellite, made visible by light it reflects from the sun; a natural satellite of any planet. ● v. behave dreamily.

moonlight n. light from the moon. ● v. (colloquial) have two paid jobs, one by day and the other in the evening.

moonlit adj. lit by the moon.

moonscape n. (a landscape resembling) the surface of the moon.

moonshine n. **1** foolish ideas. **2** illicitly distilled alcoholic liquor.

moonstone n. a pearly semi-precious stone.

Moor n. a member of a Muslim people of north-west Africa. □ **Moorish** adj.

moor¹ n. a stretch of open uncultivated land with low shrubs.

moor² v. secure (a boat etc.) to a fixed object by means of cable(s).

moorhen n. a small waterbird.

moorings n.pl. the cables or a place for mooring a boat.

moose n. (pl. **moose**) an elk of North America.

moot point n. a debatable or undecided issue.

mop n. a pad or bundle of yarn on a stick, used for cleaning things; a thick mass of hair. ● v. (**mopped**) clean with a mop.

mope v. be unhappy and listless.

moped n. a low-powered motorcycle.

moquette n. a material with raised loops or cut pile used for upholstery or carpets.

moraine n. a mass of stones etc. deposited by a glacier.

moral adj. concerned with right and wrong conduct; virtuous. ● n. 1 a moral lesson or principle. 2 (**morals**) a person's moral habits, esp. sexual conduct. □ **morally** adv.

morale n. (mŏ-rahl) n. the state of a person's or group's spirits and confidence.

moralist n. a person who expresses or teaches moral principles.

morality n. moral principles or rules; goodness or rightness.

moralize v. (also **-ise**) talk or write about the morality of something.

moral support n. encouragement.

moral victory n. a triumph though without concrete gain.

morass n. a marsh, a bog; a complex entanglement.

moratorium n. (pl. **moratoriums** or **moratoria**) a temporary ban on an activity.

morbid adj. preoccupied with gloomy or unpleasant things; unhealthy. ■ **morbidly** adv., **morbidness** n., **morbidity** n.

mordant adj. (of wit etc.) caustic. ● n. a substance that fixes a dye.

more adj. greater in quantity or intensity etc. ● n. a greater amount or number. ● adv. to a greater extent; again. □ **more or less** approximately.

moreish adj. (also **morish**) (colloquial) so tasty that you want more.

morel n. an edible mushroom.

morello n. a bitter dark cherry.

moreover adv. besides.

mores (mor-ayz) n.pl. customs or conventions.

morgue (morg) n. a mortuary.

moribund adj. in a dying state.

morish var. of **moreish**.

Mormon n. a member of a Christian sect founded in the USA.

morning n. the part of the day before noon or the midday meal.

morning sickness n. nausea felt in the morning in early pregnancy.

morocco n. goatskin leather.

moron n. (colloquial) a stupid person; an adult with a mental age of 8 to 12 years. □ **moronic** adj.

morose adj. gloomy and unsociable, sullen. □ **morosely** adv., **moroseness** n.

morphia n. morphine.

morphine n. a pain-killing drug made from opium.

morphology n. the study of forms of animals and plants or of words. □ **morphological** adj.

morris dance n. a traditional English dance performed in costume decorated with ribbons and bells.

Morse code n. a code of signals using short and long sounds or flashes of light.

morsel n. a small amount; a small piece of food.

mortal adj. subject to death; fatal; deadly. ● n. a mortal being. □ **mortally** adv.

mortality n. being mortal; loss of life on a large scale; the death rate.

mortar n. a mixture of lime or cement with sand and water for joining bricks or stones; a bowl in which substances are pounded with a pestle; a short cannon.

mortarboard n. a stiff square cap worn as part of academic dress.

mortgage (mor-gij) *n.* a loan for the purchase of property, in which the property itself is pledged as security; an agreement effecting this. ● *v.* pledge (property) as security in this way.

mortgagee (mor-gij-ee) *n.* the borrower in a mortgage.

mortgager (mor-gij-ĕr) *n.* (also **mortgagor**) the lender in a mortgage.

mortician *n.* (*Amer.*) an undertaker.

mortify *v.* humiliate greatly; (of flesh) become gangrenous. □ **mortification** *n.*

mortise *n.* (also **mortice**) a hole in one part of a framework shaped to receive the end of another part.

mortise lock *n.* a lock set in (not on) a door.

mortuary *n.* a place where dead bodies are kept temporarily.

mosaic *n.* a pattern or picture made with small pieces of coloured glass or stone.

Moslem *adj.* & *n.* = **Muslim**.

mosque *n.* a Muslim place of worship.

mosquito *n.* (*pl.* **mosquitoes**) a kind of gnat.

moss *n.* a small flowerless plant forming a dense growth in moist places. □ **mossy** *adj.*

most *adj.* greatest in quantity or intensity etc. ● *n.* the greatest amount or number. ● *adv.* to the greatest extent; very. □ **at most** not more than. **for the most part** in most cases; in most of its extent.

mostly *adv.* for the most part.

MOT *abbr.* an annual test of motor vehicles over a certain age.

motel *n.* a roadside hotel for motorists.

motet *n.* a short religious choral work.

moth *n.* an insect like a butterfly but usu. flying at night; a similar insect whose larvae feed on cloth or fur.

mothball *n.* a small ball of a pungent substance for keeping moths away from clothes.

mother *n.* a female parent; the title of the female head of a religious community. ● *v.* look after in a motherly way. □ **motherhood** *n.*

motherboard *n.* a printed circuit board containing the principal components of a microcomputer.

mother-in-law *n.* (*pl.* **mothers-in-law**) the mother of one's wife or husband.

motherland *n.* one's native country.

motherless *adj.* without a living mother.

motherly *adj.* showing a mother's kindness. □ **motherliness** *n.*

mother-of-pearl *n.* a pearly substance lining shells of oysters and mussels etc.

mother tongue *n.* one's native language.

motif *n.* a recurring design, feature, or melody.

motion *n.* **1** moving; movement. **2** a formal proposal put to a meeting for discussion. **3** emptying of the bowels, faeces. ● *v.* make a gesture directing (a person) to do something.

motionless *adj.* not moving.

motion picture *n.* a cinema film.

motivate *v.* give a motive to; stimulate the interest of, inspire. □ **motivation** *n.*

motive *n.* that which induces a person to act in a certain way. ● *adj.* producing movement or action.

mot juste (moh *zh*oost) *n.* (*pl.* **mots justes**) the most appropriate word.

motley adj. multicoloured; assorted. ● n. (old use) a jester's particoloured costume.

motocross n. a motorcycle race over rough ground.

motor n. a machine supplying motive power; a car. ● adj. producing motion; driven by a motor.

motorbike n. (colloquial) a motorcycle.

motorcade n. a procession or parade of motor vehicles.

motorcycle n. a two-wheeled motor-driven road vehicle. □ **motorcyclist** n.

motorist n. a car driver.

motorize v. (also -**ise**) equip with motor(s) or motor vehicles.

motor vehicle n. a vehicle with a motor engine, for use on ordinary roads.

motorway n. a road designed for fast long-distance traffic.

motte n. a mound forming the site of an ancient castle or camp.

mottled n. patterned with irregular patches of colour.

motto n. (pl. **mottoes**) a short sentence or phrase expressing an ideal or rule of conduct; a maxim or joke inside a paper cracker.

mould (Amer. **mold**) n. **1** a hollow container into which a liquid is poured to set in a desired shape; something made in this. **2** a furry growth of tiny fungi on a damp surface. **3** a soft fine earth rich in organic matter. ● v. shape; guide or control the development of.

moulder v. (Amer. **molder**) decay, rot away.

moulding n. (Amer. **molding**) an ornamental strip of plaster or wood.

mouldy adj. (Amer. **moldy**) (**mouldier**) **1** covered with mould; stale. **2** (slang) dull, worthless.

moult v. (Amer. **molt**) shed feathers, hair, or skin before new growth. ● n. this process.

mound n. a pile of earth or stones; a small hill.

mount v. **1** go up; get on or a horse etc.; increase; fix on or in a support or setting. **2** organize, arrange. ● n. **1** a horse for riding; a thing on which something is fixed. **2** a mountain, a hill.

mountain n. a mass of land rising to a great height; a large heap or pile.

mountain ash n. a rowan tree.

mountain bike n. a sturdy bicycle suitable for riding on rough hilly ground.

mountaineer n. a person who climbs mountains. ● v. climb mountains as a recreation.

mountainous adj. full of mountains; huge.

mourn v. feel or express sorrow about (a dead person or lost thing). □ **mourner** n.

mournful adj. sorrowful. □ **mournfully** adv.

mourning n. dark clothes worn as a symbol of bereavement.

mouse n. (pl. **mice**) a small rodent with a long tail; a quiet timid person; a small rolling device for moving the cursor on a VDU screen.

moussaka n. (also **mousaka**) a Greek dish of minced meat and aubergine.

mousse n. a frothy creamy dish; a substance of similar texture.

moustache (mus-**tahsh**) n. (Amer. **mustache**) hair on the upper lip.

mousy adj. (**mousier**) **1** dull greyish brown. **2** quiet and timid.

mouth n. (mowth) the opening in the head through which food is taken in and sounds uttered; the opening of a bag, cave, cannon, etc; a place where a river enters the sea. ● v. (mowth) form (words) soundlessly with the lips.

mouth-organ *n.* a small instrument played by blowing and sucking.

mouthpiece *n.* the part of an instrument placed between or near the lips; a spokesperson.

mouthwash *n.* a liquid for cleansing the mouth.

movable *adj.* able to be moved.

move *v.* **1** change in place, position, or attitude; change one's residence. **2** provoke an emotion in; take action. **3** put to a meeting for discussion. ● *n.* the act of moving; the moving of a piece in chess etc.; a calculated action.

movement *n.* **1** moving; a move; moving parts. **2** a group with a common cause. **3** a section of a long piece of music.

movie *n.* (*Amer. colloquial*) a cinema film.

moving *adj.* arousing pity or sympathy. □ **movingly** *adv.*

mow *v.* (**mowed, mown, mowing**) cut down (grass or grain etc.); cut grass etc. from. □ **mow down** kill or destroy by a moving force. □ **mower** *n.*

mozzarella *n.* a soft Italian cheese.

MP *abbr.* Member of Parliament.

m.p.g. *abbr.* miles per gallon.

m.p.h. *abbr.* miles per hour.

Mr *n.* (*pl.* **Messrs**) the title prefixed to a man's name.

Mrs *n.* (*pl.* **Mrs**) the title prefixed to a married woman's name.

MS *abbr.* multiple sclerosis; (*pl.* **MSS**) manuscript.

Ms *n.* the title prefixed to a married or unmarried woman's name.

M.Sc. *abbr.* Master of Science.

MS-DOS (*trade mark*) Microsoft disk operating system.

Mt. *abbr.* Mount.

much *adj.* & *n.* (existing in) great quantity. ● *adv.* in a great degree; to a great extent.

mucilage *n.* a sticky substance obtained from plants; an adhesive gum.

muck *n.* farmyard manure; (*colloquial*) dirt, mess. □ **mucky** *adj.*

muckraking *n.* seeking and exposing scandal.

mucous *adj.* like or covered with mucus.

mucus *n.* a slimy substance coating the inner surface of hollow organs of the body.

mud *n.* wet soft earth. □ **muddy** *adj.*

muddle *v.* confuse, mix up; progress in a haphazard way. ● *n.* a muddled state.

mudguard *n.* a curved cover above a wheel as a protection against spray.

muesli *n.* food of mixed crushed cereals, dried fruit, nuts, etc.

muezzin (moo-ez-in) *n.* a man who proclaims the hours of prayer for Muslims.

muff *n.* a tube-shaped furry covering for the hands. ● *v.* (*colloquial*) bungle.

muffin *n.* a light round yeast cake eaten toasted and buttered; a cup-shaped cake.

muffle *v.* wrap for warmth or protection, or to deaden sound; make less loud or less distinct.

muffler *n.* a scarf.

mufti *n.* plain clothes worn by one who usually wears a uniform.

mug *n.* **1** a large drinking vessel with a handle, for use without a saucer. **2** (*slang*) the face; a person who is easily outwitted. ● *v.* (**mugged**) rob (a person) with violence, esp. in a public place. □ **mugger** *n.*

muggy *adj.* (**muggier**) (of weather) oppressively damp and warm. □ **mugginess** *n.*

Muhammadan n. & adj. (also **Mohammedan**) Muslim.

■ **Usage** The term *Muhammadan* is not used by Muslims and is often regarded as offensive.

mulberry n. a purple or white fruit resembling a blackberry; the tree bearing this; dull purplish red.

mulch n. a mixture of wet straw, leaves, etc., spread on ground to protect plants or retain moisture. ● v. cover with mulch.

mule n. **1** an animal that is the offspring of a horse and a donkey. **2** a backless slipper.

mull v. heat (wine etc.) with sugar and spices, as a drink. □ **mull over** think over.

mullah n. a Muslim learned in Islamic law.

mullet n. a small edible sea fish.

mulligatawny n. a curry-flavoured soup.

mullion n. an upright bar between the sections of a window.

multi- comb. form many.

multicultural adj. of or involving several cultural or ethnic groups.

multifarious adj. very varied. □ **multifariously** adv.

multilateral adj. involving three or more parties.

multinational adj. & n. (a business company) operating in several countries.

multiple adj. having or affecting many parts. ● n. a quantity containing another a number of times without remainder.

multiplex adj. having many elements.

multiplicand n. a quantity to be multiplied by another.

multiplication n. multiplying.

multiplicity n. a great variety.

multiplier n. the number by which a quantity is multiplied.

multiply v. add a quantity to itself a specified number of times; increase in number.

multiracial adj. of or involving people of several races.

multitasking n. (*Computing*) the performance of different tasks simultaneously.

multitude n. a great number of things or people.

multitudinous adj. very numerous.

mum (*colloquial*) n. mother. ● adj. silent.

mumble v. speak indistinctly. ● n. indistinct speech.

mumbo-jumbo n. meaningless ritual; deliberately obscure language.

mummify v. preserve (a corpse) by embalming as in ancient Egypt. □ **mummification** n.

mummy n. **1** (*colloquial*) mother. **2** a corpse embalmed and wrapped for burial, esp. in ancient Egypt.

mumps n. a virus disease with painful swellings in the neck.

munch v. chew vigorously.

mundane adj. dull, routine; worldly.

mung bean n. a small dried pea-like bean that can be cooked or sprouted.

municipal adj. of a town or city.

municipality n. a self-governing town or district.

munificent adj. splendidly generous. □ **munificence** adv., **munificence** n.

munitions n.pl. weapons, ammunition, etc.

muntjac n. (also **muntjak**) a small deer.

muon n. an unstable elementary particle.

mural adj. of or on a wall. ● n. a painting on a wall.

murder n. intentional unlawful killing. ● v. kill intentionally and unlawfully. □ **murderer** n.

murderous *adj.* involving or capable of murder.

murk *n.* darkness, gloom. □ **murky** *adj.*

murmur *n.* a low continuous sound; softly spoken words. ● *v.* make a murmur; speak or utter softly.

murrain *n.* an infectious disease of cattle.

muscle *n.* a strip of fibrous tissue able to contract and move a part of the body; muscular power; strength. □ **muscle in** (*colloquial*) force oneself on others.

Muscovite *n.* a person from Moscow.

muscular *adj.* of muscles; having well-developed muscles. □ **muscularity** *n.*

muscular dystrophy *n.* a condition causing progressive wasting of the muscles.

muse *v.* ponder. ● *n.* a poet's source of inspiration.

museum *n.* a place where objects of historical interest are collected and displayed.

mush *n.* a soft pulp.

mushroom *n.* an edible fungus with a stem and a domed cap. ● *v.* spring up in large numbers; rise and spread in a mushroom shape.

mushy *adj.* (**mushier**) as or like mush; feebly sentimental. □ **mushiness** *n.*

music *n.* an arrangement of sounds of one or more voices or instruments; the written form of this.

musical *adj.* of or involving music; fond of or skilled in music; sweet-sounding. ● *n.* a play with songs and dancing. □ **musically** *adv.*

musician *n.* a person skilled in music.

musicology *n.* the study of the history and forms of music. □ **musicologist** *n.*

musk *n.* a substance secreted by certain animals or produced synthetically, used in perfumes.

musket *n.* a long-barrelled gun formerly used by infantry.

Muslim (also **Moslem**) *adj.* of or believing in Muhammad's teaching. ● *n.* a believer in this faith.

muslin *n.* a thin cotton cloth.

musquash *n.* a rat-like North American water animal; its fur.

mussel *n.* a bivalve mollusc.

must *v.aux.* used to express necessity or obligation, certainty, or insistence. ● *n.* (*colloquial*) something that must be done or visited etc.

■ **Usage** The negative *I must not go* means 'I am not allowed to go'. To express a lack of obligation, use *I am not obliged to go*, *I need not go*, or *I haven't got to go*.

mustache Amer. sp. of **moustache**.

mustang *n.* a wild horse of Mexico and California.

mustard *n.* a sharp-tasting yellow condiment made from the seeds of a plant; this plant.

muster *v.* gather together, assemble; summon (energy, strength).

musty *adj.* (**mustier**) smelling mouldy, stale. □ **mustiness** *n.*

mutable *adj.* liable to change, fickle. □ **mutability** *n.*

mutagen *n.* something causing genetic mutation.

mutant *adj.* & *n.* (a living thing) differing from its parents as a result of genetic change.

mutate *v.* change in form.

mutation *n.* a change in form; a mutant.

mute *adj.* silent; dumb. ● *n.* a dumb person; a device muffling the sound of a musical instrument. ● *v.* deaden or muffle the

sound of. □ **mutely** *adv.*,
muteness *n.*

mutilate *v.* injure or disfigure by
cutting off a part. □ **mutilation**
n.

mutineer *n.* a person who mutinies.

mutinous *adj.* rebellious, ready
to mutiny. ● **mutinously** *adv.*

mutiny *n.* a rebellion against
authority, esp. by members of
the armed forces. ● *v.* engage in
mutiny.

mutter *v.* speak or utter in a low
unclear tone; utter subdued
grumbles. ● *n.* muttering.

mutton *n.* the flesh of sheep as
food.

mutual *adj.* felt or done by each
to the other; (*colloquial*) common
to two or more people. □ **mutually** *adv.*

■ **Usage** The use of *mutual* to
mean 'common to two or more
people' is considered incorrect
by some people, who use *common* instead.

Muzak *n.* (*trade mark*) piped
music in public places.

muzzle *n.* the projecting nose
and jaws of certain animals; the
open end of a firearm; a strap
etc. over an animal's head to
prevent it from biting or feeding. ● *v.* put a muzzle on; prevent from expressing opinions
freely.

muzzy *adj.* (**muzzier**) dazed,
feeling stupefied. □ **muzziness**
n.

MW *abbr.* medium wave;
megawatt(s).

my *adj.* belonging to me.

myalgia *n.* muscle pain.

mycelium *n.* microscopic
threadlike parts of a fungus.

mycology *n.* the study of fungi.

myelin *n.* a substance forming a
protective sheath around nerve-
fibres.

mynah *n.* (also **myna, mina**) a
bird of the starling family that
can mimic sounds.

myopia (my-oh-piă) *n.* short-
sightedness. □ **myopic** (my-op-
ik) *adj.*

myriad *n.* a vast number.

myrrh (mer) *n.* a gum resin used
in perfumes, medicines, and
incense.

myrtle *n.* an evergreen shrub.

myself *pron.* an emphatic and
reflexive form of *I* and *me*.

mysterious *adj.* full of mystery,
puzzling. □ **mysteriously** *adv.*

mystery *n.* a matter that remains unexplained; the quality
of being unexplained or obscure;
a story dealing with a puzzling
crime.

mystic *adj.* having a hidden or
symbolic meaning, esp. in religion; inspiring a sense of mystery and awe. ● *n.* a person who
seeks to obtain union with God
by spiritual contemplation.
□ **mystical** *adj.*, **mystically**
adv., **mysticism** *n.*

mystify *v.* cause to feel puzzled.
□ **mystification** *n.*

mystique *n.* an aura of mystery
or mystical power.

myth *n.* a traditional tale containing beliefs about ancient
times or natural events; an
imaginary person or thing.
□ **mythical** *adj.*

mythology *n.* myths; the study
of myths. □ **mythological** *adj.*

myxomatosis *n.* a fatal virus
disease of rabbits.

Nn

N *abbr.* **1** north, northern. **2** newton(s). **3** (*Chess*) Knight. ● *symb.* nitrogen.

Na *symb.* sodium.

NAAFI *abbr.* Navy, Army, and Air Force Institutes; a canteen for servicemen.

naan var. of **nan** (sense 2).

nab *v.* (**nabbed**) (*slang*) catch (a wrongdoer) in the act, arrest; seize.

nadir *n.* the lowest point.

naevus (nee-vŭs) *n.* (*Amer.* **nevus**) (*pl.* **naevi**) a red birthmark.

naff *adj.* (*slang*) unfashionable; rubbishy.

nag *v.* (**nagged**) scold continually; (of pain) be felt persistently. ● *n.* (*colloquial*) a horse.

nail *n.* **1** a layer of horny substance over the outer tip of a finger or toe. **2** a small metal spike. ● *v.* **1** fasten with nail(s). **2** catch, arrest.

naive (nah-eev) *adj.* showing lack of experience or judgement. □ **naively** *adv.*, **naivety** *n.*

naked *adj.* without clothes on; without coverings. □ **nakedness** *n.*

naked eye *n.* the eye unassisted by a telescope or microscope etc.

namby-pamby *adj.* feeble or unmanly.

name *n.* the word(s) by which a person, place, or thing is known or indicated; a reputation. ● *v.* give as a name; nominate, specify.

namely *adv.* that is to say.

namesake *n.* a person or thing with the same name as another.

nan *n.* **1** (*colloquial*) a grandmother. **2** (also **naan**) a flat leavened Indian bread.

nanny *n.* a child's nurse.

nanny goat *n.* a female goat.

nano- *comb. form* one thousand millionth.

nap *n.* **1** a short sleep, esp. during the day. **2** short raised fibres on the surface of cloth or leather. ● *v.* (**napped**) have a short sleep.

napalm (nay-pahm) *n.* a jellylike petrol substance used in incendiary bombs.

nape *n.* the back part of the neck.

naphtha (naf-thǎ) *n.* an inflammable oil.

naphthalene *n.* a strongsmelling white substance obtained from coal tar, used as a moth-repellent.

napkin *n.* **1** a piece of cloth or paper used to protect clothes or for wiping one's lips at meals. **2** a nappy.

nappy *n.* a piece of absorbent material worn by a baby to absorb or retain urine and faeces.

narcissism *n.* abnormal selfadmiration. □ **narcissistic** *adj.*

narcissus *n.* (*pl.* **narcissi**) a flower of the group including the daffodil.

narcosis *n.* a state of drowsiness.

narcotic *adj.* causing sleep or drowsiness. ● *n.* a narcotic drug.

narrate *v.* tell (a story), give an account of. □ **narration** *n.*, **narrator** *n.*

narrative *n.* a spoken or written account of something. ● *adj.* in this form.

narrow *adj.* **1** small across, not wide. **2** with little margin or scope. ● *v.* make or become narrower. □ **narrowly** *adv.*, **narrowness** *n.*

narrow-minded *adj.* intolerant.

narwhal *n.* an Arctic whale with a spirally grooved tusk.

nasal adj. of the nose; sounding as if breath came out through the nose. □ **nasally** adv.

nascent adj. just coming into existence. □ **nascence** n.

nasturtium (nă-ster-shŭm) n. a plant with bright orange or yellow flowers and rounded leaves.

nasty adj. unpleasant; unkind; difficult. □ **nastily** adv., **nastiness** n.

natal adj. of or from one's birth.

nation n. people of mainly common descent and history usu. inhabiting a particular country under one government.

national adj. of a nation; common to a whole nation. ● n. a citizen of a particular country. □ **nationally** adv.

national curriculum n. a common programme of study for school pupils in England and Wales.

nationalism n. patriotic feeling; a policy of national independence. □ **nationalist** n., **nationalistic** adj.

nationality n. the condition of belonging to a particular nation.

nationalize v. (also **-ise**) convert from private to State ownership. □ **nationalization** n.

native adj. natural, inborn; belonging to a place by birth. ● n. a person born in a specified place; a local inhabitant.

nativity n. birth; **(the Nativity)** that of Christ.

NATO abbr. North Atlantic Treaty Organization; a defence association of European and North American States.

natter v. & n. (colloquial) chat.

natural adj. **1** of or produced by nature. **2** normal; not seeming artificial or affected. ● n. **1** a person or thing that seems naturally suited for something. **2** (Music) (a sign indicating) a note that is not a sharp or flat. □ **naturalness** n.

natural history n. the study of animal and plant life.

naturalism n. realism in art and literature. □ **naturalistic** adj.

naturalist n. an expert in natural history.

naturalize v. (also **-ise**) admit (a person of foreign birth) to full citizenship of a country; introduce and acclimatize (an animal or plant) into another region. □ **naturalization** n.

naturally adv. **1** in a natural manner. **2** as might be expected, of course.

natural science n. a science studying the natural or physical world.

nature n. **1** the world with all its features and living things; the physical power producing these. **2** a kind, sort, or class; innate characteristics; all that makes a thing what it is.

naturist n. a nudist. □ **naturism** n.

naught n. (literary) nothing, = **nought**.

naughty adj. (**naughtier**) **1** behaving badly, disobedient. **2** slightly indecent. □ **naughtily** adv., **naughtiness** n.

nausea n. a feeling of sickness; revulsion.

nauseate v. affect with nausea.

nauseous adj. causing nausea.

nautical adj. of sailors or seamanship.

nautical mile n. a unit of 1,852 metres (approx. 2,025 yds).

nautilus n. (pl. **nautiluses** or **nautili**) a mollusc with a spiral shell.

naval adj. of a navy.

nave n. the main part of a church.

navel n. the small hollow in the centre of the abdomen.

navigable adj. (of a river) suitable for boats to sail in; (of a boat) able to be steered and sailed.

navigate v. sail in or through (a sea, river, etc.); direct the course of (a ship, vehicle, etc.). □ **navigation** n., **navigator** n.

navvy (na-vee) n. a labourer making roads etc. where digging is necessary.

navy (nay-vee) n. **1** a country's warships; the people serving in a naval force. **2** (in full **navy blue**) very dark blue.

NB abbr. (Latin nota bene) note well.

Nb symb. niobium.

Nd symb. neodymium.

NE abbr. north-east; north-eastern.

Ne symb. neon.

Neapolitan adj. & n. (a native or inhabitant) of Naples.

neap tide n. the tide when there is least rise and fall of water.

near adv. at, to, or within a short distance or interval; nearly. ● prep. near to. ● adj. **1** with only a short distance or interval between; closely related; with little margin. **2** of the side of a vehicle that is normally nearest the kerb. ● v. draw near. □ **nearness** n.

nearby adj. & adv. near in position.

nearly adv. closely; almost.

neat adj. **1** clean and orderly in appearance or workmanship. **2** undiluted. □ **neatly** adv., **neatness** n.

neaten v. make neat.

nebula n. (pl. **nebulae**) a cloud of gas or dust in space. □ **nebular** adj.

nebulous adj. indistinct, having no definite form.

necessarily adv. as a necessary result, inevitably.

necessary adj. essential in order to achieve something; happening or existing by necessity.

necessitate v. make necessary; involve as a condition or result.

necessitous adj. poor, needy.

necessity n. the state or fact of being necessary; a necessary thing; the compelling power of circumstances; a state of need or hardship.

neck n. the narrow part connecting the head to the body; the part of a garment round this; the narrow part of a bottle, cavity, etc. **neck and neck** running level in a race.

necklace n. a piece of jewellery worn round the neck.

neckline n. the outline formed by the edge of a garment at the neck.

necromancy n. the art of predicting things by communicating with the dead. □ **necromancer** n.

necrosis n. death of bone or tissue. □ **necrotic** adj.

nectar n. a sweet fluid from plants, collected by bees; any delicious drink.

nectarine n. a kind of peach with a smooth shiny skin.

née (nay) adj. born (used in stating a married woman's maiden name).

need n. **1** a requirement. **2** a state of great difficulty or misfortune; poverty. ● v. be in need of, require; be obliged.

needful adj. necessary.

needle n. **1** a small thin pointed piece of steel used in sewing. **2** something shaped like this; a pointer of a compass or gauge; the end of a hypodermic syringe. ● v. annoy, provoke.

needlecord n. fine-ribbed corduroy fabric.

needless adj. unnecessary. □ **needlessly** adv.

needlework n. sewing or embroidery.

needy adj. (**needier**) very poor.

nefarious adj. wicked. □ **nefariously** adv.

negate v. nullify, disprove, make invalid. □ **negation** n.

negative adj. **1** expressing denial, refusal, or prohibition; not positive. **2** (of a quantity) less than zero. **3** (of a battery terminal) through which electric current leaves. ● n. **1** a negative statement or word; a negative quality or quantity. **2** a photograph with lights and shades or colours reversed, from which positive pictures can be obtained. □ **negatively** adv.

negative equity n. a situation in which the value of property falls below the debt outstanding on it.

negative pole n. the south-seeking pole of a magnet.

neglect v. pay insufficient attention to; fail to take proper care of or to do something. ● n. neglecting, being neglected. □ **neglectful** adj.

negligee (neg-li-zhay) n. a woman's light flimsy dressing gown.

negligence n. lack of proper care or attention. □ **negligent** adj., **negligently** adv.

negligible adj. too small to be worth taking into account.

negotiate v. **1** hold a discussion so as to reach agreement; arrange by such discussion. **2** get past (an obstacle) successfully. □ **negotiation** n., **negotiator** n.

Negress n. a female Negro.

■ **Usage** The term *Negress* is often considered offensive; *black* is usually preferred.

Negro n. (pl. **Negroes**) a member of the black-skinned race that originated in Africa. □ **Negroid** adj.

■ **Usage** The term *Negro* is often considered offensive; *black* is usually preferred.

neigh n. a horse's long high-pitched cry. ● v. make this cry.

neighbour n. (Amer. **neighbor**) a person living next to another; a thing situated near another.

neighbourhood n. (Amer. **neighborhood**) a district.

neighbourhood watch n. systematic vigilance by residents to deter crime in their area.

neighbouring adj. (Amer. **neighboring**) living or situated nearby.

neighbourly adj. (Amer. **neighborly**) kind and friendly towards neighbours. □ **neighbourliness** n.

neither adj. & pron. not either. ● adv. & conj. not either; also not.

nem. con. abbr. unanimously, with nobody disagreeing.

nemesis (nem-i-sis) n. the infliction of deserved and unavoidable punishment.

neo- pref. new.

neoclassical adj. of a style of art, music, etc. influenced by classical style.

neodymium n. a metallic element (symbol Nd).

neolithic adj. of the later part of the Stone Age.

neologism n. a new word.

neon n. a chemical element (symbol Ne), a gas used in illuminated signs.

neonatal adj. of the newly born.

neophyte n. a new convert; a novice.

nephew n. one's brother's or sister's son.

nephritis n. inflammation of the kidneys.

nepotism n. favouritism shown to relatives or friends in appointing them to jobs.

neptunium n. a radioactive metallic element (symbol Np).

nerd n. (also **nurd**) (slang) a foolish or uninteresting person.

nerve n. **1** a fibre carrying impulses of sensation or movement between the brain and a part of the body. **2** courage; (*colloquial*) impudence. **3** (**nerves**) nervousness, the effect of mental stress. ● *v.* give courage to.

nervous *adj.* **1** of the nerves. **2** easily alarmed; slightly afraid. □ **nervously** *adv.*, **nervousness** *n.*

nervy *adj.* (**nervier**) nervous. □ **nerviness** *n.*

nest *n.* **1** a structure or place in which a bird lays eggs and shelters its young; a breeding place, a lair; a snug place. **2** a set of articles (esp. tables) designed to fit inside each other. ● *v.* make or have a nest.

nest egg *n.* a sum of money saved for future use.

nestle *v.* settle oneself comfortably; lie sheltered.

nestling *n.* a bird too young to leave the nest.

net[1] *n.* openwork material of thread, cord, or wire etc.; a piece of this used for a particular purpose. ● *v.* (**netted**) place nets in or on; catch in a net.

net[2] *adj.* (also **nett**) remaining after all deductions; (of weight) not including wrappings etc. ● *v.* (**netted**) obtain or yield as net profit.

netball *n.* a team game in which a ball has to be thrown into a high net.

nether *adj.* lower.

netting *n.* netted fabric.

nettle *n.* a wild plant with leaves that sting when touched. ● *v.* irritate, provoke.

network *n.* an arrangement of intersecting lines; a complex system; a group of interconnected people or broadcasting stations, computers, etc.

networking *n.* an interchange of ideas between different companies, used as a business strategy.

neural *adj.* of nerves.

neuralgia *n.* a sharp pain along a nerve. □ **neuralgic** *adj.*

neuritis *n.* inflammation of a nerve.

neurology *n.* the study of nerve systems. □ **neurological** *adj.*, **neurologist** *n.*

neurosis *n.* (*pl.* **neuroses**) a mental disorder producing depression or abnormal behaviour.

neurotic *adj.* of or caused by a neurosis; subject to abnormal anxieties or obsessive behaviour. ● *n.* a neurotic person. □ **neurotically** *adv.*

neuter *adj.* neither masculine nor feminine; without male or female parts. ● *n.* a neuter word or being; a castrated animal. ● *v.* castrate.

neutral *adj.* **1** not supporting either side in a conflict. **2** without distinctive or positive characteristics. ● *n.* **1** a neutral person, country, or colour. **2** (also **neutral gear**) a position of a gear mechanism in which the engine is disconnected from driven parts. □ **neutrally** *adv.*, **neutrality** *n.*

neutralize *v.* (also **-ise**) make ineffective. □ **neutralization** *n.*

neutrino *n.* (*pl.* **neutrinos**) an elementary particle with zero electric charge and probably zero mass.

neutron *n.* an elementary particle with no electric charge.

neutron bomb *n.* a nuclear bomb that kills people but does little damage to buildings etc.

never *adv.* **1** at no time, on no occasion. **2** not; (*colloquial*) surely not. □ **never mind** do not worry.

nevermore *adv.* at no future time.

nevertheless *adv.* in spite of this.

nevus Amer. sp. of **naevus**.

new *adj.* not existing before, recently made, discovered, experienced, etc.; unfamiliar, unaccustomed. ●adv. newly, recently.

New Age n. a set of beliefs replacing Western culture with alternative approaches to religion, medicine, the environment, etc.

newcomer n. a person who has arrived recently.

newel n. the top or bottom post of the handrail of a stair; the central pillar of a winding stair.

newfangled adj. objectionably new in method or style.

newly adv. recently, freshly.

newly-wed n. a recently married person.

new man n. a man who supports women's liberation, shares household chores, etc.

new moon n. the moon seen as a crescent.

news n. new or interesting information about recent events; a broadcast report of this.

newsagent n. a shopkeeper who sells newspapers.

newscaster n. a newsreader.

newsflash n. an item of important news, broadcast as an interruption to another programme.

newsletter n. an informal printed report containing items of interest to members of a club etc.

newspaper n. a printed daily or weekly publication giving news reports; the paper forming this.

newsprint n. the type of paper on which newspapers are printed.

newsreader n. a person who reads broadcast news reports.

newsworthy adj. worth reporting as news. □ **newsworthiness** n.

newt n. a small lizard-like amphibious creature.

New Testament see **testament**.

newton n. a unit of force.

new year n. the first days of January.

New Year's Day 1 Jan.

next adj. nearest in position or time etc.; soonest come to. ●adv. in the next place or degree; on the next occasion. ●n. the next person or thing.

next door adj. in the next house or room.

next of kin n. one's closest relative(s).

nexus n. (pl. **nexus**) a connected group or series.

NHS abbr. National Health Service.

NI abbr. Northern Ireland; National Insurance.

Ni symb. nickel.

niacin n. = **nicotinic acid.**

nib n. the metal point of a pen.

nibble v. take small quick or gentle bites (at). ●n. a small quick bite; a snack. □ **nibbler** n.

nicad adj. nickel and cadmium.

Nicam n. (trade mark) a digital system used in Britain to produce high quality stereo television sound.

nice adj. 1 pleasant, satisfactory. 2 precise; fastidious. □ **nicely** adv., **niceness** n.

nicety n. precision; detail. □ **to a nicety** exactly.

niche (neesh) n. 1 a shallow recess esp. in a wall. 2 a suitable position in life or employment.

nick n. a small cut or notch. ●v. 1 make a nick in. 2 (slang) steal; arrest. □ **in good nick** (colloquial) in good condition. **in the nick of time** only just in time.

nickel n. 1 a silver-white metallic element (symbol Ni) used in alloys. 2 (Amer.) a 5-cent piece.

nickelodeon n. (Amer.) a jukebox.

nickname n. a familiar or humorous name given to a person or thing instead of the real name. ● v. give as a nickname.

nicotine n. a poisonous substance found in tobacco.

nicotinic acid n. a vitamin of the B complex.

niece n. one's brother's or sister's daughter.

niggardly adj. stingy, mean. □ **niggard** n.

nigger n. (offensive) a black person.

niggle v. fuss over details.

nigh adv. & prep. near.

night n. the dark hours between sunset and sunrise; nightfall.

nightcap n. an alcoholic or hot drink taken at bedtime.

nightclub n. a club open at night, providing refreshment and entertainment.

nightdress n. a woman's or child's loose garment worn in bed.

nightfall n. the onset of night.

nightgown n. a nightdress.

nightie n. (colloquial) a nightdress.

nightingale n. a small thrush, the male of which sings melodiously.

nightjar n. a night-flying bird with a harsh cry.

nightlife n. entertainment available in public places at night.

nightly adj. & adv. happening at night or every night.

nightmare n. an unpleasant dream or experience. □ **nightmarish** adj.

night school n. instruction or education provided in the evening.

nightshade n. a plant with poisonous berries.

nightshirt n. a long shirt worn in bed.

nightspot n. a nightclub.

nihilism n. rejection of all religious and moral principles. □ **nihilist** n., **nihilistic** adj.

nil n. nothing.

nimble adj. able to move quickly. □ **nimbly** adv.

nimbus n. (pl. **nimbi** or **nimbuses**) **1** a halo. **2** a raincloud.

Nimby n. & adj. (a person) objecting to development in one's own area (from the phrase not in my back yard).

nincompoop n. a foolish person.

nine adj. & n. one more than eight (9, IX). □ **ninth** adj. & n.

ninepins n. a game of skittles played with nine objects.

nineteen adj. & n. one more than eighteen (19, XIX). □ **nineteenth** adj. & n.

ninety adj. & n. nine times ten (90, XC). □ **ninetieth** adj. & n.

ninny n. a foolish person.

niobium n. a metallic element (symbol Nb).

nip v. (**nipped**) **1** pinch or squeeze sharply; bite quickly with the front teeth. **2** (slang) go quickly. ● n. **1** a sharp pinch, squeeze, or bite. **2** a sharp coldness. **3** a small drink of spirits.

nipple n. the small projection at the centre of a breast; a similar protuberance; the teat of a feeding bottle.

nippy adj. (**nippier**) (colloquial) **1** nimble, quick. **2** (of the weather) bitingly cold.

nirvana n. (in Buddhism and Hinduism) a state of perfect bliss achieved by the soul.

Nissen hut n. a tunnel-shaped hut of corrugated iron.

nit n. an egg of a louse or similar parasite.

nit-picking n. & adj. (colloquial) petty fault-finding.

nitrate n. a substance formed from nitric acid, esp. used as a fertilizer.

nitric acid n. a corrosive acid containing nitrogen.

nitrogen n. a chemical element (symbol N), a gas forming about four-fifths of the atmosphere.

nitroglycerine n. (also **nitroglycerin**) a powerful explosive.

nitrous oxide n. a gas used as an anaesthetic.

nitty-gritty n. (slang) the basic facts or realities of a matter.

nitwit n. (colloquial) a stupid or foolish person.

No symb. nobelium.

No., no. abbr. number.

no adj. not any; not a. ● adv. (used as a denial or refusal of something); not at all. ● n. (pl. **noes**) a negative reply, a vote against a proposal.

nobble v. (slang) get hold of; tamper with or influence dishonestly.

nobelium n. a radioactive metallic element (symbol No).

nobility n. nobleness of character or of rank; titled people.

noble adj. aristocratic; possessing excellent qualities, esp. of character; generous, not petty; imposing. ● n. a member of the nobility. □ **nobly** adv., **nobleness** n.

nobleman n. (pl. **-men**) a male member of the nobility.

noblesse oblige (noh-bless o-bleezh) privilege entails responsibility.

noblewoman n. (pl. **-women**) a female member of the nobility.

nobody pron. no person. ● n. a person of no importance.

nocturnal adj. of, happening in, or active in the night. □ **nocturnally** adv.

nocturne n. a short romantic piece of music.

nod v. (**nodded**) **1** move the head down and up quickly; indicate (agreement or casual greeting) in this way; let the head droop, be drowsy. **2** (of flowers etc.) bend and sway. ● n.

a nodding movement esp. in agreement or greeting.

node n. a knob-like swelling; a point on a stem where a leaf or bud grows out. □ **nodal** adj.

nodule n. a small rounded lump, a small node. □ **nodular** adj.

no-fly zone n. an area in which specified aircraft are forbidden to fly.

noggin n. a small measure of alcohol, usually ¼ pint.

no-go area n. an area to which entry is forbidden or restricted.

noise n. a sound, esp. a loud, harsh, or undesired one. □ **noiseless** adj.

noisome adj. harmful; evil-smelling; disgusting.

noisy adj. (**noisier**) making much noise. □ **noisily** adv., **noisiness** n.

nomad n. a member of a tribe that roams seeking pasture for its animals; a wanderer. □ **nomadic** adj.

no man's land n. an area not controlled by anyone, esp. between opposing armies.

nom de plume n. (pl. **noms de plume**) a writer's pseudonym.

nomenclature n. a system of names, e.g. in a science.

nominal adj. **1** in name only. **2** (of a fee) very small. □ **nominally** adv.

nominal value n. the face value of a coin etc.

nominate v. name as candidate for or the future holder of an office; appoint as a place or date. □ **nomination** n., **nominator** n.

nominative n. the grammatical case expressing the subject of a verb.

nominee n. a person nominated.

non- pref. not.

nonagenarian n. a person in his or her nineties.

non-aligned adj. not in alliance with another power.

nonchalant *adj.* calm and casual. □ **nonchalantly** *adv.*, **nonchalance** *n.*

noncommittal *adj.* not revealing one's opinion.

nonconformist *n.* a person not conforming to established practices; (**Nonconformist**) a member of a Protestant sect not conforming to Anglican practices.

non-contributory *adj.* not involving payment of contributions.

nondescript *adj.* lacking distinctive characteristics.

none *pron.* not any; no person. □ **none too** not in the least.

■ **Usage** The verb following *none* can be singular or plural when it means 'not any of several', e.g. *None of us knows* or *None of us know.*

nonentity *n.* a person of no importance.

non-event *n.* an event that was expected to be important but proves disappointing.

non-existent *adj.* not existing. □ **non-existence** *n.*

nonplussed *adj.* completely puzzled.

nonsense *n.* words put together in a way that does not make sense; foolish talk or behaviour. □ **nonsensical** *adj.*

non sequitur *n.* a conclusion that does not follow from the evidence given.

non-starter *n.* **1** a horse entered for a race but not running in it. **2** a person or idea etc. not worth considering for a purpose.

non-stop *adj.* & *adv.* not ceasing; (of a train etc.) not stopping at intermediate places.

noodles *n.pl.* pasta in narrow strips.

nook *n.* a secluded place; a recess.

noon *n.* twelve o'clock in the day.

no one *n.* no person, nobody.

noose *n.* a loop of rope etc. with a knot that tightens when pulled.

nor *conj.* and not.

norm *n.* a standard type; usual behaviour.

normal *adj.* conforming to what is standard or usual; free from mental or emotional disorders. □ **normally** *adv.*, **normality** *n.*

north *n.* the point or direction to the left of a person facing east; a northern part. ● *adj.* in the north; (of wind) from the north. ● *adv.* towards the north.

north-east *n.* the point or direction midway between north and east. □ **north-easterly** *adj.* & *n.*, **north-eastern** *adj.*

northerly *adj.* towards or blowing from the north.

northern *adj.* of or in the north.

northerner *n.* a native of the north.

northernmost *adj.* furthest north.

northward *adj.* towards the north. □ **northwards** *adv.*

north-west *n.* the point or direction midway between north and west. □ **north-westerly** *adj.* & *n.*, **north-western** *adj.*

Norwegian *adj.* & *n.* (a native, the language) of Norway.

Nos. *abbr.* numbers.

nose *n.* the organ at the front of the head, used in breathing and smelling; the sense of smell; the front end or projecting part. ● *v.* detect or search by use of the sense of smell; push one's nose against or into; push one's way cautiously ahead.

nosebag *n.* a bag of fodder for hanging on a horse's head.

nosedive *n.* a steep downward plunge, esp. of an aeroplane. ● *v.* make this plunge.

nosh (*slang*) *n.* food. ● *v.* eat.

nostalgia n. sentimental memory of or longing for things of the past. □ **nostalgic** adj., **nostalgically** adv.

nostril n. either of the two external openings in the nose.

nostrum n. a quack remedy; a pet scheme.

nosy adj. (**nosier**) (colloquial) inquisitive, prying. □ **nosily** adv., **nosiness** n.

not adv. expressing a negative, denial, or refusal.

notable adj. worthy of notice; remarkable, eminent. ● n. an eminent person. □ **notably** adv.

notary n. (in full **notary public**) an official authorized to witness the signing of documents, perform formal transactions, etc.

notation n. a system of signs or symbols representing numbers, quantities, musical notes, etc.

notch n. a V-shaped cut or indentation. ● v. make a notch in. □ **notch up** score, achieve.

note n. 1 a brief record written down to aid memory; a short or informal letter; a short written comment. 2 a banknote. 3 a musical tone of definite pitch; a symbol representing the pitch and duration of a musical sound; each of the keys on a piano etc. 4 eminence; notice, attention. ● v. notice, pay attention to; write down.

notebook n. a book with blank pages on which to write notes.

notecase n. a wallet for banknotes.

noted adj. famous, well known.

notelet n. a small folded card for a short informal letter.

notepaper n. paper for writing letters on.

noteworthy adj. worthy of notice, remarkable.

nothing n. no thing, not anything; no amount, nought; nonexistence; a person or thing of no importance. ● adv. not at all.

notice n. 1 attention, observation. 2 a piece of written information displayed. 3 intimation, warning; the formal announcement of the termination of an agreement or employment. 4 a review in a newspaper. ● v. perceive; take notice of; remark upon. □ **take notice** show interest. **take no notice** (**of**) pay no attention (to).

noticeable adj. easily seen or noticed. □ **noticeably** adv.

notifiable adj. that must be notified.

notify v. inform; report, make known. □ **notification** n.

notion n. a concept; an idea; a whim.

notional adj. hypothetical. □ **notionally** adv.

notorious adj. well known, esp. unfavourably. □ **notoriously** adv., **notoriety** n.

notwithstanding prep. in spite of. ● adv. nevertheless.

nougat (noo-gah) n. a chewy sweet.

nought n. the figure 0; nothing.

noun n. a word used as the name of a person, place, or thing.

nourish v. 1 keep alive and well by food. 2 encourage (a feeling).

nourishment n. nourishing; food.

nous (nowss) n. (colloquial) common sense.

nouveau riche (noo-voh **reesh**) n. (pl. **nouveaux riches**) a person who has become rich recently and makes a display of it.

nouvelle cuisine (noo-vel kwizeen) n. a light style of cooking emphasizing high quality and presentation.

Nov. abbr. November.

nova n. (pl. **novae** or **novas**) a star that suddenly becomes much brighter for a short time.

novel n. a book-length story. ● adj. of a new kind.

novelette n. a short (esp. romantic) novel.

novelist n. a writer of novels.

novelty n. a novel thing or quality; a small unusual object.

November n. the eleventh month.

novice n. an inexperienced person; a probationary member of a religious order.

now adv. **1** at the present time; immediately. **2** (with no reference to time) I wonder or am telling you; *now why didn't I think of that?* ● conj. as a consequence of or simultaneously with the fact that. ● n. the present time. □ **now and again, now and then** occasionally.

nowadays adv. in present times.

nowhere adv. not anywhere.

noxious adj. unpleasant and harmful.

nozzle n. the vent or spout of a hosepipe etc.

Np symb. neptunium.

nuance (nyoo-ahns) n. a shade of meaning.

nub n. a small lump; the central point or core of a matter or problem.

nubile adj. (of a young woman) marriageable; sexually attractive. □ **nubility** n.

nuclear adj. of a nucleus; of the nuclei of atoms; using energy released or absorbed during reactions in these.

nuclear family n. a couple and their children.

nucleic acid n. either of two complex organic molecules (DNA or RNA) present in all living cells.

nucleon n. a proton or neutron.

nucleus n. (pl. **nuclei**) the central part or thing round which others are collected; the central portion of an atom, seed, or cell.

nude adj. naked. ● n. a naked figure in a picture etc. □ **nudity** n.

nudge v. poke gently with the elbow to attract attention quietly; push slightly or gradually. ● n. this movement.

nudist n. a person who believes that going unclothed is good for the health. □ **nudism** n.

nugget n. a rough lump of gold or platinum found in the earth.

nuisance n. an annoying person or thing.

nuke v. (slang) attack with a nuclear weapon.

null adj. having no legal force. □ **nullity** n.

nullify v. make null; neutralize the effect of. □ **nullification** n.

numb adj. deprived of feeling or the power to move. ● v. make numb. □ **numbness** n.

number n. **1** a symbol or word indicating how many; a total, a quantity; a numeral assigned to a person or thing. **2** a single issue of a magazine; an item. ● v. count; amount to; mark or distinguish with a number.

■ **Usage** The phrase *a number of* is normally used with a plural verb, as in *A number of problems remain.*

numberless adj. innumerable.

number one n. (colloquial) oneself.

number plate n. a plate on a motor vehicle, bearing its registration number.

numeral n. a symbol representing a number.

numerate adj. having a good basic understanding of mathematics and science. □ **numeracy** n.

numerator n. a number above the line in a vulgar fraction.

numerical adj. of a number or series of numbers. □ **numerically** adv.

numerous adj. great in number.

numismatics n. the study of coins and medals.

nun *n.* a member of a female religious community.

nuncio *n.* (*pl.* **nuncios**) a representative of the Pope.

nunnery *n.* a residence of a community of nuns.

nuptial *adj.* of marriage or a wedding. ● *n.pl.* (**nuptials**) a wedding ceremony.

nurd var. of **nerd**.

nurse *n.* a person trained to look after sick or injured people; a person employed to take charge of young children. ● *v.* 1 work as a nurse; act as nurse (to). 2 feed at the breast or udder. 3 hold carefully; give special care to.

nursery *n.* 1 a room for young children. 2 a place where plants are reared for sale.

nurseryman *n.* (*pl.* **-men**) a person growing plants at a nursery.

nursery rhyme *n.* a traditional verse for children.

nursery school *n.* a school for children below normal school age.

nursery slopes *n.pl.* gentle slopes suitable for beginners at skiing.

nursing home *n.* a privately run hospital or home for invalids

nurture *v.* nourish, rear; bring up. ● *n.* nurturing.

nut *n.* 1 a fruit with a hard shell round an edible kernel; this kernel. 2 a small threaded metal ring for use with a bolt. 3 (*slang*) the head. ● *adj.* (**nuts**) (*slang*) crazy.

nutcase *n.* (*slang*) a crazy person.

nuthatch *n.* a small climbing bird.

nutmeg *n.* a hard fragrant tropical seed ground or grated as spice.

nutria *n.* the fur of the coypu.

nutrient *n.* a nourishing substance.

nutriment *n.* nourishing food.

nutrition *n.* nourishment; the study of nourishment. □ **nutritional** *adj.*, **nutritionally** *adv.*

nutritious *adj.* nourishing.

nutshell *n.* the hard shell of a nut. □ **in a nutshell** expressed very briefly.

nutty *adj.* (**nuttier**) 1 full of nuts; tasting like nuts. 2 (*slang*) crazy.

nuzzle *v.* press or rub gently with the nose.

NVQ *abbr.* National Vocational Qualification.

NW *abbr.* north-west; north-western.

nylon *n.* a very light strong synthetic fibre; fabric made of this.

nymph *n.* 1 a mythological semi-divine maiden. 2 a young insect.

nymphomania *n.* excessive sexual desire in a woman. □ **nymphomaniac** *n.*

NZ *abbr.* New Zealand.

Oo

O *symb.* oxygen.

oaf *n.* a clumsy person.

oak *n.* a deciduous forest tree bearing acorns; its hard wood. □ **oaken** *adj.*

oak-apple *n.* (also **oak-gall**) a growth on oak trees formed by the larvae of wasps.

OAP *abbr.* old-age pensioner.

oar *n.* a pole with a flat blade used to row a boat; a rower.

oasis *n.* (*pl.* **oases**) a fertile spot in a desert, where there is water. 2 a period of tranquillity.

oast house *n.* a building containing a kiln for drying hops.

oatcake *n.* a biscuit made of oatmeal.

oath *n.* 1 a solemn promise. 2 a swear word.

oatmeal n. **1** ground oats. **2** a greyish-fawn colour.

oats n. a hardy cereal plant; its grain.

OAU abbr. Organization of African Unity.

obdurate adj. stubborn. □ **obdurately** adv., **obduracy** n.

OBE abbr. Order of the British Empire.

obedient adj. doing what one is told to do. □ **obediently** adv., **obedience** n.

obeisance n. a bow or curtsy.

obelisk n. a tall pillar set up as a monument.

obese adj. very fat. □ **obesity** n.

obey v. do what is commanded (by).

obfuscate v. confuse, bewilder. □ **obfuscation** n.

obituary n. a printed statement of a person's death (esp. in a newspaper).

object n. (ob-jekt) **1** something solid that can be seen or touched. **2** a person or thing to which an action or feeling is directed. **3** a purpose, an intention. **4** (Grammar) a noun etc. acted upon by a verb or preposition. ● v. (ob-jekt) state that one is opposed to, protest. □ **no object** not a limiting factor. □ **objector** n.

objection n. disapproval, opposition; a statement of this; a reason for objecting.

objectionable adj. unpleasant. □ **objectionably** adv.

objective adj. **1** not influenced by personal feelings or opinions. **2** (Grammar) of the form of a word used when it is the object of a verb or preposition. ● n. the thing one is trying to achieve, reach, or capture. □ **objectively** adv., **objectiveness** n., **objectivity** n.

object lesson n. a striking practical example of a principle.

objet d'art (ob-zhay dar) n. (pl. **objets d'art**) a small artistic object.

oblation n. an offering made to God.

obligate v. oblige.

obligation n. being obliged to do something; what one must do to comply with an agreement or law.

obligatory adj. compulsory, not optional.

oblige v. **1** compel. **2** help or gratify by a small service.

obliged adj. indebted.

obliging adj. polite and helpful. □ **obligingly** adv.

oblique adj. slanting; indirect. □ **obliquely** adv., **obliqueness** n.

obliterate v. blot out, destroy. □ **obliteration** n.

oblivion n. the state of being forgotten; the state of being unaware.

oblivious adj. unaware. □ **obliviously** adv., **obliviousness** n.

oblong n. a rectangular shape with the length greater than the breadth. ● adj. having this shape.

obloquy n. **1** verbal abuse. **2** disgrace.

obnoxious adj. very unpleasant. □ **obnoxiously** adv.

oboe n. a woodwind instrument of treble pitch, with a double reed. □ **oboist** n.

obscene adj. indecent in a repulsive or offensive way. □ **obscenely** adv., **obscenity** n.

obscure adj. not clear or easily seen; not easily understood; not famous. ● v. make obscure; conceal. □ **obscurely** adv., **obscurity** n.

obsequies n.pl. funeral rites.

obsequious adj. excessively respectful. □ **obsequiously** adv., **obsequiousness** n.

observance n. the keeping of a law, custom, or festival.

observant adj. quick at noticing. □ **observantly** adv.

observation n. observing; a remark. □ **observational** adj.

observatory n. a building equipped for the observation of stars or weather.

observe v. **1** perceive; watch carefully; pay attention to. **2** remark. **3** keep or celebrate (a festival). □ **observable** adj., **observer** n.

obsess v. occupy the thoughts of continually.

obsession n. the state of being obsessed; a persistent idea. □ **obsessional** adj.

obsessive adj. of, causing, or showing obsession. □ **obsessively** adv.

obsolescent adj. becoming obsolete. □ **obsolescence** n.

obsolete adj. no longer used.

obstacle n. something that obstructs progress.

obstetrics n. the branch of medicine and surgery dealing with childbirth. □ **obstetric** adj., **obstetrician** n.

obstinate adj. not easily persuaded, influenced, or overcome. □ **obstinately** adv., **obstinacy** n.

obstreperous adj. noisy, unruly.

obstruct v. block; hinder the movement or progress of. □ **obstruction** n., **obstructive** adj.

obtain v. **1** get, come into possession of. **2** be customary.

obtrude v. force (ideas or oneself) upon others. □ **obtrusion** n.

obtrusive adj. obtruding oneself, unpleasantly noticeable. □ **obtrusively** adv., **obtrusiveness** n.

obtuse adj. **1** blunt in shape; (of an angle) more than 90° but less than 180°. **2** slow at understanding. □ **obtusely** adv., **obtuseness** n.

obverse n. the side of a coin bearing a head or the principal design.

obviate v. make unnecessary.

obvious adj. easy to perceive or understand. □ **obviously** adv., **obviousness** n.

ocarina n. an egg-shaped wind instrument.

occasion n. the time at which a particular event takes place; a special event; an opportunity. ● v. cause.

occasional adj. **1** happening sometimes but not frequently. **2** for a particular occasion. □ **occasionally** adv.

Occident n. the West, the western world.

Occidental n. a native of the Occident. ● adj. (**occidental**) of the western world.

occlude v. stop up, obstruct. □ **occlusion** n.

occluded front n. an upward movement of a mass of warm air caused by a cold front overtaking it, producing prolonged rainfall.

occult adj. **1** secret. **2** supernatural.

occupant n. a person occupying a place or dwelling. □ **occupancy** n.

occupation n. **1** an activity that keeps a person busy, a job; employment. **2** occupying.

occupational adj. of or caused by one's employment.

occupational therapy n. activities designed to assist recovery from illness or injury.

occupy v. **1** dwell in. **2** take possession of (a place) by force. **3** fill (a space or position). **4** keep busy; fill up (time). □ **occupier** n.

occur v. (**occurred**) come into being as an event or process; happen; exist.

occurrence n. occurring; an incident or event.

ocean *n.* the sea surrounding the continents of the earth. □ **oceanic** *adj.*

oceanography *n.* the study of the ocean.

ocelot *n.* a leopard-like animal of Central and South America; its fur.

oche *n.* (o-ki) the line from which a darts player throws the darts.

ochre *n.* (*Amer.* **ocher**) a yellowish clay used as pigment; pale brownish yellow.

o'clock *adv.* of the clock.

OCR *abbr.* optical character reader (or recognition).

Oct. *abbr.* October.

octagon *n.* a geometric figure with eight sides. □ **octagonal** *adj.*

octahedron *n.* a solid with eight sides. □ **octahedral** *adj.*

octane *n.* a hydrocarbon occurring in petrol.

octave *n.* the interval of eight notes between one musical note and the next note of the same name above or below it.

octavo *n.* (*pl.* **octavos**) the size of a book formed by folding a standard sheet three times to form eight leaves.

octet *n.* a group of eight voices or instruments; music for these.

October *n.* the eighth month.

octogenarian *n.* a person in his or her eighties.

octopus *n.* (*pl.* **octopuses**) a sea animal with eight tentacles.

ocular *adj.* of, for, or by the eyes.

oculist *n.* a specialist in the treatment of eye disorders and defects.

OD *n.* (*slang*) an overdose.

odd *adj.* **1** not part of a set. **3** not exactly divisible by 2. **4** exceeding a round number or amount. □ **oddly** *adv.*, **oddness** *n.*

oddity *n.* strangeness; an unusual person or thing.

oddment *n.* a thing left over; an isolated article.

odds *n.pl.* a probability; the ratio between the amounts staked by the parties to a bet. □ **at odds with** in conflict with. **odds and ends** oddments.

odds-on *adj.* with success more likely than failure.

ode *n.* a poem addressed to a person or celebrating an event.

odious *adj.* hateful. □ **odiously** *adv.*, **odiousness** *n.*

odium *n.* widespread hatred or disgust.

odometer *n.* = **milometer**.

odoriferous *adj.* having a fragrant smell.

odour *n.* (*Amer.* **odor**) a smell. □ **odorous** *adj.*

odyssey *n.* (*pl.* **odysseys**) a long adventurous journey.

OECD *abbr.* Organization for Economic Cooperation and Development.

oedema (i-dee-mă) *n.* (*Amer.* **edema**) excess fluid in tissues, causing swelling.

oesophagus (i-sof-ă-gŭs) *n.* (*Amer.* **esophagus**) the tube from the mouth to the stomach.

oestrogen *n.* (*Amer.* **estrogen**) a sex hormone responsible for controlling female bodily characteristics.

of *prep.* belonging to; from; composed or made from; concerning; for, involving.

off *adv.* **1** away; out of position, disconnected. **2** not operating, cancelled. **3** (of food) beginning to decay. ● *prep.* away from; below the normal standard of. ● *adj.* of the side of a vehicle furthest from the kerb. □ **on the off chance** in case of the remote possibility that.

offal *n.* the edible organs from an animal carcass.

offbeat *adj.* unusual, unconventional.

off colour *adj.* **1** unwell. **2** (*Amer.*) slightly indecent.

offcut *n.* a remnant of timber etc.

offence *n.* (*Amer.* **offense**) **1** an illegal act. **2** a feeling of annoyance or resentment.

offend *v.* cause offence to; do wrong. □ **offender** *n.*

offensive *adj.* **1** causing offence, insulting; disgusting. **2** used in attacking. ●*n.* an aggressive action. □ **offensively** *adv.*, **offensiveness** *n.*

offer *v.* present for acceptance or refusal, or for consideration or use; state what one is willing to do, pay, or give; show an intention. ●*n.* an expression of willingness to do, give, or pay something; an amount offered.

offering *n.* a gift; a contribution.

offertory *n.* the offering of bread and wine for consecration; a church collection.

offhand *adj.* unceremonious, casual. ●*adv.* in an offhand way.

office *n.* **1** a room or building used for clerical and similar work. **2** a position of authority or trust.

officer *n.* an official; a person holding authority in the armed services; a policeman.

official *adj.* of office or officials; authorized. ●*n.* a person holding office. □ **officially** *n.*

officialese *n.* the formal long-winded language of official documents.

officiate *v.* act in an official capacity; be in charge.

officious *adj.* asserting one's authority, bossy. □ **officiously** *adv.*

off-licence *n.* a shop with a licence to sell alcohol for consumption away from the premises.

off-line *adj.* & *adv.* (of a computer terminal or process) not directly controlled by or connected to a central processor.

offload *v.* unload.

off-putting *adj.* (*colloquial*) repellent.

offset *v.* (**offset, offsetting**) counterbalance, compensate for. ●*n.* **1** an offshoot. **2** a printing method using an inked rubber surface.

offshoot *n.* a side shoot; a subsidiary product.

offshore *adj.* at sea some distance from land; (of wind) blowing from the land to the sea.

offside *adj.* & *adv.* in a position where one may not legally play the ball (in football etc.).

offspring *n.* (*pl.* **offspring**) a person's child or children; an animal's young.

off-white *adj.* not quite pure white.

often *adv.* many times, at short intervals; in many instances.

ogee *n.* a double continuous curve as in an S.

ogle *v.* look lustfully at.

ogre *n.* a man-eating giant in fairy tales; a terrifying person.

oh *int.* an exclamation of delight or pain, or used for emphasis.

ohm *n.* a unit of electrical resistance.

OHP *abbr.* overhead projector.

oil *n.* **1** a thick slippery liquid that will not dissolve in water; petroleum, a form of this. **2** an oil painting; (**oils**) oil paints. ●*v.* lubricate or treat with oil. □ **oily** *adj.*

oilfield *n.* an area where oil is found in the ground.

oil paint *n.* paint made by mixing pigment in oil.

oil painting *n.* a picture painted in oil paints.

oil rig *n.* a structure for drilling and extracting oil.

oilfield n. an area where oil is found in the ground.

oilskin n. cloth waterproofed by treatment with oil; (**oilskins**) waterproof clothing made of this.

ointment n. a cream rubbed on skin to heal it.

OK adj. & adv. (also **okay**) (colloquial) all right.

okapi n. a giraffe-like animal of Central Africa.

okra n. an African plant with seed pods used as food.

old adj. having lived, existed, or been known for a long time; of specified age; shabby from age or wear; former, not recent or modern.

old age n. the later part of life.

old-fashioned adj. no longer fashionable.

Old Testament see testament.

old wives' tale n. a traditional but unfounded belief.

oleaginous adj. producing oil; oily.

oleander n. a flowering shrub of Mediterranean regions.

O level n. Ordinary level.

olfactory adj. concerned with smelling.

oligarch n. a member of an oligarchy.

oligarchy n. government by a small group; a country governed in this way. □ **oligarchic** adj.

olive n. 1 a small oval fruit from which an oil (**olive oil**) is obtained; a tree bearing this. 2 a greenish colour. ● adj. of this colour; (of the skin) yellowish brown.

olive branch n. something done or offered to show one's desire to make peace.

ombudsman n. (pl. **-men**) an official appointed to investigate people's complaints about maladministration by public authorities.

omega n. the last letter of the Greek alphabet (Ω, ω).

omelette n. a dish of beaten eggs cooked in a frying pan.

omen n. an event regarded as a prophetic sign.

ominous adj. seeming as if trouble is imminent. □ **ominously** adv.

omit v. (**omitted**) leave out, not include; neglect (to do something). □ **omission** n.

omnibus n. 1 (formal) a bus. 2 a publication containing several items.

omnipotent adj. having unlimited or very great power. □ **omnipotence** n.

omnipresent adj. present everywhere.

omniscient adj. knowing everything. □ **omniscience** n.

omnivorous adj. feeding on both plants and animals.

on prep. 1 supported by, attached to, covering. 2 close to, near, at; towards. 3 (of time) exactly at, during. 4 concerning, about. 5 added to. ● adv. 1 so as to be on or covering something. 2 further forward, towards something. 3 with continued movement or action; operating, taking place. □ **on and off** from time to time.

■ **Usage** See note at **onto**.

ONC abbr. Ordinary National Certificate.

once adv. 1 on one occasion or for one time only. 2 formerly. ● conj. as soon as. ● n. one time or occasion. □ **once upon a time** at some vague time in the past.

once-over n. (colloquial) a rapid inspection.

oncogene n. a gene that transforms a cell into a cancer cell.

oncology n. the study of tumours.

oncoming adj. approaching.

OND *abbr.* Ordinary National Diploma.

one *adj.* single, individual, forming a unity. ● *n.* the smallest whole number (1, I); a single thing or person. ● *pron.* a person; any person (esp. used by a speaker or writer of himself or herself as representing people in general). □ **one another** each other. **one day** at some unspecified date.

onerous *adj.* burdensome.

oneself *pron.* the emphatic and reflexive form of *one*.

one-sided *adj.* unfair, prejudiced.

one-upmanship *n.* maintaining an advantage over others.

one-way *adj.* allowing movement in one direction only.

ongoing *adj.* continuing, in progress.

onion *n.* a vegetable with a bulb that has a strong taste and smell.

online *adj. & adv.* (*Computing*) directly controlled by or connected to a central processor.

onlooker *n.* a spectator.

only *adj.* being the one specimen or all the specimens of a class; sole. ● *adv.* **1** without anything or anyone else; and that is all. **2** no longer ago than. ● *conj.* but then. □ **only too** extremely.

o.n.o. *abbr.* or near offer.

onomatopoeia (onŏ-matŏ-pee-ă) *n.* the formation of words that imitate the sound of what they stand for. □ **onomatopoeic** *adj.*

onset *n.* **1** a beginning. **2** an attack.

onshore *adj.* (of wind) blowing from the sea to the land.

onslaught *n.* a fierce attack.

onto *prep.* = on to, to a position on.

■ **Usage** Onto is much used but is still not as widely accepted as *into*. It is, however, useful in distinguishing between, e.g., *We*

drove on to the beach (i.e. towards it) and *We drove onto the beach* (i.e. into contact with it).

onus *n.* a duty or responsibility.

onward *adv. & adj.* with an advancing motion; further on. □ **onwards** *adv.*

onyx *n.* a stone like marble.

oodles *n.pl.* (*colloquial*) a great amount.

oolite (oh-ō-lyt) *n.* a granular form of limestone.

oomph *n.* (*colloquial*) energy; enthusiasm.

ooze *v.* trickle or flow out slowly; exude. ● *n.* wet mud.

op. *abbr.* opus.

opacity *n.* being opaque.

opal *n.* an iridescent quartz-like stone.

opalescent *adj.* iridescent like an opal. □ **opalescence** *n.*

opaque *adj.* not clear, not transparent.

op. cit. *abbr.* in the work already quoted.

OPEC *abbr.* (**oh**-pek) Organization of Petroleum Exporting Countries.

open *adj.* **1** able to be entered, not closed or sealed or locked; not covered or concealed or restricted. **2** unfolded. **3** frank; undisguised. **4** not yet decided. **5** (**open to**) willing to receive. ● *v.* make or become open or more open; begin, establish. □ **in the open air** not in a house or building etc. □ **openness** *n.*

opencast *adj.* (of mining) on the surface of the ground.

open-ended *adj.* with no fixed limit.

opener *n.* a device for opening tins, bottles, etc.

open-handed *adj.* giving generously.

open house *n.* hospitality to all comers.

opening *n.* **1** a gap; the place where a thing opens. **2** a beginning; an opportunity.

open letter n. one addressed to a person by name but printed in a newspaper.

openly adv. publicly, frankly.

open-plan adj. without partition walls or fences.

open prison n. a prison with few physical restraints on prisoners.

open verdict n. one not specifying whether a crime is involved.

opera n. a play in which words are sung to music.

operable adj. 1 able to be operated. 2 suitable for treatment by surgery.

opera glasses n.pl. small binoculars used in the theatre.

operate v. 1 be in action; produce an effect; control the functioning of. 2 perform a surgical operation.

operatic adj. of or like opera.

operating system n. basic software allowing a computer program to run.

operation n. 1 operating; the way a thing works; planned action. 2 a surgical treatment.

operational adj. of or used in operations; able to function.

operative adj. 1 working, functioning. 2 of surgical operations. ● n. a worker, esp. in a factory.

operator n. a person who operates a machine; one who connects lines at a telephone exchange.

operetta n. a short or light opera.

ophidian adj. & n. (a member) of the snake family.

ophthalmic adj. of or for the eyes.

ophthalmic optician n. an optician qualified to prescribe and dispense spectacles.

ophthalmology n. the study of the eye and its diseases. □ **ophthalmologist** n.

ophthalmoscope n. an instrument for examining the eye.

opiate n. a sedative containing opium.

opine v. express or hold as an opinion.

opinion n. a belief or judgement held without actual proof; what one thinks on a particular point.

opinionated adj. holding strong opinions obstinately.

opium n. a narcotic drug made from the juice of certain poppies.

opossum n. a small furry tree-living marsupial.

opponent n. one who opposes another.

opportune adj. (of time) favourable; well-timed. □ **opportunely** adv., **opportuneness** n.

opportunist n. a person who grasps opportunities. □ **opportunism** n., **opportunistic** adj.

opportunity n. circumstances suitable for a particular purpose.

oppose v. argue or fight against; place opposite; place or be in opposition to; contrast.

opposite adj. 1 facing, on the further side. 2 as different as possible from. ● n. an opposite thing or person. ● adv. & prep. in an opposite position or direction (to).

opposition n. 1 antagonism, resistance; placing or being placed opposite. 2 people opposing something; (also **the Opposition**) the chief parliamentary party opposing the one in power.

oppress v. govern or treat harshly; weigh down with cares. □ **oppression** n., **oppressor** n.

oppressive adj. oppressing; hard to endure; (of weather) sultry and tiring. □ **oppressively** adv., **oppressiveness** n.

opprobrious adj. abusive.

opprobrium n. great, disgrace from shameful conduct.

opt v. make a choice. □ **opt out** choose not to participate.

optic adj. of the eye or sight.

optical adj. of or aiding sight; visual. □ **optically** adv.

optical character reader n. (*Computing*) a scanner enabling text to be transferred to computer.

optical fibre n. a thin glass fibre used to transmit signals.

optician n. a maker or seller of spectacles.

optics n. the study of sight and of light as its medium.

optimal adj. the best or most favourable.

optimism n. a tendency to take a hopeful view of things. □ **optimist** n., **optimistic** adj., **optimistically** adv.

optimize v. (also **-ise**) make the best use of. □ **optimization** n.

optimum adj. & n. the best or most favourable (conditions, amount, etc.).

option n. 1 freedom to choose; a thing that is or may be chosen. 2 the right to buy or sell a thing within a limited time.

optional adj. not compulsory. □ **optionally** adv.

optometrist n. an ophthalmic optician.

opulent adj. 1 wealthy. 2 abundant. □ **opulently** adv., **opulence** n.

opus n. (pl. **opera**) a musical composition numbered as one of a composer's works.

or conj. as an alternative; because if not; also known as.

oracle n. a person or thing giving wise guidance. □ **oracular** adj.

oral adj. 1 spoken not written. 2 of the mouth; taken by mouth. ● n. a spoken examination. □ **orally** adv.

orange n. a round juicy citrus fruit with reddish-yellow peel; this colour. ● adj. reddish yellow.

orangeade n. an orange-flavoured soft drink.

orang-utan n. (also **orang-outang**) a large ape of Borneo and Sumatra.

oration n. a long speech, esp. of a ceremonial kind.

orator n. a person who makes public speeches; a skilful speaker.

oratorio n. (pl. **oratorios**) a musical composition for voices and orchestra, usu. with a biblical theme.

oratory n. the art of public speaking; an eloquent speech. □ **oratorical** adj.

orb n. a sphere, a globe.

orbit n. the curved path of a planet, satellite, or spacecraft round another; a sphere of influence. ● v. (**orbited**) move in an orbit (round).

orbital adj. 1 of orbits. 2 (of a road) round the outside of a city.

orchard n. a piece of land planted with fruit trees.

orchestra n. a large body of people playing various musical instruments. □ **orchestral** adj.

orchestrate v. compose or arrange (music) for an orchestra; coordinate. □ **orchestration** n.

orchid n. a showy often irregularly shaped flower.

ordain v. 1 appoint ceremonially to the Christian ministry. 2 destine; decree authoritatively.

ordeal n. a difficult experience.

order n. 1 the way things are placed in relation to each other; the proper or usual sequence. 2 an efficient or law-abiding state. 3 a command; a request to supply goods etc., the things supplied; a written instruction or permission. 4 a rank, kind, or quality; a group of plants or animals classified as similar. **5** a monastic organization. ● *v.* **1** arrange in order. **2** command; give an order for (goods etc.). □ **in order to** or **that** with the purpose of or intention that.

orderly *adj.* in due order; not unruly. ● *n.* a soldier assisting an officer; a hospital attendant. □ **orderliness** *n.*

order paper *n.* a programme of the day's business in Parliament.

ordinal *n.* (in full **ordinal number**) one defining a thing's position in a series (e.g. *first, second,* etc.).

ordinance *n.* a decree.

ordinand *n.* a candidate for ordination.

ordinary *adj.* usual, not exceptional. □ **ordinarily** *adv.*

Ordinary level *n.* (also **O level**) (*hist.*) the lowest level of GCE examination.

ordination *n.* ordaining.

ordnance *n.* military materials.

Ordnance Survey *n.* an official surveying organization preparing maps of the British Isles.

ordure *n.* dung.

ore *n.* solid rock or mineral from which metal is obtained.

oregano *n.* a herb used in cooking.

organ *n.* **1** a keyboard instrument with pipes supplied with wind by bellows. **2** a body part with a specific function. **3** a medium of communication, esp. a newspaper.

organdie *n.* a translucent cotton fabric, usu. stiffened.

organic *adj.* **1** of bodily organs. **2** of or formed from living things. **3** using no artificial fertilizers or pesticides. **4** organized as a system. □ **organically** *adv.*

organism *n.* a living being; an individual animal or plant.

organist *n.* a person who plays the organ.

organization *n.* organizing; an organized system or body of people. □ **organizational** *adj.*

organize *v.* (also **-ise**) arrange systematically; make arrangements for; form (people) into an association for a common purpose. □ **organizer** *n.*

organza *n.* a thin stiff transparent fabric.

orgasm *n.* a climax of sexual excitement.

orgy *n.* a wild party; unrestrained activity. □ **orgiastic** *adj.*

Orient *n.* the East, the eastern world.

orient *v.* place or determine the position of (a thing) with regard to points of the compass. □ **orient oneself** get one's bearings; become accustomed to a new situation. □ **orientation** *n.*

oriental *adj.* (also **Oriental**) of the Orient. ● *n.* a native of the Orient.

orientate *v.* orient.

orienteering *n.* the sport of finding one's way across country with a map and compass.

orifice *n.* an opening, esp. the mouth of a cavity.

origami *n.* the Japanese decorative art of paper folding.

origin *n.* the point, source, or cause from which a thing begins its existence; a person's ancestry or parentage.

original *adj.* **1** existing from the first; earliest. **2** being the first form of something; new in character or design; inventive, creative. ● *n.* the first form; the thing from which another is copied. □ **originally** *adv.*, **originality** *n.*

originate *v.* bring or come into being. □ **origination** *n.*, **originator** *n.*

oriole *n.* a bird with black and yellow plumage.

ormolu *n.* a gold-coloured alloy of copper; things made of this.

ornament *n.* a decorative object or detail; decoration. ● *v.* decorate with ornament(s). □ **ornamentation** *n.*

ornamental *adj.* serving as an ornament. □ **ornamentally** *adv.*

ornate *adj.* elaborately ornamented. □ **ornately** *adv.*, **ornateness** *n.*

ornithology *n.* the study of birds. □ **ornithological** *adj.*, **ornithologist** *n.*

orphan *n.* a child whose parents are dead. ● *v.* make (a child) an orphan.

orphanage *n.* an institution where orphans are cared for.

orris root *n.* a violet-scented iris root.

orthodontics *n.* correction of irregularities in teeth. □ **orthodontic** *adj.*, **orthodontist** *n.*

orthodox *adj.* of or holding conventional or currently accepted beliefs, esp. in religion. □ **orthodoxy** *n.*

Orthodox Church *n.* the Eastern or Greek Church.

orthography *n.* spelling. □ **orthographic** *adj.*

orthopaedics (orthŏ-pee-diks) *n.* (*Amer.* **orthopedics**) the surgical correction of deformities in bones or muscles. □ **orthopaedic** *adj.*, **orthopaedist** *n.*

oryx *n.* a large African antelope.

Os *symb.* osmium.

oscillate *v.* move to and fro; vary. □ **oscillation** *n.*

oscilloscope *n.* a device for recording oscillations.

osier *n.* willow with flexible twigs; a twig of this.

osmium *n.* a hard metallic element (symbol Os).

osmosis *n.* the diffusion of fluid through a porous partition from another fluid. □ **osmotic** *adj.*

osprey *n.* (*pl.* **ospreys**) a large bird preying on fish in inland waters.

osseous *adj.* like bone, bony.

ossify *v.* turn into bone; harden; make or become rigid and unprogressive. □ **ossification** *n.*

ostensible *adj.* pretended, used as a pretext. □ **ostensibly** *adv.*

ostensive *adj.* showing something directly.

ostentation *n.* a showy display intended to impress people. □ **ostentatious** *adj.*, **ostentatiously** *adv.*

osteopath *n.* a practitioner who treats certain conditions by manipulating bones and muscles. □ **osteopathic** *adj.*, **osteopathy** *n.*

ostracize *v.* (also **-ise**) refuse to associate with. □ **ostracism** *n.*

ostrich *n.* a large flightless swift-running African bird.

other *adj.* alternative, additional, being the remaining one of a set of two or more; not the same. ● *n.* & *pron.* the other person or thing. ● *adv.* otherwise. ■ **the other day** or **week** a few days or weeks ago.

otherwise *adv.* in a different way; in other respects; if circumstances are or were different.

otiose *adj.* serving no practical purpose.

OTT *abbr.* (*colloquial*) over-the-top, excessive.

otter *n.* a fish-eating water animal with thick brown fur.

ottoman *n.* a storage box with a padded top.

OU *abbr.* Open University.

ought *v.aux.* expressing duty, rightness, advisability, or strong probability.

Ouija board (wee-jǎ) *n.* (*trade mark*) a board marked with letters and signs, used in seances to indicate messages using a pointer.

ounce n. a unit of weight, one-sixteenth of a pound (about 28 grams).

our adj. of or belonging to us.

ours poss.pron. belonging to us.

ourselves pron. an emphatic and reflexive form of we and us.

oust v. drive out, eject.

out adv. **1** away from or not in a place; not at home. **2** not in effective action; in error; not possible; unconscious. **3** into the open, so as to be heard or seen. ● prep. out of (see note below). ● n. a way of escape. □ be **out** to be intending to. **out and out** thorough, extreme. **out of** from inside or among; having no more of. **out of date** no longer fashionable, current, or valid. **out of the way** no longer an obstacle; remote; unusual.

■ **Usage** The use of out as a preposition, as in He walked out the room, is not standard English. Out of should be used instead.

out- pref. more than, so as to exceed.

outback n. the remote inland areas of Australia.

outboard adj. (of a motor) attached to the outside of a boat.

outbreak n. a sudden eruption of anger, war, disease, etc.

outbuilding n. an outhouse.

outburst n. an explosion of feeling.

outcast n. a person driven out of a group or by society.

outclass v. surpass in quality.

outcome n. the result of an event.

outcrop n. part of an underlying layer of rock that projects on the surface of the ground.

outcry n. a loud cry; a strong protest.

outdistance v. get far ahead of.

outdo v. (**outdid, outdone, outdoing**) be or do better than.

outdoor adj. of or for use in the open air. □ **outdoors** adv.

outer adj. further from the centre or inside; exterior, external.

outermost adv. furthest outward.

outface v. disconcert by staring or by a confident manner.

outfit n. a set of clothes or equipment.

outfitter n. a supplier of equipment or men's clothing.

outflank v. get round the flank of (an enemy).

outgoing adj. **1** going out. **2** sociable.

outgoings n.pl. expenditure.

outgrow v. (**outgrew, outgrown, outgrowing**) **1** grow faster than; grow too large for. **2** leave aside as one develops.

outhouse n. a shed, barn, etc.

outing n. a pleasure trip.

outlandish adj. looking or sounding strange or foreign.

outlast v. last longer than.

outlaw n. (old use) a person punished by being deprived of the law's protection; a bandit. ● v. make (a person) an outlaw; declare illegal.

outlay n. money etc. spent.

outlet n. a way out; a means for giving vent to energies or feelings; a market for goods.

outline n. **1** a line showing a thing's shape or boundary. **2** a summary. ● v. draw or describe in outline; mark the outline of.

outlook n. a view, a prospect; a mental attitude; future prospects.

outlying adj. remote.

outmoded adj. no longer fashionable or accepted.

outnumber v. exceed in number.

outpace v. go faster than.

outpatient n. a person visiting a hospital for treatment but not staying overnight.

outpost n. an outlying settlement or detachment of troops.

output n. the amount of electrical power, work, etc. produced. ● v. (**output** or **outputted, outputting**) (of a computer) supply (results etc.).

outrage n. an act that shocks public opinion; a violation of rights. ● v. shock and anger greatly.

outrageous adj. greatly exceeding what is moderate or reasonable; shocking. □ **outrageously** adv.

outré (oo-tray) adj. eccentric, unseemly.

outrider n. a mounted attendant or motorcyclist riding as guard.

outrigger n. a stabilizing strip of wood fixed outside and parallel to a canoe; a canoe with this.

outright adv. **1** completely, not gradually. **2** frankly. ● adj. thorough, complete.

outrun v. (**outran, outrun, outrunning**) run faster or further than.

outset n. □ **at** or **from the outset** at or from the beginning.

outside n. the outer side, surface, or part. ● adj. of or from the outside. ● adv. on, at, or to the outside. ● prep. on, at, or to the outside of; other than.

outsider n. **1** a non-member of a group. **2** a competitor thought to have no chance in a contest.

outsize adj. much larger than average.

outskirts n.pl. the outer districts.

outspoken adj. very frank.

outstanding adj. **1** conspicuous; exceptionally good. **2** not yet paid or settled. □ **outstandingly** adv.

outstrip v. (**outstripped**) run faster or further than; surpass.

out-tray n. a tray for documents that have been dealt with.

outvote v. defeat by a majority of votes.

outward adj. on or to the outside. ● adv. outwards. □ **outwardly** adv.; outwards.

outweigh v. be of greater weight or importance than.

outwit v. (**outwitted**) defeat by one's craftiness.

outwork n. work done away from the employer's premises.

ouzel (oo-zél) n. **1** (in full **ring ouzel**) a small bird of the thrush family. **2** (in full **water ouzel**) a diving bird.

ouzo (oo-zoh) n. a Greek aniseed-flavoured spirit.

ova see **ovum.**

oval n. a rounded symmetrical shape longer than it is broad. ● adj. of this shape.

ovary n. an organ producing egg cells; that part of a pistil from which fruit is formed. □ **ovarian** adj.

ovate adj. egg-shaped.

ovation n. enthusiastic applause.

oven n. an enclosed chamber in which things are cooked or heated.

ovenware n. dishes for use in an oven.

over prep. **1** in or to a position higher than; above and across; more than. **2** throughout, during. ● adv. **1** outwards and downwards from the brink or an upright position etc. **2** from one side or end etc. to the other; across a space or distance. **3** besides, in addition. **4** with repetition. **5** at an end.

over- pref. above; excessively.

overall n. a garment worn to protect other clothing; (**overalls**) a one-piece garment of this kind covering the body and legs. ● adj. total; taking all aspects into account. ● adv. taken as a whole.

overarm *adj.* & *adv.* with the arm brought forward and down from above shoulder level.

overawe *v.* overcome with awe.

overbalance *v.* lose balance and fall; cause to do this.

overbearing *adj.* domineering.

overblown *adj.* **1** pretentious. **2** (of a flower) past its best.

overboard *adv.* from a ship into the water. □ **go overboard** (*colloquial*) show excessive enthusiasm.

overcast *adj.* covered with cloud.

overcharge *v.* charge too much.

overcoat *n.* a warm full-length outdoor coat.

overcome *v.* win a victory over; succeed in subduing or dealing with; be victorious; weaken; less.

overdo *v.* (**overdoes, overdid, overdone, overdoing**) do too much; cook for too long.

overdose *n.* too large a dose. ● *v.* give an overdose to; take an overdose.

overdraft *n.* an overdrawing of a bank account; the amount of this.

overdraw *v.* (**overdrew, overdrawn, overdrawing**) draw more from (a bank account) than the amount credited.

overdrive *n.* a mechanism providing an extra gear above top gear.

overdue *adj.* not paid or arrived etc. by the due time.

overestimate *v.* form too high an estimate of.

overflow *v.* flow over the edge or limits (of). ● *n.* what overflows; an outlet for excess liquid.

overgrown *adj.* **1** grown too large. **2** covered with weeds etc.

overhaul *v.* **1** examine and repair. **2** overtake. ● *n.* an examination and repair.

overhead *adj.* & *adv.* above the level of one's head; in the sky.

● *n.pl.* (**overheads**) the expenses involved in running a business etc.

overhead projector *n.* a projector producing an image from a transparency placed flat on a light source.

overhear *v.* (**overheard, overhearing**) hear accidentally or without the speaker's knowledge.

overjoyed *adj.* very happy.

overkill *n.* an excess of capacity to kill or destroy.

overland *adj.* & *adv.* (travelling) by land.

overlap *v.* (**overlapped**) **1** extend beyond the edge of. **2** coincide partially. ● *n.* overlapping; a part or amount that overlaps.

overleaf *adv.* on the other side of a leaf of a book etc.

overload *v.* put too great a load on or in. ● *n.* a load that is too great.

overlook *v.* **1** have a view over; oversee. **2** fail to observe or consider; ignore, not punish.

overly *adv.* excessively, too.

overman *v.* (**overmanned**) provide with too many staff or crew.

overnight *adv.* & *adj.* during or for a night.

overpass *n.* a road crossing another by means of a bridge.

overpower *v.* overcome by greater strength or numbers.

overpowering *adj.* (of heat or feelings) extremely intense.

overrate *v.* have too high an opinion of.

overreach *v.* □ **overreach oneself** fail through being too ambitious.

overreact *v.* respond more angrily etc. than is justified.

override *v.* (**overrode, overridden, overriding**) overrule; prevail over; intervene and cancel the operation of.

overrule v. set aside (a decision etc.) by using one's authority.

overrun v. (**overran, overrun, overrunning**) **1** spread over and occupy or injure. **2** exceed (a limit).

overseas adj. & adv. across or beyond the sea, abroad.

oversee v. (**oversaw, overseen, overseeing**) superintend. □ **overseer** n.

oversew v. (**oversewed, oversewn, oversewing**) sew (edges) together so that each stitch lies over the edges.

overshadow v. cast a shadow over; cause to seem unimportant in comparison.

overshoot v. (**overshot, overshooting**) pass beyond (a target or limit etc.).

oversight n. **1** an unintentional omission or mistake. **2** supervision.

overspill n. what spills over; a district's surplus population looking for homes elsewhere.

oversteer v. (of a car) tend to turn more sharply than was intended. ● n. this tendency.

overstep v. (**overstepped**) go beyond (a limit).

overt adj. done or shown openly. □ **overtly** adv.

overtake v. (**overtook, overtaken, overtaking**) **1** pass a moving person or thing. **2** (of misfortune) come suddenly upon.

overthrow v. (**overthrew, overthrown, overthrowing**) cause the downfall of. ● n. a defeat or downfall.

overtime adv. in addition to regular working hours. ● n. time worked in this way; payment for this.

overtone n. an additional quality or implication.

overture n. an orchestral composition forming a prelude to a

performance; (**overtures**) an initial approach or proposal.

overturn v. (cause to) turn or fall over; reverse or overthrow.

overview n. a general survey.

overwhelm v. **1** bury beneath a huge mass. **2** overcome completely; make helpless with emotion. □ **overwhelming** adj.

overwrought adj. in a state of nervous agitation.

oviduct n. the tube through which ova pass from the ovary.

oviparous adj. egg-laying.

ovoid adj. egg-shaped, oval.

ovulate v. produce or discharge an egg cell from an ovary. □ **ovulation** n.

ovule n. a germ cell of a plant.

ovum n. (pl. **ova**) an egg cell; a reproductive cell produced by a female.

owe v. **1** be under an obligation to pay or repay. **2** have (a thing) as the result of the action of another person or cause; I owe my life to him.

owing adj. owed and not yet paid. □ **owing to** caused by; because of.

owl n. a bird of prey with large eyes, usu. flying at night. □ **owlish** adj.

own adj. belonging to oneself or itself. ● v. have as one's own; acknowledge ownership of. □ **of one's own** belonging to oneself. **on one's own** alone; independently. **own up** (colloquial) confess. □ **owner** n., **ownership** n.

ox n. (pl. **oxen**) an animal of or related to the kind kept as domestic cattle; a fully grown bullock.

oxidation n. the process of combining with oxygen.

oxide n. a compound of oxygen and one other element.

oxidize v. (also **-ise**) combine with oxygen; coat with an oxide;

make or become rusty. □ **oxidization** n.

oxtail n. the tail of an ox, used in soups.

oxyacetylene adj. using a mixture of oxygen and acetylene, esp. in metal-cutting and welding.

oxygen n. a chemical element (symbol O), a colourless gas existing in air.

oxygenate v. supply or mix with oxygen.

oyster n. an edible shellfish.

oz. abbr. ounce(s).

ozone n. a form of oxygen; a protective layer of this in the stratosphere.

Pp

P abbr. (in road signs) parking. ● symb. phosphorus.

p abbr. **1** penny; pence (in decimal coinage). **2** (Music) piano.

PA abbr. **1** personal assistant. **2** public address system.

Pa symb. protactinium.

pa n. (colloquial) father.

p.a. abbr. per annum, yearly.

pace n. a single step in walking or running; a rate of progress. ● v. walk steadily or to and fro; measure by pacing; set the pace for.

pacemaker n. **1** a runner etc. who sets the pace for another. **2** a device regulating heart contractions.

pachyderm (pak-i-derm) n. a large thick-skinned mammal such as the elephant.

pacific adj. making or loving peace.

pacifist n. a person totally opposed to war. □ **pacifism** n.

pacify v. calm and soothe; establish peace in. □ **pacification** n.

pack n. **1** a collection of things wrapped or tied for carrying or selling. **2** a set of playing cards. **3** a group of hounds or wolves. ● v. **1** put into or fill a container. **2** press or crowd together; fill (a space) in this way. **3** cover or protect with something pressed tightly. □ **pack off** send away. **send packing** dismiss abruptly. ● **packer** n.

package n. **1** a parcel; a box etc. in which goods are packed. **2** a package deal. ● v. put together in a package. □ **packager** n.

package deal n. a set of proposals offered or accepted as a whole.

package holiday n. one with set arrangements at an inclusive price.

packet n. **1** a small package. **2** (colloquial) a large sum of money. **3** a mailboat.

pact n. an agreement, a treaty.

pad n. **1** a piece of padding. **2** a set of sheets of paper fastened together at one edge. **3** a soft fleshy part under an animal's paw. **4** a flat surface for use by helicopters or for launching rockets. ● v. (**padded**) **1** put padding on or into. **2** fill out. **3** walk softly or steadily.

padding n. soft material used to protect against jarring, add bulk, absorb fluid, etc.

paddle n. a short oar with a broad blade. ● v. **1** propel by use of a paddle or paddles; row gently. **2** walk with bare feet in shallow water.

paddock n. a small field where horses are kept; an enclosure for horses at a racecourse.

padlock n. a detachable lock with a U-shaped bar secured through the object fastened. ● v. fasten with a padlock.

padre (pah-dray) n. (colloquial) a chaplain in the army etc.

paean (pee-ăn) n. (Amer. **pean**) a song of triumph.

paediatrics (peed-i-at-riks) n. (Amer. **pediatrics**) the study of children's diseases. □ **paediatric** adj., **paediatrician** n.

paella (py-el-ă) n. a Spanish dish of rice, seafood, chicken, etc.

paeony var. of **peony**.

pagan adj. & n. heathen; nature-worshipping.

page n. **1** a sheet of paper in a book etc.; one side of this. **2** a boy attendant of a bride. ● v. summon by an announcement, messenger, or pager.

pageant n. a public show or procession, esp. with people in costume. □ **pageantry** n.

pager n. a radio device with a bleeper for summoning the wearer.

pagoda n. a Hindu temple or Buddhist tower in India, China, etc.

paid see **pay**. □ **put paid to** end (hopes or prospects).

pail n. a bucket.

pain n. **1** an unpleasant feeling caused by injury or disease; mental suffering. **2** (**pains**) careful effort. ● v. cause pain to.

painful adj. **1** causing or suffering pain. **2** laborious. □ **painfully** adv.

painkiller n. a drug for reducing pain.

painless adj. not causing pain. □ **painlessly** adv.

painstaking adj. very careful.

paint n. colouring matter for applying in liquid form to a surface; (**paints**) tubes or cakes of paint. ● v. coat with paint; portray by using paint(s) or in words; apply (liquid) to.

painter n. **1** a person who paints as an artist or decorator. **2** a rope attached to a boat's bow for tying it up.

painting n. a painted picture.

pair n. a set of two things or people; an article consisting of two parts; the other member of a pair. ● v. arrange or be arranged in a pair or pairs.

Paisley adj. having a detailed pattern of curved feather-shaped figures.

pajamas Amer. sp. of **pyjamas**.

pal n. (colloquial) a friend.

palace n. an official residence of a sovereign, archbishop, or bishop; a splendid mansion.

palaeography (pali-og-răfi) n. (Amer. **paleography**) the study of ancient writing and inscriptions. □ **palaeographer** n.

palaeolithic (pali-ŏ-lith-ik) adj. (Amer. **paleolithic**) of the early part of the Stone Age.

palaeontology (pali-ŏn-tol-ŏji) n. (Amer. **paleontology**) the study of life in the geological past. □ **palaeontologist** n.

palatable adj. pleasant to the taste or mind.

palate n. the roof of the mouth; the sense of taste.

palatial (pă-lay-shăl) adj. of or like a palace.

palaver n. (colloquial) a fuss.

pale adj. (of the face) having less colour than normal; (of colour or light) faint. ● v. turn pale. □ **beyond the pale** outside the bounds of acceptable behaviour. □ **palely** adv., **paleness** n.

paleo- Amer. sp. of words beginning with **palaeo-**.

Palestinian adj. & n. (a native) of Palestine.

palette n. a board on which an artist mixes colours.

palette knife n. a knife with a flexible blade for spreading paint or for smoothing soft substances in cookery.

palindrome n. a word or phrase that reads the same backwards as forwards.

paling n. railing(s).

pall (pawl) n. a cloth spread over a coffin; a heavy dark covering. ● v. become uninteresting.

palladium n. a rare metallic element (symbol Pd).

pall-bearer n. a person helping to carry or walking beside the coffin at a funeral.

pallet n. 1 a straw-stuffed mattress; a hard narrow or makeshift bed. 2 a tray or platform for goods being lifted or stored.

palliasse n. a straw-stuffed mattress.

palliate v. alleviate; partially excuse. □ **palliative** adj.

pallid adj. pale, esp. from illness. □ **pallidness** n., **pallor** n.

pally adj. (**pallier**) (colloquial) friendly.

palm n. 1 the inner surface of the hand. 2 a tree of warm and tropical climates, with large leaves and no branches. ● v. conceal in one's hand. □ **palm off** get (a thing) accepted fraudulently.

palmist n. a person who tells people's fortunes from lines in their palms. □ **palmistry** n.

palomino n. (pl. **palominos**) a golden or cream-coloured horse.

palpable adj. able to be touched or felt; obvious. □ **palpably** adv.

palpate v. examine medically by touch. □ **palpation** n.

palpitate v. throb rapidly; quiver with fear or excitement. □ **palpitation** n.

palsy n. paralysis, esp. with involuntary tremors. □ **palsied** adj.

paltry adj. (**paltrier**) worthless. □ **paltriness** n.

pampas n. vast grassy plains in South America.

pampas grass n. tall ornamental grass with feathery plumes.

pamper v. treat very indulgently.

pamphlet n. a leaflet or paper-covered booklet.

pamphleteer n. a writer of pamphlets.

pan n. a metal or earthenware vessel with a flat base, used in cooking; any similar vessel. ● v. (**panned**) 1 wash (gravel) in a pan in searching for gold. 2 (colloquial) criticize severely. 3 turn (the camera) horizontally in filming. □ **panful** n.

pan- comb. form all, whole.

panacea (pan-ă-see-ă) n. a remedy for all kinds of diseases or troubles.

panache n. a confident stylish manner.

panama n. a straw hat.

panatella n. a thin cigar.

pancake n. a thin round cake of fried batter.

panchromatic adj. (of film etc.) sensitive to all colours of the visible spectrum.

pancreas n. the gland near the stomach, discharging insulin into the blood. □ **pancreatic** adj.

panda n. 1 a bear-like black and white animal of south-west China. 2 a racoon-like animal of India.

pandemic adj. (of a disease) occurring over a whole country or the world.

pandemonium n. uproar.

pander v. □ **pander to** gratify by satisfying a weakness or vulgar taste.

p. & p. abbr. postage and packing.

pane n. a sheet of glass in a window or door.

panegyric n. a piece of written or spoken praise.

panel n. 1 a distinct usu. rectangular section; a strip of board etc. forming this. 2 a group assembled to discuss or decide something; a list of jurors, a jury. ● v. (**panelled**; *Amer.*

paneled) cover or decorate with panels.

panelling *n.* (*Amer.* **paneling**) a series of wooden panels in a wall; wood used for making panels.

panellist *n.* (*Amer.* **panelist**) a member of a panel.

pang *n.* a sudden sharp pain.

pangolin *n.* a scaly anteater.

panic *n.* sudden strong fear. ● *v.* (**panicked, panicking**) affect or be affected with panic. □ **panic-stricken, panic-struck** *adj.,* **panicky** *adj.*

panjandrum *n.* a mock title for an important person.

pannier *n.* a large basket carried by a donkey etc.; a bag fitted on a motorcycle or bicycle.

panoply *n.* a splendid array.

panorama *n.* a view of a wide area or set of events. □ **panoramic** *adj.*

pan pipes *n.pl.* a musical instrument made of a series of graduated pipes.

pansy *n.* **1** a garden flower of the violet family with broad petals. **2** (*offensive*) an effeminate or homosexual man.

pant *v.* breathe with short quick breaths; utter breathlessly.

pantaloons *n.pl.* baggy trousers gathered at the ankles.

pantechnicon *n.* a large van for transporting furniture etc.

pantheism *n.* the doctrine that God is in everything. □ **pantheist** *n.,* **pantheistic** *adj.*

panther *n.* a leopard.

panties *n.pl.* (*colloquial*) underpants for women or children.

pantile *n.* a curved roof tile.

pantograph *n.* a device for copying a plan etc. on any scale.

pantomime *n.* a Christmas play based on a fairy tale.

pantry *n.* a room for storing china, glass, etc.; a larder.

pants *n.pl.* (*colloquial*) underpants; knickers; trousers:

pap *n.* soft food suitable for infants or invalids; pulp.

papacy *n.* the position or authority of the pope.

papal *adj.* of the pope or papacy.

paparazzi *n.pl.* freelance photographers who pursue celebrities.

papaw, papaya *n.* = pawpaw.

paper *n.* **1** a substance manufactured in thin sheets from wood fibre, rags, etc., used for writing on, wrapping, etc. **2** a newspaper; a set of examination questions; a document; a dissertation. ● *v.* cover (walls) with wallpaper.

paperback *adj. & n.* (a book) bound in a flexible paper binding.

paperweight *n.* a small heavy object for holding loose papers down.

paperwork *n.* clerical or administrative work.

papier mâché (**mash-ay**) *n.* moulded paper pulp used for making small objects.

papoose *n.* a young North American Indian child.

paprika *n.* red pepper.

papyrus *n.* (*pl.* **papyri**) a reed-like water plant from which the ancient Egyptians made a kind of paper; this paper; a manuscript written on this.

par *n.* **1** an average or normal amount, condition, etc. **2** (*Golf*) the number of strokes needed by a first-class player for a hole or course. **3** the face value of stocks and shares.

parable *n.* a story told to illustrate a moral.

parabola *n.* a curve like the path of an object that is thrown into the air and falls back to earth. □ **parabolic** *adj.*

paracetamol *n.* a drug that relieves pain and reduces fever.

parachute *n.* a device used to slow the descent of a person or object dropping from a great height. ● *v.* descend or drop by parachute. □ **parachutist** *n.*

parade *n.* 1 a formal assembly of troops. 2 a procession. 3 an ostentatious display. 4 a public square or promenade. ● *v.* 1 assemble for parade. 2 march or walk with display; make a display of.

paradigm (pa-ră-dym) *n.* an example; a model.

paradise *n.* heaven; Eden.

paradox *n.* a statement that seems self-contradictory but contains a truth. □ **paradoxical** *adj.*, **paradoxically** *adv.*

paraffin *n.* oil from petroleum or shale, used as fuel.

paragon *n.* an apparently perfect person or thing.

paragraph *n.* one or more sentences on a single theme, beginning on a new line. ● *v.* arrange in paragraphs.

parakeet *n.* a small parrot.

parallax *n.* an apparent difference in an an object's position when viewed from different points. □ **parallactic** *adj.*

parallel *adj.* 1 (of lines or planes) going continuously at the same distance from each other. 2 similar, corresponding. ● *n.* 1 a parallel line or thing. 2 a line of latitude. 3 an analogy. ● *v.* (**paralleled, paralleling**) 1 be parallel to. 2 compare. □ **parallelism** *n.*

parallelogram *n.* a four-sided geometric figure with its opposite sides parallel to each other.

paralyse *v.* (*Amer.* **paralyze**) 1 affect with paralysis. 2 bring to a standstill.

paralysis *n.* loss of power of movement.

paralytic *adj.* 1 affected with paralysis. 2 (*slang*) very drunk.

paramedic *n.* a skilled person working in support of medical staff. □ **paramedical** *adj.*

parameter *n.* a variable quantity or quality that restricts what it characterizes.

paramilitary *adj.* organized like a military force.

paramount *adj.* chief in importance.

paranoia *n.* a mental disorder in which a person has delusions of grandeur or persecution; an abnormal tendency to mistrust others. □ **paranoiac** *adj.*, **paranoid** *adj.*

paranormal *adj.* supernatural.

parapet *n.* a low wall along the edge of a balcony or bridge.

paraphernalia *n.* numerous belongings or pieces of equipment.

paraphrase *v.* express in other words. ● *n.* a rewording in this way.

paraplegia *n.* paralysis of the legs and part or all of the trunk. □ **paraplegic** *adj.* & *n.*

parapsychology *n.* the study of mental perceptions that seem outside normal abilities.

paraquat *n.* an extremely poisonous weedkiller.

parasailing *n.* (also **parascending**) a sport in which a person wearing a parachute is towed behind a motor boat or vehicle.

parasite *n.* an animal or plant living on or in another; a person living off another or others and giving no useful return. □ **parasitic** *adj.*

parasol *n.* a light umbrella used to give shade from the sun.

paratroops *n.pl.* parachute troops. □ **paratrooper** *n.*

parboil *v.* cook partially by boiling.

parcel *n.* 1 something wrapped for carrying or post. 2 something considered as a unit. ● *v.* (**parcelled**; *Amer.* **parceled**) 1 wrap as a parcel. 2 divide into portions.

parch v. make hot and dry; make thirsty.

parchment n. writing material made from animal skins; paper resembling this.

pardon n. forgiveness. ● v. (**pardoned**) forgive or excuse. □ **pardonable** adj.

pare v. trim the edges of; peel; reduce little by little.

parent n. a father or mother; an ancestor; a source from which other things are derived. □ **parental** adj., **parenthood** n.

parentage n. ancestry.

parenthesis n. (pl. **parentheses**) a word or phrase inserted into a passage; brackets (like these) placed round this. □ **parenthetic** adj., **parenthetical** adj., **parenthetically** adv.

parenting n. being a parent.

par excellence adv. being a supreme example of its kind.

pariah (pă-ry-ă) n. an outcast.

parietal bone n. each of a pair of bones forming part of the skull.

paring n. a piece pared off.

parish n. **1** an area with its own church and clergyman. **2** a local government area within a county.

parishioner n. an inhabitant of a parish.

Parisian adj. & n. (a native) of Paris.

parity n. equality.

park n. a public garden or recreation ground; the enclosed land of a country house. ● v. place and leave (a vehicle) temporarily.

parka n. a jacket with a fur-trimmed hood.

parking ticket n. a notice of a fine for illegal parking.

Parkinson's disease n. a disease causing trembling and weakness.

Parkinson's law n. the notion that work expands to fill the time available.

parlance n. vocabulary, expressions used by people.

parley n. (pl. **parleys**) a discussion to settle a dispute. ● v. (**parleyed**, **parleying**) hold a parley.

parliament n. an assembly that makes a country's laws. □ **parliamentary** adj., **parliamentarian** n.

parlour n. (Amer. **parlor**) a sitting room.

parlous adj. difficult; dangerous.

Parmesan n. a hard Italian cheese, usually grated.

parochial adj. **1** of a church parish. **2** interested in a limited area only. □ **parochially** adv., **parochialism** n.

parody n. a comic or grotesque imitation. ● v. make a parody of.

parole n. **1** a person's word of honour. **2** the release of a prisoner before the end of his or her sentence on condition of good behaviour. ● v. release in this way.

paroxysm n. a spasm; an outburst of laughter, rage, etc.

parquet (par-kay) n. flooring of wooden blocks arranged in a pattern.

parricide n. the killing of one's own parent. □ **parricidal** adj.

parrot n. **1** a tropical bird with a short hooked bill. **2** an unintelligent imitator. ● v. (**parroted**) repeat mechanically.

parry v. ward off (a blow); evade (a question) skilfully.

parse v. analyse (a sentence) in terms of grammar.

parsec n. a unit of distance used in astronomy, about 3.25 light years.

parsimonious adj. mean, very sparing. □ **parsimoniously** adv., **parsimony** n.

parsley n. a herb with crinkled green leaves.

parsnip n. a vegetable with a large yellowish tapering root.

parson n. (colloquial) a clergyman.

parsonage n. a rectory, a vicarage.

parson's nose n. a fatty lump on the rump of a cooked fowl.

part n. **1** some but not all; a distinct portion; a portion allotted. **2** a component. **3** a character assigned to an actor in a play etc. ● adv. partly. ● v. separate, divide. □ **in good part** without taking offence. **part with** give up possession of.

partake v. (partook, partaken, partaking) **1** participate. **2** take a portion, esp. of food. □ **partaker** n.

partial adj. **1** favouring one side or person, biased. **2** not complete or total. □ **be partial to** have a strong liking for. □ **partially** adv.

partiality n. bias, favouritism; a strong liking.

participate v. have a share, take part in something. □ **participant** n., **participation** n.

participle n. (Grammar) a word formed from a verb, as a **past participle** (e.g. burnt, frightened), or a **present participle** (e.g. burning, frightening). □ **participial** adj.

particle n. a very small portion of matter; a minor part of speech.

particoloured adj. (Amer. **particolored**) coloured partly in one colour, partly in another.

particular adj. **1** relating to one person or thing and not others. **2** carefully insisting on certain standards. ● n. a detail; a piece of information. □ **in particular** specifically. □ **particularly** adv.

parting n. **1** leave-taking. **2** a line from which hair is combed in different directions.

partisan n. **1** a strong supporter. **2** a guerrilla. □ **partisanship** n.

partition n. division into parts; a structure dividing a room or space, a thin wall. ● v. divide into parts or by a partition.

partitive adj. (Grammar) (of a word) denoting part of a group or quantity.

partly adv. partially.

partner n. a person sharing with another or others in an activity; each of a pair; a husband or wife or member of an unmarried couple. ● v. be the partner of; put together as partners. □ **partnership** n.

part of speech n. a word's grammatical class (noun, verb, adjective, etc.).

partridge n. a plump brown game bird.

part-time adj. for or during only part of the working week.

parturition n. the process of giving birth to young; childbirth.

party n. **1** a social gathering. **2** a group travelling or working as a unit; a group with common aims, esp. in politics. **3** one side in an agreement or dispute.

party line n. the policy of a political party.

party wall n. a wall common to two buildings or rooms.

pascal n. a unit of pressure.

paschal (pas-kăl) adj. of the Passover; of Easter.

pass v. **1** move onward or past; go or send to another person or place; go beyond. **2** discharge from the body as excreta. **3** change from one state into another; happen; occupy (time). **4** be accepted; examine and declare satisfactory; achieve the required standard in a test. **5** (in a game) refuse one's turn. ● n.

passing. **2** a movement made with the hands or thing held. **3** a permit to enter or leave. **4** a gap in mountains, allowing passage to the other side. □ **make a pass at** (colloquial) make sexual advances to. **pass away** die. **pass out** become unconscious. **pass over** disregard. **pass up** refuse to accept. **pass water** urinate.

passable adj. **1** able to be crossed. **2** just satisfactory. □ **passably** adv.

passage n. **1** passing; the right to pass or be a passenger. **2** a way through; esp. with a wall on each side; a tube-like structure. **3** a journey by sea. **4** a section of a literary or musical work. □ **passageway** n.

passbook n. a book recording a customer's deposits and withdrawals from a bank etc.

passé adj. old-fashioned.

passenger n. **1** a person (other than the driver, pilot, or crew) travelling in a vehicle, aircraft, etc. **2** an ineffective member of a team.

passer-by n. (pl. **passers-by**) a person who happens to be going past.

passim adv. throughout a book, article, etc.

passing adj. not lasting long; casual.

passion n. **1** strong emotion; sexual love; great enthusiasm. **2** (**the Passion**) the sufferings of Christ on the Cross.

passionate adj. full of passion; intense. □ **passionately** adv.

passive adj. **1** acted upon and not active; not resisting; lacking initiative or forceful qualities. **2** (Grammar) (of a verb) of which the subject undergoes the action. □ **passively** adv., **passiveness** n., **passivity** n.

Passover n. a Jewish festival commemorating the escape of Jews from slavery in Egypt.

passport n. an official document for use by a person travelling abroad.

password n. a secret phrase used to gain admission, prove identity, etc.

past adj. belonging to the time before the present, gone by. ● n. past time or events; a person's past life. ● prep. & adv. beyond.

pasta n. dried flour paste produced in various shapes; a cooked dish made with this.

paste n. **1** a moist mixture. **2** an adhesive. **3** a glasslike substance used in imitation gems. ● v. **1** fasten or coat with paste. **2** (slang) thrash.

pasteboard n. cardboard.

pastel n. **1** a chalk-like crayon; a drawing made with this. **2** a light delicate shade of colour.

pasteurize v. (also -**ise**) sterilize by heating. □ **pasteurization** n.

pastiche n. **1** a composition of selections from various sources. **2** a work in the style of another author.

pastille n. a small flavoured sweet; a lozenge.

pastime n. something done to pass time pleasantly.

past master n. an expert.

pastor n. a clergyman in charge of a church or congregation.

pastoral adj. **1** of country life. **2** of a pastor; of spiritual guidance.

pastrami n. seasoned smoked beef.

pastry n. a dough made of flour, fat, and water, used for making pies etc.; an individual item of food made with this.

pasturage n. pasture land.

pasture n. grassy land suitable for grazing cattle. ● v. put (animals) to graze.

pasty¹ (pas-ti) n. a pastry with a sweet or savoury filling, baked without a dish.

pasty² (pays-ti) adj. (**pastier**) **1** of or like paste. **2** unhealthily pale.

pat v. (**patted**) tap gently with an open hand. ● n. **1** a patting movement. **2** a small mass of a soft substance. ● adv. & adj. known and ready.

patch n. **1** a piece put on, esp. in mending. **2** a distinct area or period. **3** a piece of ground. ● v. put patch(es) on; piece (things) together. □ **not a patch on** (colloquial) not nearly as good as. **patch up** repair; settle (a quarrel etc.).

patchouli n. a fragrant plant of the Far East.

patchwork n. needlework in which small pieces of cloth are joined decoratively; something made of assorted pieces.

patchy adj. (**patchier**) existing in patches; uneven in quality. □ **patchily** adv., **patchiness** n.

pâté (pa-tay) n. a paste of meat etc.

patella n. (pl. **patellae**) the kneecap.

patent adj. **1** obvious. **2** patented. ● v. obtain or hold a patent for. ● n. an official right to be the sole maker or user of an invention or process; an invention etc. protected by this. □ **patently** adv.

patentee n. the holder of a patent.

patent leather n. leather with a glossy varnished surface.

paternal adj. of a father; fatherly; related through one's father. □ **paternally** adv.

paternalism n. a policy of making kindly provision for people's needs but giving them no responsibility. □ **paternalistic** adj.

paternity n. fatherhood.

path n. **1** a way by which people pass on foot; a line along which a person or thing moves. **2** a course of action.

pathetic adj. arousing pity or sadness; miserably inadequate. □ **pathetically** adv.

pathogenic adj. causing disease.

pathology n. the study of disease. □ **pathological** adj., **pathologist** n.

pathos n. a pathetic quality.

patience n. **1** calm endurance. **2** a card game for one player.

patient adj. showing patience. ● n. a person receiving medical treatment. □ **patiently** adv.

patina n. a sheen on a surface produced by age or use.

patio n. (pl. **patios**) a paved courtyard.

patisserie n. fancy pastries; a shop selling these.

patois (pat-wah) n. a dialect.

patriarch n. the male head of a family or tribe; a bishop of high rank in certain Churches. □ **patriarchal** adj., **patriarchate** n.

patriarchy n. a social organization in which a male is head of the family.

patricide n. the killing of one's own father. □ **patricidal** adj.

patrimony n. heritage.

patriot n. a patriotic person.

patriotic adj. loyally supporting one's country. □ **patriotically** adv., **patriotism** n.

patrol v. (**patrolled**) walk or travel regularly through (an area or building) to see that all is well. ● n. patrolling; a person patrolling.

patron (pay-trŏn) n. **1** a person giving influential or financial support to a cause. **2** a customer.

patronage n. **1** a patron's support. **2** patronizing behaviour.

patronize v. (also **-ise**) **1** act as patron to. **2** treat in a condescending way.

patron saint n. a saint regarded as a protector.

patronymic n. a name derived from that of a father or an ancestor.

patter v. make a series of quick tapping sounds; run with short quick steps. ● n. **1** a pattering sound. **2** rapid glib speech.

pattern n. **1** a decorative design. **2** a model, design, or instructions showing how a thing is to be made; a sample of cloth etc.; an excellent example. **3** the regular manner in which things occur. □ **patterned** adj.

patty n. a small pie or pasty.

paucity n. smallness of quantity.

paunch n. a large protruding stomach.

pauper n. a very poor person.

pause n. a temporary stop. ● v. make a pause.

pave v. cover (a path, area, etc.) with a hard surface.

pavement n. a paved surface; a path at the side of a road.

pavilion n. **1** a building on a sports ground for use by players and spectators. **2** an ornamental building.

pavlova n. a meringue containing cream and fruit.

paw n. a foot of an animal that has claws. ● v. strike with a paw; scrape (the ground) with a hoof; (colloquial) touch with the hands.

pawl n. a lever with a catch that engages with the notches of a ratchet.

pawn n. **1** a chess piece of the smallest size and value. **2** a person whose actions are controlled by others. ● v. deposit with a pawnbroker as security for money borrowed.

pawnbroker n. a person licensed to lend money on the security of personal property deposited.

pawnshop n. a pawnbroker's premises.

pawpaw n. (also **papaw**, **papaya**) an edible fruit of a palm-like tropical tree.

pay v. (**paid**, **paying**) **1** give (money) in return for goods or services; give what is owed. **2** be profitable or worthwhile. **3** suffer (a penalty). ● n. payment; wages. □ **pay off** pay (a debt) in full; discharge (an employee); yield good results. □ **payer** n.

payable adj. which must or may be paid.

PAYE abbr. pay-as-you-earn.

payee n. a person to whom money is paid or is to be paid.

payload n. an aircraft's or rocket's total load.

payment n. paying; money etc. paid.

pay-off n. (slang) **1** payment. **2** a reward, retribution, or end result.

payola n. a bribe offered for dishonest use of influence to promote a commercial product.

payroll n. a list of a firm's employees receiving regular pay.

Pb symb. lead.

PC abbr. **1** police constable. **2** personal computer. **3** politically correct; political correctness.

Pd symb. palladium.

p.d.q. abbr. (slang) pretty damn quick.

PE abbr. physical education.

pea n. a plant bearing seeds in pods; its round seed used as a vegetable.

peace n. a state of freedom from war or disturbance.

peaceable adj. fond of peace, not quarrelsome; peaceful. □ **peaceably** adv.

peace dividend n. public money made available when defence spending is reduced.

peaceful adj. characterized by peace. □ **peacefully** adv., **peacefulness** n.

peacemaker n. a person who brings about peace.

peach n. a round juicy fruit with a rough stone; a tree bearing this; its yellowish-pink colour.

peacock n. a male bird with splendid plumage and a long fan-like tail.

peahen n. the female of the peacock.

peak n. a pointed top, esp. of a mountain; the projecting part of the edge of a cap; the point of highest value, intensity, etc. □ **peaked** adj.

peaky adj. (**peakier**) looking pale and sickly.

peal n. the sound of ringing bells; a set of bells with different notes; a loud burst of thunder or laughter. ● v. sound in a peal.

pean Amer. sp. of **paean**.

peanut n. **1** a plant bearing underground pods with two edible seeds; this seed. **2** (**peanuts**) a trivial sum of money.

pear n. a rounded fruit tapering towards the stalk; a tree bearing this.

pearl n. a round creamy-white gem formed inside the shell of certain oysters. □ **pearly** adj.

peasant n. a person working on the land, esp. in the Middle Ages.

peasantry n. peasants collectively.

peat n. decomposed vegetable matter from bogs etc., used in horticulture or as fuel. □ **peaty** adj.

pebble n. a small smooth round stone. □ **pebbly** adj.

pecan n. a smooth pinkish-brown nut; a tree bearing this.

peccary n. a small wild pig of Central and South America.

peck v. **1** strike, nip, or pick up with the beak. **2** kiss hastily. ● n. a pecking movement.

pecker n. (Amer., vulgar) the penis. □ **keep your pecker up** (colloquial) stay cheerful.

peckish adj. (colloquial) hungry.

pectin n. a substance found in fruits which makes jam set.

pectoral adj. of, in, or on the chest or breast. ● n. a pectoral fin or muscle.

peculiar adj. **1** strange, eccentric. **2** belonging exclusively to one person or place or thing; special. □ **peculiarly** adv., **peculiarity** n.

pecuniary adj. of or in money.

pedagogue n. a person who teaches pedantically.

pedal n. a lever operated by the foot. ● v. (**pedalled**; Amer. **pedaled**) work the pedal(s) of; operate by pedals.

pedalo n. (pl. **pedalos** or **pedaloes**) a small pedal-operated pleasure boat.

pedant n. a pedantic person. □ **pedantry** n.

pedantic adj. insisting on strict observance of rules and details. □ **pedantically** adv.

peddle v. sell (goods) as a pedlar.

peddler n. **1** a person who sells illegal drugs. **2** Amer. sp. of **pedlar**.

pedestal n. a base supporting a column or statue etc.

pedestrian n. a person walking, esp. in a street. ● adj. **1** of or for pedestrians. **2** unimaginative, dull.

pediatrics etc. Amer. sp. of **paediatrics** etc.

pedicure n. the care or treatment of the feet and toenails.

pedigree n. a line or list of (esp. distinguished) ancestors. ● adj. (of an animal) of recorded and pure breeding.

pediment n. a triangular part crowning the front of a building.

pedlar n. (*Amer.* **peddler**) a person who sells small articles from door to door.

pedometer n. a device for estimating the distance travelled on foot.

peduncle n. the stalk of a flower etc.

pee (*colloquial*) v. urinate. ● n. urine.

peek v. peep, glance. ● n. a peep.

peel n. the skin of certain fruits and vegetables. ● v. remove the peel off; strip off; come off in strips or layers, lose skin or bark etc. in this way. □ **peeler** n., **peelings** n.pl.

peep v. look through a narrow opening; look quickly or surreptitiously; show slightly. ● n. a brief or surreptitious look.

peephole n. a small hole to peep through.

peeping Tom n. a furtive voyeur.

peer[1] v. look searchingly or with difficulty or effort.

peer[2] n. **1** a duke, marquess, earl, viscount, or baron. **2** one who is the equal of another in rank, merit, age, etc.

peerage n. peers as a group; the rank of peer or peeress.

peeress n. a female peer; a peer's wife.

peerless adj. without equal, superb.

peeved adj. (*colloquial*) annoyed.

peevish adj. irritable. □ **peevishly** adv.

peewit n. (also **pewit**) a lapwing.

peg n. a wooden or metal pin or stake; a clip for holding clothes on a line. ● v. (**pegged**) **1** fix or mark by means of pegs. **2** keep (wages or prices) at a fixed level. □ **off the peg** (of clothes) ready-made.

pejorative adj. expressing disapproval.

peke n. a Pekinese dog.

Pekinese n. (also **Pekingese**) a dog of a breed with short legs, a flat face, and silky hair.

pelargonium n. a plant with showy flowers (often called *geranium*).

pelican n. a waterbird with a pouch in its long bill for storing fish.

pelican crossing n. a pedestrian crossing with lights operated by pedestrians.

pellagra n. a deficiency disease causing cracking of the skin.

pellet n. a small round mass of a substance; a piece of small shot. □ **pelleted** adj.

pell-mell adj. & adv. in a hurrying disorderly manner, headlong.

pellucid adj. very clear.

pelmet n. an ornamental strip above a window etc.

pelt[1] v. **1** throw missiles at. **2** run fast. ● n. an animal skin. □ **at full pelt** as fast as possible.

pelvis n. the framework of bones round the body below the waist. □ **pelvic** adj.

pen[1] n. **1** a device with a metal point for writing with ink. **2** a small fenced enclosure, esp. for animals. **3** a female swan. ● v. (**penned**) **1** write (a letter etc.). **2** shut in or as if in an animal pen.

penal adj. of or involving punishment.

penalize v. (also **-ise**) inflict a penalty on; put at a disadvantage. □ **penalization** n.

penalty n. a punishment for breaking a law, rule, or contract.

penance n. an act performed as an expression of penitence.

pence see **penny**.

penchant n. (pahn-shahn) a liking.

pencil n. an instrument containing graphite, used for drawing or writing. ● v. (pen-

cilled; *Amer.* **penciled)** write, draw, or mark with a pencil.

pendant *n.* an ornament hung from a chain round the neck.

pendent *adj.* (also **pendant**) hanging.

pending *adj.* waiting to be decided or settled. ● *prep.* during; until.

pendulous *adj.* hanging loosely.

pendulum *n.* a weight hung from a cord and swinging freely; a rod with a weighted end that regulates a clock's movement.

penetrate *v.* make a way into or through, pierce; see **into** or **through**. □ **penetrable** *adj.*, **penetration** *n.*

penetrating *adj.* **1** showing great insight. **2** (of sound) piercing.

penfriend *n.* a friend to whom a person writes regularly without meeting.

penguin *n.* a flightless seabird of Antarctic regions.

penicillin *n.* an antibiotic obtained from mould fungi.

peninsula *n.* a piece of land almost surrounded by water. □ **peninsular** *adj.*

penis *n.* the organ by which a male mammal copulates and urinates.

penitent *adj.* feeling or showing regret that one has done wrong. ● *n.* a penitent person. □ **penitently** *adv.*, **penitence** *n.*

penitential *adj.* of penitence or penance.

pen-name *n.* an author's pseudonym.

pennant *n.* a long tapering flag.

penniless *adj.* having no money.

pennon *n.* a flag, esp. a long triangular or forked one.

penny *n.* (*pl.* **pennies** for separate coins, **pence** for a sum of money) a British bronze coin worth one-hundredth of £1; a former coin worth one-twelfth of a shilling.

penny-pinching *adj.* niggardly, mean.

pen-pushing *n.* (*derog.*, *colloquial*) clerical work.

pension *n.* an income paid by the government, an ex-employer, or a private fund to a person who is retired, disabled, etc. ● *v.* pay a pension to. □ **pension off** dismiss with a pension.

pension (pahn-si-ahn) *n.* a guest house in Europe.

pensionable *adj.* entitled or (of a job) entitling one to a pension.

pensioner *n.* a person who receives a pension.

pensive *adj.* deep in thought. □ **pensively** *adv.*, **pensiveness** *n.*

pentagon *n.* a geometric figure with five sides. □ **pentagonal** *adj.*

pentagram *n.* a five-pointed star.

pentathlon *n.* an athletic contest involving five events.

Pentecost *n.* the Jewish harvest festival; Whit Sunday.

penthouse *n.* a flat (usu. with a terrace) on the roof of a tall building.

penultimate *adj.* last but one.

penumbra *n.* (*pl.* **penumbrae** or **penumbras**) an area of partial shadow, esp. in an eclipse.

penury *n.* poverty. □ **penurious** *adj.*

peony *n.* (also **paeony**) a plant with large showy flowers.

people *n.pl.* **1** human beings; persons; the subjects of a state. **2** parents or other relatives. ● *n.sing.* persons composing a race or nation. ● *v.* fill with people, populate.

PEP *abbr.* personal equity plan.

pep *n.* (*colloquial*) vigour. □ **pep up** (**pepped**) fill with vigour, enliven.

pepper *n.* **1** a hot-tasting seasoning powder made from the dried berries of certain plants. **2**

a capsicum. ● v. 1 sprinkle with pepper. 2 pelt; sprinkle. □ **peppery** adj.

peppercorn n. a dried black berry from which pepper is made.

peppercorn rent n. a very low rent.

peppermint n. a mint producing a strong fragrant oil; a sweet flavoured with this.

pepsin n. an enzyme in gastric juice, helping in digestion.

pep talk n. a talk urging great effort.

peptic adj. of digestion.

per prep. 1 for each. 2 in accordance with. 3 by means of.

per annum adv. for each year.

per capita adv. & adj. for each person.

per cent adv. (Amer. **percent**) in or for every hundred.

perambulate v. walk through or round (an area). □ **perambulation** n.

perceive v. become aware of, see or notice.

percentage n. a rate or proportion per hundred; a proportion, a part.

perceptible adj. able to be perceived. □ **perceptibly** adv.

perception n. perceiving, the ability to perceive.

perceptive adj. showing insight and understanding. □ **perceptively** adv., **perceptiveness** n.

perch[1] n. a bird's resting place, a rod etc. for this; a high seat. ● v. rest or place on something temporarily.

perch[2] n. (pl. **perch**) an edible freshwater fish with spiny fins.

percipient adj. perceptive. □ **percipience** n.

percolate v. filter, esp. through small holes; prepare in a percolator. □ **percolation** n.

percolator n. a coffee-making pot in which boiling water is circulated through ground coffee in a perforated drum.

percussion n. the playing of a musical instrument by striking it with a stick etc. □ **percussive** adj.

perdition n. eternal damnation.

peregrination n. travelling.

peregrine n. a falcon.

peremptory adj. imperious; insisting on obedience. □ **peremptorily** adv.

perennial adj. lasting a long time; constantly recurring; (of plants) living for several years. ● n. a perennial plant. □ **perennially** adv.

perestroika n. (in the former USSR) reform of the economic and political system.

perfect adj. (per-fekt) complete, entire; faultless, excellent; exact. ● v. (per-fekt) make perfect. □ **perfectly** adv., **perfection** n.

perfectionist n. a person who seeks perfection. □ **perfectionism** n.

perfidious adj. treacherous, disloyal. □ **perfidy** n.

perforate v. make hole(s) through. □ **perforation** n.

perforce adv. (old use) unavoidably, necessarily.

perform v. 1 carry into effect. 2 function. 3 go through (a piece of music, ceremony, etc.); act in a play, sing, do tricks etc. before an audience. □ **performance** n., **performer** n.

perfume n. a sweet smell; a fragrant liquid for applying to the body. ● v. give a sweet smell to.

perfumery n. perfumes.

perfunctory adj. done only as a duty, without much care or interest. □ **perfunctorily** adv.

pergola n. an arbour or walkway with arches covered in climbing plants.

perhaps adv. it may be, possibly.

pericardium n. (pl. **pericardia**) a membranous sac enclosing the heart.

perigee n. the point nearest to the earth in the moon's orbit.

peril n. serious danger.

perilous adj. full of risk, dangerous. □ **perilously** adv.

perimeter n. the outer edge of an area; the length of this.

perinatal adj. of the time immediately before and after birth.

period n. **1** a length or portion of time. **2** an occurrence of menstruation. **3** a complete sentence; a full stop in punctuation. ● adj. (of dress or furniture) belonging to a past age.

periodic adj. happening at intervals. □ **periodicity** n.

periodical adj. periodic. ● n. a magazine etc. published at regular intervals. □ **periodically** adv.

peripatetic adj. going or on the move from place to place.

peripheral adj. of or on the periphery; of minor but not central importance.

periphery n. a boundary, an edge; the fringes of a subject.

periphrasis n. (pl. **periphrases**) a roundabout phrase or way of speaking.

periscope n. a tube with mirrors by which a person can see things above.

perish v. suffer destruction, die; rot. □ **perishing** adj.

perishable adj. liable to decay or go bad in a short time.

peritoneum n. (pl. **peritoneums** or **peritonea**) the membrane lining the abdominal cavity.

peritonitis n. inflammation of the peritoneum.

periwinkle n. **1** a trailing plant with blue or white flowers. **2** a winkle.

perjure v. □ **perjure oneself** lie under oath.

perjury n. the deliberate giving of false evidence while under oath; this evidence.

perk n. (colloquial) a perquisite. □ **perk up** cheer up; brighten or smarten up.

perky adj. (**perkier**) lively and cheerful. □ **perkily** adv., **perkiness** n.

perm n. **1** a treatment giving a permanent artificial wave to hair. **2** a permutation. ● v. **1** treat (hair) with a perm. **2** make a permutation of.

permafrost n. permanently frozen subsoil in arctic regions.

permanent adj. lasting indefinitely. □ **permanently** adv., **permanence** n.

permeable adj. able to be permeated by fluids etc. □ **permeability** n.

permeate v. pass or flow into every part of.

permissible adj. allowable.

permission n. consent, authorization.

permissive adj. tolerant, esp. in social and sexual matters. □ **permissiveness** n.

permit v. (per-mit) (**permitted**) give permission to or for; make possible. ● n. (per-mit) a written order giving permission, esp. for entry.

permutation n. a variation of the order of or choice from a set of things.

pernicious adj. harmful.

pernickety adj. (colloquial) fastidious, scrupulous.

peroration n. the concluding part of a speech.

peroxide n. a compound of hydrogen used to bleach hair. ● v. bleach with this.

perpendicular adj. at an angle of 90° to a line or surface; upright, vertical. ● n. a perpendicular line or direction. □ **perpendicularly** adv.

perpetrate v. commit (a crime), be guilty of (a blunder). □ **perpetration** n., **perpetrator** n.

perpetual adj. lasting, not ceasing. □ **perpetually** adv.

perpetuate v. preserve from being forgotten or from going out of use. □ **perpetuation** n.

perpetuity n. □ **in perpetuity** for ever.

perplex v. bewilder, puzzle.

perplexity n. bewilderment.

per pro. see **p.p.**

perquisite n. a privilege given in addition to wages; an extra profit or right.

■ **Usage** *Perquisite* is sometimes confused with *prerequisite*, which means 'a thing required as a precondition'.

---

perry n. a drink resembling cider, made from fermented pears.

per se adv. by or in itself; intrinsically.

persecute v. treat with hostility because of race or religion; harass. □ **persecution** n., **persecutor** n.

persevere v. continue in spite of difficulties. □ **perseverance** n.

persimmon n. an edible orange tomato-like fruit.

persist v. continue firmly or obstinately; continue to exist. □ **persistent** adj., **persistently** adv., **persistence** n.

person n. 1 an individual human being; one's body. 2 (Grammar) one of the three classes of personal pronouns and verb forms, referring to the person(s) speaking, spoken to, or spoken of. □ **in person** physically present.

persona n. (pl. **personae**) a personality as perceived by others.

personable adj. attractive in appearance or manner.

personage n. an important person.

personal adj. of one's own; of or involving a person's private life; referring to a person; done in person. □ **personally** adv.

personal computer n. one designed for use by a single indivual.

personal equity plan n. a scheme for tax-free personal investment.

personality n. 1 a person's distinctive character; a person with distinctive qualities. 2 a celebrity.

personalize v. (also **-ise**) identify as belonging to a particular person. □ **personalization** n.

personal organizer n. a loose-leaf folder or pocket-sized computer for keeping details of meetings, phone numbers, addresses, etc.

personal pronoun see **pronoun**.

personify v. 1 represent in human form or as having human characteristics. 2 embody in one's behaviour. □ **personification** n.

personnel n. employees, staff.

perspective n. 1 the art of drawing so as to give an effect of solidity and relative position. 2 a mental view of the relative importance of things. □ **in perspective** drawn according to the rules of perspective; not distorting a thing's relative importance.

perspex n. (trade mark) a tough light transparent plastic.

perspicacious adj. showing great insight. □ **perspicaciously** adv., **perspicacity** n.

perspicuous adj. expressing things clearly. □ **perspicuity** n.

perspire v. sweat. □ **perspiration** n.

persuade v. cause (a person) to believe or do something by reasoning. □ **persuader** n.

persuasion n. **1** persuading; persuasiveness. **2** a belief.

persuasive adj. able or trying to persuade people. □ **persuasively** adv., **persuasiveness** n.

pert adj. **1** cheeky; saucy. **2** jaunty; lively. □ **pertly** adv., **pertness** n.

pertain v. be relevant; belong as a part.

pertinacious adj. persistent and determined. □ **pertinaciously** adv., **pertinacity** n.

pertinent adj. pertaining, relevant. □ **pertinently** adv., **pertinence** n.

perturb v. agitate, make uneasy. □ **perturbation** n.

peruse v. read carefully. □ **perusal** n.

pervade v. spread throughout (a thing). □ **pervasive** adj.

perverse adj. obstinately doing something different from what is reasonable or required. □ **perversely** adv., **perversity** n.

pervert v. (per-vert) misapply, lead astray, corrupt. ● n. (per-vert) a person who is perverted, esp. sexually. □ **perversion** n.

pervious adj. permeable; penetrable.

peseta n. a unit of money in Spain.

peso n. (pl. **pesos**) a unit of money in several South American countries.

pessary n. a vaginal suppository.

pessimism n. a tendency to take a gloomy view of things. □ **pessimist** n., **pessimistic** adj., **pessimistically** adv.

pest n. a troublesome person or thing; an insect or animal harmful to plants, stored food, etc.

pester v. annoy continually, esp. with requests or questions.

pesticide n. a substance used to destroy harmful insects etc.

pestilence n. a deadly epidemic disease. □ **pestilential** adj.

pestle n. a club-shaped instrument for pounding things to powder.

pesto n. a sauce of basil, oil, Parmesan cheese, and nuts, often used on pasta.

pet n. **1** a tame animal treated with affection. **2** a favourite. ● adj. **1** kept as a pet. **2** favourite. ● v. (**petted**) **1** treat with affection. **2** fondle.

petal n. one of the coloured outer parts of a flower head.

peter v. □ **peter out** diminish gradually.

petersham n. strong corded ribbon for stiffening waistbands.

pethidine n. a painkilling drug.

petite adj. small and dainty.

petition n. a formal written request signed by many people. ● v. make a petition to.

petrel n. a seabird.

petrify v. **1** change into a stony mass. **2** paralyse with astonishment or fear. □ **petrifaction** n.

petrochemical n. a chemical substance obtained from petroleum or gas.

petrol n. an inflammable liquid made from petroleum for use as fuel in internal-combustion engines.

petroleum n. a mineral oil found underground, refined for use as fuel or in dry-cleaning etc.

petticoat n. a dress-length undergarment worn beneath a dress or skirt.

pettifogging adj. trivial; quibbling about unimportant details.

pettish adj. peevish, irritable.

petty adj. (**pettier**) unimportant; minor, on a small scale; small-minded. □ **pettily** adv., **pettiness** n.

petty cash n. money kept by an office etc. for small payments.

petulant adj. irritable. □ **petulantly** adv., **petulance** n.

petunia n. a plant with large funnel-shaped flowers.

pew n. a long bench-like seat in a church; (colloquial) a seat.

pewit var. of **peewit**.

pewter n. a grey alloy of tin with lead or other metal.

pfennig n. a German coin, one-hundredth of a mark.

pH n. a measure of acidity or alkalinity.

PG abbr. (of a film) classified as suitable for children subject to parental guidance.

phalanx n. a compact mass esp. of people.

phallus n. (pl. **phalluses** or **phalli**) (an image of) the penis. □ **phallic** adj.

phantom n. a ghost.

Pharaoh (fair-oh) n. the title of the kings of ancient Egypt.

pharmaceutical adj. of or engaged in pharmacy.

pharmacist n. a person skilled in pharmacy.

pharmacology n. the study of the action of drugs. □ **pharmacological** adj., **pharmacologist** n.

pharmacopoeia (far-mă-kŏ-pee-ă) n. a list or stock of drugs.

pharmacy n. the preparation and dispensing of medicinal drugs; a pharmacist's shop, a dispensary.

pharynx n. the cavity at the back of the nose and throat. □ **pharyngeal** adj.

phase n. a stage of change or development. ● v. carry out (a programme etc.) in stages. □ **phase in** or **out** bring gradually into or out of use.

Ph.D. abbr. Doctor of Philosophy, a higher degree.

pheasant n. a game bird with bright feathers in the male.

phenomenal adj. extraordinary, remarkable. □ **phenomenally** adv.

phenomenon n. (pl. **phenomena**) 1 a fact, occurrence, or change perceived by the senses or the mind. 2 a remarkable person or thing.

■ **Usage** It is a mistake to use the plural form phenomena when only one phenomenon is meant.

pheromone n. a substance secreted by the body to produce a response in other animals.

phial n. a small bottle.

philander v. (of a man) flirt. □ **philanderer** n.

philanthropy n. love of mankind, esp. shown in benevolent acts. □ **philanthropist** n., **philanthropic** adj.

philately n. stamp-collecting. □ **philatelic** adj., **philatelist** n.

philistine n. an uncultured person.

philology n. the study of languages. □ **philologist** n., **philological** adj.

philosopher n. 1 a person skilled in philosophy. 2 a philosophical person.

philosophical adj. 1 of philosophy. 2 bearing misfortune calmly. □ **philosophically** adv.

philosophize v. (also **-ise**) theorize; moralize.

philosophy n. a system or study of the basic truths and principles of the universe, life, and morals, and of human understanding of these; a person's principles.

philtre n. (Amer. **philter**) a magic potion.

phlegm (flem) n. a thick mucus in the bronchial passages, ejected by coughing.

phlegmatic *adj.* not easily excited or agitated; sluggish, apathetic. □ **phlegmatically** *adv.*

phlox *n.* a flowering plant, either tall-growing or a rockery plant.

phobia *n.* a lasting abnormal fear or great dislike. □ **phobic** *adj. & n.*

phoenix (**fee-niks**) *n.* a mythical Arabian bird said to burn itself and rise young again from its ashes.

phone *n.* a telephone. ● *v.* telephone.

phonecard *n.* a card containing prepaid units for use in a cardphone.

phone-in *n.* a broadcast programme in which listeners telephone the studio.

phonetic *adj.* **1** of or representing speech sounds. **2** (of spelling) corresponding to pronunciation. ● *n.* (**phonetics**) the study or representation of speech sounds. □ **phonetically** *adv.*, **phonetician** *n.*

phoney (also **phony**) (*colloquial*) *adj.* false, sham. ● *n.* a phoney person or thing.

phonograph *n.* (*Amer.*) a record player.

phosphate *n.* a fertilizer containing phosphorus.

phosphorescent *adj.* luminous. □ **phosphorescence** *n.*

phosphorus *n.* a chemical element (symbol P); a wax-like form of this appearing luminous in the dark.

photo *n.* (*pl.* **photos**) a photograph.

photocopier *n.* a machine for photocopying documents.

photocopy *n.* a photographed copy of a document. ● *v.* make a photocopy of.

photoelectric cell *n.* an electronic device emitting an electric current when light falls on it.

photo finish *n.* a close finish where the winner is decided by a photograph.

photofit *n.* a likeness of a person made up of separate photographs of features.

photogenic *adj.* looking attractive in photographs.

photograph *n.* a picture formed by the chemical action of light or other radiation on sensitive material. ● *v.* take a photograph of; come out (well or badly) when photographed. □ **photographer** *n.*, **photography** *n.*, **photographic** *adj.*, **photographically** *adv.*

photojournalism *n.* the reporting of news by photographs.

photon *n.* an indivisible unit of electromagnetic radiation.

photosensitive *adj.* reacting to light.

photostat *n.* (*trade mark*) a photocopier; a photocopy. ● *v.* (**photostatted**) make a photocopy of.

photosynthesis *n.* the process by which green plants use sunlight to convert carbon dioxide and water into complex substances.

phrase *n.* a group of words forming a unit; a unit in a melody. ● *v.* **1** express in words. **2** divide (music) into phrases. □ **phrasal** *adj.*

phraseology *n.* the way something is worded.

phrenology *n.* the study of the shape of a person's skull.

phylum *n.* (*pl.* **phyla**) a major division of the plant or animal kingdom.

physical *adj.* **1** of the body. **2** of matter or the laws of nature. **3** of physics. □ **physically** *adv.*

physical geography *n.* the study of the earth's natural features.

physician *n.* a doctor, esp. one specializing in medicine as distinct from surgery.

physicist *n.* an expert in physics.

physics *n.* the study of the properties and interactions of matter and energy.

physiognomy *n.* the features of a person's face.

physiology *n.* the study of the bodily functions of living organisms. □ **physiological** *adj.*, **physiologist** *n.*

physiotherapy *n.* treatment of an injury etc. by massage and exercises. □ **physiotherapist** *n.*

physique *n.* a person's physical build and muscular development.

pi *n.* a Greek letter (π) used as a symbol for the ratio of a circle's circumference to its diameter (about 3.14).

pianissimo *adv.* (*Music*) very softly.

pianist *n.* a person who plays the piano.

piano *n.* (*pl.* **pianos**) a musical instrument with strings struck by hammers operated by a keyboard. ● *adv.* (*Music*) softly.

pianoforte *n.* (*old use*) a piano.

piazza (pee-at-să) *n.* a public square or market place.

picador *n.* a mounted bullfighter with a lance.

picaresque *adj.* (of fiction) dealing with the adventures of rogues.

piccalilli *n.* a pickle of chopped vegetables and hot spices.

piccaninny *n.* (*Amer.* **pickaninny**) (*offensive*) a black child; an Aboriginal child.

piccolo *n.* (*pl.* **piccolos**) a small flute.

pick *v.* **1** select. **2** use a pointed instrument or the fingers or beak etc. to make (a hole) in or remove bits from (a thing). **3** detach (flower or fruit) from the plant bearing it. ● *n.* **1** picking; a selection; the best part. **2** a pickaxe. **3** a plectrum. □ **pick a lock** open it with a tool other than a key. **pick a pocket** steal its contents. **pick a quarrel** provoke one deliberately. **pick holes in** find fault with. **pick off** pluck off; shoot or destroy one by one. **pick on** nag, find fault with; select. **pick out** select; discern. **pick up** lift or take up; collect; acquire, obtain; become acquainted with casually; recover health, improve; gather (speed). □ **picker** *n.*

pickaback var. of **piggyback**.

pickaninny *Amer.* sp. of **piccaninny**.

pickaxe *n.* (*Amer.* **pickax**) a tool with a pointed iron bar at right angles to its handle, for breaking ground etc.

picket *n.* **1** people stationed outside a workplace to dissuade others from entering during a strike; a party of sentries. **2** a pointed stake set in the ground. ● *v.* (**picketed**) **1** form a picket on (a workplace). **2** enclose with stakes.

pickings *n.pl.* odd gains or perquisites.

pickle *n.* **1** vegetables preserved in vinegar or brine; this liquid. **2** (*colloquial*) a plight, a mess. ● *v.* preserve in pickle.

pickpocket *n.* a thief who steals from people's pockets.

picnic *n.* an informal outdoor meal. ● *v.* (**picnicked**, **picnicking**) take part in a picnic. □ **picnicker** *n.*

pictograph *n.* a pictorial symbol used as a form of writing.

pictorial *adj.* of, in, or like a picture or pictures; illustrated. ● *n.* a newspaper etc. with many pictures. □ **pictorially** *adv.*

picture *n.* **1** a representation made by painting, drawing, or photography etc. **2** something that looks beautiful. **3** a scene; a description. **4** a cinema film. ● *v.* **1** depict. **2** imagine.

picturesque *adj.* forming a pleasant scene; (of words or a description) very expressive.

pick-up *n.* **1** a small open motor truck. **2** a stylus-holder in a record player.

pidgin *n.* a simplified form of English or another language with elements of a local language.

pie *n.* a baked dish of meat, fish, or fruit covered with pastry or another crust.

piebald *adj.* (of a horse) with irregular patches of white and black.

piece *n.* **1** a part, a portion. **2** something regarded as a unit. **3** a musical, literary, or artistic composition. **4** a small object used in board games. ● *v.* make by putting pieces together. □ **of a piece** of the same kind; consistent.

piecemeal *adj.* & *adv.* done piece by piece, part at a time.

pièce de résistance (pee-ess dă ray-zees-tahns) (*pl.* **pièces de résistance**) the principal dish of a meal; the most impressive item.

piecework *n.* work paid according to the quantity done.

pie chart *n.* a diagram representing quantities as sectors of a circle.

pied *adj.* particoloured.

pied-à-terre (pee-ayd ă tair) *n.* (*pl.* **pieds-à-terre**) a small house for occasional use.

pie-eyed *adj.* (*slang*) drunk.

pier *n.* **1** a structure built out into the sea. **2** a pillar supporting an arch or bridge.

pierce *v.* go into or through like a sharp-pointed instrument; make (a hole) in; force one's way into or through.

piercing *adj.* **1** (of cold or wind) penetrating sharply. **2** (of sound) shrilly audible. □ **piercingly** *adv.*

piety *n.* piousness.

piffle *n.* (*colloquial*) nonsense.

pig *n.* **1** an animal with short legs, cloven hooves, and a blunt snout. **2** (*colloquial*) a greedy or unpleasant person.

pigeon *n.* **1** a bird of the dove family. **2** (*colloquial*) a person's business or responsibility.

pigeon-hole *n.* a small compartment in a desk or cabinet. ● *v.* **1** put away for future consideration or indefinitely. **2** classify.

piggery *n.* a pig-breeding establishment; a pigsty.

piggy *adj.* like a pig.

piggyback *n.* (also **pickaback**) a ride on a person's back.

piggy bank *n.* a money box shaped like a pig.

pig-headed *adj.* obstinate.

pig-iron *n.* crude iron from a smelting furnace.

piglet *n.* a young pig.

pigment *n.* colouring matter.

pigmy var. of **pygmy**.

pigsty *n.* a covered pen for pigs.

pigtail *n.* long hair worn in a plait at the back of the head.

pike *n.* **1** a spear with a long wooden shaft. **2** (*pl.* **pike**) a large voracious freshwater fish.

pilaf *n.* (also **pilaff, pilau**) a dish of rice with meat, spices, etc.

pilaster *n.* a rectangular usu. ornamental column.

pilchard *n.* a small sea fish related to the herring.

pile *n.* **1** a number of things lying one on top of another; (*colloquial*) a large amount. **2** a lofty building. **3** a heavy beam driven vertically into the ground as a support for a building or bridge. **4** cut or uncut loops on the surface of fabric. **5** (**piles**)

haemorrhoids. ● v. heap, stack, load. □ **pile up** accumulate.

pile-up n. a collision of several vehicles.

pilfer v. steal (small items or in small quantities). □ **pilferer** n.

pilgrim n. a person who travels to a sacred place as an act of religious devotion. □ **pilgrimage** n.

pill n. a small piece of medicinal substance for swallowing whole; (**the pill**) a contraceptive pill.

pillage n. & v. plunder.

pillar n. a vertical structure used as a support or ornament.

pillar box n. a postbox.

pillbox n. 1 a small round hat. 2 a concrete gun shelter.

pillion n. a passenger seat behind the driver of a motorcycle.

pillory n. a wooden frame in which offenders used to be locked and exposed to public abuse. ● v. ridicule publicly.

pillow n. a cushion for supporting the head in bed.

pilot n. a person who operates an aircraft's flying controls; a person qualified to steer ships in or out of a harbour; a guide. ● v. (**piloted**) act as a pilot of; guide.

pilot light n. 1 a small burning jet of gas which lights a larger burner. 2 an electric indicator light.

pimento n. (pl. **pimentos**) 1 allspice. 2 (also **pimiento**) a sweet pepper.

pimp n. a man who finds clients for a prostitute or brothel.

pimple n. a small inflamed spot on the skin. □ **pimply** adj.

PIN abbr. personal identification number, a number for use with a cashcard machine.

pin n. 1 a short pointed piece of metal with a broadened head, used for fastening things together. 2 a peg or stake of wood or metal. ● v. (**pinned**) 1 fasten with pin(s); attach, fix. 2 hold down and make unable to move; transfix. □ **pin down** establish clearly; bind by a promise. **pins and needles** a tingling sensation.

pinafore n. 1 an apron. 2 (in full **pinafore dress**) a sleeveless dress worn over a blouse or jumper.

pinball n. a game in which balls are propelled across a sloping board to strike targets.

pince-nez (panss-nay) n. spectacles that clip onto the nose.

pincers n.pl. a tool with pivoted jaws for gripping and pulling things; a claw-like part of a lobster etc.

pinch v. 1 squeeze between two surfaces, esp. between the finger and thumb. 2 (slang) steal. ● n. 1 pinching. 2 the stress of circumstances. 3 a small amount. □ **at a pinch** if really necessary.

pine n. an evergreen tree with needle-shaped leaves; its wood. ● v. lose strength through grief or yearning; feel an intense longing.

pineapple n. a large juicy tropical fruit.

pine nut n. the edible seed of various pine trees.

ping n. a short sharp ringing sound. ● v. make this sound. □ **pinger** n.

ping-pong n. table tennis.

pinion n. 1 a bird's wing. 2 a small cogwheel. ● v. restrain by holding or binding the arms or legs.

pink adj. pale red. ● n. 1 a pink colour. 2 a garden plant with fragrant flowers. ● v. 1 pierce slightly. 2 cut a zigzag edge on (fabric). 3 (of an engine) make slight explosive sounds when running imperfectly. □ **in the pink** (colloquial) in very good health.

pinnacle n. a pointed ornament on a roof; a peak; the highest point.

pinpoint v. locate precisely.

pinstripe n. a very narrow stripe in cloth fabric. □ **pinstriped** adj.

pint n. a measure for liquids, one-eighth of a gallon (0.568 litre).

pin-up n. (colloquial) a picture of an attractive or famous person.

pioneer n. a person who is one of the first to explore a new region or subject. ● v. be a pioneer (in).

pious adj. devout in religion; too virtuous. □ **piously** adv., **piousness** n.

pip n. **1** a small seed in fruit. **2** a star showing rank on an army officer's uniform. **3** a short high-pitched sound. ● v. (**pipped**) (colloquial) defeat by a small margin.

pipe n. **1** a tube through which something can flow. **2** a wind instrument; (**pipes**) bagpipes. **3** a narrow tube with a bowl at one end for smoking tobacco. ● v. **1** convey through pipe(s). **2** play (music) on a pipe. **3** utter in a shrill voice. □ **pipe down** (colloquial) be quiet.

pipe dream n. an unrealistic hope or scheme.

pipeline n. a long pipe for conveying petroleum etc. to a distance; a channel of supply or information. □ **in the pipeline** on the way, in preparation.

piper n. a player of pipes.

pipette n. a slender tube for transferring or measuring small amounts of liquid.

pipistrelle n. a small bat.

pipit n. a small bird resembling a lark.

pippin n. a variety of apple.

piquant adj. pleasantly sharp in taste or smell. □ **piquancy** n.

pique (peek) v. hurt the pride of. ● n. a feeling of hurt pride.

piquet (pee-kay) n. a card game for two players.

piranha (pi-rah-nǎ) n. a fierce tropical American freshwater fish.

pirate n. **1** a person on a ship who robs another ship at sea or raids a coast. **2** one who infringes copyright or business rights, or broadcasts without authorization. ● v. reproduce (a book, video, etc.) without authorization. □ **piratical** adj., **piracy** n.

pirouette v. spin on the toe in dancing.

piss (vulgar slang) v. urinate. ● n. urine.

pissed adj. (vulgar slang) drunk.

pistachio (pis-tash-i-oh) n. (pl. **pistachios**) a nut.

piste n. a ski run.

pistil n. the seed-producing part of a flower.

pistol n. a small gun.

piston n. a sliding disc or cylinder inside a tube, esp. as part of an engine or pump.

pit n. **1** a hole in the ground; a coal mine; a sunken area. **2** a place where racing cars are refuelled etc. during a race. ● v. (**pitted**) **1** make pits or depressions in. **2** set against in competition. **3** remove stones from olives etc.

pita var. of **pitta**.

pit bull terrier n. a small stocky dog noted for its ferocity.

pitch v. **1** throw. **2** erect (a tent or camp). **3** set at a particular slope or level. **4** fall heavily; (of a ship) plunge forward and back alternately. ● n. **1** the act of pitching. **2** steepness. **3** intensity. **4** the degree of highness or lowness of a music note or voice. **5** a place where a street trader or performer is stationed;

a place where a tent is pitched. **6** a playing field. **7** a dark tarry substance. □ **pitch in** (*colloquial*) set to work vigorously.

pitch-black, **pitch-dark** *adj.* completely black; with no light.

pitchblende *n.* a mineral ore (uranium oxide) yielding radium.

pitched battle *n.* one fought from prepared positions.

pitcher *n.* **1** a baseball player who throws the ball to the batter. **2** a large usu. earthenware jug.

pitchfork *n.* a long-handled fork for lifting and tossing hay.

piteous *adj.* deserving or arousing pity. □ **piteously** *adv.*

pitfall *n.* an unsuspected danger or difficulty.

pith *n.* **1** spongy tissue in stems or fruits. **2** an essential part.

pithy *adj.* **1** like pith, containing much pith. **2** brief and full of meaning. □ **pithily** *adv.*

pitiful *adj.* deserving or arousing pity or contempt. □ **pitifully** *adv.*

piton *n.* a peg with a hole for a rope, used in rock-climbing.

pitta *n.* (also **pita**) a flat bread, hollow inside.

pittance *n.* a very small allowance or wage.

pituitary gland *n.* a gland at the base of the brain, influencing bodily growth and functions.

pity *n.* a feeling of sorrow for another's suffering; a cause for regret. ● *v.* feel pity for. □ **take pity on** pity and try to help.

pivot *n.* a central point or shaft on which a thing turns or swings. ● *v.* (**pivoted**) turn on a pivot.

pivotal *adj.* **1** of a pivot. **2** vitally important.

pixel *n.* any of the minute illuminated areas making up the image on a VDU screen.

pixie *n.* (also **pixy**) a small supernatural being in fairy tales.

pizza *n.* a layer of dough baked with a savoury topping.

pizzeria *n.* a pizza restaurant.

pizzicato *adv.* plucking the strings of a violin etc. instead of using the bow.

pl. *abbr.* **1** plural. **2** (usu. **Pl.**) Place. **3** plate.

placard *n.* a poster or similar notice. ● *v.* put up placards on.

placate *v.* conciliate. □ **placatory** *adj.*

place *n.* **1** a particular portion of space or of an area. **2** a particular town, district, building, etc. **3** a position. **4** (a duty appropriate to) one's rank. ● *v.* **1** put into a place, find a place for. **2** locate, identify. **3** put or give (an order for goods etc.). □ **be placed** (in a race) be among the first three.

placebo (pla-see-boh) *n.* (*pl.* **placebos**) a harmless substance given as medicine, esp. to humour a patient.

placement *n.* placing.

placenta *n.* (*pl.* **placentae** or **placentas**) the organ in the womb that nourishes the foetus. □ **placental** *adj.*

place setting *n.* a set of dishes or cutlery for one person at a table.

placid *adj.* calm, not easily upset. □ **placidly** *adv.*, **placidity** *n.*

placket *n.* an opening in a garment for fastenings or access to a pocket.

plagiarize (play-ji-ryz) *v.* (also **-ise**) take and use (another's writings etc.) as one's own. □ **plagiarism** *n.*, **plagiarist** *n.*

plague *n.* **1** a deadly contagious disease. **2** an infestation. ● *v.* (*colloquial*) annoy, pester.

plaice *n.* (*pl.* **plaice**) an edible flatfish.

plaid (plad) *n.* cloth with a tartan pattern; a long piece of such cloth worn as part of Scottish Highland dress.

plain *adj.* **1** unmistakable, easy to see, hear, or understand. **2** not elaborate, simple; ordinary, without affectation. **3** in exact terms, candid, frank. **4** not good-looking. ● *adv.* plainly. ● *n.* a large area of level country. □ **plainly** *adv.*, **plainness** *n.*

plain clothes *n.pl.* civilian clothes, not a uniform.

plain sailing *n.* a period or activity that is free from difficulties.

plainsong *n.* (also **plainchant**) medieval church music for voices, without regular rhythm.

plaintiff *n.* a person bringing an action in a court of law.

plaintive *adj.* sounding sad. □ **plaintively** *adv.*

plait (plat) *v.* weave (three or more strands) into one rope-like length. ● *n.* something plaited.

plan *n.* **1** a diagram showing the relative position of parts of a building, town, etc. **2** a method thought out in advance. ● *v.* (**planned**) make a plan (of). □ **planner** *n.*

plane *n.* **1** an aeroplane. **2** a level surface; a level of thought or development. **3** a tool for smoothing wood or metal by paring shavings from it. **4** a tall spreading tree with broad leaves. ● *v.* smooth or pare (wood or metal) with a plane. ● *adj.* level.

planet *n.* a celestial body orbiting round a star. □ **planetary** *adj.*

planetarium *n.* (*pl.* **planetaria** or **planetariums**) a room with a domed ceiling on which lights are projected to show the positions of the stars and planets.

plangent *adj.* **1** loud and resonant. **2** sad, mournful. □ **plangently** *adv.*, **plangency** *n.*

plank *n.* a long flat piece of timber.

plankton *n.* minute life forms floating in the sea, rivers, etc.

planning permission *n.* formal approval for construction of or changes to a building.

plant *n.* **1** a living organism with neither the power of movement nor special organs of digestion. **2** a factory; its machinery. ● *v.* place in soil for growing; place in position. □ **planter** *n.*

plantain *n.* **1** a tropical banana-like fruit; a tree bearing this. **2** a herb whose seed is used as birdseed.

plantation *n.* an area planted with trees or cultivated plants; an estate on which cotton, tobacco, tea, etc. is cultivated.

planter *n.* **1** an owner or manager of a plantation. **2** a device for planting things. **3** a large container for plants.

plaque *n.* **1** a commemorative plate fixed on a wall. **2** a film forming on teeth, encouraging harmful bacteria.

plasma *n.* **1** the colourless fluid part of blood. **2** a kind of gas. □ **plasmic** *adj.*

plaster *n.* **1** a mixture of lime, sand, water, etc. used for coating walls. **2** a sticking plaster. ● *v.* cover with plaster; coat, daub. □ **plasterer** *n.*

plasterboard *n.* board with a core of plaster, for making partitions etc.

plastered *adj.* (*slang*) drunk.

plaster of Paris *n.* a white paste made from gypsum, for making moulds or casts.

plastic *adj.* **1** able to be moulded. **2** made of plastic. ● *n.* a synthetic substance moulded to a permanent shape. □ **plasticity** *n.*

plastic bullet *n.* a solid plastic cylinder fired as a riot-control device rather than to kill.

plasticine n. (*trade mark*) a plastic substance used for modelling.

plastic money n. credit cards as opposed to cash or cheques.

plastic surgery n. an operation to replace injured or defective external tissue.

plate n. **1** an almost flat usu. circular utensil for holding food. **2** articles of gold, silver, or other metal. **3** a flat thin sheet of metal, glass, or other material. **4** an illustration on special paper in a book. **5** a denture. ● v. cover or coat with metal. □ **plateful** n.

plateau n. (*pl.* **plateaux** or **plateaus**) **1** an area of level high ground. **2** a steady state following an increase.

plate glass n. thick glass for windows etc.

platelet n. a small disc in the blood, involved in clotting.

platen n. a plate in a printing press holding the paper against the type; the roller of a typewriter or printer.

platform n. a raised level surface or area, esp. from which a speaker addresses an audience.

platinum n. an element (symbol Pt), a silver-white metal that does not tarnish.

platinum blonde n. a woman with silvery-blonde hair.

platitude n. a commonplace remark. □ **platitudinous** adj.

platonic adj. involving affection but not sexual love.

platoon n. a subdivision of a military company.

platter n. a large plate for food.

platypus n. (*pl.* **platypuses**) an Australian animal with a ducklike beak which lays eggs but suckles its young.

plaudits n.pl. applause; an expression of approval.

plausible adj. seeming probable; persuasive but deceptive. □ **plausibly** adv., **plausibility** n.

play v. **1** occupy oneself in (a game) or in other recreational activity; compete against in a game; move (a piece, a ball, etc.) in a game. **2** act the part of. **3** perform on (a musical instrument). **4** cause (a radio, recording, etc.) to produce sound. **5** move lightly; touch gently. ● n. **1** playing. **2** activity, operation. **3** a literary work for a stage or broadcast performance. **4** free movement. □ **play at** perform in a half-hearted way. **play down** minimize the importance of. **play safe** not take risks. **play the game** behave honourably. **play up** (*colloquial*) fail to work properly; cause trouble; be unruly. □ **player** n.

playboy n. a pleasure-loving usu. rich man.

playful adj. full of fun; in a mood for play, not serious. □ **playfully** adv., **playfulness** n.

playgroup n. a group of preschool children who play regularly together under supervision.

playhouse n. a theatre.

playing card n. one of a set of (usu. 52) pieces of card used in games.

playing field n. a field used for outdoor games.

playmate n. a child's companion in play.

play pen n. a portable enclosure for a young child to play in.

playwright n. a person who writes plays.

plaza n. a public square.

plc abbr. (also **PLC**) Public Limited Company.

plea n. a defendant's answer to a charge in a law court; an appeal, an entreaty; an excuse.

plead v. (**pleaded** (*Scot. & Amer.* **pled**), **pleading**) give as one's plea; put forward (a case) in a law court; make an appeal or entreaty; put forward as an excuse.

pleasant *adj.* pleasing; having an agreeable manner. □ **pleasantly** *adv.*, **pleasantness** *n.*

pleasantry *n.* a friendly or humorous remark.

please *v.* **1** give pleasure to. **2** think fit, have the desire. ● *adv.* a polite word of request. □ **please oneself** do as one chooses.

pleased *adj.* feeling or showing pleasure or satisfaction.

pleasurable *adj.* causing pleasure. □ **pleasurably** *adv.*

pleasure *n.* a feeling of satisfaction or joy; a source of this.

pleat *n.* a flat fold of cloth. ● *v.* make a pleat or pleats in.

pleb *n.* (*colloquial*) a rough uncultured person.

plebeian *adj.* of the lower social classes; uncultured, vulgar.

plebiscite (pleb-i-syt) *n.* a referendum.

plectrum *n.* (*pl.* **plectrums** or **plectra**) a small piece of plastic etc. for plucking the strings of a musical instrument.

pledge *n.* something deposited as a guarantee (e.g. that a debt will be paid); a token of something; a solemn promise. ● *v.* deposit as a pledge; promise solemnly.

plenary *adj.* entire; attended by all members.

plenipotentiary *adj.* & *n.* (an envoy) with full powers to take action.

plenitude *n.* abundance; completeness.

plentiful *adj.* existing in large amounts. □ **plentifully** *adv.*

plenty *n.* enough and more. ● *adv.* (*colloquial*) quite, fully. □ **plenteous** *adj.*

plethora *n.* an oversupply or excess.

pleurisy *n.* inflammation of the membrane round the lungs.

pliable *adj.* flexible; easily influenced. □ **pliability** *n.*

pliant *adj.* pliable. □ **pliancy** *n.*

pliers *n.pl.* pincers with flat surfaces for gripping things.

plight *n.* a predicament.

plimsoll *n.* a canvas sports shoe.

Plimsoll line *n.* a mark on a ship's side showing the legal water level when loaded.

plinth *n.* a slab forming the base of a column or statue etc.

plod *v.* (**plodded**) walk doggedly, trudge; work slowly but steadily. □ **plodder** *n.*

plonk *n.* (*colloquial*) cheap or inferior wine.

plop *n.* a sound like something small dropping into water with no splash.

plot *n.* **1** a small piece of land. **2** the story in a play, novel, or film. **3** a conspiracy, a secret plan. ● *v.* (**plotted**) **1** make a map or chart of, mark on this. **2** plan secretly. □ **plotter** *n.*

plough (plow) *n.* (*Amer.* **plow**) an implement for cutting furrows in soil and turning it up. ● *v.* **1** cut or turn up (soil etc.) with a plough. **2** make one's way laboriously. □ **ploughman** *n.*

ploughman's lunch *n.* a meal of bread, cheese, and pickle.

plover *n.* a wading bird.

ploy *n.* a cunning manoeuvre.

pluck *v.* pull at or out or off; pick (a flower etc.); strip (a bird) of its feathers. ● *n.* **1** a plucking movement. **2** courage.

plucky *adj.* (**pluckier**) brave, spirited. □ **pluckily** *adv.*

plug *n.* **1** something fitting into and stopping or filling a hole or cavity. **2** a device of this kind for making an electrical connection. ● *v.* (**plugged**) **1** put a plug into. **2** (*colloquial*) work diligently. **3** (*colloquial*) advertise by constant commendation.

□ **plug in** connect electrically by putting a plug into a socket.

plum n. **1** a fruit with sweet pulp round a pointed stone; a tree bearing this. **2** reddish purple. **3** something desirable, the best.

plumage n. a bird's feathers.

plumb (plum) n. a lead weight hung on a cord (**plumb line**), used for testing depths or verticality. ● adv. exactly; (Amer.) completely. ● v. **1** measure or test with a plumb line; reach (depths); get to the bottom of. **2** work or fit (things) as a plumber.

plumber (plum-er) n. a person who fits and repairs plumbing.

plumbing (plum-ing) n. a system of water and drainage pipes etc. in a building.

plume n. a feather, esp. as an ornament; something resembling this. □ **plumed** adj.

plummet v. (**plummeted**) fall steeply or rapidly.

plump adj. having a full rounded shape. ● v. **1** make or become plump. **2** plunge abruptly. □ **plump for** choose, decide on. □ **plumpness** n.

plunder v. rob. ● n. plundering; goods etc. stolen.

plunge v. **1** thrust or go forcefully into something. **2** go down suddenly. ● n. plunging, a dive.

plunger n. a device that works with a plunging movement.

pluperfect adj. (Grammar) of the tense used to denote some action completed before some past point of time, e.g. we had arrived.

plural n. the form of a noun or verb used in referring to more than one person or thing. ● adj. of this form; of more than one. □ **plurality** n.

plus prep. **1** with the addition of. **2** above zero. ● adj. **1** more than zero. **2** more than the amount indicated. ● n. **1** the sign (+). **2** an advantage.

■ **Usage** The use of plus as a conjunction, as in they arrived late, plus they wanted a meal, is considered incorrect except in very informal use.

plush n. cloth with a long soft nap. ● adj. **1** made of plush. **2** luxurious.

plushy adj. (**plushier**) luxurious.

plutocrat n. a wealthy and influential person. □ **plutocracy** n., **plutocratic** adj.

plutonium n. a chemical element (symbol Pu), a radioactive substance used in nuclear weapons and reactors.

pluvial adj. of or caused by rain.

ply¹ n. **1** a thickness or layer of wood, cloth, etc. **2** plywood.

ply² v. **1** use or wield (a tool etc.); work at (a trade). **2** keep offering or supplying. **3** go to and fro regularly, esp. looking for custom.

plywood n. board made by gluing layers with the grain crosswise.

PM abbr. **1** Prime Minister. **2** post-mortem.

Pm symb. promethium.

p.m. abbr. (Latin post meridiem) after noon.

PMS abbr. premenstrual syndrome.

PMT abbr. premenstrual tension.

pneumatic (new-ma-tik) adj. filled with or operated by compressed air. □ **pneumatically** adv.

pneumonia (new-moh-niă) n. inflammation of the lungs.

PO abbr. **1** Post Office. **2** postal order.

Po symb. polonium.

poach v. **1** cook (an egg without its shell) in or over boiling water; simmer in a small amount of liquid. **2** take (game or fish) illegally; trespass, encroach. □ **poacher** n.

pocket n. **1** a small bag-like part on a garment; a pouch-like compartment. **2** an isolated group or area. ● adj. suitable for carrying in one's pocket. ● v. **1** put into one's pocket. **2** take dishonestly. □ **in** or **out of pocket** having made a profit or loss. □ **pocketful** n.

pocketbook n. **1** a notebook. **2** a small folding case for money or papers.

pocket money n. money for small personal expenses; money allowed to children.

pock-marked adj. marked by scars or pits.

poco adv. (Music) a little, rather.

pod n. a long narrow seed case.

podgy adj. (**podgier**) short and fat.

podium n. (pl. **podiums** or **podia**) a pedestal or platform.

poem n. a literary composition in verse.

poet n. a person who writes poems.

poetic adj. (also **poetical**) of or like poetry. □ **poetically** adv.

poetry n. **1** poems; a poet's work. **2** a quality that pleases the mind in a poetic way.

po-faced adj. solemn.

pogo stick n. a stilt-like toy with a spring, for jumping about on.

pogrom n. an organized massacre.

poignant adj. arousing sympathy, moving; keenly felt. □ **poignantly** adv., **poignancy** n.

poinsettia n. a plant with large scarlet or cream bracts.

point n. **1** a tapered or sharp end, a tip; a promontory. **2** a particular place, moment, or stage. **3** a unit of measurement or scoring. **4** an item, a detail; a characteristic; a chief or im-

portant feature. **5** effectiveness. **6** an electrical socket. **7** a movable rail for directing a train from one line to another. ● v. **1** aim, direct (a finger or weapon etc.); have a certain direction; indicate. **2** fill in (joints of brickwork) with mortar. □ **on the point of** on the verge of (an action). **point out** draw attention to. **point up** emphasize. **to the point** relevant.

point-blank adj. **1** aimed or fired at very close range. **2** (of a remark) direct. ● adv. in a point-blank manner.

point duty n. traffic control by a policeman at a road junction.

pointed adj. **1** tapering to a point. **2** (of a remark or manner) emphasized, direct; cutting. □ **pointedly** adv.

pointer n. a thing that points to something; a dog that points towards game which it scents.

pointing n. the mortar around the edges of bricks in a wall.

pointless adj. having no purpose or meaning. □ **pointlessly** adv.

point of view n. a way of considering an issue.

poise n. **1** balance. **2** a dignified self-assured manner.

poison n. a substance that can destroy life or harm health. ● v. give poison to; put poison on or in; corrupt, fill with prejudice. □ **poisoner** n., **poisonous** adj.

poison pen letter n. a malicious unsigned letter.

poke v. **1** thrust with the end of a finger or stick etc. **2** thrust forward. **3** search, pry. ● n. a poking movement. □ **poke fun at** ridicule.

poker n. **1** a stiff metal rod for stirring up a fire. **2** a gambling card game.

poker-face n. one that does not reveal thoughts or feelings.

poky adj. (**pokier**) small and cramped. □ **pokiness** n.

polar adj. of or near the North or South Pole; of a pole of a magnet.

polar bear n. a white bear of Arctic regions.

polarize v. (also **-ise**) **1** confine similar vibrations of (light waves) to one direction or plane. **2** give magnetic poles to. **3** set at opposite extremes of opinion. □ **polarization** n.

Polaroid n. (trade mark) **1** a material that polarizes light passing through it, used in sunglasses. **2** a camera that prints a photograph as soon as it is taken.

Pole n. a Polish person.

pole n. **1** a long rod or post. **2** the north (**North Pole**) or south (**South Pole**) end of the earth's axis; one of the opposite ends of a magnet or terminals of an electric cell or battery. ● v. push along using a long rod.

polecat n. a small animal of the weasel family; (Amer.) a skunk.

polemic n. a verbal attack on a belief or opinion. □ **polemical** adj.

polenta n. porridge made from maize meal.

pole position n. the most favourable starting position in a motor race.

pole star n. a star near the North Pole in the sky.

police n. a civil force responsible for keeping public order. ● v. keep order in (a place) by means of police.

policeman n. (pl. **-men**) a male member of the police force.

police state n. a country where political police supervise and control citizens' activities.

policewoman n. (pl. **-women**) a female member of the police force.

policy n. **1** a general plan of action. **2** an insurance contract.

polio n. poliomyelitis.

poliomyelitis n. an infectious disease causing temporary or permanent paralysis.

Polish adj. & n. (the language) of Poland.

polish v. **1** make smooth and glossy by rubbing. **2** refine, perfect. ● n. **1** glossiness; polishing, a substance used for this. **2** elegance. □ **polish off** finish off. □ **polisher** n.

polished adj. (of manner or performance) elegant, perfected.

polite adj. having good manners, socially correct; refined. □ **politely** adv., **politeness** n.

politic adj. showing good judgement.

political adj. of or involving politics; of the way a country is governed. □ **politically** adv.

political correctness n. avoidance of any expressions or behaviour that may be considered discriminatory.

politician n. an MP or other political representative.

politics n. the science and art of government; political affairs or life; (as pl.) political principles.

polity n. civil government; an organized society.

polka n. a lively dance for couples.

polka dots n.pl. round evenly spaced dots on fabric.

poll n. **1** the votes cast in an election; a place for this. **2** an estimate of public opinion made by questioning people. ● v. **1** receive as votes. **2** cut off the top of (a tree etc.).

pollack n. (pl. **pollack**) an edible sea fish related to cod.

pollard v. poll (a tree) to produce a close head of young branches. ● n. a pollarded tree; a hornless animal.

pollen *n.* a fertilizing powder produced by flowers.

pollen count *n.* a measurement of the amount of pollen in the air.

pollinate *v.* fertilize with pollen. □ **pollination** *n.*

pollster *n.* a person conducting an opinion poll.

poll tax *n.* (*hist.*) a tax on each member of the population.

pollute *v.* make dirty or impure. □ **pollutant** *n.*, **pollution** *n.*

polo *n.* a game like hockey played by teams on horseback.

polonaise *n.* a slow processional dance.

polo neck *n.* a high turned-over collar.

polonium *n.* a radioactive metallic element (symbol Po).

poltergeist *n.* a spirit that throws things about noisily.

polyandry *n.* a system of having more than one husband at a time.

polyanthus *n.* (*pl.* **polyanthuses**) a cultivated primrose.

polychrome *adj.* multicoloured. □ **polychromatic** *adj.*

polyester *n.* a synthetic resin or fibre.

polyethylene *n.* = polythene.

polygamy *n.* a system of having more than one wife at a time. □ **polygamist** *n.*, **polygamous** *adj.*

polyglot *adj.* & *n.* (a person) knowing several languages.

polygon *n.* a geometric figure with many sides. □ **polygonal** *adj.*

polygraph *n.* a machine reading the pulse rate etc. used as a lie detector.

polyhedron *n.* (*pl.* **polyhedra** or **polyhedrons**) a solid with many sides. □ **polyhedral** *adj.*

polymath *n.* a person with knowledge of many subjects.

polymer *n.* a compound whose molecule is formed from a large number of simple molecules.

polymerize *v.* (also **-ise**) combine into a polymer. □ **polymerization** *n.*

polynomial *adj.* consisting of three or more terms.

polyp *n.* **1** a simple organism with a tube-shaped body. **2** an abnormal growth projecting from a mucous membrane.

polyphony *n.* a combination of melodies. □ **polyphonal** *adj.*

polystyrene *n.* a plastic, a polymer of styrene.

polytechnic *n.* a college offering courses up to degree level.

polytheism *n.* belief in or worship of more than one god. □ **polytheist** *n.*, **polytheistic** *adj.*

polythene *n.* (also **polyethylene**) a tough light plastic.

polyunsaturated *adj.* (of fat) not associated with the formation of cholesterol in the blood.

polyurethane *n.* a synthetic resin or plastic.

polyvinyl chloride *n.* a plastic used as fabric (PVC).

pomander *n.* a ball of mixed sweet-smelling substances.

pomegranate *n.* a tropical fruit with many seeds; a tree bearing this.

Pomeranian *n.* a dog of a small silky-haired breed.

pommel (pum-ĕl) *n.* a knob on the hilt of a sword; an upward projection on a saddle.

pomp *n.* stately and splendid ceremonial.

pompon *n.* (also **pompom**) a decorative tuft or ball.

pompous *adj.* full of ostentatious dignity and self-importance. □ **pompously** *adv.*, **pomposity** *n.*

ponce *n.* **1** a pimp. **2** (*offensive*) a homosexual or effeminate man.

poncho n. (pl. **ponchos**) a cloak like a blanket with a hole for the head.

pond n. a small area of still water.

ponder v. be deep in thought; think over.

ponderous adj. heavy, unwieldy; laborious. □ **ponderously** adv.

pong n. & v. (slang) stink.

pontiff n. the Pope.

pontificate v. speak pompously and at length.

pontoon n. **1** a flat-bottomed boat. **2** a floating platform. **3** a card game.

pontoon bridge n. a temporary bridge supported on pontoons.

pony n. a horse of any small breed.

ponytail n. long hair drawn back and tied to hang down.

poodle n. a dog with thick curly hair.

poof n. (also **poofter**) (offensive slang) a homosexual or effeminate man.

pooh int. an exclamation of contempt.

pooh-pooh v. dismiss (a subject) scornfully.

pool n. **1** a small area of still water; a puddle; a swimming pool. **2** a shared fund or supply. **3** a game resembling snooker. **4** (**the pools**) football pools. ● v. put into a common fund or supply.

poop n. a ship's stern; a raised deck at the stern.

pooper scooper n. a device for picking up dog faeces.

poor adj. **1** having little money or means. **2** not abundant. **3** not very good. **4** pitiable. □ **poorness** n.

poorly adv. in a poor way, badly. ● adj. unwell.

pop n. **1** a small explosive sound. **2** a fizzy drink. **3** pop music. ● v. (**popped**) **1** make or cause

to make a pop. **2** put, come, or go quickly. ● adj. in a popular modern style.

popadam var. of **poppadom**.

popcorn n. maize heated to burst and form puffy balls.

Pope n. the head of the Roman Catholic Church.

poplar n. a tall slender tree.

poplin n. a plain woven usu. cotton fabric.

pop music n. modern music appealing to young people.

poppadom n. (also **popadam**, **poppadam**) a large thin crisp savoury Indian bread.

popper n. (colloquial) a press stud.

poppy n. a plant with bright flowers on tall stems.

poppycock n. (slang) nonsense.

populace n. the general public.

popular adj. liked, enjoyed, or used by many people; of or for the general public. □ **popularly** adv., **popularity** n.

popularize v. (also **-ise**) **1** make generally liked. **2** present in a popular non-technical form.

populate v. fill with a population.

population n. the inhabitants of an area.

populous adj. thickly populated.

porcelain n. fine china.

porch n. a roofed shelter over the entrance of a building.

porcine adj. of or like a pig.

porcupine n. an animal covered with protective spines.

pore n. a tiny opening on skin or on a leaf, for giving off or taking in moisture. □ **pore over** study closely.

pork n. unsalted pig meat.

porn n. (colloquial) pornography.

pornography n. writings or pictures intended to stimulate erotic feelings by portraying sexual activity. □ **pornographer** n., **pornographic** adj.

porous adj. letting through fluid or air. □ **porosity** n.

porphyry n. a rock containing mineral crystals.

porpoise n. a small whale with a blunt rounded snout.

porridge n. a food made by boiling oatmeal or other cereal in water or milk.

port n. 1 a harbour; a town with a harbour. 2 an opening in a ship's side; a porthole. 3 the left-hand side of a ship or aircraft. 4 strong sweet wine.

portable adj. able to be carried. □ **portability** n.

Portakabin n. (trade mark) a small portable building.

portal n. a door or entrance, esp. an imposing one.

portcullis n. a vertical grating lowered to block the gateway to a castle.

portend v. foreshadow.

portent n. an omen, a significant sign. □ **portentous** adj.

porter n. 1 a person employed to carry luggage or goods. 2 a doorkeeper of a large building. 3 a dark beer.

portfolio n. (pl. **portfolios**) 1 a case for loose sheets of paper. 2 a set of investments. 3 the position of a Minister of State.

porthole n. a window in the side of a ship or aircraft.

portico n. (pl. **porticoes** or **porticos**) a roof supported by columns forming a porch or similar structure.

portion n. a part, a share; an amount of food for one person. ● v. divide into portions.

portly adj. (**portlier**) stout and dignified. □ **portliness** n.

portmanteau n. (pl. **portmanteaus** or **portmanteaux**) a two-part case for clothes.

portrait n. a picture of a person or animal; a description.

portray v. make a picture of; describe; represent in a play etc. □ **portrayal** n.

Portuguese adj. & n. (a native, the language) of Portugal.

Portuguese man-of-war n. a jellyfish.

pose v. 1 put into or take a particular attitude. 2 pretend (to be a person or thing). 3 put forward, present (a problem etc.). ● n. 1 an attitude in which someone is posed. 2 a pretence.

poser n. 1 a puzzling problem. 2 a poseur.

poseur n. a person who behaves affectedly.

posh adj. (colloquial) very smart, luxurious.

posit v. assume, postulate.

position n. 1 a place occupied by or intended for a person or thing. 2 a posture. 3 a situation. 4 status. 5 a job. ● v. place, put. □ **positional** adj.

positive adj. 1 definite; explicit. 2 constructive. 3 (of a quantity) greater than zero. 4 (of a battery terminal) through which electric current enters. 5 (of a photograph) with lights, shades, or colours as in the subject, not reversed. ● n. a positive quality, quantity, or photograph. □ **positively** adv.

positive pole n. the north-seeking pole of a magnet.

positive vetting n. an intensive enquiry into the background of an applicant.

positron n. a particle with a positive electric charge.

posse (poss-ee) n. a group of law-enforcers; a strong force or company.

possess v. 1 own; hold as belonging to oneself. 2 dominate the mind of. □ **possessor** n.

possession n. possessing; something owned. □ **take possession of** become the possessor of.

possessive *adj.* of or indicating possession; desiring to possess things. □ **possessively** *adv.*, **possessiveness** *n.*

possessive pronoun see **pro-noun.**

possible *adj.* capable of existing, happening, being done, etc. □ **possibly** *adv.*, **possibility** *n.*

possum *n.* (*colloquial*) an opos-sum. □ **play possum** pretend to be unaware.

post. 1 the official conveyance of letters etc.; the letters etc. conveyed. **2** a piece of timber, metal, etc. set upright to support or mark something. **3** a place of duty; a job; an outpost of sol-diers; a trading station. ● *v.* **1** send (letters etc.) by post. **2** put up (a notice); announce in this way. **3** place, station. □ **keep me posted** keep me informed.

post- *pref.* after.

postage *n.* a charge for sending something by post.

postal *adj.* of the post; by post.

postbox *n.* a box into which letters are put for sending by post.

postcard *n.* a card for sending messages by post without an envelope.

postcode *n.* a group of letters and figures in a postal address to assist sorting.

postdate *v.* **1** put a date on (a cheque etc.) that is later than the actual date. **2** occur later than.

poster *n.* a large picture or no-tice announcing or advertising something.

post-haste *adv.* with great haste.

post office *n.* a building where postal business is carried o:

poste restante (pohst rest-ahnt) *n.* a post office depart-ment where letters are kept un-til called for.

posterior *adj.* situated behind or at the back. ● *n.* the buttocks.

posterity *n.* future generations.

postern *n.* a small back or side entrance to a fortress etc.

postgraduate *n.* a student studying for a higher degree.

posthumous *adj.* happening, awarded, published, etc. after a person's death. □ **posthum-ously** *adv.*

postman *n.* (*pl.* **-men**) a person who delivers or collects letters etc.

postmark *n.* an official mark stamped on something sent by post, giving place and date of marking. ● *v.* mark with this.

postmaster *n.* a male official in charge of a post office.

postmistress *n.* a female official in charge of a post office.

post-mortem *n.* **1** an examina-tion of a body to determine the cause of death. **2** an analysis of something that has happened.

post-natal *adj.* after childbirth.

postpone *v.* keep (an event etc.) from occurring until a later time. □ **postponement** *n.*

postprandial *adj.* (*formal*) after lunch or dinner.

postscript *n.* an additional paragraph at the end of a letter etc.

post-traumatic stress dis-order (or **syndrome**) *n.* symp-toms that typically occur after exposure to a stressful situation.

postulant *n.* a candidate for admission to a religious order.

postulate *v.* assume to be true as a basis for reasoning. □ **postulation** *n.*

posture *n.* the way a person stands, walks, etc. ● *v.* assume a posture, esp. for effect. □ **pos-tural** *adj.*

posy *n.* a small bunch of flowers.

pot *n.* **1** a vessel for holding li-quids or solids, or for cooking in. **2** (*slang*) marijuana. ● *v.*

(**potted**) **1** put into a pot. **2** send (a ball in billiards etc.) into a pocket. □ **go to pot** (*slang*) fall into a bad state.

potable *adj.* drinkable.

potash *n.* potassium carbonate.

potassium *n.* a soft silvery-white metallic element (symbol K).

potation *n.* drinking; a drink.

potato *n.* (*pl.* **potatoes**) a plant with starchy tubers used as food; one of these tubers.

pot belly *n.* a large protuberant belly.

potboiler *n.* a book, painting, etc. produced merely to make money.

poteen (poch-**een**) *n.* illegally distilled whisky.

potent *adj.* **1** having great natural power; having a strong effect. **2** (of a male) capable of sexual intercourse. □ **potently** *adv.*, **potency** *n.*

potentate *n.* a monarch or ruler.

potential *adj.* capable of being developed or used. ● *n.* an ability or capacity for development. □ **potentially** *adv.*, **potentiality** *n.*

pothole *n.* a hole formed underground by the action of water; a hole in a road surface.

potholing *n.* caving. □ **potholer** *n.*

potluck *n.* whatever is available.

potion *n.* a liquid medicine or drug.

pot-pourri (poh poor-ee) *n.* a scented mixture of dried petals and spices; a medley or mixture.

pot roast *n.* a piece of meat cooked slowly in a covered dish.

potsherd *n.* a broken piece of earthenware, esp. in archaeology.

pot-shot *n.* a shot aimed casually.

potted *see* **pot**. *adj.* **1** preserved in a pot. **2** abridged.

potter[1] *n.* a maker of pottery.

potter[2] *v.* (*Amer.* **putter**) work on trivial tasks in a leisurely way.

pottery *n.* vessels and other objects made of baked clay; a potter's work or workshop.

potty *adj.* (**pottier**) (*slang*) trivial; crazy. ● *n.* a chamber pot, esp. for a child.

pouch *n.* a small bag or bag-like formation.

pouffe *n.* a padded stool.

poult *n.* a young domestic fowl or game bird.

poulterer *n.* a dealer in poultry.

poultice *n.* a moist usu. hot dressing applied to inflammation.

poultry *n.* domestic fowls.

pounce *v.* swoop down and grasp or attack. ● *n.* a pouncing movement.

pound[1] *n.* **1** a measure of weight, 16 oz. avoirdupois (0.454 kg) or 12 oz. troy (0.373 kg). **2** a unit of money in Britain and certain other countries.

pound[2] *n.* an enclosure where stray animals, or vehicles officially removed, are kept until claimed.

pound[3] *v.* beat or crush with heavy strokes; run heavily; (of the heart) beat heavily.

poundage *n.* a charge or commission per £ or per pound weight.

pour *v.* flow, cause to flow; rain heavily; send out freely.

pout *v.* push out one's lips. ● *n.* a pouting expression.

poverty *n.* the state of being poor; scarcity, lack; inferiority.

POW *abbr.* prisoner of war.

powder *n.* a mass of fine dry particles; a medicine or cosmetic in this form; gunpowder. ● *v.* cover with powder. □ **powdery** *adj.*

powder room *n.* a ladies' toilet.

power *n.* **1** the ability to do something. **2** vigour, strength.

control, influence, authority; an influential person or country etc. **4** a product of a number multiplied by itself a given number of times. **5** mechanical or electrical energy; the electricity supply. ●*v.* supply with mechanical or electrical power.

powerful *adj.* having great power or influence. □ **powerfully** *adv.*

powerless *adj.* without power to take action, wholly unable.

power of attorney *n.* legal authority to act for another person.

power station *n.* a building where electricity is generated for distribution.

pp *abbr.* (*Music*) pianissimo.

pp. *abbr.* pages.

p.p. *abbr.* through the agency of, by proxy (indicating that a person is signing on behalf of another).

■ **Usage** The two signatures are frequently written in the wrong order: "T. Jones, *p.p.* P. Smith" means that P. Smith is signing on behalf of T. Jones.

PPS *abbr.* **1** Parliamentary Private Secretary. **2** post-postscript, an additional postscript.

PR *abbr.* **1** public relations. **2** proportional representation.

Pr *symb.* praseodymium.

practicable *adj.* able to be done. □ **practicability** *n.*

practical *adj.* **1** involving activity as distinct from study or theory. **2** suitable for use. **3** clever at doing and making things. **4** virtual. □ **practicality** *n.*

practical joke *n.* a humorous trick played on a person.

practically *adv.* **1** in a practical way. **2** virtually, almost.

practice *n.* **1** repeated exercise to improve skill. **2** action as opposed to theory. **3** a custom or

habit. **4** a doctor's or lawyer's business.

practise *v.* (*Amer.* practice) **1** do something repeatedly or habitually. **2** (of a doctor or lawyer) perform professional work.

practised *adj.* (*Amer.* **practiced**) experienced.

practitioner *n.* a professional worker, esp. in medicine.

praesidium var. of **presidium**.

pragmatic *adj.* treating things from a practical point of view. □ **pragmatically** *adv.*, **pragmatism** *n.*, **pragmatist** *n.*

prairie *n.* a large treeless area of grassland, esp. in North America.

prairie dog *n.* a North American rodent that lives in burrows.

praise *v.* express approval or admiration of; honour (God) in words. ●*n.* praising; approval expressed in words.

praiseworthy *adj.* deserving praise.

praline *n.* a nutty sweet.

pram *n.* a four-wheeled conveyance for a baby.

prance *v.* move springily.

prang *v.* (*slang*) crash a vehicle.

prank *n.* a piece of mischief.

prankster *n.* a person playing pranks.

praseodymium *n.* a metallic element (symbol Pr).

prat *n.* (*slang*) a fool.

prattle *v.* chatter in a childish way. ●*n.* childish chatter.

prawn *n.* an edible shellfish like a large shrimp.

pray *v.* say prayers; entreat.

prayer *n.* a solemn request or thanksgiving to God; an act of praying; an entreaty.

pre- *pref.* before; beforehand.

preach *v.* deliver a sermon; expound (the Gospel etc.); advocate. □ **preacher** *n.*

preamble *n.* a preliminary statement, an introductory section.

prearrange v. arrange beforehand. □ **prearrangement** n.

precarious adj. unsafe, not secure. □ **precariously** adv., **precariousness** n.

pre-cast adj. (of concrete) cast in shape before use.

precaution n. something done in advance to avoid a risk. □ **precautionary** adj.

precede v. come or go before in time, order, etc.

precedence n. priority.

precedent n. a previous case serving as an example to be followed.

precept n. a command or rule of conduct.

precinct n. 1 an area closed to traffic in a town. 2 an enclosed area, esp. round a cathedral. 3 (**precincts**) the surrounding area.

precious adj. 1 of great value; beloved. 2 affectedly refined.

precious stone n. a small valuable piece of mineral.

precipice n. a very steep face of a cliff or rock.

precipitate v. (pri-sip-i-tayt) 1 throw headlong; cause to happen suddenly or soon. 2 cause (a substance) to be deposited. 3 condense (vapour) into drops falling as rain etc. ● n. (pri-sip-i-tăt) 1 a substance deposited from a solution. 2 moisture condensed from vapour. ● adj. (pri-sip-i-tăt) hasty, rash. □ **precipitately** adv.

precipitation n. 1 rain or snow. 2 precipitating; being precipitated.

precipitous adj. very steep.

précis (pray-see) n. (pl. **précis**) a summary. ● v. make a précis of.

precise adj. exact; correct and clearly stated. □ **precisely** adv., **precision** n.

preclude v. exclude the possibility of, prevent.

precocious adj. having developed earlier than is usual. □ **precociously** adv.

precognition n. foreknowledge, esp. supernatural.

preconceived adj. (of an idea) formed beforehand. □ **preconception** n.

precondition n. a condition that must be fulfilled beforehand.

precursor n. a forerunner.

pre-date v. exist or occur at an earlier time than.

predator n. a predatory animal.

predatory (pred-ă-ter-i) adj. preying on others.

predecease v. die earlier than (another person).

predecessor n. 1 the former holder of an office or position. 2 an ancestor.

predestination n. the doctrine that everything has been determined in advance.

predicament n. a difficult situation.

predicate n. (Grammar) the part of a sentence that says something about the subject (e.g. 'is short' in life is short). □ **predicative** adj.

predict v. foretell. □ **prediction** n., **predictive** adj., **predictor** n.

predictable adj. able to be predicted. □ **predictably** adv.

predilection n. a special liking.

predispose v. 1 influence in advance. 2 make liable or inclined. □ **predisposition** n.

predominate v. be most numerous or powerful; exert control. □ **predominant** adj., **predominantly** adv., **predominance** n.

pre-eminent adj. excelling others, outstanding. □ **pre-eminently** adv., **pre-eminence** n.

pre-empt v. obtain (a thing) before anyone else can do so;

forestall. □ pre-emption n., pre-emptive adj.

■ **Usage** Pre-empt is sometimes used to mean prevent, but this is considered incorrect in standard English.

preen v. (of a bird) smooth (feathers) with the beak. □ **preen oneself** groom oneself; show self-satisfaction.

prefabricate v. manufacture in sections for assembly on a site. □ **prefabrication** n.

preface n. an introductory statement. ● v. **1** introduce with a preface. **2** lead up to (an event).

prefect n. **1** a senior pupil authorized to maintain discipline in a school. **2** an administrative official in certain countries. □ **prefecture** n.

prefer v. (**preferred**) **1** choose as more desirable, like better. **2** put forward (an accusation).

preferable adj. more desirable. □ **preferably** adv.

preference n. **1** preferring; a thing preferred. **2** a prior right. **3** favouring.

preferential adj. giving preference. □ **preferentially** adv.

preferment n. promotion.

prefix n. a word or syllable placed in front of a word to change its meaning. ● v. add as a prefix or introduction.

pregnant adj. **1** having a child or young developing in the womb. **2** full of meaning. □ **pregnancy** n.

prehensile adj. able to grasp things.

prehistoric adj. of the ancient period before written records were made. □ **prehistorically** adv.

prejudge v. form a judgement on before knowing all the facts.

prejudice n. **1** a preconceived opinion. **2** harm to someone's

rights. ● v. **1** cause to have a prejudice. **2** harm the rights of. □ **prejudiced** adj.

prejudicial adj. harmful to rights or interests. □ **prejudicially** adv.

prelate n. a clergyman of high rank. □ **prelacy** n.

preliminary adj. preceding and preparing for a main action or event. ● n. a preliminary action or event.

prelude n. an action or event leading up to another; an introductory part or piece of music.

premarital adj. before marriage.

premature adj. coming or done before the usual or proper time. □ **prematurely** adv.

pre-medication n. medication in preparation for an operation.

premeditated adj. planned beforehand. □ **premeditation** n.

premenstrual adj. before each menstruation.

premier adj. first in importance, order, or time. ● n. a prime minister, a head of government. □ **premiership** n.

premiere (prem-i-air) n. the first public performance.

premise n. = premiss.

premises n.pl. a house or other building and its grounds.

premiss n. a statement on which reasoning is based.

premium n. **1** an amount or instalment paid for an insurance policy. **2** an extra sum of money. □ **at a premium** above the nominal or usual price; scarce and in demand.

premonition n. a feeling that something (bad) will happen. □ **premonitory** adj.

preoccupation n. being preoccupied; something that fills one's thoughts.

preoccupied adj. mentally engrossed and inattentive to other things; already occupied.

prep. *abbr.* preposition.

preparation *n.* preparing; something done to make ready; a substance prepared for use.

preparatory *adj.* preparing for something. ● *adv.* in a preparatory way.

preparatory school *n.* a private primary school or (in America) a secondary school.

prepare *v.* make or get ready. □ **prepared to** ready and willing to.

pre-pay *v.* (**pre-paid**, **pre-paying**) pay in advance or beforehand.

preponderate *v.* be greater in number, power, etc. □ **preponderant** *adj.*, **preponderance** *n.*

preposition *n.* (*Grammar*) a word used with a noun or pronoun to show position, time, or means (e.g. *at* home, *by* train). □ **prepositional** *adj.*

prepossessing *adj.* attractive.

preposterous *adj.* utterly absurd, outrageous. □ **preposterously** *adv.*

prepuce *n.* the foreskin.

prerequisite *n.* something that is required before something else can happen.

■ **Usage** *Prerequisite* is sometimes confused with *perquisite* which means 'an extra profit, right, or privilege.'

prerogative *n.* a right or privilege.

presage *n.* an omen; a presentiment. ● *v.* portend; foresee.

Presbyterian *adj.* & *n.* (a member) of a Church governed by elders of equal rank, esp. that of Scotland. □ **Presbyterianism** *n.*

pre-school *adj.* of the time before a child is old enough to go to school.

prescribe *v.* **1** advise the use of (a medicine etc.). **2** lay down as a course or rule to be followed.

■ **Usage** *Prescribe* is sometimes confused with *proscribe*, which means 'to forbid.'

prescription *n.* prescribing; a doctor's written instructions for the preparation and use of a medicine.

prescriptive *adj.* prescribing.

presence *n.* **1** being present; a person or thing that is or seems present. **2** a person's bearing. □ **presence of mind** ability to act sensibly in a crisis.

present¹ (pre-zĕnt) *adj.* **1** being in the place in question. **2** existing or being dealt with now. ● *n.* the present time, time now passing. □ **at present** now. **for the present** for now, temporarily.

present² *n.* (pre-zĕnt) a gift. ● *v.* (pri-zent) **1** give as a gift or award. **2** introduce; bring to the public. □ **presenter** *n.*

presentable *adj.* fit to be presented, of good appearance. □ **presentably** *adv.*

presentation *n.* presenting; something presented.

presentiment *n.* a feeling of something about to happen, a foreboding.

presently *adv.* **1** soon. **2** (*Scot.* & *Amer.*) now.

preservative *adj.* preserving. ● *n.* a substance that preserves perishable food.

preserve *v.* **1** keep safe, unchanged, or in existence. **2** treat (food) to prevent decay. ● *n.* **1** interests etc. regarded as one person's domain. **2** (also **preserves**) jam. □ **preservation** *n.*, **preserver** *n.*

preside *v.* be president or chairman; have the position of control.

president *n.* the head of an institution or club; the head of a republic. □ **presidency** *n.*, **presidential** *adj.*

presidium *n.* (also **praesidium**) the standing committee in a Communist organization.

press *v.* **1** apply weight or force against; squeeze; flatten; smooth; iron (clothes etc.). **2** urge. **3** throng closely. **4** bring into use as a makeshift. ● *n.* **1** the process of pressing; an instrument for pressing something. **2** newspapers and periodicals; people involved in producing these. □ **be pressed for** have barely enough of.

press conference *n.* an interview given to a number of reporters.

press cutting *n.* an article cut from a newspaper.

press-gang *v.* force into service.

pressing *adj.* urgent.

press stud *n.* a small fastener with two parts that are pressed together.

press-up *n.* an exercise of pressing on the hands to raise the body while lying face down.

pressure *n.* **1** exertion of force against a thing; this force. **2** a compelling or oppressive influence. ● *v.* pressurize (a person).

pressure-cooker *n.* a pan for cooking things quickly by steam under pressure.

pressure group *n.* an organized group seeking to exert influence by concerted action.

pressurize *v.* (also **-ise**) **1** try to compel into an action. **2** maintain a constant atmospheric pressure in (a compartment). □ **pressurization** *n.*

prestige *n.* respect resulting from good reputation or achievements.

prestigious *adj.* having or showing prestige.

presto *adv.* very quickly.

prestressed *adj.* (of concrete) strengthened by wires within it.

presumably *adv.* it may be presumed.

presume *v.* **1** suppose to be true. **2** be presumptuous. □ **presumption** *n.*

presumptuous *adj.* behaving with impudent boldness; acting beyond one's authority. □ **presumptuously** *adv.*

presuppose *v.* **1** assume beforehand. **2** assume the prior existence of. □ **presupposition** *n.*

pre-tax *adj.* before tax has been deducted.

pretence *n.* (*Amer.* **pretense**) **1** pretending, make-believe. **2** a claim (e.g. to merit or knowledge).

pretend *v.* create a false impression of (in play or deception); claim falsely that one has or is something. □ **pretender** *n.*

pretension *n.* asserting of a claim; pretentiousness.

pretentious *adj.* claiming great merit or importance. □ **pretentiously** *adv.*, **pretentiousness** *n.*

pre-term *adj. & adv.* born or occurring prematurely.

preternatural *adj.* beyond what is natural. □ **preternaturally** *adv.*

pretext *n.* a reason put forward to conceal one's true reason.

prettify *v.* make (something) look superficially attractive.

pretty *adj.* (**prettier**) attractive in a delicate way. ● *adv.* fairly, moderately. □ **prettily** *adv.*, **prettiness** *n.*

pretzel *n.* a knot-shaped salted biscuit.

prevail *v.* **1** be victorious, gain mastery. **2** be the most usual. □ **prevail on** or **upon** persuade.

prevalent adj. existing generally, widespread. ◻ **prevalence** n.

prevaricate v. speak or act evasively or misleadingly. ◻ **prevarication** n.

■ Usage *Prevaricate* is often confused with *procrastinate*, which means 'to postpone action'.

prevent v. keep from happening or doing; stop, hinder. ◻ **preventable** adj., **prevention** n.

■ Usage The use of *prevent* without 'from' as in *She prevented me going* is informal. An acceptable further alternative is *She prevented my going.*

preventative adj. & n. preventive.

preventive adj. preventing something. ● n. something that does this.

previous adj. coming before in time or order. ◻ **previously** adv.

prey n. an animal hunted or killed by another for food; a victim. ◻ **prey on** seek or take as prey; cause worry to. **bird of prey** one that kills and eats animals.

price n. **1** the amount of money for which a thing is bought or sold. **2** what must be given or done etc. to achieve something. ● v. fix, find, or estimate the price of.

priceless adj. **1** invaluable. **2** (slang) very amusing or absurd.

prick v. **1** pierce slightly; feel a pricking sensation. **2** erect (the ears). ● n. **1** an act of pricking; a sensation of being pricked. **2** (vulgar) the penis. ◻ **prick up one's ears** listen alertly.

prickle n. a small thorn or spine; a pricking sensation. ● v. feel or cause a pricking sensation.

prickly adj. (**pricklier**) **1** having prickles. **2** irritable, touchy.

pride n. **1** a feeling of pleasure or satisfaction about one's actions, qualities, or possessions; a source of this; a sense of dignity. **2** a group (of lions). ● v. ◻ **pride oneself on** be proud of.

pride of place n. the most prominent position.

priest n. a member of the clergy; an official of a non-Christian religion. ◻ **priesthood** n., **priestly** adj.

priestess n. a female priest of a non-Christian religion.

prig n. a self-righteous person. ◻ **priggish** adj., **priggishness** n.

prim adj. (**primmer**) formal and precise; prudish. ◻ **primly** adv., **primness** n.

prima ballerina n. a chief ballerina.

primacy n. pre-eminence.

prima donna n. the chief female singer in an opera.

prima facie (pry-mǎ fay-shee) adv. at first sight. ● adj. based on first impressions.

primal adj. **1** primitive, primeval. **2** fundamental.

primary adj. **1** first in time, order, or importance. **2** (of education) for children below the age of 11. ● n. (in the USA) a preliminary election to choose delegates or candidates. ◻ **primarily** adv.

primary colour n. one not made by mixing others, i.e. (for light) red, green, or blue, (for paint) red, blue, or yellow.

primary school n. a school for children below the age of 11.

primate n. **1** an archbishop. **2** a member of the highly developed order of animals that includes man, apes, and monkeys.

prime adj. **1** chief. **2** first-rate. **3** fundamental. ● n. a state of greatest perfection. ● v. prepare for use or action; provide with information in preparation for something.

prime minister *n.* the head of a parliamentary government.

prime number *n.* a number that can be divided exactly only by itself and one.

primer *n.* **1** a substance used to prime a surface for painting. **2** an elementary textbook.

primeval *adj.* of the earliest times of the world.

primitive *adj.* **1** of or at an early stage of civilization. **2** simple, crude.

primogeniture *n.* a system by which an eldest son inherits all his parents' property.

primordial *adj.*

primrose *n.* a pale yellow spring flower; its colour.

primula *n.* a perennial plant of the family that includes the primrose.

prince *n.* a male member of a royal family; a sovereign's son or grandson.

princely *adj.* **1** like a prince. **2** splendid, generous.

princess *n.* a female member of a royal family; a sovereign's daughter or granddaughter; a prince's wife.

principal *adj.* first in rank or importance. ● *n.* **1** the head of certain schools or colleges; a person with highest authority or playing the leading part. **2** a capital sum as distinct from interest or income.

principality *n.* a country ruled by a prince.

principally *adv.* mainly.

principle *n.* a general truth used as a basis of reasoning or action; a scientific law shown or used in the working of something. □ **in principle** as regards the main elements. **on principle** because of one's moral beliefs.

print *v.* **1** press (a mark) upon a surface, mark (a surface etc.) in

this way; produce by applying inked type to paper. **3** write with unjoined letters. **3** produce a positive picture from (a photographic negative). ● *n.* a mark left by pressing; printed lettering or words; a printed design, picture, or fabric.

printed circuit *n.* an electric circuit with lines of conducting material printed on a flat sheet.

printer *n.* **1** a person who prints books, newspapers, etc. **2** a machine that prints.

printout *n.* printed material produced from a computer printer or teleprinter.

prior *adj.* coming before in time, order, or importance. ● *n.* a monk who is head of a religious community, or one ranking next below an abbot.

prioress *n.* a female prior.

prioritize *v.* (also **-ise**) treat as a priority. □ **prioritization** *n.*

priority *n.* a right to be first; something that should be treated as most important.

priory *n.* a monastery or nunnery governed by a prior or prioress.

prise *v.* (*Amer.* **prize**) force out or open by leverage.

prism *n.* a solid geometric shape with ends that are equal and parallel; a transparent object of this shape with refracting surfaces.

prismatic *adj.* of or like a prism; (of colours) rainbow-like.

prison *n.* a building used to confine people convicted of crimes; a place of confinement.

prisoner *n.* a person kept in prison; a person in confinement.

prissy *adj.* (**prissier**) prim, prudish. □ **prissily** *adv.*, **prissiness** *n.*

pristine *adj.* in its original and unspoilt condition.

privacy *n.* being private.

private *adj.* **1** belonging to a person or group, not public; confidential; secluded. **2** not provided by the state. ● *n.* a soldier of the lowest rank. □ **in private** privately. □ **privately** *adv.*

privation *n.* loss, lack; hardship.

privatize *v.* (also **-ise**) transfer from state to private ownership. □ **privatization** *n.*

privet *n.* a bushy evergreen shrub much used for hedges.

privilege *n.* a special right granted to a person or group. □ **privileged** *adj.*

privy *n.* (*old use & Amer.*) a lavatory. ● **be privy to** share in the secret of, know about.

prize *n.* an award for victory or superiority; something that can be won. ● *adj.* **1** winning a prize. **2** excellent. ● *v.* **1** value highly. **2** Amer. sp. of **prise**.

pro *n.* (*pl.* **pros**) (*colloquial*) a professional. □ **pros and cons** arguments for and against something.

pro- *pref.* in favour of.

proactive *adj.* taking the initiative.

probable *adj.* likely to happen or be true. □ **probably** *adv.,* **probability** *n.*

probate *n.* the official process of proving that a will is valid; a certified copy of a will.

probation *n.* **1** testing of behaviour or abilities. **2** the supervision of an offender by an official (**probation officer**) as an alternative to imprisonment. □ **probationary** *adj.*

probationer *n.* a person undergoing a probationary period.

probe *n.* **1** a blunt surgical instrument for exploring a wound. **2** an unmanned exploratory spacecraft. **3** an investigation. ● *v.* **1** explore with a probe. **2** investigate.

probity *n.* honesty.

problem *n.* something difficult to deal with or understand; something to be solved or dealt with. □ **problematic, problematical** *adj.*

proboscis *n.* **1** a long flexible snout. **2** an insect's elongated mouthpart used for sucking things.

procedure *n.* a series of actions done to accomplish something. □ **procedural** *adj.*

proceed *v.* **1** go forward or onward; continue. **2** start a lawsuit. **3** come forth, originate.

proceedings *n.pl.* **1** what takes place. **2** a lawsuit. **3** a published report of a conference.

proceeds *n.pl.* the profit from a sale, performance, etc.

process *n.* a series of operations used in making something; a procedure; a series of changes or events. ● *v.* subject to a process; deal with.

procession *n.* a number of people or vehicles etc. going along in an orderly line.

processor *n.* a machine that processes things.

proclaim *v.* announce publicly. □ **proclamation** *n.*

proclivity *n.* a tendency.

procrastinate *v.* postpone action. □ **procrastination** *n.*

■ **Usage** *Procrastinate* is often confused with *prevaricate,* which means 'to speak or act evasively or misleadingly'.

---

procreate *v.* create or produce (offspring). □ **procreation** *n.*

procurator fiscal *n.* (in Scotland) a public prosecutor and coroner.

procure *v.* obtain by care or effort, acquire; act as procurer. □ **procurement** *n.*

procurer *n.* a person who obtains women for prostitution.

prod *v.* (**prodded**) **1** poke. **2** stimulate to action. ● *n.* **1** a prodding action; an instrument for prodding things. **2** a stimulus.

prodigal adj. wasteful, extravagant. □ **prodigally** adv., **prodigality** n.

prodigious adj. amazing; enormous. □ **prodigiously** adv.

prodigy n. a person with exceptional qualities or abilities; a wonderful thing.

produce v. (prŏ-dewss) **1** bring forward for inspection. **2** bring (a performance etc.) before the public. **3** bring into existence, cause; manufacture. ● n. (prodewss) things produced; agricultural and natural products. □ **production** n.

producer n. **1** a person who produces something. **2** a person responsible for the schedule, expenditure, and quality of a film, play, broadcast, etc.

product n. **1** a thing produced. **2** a number obtained by multiplying.

productive adj. producing things, esp. in large quantities.

productivity n. efficiency in industrial production.

profane adj. irreverent, blasphemous; not sacred. ● v. treat irreverently. □ **profanely** adv., **profanity** n.

profess v. **1** claim (a quality etc.), pretend. **2** affirm faith in (a religion).

professed adj. **1** self-acknowledged. **2** falsely claiming to be something. □ **professedly** adv.

profession n. **1** an occupation requiring advanced learning; the people engaged in this. **2** a declaration.

professional adj. **1** belonging to a profession. **2** skilful and conscientious. **3** doing something for payment, not as a pastime. ● n. a professional worker or

player. □ **professionalism** n., **professionally** adv.

professor n. a university teacher of the highest rank; (in America) a university lecturer. □ **professorial** adj.

proffer v. offer.

proficient adj. competent, skilled. □ **proficiently** adv., **proficiency** n.

profile n. **1** a side view, esp. of the face. **2** a short account of a person's character or career.

profit n. **1** money gained. **2** an advantage, a benefit. ● v. (profited) obtain a profit; benefit.

profitable adj. bringing profit. □ **profitably** adv., **profitability** n.

profiteer n. a person who makes excessive profits. □ **profiteering** n.

profiterole n. a small hollow cake of choux pastry with filling.

profligate adj. wasteful, extravagant; dissolute. ● n. a profligate person. □ **profligacy** n.

profound adj. **1** intense. **2** showing or needing great insight. □ **profoundly** adv., **profundity** n.

profuse adj. lavish; plentiful. □ **profusely** adv., **profusion** n.

progenitor n. an ancestor.

progeny n. offspring.

progesterone n. a sex hormone that maintains pregnancy.

prognosis n. (pl. **prognoses**) a forecast, esp. of the course of a disease. □ **prognostic** adj.

prognosticate v. forecast. □ **prognostication** n.

program n. **1** Amer. sp. of **programme**. **2** a series of coded instructions for a computer. ● v. (**programmed**) instruct (a computer) by means of a program. □ **programmer** n.

programme n. (Amer. **program**) **1** a plan of action. **2** a list of items in an entertainment; these items. **3** a broadcast performance.

progress n. (proh-gress) forward or onward movement; development. ● v. (prŏ-gress) make progress; develop. □ **in progress** taking place. □ **progression** n.

progressive adj. **1** favouring progress or reform. **2** (of a disease) gradually increasing in its effect. □ **progressively** adv.

prohibit v. (prohibited) forbid. □ **prohibition** n.

prohibitive adj. prohibiting; intended to prevent the use or purchase of something.

project¹ (prŏ-jekt) v. **1** extend outwards; cast, throw. **2** estimate; plan.

project² (pro-jekt) n. a plan, an undertaking; a task involving research.

projectile n. a missile.

projection n. **1** the process of projecting something. **2** something projecting from a surface. **3** an estimate of future situations based on a study of present ones.

projectionist n. a person who operates a projector.

projector n. an apparatus for projecting images on to a screen.

prolapse n. a condition in which an organ slips forward out of place.

proletariat n. working-class people. □ **proletarian** adj. & n.

pro-life adj. in favour of preserving life, esp. in opposing abortion.

proliferate v. reproduce rapidly, multiply. □ **proliferation** n.

prolific adj. producing things abundantly. □ **prolifically** adv.

prologue n. an introduction to a play, poem, etc.

prolong v. lengthen in extent or duration. □ **prolongation** n.

prolonged adj. continuing for a long time.

prom n. (colloquial) **1** a promenade concert. **2** a promenade.

promenade n. a paved public walk (esp. along a sea front).

promenade concert n. one where part of the audience is not seated and can move about.

promethium n. a radioactive metallic element (symbol Pm).

prominent adj. **1** projecting; conspicuous. **2** well known. □ **prominently** adv., **prominence** n.

promiscuous adj. **1** having sexual relations with many people. **2** indiscriminate. □ **promiscuously** adv., **promiscuity** n.

promise n. **1** a declaration that one will give or do a certain thing. **2** an indication of future results. ● v. **1** make a promise (to); say that one will do or give (a thing). **2** seem likely, produce expectation of.

promising adj. likely to turn out well.

promissory adj. conveying a promise.

promontory n. high land jutting out into the sea.

promote v. **1** raise to a higher rank or office; help the progress of. **2** publicize in order to sell. □ **promoter** n., **promotion** n., **promotional** adj.

prompt adj. done without delay; punctual. ● adv. punctually. ● v. **1** incite. **2** assist (an actor) by supplying forgotten words. □ **promptly** adv., **promptness** n.

prompter n. a person positioned off stage to prompt actors.

promulgate v. make known to the public. □ **promulgation** n., **promulgator** n.

prone adj. **1** lying face downwards. **2** likely to do or suffer something.

prong n. each of the pointed parts of a fork. □ **pronged** adj.

pronoun n. (Grammar) a word used as a substitute for a noun; demonstrative pronouns, e.g. this, that; interrogative pronouns, e.g. who?, which?; personal pronouns, e.g. I, you, her, it; possessive pronouns, e.g. my, your, her, its; reflexive pronouns, e.g. myself, oneself; relative pronouns, e.g. who, which, that. □ **pronominal** adj.

pronounce v. 1 utter (a sound or word) distinctly or in a certain way. 2 declare. □ **pronunciation** n.

pronounced adj. noticeable.

pronouncement n. a declaration.

proof n. 1 evidence that something is true or exists. 2 a copy of printed matter for correction. ● adj. able to resist penetration or damage. ● v. make (fabric) proof against something (e.g. water).

proof-read v. read and correct (printed proofs). □ **proof-reader** n.

prop n. 1 a support to prevent something from falling, sagging, or failing. 2 (colloquial) a stage property. 3 (colloquial) a propeller. ● v. (propped) support with or as if with a prop.

propaganda n. publicity intended to persuade or convince people.

propagate v. 1 breed or reproduce (a plant) from parent stock. 2 spread (news etc.); transmit. □ **propagation** n., **propagator** n.

propane n. a hydrocarbon fuel gas.

propel v. (propelled) push forwards or onwards. □ **propellant** n. & adj.

propeller n. a revolving device with blades, for propelling a ship or aircraft.

propensity n. a tendency; an inclination.

proper adj. 1 suitable; correct; conforming to social conventions. 2 (colloquial) thorough.

proper name, **proper noun** n. (Grammar) the name of an individual person or thing.

property n. 1 something owned; real estate, land. 2 a movable object used in a play or film. 3 a quality, a characteristic.

prophecy n. the power of prophesying; a statement prophesying something.

prophesy v. foretell (what will happen).

prophet n. 1 a person who foretells events. 2 a religious teacher inspired by God.

prophetic adj. prophesying.

prophylactic adj. preventing disease or misfortune. ● n. 1 a preventive medicine or action. 2 (Amer.) a condom. □ **prophylaxis** n.

propinquity n. nearness.

propitiate (prŏ-pish-i-ayt) v. win the favour of, appease. □ **propitiation** n., **propitiatory** adj.

propitious (prŏ-pish-ŭs) adj. auspicious, favourable. □ **propitiously** adv., **propitiousness** n.

proponent n. a person putting forward a proposal.

proportion n. a fraction or share of a whole; a ratio; the correct relation in size or degree; (proportions) dimensions. □ **proportional** adj., **proportionally** adv.

proportional representation n. an electoral system in which each party receives seats in proportion to the number of votes cast for its candidates.

proportionate adj. in proportion, corresponding. □ **proportionately** adv.

proposal n. 1 the proposing of something; something proposed. 2 an offer of marriage.

propose v. **1** put forward for consideration; declare as one's plan; nominate; **2** make a proposal of marriage. □ **proposer** n.

proposition n. **1** a statement; a proposal, a scheme proposed. **2** (colloquial) a problem or undertaking. ● v. (colloquial) put a proposal to, esp. of a sexual nature.

propound v. put forward for consideration.

proprietary adj. made and sold by a particular firm, usu. under a patent; of an owner or ownership.

proprietor n. the owner of a business. □ **proprietorial** adj.

propriety n. correctness of behaviour.

propulsion n. the process of propelling or being propelled.

propylene n. a gaseous hydrocarbon.

pro rata adj. & adv. proportional(ly).

prorogue (proh-rohg) v. (**prorogued, proroguing**) discontinue the meetings of (a parliament) without dissolving it.

prosaic adj. plain and ordinary, unimaginative. □ **prosaically** adv.

proscenium (prŏ-seen-i-ŭm) n. (pl. **prosceniums** or **proscenia**) the part of a theatre stage in front of the curtain.

proscribe v. forbid by law.

■ **Usage** Proscribe is sometimes confused with prescribe, which means 'to impose'.

prose n. written or spoken language not in verse form.

prosecute v. **1** take legal proceedings against (a person) for a crime. **2** carry on, conduct. □ **prosecution** n., **prosecutor** n.

proselyte n. a recent convert to a religion.

proselytize v. (also **-ise**) seek to convert.

prospect n. (pross-pekt) **1** what one is to expect. **2** a chance of advancement. ● v. (prŏ-spekt) explore in search of something. □ **prospector** n.

prospective adj. expected to be or to occur; future, possible.

prospectus n. a document giving details of a school, business, etc.

prosper v. be successful, thrive.

prosperous adj. financially successful. □ **prosperity** n.

prostate gland n. the gland round the neck of the bladder in male mammals.

prosthesis n. (pl. **prostheses**) an artificial limb or similar appliance. □ **prosthetic** adj.

prostitute n. a woman who offers sexual intercourse for payment. ● v. make a prostitute of; put (talent etc.) to an unworthy use. □ **prostitution** n.

prostrate (pross-trayt) **1** face downwards; lying horizontally. **2** overcome, exhausted. ● v. (pross-trayt) cause to be prostrate. □ **prostration** n.

protactinium n. a radioactive metallic element (symbol Pa).

protagonist n. **1** the chief person in a drama, story, etc. **2** a supporter of a cause etc.

protean adj. variable; versatile.

protect v. keep from harm or injury. □ **protection** n., **protector** n.

protectionism n. a policy of protecting home industries by tariffs etc. □ **protectionist** n.

protective adj. protecting, giving protection. □ **protectively** adv.

protectorate n. a country that is under the official protection and partial control of a stronger one.

protégé (prot-i-zhay) n. a person who is helped and protected by another.

protein n. an organic compound forming an essential part of humans' and animals' food.

pro tem adj. & adv. for the time being.

protest n. **1** a statement or action indicating disapproval. ●v. (prŏ-test) **1** express disapproval. **2** declare firmly.

Protestant n. a member of any of the western Christian Churches that are separate from the Roman Catholic Church. □ **Protestantism** n.

protestation n. a firm declaration.

protocol n. **1** etiquette applying to rank or status. **2** a draft of a treaty.

proton n. a particle of matter with a positive electric charge.

protoplasm n. the contents of a living cell.

prototype n. an original example from which others are developed.

protozoan (proh-tŏ-zoh-ăn) n. (also **protozoon**) (pl. **protozoa** or **protozoans**) a one-celled microscopic animal.

protract v. prolong in duration. □ **protraction** n.

protractor n. an instrument for measuring angles.

protrude v. project, stick out. □ **protrusion** n., **protrusive** adj.

protuberance n. a bulging part.

protuberant adj. bulging.

proud adj. **1** full of pride. **2** imposing. **3** slightly projecting. □ **proudly** adv.

prove v. (**proved** or **proven**, **proving**) **1** give or be proof of. **2** be found to be. **3** (of dough) rise because of the action of yeast.

provenance n. a place of origin.

provender n. fodder.

proverb n. a short well-known saying.

proverbial adj. **1** of or mentioned in a proverb. **2** well known.

provide v. **1** supply, make available. **2** supply the necessities of life. **3** make preparations. □ **provider** n.

provided conj. on condition (that).

providence n. **1** being provident. **2** God's or nature's protection.

provident adj. showing wise forethought for future needs, thrifty.

providential adj. happening very luckily. □ **providentially** adv.

providing conj. = **provided**.

province n. **1** an administrative division of a country. **2** (**the provinces**) all parts of a country outside its capital city. **3** a range of learning or responsibility.

provincial adj. **1** of a province or provinces. **2** having limited interests and narrow-minded views. ●n. an inhabitant of a province.

provision n. **1** the process of providing things. **2** a stipulation in a treaty or contract etc. **3** (**provisions**) food and drink.

provisional adj. arranged temporarily. □ **provisionally** adv.

proviso (prŏ-vyz-oh) n. (pl. **provisos**) a stipulation. □ **provisory** adj.

provoke v. **1** make angry. **2** rouse to action; produce as a reaction. □ **provocation** n., **provocative** adj., **provocatively** adv.

provost n. the head of a college; the head of a cathedral chapter.

prow n. a projecting front part of a ship or boat.

prowess n. great ability or daring.

prowl v. go about stealthily. ●n. prowling. □ **prowler** n.

proximate adj. nearest.

proximity n. nearness.

proxy n. a person authorized to represent or act for another; use of such a person.

prude n. a person who is extremely correct and proper, one who is easily shocked. □ **prudery** n.

prudent adj. showing care and foresight. □ **prudently** adv., **prudence** n.

prudish adj. showing prudery. □ **prudishly** adv., **prudishness** n.

prune n. a dried plum. ● v. trim by cutting away dead or unwanted parts; reduce.

prurient adj. having or exciting lustful thoughts. □ **pruriently** adv., **prurience** n.

pry v. (**pries, pried, prying**) inquire or peer impertinently (often furtively).

PS abbr. postscript.

psalm (sahm) n. a sacred song.

psalter (sawl-ter) n. a copy of the Book of Psalms.

psaltery n. an instrument like a dulcimer, played by plucking the strings.

PSBR abbr. public sector borrowing requirement; the amount of money borrowed by government.

psephology (sef-ol-ŏji) n. the study of trends in voting. □ **psephologist** n.

pseudo- (syoo-doh) comb. form false.

pseudonym (syoo-dŏ-nim) n. a fictitious name used by an author.

psoriasis (sŏ-ry-ăsis) n. a skin condition causing scaly red patches.

PSV abbr. public service vehicle.

psyche (sy-kee) n. the soul, the spirit; the mind.

psychedelic (sy-kĕ-del-ik) adj. (of a drug) producing hallucinations. **2** with vivid, abstract patterns.

psychiatry (sy-ky-ătri) n. the study and treatment of mental illness. □ **psychiatric** (sy-kee-at-rik) adj., **psychiatrist** n.

psychic (sy-kik) adj. of the soul or mind; of or having apparently supernatural powers. ● n. a person with such powers.

psycho (sy-koh) n. (pl. **psychos**) (colloquial) a psychopath.

psychoanalyse v. (Amer. **-yze**) treat by psychoanalysis. □ **psychoanalyst** n.

psychoanalysis n. a method of examining and treating mental conditions by investigating the interaction of conscious and unconscious elements.

psychology n. the study of the mind and how it works; mental characteristics. □ **psychological** adj., **psychologically** adv., **psychologist** n.

psychopath n. a person suffering from a severe mental disorder sometimes resulting in antisocial or violent behaviour. □ **psychopathic** adj.

psychosis n. (pl. **psychoses**) a severe mental disorder affecting a person's whole personality. □ **psychotic** adj. & n.

psychosomatic adj. (of illness) caused or aggravated by mental stress.

psychotherapy n. treatment of mental disorders by the use of psychological methods. □ **psychotherapist** n.

PT abbr. physical training.

Pt symb. platinum.

pt. abbr. **1** pint. **2** part. **3** point.

PTA abbr. parent-teacher association.

ptarmigan (tar-mi-găn) n. a bird of the grouse family with plumage that turns white in winter.

pterodactyl (teră-dak-til) n. an extinct reptile with wings.

PTO *abbr.* please turn over.

ptomaine (toh-mayn) *n.* a compound (often poisonous) found in rotting matter.

Pu *symb.* plutonium.

pub *n.* (*colloquial*) a public house.

puberty *n.* the stage in life when a person reaches sexual maturity and becomes capable of reproduction. □ **pubertal** *adj.*

pubic *adj.* of the abdomen at the lower front part of the pelvis.

public *adj.* of, for, or known to people in general. ● *n.* members of a community in general. □ **publicly** *adv.*

public address system *n.* a system of loudspeakers amplifying sound for an audience.

publican *n.* the keeper of a public house.

publication *n.* publishing; a published book, newspaper, etc.

public convenience *n.* a public toilet.

public house *n.* a building (other than a hotel) licensed to serve alcoholic drinks.

publicity *n.* public attention directed upon a person or thing; the process of attracting this.

publicize *v.* (also **-ise**) bring to the attention of the public. □ **publicist** *n.*

public relations *n.pl.* the promotion (by a company or political party) of a favourable public image.

public school *n.* a secondary school for fee-paying pupils; (in Scotland, USA, etc.) a school run by public authorities.

public sector *n.* the part of the economy that is owned and controlled by the state.

public servant *n.* a state official, a civil servant, etc.

public-spirited *adj.* showing readiness to do things for the benefit of people in general.

publish *v.* **1** issue copies of (a book etc.) to the public. **2** make generally known. □ **publisher** *n.*

puce *adj. & n.* brownish purple.

puck *n.* a hard rubber disc used in ice hockey.

pucker *v.* gather into wrinkles. ● *n.* a wrinkle.

pudding *n.* **1** a sweet cooked dish; the sweet course of a meal. **2** a savoury dish containing flour, suet, etc.; a kind of sausage.

puddle *n.* a small pool of rainwater or other liquid.

pudenda *n.pl.* the genitals.

puerile *adj.* childish. □ **puerility** *n.*

puerperal *adj.* of or resulting from childbirth.

puff *n.* **1** a short light blowing of breath, wind, smoke, etc. **2** a soft pad for applying powder to the skin. **3** an over-enthusiastic review. ● *v.* **1** send (air etc.) or come out in puffs; breathe hard, pant. **2** swell.

puffball *n.* a ball-shaped fungus.

puffin *n.* a seabird with a short striped bill.

puff pastry *n.* very light flaky pastry.

puffy *adj.* (**puffier**) puffed out, swollen. □ **puffiness** *n.*

pug *n.* a dog of a small breed with a flat nose and wrinkled face.

pugilist (pew-ji-list) *n.* a professional boxer. □ **pugilism** *n.*

pugnacious *adj.* eager to fight, aggressive. □ **pugnaciously** *adv.*, **pugnacity** *n.*

puke *v. & n.* (*slang*) vomit.

pukka *adj.* (*colloquial*) real, genuine.

pull *v.* **1** exert force upon (a thing) so as to move it towards the source of the force; remove by pulling. **2** exert a pulling or driving force; attract. ● *n.* **1** an act or force of pulling. **2** a means of exerting influence. **3** a deep drink. **4** a draw at a pipe etc. □ **pull in** move towards the side of the road or into a stopping place. **pull off** succeed in doing or achieving. **pull out** withdraw; move away from the side of a road or a stopping place. **pull through** come or bring successfully through an illness or difficulty. **pull up** stop.

pullet *n.* a young hen.

pulley *n.* (*pl.* **pulleys**) a wheel over which a rope etc. passes, used in lifting things.

pullover *n.* a sweater (with or without sleeves) with no fastenings.

pulmonary *adj.* of the lungs.

pulp *n.* **1** the soft moist part of fruit; any soft moist substance. **2** writing of poor quality. ● *v.* reduce to pulp. □ **pulpy** *adj.*

pulpit *n.* a raised enclosed platform from which a preacher speaks.

pulsar *n.* a source (in space) of radio pulses.

pulsate *v.* expand and contract rhythmically. □ **pulsation** *n.*

pulse *n.* **1** the rhythmical throbbing of arteries as blood is propelled along them, as felt in the wrists or temples. **2** a single beat, throb, or vibration. **3** the edible seed of beans, peas, lentils, etc. ● *v.* pulsate.

pulverize *v.* (also **-ise**) **1** crush into powder; become powder. **2** defeat thoroughly. □ **pulverization** *n.*

puma *n.* a large brown American animal of the cat family.

pumice *n.* (in full **pumice stone**) solidified lava used for scouring or polishing.

pummel *v.* (**pummelled**; *Amer.* **pummeled**) strike repeatedly, esp. with the fists.

pump *n.* a machine for moving liquid, gas, or air. ● *v.* **1** use a pump; move, inflate, or empty by using a pump. **2** move vigorously up and down. **3** pour forth. **4** question persistently.

pumpkin *n.* a large round orange-coloured fruit.

pun *n.* a humorous use of a word to suggest another that sounds the same. □ **punning** *adj.* & *n.*

punch *v.* **1** strike with the fist. **2** cut (a hole etc.) with a device. ● *n.* **1** a blow with the fist. **2** a device for cutting holes or impressing a design. **3** a drink made of wine or spirits mixed with fruit juices etc.

punch-drunk *adj.* stupefied by repeated blows.

punchline *n.* words giving the climax of a joke.

punctilious *adj.* very careful about details; conscientious. □ **punctiliously** *adv.*, **punctiliousness** *n.*

punctual *adj.* arriving or doing things at the appointed time. □ **punctually** *adv.*, **punctuality** *n.*

punctuate *v.* **1** insert the appropriate marks in written material to separate sentences etc. **2** interrupt at intervals. □ **punctuation** *n.*

puncture *n.* a small hole made by something sharp, esp. in a tyre. ● *v.* make a puncture in; suffer a puncture.

pundit *n.* an expert.

pungent *adj.* having a strong sharp taste or smell. □ **pungently** *adv.*, **pungency** *n.*

punish *v.* cause (an offender) to suffer for his or her offence; inflict a penalty for; treat roughly. □ **punishment** *n.*

punitive *adj.* inflicting or intended to inflict punishment.

punk *n.* (*slang*) **1** (in full **punk rock**) a deliberately outrageous type of rock music, popular in the 1970s; a follower of this. **2** a hooligan, a lout.

punnet *n.* a small container for fruit etc.

punt¹ *n.* a shallow flat-bottomed boat with broad square ends. ● *v.* propel (a punt) by pushing with a pole against the bottom of a river; travel in a punt.

punt² *v.* kick (a dropped football) before it touches the ground. ● *n.* this kick.

punter *n.* 1 a person who gambles. 2 (*colloquial*) a customer.

puny *adj.* (**punier**) undersized; feeble.

pup *n.* a young dog; a young wolf, rat, or seal. ● *v.* (**pupped**) give birth to pups.

pupa *n.* (*pl.* **pupae**) a chrysalis.

pupate *v.* become a pupa.

pupil *n.* 1 a person who is taught by another. 2 the opening in the centre of the iris of the eye.

puppet *n.* 1 a kind of doll made to move as an entertainment. 2 a person or state controlled by another. □ **puppetry** *n.*

puppy *n.* a young dog.

purchase *v.* buy. ● *n.* 1 buying; something bought. 2 a firm hold to pull or raise something, leverage. □ **purchaser** *n.*

purdah *n.* the Muslim or Hindu system of keeping women from the sight of men or strangers.

pure *adj.* 1 not mixed with any other substances. 2 mere; utter. 3 innocent; chaste. 4 (of mathematics or sciences) dealing with theory, not with practical applications.

purée *n.* pulped fruit or vegetables etc. ● *v.* make into a purée.

purely *adv.* 1 in a pure way. 2 entirely; only.

purgative *adj.* strongly laxative. ● *n.* a purgative substance.

purgatory *n.* a place or condition of suffering, esp. (in RC belief) in which souls undergo purification. □ **purgatorial** *adj.*

purge *v.* 1 clear the bowels of by a purgative. 2 rid of undesirable people or things. ● *n.* the process of purging.

purify *v.* make pure. □ **purification** *n.*, **purifier** *n.*

purist *n.* a stickler for correctness. □ **purism** *n.*

puritan *n.* a person who is strict in morals and regards certain pleasures as sinful. □ **puritanical** *adj.*

purity *n.* a pure state or condition.

purl *n.* a knitting stitch. ● *v.* make this stitch.

purlieu (**perl-yoo**) *n.* a person's bounds or usual haunts.

purloin *v.* steal.

purple *adj. & n.* (of) a colour made by mixing red and blue.

purport *n.* (**per-port**) meaning. ● *v.* (**per-port**) pretend; be intended to seem. □ **purportedly** *adv.*

purpose *n.* 1 the intended result of effort. 2 intention to act, determination. ● *v.* intend. □ **on purpose** by intention.

purpose-built *adj.* designed and built for a particular purpose.

purposeful *adj.* having or showing a conscious purpose, with determination. □ **purposefully** *adv.*

purposely *adv.* on purpose.

purr *v.* a low vibrant sound that a cat makes when pleased; any similar sound. ● *v.* make this sound.

purse *n.* 1 a small pouch for carrying money. 2 (*Amer.*) a handbag. 3 money, funds. ● *v.* pucker (one's lips).

purser *n.* a ship's officer in charge of accounts.

pursuance *n.* the performance (of duties etc.).

pursue *v.* (**pursued**, **pursuing**) 1 follow or chase. 2 continue, proceed along; engage in. □ **pursuer** *n.*

pursuit n. 1 pursuing. 2 an activity to which one gives time or effort.

purulent adj. of or containing pus. □ **purulence** n.

purvey v. supply (food) as a trader. □ **purveyor** n.

pus n. thick yellowish matter produced from an infected wound.

push v. 1 move away by exerting force. 2 thrust forward. 3 make demands on the abilities or tolerance of. 4 urge. 5 (colloquial) sell (drugs) illegally. ●n. 1 the act or force of pushing. 2 vigorous effort. □ **push off** (slang) go away. □ **pusher** n.

pushchair n. a folding chair on wheels, in which a child can be pushed along.

pushy adj. (**pushier**) (colloquial) self-assertive, determined to get on. □ **pushiness** n.

pusillanimous adj. cowardly.

puss n. (also **pussy**) a cat.

pussyfoot v. move stealthily; act cautiously.

pussy willow n. a willow with furry catkins.

pustule n. a pimple or blister. □ **pustular** adj.

put v. (**put**, **putting**) 1 cause to be in a certain place, position, state, or relationship. 2 express, phrase. 3 throw (the shot or weight) as an athletic exercise. ●n. a throw of the shot or weight. □ **put by** save for future use. **put down** suppress; snub; have (an animal) killed; record in writing. **put off** postpone; dissuade, repel. **put out** disconcert; inconvenience; extinguish; dislocate. **put up** construct; raise the price of; provide (money etc.); give temporary accommodation to. **put upon** (colloquial) unfairly burdened. **put up with** endure, tolerate.

putative adj. reputed, supposed. □ **putatively** adv.

putrefy v. rot. □ **putrefaction** n.

putrescent adj. rotting. □ **putrescence** n.

putrid adj. rotten; stinking.

putt v. strike (a golf ball) gently to make it roll along the ground. ●n. this stroke.

putter n. a club used for putting. ●v. (Amer.) = **potter**.

putty n. a soft paste that sets hard, used for fixing glass in frames, filling holes, etc.

put-up job n. a scheme concocted fraudulently.

puzzle n. a difficult question or problem; a problem or toy designed to test knowledge or ingenuity. ●v. (cause to) think hard. □ **puzzlement** n.

PVC abbr. polyvinyl chloride.

PW abbr. policewoman.

PWR abbr. pressurized-water reactor.

pygmy n. (also **pigmy**) a member of a dwarf black African people; a person or thing of unusually small size.

pyjamas n.pl. (Amer. **pajamas**) a loose jacket and trousers for sleeping in.

pylon n. a tall metal structure carrying electricity cables.

pyorrhoea n. (Amer. **pyorrhea**) a disease causing discharge of pus from the tooth-sockets.

pyramid n. a structure with triangular sloping sides that meet at the top. □ **pyramidal** adj.

pyre n. a pile of wood etc. for burning a dead body.

pyrethrum n. a kind of chrysanthemum; an insecticide made from its dried flowers.

pyretic adj. of or producing fever.

Pyrex n. (trade mark) a hard heat-resistant glass.

pyrites (py-ry-teez) n. a mineral sulphide of (copper and) iron.

pyromaniac n. a person with an uncontrollable impulse to set things on fire.

pyrotechnics n.pl. a firework display. □ **pyrotechnic** adj.

Pyrrhic victory (pi-rik) n. one gained at too great a cost.

python n. a large snake that crushes its prey.

Qq

QC abbr. Queen's Counsel.

qua (kway, kwah) conj. in the capacity of, as.

quack n. 1 a duck's harsh cry. 2 a person who falsely claims to have medical skill. ● v. (of a duck) make its harsh cry.

quad n. (colloquial) 1 quadrangle. 2 quadruplet.

quadrangle n. a four-sided courtyard bordered by large buildings.

quadrant n. 1 a quarter of a circle or its circumference. 2 a graduated instrument for taking angular measurements.

quadraphonic adj. (also **quadrophonic**) (of sound reproduction) using four transmission channels.

quadratic equation n. one involving the square (and no higher power) of an unknown quantity or variable.

quadrilateral n. a geometric figure with four sides.

quadrille n. a square dance.

quadriplegia n. paralysis of both arms and legs.

quadruped n. a four-footed animal.

quadruple adj. having four parts or members; four times as much as. ● v. increase by four times its amount.

quadruplet n. one of four children born at one birth.

quaff v. drink in large draughts.

quagmire n. a bog, a marsh.

quail n. a bird related to the partridge. ● v. flinch, show fear.

quaint adj. odd in a pleasing way. □ **quaintly** adv., **quaintness** n.

quake v. shake or tremble, esp. with fear.

Quaker n. a member of the Society of Friends, a Christian sect with no written creed or ordained ministers.

qualification n. 1 qualifying. 2 a thing that qualifies someone to do something. 3 something that limits a meaning.

qualify v. 1 make or become competent, eligible, or legally entitled to do something. 2 limit the meaning of. □ **qualifier** n.

qualitative adj. of or concerned with quality.

quality n. 1 a degree of excellence. 2 a characteristic, something that is special in a person or thing.

qualm (kwahm) n. a misgiving, a pang of conscience.

quandary n. a state of perplexity, a difficult situation.

quango n. (pl. quangos) an administrative body (outside the Civil Service) with senior members appointed by the government.

quantify v. express as a quantity. □ **quantifiable** adj.

quantitative adj. of or concerned with quantity.

quantity n. an amount or number of things; ability to be measured; (**quantities**) large amounts.

quantity surveyor n. a person who measures and prices building work.

quantum leap n. a sudden great advance.

quantum theory n. a theory of physics based on the assumption that energy exists in indivisible units.

quarantine *n.* isolation imposed on those who have been exposed to an infectious disease. ● *v.* put into quarantine.

quark *n.* **1** a component of elementary particles. **2** low-fat curd cheese.

quarrel *n.* an angry disagreement; a cause for complaint. ● *v.* (**quarrelled;** *Amer.* **quarreled**) engage in a quarrel.

quarrelsome *adj.* liable to quarrel.

quarry *n.* **1** an intended prey or victim; something sought or pursued. **2** an open excavation from which stone etc. is obtained. ● *v.* obtain (stone etc.) from a quarry.

quarry tile *n.* an unglazed floor tile.

quart *n.* a quarter of a gallon, two pints (1.137 litres).

quarter *n.* **1** one of four equal parts; this amount. **2** a fourth part of a year; a point of time 15 minutes before or after every hour. **3** (*Amer. & Canada*) a coin worth 25 cents. **4** a direction, a district. **5** mercy towards an opponent. **6** (**quarters**) lodgings, accommodation. ● *v.* **1** divide into quarters. **2** put into lodgings.

quarterdeck *n.* part of a ship's upper deck nearest the stern.

quarter-final *n.* a contest preceding a semi-final.

quarterly *adj. & adv.* (produced or occurring) once in every quarter of a year. ● *n.* a quarterly periodical.

quartermaster *n.* a regimental officer in charge of stores etc.; a naval petty officer in charge of steering and signals.

quartet *n.* a group of four instruments or voices; music for these.

quartz *n.* a hard mineral.

quartz clock *n.* one operated by electric vibrations of a quartz crystal.

quasar *n.* a star-like object that is the source of intense electromagnetic radiation.

quash *v.* annul; reject as not valid; suppress.

quasi- *comb. form* seeming to be but not really so.

quatrain *n.* a stanza or poem of four lines.

quaver *v.* tremble, vibrate; speak or utter in a trembling voice. ● *n.* **1** a trembling sound. **2** a note in music, half a crochet.

quay (kee) *n.* a landing place built for ships to load or unload alongside. □ **quayside** *n.*

queasy *adj.* (**queasier**) feeling slightly sick; squeamish. □ **queasiness** *n.*

queen *n.* **1** a female ruler of a country by right of birth; a king's wife; a woman or thing regarded as supreme in some way. **2** a piece in chess; a playing card bearing a picture of a queen. **3** a fertile female bee, ant, etc. **4** (*offensive slang*) a male homosexual. □ **queenly** *adj.*

queen mother *n.* a dowager queen who is the reigning sovereign's mother.

queer *adj.* **1** strange, odd, eccentric. **2** slightly ill or faint. **3** (*offensive slang*) homosexual. ● *n.* (*offensive slang*) a homosexual. ● *v.* spoil. □ **queer a person's pitch** spoil his or her chances.

quell *v.* suppress.

quench *v.* **1** extinguish (a fire or flame); cool by water. **2** satisfy (one's thirst) by drinking something.

quern *n.* a hand mill for grinding corn or pepper.

querulous *adj.* complaining peevishly. □ **querulously** *adv.*, **querulousness** *n.*

query *n.* a question; a question mark. ● *v.* ask a question or express doubt about.

quest *n.* seeking, a search.

question *n.* a sentence requesting information; a matter for discussion or solution; a doubt. ● *v.* ask or raise question(s) about. □ **in question** being discussed or disputed. **no question** of no possibility of. **out of the question** completely impracticable.

questionable *adj.* open to doubt.

question mark *n.* a punctuation mark (?) placed after a question.

questionnaire *n.* a list of questions seeking information.

queue (kew) *n.* a line of people waiting for something. ● *v.* (**queued, queuing** or **queueing**) wait in a queue.

quibble *n.* a petty objection. ● *v.* make petty objections.

quiche (keesh) *n.* an open tart with a savoury filling.

quick *adj.* **1** taking only a short time. **2** able to learn or think quickly. **3** (of temper) easily roused. ● *n.* the sensitive flesh below the nails. □ **quickly** *adv.*, **quickness** *n.*

quicken *v.* make or become quicker or livelier.

quicklime *n.* = **lime** (*sense* 1).

quicksand *n.* an area of loose wet deep sand into which heavy objects will sink.

quicksilver *n.* mercury.

quickstep *n.* a ballroom dance.

quid *n.* (*pl.* **quid**) (*slang*) £1. □ **quids in** (*slang*) in a position of profit.

quid pro quo *n.* (*pl.* **quid pro quos**) something given in return.

quiescent *adj.* inactive, quiet. □ **quiescence** *n.*

quiet *adj.* **1** with little or no sound; silent. **2** free from disturbance or vigorous activity. **3** subdued; restrained. ● *n.*

quietness. ● *v.* quieten. □ **on the quiet** unobtrusively, secretly. □ **quietly** *adv.*, **quietness** *n.*

quieten *v.* make or become quiet.

quiff *n.* an upright tuft of hair.

quill *n.* **1** a large feather; a pen made from this. **2** each of a porcupine's spines.

quilt *n.* a padded bed-covering. ● *v.* line with padding and fix with lines of stitching.

quin *n.* (*colloquial*) a quintuplet.

quince *n.* a hard yellowish fruit; a tree bearing this.

quinine *n.* a bitter-tasting drug used to treat malaria and in tonics.

quinsy *n.* an abscess on a tonsil.

quintessence *n.* essence; a perfect example of a quality. □ **quintessential** *adj.*, **quintessentially** *adv.*

quintet *n.* a group of five instruments or voices; music for these.

quintuple *adj.* having five parts or members; five times as much as. ● *v.* increase by five times its amount.

quintuplet *n.* one of five children born at one birth.

quip *n.* a witty or sarcastic remark. ● *v.* (**quipped**) utter as a quip.

quire *n.* twenty-five (formerly twenty-four) sheets of writing paper.

quirk *n.* a peculiarity of behaviour; a trick of fate.

quisling *n.* a traitor who collaborates with occupying forces.

quit *v.* (**quitted** or **quit, quitting**) **1** leave; abandon. **2** (*Amer.*) cease, stop.

quite *adv.* **1** completely. **2** somewhat. **3** really, actually. **4** (as an answer) I agree. □ **quite a few** a considerable number.

quits *adj.* on even terms after retaliation or repayment.

quiver v. shake or vibrate with a slight rapid motion. ●n. 1 a quivering movement or sound. 2 a case for holding arrows.

quixotic adj. romantically chivalrous. □ **quixotically** adv.

quiz n. (pl. **quizzes**) a series of questions testing knowledge, esp. as an entertainment. ●v. (**quizzed**) interrogate.

quizzical adj. done in a questioning way, esp. humorously. □ **quizzically** adv.

quoit (koyt) n. a ring thrown to encircle a peg in the game of **quoits**.

quorate adj. having a quorum present.

quorum n. a minimum number of people that must be present for a valid meeting.

quota n. a fixed share; a maximum number or amount that may be admitted, manufactured, etc.

quotable adj. worth quoting.

quotation n. quoting; a passage or price quoted.

quotation marks n.pl. punctuation marks (' ' or " ") enclosing words quoted.

quote v. 1 repeat words from a book or speech; mention in support of a statement. 2 state the price of, estimate.

quotidian adj. daily.

quotient (kwoh-shént) n. the result of a division sum.

q.v. abbr. which see (indicating that the reader should look at the reference given).

qwerty adj. denoting the standard layout of English-language keyboards.

Rr

R abbr. 1 Regina, Rex; Elizabeth R. 2 river. 3 registered as a trade mark. 4 (Chess) rook.

Ra symb. radium.

rabbet n. a channel cut along the edge of a piece of wood to receive a matching piece or a piece of glass.

rabbi n. a religious leader of a Jewish congregation.

rabbinical adj. of rabbis or Jewish doctrines or law.

rabbit n. a burrowing animal with long ears and a short furry tail. ●v. (**rabbited**) talk at length in a rambling way.

rabble n. a disorderly crowd.

rabid adj. furious, fanatical; affected with rabies. □ **rabidity** n.

rabies n. a contagious fatal virus disease of dogs etc., that can be transmitted to humans.

RAC abbr. Royal Automobile Club.

raccoon n. (also **racoon**) a small arboreal mammal of North America.

race n. 1 a contest of speed; (the **races**) a series of races for horses or dogs. 2 a large group of people with common ancestry and inherited physical characteristics; a genus, species, breed, or variety of animal or plant. ●v. compete in a race (with); move or operate at full or excessive speed. □ **racer** n.

racecourse n. a ground where horse races are held.

racetrack n. a racecourse; a track for motor racing.

raceme (ra-seem) n. a flower cluster with flowers attached by short stalks along a central stem.

racial adj. of or based on race. □ **racially** adv.

racialism n. racism. □ **racialist** adj. & n.

racism n. a belief in the superiority of a particular race; antagonism between races; the theory that human abilities are determined by race. □ **racist** adj. & n.

rack n. **1** a framework for keeping or placing things on. **2** a bar with teeth that engage with those of a wheel. **3** an instrument of torture on which people were tied and stretched. ● v. inflict great torment on. □ **rack one's brains** think hard about a problem. **rack and ruin** destruction.

racket n. **1** (also **racquet**) a stringed bat used in tennis and similar games. **2** a din, a noisy fuss. **3** (slang) a fraudulent business or scheme.

racketeer n. a person who operates a fraudulent business etc. □ **racketeering** n.

raconteur n. a person who is good at telling entertaining stories.

racoon var. of **raccoon**.

racquet var. of **racket** (sense 1).

racy adj. (**racier**) spirited and vigorous in style. □ **racily** adv.

rad n. a unit of absorbed dose of ionizing radiation.

RADA abbr. Royal Academy of Dramatic Art.

radar n. a system for detecting objects by means of radio waves.

radial adj. **1** of rays or radii; having spokes or lines etc. that radiate from a central point. **2** (of a tyre; also **radial-ply**) having the fabric layers parallel and the tread strengthened. ● n. a radial-ply tyre. □ **radially** adv.

radian n. an SI unit of plane angle; the angle at the centre of a circle formed by the radii of an arc equal in length to the radius.

radiant adj. **1** emitting rays of light or heat; emitted in rays. **2** looking very bright and happy. □ **radiantly** adv. **radiance** n.

radiate v. spread outwards from a central point; send or be sent out in rays.

radiation n. **1** the process of radiating. **2** the sending out of rays and atomic particles characteristic of radioactive substances; these rays and particles.

radiator n. an apparatus that radiates heat, esp. a metal case through which steam or hot water circulates; an engine-cooling apparatus.

radical adj. fundamental; drastic, thorough; holding extremist views. ● n. a person desiring radical reforms or holding radical views. □ **radically** adv.

radicchio n. (rà-dee-kioh) n. (pl. **radicchios**) a variety of chicory with reddish leaves.

radicle n. an embryo root.

radii see **radius**.

radio n. (pl. **radios**) the process of sending and receiving messages etc. by electromagnetic waves; a transmitter or receiver for this; sound broadcasting, a station for this. ● adj. of or involving radio. ● v. transmit or communicate by radio.

radioactive adj. emitting radiation caused by the decay of atomic nuclei. □ **radioactivity** n.

radiocarbon n. a radioactive form of carbon used in carbon dating.

radiography n. the production of X-ray photographs. □ **radiographer** n.

radiology n. a study of X-rays and similar radiation. □ **radiological** adj., **radiologist** n.

radiophonic adj. relating to electronically produced sound.

radiotherapy n. the treatment of disease by X-rays or similar radiation. □ **radiotherapist** n.

radish n. a plant with a crisp root that is eaten raw.

radium n. a radioactive metallic element (symbol Ra) obtained from pitchblende.

radius n. (pl. **radii** or **radiuses**) **1** a straight line from the centre to the circumference of a circle; its length. **2** the thicker long bone of the forearm.

radon n. a chemical element (symbol Rn), a radioactive gas.

RAF abbr. Royal Air Force.

raffia n. strips of fibre from the leaves of a palm tree.

raffish adj. looking vulgarly flashy or rakish. □ **raffishness** n.

raffle n. a lottery with an object as the prize. ● v. offer as the prize in a raffle.

raft n. **1** a flat floating structure of timber etc., used as a boat. **2** a large collection.

rafter n. one of the sloping beams forming the framework of a roof.

rag n. **1** a torn or worn piece of cloth; (**rags**) old and torn clothes. **2** (derog.) a newspaper. **3** a students' carnival in aid of charity. **4** a piece of ragtime music. ● v. (**ragged**) (slang) tease.

ragamuffin n. a person in ragged dirty clothes.

rage n. **1** violent anger. **2** a craze, a fashion. ● v. **1** show violent anger. **2** (of a storm or battle) continue furiously.

ragged adj. torn, frayed; wearing torn clothes; jagged.

raglan n. a sleeve that continues to the neck, joined by sloping seams.

ragout (ra-goo) n. a stew of meat and vegetables.

raid n. a brief attack to destroy or seize something; a surprise visit by police etc. to arrest suspected people or seize illicit goods. ● v. make a raid on. □ **raider** n.

rail n. **1** a horizontal bar. **2** any of the lines of metal bars on which trains or trams run; the railway system. ● v. **1** fit or protect with a rail. **2** utter angry reproaches.

railing n. a fence of rails supported on upright metal bars.

railroad n. (Amer.) a railway. ● v. force into hasty action.

railway n. a set of rails on which trains run; a system of transport using these.

raiment n. (old use) clothing.

rain n. atmospheric moisture falling as drops; a fall of this; a shower of things. ● v. send down or fall as or like rain.

rainbow n. an arch of colours formed in rain or spray by the sun's rays.

raincoat n. a rain-resistant coat.

raindrop n. a single drop of rain.

rainfall n. the total amount of rain falling in a given time.

rainforest n. dense wet tropical forest.

rainwater n. water that has fallen as rain.

rainy adj. (**rainier**) in or on which much rain falls.

raise v. **1** bring to a higher level or an upright position. **2** breed, grow; bring up (a child). **3** procure. ● n. (Amer.) an increase in salary.

raisin n. a dried grape.

raising agent n. a substance that makes bread etc. swell in cooking.

raison d'être (ray-zawn detr) (pl. **raisons d'être**) a reason for or purpose of a thing's existence.

Raj n. the period of British rule in India.

raja n. (also **rajah**) an Indian king or prince.

rake n. **1** a tool with prongs for gathering leaves, smoothing loose soil, etc. **2** a backward slope of an object. **3** a man who lives an irresponsible and immoral life. ● v. **1** gather or smooth with a rake. **2** search. **3** direct (gunfire etc.) along. **4** set at a sloping angle. □ **rake up** revive the memory of (an unpleasant incident).

rakish adj. dashing, jaunty and perhaps immoral.

rake-off n. (colloquial) a share of profits.

rallentando adv. (Music) with a gradual decrease of speed.

rally v. **1** bring or come (back) together for a united effort. **2** revive; recover strength. ● n. **1** an act of rallying, a recovery. **2** a series of strokes in tennis etc. **3** a mass meeting. **4** a driving competition over public roads.

RAM abbr. (Computing) random-access memory, a temporary working memory.

ram n. **1** an uncastrated male sheep. **2** a striking or plunging device. ● v. (**rammed**) strike or push heavily, crash against.

Ramadan n. the ninth month of the Muslim year, when Muslims fast during daylight hours.

ramble n. a walk taken for pleasure. ● v. **1** take a ramble. **2** wander, straggle; talk or write disconnectedly. □ **rambler** n.

ramekin n. a small individual baking dish.

ramification n. **1** an arrangement of branching parts. **2** a consequence, a result.

ramify v. form branches or subdivisions.

ramp n. a slope joining two levels.

rampage v. (ram-**payj**) behave or race about violently. ● n. (ram-

payj) violent behaviour. □ **on the rampage** rampaging.

rampant adj. **1** flourishing excessively, unrestrained. **2** (of an animal in heraldry) standing on one hind leg with the opposite foreleg raised.

rampart n. a broad-topped defensive wall or bank of earth.

ram-raid n. a robbery in which a vehicle is crashed into a shop etc.

ramrod n. an iron rod formerly used for ramming a charge into guns.

ramshackle adj. tumbledown, rickety.

ran see **run**.

ranch n. a cattle-breeding establishment, esp. in North America; a farm where certain other animals are bred. ● v. farm on a ranch. □ **rancher** n.

rancid adj. smelling or tasting like stale fat. □ **rancidity** n.

rancour n. (Amer. **rancor**) a feeling of bitterness or ill will. □ **rancorous** adj.

rand n. a unit of money in South Africa.

R & B abbr. rhythm and blues.

R & D abbr. research and development.

random adj. done or made etc. at random. □ **at random** without a particular aim or purpose. □ **randomness** n.

random-access adj. (of a computer memory or file) with all parts directly accessible, so that it need not be read sequentially. See also **RAM**.

randy adj. (**randier**) lustful, sexually aroused.

rang see **ring**.

range n. **1** a line or series of things. **2** the limits between which something operates or varies. **3** the distance over which a thing can travel or be effective; the distance to an objective. **4** a large open area for grazing or hunting. **5** a place with targets for shooting practice. ● *v.* **1** arrange in rows etc. **2** extend. **3** vary between limits. **4** go about a place.

rangefinder *n.* a device for calculating the distance to a target etc.

ranger *n.* an official in charge of a park or forest; a mounted warden policing a thinly populated area.

rangy *adj.* (**rangier**) tall and thin.

rank *n.* **1** a line of people or things. **2** a place in a scale of quality or value etc. **3** high social position. **4** (**the ranks**) ordinary soldiers, not officers. ● *v.* arrange in a rank; assign a rank to; have a certain rank. ● *adj.* **1** growing too thickly and coarsely; full of weeds. **2** foul-smelling; unmistakably bad. □ **the rank and file** the ordinary people of an organization. □ **rankness** *n.*

rankle *v.* cause lasting resentment.

ransack *v.* search thoroughly or roughly; rob or pillage (a place).

ransom *n.* a price demanded or paid for the release of a captive. ● *v.* demand or pay a ransom for.

rant *v.* make a violent speech.

rap *n.* **1** a quick sharp blow. **2** a knocking sound. **3** a monologue recited rhythmically to rock music. ● *v.* (**rapped**) **1** knock. **2** (*slang*) a reprimand. □ **take the rap** suffer the consequences.

rapacious *adj.* grasping, violently greedy. □ **rapacity** *adj.*

rape¹ *v.* have sexual intercourse with (esp. a woman) without consent. ● *n.* this act or crime.

rape² *n.* a plant grown as fodder and for its seeds, which yield oil.

rapid *adj.* quick, swift. ● *n.pl.* (**rapids**) a swift current where a

river bed slopes steeply. □ **rapidly** *adv.*, **rapidity** *n.*

rapier *n.* a thin light sword.

rapist *n.* a person who commits rape.

rapport (ra-por) *n.* a harmonious understanding relationship.

rapprochement (ra-prosh-mahn) *n.* a resumption of friendly relations.

rapt *adj.* very intent and absorbed, enraptured. □ **raptly** *adv.*

rapture *n.* intense delight. □ **rapturous** *adj.*, **rapturously** *adv.*

rare *adj.* **1** very uncommon. **2** exceptionally good. **3** (of the atmosphere) of low density, thin. **4** (of meat) cooked lightly not thoroughly. □ **rarely** *adv.*, **rareness** *n.*

rarebit *see* **Welsh rabbit**.

rarefied *adj.* **1** (of the atmosphere) of low density, thin. **2** (of an idea etc.) very subtle. □ **rarefaction** *n.*

raring *adj.* enthusiastic; *raring to go.*

rarity *n.* rareness; a rare thing.

rascal *n.* a dishonest or mischievous person. □ **rascally** *adv.*

rase var. of **raze**.

rash *n.* an eruption of spots or patches on the skin. ● *adj.* acting or done without due consideration of the risks. □ **rashly** *adv.*, **rashness** *n.*

rasher *n.* a slice of bacon or ham.

rasp *n.* **1** a coarse file. **2** a grating sound. ● *v.* scrape with a rasp; utter with or make a grating sound.

raspberry *n.* **1** an edible red berry; a plant bearing this. **2** (*colloquial*) a vulgar sound of disapproval.

rat *n.* **1** a rodent like a large mouse. **2** an unpleasant or treacherous person. □ **rat on**

(ratted) (*colloquial*) desert or betray.

ratable var. of **rateable**.

ratafia *n.* **1** a liqueur flavoured with fruit kernels. **2** an almond-flavoured biscuit.

ratchet *n.* a bar or wheel with notches in which a device engages to prevent backward movement.

rate *n.* **1** a standard of reckoning. **2** the ratio of one amount etc. to another. **3** rapidity. **4** (**rates**) a tax levied according to the value of buildings and land. ● *v.* **1** estimate the worth or value of; consider, regard as. **2** deserve. □ **at any rate** no matter what happens; at least.

rateable *adj.* (also **ratable**) liable to rates.

rateable value *n.* the value at which a business etc. is assessed for rates.

ratepayer *n.* a person paying rates.

rather *adv.* **1** slightly. **2** more exactly. **3** by preference. **4** emphatically yes.

ratify *v.* confirm (an agreement etc.) formally. □ **ratification** *n.*

rating *n.* **1** the level at which a thing is rated. **2** a non-commissioned sailor.

ratio *n.* (*pl.* **ratios**) the relationship between two amounts, reckoned as the number of times one contains the other.

ratiocinate *v.* reason logically. □ **ratiocination** *n.*

ration *n.* a fixed allowance of food etc. ● *v.* limit to a ration.

rational *adj.* able to reason; sane; based on reasoning. □ **rationally** *adv.*, **rationality** *n.*

rationale (rash-ŏnahl) *n.* a fundamental reason; a logical basis.

rationalism *n.* treating reason as the basis of belief and knowledge. □ **rationalist** *n.*, **rationalistic** *adj.*

rationalize *v.* (also **-ise**) **1** invent a rational explanation for. **2** make more efficient by reorganizing. □ **rationalization** *n.*

rat race *n.* a fiercely competitive struggle for success.

rattan *n.* a palm with jointed stems, used for furniture making.

rattle *v.* **1** (cause to) make a rapid series of short hard sounds. **2** (*slang*) make nervous or annoyed. ● *n.* a rattling sound; a device for making this. □ **rattle off** rattle rapidly.

rattlesnake *n.* a poisonous American snake with a rattling tail.

raucous *adj.* loud and harsh. □ **raucously** *adv.*, **raucousness** *n.*

raunchy *adj.* (**raunchier**) coarsely outspoken; sexually provocative. □ **raunchily** *adv.*, **raunchiness** *n.*

ravage *v.* do great damage to. ● *n.pl.* (**ravages**) damage.

rave *v.* talk wildly or furiously; speak with rapturous enthusiasm. ● *n.* an all-night party with loud rock music, attended by large numbers of young people.

ravel *v.* (**ravelled**; *Amer.* **raveled**) tangle.

raven *n.* a black bird with a hoarse cry. ● *adj.* (of hair) glossy black.

ravenous *adj.* very hungry. □ **ravenously** *adv.*

ravine (rǎ-veen) *n.* a deep narrow gorge.

raving *adj.* **1** delirious; (*colloquial*) mad. **2** (*colloquial*) utter; *a raving beauty.*

ravioli *n.* small square pasta cases containing a savoury filling.

ravish *v.* **1** rape. **2** enrapture.

raw *adj.* **1** not cooked; not yet processed. **2** stripped of skin, sensitive because of this. **3** (of weather) damp and chilly. □ **rawness** n.

raw deal n. unfair treatment.

rawhide n. untanned leather.

Rawlplug n. (trade mark) a thin cylindrical plug for holding a screw in masonry.

ray n. **1** a single line or narrow beam of radiation; a radiating line. **2** a large marine flatfish. **3** (Music) the second note of a major scale, or the note D.

rayon n. a synthetic fibre or fabric, made from cellulose.

raze v. (also **rase**) tear down (a building).

razor n. a sharp-edged instrument used for shaving.

razzmatazz n. excitement; extravagant publicity etc.

Rb abbr. rubidium.

RC abbr. Roman Catholic.

Rd. abbr. road.

RE abbr. **1** religious education. **2** Royal Engineers.

Re symb. rhenium.

re prep. concerning.

re- pref. again; back again.

reach v. **1** extend, go as far as; arrive at. **2** stretch out a hand in order to touch or take. **3** establish communication with. **4** achieve, attain. ● n. **1** the distance over which a person or thing can reach. **2** a section of a river.

react v. cause or undergo a reaction. □ **reactive** adj.

reaction n. a response to a stimulus or act or situation etc.; a chemical change produced by substances acting upon each other; an occurrence of one condition after a period of the opposite; a bad physical response to a drug.

reactionary adj. & n. (a person) opposed to progress and reform.

reactor n. an apparatus for the production of nuclear energy.

read v. (**read**, **reading**) **1** understand the meaning of (writ-

ten or printed words or symbols); speak (such words etc.) aloud; study or discover by reading. **2** have a certain wording. **3** (of an instrument) indicate as a measurement. **4** interpret mentally. **5** (of a computer) copy, extract, or transfer (data). ● n. a session of reading.

readable adj. **1** pleasant to read. **2** legible. □ **readably** adv.

reader n. **1** a person who reads. **2** a senior lecturer at a university. **3** a device producing a readable image from a microfilm etc.

readership n. **1** the readers of a newspaper etc. **2** the position of a reader at a university.

readily adv. **1** willingly. **2** easily.

readjust v. adjust again; adapt oneself again. □ **readjustment** n.

ready adj. **1** fit or available for action or use. **2** willing; about or inclined (to do something). **3** quick. ● adv. beforehand. □ **readiness** n.

reafforest v. (also **reforest**) replant with trees.

reagent n. a substance used to produce a chemical reaction.

real adj. existing as a thing or occurring as a fact; genuine, natural.

real estate n. immovable assets, i.e. buildings or land.

realign v. align again; regroup in politics etc.

realism n. representing or viewing things as they are in reality. □ **realist** n.

realistic adj. showing realism; practical. □ **realistically** adv.

reality n. the quality of being real; something real and not imaginary.

realize v. (also **-ise**) **1** be or become aware of; accept as a fact. **2** fulfil (a hope or plan). □ **realization** n.

really *adv.* **1** in fact. **2** thoroughly. **3** indeed; I assure you; I protest.

realm *n.* **1** a kingdom. **2** a field of activity or interest.

realty *n.* real estate.

ream *n.* a quantity of paper (usu. 500 sheets); **(reams)** a great quantity of written matter.

reap *v.* **1** cut (grain etc.) as harvest. **2** receive as the consequence of actions. □ **reaper** *n.*

reappear *v.* appear again.

rear *n.* the back part. ● *adj.* situated at the back. ● *v.* **1** bring up (children); breed and look after (animals); cultivate (crops). **2** (of a horse etc.) raise itself on its hind legs. **3** extend to a great height. □ **bring up the rear** be last.

rear admiral *n.* a naval officer next below vice admiral.

rearguard *n.* troops protecting an army's rear.

rearm *v.* arm again. □ **rearmament** *n.*

rearrange *v.* arrange in a different way. □ **rearrangement** *n.*

rearward *adj., adv., & n.* (towards or at) the rear. □ **rearwards** *adv.*

reason *n.* **1** a motive, cause, or justification. **2** the ability to think and draw conclusions; sanity; good sense or judgement. ● *v.* use one's ability to think and draw conclusions. □ **reason with** try to persuade by argument.

reasonable *adj.* **1** ready to use or listen to reason; in accordance with reason, logical. **2** moderate, not expensive. □ **reasonably** *adv.*

reassemble *v.* assemble again.

reassure *v.* restore confidence to. □ **reassurance** *n.*

rebate *n.* **1** a partial refund. **2** = **rabbet**.

rebel *n.* (re-bĕl) a person who rebels. ● *v.* (ri-bĕl) **(rebelled)**

fight against or refuse allegiance to established government or conventions; resist control. □ **rebellion** *n.*, **rebellious** *adj.*

reboot *v.* start up (a computer) again.

rebound *v.* (ri-bownd) spring back after impact. ● *n.* (ree-bownd) the act of rebounding. □ **on the rebound** while still reacting to a disappointment etc.

rebuff *n.* a snub or refusal. ● *v.* snub or refuse someone.

rebuild *v.* **(rebuilt, rebuilding)** build again after destruction.

rebuke *v.* reprove. ● *n.* a reproof.

rebus *n.* a representation of a word by pictures etc. suggesting its parts.

rebut *v.* **(rebutted)** disprove. □ **rebuttal** *n.*

recalcitrant *adj.* obstinately disobedient. □ **recalcitrance** *n.*

recall *v.* **1** summon to return. **2** remember. ● *n.* recalling, being recalled.

recant *v.* withdraw and reject (one's former statement or belief). □ **recantation** *n.*

recap (*colloquial*) *v.* **(recapped)** recapitulate. ● *n.* a recapitulation.

recapitulate *v.* state again briefly. □ **recapitulation** *n.*

recapture *v.* **1** capture again. **2** experience again. ● *n.* recapturing.

recce (re-kee) *n.* (*colloquial*) a reconnaissance.

recede *v.* **1** go or shrink back. **2** become more distant.

receipt (ri-seet) *n.* the act of receiving; a written acknowledgement that something has been received or money paid.

receive *v.* **1** acquire, accept, or take in. **2** experience, be treated with. **3** greet on arrival.

receiver *n.* **1** a person or thing that receives something. **2** a person who deals in stolen goods. **3** (in full **official receiver**) an official who handles the affairs of a bankrupt person or firm. **4** an apparatus that receives electrical signals and converts them into sound or a picture; the earpiece of a telephone.

receivership *n.* the office of official receiver; the state of being dealt with by a receiver.

recent *adj.* happening in a time shortly before the present. □ **recently** *adv.*

receptacle *n.* something for holding what is put into it.

reception *n.* **1** an act, process, or way of receiving. **2** an assembly held to receive guests. **3** a place where clients etc. are received on arrival.

receptionist *n.* a person employed to receive and direct clients etc.

receptive *adj.* quick to receive ideas. □ **receptiveness** *n.*

receptor *n.* a bodily organ able to respond to a stimulus and transmit a signal through a nerve.

recess *n.* **1** a part or space set back from the line of a wall or room etc. **2** a temporary cessation from business. ● *v.* make a recess in or of.

recession *n.* **1** receding from a point or level. **2** a temporary decline in economic activity.

recessive *adj.* **1** tending to recede. **2** (of a characteristic) remaining latent when a dominant characteristic is present.

recidivist *n.* a person who persistently relapses into crime. □ **recidivism** *n.*

recipe (ress-ipi) *n.* **1** directions for preparing a dish. **2** a way of achieving something.

recipient *n.* a person who receives something.

reciprocal *adj.* both given and received. □ **reciprocally** *adv.*, **reciprocity** *n.*

reciprocate *v.* **1** give and receive; make a return for something. **2** move backward and forward alternately. □ **reciprocation** *n.*

recital *n.* **1** reciting; a long account of events. **2** a musical entertainment.

recitation *n.* reciting; something recited.

recitative (ressi-tă-teev) *n.* a narrative part of an opera, sung in a rhythm imitating speech.

recite *v.* repeat aloud from memory; state (facts) in order.

reckless *adj.* wildly impulsive. □ **recklessly** *adv.*, **recklessness** *n.*

reckon *v.* **1** count up. **2** have as one's opinion. **3** rely. □ **reckon with** take into account.

reclaim *v.* **1** take action to recover possession of. **2** make (wasteland) usable. □ **reclamation** *n.*

recline *v.* lean (one's body), lie down.

recluse *n.* a person who avoids social life.

recognition *n.* recognizing.

recognizance *n.* a pledge made to a law court or magistrate; surety for this.

recognize *v.* (also **-ise**) **1** know again from one's previous experience. **2** acknowledge as genuine, valid, or worthy. □ **recognizable** *adj.*

recoil *v.* **1** spring back; shrink in fear or disgust. **2** have an adverse effect (on the originator). ● *n.* the act of recoiling.

recollect *v.* remember, call to mind. □ **recollection** *n.*

recommend *v.* **1** advise. **2** praise as worthy of employment or use etc. **3** (of qualities etc.) make desirable. □ **recommendation** *n.*

recompense v. repay, compensate. ● n. repayment.

reconcile v. make friendly after an estrangement; induce to tolerate something unwelcome; make compatible. □ **reconciliation** n.

recondite adj. obscure, dealing with an obscure subject.

recondition v. overhaul, repair.

reconnaissance n. a preliminary survey, esp. exploration of an area for military purposes.

reconnoitre v. (Amer. **reconnoiter**) (**reconnoitred**, **reconnoitring**) make a reconnaissance (of).

reconsider v. consider again; consider changing. □ **reconsideration** n.

reconstitute v. reconstruct; restore to its original form. □ **reconstitution** n.

reconstruct v. construct or enact again. □ **reconstruction** n.

record v. (ri-**kord**) 1 set down in writing or other permanent form. 2 preserve (sound) on a disc or magnetic tape for later reproduction. 3 (of a measuring instrument) indicate, register. ● n. (re-**kord**) 1 information set down in writing etc.; a document bearing recorded this. 2 a disc bearing recorded sound. 3 facts known about a person's past. 4 the best performance or most remarkable event etc. of its kind. ● adj. (re-**kord**) the best or most extreme hitherto recorded. □ **off the record** unofficially or not for publication.

recorder n. 1 a person or thing that records. 2 a barrister or solicitor serving as a part-time judge. 3 a kind of flute.

recount v. narrate, tell in detail.

re-count v. count again. ● n. a second or subsequent counting.

recoup v. reimburse or compensate (for); make up (a loss).

recourse n. a source of help to which one may turn. □ **have recourse to** turn to for help.

recover v. 1 regain possession or control of. 2 return to health. □ **recovery** n.

recreation n. a pastime; relaxation. □ **recreational** adj.

recriminate v. make angry accusations in retaliation. □ **recrimination** n., **recriminatory** adj.

recruit n. a new member, esp. of the armed forces. ● v. form by enlisting recruits; enlist as a recruit. □ **recruitment** n.

rectal adj. of the rectum.

rectangle n. a geometric figure with four sides and four right angles, esp. with adjacent sides unequal in length. □ **rectangular** adj.

rectify v. 1 put right. 2 purify, refine. 3 convert to direct current. □ **rectification** n., **rectifier** n.

rectilinear adj. bounded by straight lines.

rectitude n. correctness of behaviour or procedure.

rector n. 1 a clergyman in charge of a parish. 2 the head of certain schools, colleges, and universities.

rectory n. the house of a rector.

rectum n. the last section of the intestine, between the colon and the anus.

recumbent adj. lying down.

recuperate v. recover (health, strength, or losses). □ **recuperation** n., **recuperative** adj.

recur v. (**recurred**) happen again or repeatedly.

recurrent adj. recurring. □ **recurrence** n.

recusant n. a person who refuses to submit or comply.

recycle v. convert (waste material) for reuse.

red adj. (**redder**) 1 of or like the colour of blood; (of hair) reddish brown. **2** Communist, favouring Communism. ● *n.* **1** a red colour or thing. **2** a Communist. □ **in the red** having a debit balance, in debt. □ **redness** *n.*

red-brick *adj.* (of British universities) founded in the late 19th century or later.

red carpet *n.* privileged treatment for an important visitor.

redcurrant *n.* a small edible red berry; a bush bearing these.

redden *v.* make or become red.

reddish *adj.* rather red.

redeem *v.* **1** buy back; convert (tokens etc.) into goods or cash. **2** reclaim. **3** free from sin; make up for (faults). □ **redemption** *n.*, **redemptive** *adj.*

redeploy *v.* send to a new place or task. □ **redeployment** *n.*

red-handed *adj.* in the act of committing a crime.

redhead *n.* a person with red hair.

red herring *n.* a misleading clue or diversion.

red-hot *adj.* **1** glowing red from heat. **2** (of news) completely new.

redirect *v.* direct or send to another place. □ **redirection** *n.*

red-letter day *n.* a memorable day because of a success or happy event.

red light *n.* a signal to stop; a danger signal.

red-light district *n.* a district containing many brothels.

redolent *adj.* **1** smelling strongly. **2** reminiscent. □ **redolence** *n.*

redouble *v.* increase or intensify.

redoubtable *adj.* formidable.

redress *v.* set right. ● *n.* reparation, amends.

red tape *n.* excessive formalities in official transactions.

reduce *v.* **1** make or become smaller or less. **2** make lower in rank. **3** slim. **4** bring into a

specified state. **5** convert into a simpler or more general form. □ **reducible** *adj.*, **reduction** *n.*

redundant *adj.* superfluous; no longer needed. □ **redundancy** *n.*

redwood *n.* a very tall evergreen Californian tree; its wood.

reed *n.* **1** a water or marsh plant with tall hollow stems; its stem. **2** a vibrating part producing sound in certain wind instruments.

reedy *adj.* (**reedier**) (of the voice) having a thin high tone. □ **reediness** *n.*

reef *n.* **1** a ridge of rock or sand etc. reaching to or near the surface of water. **2** part of a sail that can be drawn in when there is a high wind. ● *v.* shorten (a sail).

reefer *n.* a thick double-breasted jacket.

reef knot *n.* a symmetrical double knot.

reek *n.* a strong unpleasant smell. ● *v.* smell strongly.

reel *n.* **1** a cylinder on which something is wound. **2** a lively folk or Scottish dance. ● *v.* **1** wind on or off a reel. **2** stagger. □ **reel off** say rapidly without effort.

refectory *n.* the dining room of a monastery or college etc.

refer *v.* (**referred**) □ **refer to** mention; direct to an authority or specialist; turn to for information.

referee *n.* **1** an umpire, esp. in football and boxing; a person to whom disputes are referred for decision. **2** a person willing to testify to the character or ability of one applying for a job. ● *v.* (**refereed**, **refereeing**) act as referee in (a match etc.).

reference *n.* **1** an act of referring; a mention. **2** a source of information. **3** a testimonial. **4** a person willing to testify to another's character, ability, etc.

□ **in** or **with reference to** in connection with, about.

reference book n. a book providing information.

reference library n. one containing books that can be consulted but not taken away.

referendum n. (pl. **referendums** or **referenda**) the referring of a question to the people for decision by a general vote.

referral n. referring.

refill v. (re-**fill**) fill again. ● n. (ree-fil) a second or later filling; material used for this.

refine v. **1** remove impurities or defects from. **2** make elegant or cultured. □ **refined** adj.

refinement n. **1** refining. **2** elegance of behaviour. **3** an improvement added. **4** a fine distinction.

refinery n. an establishment where crude substances are refined.

refit v. (**refitted**) renew or repair the fittings of.

reflate v. restore (a financial system) after deflation. □ **reflation** n., **reflationary** adj.

reflect v. **1** throw back (light, heat, or sound); show an image of. **2** bring (credit or discredit). **3** think deeply.

reflection n. **1** reflecting or being reflected. **2** a reflected image; reflected light etc. **3** reconsideration. **4** discredit.

reflective adj. **1** reflecting (light etc.). **2** thoughtful.

reflector n. something that reflects light or heat.

reflex n. (also **reflex action**) an involuntary or instinctive movement in response to a stimulus.

reflex angle n. an angle of more than 180°.

reflexive adj. & n. (Grammar) (a word or form) showing that the action of the verb is performed on its subject (e.g. he washed himself).

reflexology n. the massaging of points on the feet as a treatment for stress and other conditions.

reforest var. of **reafforest**

reform v. improve by removing faults; (cause to) give up bad behaviour. ● n. reforming. □ **reformation** n., **reformer** n., **reformist** n.

reformatory adj. reforming. ● n. (Amer. or hist.) an institution to which young offenders are sent to be reformed.

refract v. bend (a ray of light) where it enters water or glass etc. obliquely. □ **refraction** n., **refractive** adj., **refractor** n.

refractory adj. resisting control or discipline; resistant to treatment or heat.

refrain v. keep oneself from doing something. ● n. recurring lines of a song; music for these.

refresh v. **1** restore the vigour of by food, drink, or rest. **2** stimulate (a person's memory).

refreshing adj. **1** restoring vigour, cooling. **2** interesting because of its novelty. □ **refreshingly** adv.

refreshment n. **1** the process of refreshing; something that refreshes. **2** (**refreshments**) food and drink.

refrigerate v. make extremely cold, esp. in order to preserve. □ **refrigeration** n.

refrigerator n. a cabinet or room in which food is stored at a very low temperature.

refuge n. a shelter from pursuit or danger.

refugee n. a person who has left home and seeks refuge (e.g. from war or persecution).

refulgent adj. (literary) shining.

refund v. (ri-**fund**) pay back. ● n. (ree-fund) a repayment, money refunded.

refurbish v. make clean or bright again, redecorate. □ **refurbishment** n.

refuse[1] (ri-fewz) v. say or show that one is unwilling to accept or do (what is asked). □ **refusal** n.

refuse[2] (ref-yooss) n. waste material.

refute v. **1** prove (a statement or person) wrong. **2** deny or contradict. □ **refutation** n.

■ **Usage** The use of *refute* to mean 'deny, contradict' is considered incorrect by some people. *Repudiate* can be used instead.

regain v. obtain again after loss; reach again.

regal adj. like or fit for a king. □ **regally** adv., **regality** n.

regale v. feed or entertain well.

regalia n.pl. emblems of royalty or rank.

regard v. **1** look steadily at. **2** consider to be. ● n. **1** a steady gaze. **2** heed. **3** respect. **4** (**regards**) kindly greetings conveyed in a message. □ **as regards** regarding.

regarding prep. with reference to.

regardless adj. & adv. without consideration (for).

regatta n. boat races organized as a sporting event.

regency n. rule by a regent; a period of this.

regenerate v. give new life or vigour to. □ **regeneration** n., **regenerative** adj.

regent n. a person appointed to rule while the monarch is a minor or is unwell or absent.

reggae n. a West Indian style of music with a strong beat.

regicide n. the killing or killer of a king.

regime (ray-zheem) n. **1** a method or system of government. **2** a regimen.

regimen n. a prescribed course of treatment etc.; a way of life.

regiment n. a permanent unit of an army. ● v. organize rigidly. □ **regimentation** n.

regimental adj. of an army regiment.

Regina n. **1** a reigning queen; *Elizabeth Regina*. **2** (*Law*) the Crown; *Regina v. Jones*.

region n. a part of a surface, space, or body; an administrative division of a country. □ **in the region of** approximately. □ **regional** adj.

register n. an official list. **2** a range of a voice or musical instrument. ● v. **1** enter in a register; record in writing. **2** notice and remember. **3** indicate, record; make an impression. □ **registration** n.

register office n. a place where records of births, marriages, and deaths are kept and civil marriages are performed.

registrar n. **1** an official responsible for keeping written records. **2** a hospital doctor ranking just below specialist.

registry n. **1** registration. **2** a place where written records are kept.

registry office n. a register office.

regress v. relapse to an earlier or more primitive state. □ **regression** n., **regressive** adj.

regret n. a feeling of sorrow, annoyance, or repentance. ● v. (**regretted**) feel regret about. □ **regretful** adj., **regretfully** adv.

regrettable adj. undesirable. □ **regrettably** adv.

regular adj. **1** acting, occurring, or done in a uniform manner or at a fixed time or interval; conforming to a rule or habit; even, symmetrical. **2** forming a country's permanent armed forces. ● n. **1** a regular soldier etc. **2**

(colloquial) a regular customer, visitor, etc. □ **regularly** adv., **regularity** n.

regularize v. (also **-ise**) make regular; make lawful or correct. □ **regularization** n.

regulate v. control by rules; adjust to work correctly or according to one's requirements. □ **regulator** n.

regulation n. the process of regulating; a rule.

regulo n. a point on the temperature scale of a gas oven; *cook at regulo 6.*

regurgitate v. **1** bring (swallowed food) up again to the mouth. **2** cast out again. □ **regurgitation** n.

rehabilitate v. restore to a normal life or good condition. □ **rehabilitation** n.

rehash v. (ree-hash) put (old material) into a new form. ● n. (ree-hash) rehashing; something made of rehashed material.

rehearse v. practise beforehand; recite. □ **rehearsal** n.

rehouse v. provide with new accommodation.

reign n. a sovereign's (period of) rule. ● v. rule as king or queen; be supreme.

reimburse v. repay (a person), refund. □ **reimbursement** n.

rein n. a long strap fastened to a bridle, used to guide or check a horse; a means of control. ● v. check or control with reins.

reincarnation n. the rebirth of a soul in another body after death. □ **reincarnate** v.

reindeer n. (pl. **reindeer** or **reindeers**) a deer of Arctic regions, with large antlers.

reinforce v. strengthen with additional people, material, or quantity. □ **reinforcement** n.

reinstate v. restore to a previous position. □ **reinstatement** n.

reiterate v. say or do again or repeatedly. □ **reiteration** n.

reject v. (ri-jekt) refuse to accept. ● n. (ree-jekt) a person or thing rejected. □ **rejection** n.

rejig v. (**rejigged**) re-equip for new work; rearrange.

rejoice v. feel or show great joy; gladden.

rejoin v. **1** join again. **2** retort.

rejoinder n. a reply, a retort.

rejuvenate v. restore youthful appearance or vigour to. □ **rejuvenation** n., **rejuvenator** n.

relapse v. fall back into a previous state; become worse after improvement. ● n. relapsing.

relate v. **1** narrate. **2** establish a relation between; have a connection with; establish a successful relationship.

related adj. having a common descent or origin.

relation n. **1** a similarity connecting people or things. **2** a relative. **3** narrating. **4** (**relations**) dealings with others. **5** (**relations**) sexual intercourse. □ **relationship** n.

relative adj. considered in relation to something else; having a connection. ● n. a person related to another by descent or marriage. □ **relatively** adv.

relativity n. being relative; (*Physics*) Einstein's theory of the universe, showing that all motion is relative and treating time as a fourth dimension related to space.

relax v. **1** make or become less tight, tense, or strict. **2** rest from work. □ **relaxation** n.

relay n. (ree-lay) **1** a fresh set of workers relieving others. **2** a relay race. **3** a relayed message or transmission; a device relaying things or activating an electrical circuit. ● v. (ri-lay) receive and pass on or retransmit.

relay race n. a race between teams in which each person in turn covers part of the total distance.

release v. **1** set free; remove from a fixed position. **2** make (information, a film or recording) available to the public. ● n. **1** releasing; a handle or catch that unfastens something. **2** information or a film etc. released.

relegate v. consign to a less important position or group. □ **relegation** n.

relent v. become less severe.

relentless adj. unremitting; constant. □ **relentlessly** adv.

relevant adj. related to the matter in hand. □ **relevance** n.

reliable adj. able to be relied on; consistently good. □ **reliably** adv., **reliability** n.

reliance n. relying; trust, confidence. □ **reliant** adj.

relic n. something that survives from earlier times; (**relics**) remains.

relief n. **1** ease given by reduction or removal of pain, anxiety, etc. **2** something that breaks up monotony. **3** assistance to those in need; a person replacing one who is on duty. **4** a carving etc. in which the design projects from a surface; a similar effect given by colour or shading.

relief road n. a road by which traffic can avoid a congested area.

relieve v. give or bring relief to; release from a task or duty; raise the siege of. □ **relieve oneself** urinate or defecate.

religion n. belief in and worship of a superhuman controlling power, esp. a God; a system of this; a controlling influence.

religious adj. **1** of religion; devout, pious. **2** very conscientious. □ **religiously** adv.

relinquish v. give up, cease from. □ **relinquishment** n.

reliquary n. a receptacle for relics of a holy person.

relish n. **1** great enjoyment of something. **2** an appetizing flavour; a strong-tasting food eaten as an accompaniment. ● v. enjoy greatly.

relocate v. move to a different place. □ **relocation** n.

reluctant adj. unwilling, grudging one's consent. □ **reluctantly** adv., **reluctance** n.

rely on v. trust confidently, depend on for help etc.

REM abbr. rapid eye movement, jerky eye movements during dreaming.

remain v. stay; be left or left behind; continue in the same condition.

remainder n. the remaining people or things; a quantity left after subtraction or division. ● v. dispose of unsold copies of (a book) at a reduced price.

remains n.pl. **1** what remains, surviving parts. **2** a dead body.

remand v. send back (a prisoner) into custody while further evidence is sought. □ **on remand** remanded.

remark n. a spoken or written comment. ● v. **1** make a remark, say. **2** notice.

remarkable adj. worth noticing, unusual. □ **remarkably** adv.

remedial adj. **1** providing a remedy. **2** (of teaching) for slow or disadvantaged pupils.

remedy n. something that cures a condition or puts a matter right. ● v. be a remedy for; put right.

remember v. keep in one's mind and recall at will. □ **remembrance** n.

remind v. cause to remember.

reminder n. something that reminds someone, a letter sent as this.

reminisce v. think or talk about past events.

reminiscence n. reminiscing; an account of what one remembers.

reminiscent adj. having characteristics that remind one (of something).

remiss adj. negligent.

remission n. **1** cancellation of a debt or penalty. **2** a reduction of force or intensity.

remit v. (ri-**mit**) (**remitted**) **1** cancel (a debt or punishment). **2** make or become less intense. **3** send (money etc.). **4** refer (a matter for decision) to an authority. ● n. (**ree**-mit) terms of reference.

remittance n. the sending of money; money sent.

remnant n. a small remaining quantity; a surviving trace.

remold Amer. sp. of **remould**.

remonstrate v. make a protest. □ **remonstrance** n.

remorse n. deep regret for one's wrongdoing. □ **remorseful** adj., **remorsefully** adv.

remorseless adj. relentless. □ **remorselessly** adv.

remote adj. **1** far away in place or time; not close. **2** slight. □ **remotely** adv., **remoteness** n.

remould (Amer. **remold**) v. (ree-**mohld**) mould again; reconstruct the tread of (a tyre). ● n. (**ree**-mohld) a remoulded tyre.

remove v. take off or away; dismiss from office; get rid of. ● n. a degree of remoteness or difference. □ **remover** n., **removable** adj., **removal** n.

remunerate v. pay or reward for services. □ **remuneration** n.

remunerative adj. giving good remuneration, profitable.

Renaissance n. a revival of art and learning in Europe in the 14th–16th centuries; (**renaissance**) any similar revival.

renal adj. of the kidneys.

rend v. (**rent**, **rending**) tear.

render v. **1** give, esp. in return; submit (a bill etc.). **2** cause to become. **3** give a performance of. **4** translate. **5** melt down (fat).

rendezvous (ron-day-voo) n. (pl. **rendezvous**) a prearranged meeting or meeting place. ● v. meet at a rendezvous.

rendition n. the way something is rendered or performed.

renegade n. a person who deserts from a group, cause, etc.

renege (ri-**nayg**) v. fail to keep a promise or agreement.

renew v. restore to its original state; replace with a fresh thing or supply; get, make, or give again. □ **renewal** n.

rennet n. a substance used to curdle milk in making cheese.

renounce v. give up formally; reject. □ **renouncement** n.

renovate v. repair, restore to good condition. □ **renovation** n., **renovator** n.

renown n. fame.

renowned adj. famous.

rent¹ see **rend**.

rent² n. a tear; a gap.

rent³ n. periodical payment for use of land, rooms, machinery, etc. ● v. pay or receive rent for.

rental n. rent; renting.

renunciation n. renouncing.

reorganize v. (also **-ise**) organize in a new way. □ **reorganization** n.

reorient v. **1** change or adjust (ideas or views). **2** find one's bearings again.

rep n. (colloquial) **1** a business firm's travelling representative. **2** a varied series of theatrical performances.

repair v. put into good condition after damage or wear; make amends for. ● n. **1** process of repairing; a repaired place. **2** condition for use. □ **repairer** n.

repartee n. an exchange of witty remarks.

repast n. (formal) a meal.

repatriate v. send or bring back (a person) to his or her own country. □ **repatriation** n.

repay v. (**repaid, repaying**) pay back. □ **repayable** adj., **repayment** n.

repeal v. withdraw (a law) officially. ● n. the repealing of a law.

repeat v. say, do, produce, or occur again; tell (a thing told to oneself) to another person. ● n. repeating; something repeated.

repeatedly adv. again and again.

repel v. (**repelled**) drive away; be impossible for (a substance) to penetrate; be repulsive or distasteful to. □ **repellent** adj. & n.

repent v. feel regret about (an action or failure). □ **repentance** n., **repentant** adj.

repercussion n. 1 the recoil of something. 2 an echo. 3 an indirect effect or reaction.

repertoire (rep-er-twahr) n. a stock of songs, plays, etc., that a person or company is able to perform.

repertory n. a repertoire.

repertory company n. a theatrical company that performs a succession of plays for short periods.

repetition n. repeating; an instance of this.

repetitious adj. repetitive.

repetitive adj. characterized by repetition. □ **repetitively** adv.

repine v. fret, be discontented.

replace v. 1 put back in place. 2 provide or be a substitute for. □ **replacement** n.

replay v. (ree-play) play again. ● n. (ree-play) replaying.

replenish v. refill; renew (a supply etc.). □ **replenishment** n.

replete adj. full; well-supplied.

replica n. an exact copy.

replicate v. make a replica of. □ **replication** n.

reply v. answer. ● n. an answer.

report v. 1 give an account of; tell as news. 2 make a formal complaint about. 3 present oneself on arrival. ● n. 1 a spoken or written account; a written statement about a pupil's work etc.; a rumour. 2 an explosive sound.

reportage (re-por-tahzh) n. the reporting of news.

reported speech n. a speaker's words as reported by another person.

reporter n. a person employed to report news for publication or broadcasting.

repose n. rest, sleep; tranquillity. ● v. rest, lie.

repository n. a storage place.

repossess v. take back something for which payments have not been made. □ **repossession** n.

reprehend v. rebuke.

reprehensible adj. deserving rebuke. □ **reprehensibly** adv.

represent v. 1 show in a picture or play etc. 2 describe or declare (to be). 3 be an example or embodiment of. 4 act on behalf of.

representation n. 1 representing, being represented; a picture, diagram, etc. 2 (**representations**) statements made as an appeal, protest, or allegation.

representative adj. typical of a group or class. ● n. a person's or firm's agent; a person chosen to represent others.

repress v. suppress, keep (emotions) from finding an outlet. □ **repression** n., **repressive** adj.

reprieve n. a postponement or cancellation of punishment (esp. a death sentence); a temporary relief from trouble. ● v. give a reprieve to.

reprimand v. rebuke. ● n. a rebuke.

reprint v. (ree-print) print again. ● n. (ree-print) a book reprinted.

reprisal n. an act of retaliation.

reproach v. express disapproval to (a person) for a fault or offence. ● n. an act or instance of reproaching; (a cause of) discredit. □ reproachful adj., reproachfully adv.

reprobate n. an immoral or unprincipled person.

reproduce v. produce again; produce a copy of; produce further members of the same species. □ reproduction n.

reproductive adj. of reproduction.

reproof n. an expression of condemnation for a fault.

reprove v. give a reproof to.

reptile n. a member of the class of cold-blooded animals with a backbone and rough or scaly skin. □ reptilian adj. & n.

republic n. a country in which the supreme power is held by the people's representatives, not by a monarch.

republican adj. of or advocating a republic. ● n. a person advocating republican government.

repudiate v. reject or disown utterly, deny. □ repudiation n.

repugnant adj. distasteful, objectionable. □ repugnance n.

repulse v. drive back (an attacking force); reject, rebuff. ● n. a driving back; a rejection, a rebuff.

repulsion n. repelling; a strong feeling of distaste, revulsion.

repulsive adj. 1 arousing disgust. 2 able to repel. □ repulsively adv.

reputable adj. having a good reputation, respected.

reputation n. what is generally believed about a person or thing.

repute n. reputation.

reputed adj. said or thought to be.

reputedly adv. by repute.

request n. asking for something; something asked for. ● v. make a request.

requiem (rek-wi-em) n. a special Mass for the repose of the souls of the dead; music for this.

require v. 1 need, depend on for success or fulfilment. 2 order, oblige.

requirement n. a need.

requisite adj. required, necessary. ● n. something needed.

requisition n. a formal written demand, an order laying claim to the use of property or materials. ● v. demand or order by this.

requite v. make return for (a service).

resale n. a sale to another person of something one has bought.

rescind v. repeal or cancel (a law etc.).

rescue v. save from danger or capture etc. ● n. rescuing. □ rescuer n.

research n. study and investigation to discover facts. ● v. perform research (into). □ researcher n.

resemble v. be like. □ resemblance n.

resent v. feel displeased and indignant about. □ resentful adj., resentfully adv., resentment n.

reservation n. 1 reserving; reserved accommodation etc. 2 doubt. 3 an area of land set aside for a purpose.

reserve v. put aside for future or special use; order or set aside for a particular person; retain. ● n. 1 something reserved, extra stock available. 2 (also reserves) forces outside the regular armed services. 3 land set aside for special use. 4 a substitute player in team games.

5 the lowest acceptable price for an item to be auctioned. **6** a tendency to avoid showing feelings or friendliness. □ **in reserve** unused and available.

reserved adj. (of a person) showing reserve of manner.

reservist n. a member of a reserve force.

reservoir (rez-er-vwahr) n. a lake used as a store for a water supply; a container for a supply of fluid.

reshuffle v. interchange; reorganize. ● n. reshuffling.

reside v. dwell permanently.

residence n. residing; the place where a person lives. □ **in residence** living in a specified place to perform one's work.

resident adj. residing, in residence. ● n. a permanent inhabitant; (at a hotel) a person staying overnight.

residential adj. containing houses; providing accommodation.

residue n. what is left over. □ **residual** adj.

resign v. give up (one's job, claim, etc.). □ **resign oneself** be ready to accept and endure. □ **resignation** n.

resigned adj. having resigned oneself. □ **resignedly** adv.

resilient adj. **1** springy. **2** readily recovering from shock etc. □ **resilience** n.

resin n. a sticky substance from plants and certain trees; a similar substance made synthetically, used in plastics. □ **resinous** adj.

resist v. oppose strongly or forcibly; withstand; refrain from accepting or yielding to. □ **resistance** n., **resistant** adj.

resistible adj. able to be resisted.

resistivity n. resistance to the passage of an electric current.

resistor n. a device having resistance to the passage of an electric current.

resolute adj. showing great determination. □ **resolutely** adv.

resolution n. **1** resolving; great determination. **2** a formal statement of a committee's opinion.

resolve v. **1** decide firmly. **2** solve or settle (a problem or doubt). **3** separate into constituent parts. ● n. great determination.

resonant adj. resounding, echoing; reinforcing sound, esp. by vibration. □ **resonance** n.

resonate v. produce or show resonance. □ **resonator** n.

resort n. **1** a popular holiday place. **2** recourse. □ **resort to** turn to for help; adopt as a measure.

resound v. fill a place or be filled with sound; echo.

resource n. something to which one can turn for help; ingenuity; (**resources**) available assets.

resourceful adj. clever at finding ways of doing things. □ **resourcefully** adv., **resourcefulness** n.

respect n. **1** admiration or esteem; politeness arising from this; consideration. **2** relation, reference; a particular aspect. ● v. feel or show respect for.

respectable adj. **1** worthy of respect. **2** considerable. □ **respectably** adv., **respectability** n.

respective adj. belonging to each as an individual.

respectively adv. for each separately in the order mentioned.

respiration n. breathing.

respirator n. a device worn over the nose and mouth to purify air before it is inhaled; a device for giving artificial respiration.

respiratory adj. of respiration.

respire v. breathe.

respite n. an interval of rest or relief; a permitted delay.

resplendent adj. brilliant with colour or decorations.

respond v. answer; react.

respondent n. a defendant in a lawsuit, esp. in a divorce case.

response n. 1 an answer. 2 an act, feeling, or movement produced by a stimulus or another's action.

responsibility n. being responsible; something for which one is responsible.

responsible adj. 1 liable to be blamed for loss or failure etc.; being the cause of something. 2 having to account for one's actions. 3 sensible; trustworthy. 4 involving important duties. □ **responsibly** adv.

responsive adj. responding well to an influence. □ **responsiveness** n.

rest v. 1 be still; (cause or allow to) cease from tiring activity. 2 (of a matter) be left without further discussion. 3 place or be placed for support. 4 rely. 5 remain in a specified state. ● n. 1 (a period of) inactivity or sleep. 2 a prop or support for an object. □ **the rest** the remaining part; the others.

restaurant n. a place where meals can be bought and eaten.

restaurateur n. a restaurant-keeper.

restful adj. giving rest; relaxing.

restitution n. the restoring of a thing to its proper owner or original state. 2 compensation.

restive adj. restless; impatient.

restless adj. unable to rest or be still. □ **restlessly** adv., **restlessness** n.

restoration n. restoring; something restored.

restorative adj. restoring health or strength. ● n. a restorative food, medicine, or treatment.

restore v. bring back to its original state (e.g. by repairing), or to good health or vigour; put back in a former position. □ **restorer** n.

restrain v. hold back from movement or action, keep under control. □ **restraint** n.

restrict v. put a limit on, subject to limitations. □ **restriction** n., **restrictive** adj.

result n. 1 the product of an activity, operation, or calculation. 2 the score, marks, or name of the winner in a contest. ● v. occur or have as a result.

resultant adj. occurring as a result.

resume v. get or take again; begin again after stopping. □ **resumption** n., **resumptive** adj.

résumé (rez-yoo-may) n. 1 a summary. 2 (Amer.) = **curriculum vitae.**

resurface v. 1 put a new surface on. 2 return to the surface.

resurgent adj. rising or arising again. □ **resurgence** n.

resurrect v. bring back into use.

resurrection n. 1 rising from the dead. 2 revival after disuse.

resuscitate v. revive. □ **resuscitation** n.

retail n. selling of goods to the public. ● adj. & adv. by retail. ● v. sell or be sold in the retail trade; relate details of. □ **retailer** n.

retain v. 1 keep, esp. in one's possession or in use. 2 hold in place.

retainer n. a fee paid to retain services.

retaliate v. repay an injury, insult, etc. by inflicting one in return. □ **retaliation** n., **retaliatory** adj.

retard v. cause delay to. □ **retardation** n.

retarded adj. backward in mental or physical development.

retch v. strain one's throat as if vomiting.

retention n. retaining.

retentive adj. able to retain things.

rethink v. (**rethought, rethinking**) reconsider; plan again and differently.

reticent adj. not revealing one's thoughts; reserved. □ **reticence** n.

retina n. (pl. **retinas** or **retinae**) a membrane at the back of the eyeball, sensitive to light.

retinue n. attendants accompanying an important person.

retire v. 1 give up one's regular work because of age; cause (an employee) to do this. 2 withdraw, retreat; go to bed. □ **retirement** n.

retiring adj. shy, avoiding society.

retort v. make (as) a witty or angry reply. ● n. 1 retorting; a reply of this kind. 2 a vessel with a bent neck, used in distilling; a vessel used in making gas or steel.

retouch v. touch up (a picture or photograph).

retrace v. go back over or repeat (a route).

retract v. withdraw. □ **retractable** adj., **retraction** n., **retractor** n.

retractile adj. able to be retracted.

retreat v. withdraw, esp. after defeat or when faced with difficulty. ● n. 1 retreating, withdrawal; a military signal for this. 2 a place of shelter or seclusion.

retrench v. reduce (expenditure or operations). □ **retrenchment** n.

retrial n. a trial of a lawsuit or defendant again.

retribution n. deserved punishment.

retrieve v. 1 regain possession of; extract; bring back. 2 set right (an error etc.). □ **retrieval** n.

retriever n. a dog of a breed used to retrieve game.

retroactive adj. operating retrospectively. □ **retroactively** adv.

retrograde adj. going backwards; reverting to an inferior state.

retrogress v. move backwards, deteriorate. □ **retrogression** n., **retrogressive** adj.

retrospect n. □ **in retrospect** when one looks back on a past event.

retrospective adj. looking back on the past; (of a law etc.) made to apply to the past as well as the future. □ **retrospectively** adv.

retroussé adj. (of the nose) turned up.

retroverted adj. turned backwards. □ **retroversion** n.

retry v. try (a lawsuit or defendant) again.

retsina n. a Greek resin-flavoured wine.

return v. come or go back; bring, give, put, or send back. ● n. 1 returning. 2 a profit; a return ticket; a return match. 3 a formal report submitted by order.

return match n. a second match between the same opponents.

return ticket n. a ticket for a journey to a place and back again.

reunion n. a gathering of people who were formerly associated.

reunite v. bring or come together again.

reusable adj. able to be used again.

rev n. (colloquial) a revolution of an engine. ● v. (**revved**) (colloquial) cause (an engine) to run quickly; (of an engine) revolve.

Rev. abbr. Reverend.

revalue v. put a new value on. □ **revaluation** n.

revamp v. renovate, give a new appearance to.

Revd abbr. Reverend.

reveal v. make visible by uncovering; make known.

reveille (ri-va-li) n. a military waking-signal.

revel v. (**revelled**; Amer. **reveled**) take great delight; hold revels. ● n.pl. (**revels**) lively festivities, merrymaking. □ **reveller** n., **revelry** n.

revelation n. revealing; a surprising thing revealed.

revenge n. injury inflicted in return for what one has suffered; an opportunity to defeat a victorious opponent. ● v. avenge.

revenue n. a country's income from taxes etc.; the department collecting this.

reverberate v. echo, resound. □ **reverberation** n.

revere v. feel deep respect or religious veneration for.

reverence n. a feeling of awe and respect or veneration.

reverend adj. deserving to be treated with respect. **2** (**Reverend**) the title of a member of the clergy.

reverent adj. feeling or showing reverence. □ **reverently** adv.

reverie n. a daydream.

revers (ri-veer) n. (pl. **revers**) a turned-back front edge at the neck of a jacket.

reversal n. reversing.

reverse adj. opposite in character or order; upside down. ● v. turn the other way round, upside down, or inside out; move backwards or in the opposite direction. **2** convert to the opposite; annul (a decree etc.). ● n. **1** the reverse side or effect. **2** a piece of misfortune. □ **reversely** adv., **reversible** adj.

revert v. return to a former condition or habit; return to a subject in talk etc.; (of property etc.) return or pass to another holder. □ **reversion** n.

review n. **1** a general survey of events or a subject; revision or reconsideration. **2** a ceremonial inspection of troops etc. **3** a critical report of a book, play, etc. ● v. make or write a review of. □ **reviewer** n.

revile v. criticize angrily in abusive language.

revise v. **1** re-examine and alter or correct. **2** study again (work learnt or done) in preparation for an examination. □ **reviser** n., **revision** n., **revisory** adj.

revivalist n. person who seeks to promote religious fervour. **revivalism** n.

revive v. come or bring back to life or consciousness or strength, or into use. □ **revival** n.

revivify v. restore to vigour; bring back to life.

revocable adj. able to be revoked.

revoke v. withdraw (a decree, licence, etc.). □ **revocation** n.

revolt v. **1** take part in a rebellion; be in a mood of protest or defiance. **2** cause strong disgust in. ● n. **1** an act or state of rebelling. **2** a sense of disgust.

revolting adj. **1** causing disgust. **2** in revolt.

revolution n. **1** revolving, a single complete orbit or rotation. **2** the substitution of a new system of government, esp. by force. **3** a complete change of method or conditions.

revolutionary adj. **1** involving a great change. **2** of political revolution. ● n. a person who begins or supports a political revolution.

revolutionize v. (also **-ise**) alter completely.

revolve v. turn round; move in an orbit.

revolver n. a pistol.

revue n. an entertainment consisting of a series of items.

revulsion n. **1** strong disgust. **2** a sudden violent change of feeling.

reward n. something given or received in return for service or merit. ● v. give a reward to.

rewire v. renew the electrical wiring of.

rewrite v. (**rewrote, rewritten, rewriting**) write again in a different form or style.

Rex n. a reigning king.

RFC abbr. Rugby Football Club.

Rh abbr. rhesus. ● symb. rhodium.

rhapsodize v. (also **-ise**) talk or write about something ecstatically.

rhapsody n. **1** an ecstatic statement. **2** a romantic musical composition. □ **rhapsodic** adj.

rhenium n. a rare metallic element (symbol Re).

rheostat n. a device for varying the resistance to electric current.

rhesus n. a small monkey common in northern India.

rhesus factor n. a substance usu. present in human blood. □ **rhesus-positive** having this substance. **rhesus-negative** not having this substance.

rhetoric n. the art of using words impressively; impressive language.

rhetorical adj. expressed so as to sound impressive. □ **rhetorically** adv.

rhetorical question n. one used for dramatic effect, not seeking an answer.

rheumatic adj. of or affected with rheumatism. □ **rheumaticky** adj.

rheumatism n. a disease causing pain in the joints, muscles, or fibrous tissue.

rheumatoid adj. having the character of rheumatism.

rhinestone n. an imitation diamond.

rhino n. (pl. **rhino** or **rhinos**) (colloquial) a rhinoceros.

rhinoceros n. (pl. **rhinoceroses** or **rhinoceros**) a large thick-skinned animal with one horn (or occasionally two) on its nose.

rhizome n. a root-like stem producing roots and shoots.

rhodium n. a metallic element (symbol Rh) used in alloys to increase hardness.

rhododendron n. an evergreen shrub with large clusters of flowers.

rhomboid adj. like a rhombus. ● n. a rhomboid figure.

rhombus n. a quadrilateral with opposite sides and angles equal (and not right angles).

rhubarb n. a plant with red leaf-stalks used like fruit.

rhyme n. a similarity of sound between words or syllables; a word providing a rhyme to another; a poem with line-endings that rhyme. ● v. form a rhyme.

rhythm n. a pattern produced by emphasis and duration of notes in music or of syllables in words, or by regular movements or events. □ **rhythmic, rhythmical** adj., **rhythmically** adv.

rhythm method n. contraception by avoiding sexual intercourse near the time of ovulation.

rib n. **1** one of the curved bones round the chest; a structural part resembling this. **2** a pattern of raised lines in knitting. ● v. (**ribbed**) **1** support with ribs. **2** knit as rib. **3** (colloquial) tease.

ribald adj. humorous in a cheerful but vulgar or disrespectful way. □ **ribaldry** n.

riband n. a ribbon.

ribbon n. a narrow band of silky material; a strip resembling this.

riboflavin n. (also **riboflavine**) a B vitamin found in liver, milk, and eggs.

ribonucleic acid n. a substance controlling protein synthesis in cells.

rice n. a cereal plant grown in marshes in hot countries; its seeds used as food.

rich adj. **1** having much wealth. **2** abundant; containing a large proportion of something (e.g. fat or fuel). **3** (of soil) fertile. **4** (of colour, sound, or smell) pleasantly deep and strong. ● n.pl. (**riches**) wealth. □ **richness** n.

richly adv. in a rich way; fully, thoroughly.

rick n. **1** a large stack of hay etc. **2** a slight sprain or strain. ● v. sprain or strain slightly.

rickets n. a bone disease caused by vitamin D deficiency.

rickety adj. shaky, insecure.

rickrack var. of **ricrac**.

rickshaw n. a two-wheeled vehicle used in the Far East, pulled along by a person.

ricochet (rik-ŏ-shay) v. (**ricocheted**) rebound from a surface after striking it with a glancing blow. ● n. a rebound of this kind.

ricotta n. soft Italian cheese.

ricrac n. (also **rickrack**) a zigzag braid trimming.

rid v. (**rid, ridding**) free from something unpleasant or unwanted. □ **get rid of** cause to go away; free oneself of.

ridden see **ride**.

riddle n. **1** a question etc. designed to test ingenuity esp. for amusement; something puzzling or mysterious. **2** a coarse sieve. ● v. pass through a coarse sieve; permeate thoroughly.

ride v. (**rode, ridden, riding**) sit on and be carried by (a horse or bicycle etc.); travel in a vehicle; float (on). ● n. a spell of riding; a journey in a vehicle.

rider n. **1** a person who rides a horse etc. **2** an additional statement.

ridge n. a narrow raised strip; a line where two upward slopes meet; an elongated region of high barometric pressure. □ **ridged** adj.

ridicule n. making or being made to seem ridiculous. ● v. subject to ridicule, make fun of.

ridiculous adj. deserving to be laughed at; not worth serious consideration. □ **ridiculously** adv.

rife adj. occurring frequently, widespread. □ **rife with** full of.

riff n. a short repeated phrase in jazz etc.

riffle v. flick through (pages etc.).

riff-raff n. rabble; disreputable people.

rifle n. a gun with a long barrel. ● v. **1** search and rob. **2** cut spiral grooves in (a gun barrel).

rift n. a cleft in earth or rock; a crack, a split; a breach in friendly relations.

rift valley n. a steep-sided valley formed by subsidence.

rig v. (**rigged**) **1** provide with clothes or equipment; fit (a ship) with spars, ropes, etc.; set up, esp. in a makeshift way. **2** manage or control fraudulently. ● n. **1** the way a ship's masts and sails etc. are arranged. **2** an apparatus for drilling an oil well etc.

rigging n. ropes etc. used to support a ship's masts and sails.

right adj. **1** morally good; in accordance with justice; proper, correct, true. **2** in a good condition. **3** of or on the side of the body which is on the east when one is facing north. ● n. **1** what is just; something one is entitled to. **2** the right side or region; the right hand or foot. **3** people supporting more conservative policies than others in their group. ● *v.* **1** restore to a correct or upright position. **2** set right. ● *adv.* **1** on or towards the right-hand side. **2** directly. □ **in the right** having truth or justice on one's side. **right away** immediately. □ **rightly** *adv.*

right angle *n.* an angle of 90°.

righteous *adj.* doing what is morally right, making a show of this; morally justifiable. □ **righteously** *adv.*, **righteousness** *n.*

rightful *adj.* just, proper, legal. □ **rightfully** *adv.*, **rightfulness** *n.*

right-handed *adj.* using the right hand.

right-hand man *n.* an indispensable assistant.

right of way *n.* the right to pass over another's land; a path subject to this; the right to proceed while another vehicle must wait.

rigid *adj.* stiff; strict, inflexible. □ **rigidly** *adv.*, **rigidity** *n.*

rigmarole *n.* a long rambling statement; a complicated formal procedure.

rigor mortis *n.* stiffening of the body after death.

rigour *n.* (*Amer.* **rigor**) strictness, severity; harshness of weather or conditions. □ **rigorous** *adj.*, **rigorously** *adv.*, **rigorousness** *n.*

rile *v.* (*colloquial*) annoy.

rill *n.* a small stream.

rim *n.* an edge or border, esp. of something circular. □ **rimmed** *adj.*

rime *n.* frost.

rind *n.* a tough outer layer on fruit, cheese, bacon, etc.

ring¹ *n.* **1** an outline of a circle; a circular metal band usu. worn on a finger. **2** an enclosure where a performance or activity takes place. **3** a group of people acting together dishonestly. ● *v.* put a ring on; surround.

ring² *v.* (**rang, rung, ringing**) **1** give out a loud clear resonant sound; cause (a bell) to do this; signal by ringing; be filled with sound. **2** telephone. ● *n.* **1** an act or sound of ringing. **2** a specified tone or feeling of a statement etc. **3** (*colloquial*) a telephone call. □ **ring off** end a telephone call. **ring the changes** vary things. **ring up** make a telephone call (to).

ringleader *n.* a person who leads others in wrongdoing.

ringlet *n.* a long tubular curl of hair.

ringside *n.* the area beside a boxing ring.

ringside seat *n.* a position from which one has a clear view of the scene of action.

ringworm *n.* a skin condition producing round scaly patches on the skin.

rink *n.* an area of ice for skating; an enclosed area for roller skating.

rinse *v.* wash lightly; wash out soap etc. from. ● *n.* **1** the process of rinsing. **2** a solution washed through hair to tint or condition it.

riot *n.* **1** a wild disturbance by a crowd of people. **2** a profuse display. **3** (*colloquial*) a very amusing person or thing. ● *v.* take part in a riot. □ **run riot** behave in an unruly way; grow in an uncontrolled way.

riotous *adj.* disorderly, unruly; boisterous. □ **riotously** *adv.*

RIP *abbr.* rest in peace.

rip *v.* (**ripped**) **1** tear apart; remove by pulling roughly; become torn. **2** rush along. ● *n.* an act of ripping; a torn place. □ **rip off** (*colloquial*) defraud; steal.

ripcord n. a cord for pulling to release a parachute.

ripe adj. ready to be gathered and used; matured; (of age) advanced; ready. □ **ripeness** n.

ripen v. make or become ripe.

rip-off n. (colloquial) a swindle.

riposte (ri-posst) n. a quick counterstroke or retort. ● v. deliver a riposte.

ripple n. a small wave; a gentle sound that rises and falls. ● v. form ripples.

rise v. (**rose, risen, rising**) 1 come, go, or extend upwards; get up from lying or sitting, get out of bed. 2 become higher; increase. 3 rebel. 4 have its origin or source. ● n. 1 an act or amount of rising; an increase. 2 an upward slope. 3 an increase in wages. □ **give rise to** cause.

risible adj. ridiculous; inclined to laugh. □ **risibly** adv., **risibility** n.

risk n. a possibility of meeting danger or suffering harm; a person or thing representing a source of danger or harm. ● v. expose to the chance of injury or loss; accept the risk of.

risky adj. (**riskier**) full of risk. □ **riskily** adv., **riskiness** n.

risotto n. (pl. **risottos**) a dish of rice containing chopped meat, vegetables, etc.

risqué (riss-kay) adj. slightly indecent.

rissole n. a mixture of minced meat formed into a flat shape and fried.

rite n. a ritual.

ritual n. a series of actions used in a religious or other ceremony. ● adj. of or done as a ritual. □ **ritually** adv., **ritualistic** adj., **ritualism** n.

rival n. a person or thing that competes with or can equal another. ● adj. being a rival or rivals. ● v. (**rivalled**; Amer. ri-

valed) be a rival of; seem as good as. □ **rivalry** n.

river n. a large natural stream of water; a great flow.

rivet n. a bolt for holding pieces of metal together, with its end beaten to form a head when in place. ● v. (**riveted**) 1 fasten with a rivet. 2 attract and hold the attention of, engross.

rivulet n. a small stream.

RN abbr. Royal Navy.

Rn symb. radon.

RNA abbr. ribonucleic acid.

RNLI abbr. Royal National Lifeboat Institution.

roach n. (pl. **roach**) a small freshwater fish of the carp family.

road n. a prepared track along which people and vehicles may travel; a way of reaching something. □ **on the road** travelling.

road hog n. a reckless or inconsiderate driver.

roadie n. (colloquial) a person helping a touring band with their equipment.

road metal n. broken stone for making the foundation of a road or railway.

roadster n. an open car with no rear seats.

roadway n. a road, esp. as distinct from a footpath beside it.

roadworks n.pl. construction or repair of roads.

roadworthy adj. (of a vehicle) fit to be used on a road. □ **roadworthiness** n.

roam v. wander.

roan n. a horse with a dark coat sprinkled with white hairs.

roar n. a long deep sound like that made by a lion; loud laughter. ● v. give a roar; express in this way.

roaring adj. 1 noisy. 2 briskly active.

roast v. cook (meat etc.) with fat in an oven; expose to great heat;

undergo roasting. ● n. a roast joint of meat.

rob v. (**robbed**) steal from; deprive. □ **robber** n., **robbery** n.

robe n. a long loose esp. ceremonial garment. ● v. dress in a robe.

robin n. a red-breasted bird.

robot n. a machine programmed to move and perform certain tasks. □ **robotic** adj.

robotics n. the study of robots and their design and operation.

robust adj. strong, vigorous. □ **robustly** adv., **robustness** n.

rock n. **1** the hard part of the earth's crust, below the soil; a mass of this, a large stone. **2** a hard sugar sweet made in sticks. **3** modern popular music with a heavy beat. **4** a rocking movement. ● v. **1** move to and fro while supported. **2** disturb greatly by shock. □ **on the rocks** (of a drink) served with ice cubes.

rock and roll n. rock music with elements of blues.

rock-bottom adj. (colloquial) very low.

rock cake n. a small fruit cake with a rough surface.

rocker n. something that rocks; a pivoting switch.

rockery n. a bank containing large stones planted with rock plants.

rocket n. **1** a structure that flies by expelling burning gases; a spacecraft propelled in this way. **2** a firework that rises into the air and explodes. ● v. (**rocketed**) move rapidly upwards or away.

rocketry n. the science or practice of rocket propulsion.

rock plant n. an alpine plant, a plant that grows among rocks.

rocky adj. (**rockier**) **1** of or like rock; full of rock. **2** (colloquial) unsteady. □ **rockily** adv., **rockiness** n.

rococo adj. & n. (of or in) an ornate style of decoration in Europe in the 18th century.

rod n. a slender straight round stick or metal bar; a fishing rod.

rode see **ride**.

rodent n. an animal with strong front teeth for gnawing things.

rodeo n. (pl. **rodeos**) a competition or exhibition of cowboys' skill.

roe¹ n. a mass of eggs in a female fish's ovary.

roe² n. (pl. **roe** or **roes**) a small deer.

roebuck n. a male roe deer.

roentgen (runt-yĕn) n. a unit of ionizing radiation.

roger int. (in signalling) message received and understood.

rogue n. **1** a dishonest or mischievous person. **2** a wild animal living apart from the herd.

roguish adj. mischievous; playful. □ **roguishly** adv., **roguishness** n.

roister v. make merry noisily.

role n. an actor's part; a person's or thing's function.

roll v. **1** move (on a surface) on wheels or by turning over and over; turn on an axis or over and over; form into a cylindrical or spherical shape; flatten with a roller. **2** rock from side to side; undulate; move or pass steadily. ● n. **1** a cylinder of flexible material turned over and over upon itself. **2** a small individual loaf of bread. **3** undulation. **4** an official list or register. **5** a long deep sound. □ **be rolling (in money)** (colloquial) be wealthy. **roll-on roll-off** (of a ferry) that vehicles can be driven on to and off.

roll-call n. the calling of a list of names to check that all are present.

rolled gold n. a thin coating of gold on another metal.

roller n. **1** a cylinder rolled over things to flatten or spread them, or on which something is wound. **2** a long swelling wave.

roller coaster n. a switchback at a fair.

roller skate n. a boot or frame with wheels, strapped to the foot for gliding across a hard surface. ● v. move on roller skates.

rollicking adj. full of boisterous high spirits.

rolling pin n. a roller for flattening dough.

rolling stock n. railway engines and carriages, wagons, etc.

rolling stone n. a person who does not settle in one place.

rollmop n. a rolled uncooked pickled herring fillet.

roly-poly n. a pudding of suet pastry spread with jam, rolled up, and boiled. ● adj. plump, podgy.

ROM abbr. read-only memory.

Roman adj. & n. (a native, an inhabitant) of Rome or of the ancient Roman republic or empire.

roman n. plain upright type.

Roman Catholic adj. & n. (a member) of the Church that acknowledges the Pope as its head.

romance n. an imaginative story; a romantic situation, event, or atmosphere; a love story, a love affair; picturesque exaggeration. ● v. distort the truth or invent imaginatively.

Romance language n. one descended from Latin.

Romanesque adj. & n. (of or in) a style of architecture in Europe about 900–1200.

Roman numerals n.pl. letters representing numbers (I = 1, V = 5, etc.).

romantic adj. appealing to the emotions by its imaginative, heroic, or picturesque quality; involving a love affair; enjoying romantic situations etc. ● n. a romantic person. □ **romantically** adv.

romanticism n. romantic style.

romanticize v. (also **-ise**) make romantic; indulge in romance.

Romany n. a gypsy; the gypsy language.

romp v. **1** play about in a lively way. **2** (colloquial) go along easily.

rondeau n. (pl. **rondeaux**) a short poem with the opening words used as a refrain.

rondo n. (pl. **rondos**) a piece of music with a recurring theme.

rood-screen n. a screen separating the nave from the chancel in a church.

roof n. (pl. **roofs**) the upper covering of a building, car, cavity, etc. ● v. cover with a roof; be the roof of.

rook n. **1** a bird of the crow family. **2** a chess piece with a top shaped like battlements.

rookery n. a colony of rooks.

room n. an enclosed part of a building; space that is or could be occupied; scope to allow something.

roomy adj. (**roomier**) having plenty of space.

roost n. a place where birds perch or rest. ● v. perch, esp. for sleep.

root n. **1** the part of a plant that grows into the earth and absorbs nourishment from the soil; the embedded part of a hair, tooth, etc. **2** a source, a basis. **3** a number in relation to another which it produces when multiplied by itself a specified number of times. **4** (**roots**) emotional attachment to a place. ● v. **1** (cause to) take root; cause to stand fixed and unmoving. **2** (of an animal) turn up ground with the snout or beak in search of food; rummage, extract. □ **root out** or **up** drag or dig up by the roots; get rid of. **take root** send down roots; become established. ▫ **rootless** *adj.*

rope *n.* a strong thick cord. ● *v.* fasten or secure with rope; fence off with rope. ▫ **know** or **show the ropes** know or show the procedure. **rope in** persuade to take part.

rosary *n.* **1** a set series of prayers; a string of beads for keeping count in this. **2** a rose garden.

rose[1] *n.* **1** an ornamental usu. fragrant flower; a bush bearing this. **2** a reddish-pink colour.

rose[2] *see* **rise.**

rosé (roh-zay) *n.* a light pink wine.

rosemary *n.* a shrub with fragrant leaves used as a herb to flavour food.

rosette *n.* a round badge or ornament made of ribbons.

rosewood *n.* a dark fragrant wood used for making furniture.

rosin *n.* a resin.

roster *n.* a list showing people's turns of duty etc.

rostrum *n.* (*pl.* **rostra** or **rostrums**) a platform.

rosy (**rosier**) **1** deep pink. **2** promising, hopeful.

rot *v.* (**rotted**) lose its original form by chemical action caused by bacteria; decay; perish through lack of use. ● *n.* **1** rotting, rottenness. **2** (*colloquial*) nonsense.

rota *n.* a list of duties to be done or people to do them in rotation.

rotary *adj.* acting by rotating.

rotate *v.* revolve; arrange, occur, or deal with in a recurrent series. ▫ **rotation** *n.,* **rotatory** *adj.*

Rotavator *n.* (also **Rotovator**) (*trade mark*) a machine with rotating blades for breaking up uncultivated ground.

rote *n.* ▫ **by rote** by memory without thought of the meaning; by a fixed procedure.

rotisserie *n.* a revolving spit for roasting.

rotor *n.* a rotating part of a machine.

rotten *adj.* **1** decayed; breaking easily from age or use. **2** (*colloquial*) worthless, unpleasant. ▫ **rottenness** *n.*

Rottweiler *n.* a dog of a large powerful black and tan breed.

rotund *adj.* rounded, plump. ▫ **rotundity** *n.*

rotunda *n.* a circular domed building or hall.

rouble *n.* (also **ruble**) a unit of money in Russia.

rouge *n.* a reddish cosmetic colouring for the cheeks. ● *v.* colour with rouge.

rough *adj.* **1** having an uneven or irregular surface. **2** not gentle or careful, violent; (of weather) stormy. **3** not perfected or detailed; approximate. ● *n.* a rough thing or state; rough ground. ● *v.* make rough. ▫ **rough it** do without ordinary comforts. **rough out** plan or sketch roughly. ▫ **roughly** *adv.,* **roughness** *n.*

roughage *n.* dietary fibre.

rough-and-ready *adj.* rough but effective.

rough-and-tumble *n.* a haphazard struggle.

rough diamond *n.* a person of good nature but lacking polished manners.

roughen *v.* make or become rough.

roughshod ▫ **ride roughshod over** treat inconsiderately or arrogantly.

roulette *n.* a gambling game played with a small ball on a revolving disc.

round *adj.* **1** curved, circular, spherical, or cylindrical. **2** complete. ● *n.* **1** a round object;

a circular or recurring course or series. **2** a song for two or more voices that start the same tune at different times. **3** a shot from a firearm, ammunition for this. **4** one section of a competition. ● *prep.* **1** so as to circle or enclose. **2** visiting in a series; to all points of interest in. ● *adv.* **1** in a circle or curve; by a circuitous route. **2** so as to face in a different direction. **3** round a place or group. **4** to a person's house etc. ● *v.* **1** make or become round. **2** make into a round figure. **3** travel round. □ **in the round** with all sides visible. **round about** nearby; approximately. **round off** complete. **round the clock** continuously through day and night. **round up** gather into one place. □ **roundness** *n.*

roundabout *n.* **1** a revolving platform at a funfair, with model horses etc. to ride on. **2** a road junction with a circular island round which traffic has to pass in one direction. ● *adj.* indirect.

roundel *n.* a small disc.

rounders *n.* a team game played with bat and ball, in which players have to run round a circuit.

round figure *n.* (also **round number**) an approximation without odd units.

Roundhead *n.* (*hist.*) a supporter of Parliament against Charles I in the English Civil War.

roundly *adv.* **1** thoroughly, severely. **2** in a rounded shape.

round robin *n.* a petition signed with names in a circle to conceal the order of writing.

round trip *n.* a circular tour; an outward and return journey.

round-up *n.* a systematic rounding up; a summary (of news etc.).

roundworm *n.* a parasitic worm with a rounded body.

rouse *v.* wake; cause to become active or excited.

rousing *adj.* vigorous, stirring.

rout *n.* a complete defeat; a disorderly retreat. ● *v.* **1** defeat completely; put to flight. **2** fetch (out); rummage.

route *n.* a course or way from starting point to finishing point.

route march *n.* a training march for troops.

routine *n.* a standard procedure; a set sequence of movements. ● *adj.* in accordance with routine. □ **routinely** *adv.*

roux (roo) *n.* a mixture of heated fat and flour as a basis for a sauce.

rove *v.* wander.

row[1] (roh) *n.* people or things in a line.

row[2] (roh) *v.* propel (a boat) using oars; carry in a boat that one rows. ● *n.* a spell of rowing. □ **rowboat, rowing boat** *n.*

row[3] (*rhymes with* cow) (*colloquial*) *n.* a loud noise; an angry argument. ● *v.* quarrel, argue angrily.

rowan *n.* a tree bearing clusters of red berries.

rowdy *adj.* (**rowdier**) noisy and disorderly. ● *n.* a rowdy person. □ **rowdily** *adv.*, **rowdiness** *n.*

rowlock (rol-ŏk) *n.* a device on the side of a boat for holding an oar.

royal *adj.* **1** of or suited to a king or queen; of the family or in the service of royalty. **2** splendid. ● *n.* (*colloquial*) a member of a royal family. □ **royally** *adv.*

royal blue *n.* & *adj.* bright blue.

royalist *n.* a person supporting or advocating monarchy.

royalty *n.* **1** being royal; a royal person. **2** payment to an author, patentee, etc. for each copy, performance, or use of his or her work.

RPI *abbr.* retail price index, a list showing monthly variation in the prices of basic goods.

rpm *abbr.* revolutions per minute.

RSJ *abbr.* rolled steel joint, a load-bearing beam.

RSPB *abbr.* Royal Society for the Protection of Birds.

RSPCA *abbr.* Royal Society for the Prevention of Cruelty to Animals.

RSVP *abbr.* (French *répondez s'il vous plaît*) please reply.

Ru *symb.* ruthenium.

rub *v.* (**rubbed**) press against a surface and slide to and fro; polish, clean, dry, or make sore by rubbing. ● *n.* **1** an act or process of rubbing. **2** a difficulty. □ **rub it in** emphasize or remind a person constantly of an unpleasant fact. **rub out** remove (marks etc.) by using a rubber.

rubber *n.* **1** a tough elastic substance made from the juice of certain plants or synthetically; a piece of this for rubbing out pencil or ink marks. **2** a device for rubbing things. □ **rubberstamp** *v.* approve automatically without consideration. □ **rubbery** *adj.*

rubberize *v.* (also **-ise**) treat or coat with rubber.

rubbish *n.* waste or worthless material; nonsense.

rubble *n.* waste or rough fragments of stone, brick, etc.

rubella *n.* German measles.

rubidium *n.* a soft silvery metallic element (symbol Rb).

ruble var. of **rouble**.

rubric *n.* words put as a heading or note of explanation.

ruby *n.* a red gem; a deep red colour. ● *adj.* deep red.

ruby wedding *n.* a 40th wedding anniversary.

RUC *abbr.* Royal Ulster Constabulary.

ruche (roosh) *n.* fabric gathered as a trimming.

ruck *v.* crease, wrinkle. ● *n.* **1** a crease or wrinkle. **2** an undistinguished crowd; (in rugby) a loose scrum.

rucksack *n.* a bag carried on the back, used by walkers.

ructions *n.pl.* (*colloquial*) unpleasant arguments or protests.

rudder *n.* a vertical piece of metal or wood hinged to the stern of a boat or aircraft, used for steering.

ruddy *adj.* (**ruddier**) reddish.

rude *adj.* **1** impolite, showing no respect. **2** primitive, roughly made. □ **rudely** *adv.*, **rudeness** *n.*

rudiment *n.* a rudimentary part; (also **rudiments**) elementary principles.

rudimentary *adj.* incompletely developed; basic, elementary.

rue *n.* a shrub with bitter leaves formerly used in medicine. ● *v.* repent, regret.

rueful *adj.* showing or feeling good-humoured regret. □ **ruefully** *adv.*

ruff *n.* **1** a pleated frill worn round the neck; a projecting or coloured ring of feathers or fur round a bird's or animal's neck. **2** a bird of the sandpiper family.

ruffian *n.* a violent lawless person.

ruffle *v.* disturb the calmness or smoothness of; annoy. ● *n.* a gathered frill.

rug *n.* a thick floor-mat; a piece of thick warm fabric used as a covering.

rugby *n.* (in full **rugby football**) a kind of football game played with an oval ball which may be kicked or carried.

rugged *adj.* **1** uneven, irregular, craggy. **2** rough but kindly. □ **ruggedly** *adv.*, **ruggedness** *n.*

rugger n. (colloquial) rugby football.

ruin n. destruction; complete loss of one's fortune or prospects; broken remains; a cause of ruin. ● v. cause ruin to; reduce to ruins. □ **ruination** n.

ruinous adj. bringing ruin; in ruins, ruined. □ **ruinously** adv.

rule n. 1 a statement of what can or should be done; a dominant custom; governing, control. 2 a ruler used by carpenters etc. ● v. 1 govern; keep under control; give an authoritative decision. 2 draw (a line) using a ruler. □ **as a rule** usually. **rule of thumb** a rough practical method of procedure. **rule out** exclude.

ruler n. 1 a person who rules. 2 a straight strip used in measuring or for drawing straight lines.

ruling n. an authoritative decision.

rum n. an alcoholic spirit distilled from sugar cane or molasses. ● adj. (colloquial) strange, odd.

rumba n. a ballroom dance.

rumble v. 1 make a low continuous sound. 2 (slang) detect the true character of. ● n. a rumbling sound.

rumbustious adj. (colloquial) boisterous, uproarious.

ruminant n. an animal that chews the cud. ● adj. ruminating.

ruminate v. chew the cud; meditate, ponder. □ **rumination** n., **ruminative** adj.

rummage v. search by disarranging things. ● n. an untidy search through a number of things.

rummage sale n. (esp. Amer.) a jumble sale.

rummy n. a card game.

rumour n. (Amer. **rumor**) information spread by talking but not certainly true. □ **be rumoured** be spread as a rumour.

rump n. the buttocks; a bird's back near the tail.

rumple v. make or become crumpled; make untidy.

rumpus n. (colloquial) an uproar, an angry dispute.

run v. (**ran**, **run**, **running**) 1 move with quick steps with always at least one foot of the ground; go smoothly or swiftly; compete in a race. 2 spread, flow, exude liquid. 3 function. 4 travel or convey from one point to another; extend. 5 be current or valid. 6 manage, organize; own and use (a vehicle etc.). ● n. 1 a spell of running. 2 a point scored in cricket or baseball. 3 a ladder in fabric. 4 a continuous stretch or sequence. 5 an enclosure where domestic animals can range. 6 permission to make unrestricted use of something. □ **in** or **out of the running** with a good or with no chance of winning. **in the long run** in the end, over a long period. **on the run** fleeing. **run across** happen to meet or find. **run a risk** take a risk. **run a temperature** be feverish. **run away** flee; leave secretly. **run down** reduce the numbers of; speak of in a slighting way. **run into** collide with; happen to meet. **run out** become used up. **run over** knock down or crush with a vehicle. **run up** allow (a bill) to mount.

rundown n. a detailed analysis. ● adj. (**run-down**) weak or exhausted.

rune n. any of the letters in an early Germanic alphabet. □ **runic** adj.

rung¹ n. a crosspiece of a ladder etc.

rung² see **ring²**.

runner n. 1 a person or animal that runs; a messenger. 2 a creeping stem that roots. **3** a groove, strip, or roller etc. for a thing to move on; a long narrow strip of carpet or ornamental cloth.

runner-up n. one who finishes second in a competition.

runny adj. (**runnier**) semi-liquid; tending to flow or exude fluid.

run-of-the-mill adj. ordinary.

runt n. an undersized person or animal.

run-up n. the period leading up to an event.

runway n. a prepared surface on which aircraft may take off and land.

rupee n. a unit of money in India, Pakistan, etc.

rupture n. a breaking, a breach; an abdominal hernia. ● v. burst, break; cause a hernia in.

rural adj. of, in, or like the countryside.

ruse n. a deception, a trick.

rush v. go or convey with great speed; act hastily; force into hasty action; attack with a sudden assault. ● n. **1** rushing, an instance of this; a period of great activity. **2** a marsh plant with a slender pithy stem.

rush hour n. one of the times of day when traffic is busiest.

rusk n. a biscuit, esp. for babies.

russet adj. soft reddish brown. ● n. **1** a reddish-brown colour. **2** a variety of apple with a rough skin.

rust n. a brownish corrosive coating formed on iron exposed to moisture. ● adj. reddish brown. ● v. make or become rusty. □ **rustproof** adj. & v.

rustic adj. **1** of or like country life or people. **2** made of rough timber or untrimmed branches.

rustle v. **1** (cause to) make a sound like paper being crumpled. **2** steal (horses or cattle). ● n. a rustling sound. □ **rustler** n.

rusty adj. (**rustier**) **1** affected with rust; rust-coloured. **2** having lost quality by lack of use.

rut¹ n. **1** a deep track made by wheels. **2** a habitual dull course of life.

rut² n. the periodic sexual excitement of a male deer, goat, etc. ● v. (**rutted**) be affected with this.

ruthenium n. a rare metallic element (symbol Ru).

ruthless adj. having no pity. □ **ruthlessly** adv., **ruthlessness** n.

rye n. a cereal; whisky made from this.

Ss

S. abbr. **1** south; southern. **2** siemens. ● symb. sulphur.

s. abbr. **1** second(s). **2** (hist.) shilling(s).

SA abbr. **1** South Africa. **2** South Australia. **3** Salvation Army.

sabbath n. a day of worship and rest from work (Saturday for Jews, Sunday for Christians).

sabbatical n. leave granted to a university teacher for study and travel.

sable n. a small Arctic mammal with dark fur; its fur. ● adj. black.

sabotage (sab-ŏ-tahzh) n. wilful damage to machinery or materials, or disruption of work. ● v. commit sabotage on. □ **saboteur** n.

sabre n. (Amer. **saber**) a curved sword.

sac n. a bag-like part in an animal or plant.

saccharin n. a very sweet substance used instead of sugar.

saccharine adj. intensely and unpleasantly sweet.

sachet (sa-shay) n. a small bag or sealed pack.

sack n. **1** a large bag made of a strong coarse fabric. **2 (the sack)** (colloquial) dismissal from one's employment. ● v. **1** (colloquial) dismiss. **2** plunder (a captured town). □ **sackful** n.

sackcloth n. (also **sacking**) coarse fabric for making sacks.

sacral (say-krăl) adj. of the sacrum.

sacrament n. any of the symbolic Christian religious ceremonies. □ **sacramental** adj.

sacred adj. holy; connected with religion; sacrosanct; dedicated (to a person or purpose).

sacred cow n. (colloquial) an idea etc. which its supporters will not allow to be criticized.

sacrifice n. the slaughter of a victim or presenting of a gift to win a god's favour; this victim or gift; the giving up of a valued thing for the sake of something else; the thing given up, a loss entailed. ● v. offer, kill, or give up as a sacrifice. □ **sacrificial** adj.

sacrilege n. disrespect to a sacred thing. □ **sacrilegious** adj.

sacristan n. the person in charge of the sacred vessels etc. in a church.

sacristy n. the place where sacred vessels etc. are kept in a church.

sacrosanct adj. most sacred; respected and not to be harmed.

sacrum (say-krum) n. (pl. **sacrums** or **sacra**) the triangular bone at the base of the spine.

sad adj. (**sadder**) showing or causing sorrow; regrettable. □ **sadly** adv., **sadness** n.

sadden v. make sad.

saddle n. **1** a seat for a rider. **2** a joint of meat consisting of the two loins. ● v. put a saddle on (an animal); burden with a task.

saddler n. a person who makes or deals in saddles and harness.

sadism n. pleasure from inflicting or watching cruelty. □ **sadist** n., **sadistic** adj., **sadistically** adv.

s.a.e. abbr. stamped addressed envelope.

safari n. an expedition to observe or hunt wild animals.

safari park n. a park where exotic wild animals are kept in the open for visitors to see.

safe adj. free from risk or danger; providing security. ● adv. safely. ● n. a strong lockable cupboard for valuables. □ **safely** adv.

safe conduct n. immunity or protection from arrest or harm.

safe deposit n. a building containing safes and strongrooms for hire.

safeguard n. a means of protection. ● v. protect.

safe sex n. the use of condoms during sexual activity as a precaution against Aids etc.

safety n. being safe, freedom from risk or danger.

safety belt n. a seat belt.

safety pin n. a brooch-like pin with a guard enclosing the point.

safety valve n. **1** a valve that opens automatically to relieve excessive pressure in a steam boiler. **2** a harmless outlet for emotion.

safflower n. a thistle-like plant yielding red dye and edible oil.

saffron n. orange-coloured stigmas of a crocus, used to colour and flavour food; the colour of these.

sag v. (**sagged**) droop or curve down in the middle. ● n. sagging.

saga n. a long story.

sagacious adj. wise. □ **sagaciously** adv., **sagacity** n.

sage n. **1** a herb with fragrant grey-green leaves used to flavour food. **2** an old and wise man. ● adj. wise. □ **sagely** adv.

sago n. the starchy pith of the sago palm, used in puddings.

said see **say**.

sail n. **1** a piece of fabric spread to catch the wind and drive a boat along; a journey by boat. **2** the arm of a windmill. ● v. **1** travel by water; start on a sea voyage; control (a boat). **2** move smoothly; do something easily. □ **sailboat**, **sailing ship** n.

sailboard n. a board with a mast and sail, used in windsurfing. □ **sailboarder** n., **sailboarding** n.

sailcloth n. canvas for sails; a canvas-like dress material.

sailor n. a member of a ship's crew.

saint n. a holy person, esp. one venerated by the RC or Orthodox Church; a very good, patient, or unselfish person. □ **sainthood** n., **saintly** adj., **saintliness** n.

sake¹ n. □ **for the sake of** in order to please or honour (a person) or to get or keep (a thing).

sake² (sah-ki) n. a Japanese fermented liquor made from rice.

salaam n. an Oriental greeting meaning 'Peace'; a Muslim greeting consisting of a low bow.

salacious adj. lewd, erotic. □ **salaciously** adv., **salaciousness** n., **salacity** n.

salad n. a cold dish of (usu. raw) vegetables etc.

salamander n. a lizard-like animal.

salami n. a strongly flavoured sausage, eaten cold.

salaried adj. receiving a salary.

salary n. a fixed regular (usu. monthly) payment to an employee.

sale n. selling, the exchange of a commodity for money; an event at which goods are sold; the disposal of stock at reduced prices. □ **for** or **on sale** offered for purchase.

saleable adj. fit to be sold, likely to find a purchaser.

salesman, **salesperson**, **saleswoman** n. (pl. **-men**, **-persons** or **-people**, **-women**) a person employed to sell goods.

salesmanship n. skill at selling.

salient adj. prominent; most noticeable. ● n. a projecting part.

saline adj. salty, containing salt(s). □ **salinity** n.

saliva n. the colourless liquid that forms in the mouth. □ **salivary** adj.

salivate v. produce saliva. □ **salivation** n.

sallow adj. (of the complexion) yellowish. ● n. a low-growing willow tree.

sally n. **1** a sudden swift attack. **2** a lively or witty remark. □ **sally forth** rush out in attack; set out on a journey.

salmon (sa-mŏn) n. (pl. **salmon**) **1** a large fish with pinkish flesh. **2** salmon pink.

salmon pink adj. & n. yellowish pink.

salmonella n. a bacterium causing food poisoning.

salon n. a place where a hairdresser, couturier, etc. receives clients; an elegant room for receiving guests.

saloon n. **1** a public room, esp. on board ship. **2** a saloon car.

saloon car n. a car with a closed body and no partition behind the driver; a car with a separate boot, not a hatchback.

salsa n. **1** dance music of Cuban origin with jazz and rock elements. **2** a spicy sauce.

salsify n. a plant with a long fleshy root used as a vegetable.

SALT *abbr.* Strategic Arms Limitation Treaty.

salt *n.* **1** sodium chloride used to season and preserve food. **2** (**salts**) a substance resembling salt in form, esp. a laxative. **3** a chemical compound of a metal and an acid. ● *adj.* tasting of salt; impregnated with salt. ● *v.* season with salt; preserve in salt. □ **take with a grain (or pinch) of salt** regard sceptically. **worth one's salt** competent. □ **salty** *adj.*, **saltiness** *n.*

salt cellar *n.* a small container for salt used at meals.

saltpetre *n.* (*Amer.* **saltpeter**) a salty white powder used in gunpowder, in medicine, and in preserving meat.

salubrious *adj.* health-giving.

saluki *n.* a dog of a tall silky-coated breed.

salutary *adj.* producing a beneficial or wholesome effect.

salutation *n.* a greeting; an expression of respect.

salute *n.* a gesture of respect or greeting. ● *v.* make a salute to.

salvage *n.* the recovery of a ship or its cargo from loss at sea, or of property from fire etc.; the saving and use of waste material; items saved in this way. ● *v.* save from loss or for use as salvage. □ **salvageable** *adj.*

salvation *n.* saving from disaster, esp. from the consequences of sin.

salve *n.* a soothing ointment; something that soothes. ● *v.* soothe (conscience etc.).

salver *n.* a small tray.

salvo *n.* (*pl.* **salvoes** or **salvos**) **1** the firing of guns simultaneously. **2** a round of applause.

sal volatile *n.* a solution of ammonium carbonate used as a remedy for faintness.

Samaritan *n.* a charitable or helpful person.

samarium *n.* a metallic element (symbol Sm).

samba *n.* a ballroom dance of Brazilian origin.

same *adj.* **1** being of one kind, not changed or different. **2** previously mentioned. □ **sameness** *n.*

samosa *n.* a fried triangular pastry containing spiced vegetables or meat.

sampan *n.* a small flat-bottomed Chinese boat.

samphire *n.* a plant with edible fleshy leaves, growing by the sea.

sample *n.* a small part showing the quality of the whole; a specimen. ● *v.* test by taking a sample of.

sampler *n.* a piece of embroidery worked in various stitches to show one's skill.

samurai *n.* a Japanese army officer.

sanatorium *n.* (*pl.* **sanatoriums** or **sanatoria**) an establishment for treating chronic diseases or convalescents; a room for sick people in a school.

sanctify *v.* make holy or sacred. □ **sanctification** *n.*

sanctimonious *adj.* too righteous or pious. □ **sanctimoniously** *adv.*, **sanctimoniousness** *n.*

sanction *n.* **1** permission, approval. **2** a penalty imposed on a country or organization. ● *v.* give sanction to, authorize.

sanctity *n.* sacredness, holiness.

sanctuary *n.* **1** a sacred place. **2** a place where birds or wild animals are protected. **3** refuge.

sanctum *n.* **1** a holy place. **2** (*colloquial*) a person's private room.

sand *n.* very fine loose fragments of crushed rock; (**sands**) an expanse of sand, a sandbank. ● *v.* **1** sprinkle with sand. **2**

smooth with sandpaper or a sander.

sandal n. a light shoe with straps.

sandalwood n. a scented wood.

sandbag n. a bag filled with sand, used to protect a wall or building. ● v. (**sandbagged**) protect with sandbags.

sandbank n. an underwater deposit of sand.

sandblast v. treat or clean with a jet of sand driven by compressed air or steam.

sandcastle n. a structure of sand, made for fun.

sander n. a mechanical tool for smoothing surfaces.

sandpaper n. paper with a coating of sand or other abrasive substance, used for smoothing surfaces. ● v. smooth with this.

sandstone n. rock formed of compressed sand.

sandstorm n. a desert storm of wind with blown sand.

sandwich n. two or more slices of bread with a layer of filling between; something arranged like this. ● v. put between two others.

sandy adj. (**sandier**) 1 like sand; covered with sand. 2 (of hair) yellowish red.

sane adj. having a sound mind; rational. □ **sanely** adv.

sang see **sing**.

sangria n. a Spanish drink of red wine, lemonade, and fruit.

sanguinary adj. full of bloodshed; bloodthirsty.

sanguine adj. optimistic.

sanitarium n. (pl. **sanitariums** or **sanitaria**) (Amer.) a sanatorium.

sanitary adj. of hygiene; hygienic; of sanitation.

sanitary towel n. (Amer. **sanitary napkin**) an absorbent pad worn during menstruation.

sanitation n. arrangements to protect public health, esp. drainage and disposal of sewage.

sanitize v. (also **-ise**) make sanitary.

sanity n. the condition of being sane.

sank see **sink**.

sap n. 1 the food-carrying liquid in plants. 2 (slang) a foolish person. ● v. (**sapped**) exhaust gradually. □ **sappy** adj.

sapling n. a young tree.

sapphire n. a blue precious stone; its colour. ● adj. bright blue.

saprophyte n. a fungus or related plant living on decayed matter.

sarcasm n. ironically scornful language. □ **sarcastic** adj., **sarcastically** adv.

sarcophagus n. (pl. **sarcophagi**) a stone coffin.

sardine n. a young pilchard or similar small fish.

sardonic adj. humorous in a grim or sarcastic way. □ **sardonically** adv.

sargasso n. (also **sargassum**) a seaweed with berry-like air-vessels.

sari n. a length of cloth draped round the body, worn by Indian women.

sarong n. a strip of cloth worn round the body, esp. in Malaya.

sarsen n. a sandstone boulder.

sartorial adj. of tailoring; of men's clothing.

SAS abbr. Special Air Service; an armed regiment trained in commando techniques.

sash n. 1 a strip of cloth worn round the waist or over one shoulder. 2 a frame holding a pane of a window and sliding up and down in grooves.

sash window n. a window sliding up and down in grooves.

Sat. abbr. Saturday.

sat see **sit**.

satanic adj. **1** of Satan. **2** devilish, hellish.

Satanism n. the worship of Satan.

satchel n. a bag for school books, hung over the shoulder.

sateen n. a closely woven cotton fabric resembling satin.

satellite n. **1** a heavenly or artificial body revolving round a planet. **2** a country that is subservient to another.

satellite dish n. a dish-shaped aerial for receiving broadcasts transmitted by satellite.

satiate (say-shi-ayt) v. satisfy fully, glut. □ **satiation** n. **satiety** n.

satin n. a silky material that is glossy on one side. ● adj. smooth as satin.

satinwood n. a yellow glossy wood used for furniture-making.

satire n. the use of ridicule, irony, or sarcasm; a novel or play etc. that ridicules something. □ **satirical** adj., **satirically** adv.

satirize v. (also **-ise**) attack with satire; describe satirically. □ **satirist** n.

satisfactory adj. satisfying; adequate. □ **satisfactorily** adv.

satisfy v. give (a person) what he or she wants or needs; make pleased or contented; end (a demand etc.) by giving what is required; convince. □ **satisfaction** n.

satsuma n. a small variety of orange.

saturate v. make thoroughly wet; cause to absorb or accept as much as possible. □ **saturation** n.

Saturday n. the day following Friday.

saturnine adj. having a gloomy temperament or appearance.

satyr (sat-er) n. a woodland god in classical mythology, with a goat's ears, tail, and legs.

sauce n. **1** a liquid food added for flavour. **2** (colloquial) impudence.

saucepan n. a metal cooking pot with a long handle.

saucer n. a curved dish on which a cup stands; something shaped like this.

saucy adj. (**saucier**) impudent; jaunty. □ **saucily** adv.

sauerkraut (sowr-krowt) n. chopped pickled cabbage.

sauna n. a specially designed hot room for cleaning and refreshing the body.

saunter v. stroll. ● n. a stroll.

sausage n. minced seasoned meat in a tubular case of thin skin.

sauté v. fry quickly in shallow oil.

savage adj. uncivilized; wild and fierce; cruel and hostile. ● n. an uncivilized person. ● v. maul savagely. □ **savagely** adv., **savagery** n.

savannah n. a grassy plain in hot regions.

save v. **1** rescue; protect; prevent the scoring of (a goal etc.). **2** avoid wasting; keep for future use, put aside money in this way. ● n. an act of saving in football etc. □ **saver** n.

saveloy n. a highly seasoned sausage.

savings n.pl. money put aside for future use.

saviour n. (Amer. **savior**) a person who rescues people from harm.

savoir faire (sav-wahr fair) n. knowledge of how to behave; social tact.

savory n. a herb.

savour (Amer. **savor**) n. flavour; smell. ● v. have a certain savour; taste or smell with enjoyment.

savoury (Amer. **savory**) adj. having an appetizing taste or smell; salty or piquant, not sweet. ● n. a savoury dish, esp. at the end of a meal.

savoy n. a variety of cabbage.

saw¹ see **see**.

saw² n. a cutting tool with a zigzag edge. ● v. (**sawed, sawn, sawing**) cut with a saw; make a to-and-fro movement.

sawdust n. powdery fragments of wood, made in sawing timber.

sawfish n. a large sea fish with a jagged blade-like snout.

sawmill n. a mill where timber is cut.

sawn see **saw²**.

sax n. (colloquial) a saxophone.

saxe blue adj. & n. greyish blue.

saxifrage n. a rock plant.

saxophone n. a brass wind instrument with finger-operated keys. □ **saxophonist** n.

say v. (**said, saying**) utter, recite; express in words, state; give as an opinion; suppose as a possibility etc. ● n. the right to have one's opinion considered.

SAYE abbr. save-as-you-earn.

saying n. a well-known phrase or proverb.

Sb symb. antimony.

Sc symb. scandium.

scab n. 1 a crust forming over a sore; a skin disease or plant disease causing similar roughness. 2 (colloquial, derog.) a blackleg. □ **scabby** adj.

scabbard n. the sheath of a sword etc.

scabies n. a contagious skin disease causing itching.

scabious n. a herbaceous plant with clustered flowers.

scabrous adj. 1 rough-surfaced. 2 indecent.

scaffold n. 1 a platform for the execution of criminals. 2 scaffolding.

scaffolding n. poles and planks providing platforms for people working on buildings etc.

scald v. 1 injure with hot liquid or steam. 2 cleanse with boiling water. ● n. an injury by scalding.

scale n. 1 an ordered series of units or qualities etc. for measuring or classifying things; a fixed series of notes in a system of music. 2 relative size or extent. 3 (**scales**) an instrument for weighing things. 4 each of the overlapping plates of horny membrane protecting the skin of many fishes and reptiles; something resembling this. 5 an incrustation caused by hard water or forming on teeth. ● v. 1 climb. 2 represent in proportion to the size of the original. 3 remove scale from. □ **scaly** adj.

scalene adj. (of a triangle) having unequal sides.

scallion n. a shallot or spring onion.

scallop n. (also **scollop**) 1 a shellfish with a hinged fan-shaped shell. 2 (**scallops**) semicircular curves as an ornamental edging. □ **scalloped** adj.

scallywag n. a rascal.

scalp n. the skin of the head excluding the face. ● v. cut the scalp from.

scalpel n. a surgeon's or painter's small straight knife.

scam n. (slang) 1 a trick; a swindle. 2 a story or rumour.

scamp n. a rascal.

scamper v. run hastily or in play. ● n. a scampering run.

scampi n.pl. large prawns.

scan v. (**scanned**) 1 look at all parts of, esp. quickly. 2 pass a radar or electronic beam over; resolve a picture into elements of light and shade for transmission or reproduction. 3 (of verse) have a regular rhythm. ● n. scanning. □ **scanner** n.

scandal n. something disgraceful; gossip about a wrongdoing. □ **scandalous** adj., **scandalously** adv.

scandalize v. (also **-ise**) shock; outrage.

scandalmonger n. a person who spreads scandal.

Scandinavian adj. & n. (a native) of Scandinavia.

scandium n. a metallic element (symbol Sc).

scansion n. scanning of verse.

scant adj. barely enough.

scanty adj. (**scantier**) small in amount or extent; barely enough. □ **scantily** adv., **scantiness** n.

scapegoat n. a person made to bear blame that should fall on others.

scapula n. (pl. **scapulae** or **scapulas**) the shoulder blade. □ **scapular** adj.

scar n. the mark where a wound has healed. ● v. (**scarred**) mark with a scar; form scar(s).

scarab n. a sacred beetle of ancient Egypt.

scarce adj. not enough to supply a demand, rare.

scarcely adv. only just, almost not; surely not.

scarcity n. a shortage.

scare v. frighten; be frightened. ● n. fright, alarm.

scarecrow n. a figure dressed in old clothes and set up to scare birds away from crops.

scaremonger n. a person who spreads alarming rumours. □ **scaremongering** n.

scarf n. (pl. **scarves** or **scarfs**) a piece or strip of material worn round the neck or tied over a woman's head.

scarify v. **1** drag a rake through the surface of (grass or soil); make slight cuts in. **2** criticize harshly.

scarlet adj. & n. brilliant red.

scarlet fever n. an infectious fever producing a scarlet rash.

scarp n. a steep slope on a hillside.

scary adj. (**scarier**) (colloquial) frightening.

scathing adj. (of criticism) very severe.

scatology n. an excessive interest in excrement or obscenity. □ **scatological** adj.

scatter v. throw or put here and there; go or send in different directions.

scatterbrain n. a careless or forgetful person. □ **scatter-brained** adj.

scatty adj. (**scattier**) (colloquial) scatterbrained, disorganized.

scavenge v. search for (usable objects) among rubbish etc.; (of animals) search for decaying flesh as food. □ **scavenger** n.

scenario n. (pl. **scenarios**) **1** the script or summary of a film or play. **2** a suggested sequence of events.

■ **Usage** Scenario should not be used in standard English to mean 'situation', as in It was an unpleasant scenario.

scene n. **1** the place of an event; a piece of continuous action in a play or film; stage scenery; a view of a place or incident. **2** a display of temper or emotion. **3** (colloquial) an area of activity or interest. □ **behind the scenes** hidden from public view.

scenery n. the general (esp. picturesque) appearance of a landscape; structures used on a theatre stage to represent the scene of action.

scenic adj. picturesque.

scent n. a pleasant smell; liquid perfume; an animal's trail perceptible to a hound's sense of smell. ● v. discover by smell; suspect the presence or existence of; apply scent to, make fragrant.

sceptic (**skep**-tik) n. (Amer. **skeptic**) a sceptical person.

sceptical (skep-tik-ǎl) *adj.* (*Amer.* **skeptical**) unwilling to believe things. □ **sceptically** *adv.*, **scepticism** *n.*

sceptre *n.* (*Amer.* **scepter**) an ornamental rod carried as a symbol of sovereignty.

schadenfreude (shah-děn-froi-dě) *n.* enjoyment of others' misfortunes.

schedule *n.* a programme or timetable of events. ● *v.* **1** include in a schedule. **2** list a building) for preservation.

scheduled flight *n.* a regular public flight, not a chartered flight.

schema *n.* (*pl.* **schemata** or **schemas**) a summary, outline, or diagram.

schematic *adj.* in the form of a diagram. □ **schematically** *adv.*

schematize *v.* (also **-ise**) put into schematic form. □ **schematization** *n.*

scheme *n.* a plan of work or action. ● *v.* make plans, plot. □ **schemer** *n.*

scherzo (skair-tsoh) *n.* (*pl.* **scherzos**) a lively piece of music.

schism (skizm) *n.* division into opposing groups through a difference in belief or opinion. □ **schismatic** *adj.* & *n.*

schist (shist) *n.* a rock formed in layers.

schizoid *adj.* like or suffering from schizophrenia. ● *n.* a schizoid person.

schizophrenia *n.* a mental disorder in which a person is unable to act or reason rationally. □ **schizophrenic** *adj.* & *n.*

schmaltz (shmawlts) *n.* sugary sentimentality.

schmuck *n.* (*Amer. slang*) a foolish person.

schnapps *n.* a strong alcoholic spirit.

schnitzel (shnit-sěl) *n.* a fried veal cutlet.

scholar *n.* a learned person; a pupil; the holder of a scholarship. □ **scholarly** *adj.*, **scholarliness** *n.*

scholarship *n.* a grant of money towards education; scholars' methods and achievements.

scholastic *adj.* of schools or education; academic.

school *n.* **1** an educational institution. **2** a group of artists etc. following the same principles. **3** a shoal of whales. ● *v.* train, discipline. □ **schoolboy**, **schoolchild**, **schoolgirl** *n.*

schoolteacher *n.* a teacher in a school.

schooner *n.* **1** a sailing ship. **2** a measure for sherry.

sciatica *n.* a condition causing pain in the hip and thigh.

science *n.* a branch of knowledge requiring systematic study and method, esp. dealing with substances, life, and natural laws. □ **scientific** *adj.*, **scientifically** *adv.*

scientist *n.* an expert in a science.

scimitar *n.* a short curved oriental sword.

scintilla (sin-til-ǎ) *n.* a trace.

scintillating *adj.* lively; witty.

scion (sy-ǒn) *n.* **1** a plant shoot cut for grafting. **2** a descendant.

scissors *n.pl.* a cutting instrument with two pivoted blades.

sclerosis *n.* abnormal hardening of body tissue.

scoff *v.* **1** speak contemptuously, jeer. **2** (*colloquial*) eat quickly.

scold *v.* rebuke (esp. a child). □ **scolding** *n.*

scollop var. of **scallop**.

sconce *n.* an ornamental bracket on a wall, holding a light.

scone (skon, skohn) *n.* a soft flat cake eaten buttered.

scoop *n.* **1** a deep shovel-like tool; a ladle. **2** a piece of news published by one newspaper before its rivals. ● *v.* **1** lift or hollow with (or as if with) a scoop. **2** forestall with a news scoop.

scoot *v.* run, dart.

scooter *n.* **1** a child's toy vehicle with a footboard and long steering-handle. **2** a lightweight motorcycle.

scope *n.* the range of a subject, activity, etc.; opportunity.

scorch *v.* burn or become burnt on the surface.

scorching *adj.* very hot.

score *n.* **1** the number of points gained in a contest. **2** a set of twenty. **3** a line or mark cut into something. **4** written or printed music; music for a film or play. ● *v.* **1** gain (points etc.) in a contest; keep a record of the score; achieve a success. **2** cut a line or mark into. **3** write or compose as a musical score. □ **score out** cross out. □ **scorer** *n.*

scorn *n.* strong contempt. ● *v.* feel or show scorn for; reject with scorn. □ **scornful** *adj.*, **scornfully** *adv.*

scorpion *n.* a small animal of the spider group with lobster-like claws and a sting in its long tail.

Scot *n.* a native of Scotland.

Scotch *adj.* Scottish. ● *n.* Scotch whisky.

■ *Usage* Scots or Scottish is preferred to the adjective Scotch in Scotland.

scotch *v.* put an end to (a rumour).

scot-free *adv.* unharmed, not punished.

Scots *adj.* Scottish. ● *n.* the Scottish form of the English language. □ **Scotsman**, **Scotswoman** *n.*

Scottish *adj.* of Scotland or its people.

scoundrel *n.* a dishonest person.

scour *v.* **1** cleanse by rubbing; clear out (a channel etc.) by flowing water; purge drastically. **2** search thoroughly. ● *n.* scouring; action of water on a channel etc. □ **scourer** *n.*

scourge (skurj) *n.* **1** a whip. **2** a great affliction. ● *v.* flog; afflict greatly.

scout *n.* a person sent to gather information, esp. about enemy movements. ● *v.* act as a scout, search.

scowl *n.* a sullen or angry frown. ● *v.* make a scowl.

scrabble *v.* scratch or search busily with the hands, paws, etc.

scraggy *adj.* (**scraggier**) thin and bony. □ **scragginess** *n.*

scram *v.* (**scrammed**) (*colloquial*) go away.

scramble *v.* **1** clamber; move hastily or awkwardly; (of aircraft or crew) hurry to take off quickly. **2** mix indiscriminately; make a (transmission) unintelligible except by means of a special receiver. **3** cook (eggs) by heating and stirring. ● *n.* **1** a scrambling walk or movement; an eager struggle. **2** a motorcycle race over rough ground. □ **scrambler** *n.*

scrap *n.* **1** a fragment. **2** waste material; discarded metal suitable for reprocessing. **3** (*colloquial*) a fight, a quarrel. ● *v.* (**scrapped**) **1** discard as useless. **2** fight, quarrel.

scrapbook *n.* a book for newspaper cuttings or similar souvenirs.

scrape *v.* **1** clean, smooth, or damage by passing a hard edge across a surface; make the sound of scraping. **2** get along or through etc. with difficulty; be very economical. ● *n.* **1** a scraping movement or sound; a scraped place. **2** an awkward situation. □ **scraper** *n.*

scrapie n. a disease of sheep causing loss of coordination.

scraping n. a fragment produced by scraping.

scrappy adj. (**scrappier**) made up of scraps or disconnected elements. □ **scrappily** adv., **scrappiness** n.

scratch v. 1 scrape with the fingernails; make a thin scraping sound. 2 obtain with difficulty. 3 withdraw from a race or competition. ● n. a mark, wound, or sound made by scratching; a spell of scratching. ● adj. collected from what is available. □ **from scratch** from the very beginning or with no preparation. **up to scratch** up to the required standard. □ **scratchy** adj.

scratchings n.pl. crisp residue of pork fat left after rendering lard.

scrawl v. write in a hurried untidy way. ● n. bad handwriting; a scrawled note.

scrawny adj. (**scrawnier**) scraggy.

scream v. make a long piercing cry or sound. ● n. 1 a screaming cry or sound. 2 (slang) a very amusing person or thing.

scree n. a mass of loose stones on a mountain side.

screech n. a harsh high-pitched scream or sound. ● v. make a screech.

screed n. 1 a tiresomely long piece of writing or speech. 2 a thin layer of cement.

screen n. 1 a structure used to conceal, protect, or divide something. 2 a windscreen. 3 a blank surface on which pictures, cinema films, or television transmissions are projected. ● v. 1 shelter, conceal. 2 show (images etc.) on a screen. 3 examine for the presence or absence of a disease, quality, etc.

screenplay n. the script of a film.

screw n. 1 a metal pin with a spiral ridge round its length, fastened by turning; a thing twisted to tighten or press something; an act of twisting or tightening. 2 a propeller. 3 (slang) a prison officer. 4 (slang) sexual intercourse. ● v. 1 fasten or tighten with screw(s); turn (a screw); twist, become twisted. 2 oppress, extort. 3 (slang) have sexual intercourse with (a person). □ **screw up** summon up (one's courage); (slang) bungle.

screwdriver n. a tool for turning screws.

scribble v. write hurriedly or carelessly; make meaningless marks. ● n. something scribbled.

scribe n. a person who (before the invention of printing) made copies of writings; (in New Testament times) a professional religious scholar.

scrimmage n. a confused struggle.

scrimp v. skimp.

script n. 1 handwriting; a style of printed characters resembling this. 2 the text of a play, broadcast talk, etc.

scripture n. sacred writings; (**the Scriptures**) those of the Christians or the Jews.

scroll n. a roll of paper or parchment; an ornamental design in this shape. ● v. move (a display on a VDU screen) up or down as the screen is filled.

scrotum n. (pl. **scrota** or **scrotums**) the pouch of skin enclosing the testicles.

scrounge v. cadge; borrow; collect by foraging. □ **scrounger** n.

scrub v. (**scrubbed**) rub hard, esp. with something coarse or bristly. ● n. 1 the process of scrubbing. 2 vegetation consisting of stunted trees and shrubs; land covered with this.

scruff *n.* the back of the neck.

scruffy *adj.* (**scruffier**) (*colloquial*) shabby and untidy. □ **scruffily** *adv.*, **scruffiness** *n.*

scrum *n.* a scrummage; a confused struggle.

scrummage *n.* a grouping of forwards in rugby football to struggle for possession of the ball by pushing.

scrumptious *adj.* (*colloquial*) delicious.

scrunch *v.* crunch, crush. ● *n.* an act or sound of scrunching.

scruple *n.* a doubt about doing something, produced by one's conscience. ● *v.* hesitate because of scruples.

scrupulous *adj.* very conscientious or careful. □ **scrupulously** *adv.*, **scrupulousness** *n.*

scrutinize *v.* (also **-ise**) examine carefully.

scrutiny *n.* a careful look or examination.

scuba *n.* an aqualung (acronym from *self*-contained *u*nderwater *b*reathing *a*pparatus). ● **scuba-diving** *n.*

scud *v.* (**scudded**) move along fast and smoothly.

scuff *v.* scrape or drag (one's feet) in walking; mark or scrape by doing this.

scuffle *n.* a confused struggle or fight. ● *v.* take part in a scuffle.

scull *n.* one of a pair of small oars; an oar that rests on a boat's stern, worked with a screw-like movement. ● *v.* row with sculls.

scullery *n.* a room where dishes are washed.

sculpt *v.* sculpture.

sculptor *n.* a maker of sculptures.

sculpture *n.* the art of carving or modelling; work made in this way. ● *v.* represent in or decorate with sculpture; be a sculptor. □ **sculptural** *adj.*

scum *n.* **1** a layer of impurities or froth etc. on the surface of a liquid. **2** a worthless person.

scupper *n.* an opening in a ship's side to drain water from the deck. ● *v.* **1** sink (a ship) deliberately. **2** (*slang*) ruin.

scurf *n.* flakes of dry skin, esp. from the scalp; similar scaly matter.

scurrilous *adj.* abusive and insulting; coarsely humorous. □ **scurrilously** *adv.*, **scurrility** *n.*

scurry *v.* run hurriedly, scamper. ● *n.* scurrying, a rush.

scurvy *n.* a disease caused by lack of vitamin C.

scuttle *n.* a box or bucket for holding coal in a room. ● *v.* **1** scurry. **2** sink (a ship) by letting in water.

scythe (*syth*) *n.* an implement with a curved blade on a long handle, for cutting long grass.

SE *abbr.* south-east; southeastern.

Se *symb.* selenium.

sea *n.* the expanse of salt water surrounding the continents; a section of this; a large inland lake; the waves of the sea; a vast expanse. □ **at sea** in a ship on the sea; perplexed.

seaboard *n.* the coast.

seafaring *adj.* & *n.* working or travelling on the sea. □ **seafarer** *n.*

seafood *n.* fish or shellfish from the sea eaten as food.

seagoing *adj.* for sea voyages.

seagull *n.* a gull.

sea horse *n.* a small fish with a horse-like head.

seal *n.* **1** an amphibious sea animal with thick fur or bristles. **2** an engraved piece of metal used to stamp a design; its impression. **3** an action etc. serving to confirm or guarantee something. **4** a paper sticker. **5** something used to close an opening very tightly. ● v. 1 close or coat so as to prevent penetration; stick down; affix a seal to. 2 settle (e.g. a bargain).

sealant n. a substance for coating a surface to seal it.

sea lion n. a large seal.

sealskin n. a seal's skin or fur used as a clothing material.

seam n. 1 a line where two edges join. 2 a layer of coal etc. in the ground. ● v. join by a seam. □ **seamless** adj.

seaman n. (pl. **seamen**) a sailor; a person skilled in seafaring. □ **seamanship** n.

seamstress n. a woman who sews, esp. for a living.

seance (say-ahns) n. a spiritualist meeting.

seaplane n. an aeroplane designed to take off from and land on water.

sear v. scorch, burn.

search v. look, feel, or go over (a person or place) in order to find something. ● n. the process of searching.

searching adj. thorough.

searchlight n. an outdoor lamp with a powerful beam; its beam.

seascape n. a picture or view of the sea.

seasick adj. made sick by the motion of a ship. □ **seasickness** n.

seaside n. the coast as a place for holidays.

season n. a section of the year associated with a type of weather; the time when something takes place or is plentiful. ● v. 1 give extra flavour to (food). 2 dry or treat until ready for use.

seasonable adj. suitable for the season; timely.

■ **Usage** Seasonable is sometimes confused with seasonal.

seasonal adj. of a season or seasons; varying with the seasons. □ **seasonally** adv.

■ **Usage** Seasonal is sometimes confused with seasonable.

seasoned adj. experienced.

seasoning n. a substance used to season food.

season ticket n. a ticket valid for any number of journeys or performances in a specified period.

seat n. 1 a thing made or used for sitting on; the buttocks, the part of a garment covering these. 2 a place as member of a committee or parliament etc. 3 the place where something is based; a country mansion. ● v. cause to sit; have seats for.

seat belt n. a strap securing a person to a seat in a vehicle or aircraft.

sea urchin n. a sea animal with a round spiky shell.

seaward adj. & adv. towards the sea. □ **seawards** adv.

seaweed n. any plant that grows in the sea.

seaworthy adj. (of ships) fit for a sea voyage. □ **seaworthiness** n.

sebaceous adj. secreting an oily or greasy substance.

secateurs n.pl. clippers for pruning plants.

secede v. withdraw from membership. □ **secession** n.

seclude v. keep (a person) apart from others.

secluded adj. (of a place) screened from view.

seclusion n. secluding, being secluded; privacy.

second¹ (sek-ŏnd) adj. next after the first; secondary; inferior. ● n. 1 a second thing, class, etc. 2 an attendant in a duel or boxing match. 3 a sixtieth part of a minute of time or (in measuring angles) degree. ● v.

state one's support of (a proposal) formally. □ **secondly** adv.

second² (si-kond) v. transfer temporarily to another job or department. □ **secondment** n.

secondary adj. 1 coming after or derived from what is primary. 2 (of education) for children who have had primary education. ● n. a secondary thing. □ **secondarily** adv.

secondary colours n.pl. those obtained by mixing two primary colours.

secondary school n. a school for children who have received primary education, usu. between the ages of 11 and 18.

second-best adj. next to the best in quality; inferior.

second-class adj. & adv. next or inferior to first-class in quality etc.

second cousin see **cousin**.

second-hand adj. bought after use by a previous owner.

second nature n. a habit or characteristic that has become automatic.

second-rate adj. inferior in quality.

second sight n. the supposed power to foresee future events.

second thoughts n.pl. a change of mind after reconsideration.

second wind n. renewed capacity for effort.

secret adj. kept from the knowledge of most people. ● n. something secret; a mystery. □ **in secret** secretly. □ **secretly** adv., **secrecy** n.

secretariat n. an administrative office or department.

secretary n. a person employed to deal with correspondence and routine office work; an official in charge of an organization's correspondence; an ambassador's or government minister's chief assistant. □ **secretarial** adj.

Secretary-General n. a principal administrative officer.

secrete v. 1 put into a hiding place. 2 produce by secretion. □ **secretor** n.

secretion adj. the process of secreting; the production of a substance within the body; this substance.

secretive adj. making a secret of things. □ **secretively** adv., **secretiveness** n.

secretory adj. of physiological secretion.

sect n. a group with beliefs that differ from those generally accepted.

sectarian adj. 1 of a sect or sects. 2 narrow-mindedly promoting the interests of one's sect.

section n. a distinct part; a cross-section; a subdivision. ● v. 1 divide into sections. 2 commit (a person) to a psychiatric hospital.

sectional adj. of a section or sections; made in sections.

sector n. a part of an area; a branch of an activity; a section of a circular area between two lines drawn from centre to circumference.

secular adj. of worldly (not religious or spiritual) matters.

secure adj. safe; certain not to slip or fail. ● v. 1 make secure; fasten securely. 2 obtain; guarantee by pledging something as security. □ **securely** adv.

security n. 1 safety. 2 precaution or protection against espionage, theft, or other danger. 3 something serving as a pledge. 4 a certificate showing ownership of financial stocks etc.

sedate¹ adj. calm and dignified. □ **sedately** adv., **sedateness** n.

sedate² v. treat with sedatives. □ **sedation** n.

sedative adj. having a calming effect. ● n. a sedative drug.

sedentary adj. seated; (of work) done while sitting.

sedge n. a grasslike plant growing by water.

sediment n. particles of solid matter in a liquid or deposited by water or wind. □ **sedimentary** adj., **sedimentation** n.

sedition n. words or actions inciting rebellion. □ **seditious** adj., **seditiously** adv.

seduce v. tempt (esp. into wrongdoing); persuade into sexual intercourse. □ **seducer** n., **seductress** n., **seduction** n., **seductive** adj.

sedulous adj. diligent and persevering. □ **sedulously** adv.

see v. (**saw, seen, seeing**) **1** perceive with the eye(s) or mind; watch; understand. **2** consider. **3** experience. **4** meet; interview. **5** escort. **6** make sure. ● n. the district of a bishop or archbishop. □ **see about** attend to. **seeing that** in view of the fact that. **see to** attend to.

seed n. **1** a plant's fertilized ovule. **2** semen. **3** the origin of something. ● v. produce seed; sprinkle with seeds; remove seeds from. □ **go** or **run to seed** cease flowering as seed develops; become shabby or less efficient.

seedless adj. not containing seeds.

seedling n. a very young plant.

seedy adj. (**seedier**) **1** looking shabby and disreputable. **2** (colloquial) unwell.

seek v. (**sought, seeking**) try to find or obtain; try (to do something).

seem v. appear to be or exist or be true. □ **seemingly** adv.

seen see **see**.

seep v. ooze slowly out or through. □ **seepage** n.

seer n. a prophet.

seersucker n. a fabric woven with a puckered surface.

see-saw n. a long board balanced on a central support so that children sitting on each end can ride up and down; a constantly repeated up-and-down change. ● v. make this movement or change.

seethe v. bubble as if boiling; be very agitated or excited.

segment n. a part cut off, marked off, or separable from others; a part of a circle or sphere cut off by a straight line or plane. □ **segmented** adj.

segregate v. put apart from others. □ **segregation** n.

seine (sayn) n. a fishing net that hangs from floats.

seismic (syz-mik) adj. of earthquakes.

seismograph n. an instrument for recording earthquakes.

seismology n. the study of earthquakes. □ **seismological** adj., **seismologist** n.

seize v. **1** take hold of forcibly or suddenly; take possession of by force or legal right. **2** affect suddenly. □ **seize on** make use of eagerly. **seize up** (of a mechanism) become stuck, esp. through overheating.

seizure n. seizing; a sudden violent attack of an illness.

seldom adv. rarely, not often.

select v. pick out as the best or most suitable. ● adj. carefully chosen; exclusive. □ **selector** n.

selection n. selecting; things selected; things from which to choose.

selective adj. chosen or choosing carefully. □ **selectively** adv., **selectivity** n.

selenium n. a chemical element (symbol Se).

self n. (pl. **selves**) a person as an individual; a person's special nature; a person or thing as the object of reflexive action; one's own advantage or interests.

self- *comb. form* of or to or done by oneself or itself. □ **self-assured** *adj.* confident. **self-centred** *adj.* thinking chiefly of oneself or one's own affairs. **self-confidence** *n.*, **self-confident** *adj.* being confident of one's own abilities. **self-conscious** *adj.* embarrassed from knowing that one is observed. **self-contained** *adj.* complete in itself, having all the necessary facilities. **self-control** *n.* ability to control one's behaviour and not act emotionally. **self-denial** *n.* deliberately going without things one would like to have. **self-determination** *n.* free will; a nation's own choice of its form of government or allegiance. **self-evident** *adj.* clear without proof; obvious. **self-interest** *n.* one's own advantage. **self-made** *adj.* having risen from poverty or obscurity by one's own efforts. **self-possessed** *adj.* calm and dignified. **self-respect** *n.* proper regard for oneself and one's own dignity and principles etc. **self-righteous** *adj.* smugly sure of one's own righteousness. **self-satisfied** *adj.* smugly pleased with oneself and one's achievements. **self-seeking** *adj.* & *n.* seeking to promote one's own interests rather than those of others. **self-service** *adj.* at which customers help themselves and pay a cashier for goods taken. **self-styled** *adj.* using a name or description one has adopted without right. **self-sufficient** *adj.* able to provide what one needs without outside help. **self-willed** *adj.* obstinately doing what one wishes.

selfish *adj.* acting or done according to one's own interests without regard to those of others; keeping good things for oneself. □ **selfishly** *adv.*, **selfishness** *n.*

selfless *adj.* unselfish. □ **selflessly** *adv.*

selfsame *adj.* the very same.

sell *v.* (**sold, selling**) exchange (goods etc.) for money; keep (goods) for sale; promote sales of; (of goods) be sold; have a specified price; persuade into accepting (an idea etc.). ● *n.* a manner of selling. □ **sell off** dispose of by selling, esp. at a reduced price. **sell out** dispose of all one's stock etc. by selling; betray. **sell up** sell one's house or business. □ **seller** *n.*

Sellotape *n.* (*trade mark*) an adhesive usu. transparent tape.

selvedge *n.* (also **selvage**) an edge of cloth woven so that it does not unravel.

semantic *adj.* of meaning in language. ● *n.* (**semantics**) the study of meaning. □ **semantically** *adv.*

semaphore *n.* a system of signalling with the arms.

semblance *n.* an outward appearance, a show; a resemblance.

semen *n.* the sperm-bearing fluid produced by male animals.

semester *n.* a half-year course or university term.

semi- *pref.* half; partly.

semibreve *n.* a note in music, equal to two minims.

semicircle *n.* half of a circle. □ **semicircular** *adj.*

semicolon *n.* a punctuation mark (;).

semiconductor *n.* a substance that conducts electricity in certain conditions.

semi-detached *adj.* (of a house) joined to another on one side.

semi-final *n.* a match or round preceding the final. □ **semi-finalist** *n.*

seminal adj. **1** of seed or semen. **2** giving rise to new developments. □ **seminally** adv.

seminar n. a small class for discussion and research.

seminary n. a training college for priests or rabbis.

semi-precious adj. (of gems) less valuable than those called precious.

semiquaver n. a note in music, equal to half a quaver.

Semite n. a member of the group of races that includes Jews and Arabs. □ **Semitic** adj.

semitone n. half a tone in music.

semolina n. hard particles left when wheat is ground and sifted, used to make puddings.

SEN abbr. State Enrolled Nurse.

senate n. the upper house of certain parliaments; the governing body of certain universities.

senator n. a member of a senate.

send v. (**sent, sending**) order or cause to go to a certain destination; send a message; cause to move or go or become. □ **send for** order to come or be brought. **send up** make fun of by imitating.

senile adj. weak in body or mind because of old age. □ **senility** n.

senior adj. older; higher in rank or authority; for older children. ● n. a senior person; a member of a senior school. □ **seniority** n.

senior citizen n. an elderly person.

senna n. dried pods or leaves of a tropical tree, used as a laxative.

sensation n. a feeling produced by stimulation of a sense organ or of the mind; excited interest, a person or thing producing this.

sensational adj. causing great excitement or admiration. □ **sensationally** adv.

sensationalism n. the use of or interest in sensational matters. □ **sensationalist** n.

sense n. **1** any of the special powers (sight, hearing, smell, taste, touch) by which a living thing becomes aware of the external world; the ability to perceive or be conscious of a thing. **2** practical wisdom. **3** meaning. **4** (**senses**) consciousness, sanity. ● v. perceive by a sense or by a mental impression. □ **make sense** have a meaning; be a sensible idea. **make sense of** find a meaning in.

senseless adj. **1** foolish. **2** unconscious.

sensibility n. sensitiveness.

sensible adj. **1** having or showing good sense. **2** aware. □ **sensibly** adv.

sensitive adj. **1** receiving impressions or responding to stimuli easily. **2** easily hurt or offended; requiring tact. □ **sensitively** adv., **sensitivity** n.

sensitize v. (also **-ise**) make sensitive. □ **sensitization** n., **sensitizer** n.

sensor n. a device that responds to a certain stimulus.

sensory adj. of the senses; receiving and transmitting sensations.

sensual adj. gratifying to the body; indulging oneself with physical pleasures. □ **sensualism** n., **sensuality** n., **sensually** adv.

■ Usage *Sensual* is sometimes confused with *sensuous*.

sensuous *adj.* affecting the senses pleasantly. □ **sensuously** *adv.*, **sensuousness** *n.*

■ **Usage** *Sensuous* is sometimes confused with *sensual.*

sent *see* send.

sentence *n.* **1** a series of words making a single complete statement. **2** a punishment decided by a law court. ● *v.* pass sentence on (a person).

sententious *adj.* dull and moralizing. □ **sententiously** *adv.*, **sententiousness** *n.*

sentient *adj.* capable of perceiving and feeling things. □ **sentiently** *adv.*, **sentience** *n.*

sentiment *n.* **1** mental feeling; opinion. **2** sentimentality.

sentimental *adj.* full of romantic or nostalgic feeling. □ **sentimentalism** *n.*, **sentimentality** *n.*, **sentimentally** *adv.*

sentinel *n.* a sentry.

sentry *n.* a soldier posted to keep watch and guard something.

sepal *n.* each of the leaf-like parts forming a calyx.

separable *adj.* able to be separated. □ **separability** *n.*

separate *adj.* (sep-er-ăt) not joined or united with others. ● *v.* (sep-er-ayt) divide; set, move, or keep apart; cease to live together as a married couple. □ **separately** *adv.*, **separation** *n.*, **separator** *n.*

■ **Usage** *Separate, separation,* etc. are not spelt with an *e* in the middle.

separatist *n.* a person who favours separation from a larger (esp. political) unit. □ **separatism** *n.*

sepia *n.* a brown colouring matter; a rich reddish-brown colour.

sepsis *n.* septic condition.

Sept. *abbr.* September.

September *n.* the ninth month.

septet *n.* a group of seven instruments or voices; music for these.

septic *adj.* infected with harmful micro-organisms.

septicaemia (sep-ti-see-mia) *n.* (*Amer.* **septicemia**) blood poisoning.

septic tank *n.* a tank in which sewage is liquefied by bacterial activity.

septuagenarian *n.* a person in his or her seventies.

sepulchre (sep-ŭl-ker) *n.* (*Amer.* **sepulcher**) a tomb.

sequel *n.* what follows, esp. as a result; a novel or film etc. continuing the story of an earlier one.

sequence *n.* **1** the following of one thing after another; a set of things belonging next to each other in a particular order. **2** a section dealing with one topic in a film.

sequential *adj.* forming a sequence; occurring as a result; serial. □ **sequentially** *adv.*

sequester *v.* seclude; confiscate.

sequestrate *v.* confiscate; take temporary possession of. □ **sequestration** *n.*

sequin *n.* a circular spangle for decorating clothes. □ **sequinned** *adj.*

sequoia (si-kwoi-ă) *n.* a Californian tree growing to a great height.

seraglio (si-rahl-yoh) *n.* (*pl.* **seraglios**) a harem.

seraph *n.* (*pl.* **seraphim** or **seraphs**) a member of the highest order of angels. □ **seraphic** *adj.*

serenade *n.* music played for a lover, or suitable for this. ● *v.* sing or play a serenade to.

serendipity *n.* the making of pleasant discoveries by accident.

serene *adj.* calm and cheerful. □ **serenely** *adv.*, **serenity** *n.*

serf *n.* a medieval farm labourer forced to work for his landowner; an oppressed labourer. □ **serfdom** n.

serge n. strong twilled fabric.

sergeant (sar-jĕnt) n. a non-commissioned army officer ranking just above corporal; a police officer ranking just below inspector.

serial n. a story presented in a series of instalments. ● adj. of or forming a series. □ **serially** adv.

serialize v. (also **-ise**) produce as a serial. □ **serialization** n.

serial killer n. a person who murders repeatedly.

serial number n. an identification number.

series n. (pl. **series**) a number of similar things occurring, arranged, or produced in order.

serious adj. 1 solemn; sincere. 2 important; not slight. □ **seriously** adv., **seriousness** n.

sermon n. a talk on a religious or moral subject, esp. during a religious service.

sermonize v. (also **-ise**) give a long moralizing talk.

serpent n. a large snake.

serpentine adj. twisting like a snake.

SERPS abbr. state earnings related pension scheme.

serrated adj. having a series of small projections. □ **serration** n.

serried adj. arranged in a close series.

serum n. (pl. **sera** or **serums**) fluid left when blood has clotted; this used for inoculation; a watery fluid from animal tissue. □ **serous** adj.

servant n. a person employed to do domestic work in a household or as an attendant; an employee.

serve v. 1 perform or provide services for; be employed (in the army etc.). 2 be suitable (for). 3 present (food etc.) for others to consume; (of food) be enough for. 4 attend to (customers). 5 set the ball in play at tennis etc. 6 deliver (a legal writ etc.) to (a person). ● n. a service in tennis etc. □ **server** n.

service n. 1 the act of serving; being a servant; working for an employer; a department of people employed by a public organization; a system that performs work for customers or supplies public needs. 2 (the **services**) the armed forces. 3 a religious ceremony or meeting. 4 a set of dishes etc. for serving a meal. 5 a game in which one serves in tennis etc. 6 maintenance and repair of machinery. ● v. 1 maintain and repair (machinery). 2 supply with services. 3 pay the interest on (a loan).

serviceable adj. 1 useful or usable. 2 hard-wearing.

service area n. an area beside a motorway where petrol and refreshments etc. are available.

service flat n. a flat where domestic service is provided.

serviceman n. (pl. **-men**) 1 a member of the armed services. 2 a man providing a maintenance service.

service road n. a road giving access to houses etc. but not for use by through traffic.

service station n. a place beside a road selling petrol etc.

servicewoman n. (pl. **-women**) a woman in the armed services.

serviette n. a table napkin.

servile adj. menial; excessively submissive. □ **servility** n.

servitude n. the condition of being forced to work for others.

sesame (sess-ămi) n. a tropical plant with seeds that yield oil or are used as food.

session n. a meeting or meetings for discussing something; a period spent in an activity; an academic year or term.

set v. (**set**, **setting**) **1** put, place, or fix in position or readiness. **2** make or become hard, firm, or established. **3** fix or appoint (a date etc.); assign as something to be done; put into a specified state. **4** have a certain movement. **5** be brought below the horizon by the earth's movement. ●n. **1** the way a thing sets or is set. **2** people or things grouped as similar or forming a unit. **3** games forming part of a match in tennis etc. **4** a radio or television receiver. **5** the process of setting hair. **6** scenery or stage for a play or film. **7** var. of **sett**. □ **be set on** be determined about. **set about** begin (a task); attack. **set back** halt or slow the progress of; (*slang*) cost (a person) a specified amount. **set eyes on** catch sight of. **set fire to** cause to burn. **set forth** set out. **set in** become established. **set off** begin a journey; cause to explode; improve the appearance of by contrast. **set out** declare, make known; begin a journey. **set sail** begin a sea voyage.

setback n. a delay in progress; a problem.

set piece n. a formal or elaborate construction.

set square n. a right-angled triangular drawing instrument.

set-up n. an organization or arrangement.

sett n. (also **set**) **1** a badger's burrow. **2** a paving block.

settee n. a sofa.

setter n. **1** a dog of a long-haired breed. **2** a person or thing that sets.

setting n. **1** the way or place in which a thing is set. **2** a set of cutlery or crockery for one person.

settle v. **1** place so as to stay in position; establish, become established; make one's home; occupy (a previously unoccupied area); sink, come to rest. **2** arrange as desired or conclusively; deal with. **3** make or become calm or orderly. **4** pay (a bill etc.). **5** bestow legally. ●n. a wooden seat with a high back and arms. □ **settle up** pay what is owing. □ **settler** n.

settlement n. **1** settling. **2** a business or financial arrangement. **3** an amount or property settled legally on a person. **4** a place occupied by settlers.

seven adj. & n. one more than six (7, VII). □ **seventh** adj. & n.

seventeen adj. & n. one more than sixteen (17, XVII). □ **seventeenth** adj. & n.

seventy adj. & n. seven times ten (70, LXX). □ **seventieth** adj. & n.

sever v. cut or break off. □ **severance** n.

several adj. **1** a few, more than two but not many. **2** separate, individual. ●pron. several people or things.

severally adv. separately.

severe adj. **1** strict; harsh; intense. **2** (of style) plain, without decoration. □ **severely** adv., **severity** n.

sew (soh) v. (**sewed**, **sewn** or **sewed**, **sewing**) fasten by passing thread through material, using a needle etc.; make or fasten (a thing) by sewing. □ **sewer** n., **sewing** n.

sewage (soo-ij) n. liquid waste drained from houses etc. for disposal.

sewer (soo-er) n. a drain for carrying sewage.

sewerage n. a system of sewers.

sewn see **sew**.

sex n. **1** either of the two main groups (*male* and *female*) into which living things are placed according to their reproductive functions; the fact of belonging to one of these. **2** sexual feelings or impulses; sexual intercourse. ● *v.* judge the sex of.

sexagenarian *n.* a person in his or her sixties.

sexist *adj.* discriminating in favour of members of one sex; assuming a person's abilities and social functions are predetermined by his or her sex. ● *n.* a person who does this. □ **sexism** *n.*

sexless *adj.* **1** lacking a sex, neuter. **2** not involving sexual feelings.

sextant *n.* an instrument for finding one's position by measuring the height of the sun etc.

sextet *n.* a group of six instruments or voices; music for these.

sexton *n.* an official in charge of a church and churchyard.

sextuplet *n.* one of six children born at one birth.

sexual *adj.* of sex or the sexes; (of reproduction) occurring by fusion of male and female cells. □ **sexually** *adv.*, **sexuality** *n.*

sexual intercourse *n.* copulation, insertion of the penis into the vagina.

sexy *adj.* (**sexier**) sexually attractive or stimulating. □ **sexily** *adv.*, **sexiness** *n.*

Sgt. *abbr.* sergeant.

shabby *adj.* (**shabbier**) **1** worn; dilapidated; poorly dressed. **2** unfair, dishonourable. □ **shabbily** *adv.*, **shabbiness** *n.*

shack *n.* a roughly built hut.

shackle *n.* one of a pair of metal rings joined by a chain, for fastening a prisoner's wrists or ankles. ● *v.* put shackles on; impede, restrict.

shade *n.* **1** comparative darkness; a place sheltered from the sun; a screen or cover used to block or moderate light; (**shades**) sunglasses. **2** a degree or depth of colour. **3** a differing variety. **4** a small amount. ● *v.* **1** block the rays of; give shade to;

darken (parts of a drawing etc.). **2** pass gradually into another colour or variety.

shadow *n.* **1** shade; a patch of this where a body blocks light rays; a person's inseparable companion. **2** a slight trace. **3** gloom. ● *v.* **1** cast shadow over. **2** follow and watch secretly. □ **shadower** *n.*, **shadowy** *adj.*

shadow-boxing *n.* boxing against an imaginary opponent as a form of training.

Shadow Cabinet *n.* members of the main opposition party in Parliament holding posts parallel to those of the government Cabinet.

shady *adj.* (**shadier**) **1** giving shade; situated in shade. **2** disreputable, not completely honest. □ **shadily** *adv.*, **shadiness** *n.*

shaft *n.* **1** an arrow, a spear; a long slender straight part of a thing; a long bar; a large axle. **2** a vertical or sloping passage or opening.

shag *n.* **1** a shaggy mass. **2** a strong coarse tobacco. **3** a cormorant. ● *adj.* (of a carpet) with a long rough pile.

shaggy *adj.* (**shaggier**) having long rough hair or fibre; (of hair etc.) rough and thick. □ **shagginess** *n.*

shaggy-dog story *n.* a lengthy anecdote with a twist of humour at the end.

shagreen *n.* **1** untanned leather with a granulated surface. **2** sharkskin.

shah *n.* a title of the former ruler of Iran.

shake *v.* (**shook**, **shaken**, **shaking**) **1** move quickly up and down or to and fro; dislodge by doing this. **2** shock. **3** make less firm. **4** (of the voice) become uneven. **5** shake hands. ● *n.* **1** shaking, being shaken. **2** a shock. □ **shake hands** clasp right hands in greeting, parting, or agreement. **shake up** mix by shaking; rouse from lethargy, shock. □ **shaker** n.

shake-up n. an upheaval, a re-organization.

shaky adj. (**shakier**) shaking, unsteady; unreliable. □ **shakily** adv., **shakiness** n.

shale n. stone that splits easily. □ **shaly** adj.

shall v.aux. used with I and we to express future tense, and with other words in promises or statements of obligation.

shallot n. a small onion-like plant.

shallow adj. of little depth; superficial. ● n. a shallow place. ● v. make or become shallow. □ **shallowness** n.

shalom n. a Jewish expression of greeting or leave-taking.

sham n. a pretence; something that is not genuine. ● adj. pretended; not genuine. ● v. (**shammed**) pretend; pretend to be.

shamble v. walk or run in a shuffling or lazy way.

shambles n. a scene or condition of great disorder.

shame n. a painful mental feeling aroused by having done something dishonourable or ridiculous; the ability to feel this; a person or thing causing shame; something regrettable. ● v. bring shame on; make ashamed; compel by arousing shame. □ **shameful** adj., **shamefully** adv., **shameless** adj., **shamelessly** adv.

shamefaced adj. looking ashamed.

shammy n. a chamois leather.

shampoo n. a liquid used to wash hair; a preparation for cleaning upholstery etc.; the process of shampooing. ● v. wash or clean with shampoo.

shamrock n. a clover-like plant.

shandy n. a mixed drink of beer and lemonade or ginger beer.

shank n. a leg, esp. from knee to ankle; a thing's shaft or stem.

shantung n. soft Chinese silk.

shanty n. **1** a shack. **2** a sailors' traditional song.

shanty town n. an area of makeshift housing of rough shacks.

shape n. **1** an area or form with a definite outline. **2** the condition of something. **3** a state of orderly arrangement. ● v. give shape to; develop into a certain condition. □ **shapeless** adj.

shapely adj. (**shapelier**) having a pleasant shape. □ **shapeliness** n.

shard n. a broken piece of pottery.

share n. **1** part of something that one is entitled to have or do. **2** one of the equal parts forming a business company's capital and entitling the holder to a proportion of the profits. ● v. give or have a share (of). □ **shareholder** n., **sharer** n.

shareware n. computer programs freely available for trial, paid for by a fee to the author if used regularly.

shark n. **1** a large voracious sea fish. **2** (colloquial) a person who ruthlessly extorts money.

sharkskin n. a fabric with a slightly lustrous textured weave.

sharp adj. **1** having a fine edge or point capable of cutting. **2** abrupt, not gradual. **3** well-defined. **4** (of tastes or smells) causing a smarting sensation. **5** mentally alert. **6** unscrupulous. **7** above the correct pitch in music. ● adv. **1** punctually. **2** suddenly. **3** at a sharp angle. ● n. (Music) (a sign indicating) a note raised by a semitone. □ **sharply** adv., **sharpness** n.

sharpen v. make or become sharp or sharper. □ **sharpener** n.

sharp practice n. barely honest dealings.

sharpshooter n. a skilled marksman.

shatter v. break violently into small pieces; destroy utterly; upset the calmness of.

shave v. remove (growing hair) off the face etc. with a razor; cut thin slices from; graze gently in passing. ● n. an act of shaving hair from the face. □ **shaver** n.

shaven adj. shaved.

shaving n. a thin strip of wood etc. shaved off.

shawl n. a large piece of soft fabric worn round the shoulders or wrapped round a baby.

she pron. the female previously mentioned. ● n. a female.

s/he pron. a written representation of 'he or she'.

sheaf n. (pl. **sheaves**) a bundle of papers; a tied bundle of cornstalks.

shear v. (**sheared, shorn** or **sheared, shearing**) 1 cut or trim with shears or some other sharp device. 2 break because of strain. □ **shearer** n.

shears n.pl. a large cutting instrument shaped like scissors.

sheath (sheeth) n. a close-fitting cover, esp. for a blade or tool; a condom.

sheathe (sheeth) v. put into a case; encase in a covering.

shed n. a building for storing things, or for use as a workshop. ● v. (**shed, shedding**) lose by a natural falling off; take off; allow to fall off or flow.

sheen n. gloss, lustre. □ **sheeny** adj.

sheep n. (pl. **sheep**) a grass-eating animal with a thick fleecy coat.

sheepdog n. a dog trained to guard and herd sheep.

sheepish adj. bashful, embarrassed. □ **sheepishly** adv., **sheepishness** n.

sheepskin n. a sheep's skin with the fleece on.

sheer adj. 1 pure, not mixed or qualified. 2 very steep. 3 (of fabric) very thin, transparent. ● adv. directly, straight up or down. ● v. swerve from a course. □ **sheerly** adv., **sheerness** n.

sheet n. 1 a piece of cotton used as part of the bedclothes. 2 a large thin piece of glass, metal, paper, etc. 3 an expanse of water, flame, etc. 4 a rope securing the lower corner of a sail.

sheet anchor n. a large anchor for emergency use; a person or thing on which one relies.

sheikh (shayk) n. an Arab ruler. □ **sheikhdom** n.

shekel n. a unit of money in Israel; (**shekels**) (colloquial) money, riches.

shelf n. (pl. **shelves**) a board or slab fastened horizontally for things to be placed on; something resembling this, a ledge.

shelf-life n. the time for which a stored thing remains usable.

shell n. 1 the hard outer covering of eggs, nut kernels, and of animals such as snails and tortoises; a firm framework or covering. 2 a metal case filled with explosive, fired from a large gun. ● v. 1 remove the shell(s) of. 2 fire explosive shells at.

shellac n. a resinous substance used in varnish. ● v. (**shellacked**) coat with this.

shellfish n. (pl. **shellfish**) an edible water animal that has a shell.

shell-shock n. psychological disturbance from exposure to battle conditions.

shelter n. a structure that shields against danger, wind, rain, etc.; refuge, a shielded condition. ● *v.* provide with shelter; find or take shelter.

shelve *v.* **1** put aside for later consideration or permanently. **2** slope.

shelving *n.* shelves; material for these.

shepherd *n.* a person who tends sheep. ● *v.* guide (people).

shepherd's pie *n.* a pie of minced meat topped with mashed potato.

sherbet *n.* a fizzy sweet drink, the powder from which this is made.

sheriff *n.* the Crown's chief executive officer in a county; a chief judge of a district in Scotland; (*Amer.*) the chief law-enforcing officer of a county.

Sherpa *n.* a member of a Himalayan people of Nepal and Tibet.

sherry *n.* a strong wine originally from southern Spain.

shiatsu *n.* a Japanese therapy involving the application of pressure to the body.

shibboleth *n.* an old slogan or principle still considered essential by some people.

shield *n.* a protective piece of armour carried on the arm; a trophy in the form of this; any protective structure. ● *v.* protect, screen.

shift *v.* change or move from one position or form to another; transfer (blame etc.); (*slang*) move quickly. ● *n.* **1** a change of place or form etc. **2** a set of workers who start work when another set finishes; the time for which they work.

shiftless *adj.* lazy and inefficient.

shifty *adj.* (**shiftier**) evasive; untrustworthy. □ **shiftily** *adv.*, **shiftiness** *n.*

Shi'ite (shee-It) *n.* & *adj.* (a member) of a Muslim sect opposed to the Sunni.

shilly-shally *v.* be unable to make up one's mind firmly.

shim *n.* a thin wedge used to make parts of machinery fit together.

shimmer *v.* & *n.* (shine with) a soft quivering light.

shin *n.* the front of the leg below the knee; the lower foreleg. □ **shin up** (**shinned**) climb quickly.

shindig *n.* (*colloquial*) a din, a brawl.

shine *v.* **1** (**shone, shining**) give out or reflect light, be bright; excel; cause to shine. **2** (**shined, shining**) polish. ● *n.* brightness; a high polish.

shingle *n.* **1** small rounded pebbles; a stretch of these, esp. on a shore. **2** a wooden roof tile. **3** (**shingles**) a disease with a rash of small blisters. □ **shingly** *adj.*

Shinto *n.* a Japanese religion revering ancestors and nature spirits. □ **Shintoism** *n.*

shiny *adj.* (**shinier**) shining, glossy. □ **shininess** *n.*

ship *n.* a large seagoing vessel. ● *v.* (**shipped**) put or take on board a ship; transport. □ **shipper** *n.*

shipmate *n.* a person travelling or working on the same ship as another.

shipment *n.* the shipping of goods; a consignment shipped.

shipping *n.* ships collectively.

shipshape *adv.* & *adj.* in good order, tidy.

shipwreck *n.* the destruction of a ship by storm or striking rocks etc. □ **shipwrecked** *adj.*

shipyard *n.* an establishment where ships are built.

shire *n.* a county.

shire-horse *n.* a horse of a heavy powerful breed.

shirk *v.* avoid (duty or work etc.) selfishly. □ **shirker** *n.*

shirt n. a lightweight garment for the upper part of the body.

shirty adj. (**shirtier**) (colloquial) annoyed, angry.

shish kebab n. pieces of meat and vegetable grilled on skewers.

shit (vulgar slang) n. **1** faeces. **2** a contemptible person. ● v. (**shitted** or **shat** or **shit**, **shitting**) empty the bowels.

shiver v. **1** tremble slightly, esp. with cold or fear. **2** shatter. ● n. a shivering movement. □ **shivery** adj.

shoal n. **1** a great number of fish swimming together. **2** a shallow place; an underwater sandbank. ● v. form shoals.

shock n. **1** the effect of a violent impact or shake; a sudden violent effect on the mind or emotions; acute weakness caused by injury, pain, or mental shock; the effect of a sudden discharge of electricity through the body. **2** a bushy mass of hair. ● v. cause to suffer shock or a shock; horrify, disgust, seem scandalous to.

shocker n. (colloquial) a shocking person or thing.

shocking adj. causing great astonishment, indignation, or disgust; scandalous; (colloquial) very bad.

shod see **shoe**.

shoddy adj. (**shoddier**) of very poor quality. □ **shoddily** adv., **shoddiness** n.

shoe n. **1** an outer covering for a person's foot. **2** a horseshoe. **3** the part of a brake that presses against a wheel. ● v. (**shod**, **shoeing**) fit with a shoe or shoes.

shoehorn n. a curved implement for easing one's heel into a shoe.

shoelace n. a cord for lacing up shoes.

shoeshine n. (Amer.) a polish given to shoes.

shoestring n. **1** a shoelace. **2** (colloquial) a barely adequate amount of money.

shoe-tree n. a shaped block for keeping a shoe in shape.

shone see **shine**.

shoo int. a sound uttered to frighten animals away. ● v. (**shooed**) drive away by this.

shook see **shake**.

shoot v. (**shot**, **shooting**) **1** fire (a gun or missile); kill or wound with a missile from a gun etc.; hunt with a gun for sport. **2** send out or move swiftly. **3** (of a plant) put forth buds or shoots. **4** take a shot at goal. **5** photograph, film. ● n. a young branch or new growth of a plant. □ **shoot up** rise suddenly; grow rapidly. □ **shooter** n.

shooting star n. a small meteor seen to move quickly.

shooting stick n. a walking stick with a small folding seat in the handle.

shop n. **1** a building where goods or services are sold to the public. **2** a workshop. **3** one's own work as a subject of conversation. ● v. (**shopped**) **1** buy things from shops. **2** (slang) inform against. □ **shop around** look for the best bargain. □ **shopper** n.

shop floor n. workers as distinct from management.

shoplifter n. a person who steals goods from a shop. □ **shoplifting** n.

shopping n. buying goods in shops; the goods bought.

shop-soiled adj. soiled from being on display in a shop.

shop steward n. a trade union official elected by workers as their spokesperson.

shore n. the land along the edge of the sea or a lake. ● v. prop or support with a length of timber.

shorn see **shear**.

short *adj.* **1** measuring little from end to end in space or time, or from head to foot; insufficient; having insufficient; concise, brief. **2** curt. **3** (of pastry) crisp and easily crumbled. ●*adv.* abruptly. ●*n.* **1** a drink of spirits. **2** a short circuit. **3** (**shorts**) trousers that do not reach the knee; (*Amer.*) underpants. ●*v.* short-circuit.

shortage *n.* a lack, an insufficiency.

shortbread *n.* a rich sweet biscuit.

shortcake *n.* shortbread.

short-change *v.* cheat, esp. by giving insufficient change.

short circuit *n.* a fault in an electrical circuit when current flows by a shorter route than the normal one. ●*v.* (**short-circuit**) cause a short circuit in; bypass.

shortcoming *n.* failure to reach a required standard; a fault.

short cut *n.* a quicker route or method.

shorten *v.* make or become shorter.

shortening *n.* fat used to make pastry etc.

shortfall *n.* a deficit.

shorthand *n.* a method of writing rapidly with quickly made symbols.

short-handed *adj.* having insufficient workers.

shortlist *n.* a list of selected candidates from which a final choice will be made. ●*v.* put on a shortlist.

shortly *adv.* **1** after a short time. **2** in a few words; curtly.

short-sighted *adj.* able to see clearly only what is close; lacking foresight.

short ton *see* **ton**.

short wave *n.* a radio wave of frequency greater than 3 MHz.

shot *see* **shoot**. *n.* **1** a firing of a gun etc.; the sound of this; a person of specified skill in shooting; missiles for a cannon, gun, etc. **2** an attempt to hit something or reach a target; a stroke in certain ball games; an attempt. **3** a heavy ball thrown as a sport. **4** a photograph. **5** an injection. **6** (*colloquial*) a measure of spirits. □**like a shot** (*colloquial*) without hesitation.

shotgun *n.* a gun for firing small shot at close range.

should *v.aux.* used to express duty or obligation, a possible or expected future event, or (with *I* and *we*) a polite statement or a conditional or indefinite clause.

shoulder *n.* the part of the body where the arm or foreleg is attached; an animal's upper foreleg as a joint of meat. ●*v.* **1** push with one's shoulder. **2** take (blame or responsibility) on oneself.

shoulder blade *n.* the large flat bone of the shoulder.

shout *n.* a loud cry or utterance. ●*v.* utter a shout; call loudly. □**shout down** silence by shouting.

shove *n.* a rough push. ●*v.* push roughly; (*colloquial*) put.

shovel *n.* a spade-like tool for moving sand, snow, etc.; a mechanical scoop. ●*v.* (**shovelled**; *Amer.* **shoveled**) shift or clear with or as if with a shovel; scoop or thrust roughly.

shoveller *n.* a duck with a broad shovel-like beak.

show *v.* (**showed**, **shown**, **showing**) **1** allow or cause to be seen, offer for inspection or viewing; be able to be seen; present an image of. **2** demonstrate, point out, prove; cause to understand. **3** conduct. ●*n.* a process of showing; a display, a public exhibition or performance; an outward appearance. □**show off** display well, proudly, or ostentatiously; try to impress people. **show up** make or be clearly visible; reveal (a fault etc.); (*colloquial*) arrive.

show business *n.* the entertainment profession.

showdown *n.* a confrontation that settles an argument.

shower *n.* **1** a brief fall of rain or of snow, stones, etc. **2** a device or cabinet in which water is sprayed on a person's body; a wash in this. **3** a sudden influx of letters or gifts etc. **4** (*Amer.*) a party for giving presents. ● *v.* **1** send or come in a shower. **2** take a shower.

showerproof *adj.* (of fabric) able to keep out slight rain. ● *v.* make showerproof.

showery *adj.* with showers of rain.

showjumping *n.* the competitive sport of riding horses to jump over obstacles. □ **showjumper** *n.*

showman *n.* (*pl.* **-men**) an organizer of circuses or theatrical entertainments.

showmanship *n.* skill in presenting entertainment or goods etc. well.

shown see **show**.

show of hands *n.* the raising of hands in voting.

showpiece *n.* an excellent specimen used for exhibition.

showroom *n.* a room where goods are displayed for inspection.

showy *adj.* (**showier**) making a good display; brilliant, gaudy. □ **showily** *adv.*, **showiness** *n.*

shrank see **shrink**.

shrapnel *n.* pieces of metal scattered from an exploding bomb.

shred *n.* a small strip torn or cut from something; a small amount. ● *v.* (**shredded**) tear or cut into shreds. □ **shredder** *n.*

shrew *n.* a small mouse-like animal.

shrewd *adj.* showing sound judgement, clever. □ **shrewdly** *adv.*, **shrewdness** *n.*

shriek *n.* a shrill cry or scream. ● *v.* utter (with) a shriek.

shrike *n.* a bird with a strong hooked beak.

shrill *adj.* piercing and high-pitched in sound. □ **shrilly** *adv.*, **shrillness** *n.*

shrimp *n.* **1** a small edible shellfish. **2** (*colloquial*) a very small person.

shrimping *n.* catching shrimps.

shrine *n.* a sacred or revered place.

shrink *v.* (**shrank**, **shrunk**, **shrinking**) make or become smaller; draw back to avoid something. ● *n.* (*slang*) a psychiatrist.

shrinkage *n.* shrinking of textile fabric; loss by theft or wastage.

shrive *v.* (**shrove**, **shriven**, **shriving**) (*old use*) hear the confession of (a person) and give absolution.

shrivel *v.* (**shrivelled**; *Amer.* **shriveled**) shrink and wrinkle from heat or cold or lack of moisture.

shroud *n.* **1** a cloth wrapping a dead body for burial; something that conceals. **2** one of the ropes supporting a ship's mast. ● *v.* wrap in a shroud; conceal.

shrub *n.* a woody plant smaller than a tree. □ **shrubby** *adj.*

shrubbery *n.* an area planted with shrubs.

shrug *v.* (**shrugged**) raise (one's shoulders) as a gesture of indifference, doubt, or helplessness. ● *n.* this movement.

shrunk see **shrink**.

shrunken *adj.* having shrunk.

shudder *v.* shiver or shake violently. ● *n.* this movement.

shuffle *v.* **1** walk without lifting one's feet clear of the ground. **2** rearrange, jumble. ● *n.* **1** a shuffling movement or walk. **2** rearrangement.

shun v. (**shunned**) avoid.

shunt v. move (a train) to a side track; divert.

shush int. & v. (colloquial) hush.

shut v. (**shut**, **shutting**) move (a door or window etc.) into position to block an opening; be moved in this way; prevent access to (a place); bring or fold parts of (a thing) together; trap or exclude by shutting something. □ **shut down** stop or cease working or business. **shut up** shut securely; (colloquial) stop talking or making a noise.

shutter n. a screen that can be closed over a window; a device that opens and closes the aperture of a camera. □ **shuttered** adj.

shuttle n. **1** a device carrying the weft-thread in weaving. **2** a vehicle used in a shuttle service. **3** a spacecraft for repeated use. ● v. move, travel, or send to and fro.

shuttlecock n. a small cone-shaped feathered object struck to and fro in badminton.

shuttle service n. a transport service making frequent journeys to and fro.

shut-eye n. (colloquial) sleep.

shy adj. timid and lacking self-confidence. ● v. **1** jump or move suddenly in alarm. **2** (colloquial) throw. □ **shyly** adv., **shyness** n.

SI abbr. Système International, the international system of units of measurement.

Si symb. silicon.

Siamese adj. of Siam, the former name of Thailand.

Siamese cat n. a cat with pale fur and darker face, paws, and tail.

Siamese twins n.pl. twins whose bodies are joined at birth.

sibling n. a brother or sister.

sibyl n. a prophetess.

sic adv. used or spelt in the way quoted.

Sicilian adj. & n. (a native) of Sicily.

sick adj. **1** unwell; vomiting; likely to vomit. **2** distressed, disgusted. **3** finding amusement in misfortune or morbid subjects.

sicken v. **1** become ill. **2** distress; disgust. □ **be sickening for** be in the first stages of (a disease).

sickle n. a curved blade used for cutting corn etc.

sickly adj. (**sicklier**) unhealthy; causing sickness or distaste; weak. □ **sickliness** n.

sickness n. illness; vomiting.

side n. **1** a surface of an object, esp. one that is not the top, bottom, front, back, or end; a bounding line of a plane figure; a slope of a hill or ridge. **2** a part near an edge; a region next to a person or thing. **3** either of the two halves into which something is divided; an aspect of a problem etc.; one of two opposing groups or teams etc. ● adj. at or on the side. ● v. join forces with (a person) in a dispute. □ **on the side** as a sideline; as a surreptitious activity. **side by side** close together.

sideboard n. **1** a piece of dining-room furniture with drawers and cupboards for china etc. **2** (**sideboards**) (colloquial) sideburns.

sideburns n.pl. short whiskers on the cheeks.

side effect n. a secondary (usu. bad) effect.

sidelight n. one of two small lights on the front of a vehicle.

sideline n. **1** something done in addition to one's main activity. **2** (**sidelines**) lines bounding the sides of a football pitch etc.; a place for spectators; a position etc. apart from the main action.

sidelong adj. & adv. sideways.

sidereal (sy-deer-iăl) adj. of or measured by the stars.

side-saddle n. a saddle on which a woman rider sits with both legs on the same side of the horse. ● adv. sitting in this way.

sideshow n. a small show forming part of a large one.

sidestep v. (**sidestepped**) avoid by stepping sideways; evade.

sidetrack v. divert.

sidewalk n. (Amer.) a pavement.

sideways adv. & adj. to or from one side; with one side forward.

siding n. a short track by the side of a railway, used in shunting.

sidle v. advance in a timid, furtive, or cringing way.

SIDS abbr. sudden infant death syndrome; cot death.

siege n. the surrounding and blockading of a place by armed forces in order to capture it.

sienna n. a brownish clay used as colouring matter.

sierra n. a chain of mountains with jagged peaks, esp. in Spain or Spanish America.

siesta n. an afternoon nap or rest, esp. in hot countries.

sieve (siv) n. a utensil with a mesh through which liquids or fine particles can pass. ● v. put through a sieve.

sift v. 1 sieve. 2 examine carefully and select or analyse.

sigh n. a long deep breath given out audibly in sadness, tiredness, relief, etc. ● v. give or express with a sigh.

sight n. 1 the ability to see; seeing, being seen. 2 something seen or worth seeing; an unsightly thing. 3 a device looked through to aim or observe with a gun or telescope etc. ● v. get a sight of; aim or observe with a sight etc. □ at or on sight as soon as seen.

sightless adj. blind.

sight-read v. play or sing (music) without preliminary study of the score.

sightseeing n. visiting places of interest. □ **sightseer** n.

sign n. 1 something perceived that suggests the existence of a quality, future occurrence, etc. 2 a board bearing the name of a shop etc.; a notice displayed. 3 an action or gesture conveying information etc. 4 any of the twelve divisions of the zodiac. ● v. 1 write (one's name) on a document; convey, engage, or acknowledge by this. 2 make a sign.

signal n. 1 a sign or gesture giving information or a command; an object placed to give notice or warning. 2 a sequence of electrical impulses or radio waves transmitted or received. ● v. (**signalled**; Amer. **signaled**) make a signal or signals; communicate with or announce in this way. ● adj. noteworthy. □ **signaller** n., **signally** adv.

signal box n. a small railway building with signalling apparatus.

signalman n. (pl. -men) a person responsible for operating railway signals.

signatory n. one of the parties who sign an agreement.

signature n. 1 a person's name or initials written by himself or herself in signing something. 2 an indication of key or tempo at the beginning of a musical score.

signature tune n. a tune used to announce a particular performer or programme.

signet ring n. a finger ring with an engraved design.

significance n. meaning; importance. □ **significant** adj., **significantly** adv.

signification n. meaning.

signify v. be a sign or symbol of; have as a meaning; make known; matter.

signpost n. a post with arms showing the direction of certain places.

Sikh (seek) n. a member of an Indian religious sect. □ **Sikhism** n.

silage n. green fodder stored and fermented in a silo.

silence n. the absence of sound or of speaking. ● v. make silent.

silencer n. a device for reducing sound.

silent adj. without sound; not speaking. **silently** adv.

silhouette (sil-oo-et) n. a dark shadow or outline seen against a light background. ● v. show as a silhouette.

silica n. a compound of silicon occurring as quartz and in sandstone etc. □ **siliceous** adj.

silicate n. a compound of silicon.

silicon n. a chemical element (symbol Si) found in the earth's crust in its compound forms.

silicon chip n. a silicon microchip.

silicone n. an organic compound of silicon, used in paint, varnish, and lubricants.

silicosis n. a lung disease caused by inhaling dust that contains silica.

silk n. a fine strong soft fibre produced by silkworms; thread or cloth made from this. □ **silky** adj.

silken adj. like silk.

silkworm n. a caterpillar which spins its cocoon of silk.

sill n. a strip of stone, wood, or metal at the base of a doorway or window opening.

sillabub var. of **syllabub**.

silly adj. (**sillier**) lacking good sense, foolish, unwise; feeble-minded. □ **silliness** n.

silo n. (pl. **silos**) a pit or airtight structure for holding silage; a pit or tower for storing grain, cement, or radioactive waste; an underground place where a missile is kept ready for firing.

silt n. sediment deposited by water in a channel or harbour etc. ● v. block or become blocked with silt.

silvan var. of **sylvan**.

silver n. a chemical element (symbol Ag), a white precious metal; articles made of this; coins made of an alloy resembling it; household cutlery; the colour of silver. ● adj. made of or coloured like silver.

silverfish n. (pl. **silverfish**) a small wingless insect with a fish-like body.

silver jubilee n. the 25th anniversary of a sovereign's reign.

silverside n. a joint of beef cut from the haunch, below topside.

silver wedding n. the 25th anniversary of a wedding.

silvery adj. 1 like silver. 2 having a clear gentle ringing sound.

simian adj. monkey-like, ape-like.

similar adj. alike but not the same; of the same kind or amount. □ **similarity** n., **similarly** adv.

simile (sim-i-lee) n. a figure of speech in which one thing is compared to another.

similitude n. similarity.

simmer v. 1 boil very gently. 2 be in a state of barely suppressed anger or excitement. □ **simmer down** become less excited.

simper v. smile in an affected way. ● n. an affected smile.

simple adj. 1 of one element or kind; not complicated or showy. 2 foolish; feeble-minded. □ **simply** adv., **simplicity** n.

simpleton n. a foolish or feeble-minded person.

simplify v. make simple; make easy to do or understand. □ **simplification** n.

simplistic adj. over-simplified. □ **simplistically** adv.

simulate v. pretend; imitate the form or condition of. □ **simulation** n., **simulator** n.

simultaneous adj. occurring at the same time. □ **simultaneously** adv., **simultaneity** n.

sin n. the breaking of a religious or moral law; an act which does this. ● v. (**sinned**) commit a sin. ● abbr. sine.

since prep. after; from (a specified time) until now, within that period. ● conj. 1 from the time that. 2 because. ● adv. since that time or event.

sincere adj. without pretence or deceit. □ **sincerely** adv., **sincerity** n.

sine n. (Maths) the ratio of the side opposite an angle (in a right-angled triangle) to the hypotenuse.

sinecure n. a position of profit with no work attached.

sine qua non (sinay kwah nohn) n. an indispensable condition.

sinew n. tough fibrous tissue joining muscle to bone; a tendon; (**sinews**) muscles, strength. □ **sinewy** adj.

sinful adj. full of sin, wicked. □ **sinfully** adv., **sinfulness** n.

sing v. (**sang, sung, singing**) make musical sounds with the voice; perform (a song); make a humming sound. □ **singer** n.

singe (sinj) v. (**singeing**) burn slightly; burn the ends or edges of. ● n. a slight burn.

single adj. one only, not double or multiple; designed for one person or thing; unmarried; (of a ticket) valid for an outward journey only. ● n. one person or thing; a room etc. for one person; a single ticket; a pop record

with one piece of music on each side; (**singles**) a game with one player on each side. ● v. choose or distinguish from others. □ **singly** adv.

single figures n.pl. numbers from 1 to 9.

single-handed adj. without help.

single market n. an association of countries trading without restrictions.

single-minded adj. with one's mind set on a single purpose.

single parent n. a person bringing up a child or children without a partner.

singlet n. a sleeveless vest.

singleton n. a thing that occurs singly.

singsong adj. with a monotonous rise and fall of the voice. ● n. informal singing by a group of people.

singular n. the form of a noun or verb used in referring to one person or thing. ● adj. 1 of this form. 2 uncommon, extraordinary. □ **singularly** adv., **singularity** n.

sinister adj. suggestive of evil; involving wickedness.

sink v. (**sank, sunk, sinking**) 1 fall or come gradually downwards; make or become submerged; lose value or strength. 2 send (a ball) into a pocket or hole. 3 invest (money). ● n. a fixed basin with a drainage pipe. □ **sink in** become understood.

sinker n. a weight used to sink a fishing line etc.

sinking fund n. money set aside regularly for repayment of a debt etc.

sinner n. a person who sins.

sinuous adj. curving, undulating.

sinus (sy-nǔs) n. a cavity in bone or tissue, esp. that connecting with the nostrils.

sinusitis *n.* inflammation of the sinus.

sip *v.* (**sipped**) drink in small mouthfuls. ● *n.* an amount sipped.

siphon *n.* a bent pipe or tube used for transferring liquid using atmospheric pressure; a bottle from which soda water etc. is forced out by pressure of gas. ● *v.* flow or draw out through a siphon; take from a source.

sir *n.* a polite form of address to a man; (**Sir**) the title of a knight or baronet.

sire *n.* an animal's male parent. ● *v.* be the sire of.

siren *n.* **1** a device that makes a loud prolonged sound as a signal. **2** a dangerously fascinating woman.

sirloin *n.* the upper (best) part of loin of beef.

sirocco *n.* (*pl.* **siroccos**) a hot wind that reaches Italy from Africa.

sisal (sy-sǎl) *n.* rope-fibre made from a tropical plant; this plant.

sissy *n.* a weak or timid person.

sister *n.* **1** a daughter of the same parents as another person. **2** a woman who is a fellow member of a group. **3** a nun. **4** a senior female nurse. □ **sisterly** *adj.*

sisterhood *n.* **1** the relationship of sisters. **2** an order of nuns. **3** a group of women with common aims.

sister-in-law *n.* (*pl.* **sisters-in-law**) a sister of one's husband or wife; the wife of one's brother.

sit *v.* (**sat, sitting**) **1** take or be in a position with the body resting on the buttocks; cause to sit; pose for a portrait; perch; (of animals) rest with legs bent and body on the ground; (of birds) remain on the nest to hatch eggs. **2** be situated. **3** be a candidate (for). **4** (of a committee etc.) hold a session.

sitar *n.* a guitar-like Indian musical instrument.

sitcom *n.* (*colloquial*) a situation comedy.

site *n.* the place where something is, was, or is to be located. ● *v.* locate, provide with a site.

sitter *n.* a person sitting; a babysitter.

sitting *see* **sit**. *n.* the time during which a person or assembly etc. sits.

sitting room *n.* a room in a house used for relaxed sitting in.

sitting tenant *n.* one already in occupation.

situate *v.* place or put in a certain position.

situation *n.* the place (with its surroundings) occupied by something; a set of circumstances; a position of employment. □ **situational** *adj.*

situation comedy *n.* a broadcast comedy involving the same characters in a series of episodes.

six *adj.* & *n.* one more than five (6, VI). □ **sixth** *adj.* & *n.*

sixteen *n.* one more than fifteen (16, XVI). □ **sixteenth** *adj.* & *n.*

sixty *adj.* & *n.* six times ten (60, LX). □ **sixtieth** *adj.* & *n.*

size *n.* **1** relative bigness; extent; one of a series of standard measurements in which things are made and sold. **2** a gluey solution used to glaze paper or stiffen textiles. ● *v.* group according to size. □ **size up** estimate the size of; (*colloquial*) form a judgement of.

sizeable *adj.* (also **sizable**) large; fairly large.

sizzle *v.* make a hissing sound like that of frying.

skate *n.* **1** a boot with a blade or wheels attached, for gliding over ice or a hard surface. **2** a marine flatfish used as food. ● v. move on skates. □ **skate over** make only a passing reference to. □ **skater** n.

skateboard n. a small board with wheels for riding on while standing. ● v. ride on a skateboard.

skedaddle v. (colloquial) run away.

skein (skayn) n. 1 a loosely coiled bundle of yarn. 2 a flock of wild geese etc. in flight.

skeletal adj. of or like a skeleton.

skeleton n. the hard supporting structure of an animal body; any supporting structure; a framework.

skeleton service, **skeleton staff** n. one reduced to a minimum.

skeleton key n. a key made so as to fit many locks.

skeptic Amer. sp. of **sceptic**.

sketch n. a rough drawing or painting; a brief account; a short usu. comic play. ● v. make a sketch or sketches (of).

sketchy adj. (**sketchier**) rough and not detailed or substantial. □ **sketchily** adv., **sketchiness** n.

skew adj. slanting, askew. ● v. make skew; turn or twist round.

skewbald adj. (of an animal) with irregular patches of white and another colour.

skewer n. a pin to hold pieces of food together while cooking. ● v. pierce with a skewer.

ski n. one of a pair of long narrow strips of wood etc. fixed under the feet for travelling over snow. ● v. (**skis**, **skied**, **skiing**) travel on skis. □ **skier** n.

skid v. (**skidded**) (of a vehicle) slide uncontrollably. ● n. a skidding movement.

skid-pan n. a surface used for practising control of skidding vehicles.

skiff n. a small light rowing boat.

skilful adj. (Amer. **skillful**) having or showing great skill. □ **skilfully** adv.

skill n. ability to do something well. □ **skilled** adj.

skillet n. a long-handled cooking pot; (Amer.) a frying pan.

skim v. (**skimmed**) 1 take (matter) from the surface of (liquid). 2 glide. 3 read quickly.

skim milk n. (also **skimmed milk**) milk from which the cream has been skimmed.

skimp v. supply or use rather less than what is necessary.

skimpy adj. (**skimpier**) scanty. □ **skimpily** adv., **skimpiness** n.

skin n. an outer covering of the human or other animal body; material made from animal skin; complexion; an outer layer; a skin-like film on liquid. ● v. (**skinned**) strip skin from.

skin diving n. the sport of swimming under water with flippers and breathing apparatus. □ **skin diver** n.

skinflint n. a miserly person.

skinny adj. (**skinnier**) very thin.

skint adj. (slang) having no money left.

skip v. (**skipped**) 1 move lightly, taking two steps with each foot in turn; jump with a skipping rope. 2 (colloquial) omit. 3 (colloquial) leave hastily or secretly. ● n. 1 a skipping movement. 2 a large container for builders' rubbish etc.

skipper n. a captain.

skipping rope n. a rope turned over the head and under the feet while jumping.

skirmish n. a minor fight or conflict. ● v. take part in a skirmish.

skirt n. 1 a woman's garment hanging from the waist; this part of a garment; any similar part. 2 a cut of beef from the lower flank. ● v. go or be along the edge of.

skirting n. (in full **skirting board**) a narrow board round the bottom of the wall of a room.

skit n. a humorous imitation in the form of a short play or piece of writing.

skittish adj. frisky.

skittle n. one of the wooden pins set up to be bowled down with a ball in the game of **skittles**.

skive v. (colloquial) dodge a duty; play truant.

skua n. a large seagull.

skulduggery n. trickery.

skulk v. loiter stealthily.

skull n. the bony framework of the head.

skullcap n. a small cap with no peak, for the crown of the head.

skunk n. **1** a black bushy-tailed American animal able to spray an evil-smelling liquid. **2** (colloquial) a contemptible person.

sky n. the region of the clouds or upper air.

sky blue adj. & n. bright clear blue.

skydiving n. the sport of jumping from an aircraft and performing acrobatic movements in the sky before opening one's parachute.

skylark n. a lark that soars while singing. ● v. play mischievously.

skylight n. a window set in a roof or ceiling.

skyscraper n. a very tall building.

slab n. a broad flat piece of something solid.

slack adj. **1** not tight. **2** not busy. **3** negligent. ● n. **1** a slack part. **2** coal dust. ● v. **1** slacken. **2** be lazy about work. □ **slacker** n., **slackly** adv., **slackness** n.

slacken v. make or become slack.

slacks n.pl. trousers for casual wear.

slag n. **1** solid waste left when metal has been smelted. **2** (slang) an immoral woman. ● v. (also **slag off**) (**slagged**) (slang) criticize, insult.

slag heap n. a mound of waste matter.

slain see **slay**.

slake v. **1** satisfy (thirst). **2** combine (lime) with water.

slalom n. a ski race down a zig-zag course; an obstacle race in canoes etc.

slam v. (**slammed**) **1** shut forcefully and noisily; put or hit forcefully. **2** (colloquial) criticize severely. ● n. a slamming noise.

slander n. a false statement uttered maliciously that damages a person's reputation; the crime of uttering this. ● v. utter slander about. □ **slanderer** n., **slanderous** adj.

slang n. words or phrases or particular meanings of these used very informally for vividness or novelty.

slant v. **1** slope. **2** present (news etc.) from a particular point of view. ● n. **1** a slope. **2** the way news etc. is slanted; bias. □ **slantwise** adv.

slap v. (**slapped**) strike with the open hand or with something flat; place forcefully or carelessly. ● n. a slapping blow. ● adv. with a slap, directly.

slapdash adj. hasty and careless.

slap-happy adj. (colloquial) cheerfully casual.

slap-up adj. (colloquial) first-class, lavish.

slapstick n. boisterous comedy.

slash v. **1** gash; make a sweeping stroke; strike in this way. **2** reduce drastically. ● n. a slashing stroke; a cut.

slat n. a narrow strip of wood, metal, etc.

slate n. rock that splits easily into flat greyish plates; a piece of this used as roofing-material or (formerly) for writing on. ● *v.* **1** cover with slates. **2** (*colloquial*) criticize severely.

slaughter *v.* kill (animals) for food; kill ruthlessly or in great numbers. ● *n.* this process.

slaughterhouse *n.* a place where animals are killed for food.

Slav *adj. & n.* (a member) of any of the peoples of Europe who speak a Slavonic language.

slave *n.* a person who is owned by and must work for another; a victim of or to a dominating influence; a mechanism directly controlled by another. ● *v.* work very hard.

slave-driver *n.* a person who makes others work very hard. □ **slave-driving** *n.*

slaver *v.* have saliva flowing from the mouth.

slavery *n.* the existence or condition of slaves; very hard work.

slavish *adj.* excessively submissive or imitative. □ **slavishly** *adv.*

Slavonic *adj. & n.* (of) the group of languages including Russian and Polish.

slay *v.* (**slew, slain, slaying**) kill.

sleazy *adj.* (**sleazier**) disreputable; squalid; dirty. □ **sleaze** *n.,* **sleaziness** *n.*

sled *n.* (*Amer.*) a sledge.

sledge *n.* a cart with runners instead of wheels, used on snow. ● *v.* travel or convey in a sledge.

sledgehammer *n.* a large heavy hammer used with both hands.

sleek *adj.* smooth and glossy; looking well fed and thriving. □ **sleekness** *n.*

sleep *n.* the natural condition of rest with unconsciousness and relaxation of muscles; a spell of this. ● *v.* (**slept, sleeping**) **1** be or spend (time) in a state of sleep. **2** provide with sleeping accommodation.

sleeper *n.* **1** one who sleeps. **2** a railway coach fitted for sleeping in. **3** a beam on which the rails of a railway rest. **4** a ring worn in a pierced ear to keep the hole from closing.

sleeping bag *n.* a padded bag for sleeping in.

sleepwalk *v.* walk about while asleep. □ **sleepwalker** *n.*

sleepy *adj.* (**sleepier**) feeling a desire to sleep; quiet, without stir or bustle. □ **sleepily** *adv.,* **sleepiness** *n.*

sleet *n.* hail or snow and rain falling together. ● *v.* fall as sleet.

sleeve *n.* the part of a garment covering the arm; a tube-like cover; the cover for a record. □ **up one's sleeve** concealed but available.

sleigh (slay) *n.* a sledge, esp. drawn by horses.

sleight of hand (slyt) *n.* skill in using the hands to perform conjuring tricks etc.

slender *adj.* slim and graceful; small in amount. □ **slenderness** *n.*

slept *see* sleep.

sleuth (slooth) *n.* a detective.

slew *see* slay. *v.* (also **slue**) turn or swing round.

slice *n.* a thin flat piece (or a wedge) cut from something; a portion; an implement for cutting and serving food; a slicing stroke. ● *v.* **1** cut, esp. into slices. **2** strike (a ball) so that it spins away from the direction intended. □ **slicer** *n.*

slick *adj.* quick and cunning; smooth in manner. ● *n.* a patch of oil on the sea; a slippery place. ● *v.* make sleek.

slide *v.* (**slid, sliding**) (cause to) move along a smooth surface touching it always; move or pass smoothly. ● *n.* **1** the act of sliding; a smooth slope or surface for sliding; a sliding part. **2** a piece of glass for holding an object under a microscope. **3** a picture for projecting onto a screen. **4** a hinged clip for holding hair in place.

sliding scale n. a scale of fees or taxes etc. that varies according to the variation of some standard.

slight adj. not much, not great, not thorough; slender. ● v. insult by treating with lack of respect; snub. ● n. a snub. □ **slightly** adv., **slightness** n.

slim adj. (**slimmer**) of small girth or thickness; small, slight, insufficient. ● v. (**slimmed**) make (oneself) slimmer by dieting, exercise, etc. □ **slimness** n.

slime n. an unpleasant thick liquid substance.

slimline adj. of slender design.

slimy adj. **1** like slime; covered with slime. **2** disgustingly dishonest, meek, or flattering. □ **slimily** adv., **sliminess** n.

sling n. a belt, chain, or bandage etc. looped round an object to support or lift it; a looped strap used to throw a stone etc. ● v. (**slung**, **slinging**) suspend, lift, or hurl with a sling; (colloquial) throw.

slink v. (**slunk**, **slinking**) move in a stealthy or shamefaced way.

slinky adj. (**slinkier**) smooth and sinuous.

slip v. (**slipped**) slide accidentally; lose one's balance in this way; go or put smoothly; escape hold or capture; detach, release; become detached from. ● n. **1** an act of slipping. **2** a slight mistake. **3** a small piece of paper. **4** a petticoat. **5** a liquid containing clay used in pottery. □ **give a person the slip** escape from or evade him or her. **slip up** (colloquial) make a mistake.

slipped disc n. a disc of cartilage between vertebrae that has

become displaced and causes pain.

slipper n. a light loose shoe for indoor wear.

slippery adj. **1** smooth or wet and difficult to hold or causing slipping. **2** (of a person) not trustworthy. □ **slipperiness** n.

slip road n. a road for entering or leaving a motorway.

slipshod adj. done or doing things carelessly.

slipstream n. a current of air driven backward as something is propelled forward.

slipway n. a sloping structure on which boats are landed or ships built or repaired.

slit n. a narrow straight cut or opening. ● v. (**slit**, **slitting**) cut a slit in; cut into strips.

slither v. slide unsteadily.

sliver n. a small thin strip.

slob n. (slang) a lazy, untidy person.

slobber v. slaver, dribble.

sloe n. a blackthorn; its small dark plum-like fruit.

slog v. (**slogged**) **1** work or walk hard and steadily. **2** hit hard. ● n. **1** a spell of hard steady work or walking. **2** a hard hit.

slogan n. a word or phrase adopted as a motto or in advertising.

sloop n. a small ship with one mast.

slop v. (**slopped**) spill; splash. ● n. an unappetizing liquid; slopped liquid; (**slops**) liquid refuse.

slope v. lie or put at an angle from the horizontal or vertical. ● n. a sloping surface; the amount by which a thing slopes.

sloppy adj. (**sloppier**) **1** wet, slushy. **2** slipshod. **3** weakly sentimental. □ **sloppily** adv., **sloppiness** n.

slosh v. **1** splash, pour clumsily. **2** (slang) hit. ● n. **1** a splashing sound. **2** (slang) a heavy blow.

sloshed adj. (slang) drunk.

slot n. **1** a narrow opening into or through which something is to be put. **2** a position in a series or scheme. ● v. (**slotted**) make slot(s) in; put or fit into a slot.

sloth (slohth) n. **1** laziness. **2** a slow-moving animal of tropical America.

slothful adj. lazy. □ **slothfully** adv.

slot machine n. a machine operated by inserting a coin into a slot.

slouch v. stand, sit, or move in a lazy way. ● n. a slouching movement or posture.

slough¹ (rhymes with cow) n. a swamp, a marsh.

slough² (sluf) v. shed (skin); be shed in this way.

slovenly adj. careless and untidy. □ **slovenliness** n.

slow adj. **1** not quick or fast; showing an earlier time than the correct one. **2** not able to learn easily. ● adv. slowly. ● v. reduce the speed (of). □ **slowly** adv., **slowness** n.

slowcoach n. a slow or lazy person.

slow-worm n. a small lizard with no legs.

sludge n. thick mud.

slue var. of **slew**.

slug n. **1** a small slimy creature like a snail without a shell. **2** a small lump of metal; a bullet. ● v. (**slugged**) hit hard.

sluggard n. a slow or lazy person.

sluggish adj. slow-moving, not lively. □ **sluggishly** adv., **sluggishness** n.

sluice (slooss) n. a sliding gate controlling a flow of water; this water; a channel carrying off water.

slum n. a squalid house or district.

slumber v. & n. sleep.

slump n. a sudden great fall in prices or demand. ● v. undergo a slump; sit or flop down slackly.

slung see **sling**.

slunk see **slink**.

slur v. (**slurred**) **1** utter with each letter or sound running into the next. **2** pass lightly over (a fact). ● n. **1** a slurred sound; a curved line marking notes to be slurred in music. **2** a discredit.

slurp v. & n. (make) a noisy sucking sound.

slurry n. thin mud; thin liquid cement; fluid manure.

slush n. **1** partly melted snow on the ground. **2** silly sentimental talk or writing. □ **slushy** adj.

slush fund n. a fund of money for bribes etc.

slut n. a slovenly or immoral woman. □ **sluttish** adj.

sly adj. unpleasantly cunning and secretive; mischievous and knowing. □ **on the sly** secretly. □ **slyly** adv., **slyness** n.

Sm symb. samarium.

smack n. **1** a slap; a hard hit. **2** a loud kiss. **3** a slight flavour or trace. **4** a single-masted boat. **5** a hard drug, esp. heroin. ● v. **1** slap, hit hard. **2** close and part (lips) noisily. **3** have a slight flavour or trace.

small adj. not large or great; unimportant; petty. ● n. **1** the narrowest part (of the back). **2** (**smalls**) (colloquial) underwear. ● adv. in a small way. □ **smallness** n.

smallholding n. a small farm. □ **smallholder** n.

small hours n.pl. the period soon after midnight.

small-minded adj. narrow or selfish in outlook.

smallpox n. a disease with pustules that often leave bad scars.

small talk n. social conversation on unimportant subjects.

small-time adj. of an unimportant level.

smarmy adj. (**smarmier**) (colloquial) ingratiating, fulsome. □ **smarmily** adv., **smarminess** n.

smart adj. **1** neat and elegant. **2** clever. **3** forceful, brisk. ● v. feel a stinging pain. □ **smartly** adv., **smartness** n.

smarten v. make or become smarter.

smash v. break noisily into pieces; strike forcefully; crash; ruin. ● n. the act or sound of smashing; a collision; disaster; ruin.

smashing adj. (colloquial) excellent.

smattering n. a slight knowledge.

smear v. **1** spread with a greasy or dirty substance. **2** try to damage the reputation of. ● n. a mark made by smearing; a slander.

smell n. the ability to perceive things with the sense organs of the nose; a quality perceived in this way; an act of smelling. ● v. (**smelt** or **smelled**, **smelling**) perceive the smell of; give off a smell. □ **smelly** adj.

smelt v. heat and melt (ore) to extract metal; obtain (metal) in this way. ● n. a small fish related to the salmon.

smidgen n. (also **smidgin**) (colloquial) a very small amount.

smile n. a facial expression indicating pleasure or amusement, with lips upturned. ● v. give a smile; look favourable.

smirch v. & n. discredit, disgrace.

smirk n. a self-satisfied smile. ● v. give a smirk.

smite v. (**smote**, **smitten**, **smiting**) (literary) hit hard; affect suddenly.

smith n. a person who makes things in metal; a blacksmith.

smithereens n.pl. small fragments.

smithy n. a blacksmith's workshop.

smitten see **smite**.

smock n. a loose shirt-like garment; a loose overall.

smog n. dense smoky fog.

smoke n. visible vapour given off by a burning substance; the act of smoking tobacco; (colloquial) a cigarette, a cigar. ● v. give out smoke or steam; preserve with smoke; draw smoke from (a cigarette or cigar or pipe) into the mouth; do this as a habit. □ **smoky** adj.

smokeless adj. with little or no smoke.

smoker n. a person who smokes tobacco as a habit.

smokescreen n. something intended to disguise or conceal activities.

smolder Amer. sp. of **smoulder**.

smooch v. (colloquial) dance slowly, holding one's partner closely.

smooth adj. **1** having an even surface with no projections; not harsh in sound or taste; moving evenly without bumping. **2** polite but perhaps insincere. ● v. make smooth. □ **smoothly** adv., **smoothness** n.

smorgasbord n. a buffet meal with a variety of dishes.

smote see **smite**.

smother v. suffocate, stifle; cover thickly; suppress.

smoulder v. (Amer. **smolder**) **1** burn slowly with smoke but no flame. **2** show silent or suppressed anger etc.

smudge n. a dirty or blurred mark. ● v. make a smudge on or with; become smudged; blur. □ **smudgy** adj.

smug adj. (**smugger**) self-satisfied. □ **smugly** adv., **smugness** n.

smuggle v. convey secretly; bring (goods) illegally into or out of a country, esp. without paying customs duties. □ **smuggler** n.

smut n. **1** a small flake of soot; a small black mark. **2** indecent pictures, stories, etc. □ **smutty** adj.

Sn symb. tin.

snack n. a small or casual meal.

snaffle n. a horse's bit without a curb. ● v. (colloquial) take for oneself.

snag n. a jagged projection; a tear caused by this. ● v. (**snagged**) catch or tear on a snag.

snail n. a soft-bodied creature with a shell.

snake n. a reptile with a long narrow body and no legs. ● v. move in a winding course. □ **snaky** adj., **snakeskin** n.

snap v. (**snapped**) **1** (cause to) make a sharp cracking sound; break suddenly; bite at suddenly. **2** speak with sudden irritation. **3** move smartly. **4** take a snapshot of. ● n. **1** an act or sound of snapping. **2** a snapshot. ● adj. sudden, done or arranged at short notice. □ **snap up** take eagerly.

snapdragon n. a plant with flowers that have a mouth-like opening.

snapper n. any of several sea fish used as food.

snappy adj. (**snappier**) (colloquial) **1** irritable. **2** brisk. **3** neat and elegant. □ **snappily** adv., **snappiness** n.

snapshot n. an informal photograph.

snare n. a trap, usu. with a noose. ● v. trap in a snare.

snarl v. **1** growl angrily with bared teeth; speak or utter in a bad-tempered way. **2** become entangled. ● n. **1** an act or sound of snarling. **2** a tangle.

snarl-up n. a traffic jam; a muddle.

snatch v. seize quickly or eagerly. ● n. an act of snatching; a short or brief part.

snazzy adj. (**snazzier**) (slang) stylish.

sneak v. **1** go, convey, or (colloquial) steal furtively. **2** (slang) tell tales. ● n. (slang) a tell-tale.

sneaking adj. (of a feeling) persistent but not openly acknowledged.

sneer n. a scornful expression or remark. ● v. show contempt by a sneer.

sneeze n. a sudden audible involuntary expulsion of air through the nose. ● v. give a sneeze.

snicker v. & n. = **snigger**.

snide adj. sneering slyly.

sniff v. draw air audibly through the nose; draw in as one breathes; try the smell of. ● n. the act or sound of sniffing. □ **sniffer** n.

sniffle v. sniff slightly or repeatedly. ● n. this act or sound.

snifter n. (slang) a small alcoholic drink.

snigger v. & n. (give) a sly giggle.

snip v. (**snipped**) cut with scissors or shears in small quick strokes. ● n. **1** the act or sound of snipping. **2** (colloquial) a bargain, a certainty.

snipe n. (pl. **snipe** or **snipes**) a wading bird with a long straight bill. ● v. **1** fire shots from a hiding place. **2** make sly critical remarks. □ **sniper** n.

snippet n. a small piece.

snivel v. (**snivelled**; Amer. **sniveled**) cry in a miserable whining way.

snob n. a person with an exaggerated respect for social position or wealth, despising those he or she considers inferior. □ **snobbery** n., **snobbish** adj.

snood n. a loose bag-like ornamental net, holding a woman's hair at the back.

snooker n. a game played on a baize-covered table with 15 red and 6 other coloured balls.

snoop v. (colloquial) pry. □ **snooper** n.

snooty adj. (**snootier**) (colloquial) haughty and contemptuous. □ **snootily** adv.

snooze (colloquial) n. a nap. ● v. take a snooze.

snore n. a snorting or grunting sound made during sleep. ● v. make such sounds. ● **snorer** n.

snorkel n. a tube by which an underwater swimmer can breathe. ● v. (**snorkelled**; Amer. **snorkeled**) swim with a snorkel.

snort n. a sound made by forcing breath through the nose, esp. in indignation. ● v. make a snort; (slang) inhale (a powdered drug).

snout n. an animal's long projecting nose or nose and jaws.

snow n. frozen atmospheric vapour falling to earth in white flakes; a fall or layer of snow. ● v. fall as or like snow. □ **snowed under** covered with snow; overwhelmed with work etc. □ **snowstorm** n., **snowy** adj.

snowball n. snow pressed into a compact mass for throwing. ● v. increase in size or intensity.

snowblower n. a machine that clears snow by blowing it to the side of the road.

snowdrift n. a mass of snow piled up by the wind.

snowdrop n. a plant with white flowers blooming in spring.

snowman n. (pl. **-men**) a figure made of snow.

snowplough n. (Amer. **snowplow**) a device for clearing roads by pushing snow aside.

snub v. (**snubbed**) reject (a person) contemptuously. ● n. treatment of this kind. ● adj. (of the nose) short and stumpy. □ **snub-nosed** adj.

snuff n. powdered tobacco for sniffing up the nostrils. ● v. put out (a candle). □ **snuff it** (slang) die. □ **snuffer** n.

snuffle v. breathe with a noisy sniff. ● n. a snuffling sound.

snug adj. (**snugger**) cosy; close-fitting. ● n. a small comfortable room in a pub. □ **snugly** adj.

snuggle v. nestle, cuddle.

so adv. **1** to the extent or in the manner or with the result indicated. **2** very. **3** also. ● conj. **1** for that reason. **2** in order that, with the result that. □ **so that** in order that.

soak v. place or lie in liquid so as to become thoroughly wet; (of liquid) penetrate; absorb. ● n. **1** a soaking. **2** (slang) a heavy drinker.

so-and-so n. **1** a person or thing that need not be named. **2** (colloquial) a disliked person.

soap n. **1** a substance used in washing things, made of fat or oil and an alkali. **2** (colloquial) a soap opera. ● v. apply soap to.

soap opera a television or radio serial usu. dealing with domestic issues.

soapstone n. steatite.

soapsuds n.pl. froth of soapy water.

soapy adj. **1** of or like soap; containing or smeared with soap. **2** trying to win favour by flattery. □ **soapiness** n.

soar v. rise high, esp. in flight.

sob n. an uneven drawing of breath when weeping or gasping. ● v. (**sobbed**) weep, breathe, or utter with sobs.

sober adj. **1** not drunk; serious. **2** (of colour) not bright. ● v. make or become sober. □ **soberly** adv., **sobriety** n.

sobriquet n. (also **soubriquet**) a nickname.

so-called adj. called by that name, but perhaps wrongly.

soccer n. football, Association football.

sociable adj. fond of company; characterized by friendly companionship. □ **sociably** adv., **sociability** n.

social adj. living in a community; of society or its organization; sociable. ● n. a social gathering. □ **socially** adv.

socialism n. a political and economic theory that resources, industries, and transport should be owned and managed by the state. □ **socialist** n., **socialistic** adj.

socialite n. a person prominent in fashionable society.

socialize v. (also **-ise**) behave sociably. □ **socialization** n.

social security n. state assistance for those who lack economic security.

social services n. welfare services provided by the State.

social worker n. a person trained to help people with social problems.

society n. **1** an organized community; a system of living in this. **2** people of the higher social classes. **3** company, companionship. **4** a group organized for a common purpose.

sociology n. the study of human society or of social problems. □ **sociological** adj., **sociologist** n.

sock n. **1** a covering for the foot; a loose insole. **2** (colloquial) a forceful blow. ● v. (colloquial) hit forcefully.

socket n. a hollow into which something fits.

sod n. **1** turf; a piece of this. **2** (vulgar) an unpleasant or awkward person or thing.

soda n. **1** a compound of sodium. **2** soda water.

soda water n. water made fizzy by being charged with carbon dioxide under pressure.

sodden adj. made very wet.

sodium n. a soft silver-white metallic element (symbol Na).

sodomy n. anal intercourse; bestiality. □ **sodomite** n.

sofa n. a long upholstered seat with a back.

sofa bed n. a sofa that can be converted into a bed.

soft adj. **1** not hard or firm or rough. **2** not loud. **3** gentle; tender-hearted; flabby, feeble. **4** (of drinks) non-alcoholic; (of drugs) not likely to cause addiction. **5** (of currency) likely to fall suddenly in value. □ **softly** adv., **softness** n. **soft-hearted** adj.

softball n. a form of baseball using a large soft ball.

soften v. make or become soft or softer. □ **softener** n.

soft fruit n. any small stoneless fruit (e.g. raspberry).

soft furnishings n.pl. cushions, curtains, rugs, etc.

softie n. (also **softy**) a person who is weak or tender-hearted.

soft option n. an easy alternative.

soft-pedal v. (**soft-pedalled**; Amer. **soft-pedaled**) refrain from emphasizing; be restrained.

soft spot n. a feeling of affection.

software n. computer programs; disks containing these.

softwood n. the soft wood of coniferous trees.

softy var. of **softie**.

soggy adj. (**soggier**) very wet; moist and heavy. □ **sogginess** n.

soh | solution

soh n. (*Music*) the fifth note of a major scale, or the note G.

soigné adj. well-groomed and sophisticated.

soil n. loose earth; ground as territory. ● v. make or become dirty.

soirée n. a social gathering in the evening for music etc.

sojourn n. a temporary stay. ● v. stay temporarily.

solace v. & n. comfort in distress.

solar adj. of or from the sun; reckoned by the sun.

solar cell n. a device converting solar radiation into electricity.

solar plexus n. a network of nerves at the pit of the stomach.

solar system n. the sun with the planets etc. that revolve round it.

sold see **sell**.

solder n. a soft alloy used to cement metal parts together. ● v. join with solder.

soldering iron n. a tool for melting and applying solder.

soldier n. a member of an army. ● v. serve as a soldier. □ **soldier on** (*colloquial*) persevere doggedly.

sole n. **1** the undersurface of a foot; the part of a shoe etc. covering this. **2** a flatfish used as food. ● v. put a sole on a shoe etc. ● adj. one and only; belonging exclusively to one person or group. □ **solely** adv.

solemn adj. not smiling or cheerful; formal and dignified. □ **solemnly** adv., **solemnity** n.

solemnize v. (also **-ise**) celebrate (a festival etc.); perform with formal rites. □ **solemnization** n.

solenoid n. a coil of wire magnetized by electric current.

sol-fa n. the system of syllables (doh, ray, me, etc.) representing the notes of a musical scale.

solicit v. seek to obtain by asking (for); (of a prostitute) offer sexual services. □ **solicitation** n.

solicitor n. a lawyer who advises clients and instructs barristers.

solicitous adj. anxious about a person's welfare or comfort. □ **solicitously** adv., **solicitude** n.

solid adj. **1** keeping its shape, firm; not liquid or gas. **2** not hollow. **3** of the same substance throughout; continuous. **4** of solids. **5** three-dimensional. **6** sound and reliable. ● n. a solid substance, body, or food. □ **solidly** adv., **solidity** n.

solidarity n. unity resulting from common aims or interests etc.

solidify v. make or become solid. □ **solidification** n.

solidus n. (pl. **solidi**) an oblique stroke (/).

soliloquize v. (also **-ise**) utter a soliloquy.

soliloquy n. a speech made aloud to oneself.

solitaire n. **1** a gem set by itself. **2** a game for one person played on a board with pegs.

solitary adj. alone; single; not frequented, lonely. ● n. a recluse.

solitude n. being solitary.

solo n. (pl. **solos**) music for a single performer; an unaccompanied performance etc. ● adj. & adv. unaccompanied, alone.

soloist n. the performer of a solo.

solstice n. either of the times (about 21 June and 22 Dec.) or points reached when the sun is furthest from the equator.

soluble adj. **1** able to be dissolved. **2** able to be solved. □ **solubility** n.

solution n. **1** a liquid containing something dissolved; the process of dissolving. **2** the process of solving a problem etc.; the answer found.

solve v. find the answer to. □ **solvable** adj.

solvent adj. **1** having enough money to pay one's debts etc. **2** able to dissolve another substance. ● n. a liquid used for dissolving something. □ **solvency** n.

somatic adj. of the body as distinct from the mind or spirit.

sombre adj. (Amer. **somber**) dark, gloomy.

sombrero n. (pl. **sombreros**) a hat with a very wide brim.

some adj. **1** an unspecified quantity or number of. **2** unknown, unnamed. **3** a considerable quantity. **4** approximately. **5** (colloquial) remarkable. ● pron. some people or things.

somebody n. & pron. **1** an unspecified person. **2** a person of importance.

somehow adv. in an unspecified or unexplained manner.

someone n. & pron. somebody.

somersault n. a leap or roll turning one's body upside down and over. ● v. move in this way.

something n. & pron. an unspecified thing or extent; an important or praiseworthy thing. □ **something like** rather like; approximately.

sometime adj. former. ● adv. at some unspecified time.

sometimes adv. at some times but not all the time.

somewhat adv. to some extent.

somewhere adv. at, in, or to an unspecified place.

somnambulist n. a sleepwalker. □ **somnambulism** n.

somnolent adj. sleepy; asleep. □ **somnolence** n.

son n. a male in relation to his parents.

sonar n. a device for detecting objects under water by reflection of sound waves.

sonata n. a musical composition for one instrument or two, usu. in several movements.

son et lumière (son ay loom-yair) n. a night-time entertainment dramatizing a historical event with lighting and sound effects.

song n. singing; music for singing. □ **going for a song** (colloquial) being sold very cheaply.

songbird n. a bird with a musical cry.

songster n. a singer; a songbird.

sonic adj. of sound waves.

sonic boom n. (also **sonic bang**) a loud noise caused by an aircraft travelling faster than the speed of sound.

son-in-law n. (pl. **sons-in-law**) a daughter's husband.

sonnet n. a poem of 14 lines.

sonorous adj. resonant, with a deep powerful sound. □ **sonorously** adv., **sonority** n.

soon adv. in a short time; early; readily. □ **sooner or later** at some time, eventually.

soot n. a black powdery substance in smoke. □ **sooty** adj.

soothe v. calm, ease (pain etc.). □ **soothing** adj.

soothsayer n. a prophet.

sop n. a concession to pacify a troublesome person. ● v. (**sopped**) dip in liquid; soak up (liquid).

sophisticated adj. **1** characteristic of or experienced in fashionable life and its ways. **2** complicated, elaborate. □ **sophistication** n.

sophistry n. (also **sophism**) clever and subtle but perhaps misleading reasoning. □ **sophist** n.

soporific adj. tending to cause sleep. ● n. a soporific drug etc.

sopping adj. very wet, drenched.

soppy adj. (**soppier**) (colloquial) sentimental in a sickly way. □ **soppiness** n.

soprano n. (pl. **sopranos**) the highest female or boy's singing voice.

sorbet (sor-bay) n. a flavoured water ice.

sorcerer n. a magician. □ **sorcery** n.

sordid adj. dirty, squalid; (of motives etc.) not honourable. □ **sordidly** adv., **sordidness** n.

sore adj. **1** causing or suffering pain from injury or disease. **2** distressed, vexed. ● n. a sore place; a source of distress or annoyance. □ **soreness** n.

sorely adv. very much, severely.

sorghum n. a tropical cereal plant.

sorrel n. a sharp-tasting herb. ● adj. reddish brown.

sorrow n. mental suffering caused by loss, disappointment, etc.; something causing this. ● v. feel sorrow, grieve. □ **sorrowful** adj., **sorrowfully** adv.

sorry adj. (**sorrier**) feeling pity or regret or sympathy; wretched.

sort n. a particular kind or variety. ● v. arrange according to type, size, destination, etc.

sortie n. a sally by troops from a besieged place; a flight of an aircraft on a military operation.

SOS n. an international distress signal; an urgent appeal for help.

sot n. a habitual drunkard.

sotto voce (sott-oh vo-chi) adv. in an undertone.

sou (soo) n. a former French coin of low value.

soubriquet var. of **sobriquet**.

soufflé n. a light dish made with beaten egg white.

sough ((suf) or rhymes with cow) n. & v. (make) a moaning or whispering sound as of wind in trees.

sought see **seek**.

souk n. a market place in Muslim countries.

soul n. **1** the spiritual or immortal element in a person; a person's mental or emotional nature. **2** a personification; she's the soul of discretion. **3** a person. **4** (also **soul music**) black American music with elements of rhythm and blues, rock, and gospel.

soulful adj. showing deep feeling, emotional. □ **soulfully** adv.

soulless adj. lacking sensitivity or noble qualities; dull.

sound n. **1** vibrations of air detectable by the ear; the sensation produced by these; what is or may be heard. **2** a strait. ● v. **1** produce or cause to produce sound; utter, pronounce; seem when heard. **2** test the depth of (a river or sea etc.); examine with a probe. ● adj. **1** healthy; not diseased or damaged; secure. **2** correct, well-founded; thorough. □ **sound off** express one's opinions loudly. □ **sounder** n., **soundly** adv., **soundness** n.

sound barrier n. the high resistance of air to objects moving at speeds near that of sound.

sounding board n. **1** a board to reflect sound or increase resonance. **2** a person used to test opinion.

soundproof adj. not able to be penetrated by sound. ● v. make soundproof.

soup n. liquid food made from meat, vegetables, etc. □ **soup up** (colloquial) increase the power of (an engine etc.). □ **soupy** adj.

soupçon (soop-son) n. a very small quantity.

soup kitchen n. a place where soup etc. is served free to the poor.

sour adj. **1** tasting sharp; not fresh, tasting or smelling stale. **2** bad-tempered. ● v. make or become sour. □ **sourly** adv., **sourness** n.

source *n.* the place from which something comes or is obtained; a river's starting point; a person or book etc. supplying information.

sourpuss *n.* (*colloquial*) a bad-tempered person.

souse *v.* steep in pickle; drench.

south *n.* the point or direction to the right of a person facing east; a southern part. ● *adj.* in the south; (of wind) from the south. ● *adv.* towards the south.

south-east *n.* the point or direction midway between south and east. □ **south-easterly** *adj.* & *n.*, **south-eastern** *adj.*

southerly *adj.* towards or blowing from the south.

southern *adj.* of or in the south.

southerner *n.* a native of the south.

southernmost *adj.* furthest south.

southpaw *n.* (*colloquial*) a left-handed person.

southward *adj.* towards the south. □ **southwards** *adv.*

south-west *n.* the point or direction midway between south and west. □ **south-westerly** *adj.* & *n.*, **south-western** *adj.*

souvenir *n.* something serving as a reminder of an incident or place visited.

sou'wester *n.* a waterproof hat with a broad flap at the back.

sovereign *n.* a king or queen who is the supreme ruler of a country. ● *adj.* supreme; (of a state) independent. □ **sovereignty** *n.*

Soviet *n.* & *adj.* (a citizen) of the former USSR.

sow¹ (soh) *v.* (**sowed, sown** or **sowed, sowing**) plant or scatter (seed) for growth; plant seed in; implant (ideas etc.). □ **sower** *n.*

sow² (*rhymes with* cow) *n.* an adult female pig.

soy *n.* = **soya**.

soya *n.* (in full **soya bean**) a plant from whose seed an edible oil and flour are obtained.

sozzled *adj.* (*colloquial*) drunk.

spa *n.* a place with a curative mineral spring.

space *n.* the boundless expanse in which all objects exist and move; a portion of this; an empty area or extent; the universe beyond the earth's atmosphere; an interval. ● *v.* arrange with spaces between.

spacecraft *n.* a vehicle for travelling in outer space.

spaceship *n.* a spacecraft.

spacious *adj.* providing much space, roomy. □ **spaciousness** *n.*

spade *n.* **1** a tool for digging, with a broad metal blade on a handle. **2** a playing card of the suit marked with black figures shaped like an inverted heart with a small stem.

spadework *n.* hard preparatory work.

spaghetti *n.* pasta made in long strings.

span *n.* the extent from end to end; the distance or part between the uprights of an arch or bridge. ● *v.* (**spanned**) extend across.

spangle *n.* a small piece of glittering material decorating a dress etc. ● *v.* cover with spangles or sparkling objects.

Spaniard *n.* a native of Spain.

spaniel *n.* a dog with drooping ears and a silky coat.

Spanish *adj.* & *n.* (the language) of Spain.

spank *v.* slap on the buttocks.

spanking *adj.* **1** brisk, lively. **2** (*colloquial*) striking; excellent.

spanner *n.* a tool for gripping and turning the nut on a screw etc.

spar *n.* a strong pole used as a ship's mast, yard, or boom. ● *v.*

(**sparred**) box, esp. for practice; quarrel, argue.

spare v. **1** afford to give. **2** refrain from hurting or harming. ● adj. **1** additional to what is usually needed or used, kept in reserve. **2** thin. ● n. an extra thing kept in reserve. □ **to spare** additional to what is needed.

sparing adj. economical, not generous or wasteful.

spark n. a fiery particle; a flash of light produced by an electrical discharge; a particle of energy, genius, etc.). ● v. give off spark(s).

sparkle v. shine with flashes of light; be lively or merry. ● n. a sparkling light.

sparkler n. a hand-held sparking firework.

sparkling adj. (of wine or mineral water) effervescent, fizzy.

spark plug n. (also **sparking plug**) a device for making a spark in an internal-combustion engine.

sparrow n. a small brownish-grey bird.

sparrowhawk n. a small hawk.

sparse adj. thinly scattered. □ **sparsely** adv., **sparseness** n.

spartan adj. (of conditions) simple and sometimes harsh.

spasm n. a strong involuntary contraction of a muscle; a sudden brief spell of activity or emotion etc.

spasmodic adj. of or occurring in spasms. □ **spasmodically** adv.

spastic adj. affected by cerebral palsy which causes jerky, involuntary movements. ● n. a person with this condition. □ **spasticity** n.

spat n. **1** a short gaiter. **2** a slight quarrel. See also **spit**.

spate n. a sudden flood.

spatial adj. of or existing in space. □ **spatially** adv.

spatter v. scatter or fall in small drops (on). ● n. a splash; the sound of spattering.

spatula n. a knife-like tool with a broad blunt blade; a medical instrument for pressing down the tongue.

spawn n. the eggs of fish, frogs, or shellfish. ● v. deposit spawn; generate.

spay v. sterilize (a female animal) by removing the ovaries.

speak v. (**spoke**, **spoken**, **speaking**) utter (words) in an ordinary voice; say something; converse; express by speaking.

-speak comb. form jargon; computerspeak.

speaker n. **1** a person who speaks, one who makes a speech. **2** a loudspeaker.

spear n. **1** a weapon for hurling, with a long shaft and pointed tip. **2** a pointed stem. ● v. pierce with or as if with a spear.

spearhead n. the foremost part of an advancing force. ● v. be the spearhead of.

spearmint n. a variety of mint.

spec n. □ **on spec** (colloquial) as a gamble, on the off chance.

special adj. **1** of a particular kind; for a particular purpose. **2** exceptional. □ **specially** adv.

specialist n. an expert in a particular branch of a subject.

speciality n. a special quality, product, or activity.

specialize v. (also **-ise**) **1** be or become a specialist. **2** adapt for a particular purpose. □ **specialization** n.

species n. (pl. **species**) a group of similar animals or plants which can interbreed.

specific adj. particular; exact, not vague. ● n. a specific aspect. □ **specifically** adv.

specification n. specifying; details describing a thing to be made or done.

specify v. mention definitely; include in specifications.

specimen n. a part or individual taken as an example or for examination or testing.

specious adj. seeming good or sound but lacking real merit. □ **speciously** adv., **speciousness** n.

speck n. a small spot or particle.

speckle n. a small spot, esp. as a natural marking. □ **speckled** adj.

specs n.pl. (colloquial) a pair of spectacles.

spectacle n. 1 an impressive sight; a lavish public show; a ridiculous sight. 2 (**spectacles**) a pair of lenses in a frame, worn in front of the eyes.

spectacular adj. impressive. ●n. a spectacular performance or production. □ **spectacularly** adv.

spectator n. a person who watches a game, incident, etc.

spectral adj. 1 of or like a spectre. 2 of the spectrum.

spectre n. (Amer. **specter**) a ghost; a haunting fear.

spectrum n. (pl. **spectra**) bands of colour or sound forming a series according to their wavelengths; an entire range of ideas etc.

speculate v. 1 form opinions by guessing. 2 buy in the hope of making a profit. □ **speculation** n., **speculative** adj., **speculator** n.

speculum n. (pl. **specula**) a medical instrument for looking into bodily cavities.

speech n. the act, power, or manner of speaking; spoken communication, esp. to an audience; language; dialect.

speechless adj. unable to speak because of emotion or shock.

speed n. the rate of time at which something moves or operates; rapidity. ●v. 1 (**sped,** speeding) move, pass, or send quickly. 2 (**speeded, speeding**) travel at an illegal speed. □ **speed up** accelerate.

speedboat n. a fast motor boat.

speedometer n. a device in a vehicle, showing its speed.

speedway n. 1 an arena for motorcycle racing. 2 (Amer.) a road for fast traffic.

speedy adj. (**speedier**) rapid. □ **speedily** adv., **speediness** n.

speleology n. the exploration and study of caves. □ **speleologist** n.

spell v. (**spelt** or **spelled,** spelling) 1 give in correct order the letters that form (a word). 2 produce as a result. ●n. 1 words supposed to have magic power; their influence; a fascination, an attraction. 2 a period of time, weather, or activity. □ **spell out** state explicitly. □ **speller** n.

spellbound adj. entranced.

spend v. (**spent, spending**) 1 pay out (money) in buying something. 2 use up; pass (time etc.). □ **spender** n.

spendthrift n. a wasteful spender.

sperm n. (pl. **sperms** or **sperm**) a male reproductive cell; semen.

spermatozoon (sper-mǎ-tŏ-zoh-ŏn) n. (pl. **spermatozoa**) the fertilizing cell of a male organism.

spermicidal adj. killing sperm.

sperm whale n. a large whale.

spew v. (also **spue**) vomit; cast out in a stream.

sphagnum n. moss growing on bogs.

sphere n. 1 a perfectly round solid geometric figure or object. 2 a field of action or influence etc.

spherical adj. shaped like a sphere.

sphincter n. a ring of muscle controlling an opening in the body.

sphinx n. **1** an ancient Egyptian statue with a lion's body and human or ram's head. **2** an enigmatic person.

spice n. a flavouring substance with a strong taste or smell; something that adds zest. ● v. flavour with spice. □ **spicy** adj.

spick and span adj. neat and clean.

spider n. a small creature with a segmented body and eight legs. □ **spidery** adj.

spiel (shpeel) n. (slang) a glib persuasive speech.

spigot n. a plug stopping the vent-hole of a cask or controlling the flow of a tap.

spike n. a pointed piece; a pointed piece of metal. ● v. **1** put spikes on; pierce or fasten with a spike. **2** (colloquial) add alcohol to (drink). □ **spiky** adj.

spill v. (**spilt** or **spilled**, **spilling**) cause or allow to run over the edge of a container; become spilt. ● n. **1** a thin strip of wood or paper for transferring flame. **2** a fall. □ **spillage** n.

spin v. (**spun**, **spinning**) **1** turn rapidly on its axis. **2** draw out and twist into threads; make (yarn etc.) in this way. ● n. **1** a spinning movement. **2** a short drive for pleasure. □ **spin out** prolong. □ **spinner** n.

spina bifida n. an abnormal condition in which the spine is not properly developed and is protruding.

spinach n. a vegetable with green leaves.

spinal adj. of the spine.

spindle n. a rod on which thread is wound in spinning; a revolving pin or axis.

spindly adj. (**spindlier**) long or tall and thin.

spindrift n. sea spray.

spine n. **1** the backbone. **2** a needle-like projection. **3** part of a book where the pages are hinged.

spineless adj. **1** having no backbone. **2** lacking determination.

spinet n. a small harpsichord.

spinnaker n. a large extra sail on a racing yacht.

spinneret n. the thread-producing organ in a spider, silkworm, etc.

spinney n. (pl. **spinneys**) a thicket.

spin-off n. an incidental benefit.

spinster n. an unmarried woman.

spiny adj. (**spinier**) full of spines, prickly.

spiracle n. an opening through which an insect breathes; the blowhole of a whale etc.

spiral adj. forming a continuous curve round a central point or axis. ● n. a spiral line or thing; a continuous increase or decrease in two or more quantities alternately. ● v. (**spiralled**; Amer. **spiraled**) move in a spiral course. □ **spirally** adv.

spire n. a tall pointed structure esp. on a church tower.

spirit n. **1** the mind as distinct from the body; the soul; a person's nature. **2** a ghost. **3** a characteristic quality; the intended meaning of a law etc. **4** liveliness, boldness. **5** (**spirits**) a feeling of cheerfulness or depression. **6** a strong distilled alcoholic drink. ● v. carry off swiftly and mysteriously.

spirited adj. lively, bold. □ **spiritedly** adv.

spirit level n. a sealed glass tube containing a bubble in liquid, used to test that things are horizontal.

spiritual adj. **1** of the human spirit or soul. **2** of the Church or religion. ● n. a religious folk

song of American blacks.
□ **spiritually** adv., **spirituality** n.

spiritualism n. attempted communication with spirits of the dead. □ **spiritualist** n., **spiritualistic** adj.

spirituous adj. strongly alcoholic.

spirogyra n. a freshwater alga with spiral bands of chlorophyll.

spit v. (**spat** or **spit**, **spitting**) 1 eject from the mouth; eject saliva. 2 (of rain) fall lightly. ● n. 1 spittle; the act of spitting. 2 a metal spike holding meat while it is roasted. 3 a narrow strip of land projecting into the sea.

spite n. malicious desire to hurt or annoy someone. ● v. hurt or annoy from spite. □ **in spite of** not being prevented by. □ **spiteful** adj., **spitefully** adv., **spitefulness** n.

spitfire n. a fiery-tempered person.

spittle n. saliva.

splash v. 1 cause (liquid) to fly about in drops; move, fall, or wet with such drops. 2 decorate with irregular patches of colour. 3 display in large print. 4 spend (money) freely. ● n. 1 splashing; an act or mark made by this. 2 a patch of colour or light. 3 a striking display.

splatter v. splash, spatter. ● n. a splash.

splay v. spread apart; slant outwards or inwards.

spleen n. 1 an abdominal organ involved in maintaining the proper condition of the blood. 2 bad temper; peevishness.

splendid adj. brilliant, very impressive; excellent. □ **splendidly** adv.

splendour n. (Amer. **splendor**) a splendid appearance.

splenetic adj. bad-tempered, peevish.

splice v. join by interweaving or overlapping the ends.

splint n. a rigid framework preventing a limb etc. from movement, e.g. while a broken bone heals. ● v. secure with a splint.

splinter n. a thin sharp piece of broken wood etc. ● v. break into splinters.

splinter group n. a small group that has broken away from a larger one.

split v. (**split**, **splitting**) break or come apart, esp. lengthwise; divide, share. ● n. splitting; a split thing or place; (**splits**) an acrobatic position with legs stretched fully apart.

split infinitive n. an infinitive with a word placed between to and the verb.

■ **Usage** Split infinitives, as in I want to quickly sum up and Your job is to really get to know everybody, are common in informal English, but many people consider them incorrect and prefer I want quickly to sum up or I want to sum up quickly. They should therefore be avoided in formal English, but note that just changing the order of words can alter the meaning, e.g. Your job is really to get to know everybody.

split second n. a very brief moment.

splodge n. a splash, a blotch. ● v. make a splodge.

splotch v. & n. = **splodge**.

splurge n. a spending spree; an ostentatious display. ● v. make a splurge, spend money freely.

splutter v. make a rapid series of spitting sounds; speak or utter incoherently. ● n. a spluttering sound.

spoil v. (**spoilt** or **spoiled**, **spoiling**) 1 make useless or unsatisfactory; become unfit for use. **2** harm the character of (esp. a child) by being indulgent. ● *n.* (also **spoils**) plunder. □ **be spoiling for** desire (a fight etc.) eagerly.

spoiler *n.* a device that slows down an aircraft by interrupting the air flow; a similar device on a vehicle, preventing it from being lifted off the road at speed.

spoilsport *n.* a person who spoils others' enjoyment.

spoke[1] *n.* any of the bars connecting the hub to the rim of a wheel.

spoke[2], **spoken** see **speak**.

spokesman *n.* (pl. **-men**) a person who speaks on behalf of a group.

spokesperson *n.* (pl. **-persons** or **-people**) a person who speaks on behalf of a group.

spokeswoman *n.* (pl. **-women**) a woman who speaks on behalf of a group.

sponge *n.* **1** a water animal with a porous structure; its skeleton, or a similar substance, esp. used for washing, cleaning, or padding. **2** a sponge cake. ● *v.* **1** wipe or wash with a sponge. **2** live off the generosity of others. □ **spongeable** *adj.*, **spongy** *adj.*

sponge cake *n.* a cake with a light open texture.

sponger *n.* a person who sponges on others.

spongiform *adj.* of or like a sponge.

sponsor *n.* **1** a person who provides funds for a broadcast, sporting event, etc.; a person who gives to charity in return for another's activity. **2** a person who puts forward a proposal. **3** a godparent. ● *v.* act as sponsor for. □ **sponsorship** *n.*

spontaneous *adj.* resulting from natural impulse; not caused or suggested from outside. □ **spontaneously** *adv.*, **spontaneity** *n.*

spoof *n.* (*colloquial*) a hoax; a parody.

spook *n.* (*colloquial*) a ghost. □ **spooky** *adj.*

spool *n.* a reel on which something is wound. ● *v.* wind on a spool.

spoon *n.* a food utensil with a rounded bowl and a handle; the amount it contains. ● *v.* take or lift with a spoon. □ **spoonful** *n.*

spoonerism *n.* the interchange of the initial sounds of two words, e.g. *he's a boiled sprat.*

spoon-feed *v.* (**spoon-fed**, **spoon-feeding**) **1** feed from a spoon. **2** give excessive help to.

spoor *n.* a track or scent left by an animal.

sporadic *adj.* occurring here and there or now and again. □ **sporadically** *adv.*

spore *n.* one of the tiny reproductive cells of fungi, ferns, etc.

sporran *n.* a pouch worn in front of a kilt.

sport *n.* **1** an athletic (esp. outdoor) activity; game(s), pastime(s). **2** (*colloquial*) a sportsmanlike person. ● *v.* **1** play, amuse oneself. **2** wear.

sporting *adj.* **1** concerning or interested in sport. **2** sportsmanlike, generous.

sporting chance *n.* a reasonable chance of success.

sportive *adj.* playful. □ **sportively** *adv.*

sports car *n.* an open low-built fast car.

sports jacket *n.* a man's jacket for informal wear.

sportsman *n.* (pl. **-men**) **1** a man who takes part in sports. **2** a fair and generous person. □ **sportsmanlike** *adj.*, **sportsmanship** *n.*

sportswoman *n.* (pl. **-women**) a woman who takes part in sports.

spot *n.* **1** a round mark or stain; a pimple; a drop. **2** a place. **3** (*colloquial*) a small amount. **4** a spotlight. ● v. (**spotted**) **1** mark with a spot or spots. **2** (colloquial) notice; watch for and take note of. □ **on the spot** without delay or change of place; at the scene of action. **put on the spot** (colloquial) compel (a person) to take action or justify himself or herself. □ **spotter** n.

spot check n. a random check.

spotless adj. free from stain or blemish, perfectly clean. □ **spotlessly** adj.

spotlight n. a lamp or its beam directed on a small area. ● v. (**spotlit** or **spotlighted**, **spotlighting**) direct a spotlight on; draw attention to.

spotty adj. (**spottier**) marked with spots.

spouse n. a husband or wife.

spout n. a projecting tube through which liquid is poured or conveyed; a jet of liquid. ● v. **1** come or send out forcefully as a jet of liquid. **2** utter or speak lengthily.

sprain v. injure by wrenching violently. ● n. this injury.

sprang see spring.

sprat n. a small herring-like fish.

sprawl v. sit, lie, or fall with arms and legs spread loosely; spread out irregularly. ● n. a sprawling attitude or arrangement.

spray n. **1** liquid dispersed in very small drops; a liquid for spraying; a device for spraying liquid. **2** a bunch of cut flowers; a branch with its leaves and flowers; an ornament in similar form. ● v. come or send out as spray; wet with liquid in this way. □ **sprayer** n.

spray-gun n. a device for spraying paint etc.

spread v. (**spread**, **spreading**) **1** open out; become longer or wider. **2** apply as a layer. **3** be able to be spread. **4** make or become widely known or felt or

suffered; distribute, become distributed. ● n. **1** spreading. **2** (colloquial) a lavish meal. **3** a thing's range. **4** a paste for spreading on bread.

spreadeagled adj. with arms and legs spread.

spreadsheet n. a computer program that manipulates figures in tables.

spree n. (colloquial) a lively outing, some fun.

sprig n. a twig, a shoot.

sprightly adj. (**sprightlier**) lively, full of energy. □ **sprightliness** n.

spring v. (**sprang**, **sprung**, **springing**) **1** jump; move rapidly. **2** issue, arise; produce or cause to operate suddenly. ● n. **1** an act of springing, a jump. **2** a device that reverts to its original position after being compressed, tightened, or stretched; elasticity. **3** a place where water or oil flows naturally from the ground. **4** the season between winter and summer.

springboard n. a flexible board giving impetus to a gymnast or diver.

springbok n. a South African gazelle.

spring-clean v. clean (one's home etc.) thoroughly.

spring roll n. a fried pancake filled with vegetables.

spring tide n. a tide when there is the largest rise and fall of water.

springtime n. the season of spring.

springy adj. (**springier**) able to spring back easily after being squeezed, tightened, or stretched. □ **springiness** n.

sprinkle v. scatter or fall in drops or particles on (a surface). ● n. a light shower. □ **sprinkler** n.

sprinkling n. something sprinkled; a few here and there.

sprint v. run or swim etc. at full speed. ● n. a brief run, swim, etc. of this kind. □ **sprinter** n.

sprite n. an elf, fairy, or goblin.

spritzer n. a drink of white wine and soda water.

sprocket n. a projection engaging with links on a chain etc.

sprout v. begin to grow or appear; put forth. ● n. 1 a plant's shoot. 2 a Brussels sprout.

spruce adj. neat, smart. ● v. smarten. ● n. a fir tree. □ **sprucely** adv., **spruceness** n.

sprung see **spring**. adj. fitted with springs.

spry adj. active, lively. □ **spryly** adv., **spryness** n.

spud n. 1 (slang) a potato. 2 a narrow spade.

spue var. of **spew**.

spume n. froth.

spun see **spin**.

spur n. 1 a pricking device worn on a horse-rider's heel; a stimulus, an incentive. 2 a projection. ● v. (**spurred**) urge on (a horse) with one's spurs; urge on, incite; stimulate. □ **on the spur of the moment** on impulse. **win one's spurs** prove one's ability.

spurious adj. not genuine or authentic. □ **spuriously** adv., **spuriousness** n.

spurn v. reject contemptuously.

spurt v. gush; send out (liquid) suddenly; increase speed suddenly. ● n. a sudden gush; a short burst of activity; a sudden increase in speed.

sputter v. splutter. ● n. a spluttering sound.

sputum n. spittle; matter that is spat out.

spy n. a person who secretly watches or gathers information. ● v. 1 catch sight of. 2 be a spy, watch secretly.

sq. abbr. square.

squabble v. quarrel pettily or noisily. ● n. a quarrel of this kind.

squad n. a small group working together.

squadron n. a division of a cavalry unit or an airforce; a detachment of warships.

squalid adj. dirty and unpleasant; morally degrading. □ **squalidly** adv., **squalor** n.

squall n. a sudden storm or wind. □ **squally** adj.

squander v. spend wastefully.

square n. 1 a geometric figure with four equal sides and four right angles; an area or object shaped like this. 2 the product of a number multiplied by itself. ● adj. 1 of square shape. 2 right-angled. 3 of or using units expressing the measure of an area. 4 properly arranged; equal, not owed or owing anything; straightforward; honest. ● adv. squarely, directly. ● v. 1 make right-angled. 2 mark with squares. 3 place evenly. 4 multiply by itself. 5 settle (an account etc.). 6 make or be consistent. 7 (colloquial) bribe. □ **square up to** face in a fighting attitude; face resolutely. □ **squarely** adv., **squareness** n.

square dance n. a dance in which four couples face inwards from four sides.

square meal n. a substantial meal.

square root n. the number of which a given number is the square.

squash v. 1 crush; squeeze or become squeezed flat or into pulp. 2 suppress; silence with a crushing reply. ● n. 1 a crowd of people squashed together. 2 a fruit-flavoured soft drink. 3 (in full **squash rackets**) a game played with rackets and a small ball in a closed court. 4 a vegetable gourd. □ **squashy** adj.

squat v. (**squatted**) **1** sit on one's heels; crouch; (*colloquial*) sit down. **2** be a squatter. ● n. **1** a squatting posture. **2** a place occupied by squatters. ● adj. dumpy.

squatter n. a person who takes unauthorized possession of an unoccupied building.

squawk n. a loud harsh cry. ● v. make a squawk.

squeak n. a short high-pitched cry or sound. ● v. make or utter with a squeak. □ **narrow squeak** a narrow escape. □ **squeaky** adj.

squeal n. a long shrill cry or sound. ● v. **1** make or utter with a squeal. **2** (*slang*) become an informer.

squeamish adj. **1** easily sickened or disgusted. **2** overscrupulous. □ **squeamishly** adv., **squeamishness** n.

squeeze v. exert pressure on; treat in this way to extract moisture; force into or through, force one's way; crowd; produce by pressure, effort, or compulsion; extort money etc. from. ● n. squeezing; an affectionate clasp or hug; drops produced by squeezing; a crowd, a crush; restrictions on borrowing. □ **squeezer** n.

squelch v. & n. (make) a sound like someone treading in thick mud.

squib n. a small exploding firework.

squid n. a sea creature with ten arms round its mouth.

squiffy adj. (**squiffier**) (*slang*) slightly drunk.

squiggle n. a short curly line. □ **squiggly** adj.

squint v. have one eye turned abnormally from the line of gaze of the other; look sideways or through a small opening. ● n. a squinting condition of one eye; a sideways glance.

squire n. a country gentleman, esp. a landowner.

squirm v. wriggle; feel embarrassed. ● n. wriggle.

squirrel n. a small tree-climbing animal with a bushy tail.

squirt v. send out (liquid) or be sent out in a jet; wet in this way. ● n. a jet of liquid.

squish v. & n. (move with) a soft squelching sound. □ **squishy** adj.

Sr symb. strontium.

SRN abbr. State Registered Nurse.

SS abbr. **1** steamship. **2** (*hist.*) the Nazi special police force.

St abbr. Saint.

St. abbr. Street.

st. abbr. stone (in weight).

stab v. (**stabbed**) pierce, wound or kill with something pointed; poke. ● n. **1** an act of stabbing; a sensation of being stabbed. **2** (*colloquial*) an attempt.

stabilize v. (also **-ise**) make or become stable. □ **stabilization** n., **stabilizer** n.

stabilizer n. (also **-iser**) a device to prevent a ship from rolling.

stable n. a building in which horses are kept; an establishment for training racehorses; the horses, people, etc. from the same establishment. ● v. put or keep in a stable. ● adj. firmly fixed or established, not easily shaken or destroyed. □ **stably** adv., **stability** n.

staccato adv. (*Music*) with each sound sharply distinct.

stack n. **1** an orderly pile or heap; (*colloquial*) a large quantity. **2** a tall chimney. **3** a storage section of a library. ● v. arrange in a stack or stacks; arrange (cards) secretly for cheating; cause (aircraft) to fly at different levels while waiting to land.

stadium n. a sports ground surrounded by tiers of seats for spectators.

staff *n.* **1** a stick used as a weapon, support, or symbol of authority. **2** the people employed by an organization, esp. those doing administrative work. **3** (*pl.* **staffs** or **staves**) a set of five horizontal lines on which music is written. ●*v.* provide with a staff of people.

stag *n.* a fully grown male deer.

stag beetle *n.* a beetle with branched projecting mouthparts.

stage *n.* **1** a raised floor or platform; one on which plays etc. are performed; (**the stage**) the theatrical profession. **2** a division of point reached in a process or journey. ●*v.* **1** present on the stage. **2** arrange and carry out. □ **go on the stage** become an actor or actress.

stagecoach *n.* (*hist.*) a large horse-drawn coach running on a regular route by stages.

stage fright *n.* nervousness on facing an audience.

stagflation *n.* a state of inflation without an increase in demand and employment.

stagger *v.* **1** move or go unsteadily. **2** shock deeply. **3** arrange so as not to coincide exactly. ●*n.* a staggering movement.

staggering *adj.* astonishing.

staging *n.* scaffolding; a temporary platform; a platform for plants to stand on in a greenhouse.

stagnant *adj.* not flowing, still and stale; not developing; without activity. □ **stagnancy** *n.*

stagnate *v.* be or become stagnant. □ **stagnation** *n.*

stag-night *n.* an all-male party for a man about to marry.

staid *adj.* steady and serious.

stain *v.* be or become discoloured; blemish; colour with a penetrating pigment. ●*n.* a mark caused by staining; a blemish; a liquid for staining things.

stainless *adj.* free from stains.

stainless steel *n.* a steel alloy not liable to rust or tarnish.

stair *n.* one of a flight of fixed indoor steps; (**stairs**) a flight of these.

staircase *n.* stairs and their supporting structure.

stairway *n.* a staircase.

stairwell *n.* the space for a staircase.

stake *n.* **1** a pointed stick or post for driving into the ground. **2** money etc. wagered; a share or interest in an enterprise etc. ●*v.* **1** fasten, support, or mark with a stake or stakes. **2** wager. □ **at stake** being risked. **stake a claim** claim a right to something. **stake out** place under surveillance.

stalactite *n.* a deposit of calcium carbonate hanging like an icicle.

stalagmite *n.* a deposit of calcium carbonate standing like a pillar.

stale *adj.* not fresh; unpleasant or uninteresting from lack of freshness; spoilt by too much practice. ●*v.* make or become stale. □ **staleness** *n.*

stalemate *n.* a drawn position in chess; a drawn contest; deadlock. ●*v.* bring to such a state.

stalk *n.* a stem or similar supporting part. ●*v.* walk in a stately or imposing manner; track or pursue stealthily. □ **stalker** *n.*

stalking-horse *n.* a pretext concealing one's real intentions.

stall *n.* **1** a stable, a cowhouse; a compartment in this. **2** a ground floor seat in a theatre. **3** a booth or stand where goods are displayed for sale. **4** a stalling of an aircraft. ●*v.* **1** place or keep in a stall. **2** (of an engine) stop suddenly through lack of power; (of an aircraft) begin to drop because the speed is too low; cause to stall. **3** play for time when being questioned.

stallion n. an uncastrated male horse.

stalwart adj. sturdy; strong and faithful. ● n. a stalwart person.

stamen n. the pollen-bearing part of a flower.

stamina n. the ability to withstand long physical or mental strain.

stammer v. speak with involuntary pauses or repetitions of a syllable. ● n. this act or tendency.

stamp v. **1** bring (one's foot) down heavily on the ground. **2** press so as to cut or leave a mark or pattern. **3** fix a postage stamp to. **4** give a specified character to. ● n. **1** the act or sound of stamping. **2** an instrument for stamping a mark etc., this mark. **3** a small adhesive label for affixing to an envelope or document to show the amount paid as postage or a fee etc. **4** a characteristic feature. □ **stamp out** suppress by force.

stampede n. a sudden rush of animals or people. ● v. (cause to) take part in a stampede.

stamping ground n. a person's usual haunt.

stance n. a manner of standing.

stanch v. (also **staunch**) stop or slow down the flow (of blood) from a wound.

stanchion n. an upright post or support.

stand v. (**stood, standing**) **1** have, take, or keep a stationary upright position; be situated; place, set upright. **2** stay firm or valid. **3** offer oneself for election. **4** endure. **5** provide at one's own expense. ● n. **1** a sta-

tionary condition; a position taken up. **2** resistance to attack. **3** a rack, a pedestal. **4** a raised structure with seats at a sports ground etc. **5** a stall for goods. □ **stand a chance** have a chance of success. **stand by** look on without interfering; stand ready for action; support in a difficulty; keep to (a promise etc.). **stand down** withdraw. **stand for** represent; (colloquial) tolerate. **stand in** deputize. **stand one's ground** not yield. **stand to reason** be logical. **stand up** come to or place in a standing position; be valid; (colloquial) fail to keep an appointment with. **stand up for** speak in defence of. **stand up to** resist courageously; be strong enough to endure.

standard n. **1** a thing against which something may be compared for testing or measurement; an average quality; a required level of quality or proficiency. **2** a distinctive flag. **3** a moral principle. ● adj. serving as or conforming to a standard; of average or usual quality.

standardize v. (also **-ise**) cause to conform to a standard. □ **standardization** n.

standard lamp n. a household lamp set on a tall support.

stand-by adj. & n. (a person or thing) available as a substitute.

stand-in n. a deputy, a substitute.

standing n. **1** status. **2** duration.

standing order n. an instruction to a bank to make regular payments.

stand-offish adj. cold or distant in manner.

standpipe n. a vertical pipe for fluid to rise in, esp. for attachment to a water main.

standpoint n. a point of view.

standstill n. inability to proceed.

stank *see* **stink**.

stanza *n.* a verse of poetry.

staphylococcus *n.* (*pl.* **staphylococci**) a pus-producing bacterium.

staple *n.* **1** a U-shaped spike for holding something in place; a piece of wire driven into papers and clenched to fasten them. **2** a principal or standard food or product etc. ● *adj.* principal, most important. ● *v.* secure with staple(s). □ **stapler** *n.*

star *n.* **1** a heavenly body appearing as a point of light. **2** an asterisk; a star-shaped mark indicating a category of excellence. **3** a famous actor, performer, etc. ● *v.* (**starred**) **1** put an asterisk beside (an item). **2** present or perform as a star actor.

starboard *n.* the right-hand side of a ship or aircraft.

starch *n.* **1** a white carbohydrate. **2** a preparation for stiffening fabrics; stiffness of manner. ● *v.* stiffen with starch. □ **starchy** *adj.*

stardom *n.* being a star actor etc.

stare *v.* gaze fixedly esp. in astonishment. ● *n.* a staring gaze.

starfish *n.* a star-shaped sea creature.

stark *n.* **1** desolate, bare. **2** sharply evident; downright. ● *adv.* completely. □ **starkly** *adv.*, **starkness** *n.*

starling *n.* a noisy bird with glossy black speckled feathers.

starry *adj.* (**starrier**) set with stars; shining like stars.

starry-eyed *adj.* (*colloquial*) romantically enthusiastic.

START *abbr.* Strategic Arms Reduction Treaty.

start *v.* **1** begin, cause to begin; begin operating; begin a journey. **2** make a sudden movement, esp. from pain or surprise. ● *n.* **1** a beginning; the place where a race etc. starts; an advantage gained or allowed in starting. **2** a sudden movement of pain or surprise. ● **starter** *n.*

startle *v.* shock, surprise.

starve *v.* die or suffer acutely from lack of food; cause to do this; force by starvation; (*colloquial*) feel very hungry. □ **starvation** *n.*

stash *v.* (*colloquial*) hide away.

state *n.* **1** a mode of being, with regard to characteristics or circumstances; (*colloquial*) an excited or agitated condition of mind. **2** a grand imposing style. **3** (often **State**) a political community under one government or forming part of a federation; civil government. ● *adj.* of or involving the State; ceremonial. ● *v.* express in words; specify.

stateless *adj.* not a citizen or subject of any country.

stately *adj.* (**statelier**) dignified, grand. □ **stateliness** *n.*

statement *n.* the process of stating; a thing stated; a formal account of facts; a written report of a financial account.

stateroom *n.* a room used on ceremonial occasions; a passenger's private compartment on a ship.

statesman *n.* (*pl.* **-men**) an experienced and respected political leader. □ **statesmanship** *n.*

stateswoman *n.* (*pl.* **-women**) a woman who is an experienced and respected political leader.

static *adj.* of force acting by weight without motion; stationary; not changing. ● *n.* electrical disturbances in the air, causing interference in telecommunications; (in full **static electricity**) electricity present in a body, not flowing as current.

statics n. the branch of mechanics concerned with bodies at rest or forces in equilibrium.

station n. **1** a place where a person or thing stands or is stationed; a place where a public service or specialized activity is based. **2** a broadcasting channel. **3** a stopping place on a railway. **4** status. ● v. put at or in a certain place for a purpose.

stationary adj. not moving; not movable.

stationer n. a dealer in stationery.

stationery n. writing paper, envelopes, labels, etc.

station wagon n. (Amer.) an estate car.

statistic n. an item of information expressed in numbers; (**statistics**) the science of collecting and interpreting numerical information. □ **statistical** adj., **statistically** adv.

statistician n. an expert in statistics.

statue n. a sculptured, cast, or moulded figure.

statuesque adj. like a statue in size, dignity, or stillness.

statuette n. a small statue.

stature n. bodily height; greatness gained by ability or achievement.

status n. a person's position or rank in relation to others; high rank or prestige.

status quo n. the existing state of affairs.

statute n. a law passed by Parliament or a similar body.

statutory adj. fixed, done, or required by statute.

staunch adj. firm in opinion or loyalty. ● v. var. of **stanch**. □ **staunchly** adv.

stave n. **1** one of the strips of wood forming the side of a cask or tub. **2** a staff in music. ● v. (**stove** or **staved**, **staving**) dent, break a hole in. □ **stave off** (**staved**) ward off.

stay v. **1** continue in the same place or state; dwell temporarily. **2** postpone. **3** show endurance. ● n. **1** a period of staying. **2** a postponement. □ **stay the course** be able to reach the end of it.

STD abbr. subscriber trunk dialling.

stead (sted) n. □ **in a person's** or **thing's stead** instead of him, her, or it. **stand in good stead** be of great service to.

steadfast adj. firm and not changing or yielding. □ **steadfastly** adv.

steady adj. (**steadier**) **1** not shaking. **2** regular, uniform. **3** dependable, not excitable. ● adv. steadily. ● v. make or become steady. □ **steadily** adv., **steadiness** n.

steak n. a thick slice of meat (esp. beef) or fish, usu. grilled or fried.

steal v. (**stole, stolen, stealing**) **1** take dishonestly. **2** move stealthily. ● n. (colloquial) **1** stealing. **2** a bargain, an easy task. □ **steal the show** outshine other performers.

stealth n. secrecy.

stealthy adj. (**stealthier**) quiet so as to avoid notice. □ **stealthily** adv., **stealthiness** n.

steam n. gas into which water is changed by boiling; this as motive power; energy, power. ● v. give out steam; cook or treat by steam; move by the power of steam; cover or become covered by steam. □ **steamy** adj.

steamer n. **1** a steam-driven ship. **2** a container in which things are cooked or heated by steam.

steamroller n. a heavy engine with a large roller, used in roadmaking.

steatite n. a greyish talc that feels smooth and soapy.

steed n. (*poetic*) a horse.

steel n. a very strong alloy of iron and carbon; a tapered steel rod for sharpening knives. ● v. make resolute. □ **steely** adj.
steeliness n.

steel wool n. a mass of fine shavings of steel used as an abrasive.

steep adj. **1** sloping sharply not gradually. **2** (*colloquial*) (of prices) unreasonably high. ● v. soak in liquid; permeate thoroughly. □ **steeply** adv., **steepness** n.

steeple n. a tall tower with a spire, rising above a church roof.

steeplechase n. a race for horses or athletes with fences to jump. □ **steeplechaser** n., **steeplechasing** n.

steeplejack n. a person who climbs tall chimneys etc. to do repairs.

steer v. direct the course of, guide by a mechanism; be able to be steered. ● n. a bullock.

stellar adj. of a star or stars.

stem n. **1** the supporting usu. cylindrical part, esp. of a plant. **2** (*Grammar*) the main usu. unchanging part of a noun or verb. ● v. (**stemmed**) restrain the flow of; dam. □ **stem from** have as its source.

stench n. a foul smell.

stencil n. a sheet of card etc. with a cut-out design, painted over to produce this on the surface below; the design produced. ● v. (**stencilled**; *Amer.* **stenciled**) produce or ornament in this way.

stenography n. shorthand. □ **stenographer** n.

stentorian adj. (of a voice) extremely loud.

step v. (**stepped**) lift and set down a foot or alternate feet; move a short distance in this way; progress. ● n. **1** a movement of a foot and leg in stepping; the distance covered in this way; a short distance; a pattern of steps in dancing. **2** one of a series of actions; a stage in a scale. **3** a level surface for placing the foot on in climbing; (**steps**) a stepladder. □ **in step** stepping in time with others; conforming. **mind** or **watch one's step** take care. **step in** intervene; enter. **step up** increase.

step- *comb. form* related by remarriage of a parent, as *stepfather, stepmother, stepson*, etc.

stepladder n. a short ladder with a supporting framework.

steppe n. a grassy plain, esp. in south-east Europe and Siberia.

stepping stone n. a raised stone for stepping on in crossing a stream etc; a means of progress.

stereo n. (*pl.* **stereos**) stereophonic sound; a stereophonic hi-fi system; a stereoscope.

stereophonic adj. using two transmission channels so as to give the effect of naturally distributed sound. □ **stereophony** n.

stereoscope n. a device for giving a three-dimensional effect. □ **stereoscopic** adj.

stereotype n. a standardized conventional idea or character etc. ● v. standardize; cause to conform to a preconceived type. □ **stereotyped** adj.

sterile adj. **1** unable to produce fruit or young. **2** free from micro-organisms. □ **sterility** n.

sterilize v. (also **-ise**) make sterile. □ **sterilization** n., **sterilizer** n.

sterling n. British money. ● adj. of standard purity; excellent.

stern adj. strict, severe. ● n. the rear of a ship or aircraft. □ **sternly** adv., **sternness** n.

sternum n. (pl. **sternums** or **sterna**) the breastbone.

steroid n. any of a group of organic compounds that includes certain hormones.

stertorous adj. making a snoring or rasping sound.

stet v. (placed by a word that has been crossed out etc.) ignore the alteration.

stethoscope n. an instrument for listening to sounds within the body.

stetson n. a hat with a wide brim and a high crown.

stevedore n. a docker.

stew v. 1 cook by simmering in a closed pot. 2 (colloquial) swelter. ● n. 1 a dish made by stewing meat etc. 2 (colloquial) a state of great anxiety.

steward n. 1 a person employed to manage an estate etc. 2 a passengers' attendant on a ship, aircraft, or train. 3 an official at a race meeting or show etc.

stewardess n. a female attendant on an aircraft, ship, etc.

stick v. (**stuck**, **sticking**) 1 thrust (a thing) into something; (colloquial) put. 2 fix or be fixed by glue or suction etc.; jam; (colloquial) remain in a specified place, not progress. 3 (colloquial) endure. ● n. 1 a thin piece of wood; something shaped like this; a walking stick; an implement used to propel the ball in hockey, polo, etc. 2 (colloquial) criticism. □ **stick out** stand above the surrounding surface; be conspicuous. **stick to** remain faithful to; keep to (a subject, position, etc.). **stick together** (colloquial) remain united or loyal. **stick to one's guns** not yield. **stick up for** (colloquial) stand up for.

sticker n. an adhesive label or sign.

sticking plaster n. an adhesive fabric for covering small cuts.

stick-in-the-mud n. (colloquial) a person who will not adopt new ideas etc.

stickleback n. a small fish with sharp spines on its back.

stickler n. □ **a stickler for** a person who insists on (something).

sticky adj. (**stickier**) 1 sticking to what is touched; humid. 2 (colloquial) unpleasant, difficult. □ **stickily** adv., **stickiness** n.

stiff adj. 1 not bending or moving easily. 2 difficult. 3 formal in manner. 4 (of wind) blowing briskly; (of a drink etc.) strong; (of a price or penalty) severe. ● n. (slang) a corpse. □ **stiffly** adv., **stiffness** n.

stiffen v. make or become stiff. □ **stiffener** n.

stiff-necked adj. obstinate; haughty.

stifle v. feel or cause to feel unable to breathe; suppress.

stigma n. 1 a mark of shame. 2 part of a flower pistil.

stigmata n.pl. marks corresponding to the Crucifixion marks on Christ's body.

stigmatize v. (also **-ise**) brand as something disgraceful. □ **stigmatization** n.

stile n. steps or bars for people to climb over a fence.

stiletto n. (pl. **stilettos**) a dagger with a narrow blade.

stiletto heel n. a long tapering heel of a shoe.

still adj. 1 with little or no motion or sound. 2 (of drinks) not fizzy. ● n. 1 silence and calm. 2 a photograph taken from a cinema film. 3 a distilling apparatus. ● adv. 1 without moving. 2 then or now as before. 3 nevertheless. 4 in a greater amount or degree. □ **stillness** n.

stillborn adj. born dead.

still life n. a picture of inanimate objects.

stilted adj. stiffly formal.

stilts *n.pl.* a pair of poles with footrests, enabling the user to walk above the ground; piles or posts on which a building stands.

stimulant *adj.* stimulating. ● *n.* a stimulating drug or drink.

stimulate *v.* make more active; apply a stimulus to. □ **stimulation** *n.*, **stimulative** *adj.*, **stimulator** *n.*

stimulus *n.* (*pl.* **stimuli**) something that rouses a person or thing to activity or energy.

stimy var. of **stymie**.

sting *n.* a sharp wounding part of an insect, nettle, etc.; a wound made by this; sharp bodily or mental pain. ● *v.* (**stung**, **stinging**) 1 wound or affect with a sting; feel or cause sharp pain. 2 (*slang*) overcharge; extort money from.

stingy *adj.* (**stingier**) spending or given grudgingly or in small amounts. □ **stingily** *adv.*, **stinginess** *n.*

stink *n.* 1 an offensive smell. 2 (*colloquial*) a row or fuss. ● *v.* (**stank** or **stunk**, **stinking**) 1 give off a stink. 2 (*colloquial*) seem very unpleasant or dishonest.

stinker *n.* (*slang*) an objectionable person; a difficult task.

stinking *adj.* smelling horrible; (*slang*) objectionable.

stint *v.* restrict to a small allowance. ● *n.* an allotted amount of work.

stipend (sty-pend) *n.* a salary.

stipendiary *adj.* receiving a stipend.

stipple *v.* paint, draw, or engrave in small dots. ● *n.* this process or effect.

stipulate *v.* demand or insist (on) as part of an agreement. □ **stipulation** *n.*

stir *v.* (**stirred**) 1 move. 2 mix (a substance) by moving a spoon etc. round in it. 3 stimulate, ex-cite. ● *n.* 1 the act or process of stirring. 2 a commotion, excitement.

stirrup *n.* a support for a rider's foot, hanging from the saddle.

stitch *n.* 1 a single movement of a thread in and out of fabric in sewing, or of a needle in knitting or crochet; the loop made in this way; a method of making a stitch. 2 a sudden pain in the side. ● *v.* sew; join or close with stitches. □ **in stitches** (*colloquial*) laughing uncontrollably.

stoat *n.* the ermine, especially when its fur is brown.

stock *n.* 1 the amount of something available. 2 livestock. 3 lineage; standing or status. 4 a business company's capital; a portion of this held by an investor. 5 liquid made by stewing bones, vegetables, etc. 6 a plant with fragrant flowers. 7 a plant into which a graft is inserted. 8 the handle of a rifle. 9 a cravat. 10 (**stocks**) a wooden frame with holes in which people had their feet locked as a punishment; a framework on which a ship rests during construction. ● *adj.* stocked and regularly available; commonly used. ● *v.* keep in stock; provide with a supply.

stockade *n.* a protective fence.

stockbroker *n.* a person who buys and sells shares for clients.

stock car *n.* a car used in racing where deliberate bumping is allowed.

stock exchange *n.* the stock market.

stockinet *n.* a fine machine-knitted fabric.

stocking *n.* a close-fitting covering for the foot and leg.

stock-in-trade *n.* all the stock and requisites for carrying on a business.

stockist *n.* a firm that stocks certain goods.

stock market *n.* an institution for buying and selling stocks and shares; the transactions of this.

stockpile *n.* an accumulated stock of goods etc. kept in reserve. ● *v.* accumulate a stockpile of.

stock-still *adj.* motionless.

stocktaking *n.* making an inventory of stock.

stocky *adj.* (**stockier**) short and solidly built. □ **stockily** *adv.*, **stockiness** *n.*

stodge *n.* (*colloquial*) stodgy food.

stodgy *adj.* (**stodgier**) 1 (of food) heavy and filling. 2 dull.

stoic (stoh-ik) *n.* a stoical person.

stoical *adj.* calm and uncomplaining. □ **stoically** *adv.*, **stoicism** *n.*

stoke *v.* tend and put fuel on (a fire etc.). □ **stoker** *n.*

STOL *abbr.* short take-off and landing.

stole[1] *n.* a woman's wide scarf-like garment.

stole[2], **stolen** *see* **steal**.

stolid *adj.* not excitable. □ **stolidly** *adv.*, **stolidity** *n.*

stomach *n.* the integral organ in which the first part of digestion occurs; the abdomen; appetite. ● *v.* endure, tolerate.

stomp *v.* tread heavily.

stone *n.* 1 a piece of rock; stones or rock as a substance or material. 2 a gem. 3 a hard substance formed in the bladder or kidney etc. 4 the hard case round the kernel of certain fruits. 5 (*pl.* **stone**) a unit of weight, 14 lb. ● *adj.* made of stone. ● *v.* 1 pelt with stones. 2 remove stones from (fruit).

Stone Age *n.* a prehistoric period when weapons and tools were made of stone.

stoneground *adj.* (of flour) ground with millstones.

stonemason *n.* a person who shapes or builds in stone.

stonewall *v.* give noncommittal replies.

stoneware *n.* heavy heat-resistant pottery.

stonewashed *adj.* washed with abrasives to give a worn faded look.

stony *adj.* (**stonier**) 1 full of stones. 2 hard, unfeeling; unresponsive. □ **stonily** *adv.*

stood *see* **stand**.

stooge *n.* a comedian's assistant; a person who is another's puppet.

stool *n.* 1 a movable seat without arms or raised back. 2 (**stools**) faeces.

stool-pigeon *n.* a person acting as a decoy, esp. to trap a criminal.

stoop *v.* bend forwards and down; condescend; lower oneself morally. ● *n.* a stooping posture.

stop *v.* (**stopped**) 1 put an end to the movement, progress, or operation (of); refuse to give or allow. 2 close by plugging or obstructing. ● *n.* 1 stopping; a place where a train or bus etc. stops regularly; something that stops or regulates motion. 2 a row of organ pipes providing tones of one quality; the knob etc. controlling these.

stopcock *n.* a valve regulating the flow in a pipe etc.

stopgap *n.* a temporary substitute.

stopover *n.* an overnight break in a journey.

stoppage *n.* stopping; an obstruction.

stopper *n.* a plug for closing a bottle etc. ● *v.* close with a stopper.

stop press *n.* late news inserted in a newspaper after printing has begun.

stopwatch *n.* a watch that can be started and stopped, used for timing races etc.

storage *n.* storing; a space for this.

storage heater *n.* an electric radiator accumulating heat in off-peak periods.

store *n.* **1** a supply of something available for use; a storehouse. **2** a large shop. ● *v.* collect and keep for future use; deposit in a warehouse. □ **in store** being stored; destined to happen, imminent. **set store by** value greatly.

storehouse *n.* a place where things are stored.

storeroom *n.* a room used for storing things.

storey *n.* (*pl.* **storeys** or **stories**) each horizontal section of a building. □ **storeyed** *adj.*

stork *n.* a large wading bird.

storm *n.* a disturbance of the atmosphere with strong winds and usu. rain or snow; a violent shower (of missiles etc.); a great outbreak (of anger or abuse etc.). ● *v.* rage, be violent; attack or capture suddenly. □ **stormy** *adj.*

story *n.* an account of an incident (true or invented).

stout *adj.* thick and strong; fat; brave and resolute. ● *n.* a strong dark beer. □ **stoutly** *adv.*, **stoutness** *n.*

stove see **stave**. *n.* an apparatus containing an oven; a closed apparatus used for heating rooms etc.

stow *v.* a place for storage. □ **stow away** put away, store in reserve; conceal oneself as a stowaway.

stowaway *n.* a person who hides on a ship or aircraft so as to travel free of charge.

straddle *v.* sit or stand (across) with legs wide apart; stand or place (things) in a line across.

strafe *v.* attack with gunfire from the air.

straggle *v.* grow or spread untidily; wander separately; lag behind others. □ **straggler** *n.*, **straggly** *adj.*

straight *adj.* **1** extending or moving in one direction, not curved or bent; correctly or tidily arranged. **2** in unbroken succession; not modified or elaborate; without additions. **3** honest, frank. ● *adv.* **1** in a straight line; direct; without delay. **2** frankly. ● *n.* the straight part of something, esp. of a racecourse. □ **go straight** live honestly after being a criminal. **straight away** without delay. **straightness** *n.*

straighten *v.* make or become straight.

straight face *n.* one not smiling.

straight fight *n.* a contest between only two opponents.

straightforward *adj.* honest, frank; without complications. □ **straightforwardly** *adv.*

straightjacket var. of **straitjacket**.

straight off *adv.* (*colloquial*) immediately, without hesitation.

strain *v.* **1** make taut; injure by excessive stretching or overexertion; make an intense effort (with). **2** sieve to separate solids from liquid. ● *n.* **1** straining, the force exerted; injury or exhaustion caused by straining; severe demand on strength or resources. **2** lineage; a variety or breed of animal etc.; a slight or inherited tendency. **3** a passage from a tune. □ **strainer** *n.*

strained *adj.* (of manner etc.) tense, not natural or relaxed.

strait *n.* **1** (also **straits**) a narrow stretch of water connecting two seas. **2** (**straits**) a difficult state of affairs.

straitened *adj.* (of conditions) poverty-stricken.

straitjacket n. (also **straightjacket**) a strong garment put round a violent person to restrain his or her arms. ● v. restrict severely.

strait-laced adj. very prim and proper.

strand n. **1** a single thread; each of these twisted to form a cable or yarn etc.; a lock of hair. **2** a shore. ● v. run aground; leave in difficulties.

strange adj. not familiar; alien; unusual; odd; unaccustomed. □ **strangely** adv., **strangeness** n.

stranger n. one who is strange to a place, company, or experience.

strangle v. **1** kill or be killed by squeezing the throat. **2** restrict the growth or utterance of. □ **strangler** n.

stranglehold n. a strangling grip.

strangulation n. strangling.

strap n. a strip of leather or other flexible material for holding things together or in place, or supporting something. ● v. (**strapped**) secure with strap(s). □ **strapped for** (colloquial) short of.

straphanger n. (slang) a standing passenger.

strapping adj. tall and robust.

strata see **stratum**.

stratagem n. a cunning plan or scheme; a trick.

strategic adj. **1** of strategy. **2** (of weapons) very long-range. □ **strategically** adv.

strategist n. an expert in strategy.

strategy n. the planning and directing of the whole operation of a campaign or war; a plan, a policy.

stratify v. arrange in strata. □ **stratification** n.

stratosphere n. a layer of the atmosphere about 10–60 km above the earth's surface.

stratum n. (pl. **strata**) one of a series of layers or levels.

■ **Usage** It is a mistake to use the plural form **strata** when only one stratum is meant.

straw n. **1** dry cut stalks of corn etc.; a single piece of this. **2** a narrow tube for sucking up liquid to drink.

strawberry n. a soft edible red fruit with seeds on the surface.

strawberry mark n. a red birthmark.

straw poll n. an unofficial poll as a test of general feeling.

stray v. leave one's group or proper place aimlessly; wander; deviate from a subject. ● adj. having strayed; isolated. ● n. a stray domestic animal.

streak n. **1** a thin line or band of a colour or substance different from its surroundings. **2** an element, a trait. **3** a series, a spell of something. ● v. **1** mark with streaks. **2** move very rapidly; (colloquial) run naked in a public place. □ **streaker** n., **streaky** adj.

stream n. **1** a small river; a flow of liquid, things, or people; the direction of this. **2** a section into which schoolchildren of the same level of ability are placed. ● v. **1** flow; run with liquid; float or wave at full length. **2** arrange (schoolchildren) in streams. □ **on stream** in active operation or production.

streamer n. a long narrow flag; a strip of ribbon or paper etc. attached at one or both ends.

streamline v. **1** give a smooth even shape that offers least resistance to movement through water or air. **2** make more efficient by simplifying.

street n. a public road lined with buildings.

streetcar n. (*Amer.*) a tram.

street credibility n. (also **street cred**) a personal image of being fashionable and successful in city life.

streetwise adj. knowing how to survive in modern city life.

strength n. **1** the quality of being strong; its intensity. **2** an advantageous skill or quality. **3** the total number of people making up a group. □ **on the strength of** relying on as a basis or support.

strengthen v. make or become stronger.

strenuous adj. making or requiring great effort. □ **strenuously** adv.

streptococcus n. (*pl.* **streptococci**) a bacterium causing serious infections.

stress n. emphasis; pressure, tension, strain; extra force used on a sound in speech or music. ● v. put stress on. □ **stressful** adj.

stretch v. **1** pull out tightly or to a greater extent; be able or tend to become stretched. **2** be continuous. **3** thrust out one's limbs. **4** strain; exaggerate. ● n. **1** stretching; the ability to be stretched. **2** a continuous expanse or period. □ **stretch a point** agree to something not normally allowed. □ **stretchy** adj.

stretcher n. a framework for carrying a sick or injured person in a lying position.

strew v. (**strewed**, **strewn** or **strewed**, **strewing**) scatter over a surface; cover with scattered things.

striation n. each of a series of lines or grooves.

stricken adj. afflicted by an illness, shock, or grief.

strict adj. **1** precisely limited or defined; without exception or deviation. **2** requiring or giving complete obedience or exactitude. □ **strictly** adv., **strictness** n.

stricture n. **1** severe criticism. **2** abnormal constriction.

stride v. (**strode**, **stridden**, **striding**) walk with long steps. ● n. a single long step; a manner of striding; progress.

strident adj. loud and harsh. □ **stridently** adv., **stridency** n.

strife n. quarrelling, conflict.

strike v. (**struck**, **striking**) **1** hit; knock. **2** attack suddenly; afflict. **3** ignite (a match) by friction. **4** agree on (a bargain). **5** indicate (the hour) or be indicated by a sound. **6** find (gold or mineral oil etc.) **7** occur to the mind of; produce a mental impression on. **8** stop work in protest. **9** assume (an attitude) dramatically. **10** take down (a flag or tent etc.). ● n. **1** an act or instance of striking. **2** an attack. **3** a workers' refusal to work as a protest. □ **strike home** deal an effective blow. **strike off** or **out** cross out. **strike up** begin playing or singing; start (a friendship etc.) casually.

strikebound adj. immobilized by a workers' strike.

striker n. **1** a person or thing that strikes; a worker who is on strike. **2** a football player whose main function is sure to try to score goals.

striking adj. sure to be noticed; impressive. □ **strikingly** adv.

strimmer n. a long-handled machine for cutting rough grass etc.

string n. **1** a narrow cord; a stretched piece of catgut or wire etc. in a musical instrument, vibrated to produce tones; (**strings**) stringed instruments. **2** a set of objects strung together; a series. **3** (**strings**) conditions insisted upon. ● v. (**strung, stringing**) fit or fasten with string(s); thread on a string. □ **pull strings** use one's influence. **string along** (colloquial) deceive; keep company (with). **string out** spread out on a line. **string up** hang up on strings; kill by hanging.

stringent adj. strict, with firm restrictions. □ **stringently** adv., **stringency** n.

stringy adj. (**stringier**) like string; fibrous.

strip v. (**stripped**) remove (clothes, coverings, or parts etc.); pull or tear away (from); undress; deprive, e.g. of property or titles. ● n. a long narrow piece or area. □ **stripper** n.

strip cartoon n. = **comic strip**.

stripe n. a long narrow band on a surface, differing in colour or texture from its surroundings; a chevron on a sleeve, indicating rank. □ **striped** adj., **stripy** adj.

strip light n. a tubular fluorescent lamp.

stripling n. a youth.

striptease n. an entertainment in which a performer gradually undresses.

strive v. (**strove, striven, striving**) **1** make great efforts. **2** carry on a conflict.

stroboscope n. (also **strobe**) an apparatus for producing a rapidly flashing light. □ **stroboscopic** adj.

strode see **stride**.

Stroganoff n. meat etc. cut in strips and cooked in a sour-cream sauce.

stroke v. pass the hand gently along the surface of. ● n. **1** an act of stroking. **2** an act of striking something. **3** a single movement, action, or effort. **4** a particular sequence of movements (e.g. in swimming). **5** a mark made by a movement of a

pen or paintbrush etc. **6** a sound made by a clock striking. **7** an attack of apoplexy or paralysis.

stroll v. walk in a leisurely way. ● n. a leisurely walk.

stroller n. (Amer.) a pushchair.

strong adj. **1** capable of exerting or resisting great power; powerful through numbers, resources, or quality. **2** concentrated; containing much alcohol; having a considerable effect. **3** having a specified number of members. ● adv. strongly. □ **strongly** adv.

stronghold n. a fortified place; the centre of support for a cause.

strong language n. forceful language; swearing.

strong-minded adj. determined.

strongroom n. a room designed for safe storage of valuables.

strontium n. a silver-white metallic element (symbol Sr).

strontium 90 n. a radioactive isotope of strontium.

strop n. a leather strip on which a razor is sharpened.

stroppy adj. (**stroppier**) (colloquial) bad-tempered, awkward.

strove see **strive**.

struck see **strike**. □ **struck on** (colloquial) impressed with, liking.

structure n. the way a thing is constructed or organized; a thing's supporting framework or essential parts; a constructed thing, a complex whole. □ **structural** adj., **structurally** adv.

strudel n. flaky pastry filled with apple etc.

struggle v. move in a vigorous effort to get free; make one's way or a living etc. with difficulty; make a vigorous effort. ● n. a spell of struggling; a vigorous effort; a hard contest.

strum v. (**strummed**) play on (a stringed or keyboard instrument), esp. unskilfully or monotonously . ●n. a sound made by strumming.

strumpet n. (old use) a prostitute.

strung see **string**.

strut n. 1 a bar of wood or metal supporting something. 2 a strutting walk. ●v. (**strutted**) walk in a pompous self-satisfied way.

strychnine (strik-neen) n. a bitter highly poisonous substance.

stub n. 1 a short stump. 2 a counterfoil of a cheque or receipt etc. ●v. (**stubbed**) 1 strike (one's toe) against a hard object. 2 extinguish (a cigarette) by pressure.

stubble n. 1 the lower ends of cornstalks left in the ground after harvest. 2 short stiff hair or bristles growing after shaving. □ **stubbly** adj.

stubborn adj. obstinate. □ **stubbornly** adv., **stubbornness** n.

stubby adj. (**stubbier**) short and thick. □ **stubbiness** n.

stucco n. plaster or cement used for coating walls or moulding into decorations. ● **stuccoed** adj.

stuck adj. unable to move. See also **stick**.

stuck-up adj. (colloquial) conceited; snobbish.

stud n. 1 a projecting nail-head or similar knob on a surface; a device for fastening two parts, a detachable shirt-collar. 2 horses kept for breeding; an establishment keeping these. ●v. (**studded**) decorate with studs or precious stones; strengthen with studs.

student n. a person who is studying at a college or university.

studied adj. deliberate and artificial.

studio n. (pl. **studios**) the workroom of a painter, photographer, etc.; premises where cinema films are made; a room from which broadcasts are transmitted or where recordings are made.

studio flat n. a one-room flat with a kitchen and bathroom.

studious adj. spending much time in study; deliberate and careful. □ **studiously** adv., **studiousness** n.

study n. 1 the process of studying; its subject; work presenting the results of studying; a room used for studying. 2 a musical composition designed to develop a player's skill. 3 a preliminary drawing. ●v. give one's attention to acquiring knowledge of (a subject); examine attentively.

stuff n. material; unnamed things, belongings, subjects, etc. ●v. pack tightly; fill with padding or stuffing; eat greedily.

stuffing n. padding used to fill something; a savoury mixture put inside poultry, vegetables, etc. before cooking.

stuffy adj. (**stuffier**) 1 lacking fresh air or ventilation. 2 dull; (colloquial) old-fashioned, prim, narrow-minded. □ **stuffily** adv., **stuffiness** n.

stultify v. impair, make ineffective. □ **stultification** n.

stumble v. trip and lose one's balance; walk with frequent stumbles; make mistakes in speaking etc. ●n. an act of stumbling.

stumbling block n. an obstacle, a difficulty.

stump n. 1 the base of a tree left in the ground when the rest has gone; a similar remnant of something cut, broken, or worn down. 2 one of the uprights of a wicket in cricket. ●v. 1 walk stiffly or noisily. 2 (colloquial) baffle. □ **stump up** (*colloquial*) pay over (money required).

stumpy *adj.* (**stumpier**) short and thick. □ **stumpiness** *n.*

stun *v.* (**stunned**) knock senseless; astound.

stung *see* **sting.**

stunk *see* **stink.**

stunning *adj.* (*colloquial*) very attractive. □ **stunningly** *adv.*

stunt *n.* something unusual or difficult done as a performance. ● *v.* hinder the growth or development of.

stupefy *v.* dull the wits or senses of; stun. □ **stupefaction** *n.*

stupendous *adj.* amazing; exceedingly great. □ **stupendously** *adv.*

stupid *adj.* not clever; slow at learning or understanding; in a stupor. □ **stupidly** *adv.*, **stupidity** *n.*

stupor *n.* a dazed condition.

sturdy *adj.* (**sturdier**) strongly built, hardy, vigorous. □ **sturdily** *adv.*, **sturdiness** *n.*

sturgeon *n.* a large sharklike fish yielding caviar.

stutter *v.* stammer, esp. repeating consonants. ● *n.* a stammer.

sty *n.* (*pl.* **sties**) 1 a pigsty. 2 (also **stye**) an inflamed swelling on the edge of the eyelid.

style *n.* 1 a manner of writing, speaking, or doing something; a design. 2 elegance. ● *v.* design, shape, or arrange, esp. fashionably. □ **in style** elegantly, luxuriously.

stylish *adj.* fashionable, elegant. □ **stylishly** *adv.*, **stylishness** *n.*

stylist *n.* a person who has a good style; a person who styles things.

stylistic *adj.* of literary or artistic style. □ **stylistically** *adv.*

stylized *adj.* (also **-ised**) made to conform to a conventional style.

stylus *n.* (*pl.* **styluses** or **styli**) a needle-like device for cutting or following a groove in a record.

stymie *v.* (also **stimy**) (**stymied**, **stymieing** or **stymying**) obstruct, thwart.

styptic *adj.* checking bleeding by causing blood vessels to contract.

styrene *n.* a liquid hydrocarbon used in plastics.

suave (swahv) *adj.* smooth-mannered. □ **suavely** *adv.*, **suavity** *n.*

sub *n.* (*colloquial*) 1 a submarine. 2 a subscription. 3 a substitute. 4 a sub-editor.

sub- *pref.* under; subordinate.

subaltern *n.* an army officer below the rank of captain.

sub-aqua *adj.* underwater.

subatomic *adj.* smaller than an atom; occurring in an atom.

subcommittee *n.* a committee formed from some members of a main committee.

subconscious *adj.* & *n.* (of) our own mental activities of which we are not aware. □ **subconsciously** *adv.*

subcontinent *n.* a large land mass forming part of a continent.

subcontract *v.* give or accept a contract to carry out all or part of another contract. □ **subcontractor** *n.*

subculture *n.* a culture within a larger one.

subcutaneous *adj.* under the skin.

subdivide *v.* divide (a part) into smaller parts. □ **subdivision** *n.*

subdue *v.* bring under control; make quieter or less intense.

sub-editor *n.* a person who prepares material for printing; an assistant editor.

subhuman *adj.* less than human; not fully human.

subject *adj.* (**sub**-jekt) not politically independent. ● *n.* (**sub**-jekt) 1 a person subject to a particular political rule or ruler. 2 a person or thing being discussed or studied. **3** (*Grammar*) the words in a sentence that name who or what does the action of the verb. ● *v.* (sub-**jekt**) subjugate; cause to undergo. □ **subject to** depending upon as a condition. □ **subjection** *n.*

subjective *adj.* **1** dependent on personal taste or views etc. **2** (*Grammar*) of the form of a word used when it is the subject of a sentence. □ **subjectively** *adv.*

subject-matter *n.* the matter treated in a book or speech etc.

sub judice (sub joo-di-si) *adj.* under judicial consideration, not yet decided.

subjugate *v.* bring (a country) into subjection. □ **subjugation** *n.*

subjunctive *n.* (*Grammar*) of the form of a verb expressing what is imagined, wished, or possible; e.g. *were* in *if I were you*.

sub-let *v.* (**sub-let, sub-letting**) let (rooms etc. that one holds by lease) to a tenant.

sublimate *v.* divert the energy of (an emotion or impulse) into a culturally higher activity. □ **sublimation** *n.*

sublime *adj.* most exalted. □ **sublimely** *adv.*

subliminal *adj.* below the level of conscious awareness.

sub-machine gun *n.* a lightweight machine-gun.

submarine *adj.* under the surface of the sea. ● *n.* a vessel that can operate under water.

submerge *v.* put or go below the surface of water or other liquid; flood. □ **submersion** *n.*

submersible *adj.* able to submerge. ● *n.* a submersible craft.

submission *n.* submitting; a statement etc. submitted; obedience.

submissive *adj.* submitting to authority. □ **submissively** *adv.*, **submissiveness** *n.*

submit *v.* (**submitted**) **1** yield to authority or control; surrender; subject to a process. **2** present for consideration.

subordinate *adj.* (sub-**or**-di-nát) of lesser importance or rank; working under another's authority. ● *n.* (sub-**or**-di-nát) a subordinate person. ● *v.* (sub-**or**-di-nayt) make or treat as subordinate. □ **subordination** *n.*

suborn *v.* induce by bribery to commit perjury or another unlawful act. □ **subornation** *n.*

subpoena (sŭ-pee-nă) *n.* a writ commanding a person to appear in a law court. ● *v.* (sub-**poenaed**, **subpoenaing**) summon with a subpoena.

subscribe *v.* **1** pay (a subscription). **2** sign. □ **subscribe to** express agreement with (a theory etc.). □ **subscriber** *n.*

subscription *n.* a sum of money contributed; a fee for membership etc.; a process of subscribing.

subsequent *adj.* occurring after. □ **subsequently** *adv.*

subservient *adj.* subordinate; servile. □ **subserviently** *adv.*, **subservience** *n.*

subset *n.* a set of which all the elements are contained in another set.

subside *v.* sink to a lower or normal level; become less intense. □ **subsidence** *n.*

subsidiary *adj.* of secondary importance; (of a business company) controlled by another. ● *n.* a subsidiary thing.

subsidize *v.* (also **-ise**) pay a subsidy to or for.

subsidy *n.* money given to support an industry etc. or to keep prices down.

subsist *v.* keep oneself alive, exist. □ **subsistence** *n.*

subsoil n. the soil lying immediately below the surface layer.

subsonic adj. of or flying at speeds less than that of sound.

substance n. **1** matter with more or less uniform properties; a particular kind of this. **2** the essence of something spoken or written. **3** reality, solidity.

substantial adj. **1** of solid material or structure. **2** of considerable amount, intensity, or validity. **3** wealthy. **4** in essentials. □ **substantially** adv.

substantiate v. support with evidence. □ **substantiation** n.

substantive adj. genuine, real; substantial.

substitute n. a person or thing that acts or serves in place of another. ● v. use or serve as a substitute. □ **substitution** n.

subsume v. bring or include under a particular classification.

subtenant n. a person to whom a room etc. is sub-let.

subterfuge n. a trick used to avoid blame or defeat etc.

subterranean adj. underground.

subtext n. an underlying theme.

subtitle n. a subordinate title; a caption on a cinema film. ● v. provide with subtitle(s).

subtle (sutt-ĕl) adj. slight and difficult to detect or identify; making fine distinctions; ingenious. □ **subtly** adv., **subtlety** n.

subtotal n. the total of part of a group of figures.

subtract v. remove (a part or quantity or number) from a greater one. □ **subtraction** n.

subtropical adj. of regions bordering on the tropics.

suburb n. a residential area outside the central part of a town. □ **suburban** adj., **suburbanite** n.

suburbia n. suburbs and their inhabitants.

subvention n. a subsidy.

subvert v. overthrow the authority of, esp. by weakening people's trust. □ **subversion** n., **subversive** adj.

subway n. an underground passage; (Amer.) an underground railway.

succeed v. **1** be successful. **2** take the place previously filled by; come next in order.

success n. a favourable outcome; the attainment of one's aims, or of wealth, fame, or position; a successful person or thing.

successful adj. having success. □ **successfully** adv.

succession n. following in order; a series of people or things following each other; succeeding to a throne or other position. □ **in succession** one after another.

successive adj. following in succession. □ **successively** adv.

successor n. a person who succeeds another.

succinct (suk-sinkt) adj. concise and clear. □ **succinctly** adv.

succour v. & n. (Amer. **succor**) help.

succulent adj. juicy; (of plants) having thick fleshy leaves. ● n. a succulent plant. □ **succulence** n.

succumb v. give way to something overpowering.

such adj. of the same or that kind or degree; so great or intense. ● pron. that.

such-and-such adj. particular but not needing to be specified.

suchlike adj. of the same kind.

suck v. draw (liquid or air etc.) into the mouth; draw liquid from; squeeze in the mouth by using the tongue; draw in. ● n. the act or process of sucking. □ **suck up to** (colloquial) flatter or please (a person) to achieve advantage.

sucker n. **1** an organ or device that can adhere to a surface by suction. **2** (slang) a person who is easily deceived.

suckle v. feed at the breast.

suckling n. an unweaned child or animal.

sucrose n. sugar.

suction n. sucking; the production of a partial vacuum so that external atmospheric pressure forces fluid etc. into the vacant space or causes adhesion.

sudden adj. happening or done quickly or without warning. □ **all of a sudden** suddenly. □ **suddenly** adv., **suddenness** n.

sudorific adj. causing sweating.

suds n.pl. soapsuds.

sue v. (**sued, suing**) take legal proceedings against.

suede (swayd) n. leather with a velvety nap on one side.

suet n. hard white fat from round an animal's kidneys, used in cooking.

suffer v. undergo or be subjected to (pain, loss, damage, etc.); tolerate. □ **suffering** n.

sufferance n. □ **on sufferance** tolerated but only grudgingly.

suffice v. be enough (for).

sufficient adj. enough. □ **sufficiently** adv., **sufficiency** n.

suffix n. letters added at the end of a word to make another word.

suffocate v. kill by stopping the breathing; cause discomfort to by making breathing difficult; be suffocated. □ **suffocation** n.

suffrage n. the right to vote in political elections.

suffragette n. a woman who campaigned for the right to vote.

suffuse v. spread throughout or over. □ **suffusion** n.

sugar n. a sweet crystalline substance obtained from the juices of various plants. □ **sugary** adj.

sugar beet n. white beet from which sugar is obtained.

sugar cane n. a tall tropical plant from which sugar is obtained.

sugar soap n. an abrasive cleaning compound.

suggest v. bring to mind; propose for acceptance or rejection.

suggestible adj. easily influenced. □ **suggestibility** n.

suggestion n. **1** suggesting; something suggested. **2** a slight trace.

suggestive adj. conveying a suggestion; suggesting something indecent. □ **suggestively** adv.

suicidal adj. of or involving suicide; liable to commit suicide. □ **suicidally** adv.

suicide n. the intentional killing of oneself; a person who commits suicide; an act destructive to one's own interests. □ **commit suicide** kill oneself intentionally.

sui generis adj. (the only one) of its own kind; unique.

suit n. **1** a set of clothing, esp. jacket and trousers or skirt. **2** any of the four sets which a pack of cards is divided. **3** a lawsuit. ● v. **1** make or be suitable or convenient for. **2** give a pleasing appearance to.

suitable adj. right for the purpose or occasion. □ **suitably** adv., **suitability** n.

suitcase n. a rectangular case for carrying clothes.

suite (sweet) n. **1** a set of rooms or furniture. **2** a retinue. **3** a set of musical pieces.

suitor n. a man who is courting a woman.

sulk v. be sullen because of resentment or bad temper. ● n.pl. (**the sulks**) a fit of sulking. □ **sulky** adj., **sulkily** adv., **sulkiness** n.

sullen adj. gloomy and unresponsive; dark and dismal. □ **sullenly** adv., **sullenness** n.

sully v. stain, blemish.

sulphate n. (Amer. **sulfate**) a salt of sulphuric acid.

sulphide n. (Amer. **sulfide**) a compound of sulphur and an element or radical.

sulphite n. (Amer. **sulfite**) a salt of sulphurous acid.

sulphur n. (Amer. **sulfur**) a pale yellow chemical element (symbol S). □ **sulphurous** adj.

sulphuric acid n. (Amer. **sulfuric acid**) a strong corrosive acid.

sultan n. a ruler of certain Muslim countries.

sultana n. **1** a seedless raisin. **2** a sultan's wife, mother, or daughter.

sultanate n. a sultan's territory.

sultry adj. (**sultrier**) hot and humid; (of a woman) passionate and sensual. □ **sultriness** n.

sum n. a total; an amount of money; a problem in arithmetic. □ **sum up** give the total of; summarize; form an opinion of.

summarize v. (also **-ise**) make or be a summary of. □ **summarization** n.

summary n. a statement giving the main points of something. ● adj. giving the main points only; without attention to details or formalities. □ **summarily** adv.

summation n. adding up; summarizing.

summer n. the warmest season of the year. □ **summery** adj.

summertime n. summer.

summer time n. the time shown by clocks put forward in summer to give longer light evenings.

summit n. **1** the highest point; the top of a mountain. **2** a conference between heads of states.

summon v. **1** send for (a person); order to appear in a law court; call upon to do something. **2** gather together (one's courage etc.).

summons n. a command summoning a person; a written order to appear in a law court. ● v. serve with a summons.

sump n. a reservoir of oil in a petrol engine; a hole or low area into which liquid drains.

sumptuous adj. splendid, lavish, costly. □ **sumptuously** adv., **sumptuousness** n.

sumo n. Japanese wrestling.

Sun. abbr. Sunday.

sun n. the star around which the earth travels; the light or warmth from this; any fixed star. ● v. (**sunned**) expose to the sun.

sunbathe v. expose one's body to the sun. □ **sunbather** n.

sunbeam n. a ray of sun.

sunbed n. a bed for artificial sunbathing under a lamp.

sunblock n. a cream protecting the skin against the sun.

sunburn n. inflammation caused by exposure to sun. ● v. suffer sunburn. □ **sunburnt** adj.

sundae (sun-day) n. a dish of ice cream and fruit, nuts, syrup, etc.

Sunday n. the day after Saturday.

Sunday school n. a school for the religious instruction of Christian children, held on Sundays.

sunder v. (old use) break or tear apart.

sundew n. a bog plant that traps insects.

sundial n. a device that shows the time by means of a shadow on a dial.

sundown n. sunset.

sundry adj. various. ● n.pl. (**sundries**) various small items. □ **all and sundry** everyone.

sunflower n. a tall plant producing large yellow flowers and edible seeds.

sung see **sing**.

sunk see **sink**.

sunken adj. lying below the level of the surrounding surface.

Sunni n. & adj. (pl. **Sunni** or **Sunnis**) (a member) of a Muslim sect opposed to the Shiites.

sunny adj. (**sunnier**) 1 full of sunshine. 2 cheerful. □ **sunnily** adv.

sunrise n. the rising of the sun.

sunset n. the setting of the sun; the sky full of colour at sunset.

sunshade n. a parasol; an awning.

sunshine n. direct sunlight.

sunspot n. 1 a dark patch observed on the sun's surface. 2 (colloquial) a place with a sunny climate.

sunstroke n. illness caused by too much exposure to sun.

super adj. (colloquial) excellent, superb.

superannuation n. an employee's pension.

superb adj. of the most impressive or splendid kind. □ **superbly** adv.

supercharge v. increase the power of (an engine) by a device that forces extra air or fuel into it. □ **supercharger** n.

supercilious adj. haughty and superior. □ **superciliously** adv., **superciliousness** n.

supercomputer n. a very powerful computer.

superficial adj. of or on the surface, not deep or penetrating. □ **superficially** adv., **superficiality** n.

superfluous adj. more than is required. □ **superfluously** adv., **superfluity** n.

superhuman adj. beyond ordinary human capacity or power; higher than humanity, divine.

superimpose v. place on top of something else.

superintend v. supervise. □ **superintendence** n.

superintendent n. 1 a supervisor. 2 a police officer next above inspector.

superior adj. higher in position or rank; better, greater; showing that one feels wiser or better etc. than others. ● n. a person or thing of higher rank or ability or quality. □ **superiority** n.

superlative adj. 1 of the highest quality. 2 of the grammatical form expressing 'most'. ● n. a superlative form of a word. □ **superlatively** adv.

supermarket n. a large self-service store selling food and household goods.

supernatural adj. of or involving a power above the forces of nature. □ **supernaturally** adv.

supernova n. (pl. **supernovas** or **supernovae**) a star that suddenly increases in brightness because of an explosion.

supernumerary adj. extra.

superphosphate n. a fertilizer containing soluble phosphates.

superpower n. an extremely powerful nation.

superscript adj. written just above and to the right of a word etc.

supersede v. take the place of; put or use in place of.

supersonic adj. of or flying at speeds greater than that of sound. □ **supersonically** adv.

superstition n. 1 a belief in magical and similar influences; an idea or practice based on this. 2 a widely held but wrong idea. □ **superstitious** adj.

superstore n. a large supermarket.

superstructure n. a structure that rests on something else.

supervene v. occur as an interruption or a change. □ **supervention** n.

supervise v. direct and inspect (workers etc.). □ **supervision** n., **supervisor** n., **supervisory** adj.

supine adj. **1** lying face upwards. **2** not inclined to take action.

supper n. an evening meal, esp. a light or informal one.

supplant v. oust and take the place of.

supple adj. bending easily. □ **supply** adv., **suppleness** n.

supplement n. something added as an extra part or to make up for a deficiency. ● v. provide or be a supplement to.

supplementary adj. serving as a supplement.

suppliant (sup-li-ǎnt) n. & adj. (a person) asking humbly for something.

supplicate v. ask humbly for; beseech. □ **supplication** n.

supply v. give or provide with, make available; satisfy (a need). ● n. supplying; a stock, an amount provided or available.

support v. **1** bear the weight of; strengthen. **2** supply with necessaries; help, encourage. ● n. the act of supporting; a person or thing that supports. □ **supporter** n., **supportive** adj.

suppose v. be inclined to think; assume to be true; consider as a proposal. □ **be supposed to** expected to; have as a duty.

supposedly adv. according to what is generally thought or believed.

supposition n. the process of supposing; what is supposed.

suppository n. a solid medicinal substance placed in the rectum or vagina and left to melt.

suppress v. put an end to the activity or existence of. **2** keep from being known. □ **suppression** n., **suppressor** n.

suppurate v. form pus, fester. □ **suppuration** n.

supra- pref. above, over.

supreme adj. highest in authority, rank, or quality. □ **supremely** adv., **supremacy** n.

supremo n. (pl. **supremos**) a supreme leader.

surcharge n. an additional charge. ● v. make a surcharge on or to.

sure adj. without doubt or uncertainty; reliable, unfailing. ● adv. (colloquial) certainly. □ **make sure** act so as to be certain. □ **sureness** n.

sure-footed adj. never slipping or stumbling.

surely adv. in a sure manner; (used for emphasis) that must be right; (as an answer) certainly.

surety n. a guarantee; a guarantor of a person's promise.

surf n. white foam of breaking waves.

surface n. the outside or outward appearance of something; any side of an object; the uppermost area, the top. ● adj. of or on the surface. ● v. put a specified surface on; come or bring to the surface.

surfboard n. a narrow board for riding over surf.

surfeit (ser-fit) n. too much, esp. of food or drink. ● v. cause to take too much of something; satiate.

surfing n. the sport of riding on a surfboard.

surge v. move forward in or like waves; increase in volume or intensity. ● n. a surging movement or increase.

surgeon n. a doctor qualified to perform surgical operations.

surgery n. **1** treatment by cutting or manipulation of affected parts of the body. **2** a place where (or times when) a doctor or dentist or an MP etc. is available for consultation. □ **surgical** adj., **surgically** adv.

surly adj. (**surlier**) bad-tempered and unfriendly. □ **surliness** n.

surmise v. guess, suppose. ● n. a guess, a supposition.

surmount v. overcome (a difficulty); get over (an obstacle); be on the top of. □ **surmountable** adj.

surname n. a family name.

surpass v. outdo; excel.

surplice n. a loose white garment with full sleeves worn by clergy and choir.

surplus n. an amount left over after what is needed has been used.

surprise n. an emotion aroused by something sudden or unexpected; something causing this. ● v. cause to feel surprise; come upon or attack unexpectedly.

surreal adj. bizarre; dreamlike.

surrealism n. a style of art and literature seeking to express what is in the subconscious mind, characterized by unusual images. □ **surrealist** n., **surrealistic** adj.

surrender v. hand over, give into another's power or control, esp. under compulsion; give oneself up. ● n. surrendering.

surreptitious adj. acting or done stealthily. □ **surreptitiously** adv.

surrogate n. a deputy. □ **surrogacy** n.

surrogate mother n. a woman who bears a child on behalf of another.

surround v. come, place, or be all round, encircle. ● n. a border.

surroundings n.pl. things or conditions around a person or place.

surtax n. additional tax.

surveillance n. supervision; a close watch.

survey v. (ser-**vay**) look at and take a general view of; examine the condition of (a building); measure and map out. ● n. (**ser-vay**) a general look at or examination of something; a report or map produced by surveying.

surveyor n. a person whose job is to survey land or buildings.

survival n. surviving; something that has survived from an earlier time.

survive v. continue to live or exist; remain alive or in existence after. □ **survivable** adj., **survivability** n., **survivor** n.

sus var. of **suss**.

susceptible adj. easily affected or influenced. □ **susceptibility** n.

sushi n. a Japanese dish of flavoured balls of cold rice usu. garnished with fish.

suspect v. (sŭ-**spekt**) **1** feel that something may exist or be true. **2** mistrust; feel to be guilty but have no proof. ● n. (**sus-pekt**) a person suspected of a crime etc. ● adj. (**sus-pekt**) suspected, open to suspicion.

suspend v. **1** hang up; keep from falling or sinking in air or liquid. **2** stop temporarily; deprive temporarily of a position or right.

suspender n. an attachment to hold up a sock or stocking by its top; (**suspenders**) (Amer.) braces.

suspense n. anxious uncertainty while awaiting an event etc.

suspension n. suspending; a means by which a vehicle is supported on its axles.

suspension bridge n. a bridge suspended from cables that pass over supports at each end.

suspicion n. **1** suspecting; an unconfirmed belief. **2** a slight trace.

suspicious adj. feeling or causing suspicion. □ **suspiciously** adv.

suss v. (also **sus**) (slang) **1** investigate. **2** work out, realize.

sustain v. support; keep alive; keep (a sound or effort) going continuously; undergo; endure; uphold the validity of.

sustenance n. food, nourishment.

suture (soo-cher) n. surgical stitching of a wound; a stitch or thread used in this. ● v. stitch (a wound).

suzerain n. a country or ruler with some authority over a self-governing country; an overlord.

svelte adj. slender and graceful.

SW abbr. south-west; south-western.

swab n. a mop or pad for cleansing, drying, or absorbing things; a specimen of a secretion taken with this. ● v. (**swabbed**) cleanse with a swab.

swaddle v. swathe in wraps or warm garments.

swag n. **1** (slang) loot. **2** a decorative festoon of flowers, drapery, etc.

swagger v. walk or behave with aggressive pride. ● n. this gait or manner.

Swahili n. a Bantu language widely used in East Africa.

swallow v. **1** cause or allow to go down one's throat; work the throat muscles in doing this. **2** take in and engulf or absorb; accept. ● n. **1** an act of swallowing; an amount swallowed. **2** a small migratory bird with a forked tail.

swam see **swim**.

swami n. a Hindu male religious teacher.

swamp n. a marsh. ● v. flood, drench or submerge in water; overwhelm with a mass or number of things. □ **swampy** adj.

swan n. a large usu. white waterbird with a long curving neck.

swank (colloquial) n. boastful or showy behaviour. ● v. behave in this way, show off.

swansong n. a person's last performance or achievement etc.

swap (also **swop**) v. (**swapped**) exchange. ● n. an exchange; a thing exchanged.

swarm n. a large cluster of people, insects, etc. ● v. move in a swarm; be crowded. □ **swarm up** climb by gripping with arms and legs.

swarthy adj. (**swarthier**) having a dark complexion. □ **swarthiness** n.

swashbuckling adj. & n. swaggering boldly. □ **swashbuckler** n.

swastika n. a symbol formed by a cross with ends bent at right angles.

swat v. (**swatted**) hit hard with something flat. □ **swatter** n.

swatch n. a sample of cloth etc.

swath (swawth) n. (also **swathe**) a strip cut in one sweep or passage by a scythe or mower.

swathe v. wrap with layers of coverings. ● n. var. of **swath**.

sway v. **1** swing gently, lean to and fro. **2** influence the opinions of; waver in one's opinion. ● n. **1** a swaying movement. **2** influence.

swear v. (**swore**, **sworn**, **swearing**) **1** state or promise on oath; use emphatically. **2** use a swear word. □ **swear by** have great confidence in.

swear word n. a profane or indecent word used in anger etc.

sweat n. moisture given off by the body through the pores; a state of sweating; moisture forming in drops on a surface. ● v. **1** give off sweat or as sweat. **2** be in a state of great anxiety. **3** work long and hard. □ **sweaty** adj.

sweatband n. a band of material worn to absorb or wipe away sweat.

sweated labour n. labour of workers with poor pay and conditions.

sweater n. a jumper, a pullover.

sweatshirt n. a long-sleeved cotton sweater with a fleecy lining.

sweatshop n. a place employing sweated labour.

Swede n. a native of Sweden.

swede n. a large variety of turnip.

Swedish adj. & n. (the language) of Sweden.

sweep v. (**swept, sweeping**) 1 clear away with a broom or brush; clean or clear (a surface) in this way. 2 go smoothly and swiftly or majestically. 3 extend in a continuous line. ● n. 1 a sweeping movement or line; an act of sweeping. 2 a chimney sweep. 3 a sweepstake. □ **sweep the board** win all the prizes. □ **sweeper** n.

sweeping adj. comprehensive; making no exceptions.

sweepstake n. a form of gambling in which the money staked is divided among those who have drawn numbered tickets for the winners.

sweet adj. 1 tasting as if containing sugar. 2 fragrant; melodious; pleasant; (colloquial) charming. ● n. a small shaped piece of sweet substance; a sweet dish forming one course of a meal. □ **sweetly** adv., **sweetness** n.

sweetbread n. an animal's thymus gland or pancreas used as food.

sweetcorn n. maize with yellow kernels.

sweeten v. make or become sweet or sweeter.

sweetener n. 1 a sweetening substance. 2 a bribe.

sweetheart n. a girlfriend or boyfriend; a term of affection.

sweetmeal adj. of sweetened wholemeal.

sweet pea n. a climbing plant with fragrant flowers.

sweet tooth n. a liking for sweet things.

swell v. (**swelled, swollen** or **swelled, swelling**) make or become larger from pressure within; curve outwards; make or become greater in amount or intensity. ● n. 1 the act or state of swelling. 2 the heaving movement of the sea. 3 a gradual increase in loudness. 4 (colloquial, old use) a person of high social position.

swelling n. a swollen place on the body.

swelter v. be uncomfortably hot.

swept see **sweep**.

swerve v. turn aside from a straight course. ● n. a swerving movement or direction.

swift adj. quick, rapid. ● n. a swiftly flying bird with narrow wings. □ **swiftly** adv., **swiftness** n.

swill v. 1 wash, rinse; pour. 2 drink greedily. ● n. 1 rinse. 2 sloppy food fed to pigs.

swim v. (**swam, swum, swimming**) 1 travel through water by movements of the body. 2 be covered with liquid. 3 seem to be whirling or waving; be dizzy. ● n. an act or period of swimming. □ **swimmer** n.

swimming bath n. a building containing a public swimming pool.

swimmingly adv. with easy unobstructed progress.

swimming pool n. an artificial pool for swimming in.

swindle v. cheat in a business transaction. ● n. an act of swindling. □ **swindler** n.

swine n. 1 (as pl.) pigs. 2 (pl. **swine** or **swines**) (colloquial) a hated person or thing.

swing v. (**swung, swinging**) 1 move to and fro while supported; turn in a curve. 2 change from one mood or opinion to another. 3 influence decisively. ● n. 1 the act, movement, or extent of swinging; a hanging seat for swinging in. 2 jazz with the time of the melody varied. □ **in full swing** with activity at its greatest. □ **swinger** n.

swing-bridge n. a bridge that can be swung aside for boats to pass.

swingeing adj. forcible; huge in amount or scope.

swing-wing n. an aircraft wing that can be moved to slant backwards.

swipe (colloquial) v. 1 hit with a swinging blow. 2 snatch, steal. ● n. a swinging blow.

swirl v. whirl, flow with a whirling movement.

swish v. move with a hissing sound. ● n. a swishing sound. ● adj. (colloquial) smart, fashionable.

Swiss adj. & n. (a native) of Switzerland.

Swiss roll n. a thin flat sponge cake spread with jam etc. and rolled up.

switch n. 1 a device operated to turn electric current on or off. 2 a shift in opinion or method etc. 3 a flexible stick or rod, a whip; a tress of hair tied at one end. ● v. 1 turn (on or off) by means of a switch. 2 transfer, divert; change. 3 swing round quickly.

switchback n. a railway used for amusement at a fair etc., with alternate steep ascents and descents; a road with similar slopes.

switchboard n. a panel of switches for making telephone connections or operating electric circuits.

swivel n. a link or pivot enabling one part to revolve without turning another. ● v. (**swivelled**; Amer. **swiveled**) turn on or as if on a swivel.

swizz n. (also **swiz**) (colloquial) a swindle, a disappointment.

swollen see **swell**.

swoop v. make a sudden downward rush, attack suddenly. ● n. a swooping movement or attack.

swop var. of **swap**.

sword (sord) n. a weapon with a long blade and a hilt.

swordfish n. (pl. **swordfish**) an edible sea fish with a long swordlike upper jaw.

swore see **swear**.

sworn see **swear**. adj. open and determined, esp. in enmity.

swot (colloquial) v. (**swotted**) study hard. ● n. a person who studies hard.

swum see **swim**.

swung see **swing**.

sybarite n. a person who is excessively fond of comfort and luxury. □ **sybaritic** adj.

sycamore n. a large tree of the maple family.

sycophant n. a person who tries to win favour by flattery. □ **sycophantic** adj., **sycophantically** adv.

syllable n. a unit of sound in a word. □ **syllabic** adj.

syllabub n. (also **sillabub**) a dish of whipped cream flavoured esp. with wine.

syllabus n. (pl. **syllabuses** or **syllabi**) a statement of the subjects to be covered by a course of study.

syllogism n. a form of reasoning involving a conclusion drawn from two statements.

sylph n. 1 a slender girl or woman. 2 a spirit of the air.

sylvan *adj.* (also **silvan**) of the woods; having woods, rural.

symbiosis *n.* (*pl.* **symbioses**) a relationship of different organisms living in close association. □ **symbiotic** *adj.*

symbol *n.* something regarded as suggesting something; a mark or sign with a special meaning.

symbolic *adj.* (also **symbolical**) of, using, or used as a symbol. □ **symbolically** *adv.*

symbolism *n.* the use of symbols to express things. □ **symbolist** *n.*

symbolize *v.* (also **-ise**) be a symbol of; represent by means of a symbol.

symmetry *n.* the state of having parts that correspond in size, shape, and position on either side of a dividing line or round a centre. □ **symmetrical** *adj.*, **symmetrically** *adv.*

sympathetic *adj.* feeling, showing, or resulting from sympathy; likeable. □ **sympathetically** *adv.*

sympathize *v.* (also **-ise**) feel or express sympathy. □ **sympathizer** *n.*

sympathy *n.* the ability to share another's emotions or sensations; pity or tenderness towards a sufferer; liking for each other.

symphony *n.* a long elaborate musical composition for a full orchestra. □ **symphonic** *adj.*

symposium *n.* (*pl.* **symposia**) a meeting for discussing a particular subject.

symptom *n.* a sign of the existence of a condition.

symptomatic *adj.* serving as a symptom.

synagogue *n.* a building for public Jewish worship.

synapse *n.* a junction of two nerve cells.

synchromesh *n.* a device that makes gear wheels revolve at the same speed.

synchronic *adj.* concerned with something as it exists at a particular time, not with its history.

synchronize *v.* (also **-ise**) (cause to) occur or operate at the same time; cause (clocks etc.) to show the same time. □ **synchronization** *n.*

synchronous *adj.* occurring or existing at the same time.

syncopate *v.* change the beats or accents in (music). □ **syncopation** *n.*

syndicate *n.* (sin-di-kăt) an association of people or firms to carry out a business undertaking. ● *v.* (sin-di-kayt) combine into a syndicate; arrange publication in many newspapers etc. simultaneously. □ **syndication** *n.*

syndrome *n.* a combination of signs, symptoms, etc. characteristic of a specified condition.

synod *n.* a council of clergy and officials to discuss church policy, teaching, etc.

synonym *n.* a word or phrase meaning the same as another in the same language. □ **synonymous** *adj.*

synopsis *n.* (*pl.* **synopses**) a summary, a brief general survey.

synovia *n.* a thick sticky fluid lubricating body parts. □ **synovial** *adj.*

syntax *n.* the way words are arranged to form phrases and sentences. □ **syntactic** *adj.*, **syntactically** *adv.*

synthesis *n.* (*pl.* **syntheses**) **1** combining. **2** the artificial production of a substance that occurs naturally.

synthesize *v.* (also **-ise**) make by synthesis.

synthesizer n. (also **-iser**) an electronic musical instrument able to produce a great variety of sounds.

synthetic adj. made by synthesis. ● n. a synthetic substance or fabric. □ **synthetically** adv.

syphilis n. a venereal disease. □ **syphilitic** adj.

syringe n. a device for drawing and injecting liquid. ● v. wash out or spray with a syringe.

syrup n. a thick sweet liquid; water sweetened with sugar. □ **syrupy** adj.

system n. **1** a set of connected things that form a whole or work together; the animal body as a whole; a set of rules or practices used together. **2** a method of classification, notation, or measurement. **3** orderliness.

systematic adj. methodical. □ **systematically** adv.

systematize v. (also **-ise**) arrange according to a system.

systemic adj. of or affecting the body as a whole.

....

Tt

T symb. tritium. □ **to a T** exactly; in every respect.

Ta int. tantalum.

ta int. (colloquial) thank you.

tab n. a small projecting flap or strip. □ **keep tabs on** (colloquial) keep under observation.

tabard n. a short sleeveless tunic-like garment.

Tabasco n. (trade mark) a hot-tasting pepper sauce.

tabby n. a cat with grey or brown fur and dark stripes.

tabernacle n. (Bible) a portable shrine used by the Israelites; (RC Church) a receptacle for the Eucharist; a meeting place for worship used by Nonconformists or Mormons.

table n. **1** a piece of furniture with a flat top supported on one or more legs. **2** food provided. **3** a list of facts or figures arranged in columns. ● v. submit (a motion or report) for discussion.

tableau (tab-loh) n. (pl. **tableaux**) a silent motionless group arranged to represent a scene.

table d'hôte (tahbl **doht**) n. a meal in a hotel etc. served at a fixed inclusive price.

tableland n. a plateau of land.

tablespoon n. a large spoon for serving food; the amount held by this. □ **tablespoonful** n.

tablet n. **1** a slab bearing an inscription etc. **2** a measured amount of a drug compressed into a solid form.

table tennis n. a game played with bats and a light hollow ball on a table.

tabloid n. a small-sized newspaper, often sensational in style.

taboo n. (also **tabu**) a ban or prohibition made by religion or social custom. ● adj. prohibited by a taboo.

tabular adj. arranged in a table or list.

tabulate v. arrange in tabular form. □ **tabulation** n.

tachograph n. a device in a vehicle to record speed and travel time.

tachometer n. an instrument measuring the speed of a vehicle or the rotation of its engine.

tacit adj. implied or understood without being put into words. □ **tacitly** adv.

taciturn adj. saying very little. □ **taciturnity** n.

tack n. **1** a small broad-headed nail. **2** a long temporary stitch. **3** a sailing ship's oblique course. **4** riding saddles, bridles, etc. ● v. **1** nail with tacks. **2** stitch with tacks. **3** add as an extra thing. **4** sail a zigzag course.

tackle *n.* **1** a set of ropes and pulleys for lifting etc. **2** equipment for a task or sport. **3** the act of tackling in football etc. ● *v.* **1** try to deal with or overcome (an opponent or problem). **2** intercept (an opponent who has the ball in football etc.).

tacky *adj.* (**tackier**) **1** (of paint etc.) sticky, not quite dry. **2** (*colloquial*) cheap; shabby. □ **tackiness** *n.*

taco *n.* (*pl.* **tacos**) a folded tortilla with a savoury filling.

tact *n.* skill in avoiding offence or in winning goodwill. □ **tactful** *adj.*, **tactfully** *adv.*

tactic *n.* a piece of tactics; (**tactics**) skilful use of the available means to achieve an (esp. military) objective.

tactical *adj.* of tactics; (of weapons) for use in a battle or at close quarters. □ **tactically** *adv.*

tactical voting *n.* voting for the candidate most likely to defeat the leading candidate.

tactician *n.* an expert in tactics.

tactile *adj.* of or using the sense of touch. □ **tactility** *n.*

tactless *adj.* lacking in tact. □ **tactlessly** *adv.*, **tactlessness** *n.*

tadpole *n.* the larva of a frog or toad etc. at the stage when it has gills and a tail.

taffeta *n.* a shiny silklike fabric.

tag *n.* **1** a label. **2** a metal point on a shoelace etc. **3** a much-used phrase or quotation. ● *v.* (**tagged**) **1** label. **2** attach, add.

tagliatelle *n.* pasta in ribbon-shaped strips.

t'ai chi (ty chee) *n.* a Chinese martial art and system of exercises.

tail *n.* **1** an animal's hindmost part, esp. when extending beyond its body; a rear, hanging, or inferior part. **2** (*colloquial*) a person tailing another. **3** (**tails**) the reverse of a coin as a choice when tossing. **4** (**tails**) a tailcoat. ● *v.* (*colloquial*) follow closely, shadow. □ **tail off** become fewer, smaller, or slighter; end inconclusively.

tailback *n.* a queue of traffic extending back from an obstruction.

tailboard *n.* a hinged or removable back of a lorry etc.

tailcoat *n.* a man's coat with a long divided flap at the back.

tailgate *n.* a rear door in a motor vehicle.

tail light *n.* a light at the back of a motor vehicle or train etc.

tailor *n.* a maker of men's clothes, esp. to order. ● *v.* **1** make (clothes) as a tailor. **2** make or adapt for a special purpose. □ **tailor-made** *adj.*

tailplane *n.* the horizontal part of an aeroplane's tail.

tailspin *n.* an aircraft's spinning dive.

taint *n.* a trace of decay, infection, or other bad quality. ● *v.* affect with a taint.

take *v.* (**took**, **taken**, **taking**) **1** get possession of, capture. **2** cause to come or go with one; carry, remove. **3** be effective. **4** make use of. **5** accept, endure. **6** study or teach (a subject). **7** make a photograph of. ● *n.* **1** an amount taken or caught. **2** an instance of photographing a scene for a cinema film. □ **be taken by** or **with** find attractive. **be taken ill** become ill. **take after** resemble (a parent etc.). **take back** withdraw (a statement). **take in** include; make (a garment etc.) smaller; understand; deceive, cheat. **take off** mimic humorously; become airborne. **take on** acquire; undertake; engage (an employee); accept as an opponent. **take**

one's time not hurry. **take out** remove. **take over** take control of. **take part** share in an activity. **take place** occur. **take sides** support one side or another. **take to** adopt as a habit or custom; go to as a refuge; develop a liking or ability for. **take up** take as a hobby, business, etc.; occupy (time or space); accept (an offer). **take up with** begin to associate with. □ **taker** n.

takeaway n. a cooked meal bought at a restaurant etc. for eating elsewhere; (a place) selling this.

take-off n. **1** a piece of humorous mimicry. **2** the process of becoming airborne.

takeover n. the gaining of control of a business etc.

takings n.pl. money taken in business.

talc n. talcum powder; magnesium silicate used as a lubricator.

talcum n. talc.

talcum powder n. talc powdered and usu. perfumed for use on the skin.

tale n. a narrative, a story; a report spread by gossip.

talent n. a special ability.

talented adj. having talent.

talisman n. (pl. **talismans**) an object supposed to bring good luck.

talk v. convey or exchange ideas by spoken words; use (a specified language) in talking. ● n. talking, conversation; a style of speech; an informal lecture; rumour. □ **talker** n.

talkative adj. talking very much.

talking book n. a recorded reading of a book.

talking shop n. a place for empty talk; an institution without power.

talking-to n. (colloquial) a reproof or reprimand.

tall adj. of great or specified height. □ **tallness** n.

tallboy n. a tall chest of drawers.

tall order n. a difficult task.

tallow n. animal fat used to make candles, lubricants, etc.

tall story n. (colloquial) one that is hard to believe.

tally n. the total of a debt or score. ● v. correspond.

Talmud n. the ancient writings on Jewish law and tradition. □ **Talmudic** adj.

talon n. a bird's large claw.

tambourine n. a percussion instrument with jingling metal discs.

tame adj. (of animals) gentle and not afraid of human beings; docile; not exciting. ● v. make tame or manageable. □ **tamely** adv., **tameness** n.

Tamil n. & adj. (a member, the language) of a people of south India and Sri Lanka.

tamp v. pack down tightly.

tamper v. □ **tamper with** meddle or interfere with.

tampon n. a plug of absorbent material inserted into the body, esp. to absorb menstrual blood.

tan v. (**tanned**) **1** make or become brown by exposure to sun. **2** convert (hide) into leather. **3** (slang) thrash. ● n. yellowish brown; the brown colour of suntanned skin. ● adj. yellowish brown. ● abbr. tangent.

tandem n. a bicycle for two people one behind another. ● adv. one behind another. □ **in tandem** arranged one behind another; together; alongside each other.

tandoor n. a clay oven.

tandoori n. food cooked in a tandoor.

tang n. a strong taste or smell. □ **tangy** adj.

tangent n. (Maths) **1** a straight line that touches the outside of a curve without intersecting it. **2**

(in a right-angled triangle) the ratio of the sides (other than the hypotenuse) opposite and adjacent to an angle. □ **go off at a tangent** diverge suddenly from a line of thought etc. □ **tangential** adj.

tangerine n. a small orange; its colour.

tangible adj. **1** able to be perceived by touch. **2** clear and definite, real. □ **tangibly** adv., **tangibility** n.

tangle v. **1** twist into a confused mass; entangle. **2** become involved in conflict with. ● n. a tangled mass or condition.

tango n. (pl. **tangos**) a ballroom dance with gliding steps. ● v. dance a tango.

tank n. **1** a large container for liquid or gas. **2** an armoured fighting vehicle moving on a tracked carriage.

tankard n. a one-handled usu. metal drinking vessel.

tanker n. a ship, aircraft, or vehicle for carrying liquid in bulk.

tanner n. a person who tans hides.

tannery n. a place where hides are tanned into leather.

tannic acid n. tannin.

tannin n. a substance (found in tree-barks and also in tea) used in tanning and dyeing.

tannoy n. (trade mark) a public address system.

tantalize v. (also **-ise**) torment by the sight of something desired but kept out of reach or withheld.

tantalum n. a hard white metallic element (symbol Ta). □ **tantalic** adj.

tantamount adj. equivalent.

tantra n. any of a class of Hindu or Buddhist mystical or magical writings.

tantrum n. an outburst of bad temper.

tap n. **1** a device for drawing liquid in a controlled flow. **2** a light blow; the sound of this. **3** a connection for tapping a telephone. ● v. (**tapped**) **1** knock gently. **2** fit a tap into; draw off through a tap or incision. **3** obtain supplies or information from. **4** cut a screw-thread in (a cavity). **5** fit a listening device in (a telephone circuit). □ **on tap** (colloquial) available for use.

tapas n.pl. small Spanish-style savoury dishes.

tap-dance n. a dance in which the feet tap an elaborate rhythm.

tape n. **1** a narrow strip of material for tying, fastening, or labelling things. **2** magnetic tape; a tape recording. **3** a tape-measure. ● v. **1** tie or fasten with tape. **2** record on magnetic tape. □ **have a thing taped** (colloquial) understand and be able to deal with it.

tape deck n. a machine for playing and recording audiotapes.

tape-measure n. a strip of tape or flexible metal marked for measuring length.

taper n. a thin candle. ● v. make or become gradually narrower. □ **taper off** become gradually less.

tape recorder n. an apparatus for recording sounds or data on magnetic tape. □ **tape recording** n.

tapestry n. a textile fabric woven or embroidered ornamentally.

tapeworm n. a ribbon-like worm living as a parasite in intestines.

tapioca n. starchy grains obtained from cassava, used in making puddings.

tapir (tay-peer) n. a small pig-like animal with a long snout.

tappet n. a projection used in machinery to tap against something.

taproom n. a room in a pub with alcoholic drinks (esp. beer) on tap.

tap root n. a plant's chief root.

tar n. a thick dark liquid distilled from coal etc.; a similar substance formed by burning tobacco. ● v. (**tarred**) coat with tar.

taramasalata n. a pâté made from the roe of mullet or smoked cod.

tarantella n. a whirling Italian dance.

tarantula n. a large black hairy spider.

tardy adj. (**tardier**) slow; late. □ **tardily** adv., **tardiness** n.

tare n. 1 a vetch, as a cornfield weed or fodder. 2 an allowance for the weight of the container or vehicle weighed with the goods it holds.

target n. an object or mark to be hit in shooting etc.; the object of criticism; an objective. ● v. (**targeted**) aim at (as) a target.

tariff n. a list of fixed charges; a duty to be paid.

tarmac n. (trade mark) broken stone or slag mixed with tar; an area surfaced with this, e.g. a runway. □ **tarmacked** adj.

tarn n. a small mountain lake.

tarnish v. 1 lose or cause (metal) to lose lustre. 2 blemish (a reputation). ● n. 1 loss of lustre. 2 a blemish.

tarot (ta-roh) n. a pack of 78 cards mainly used for fortune-telling.

tarpaulin n. a waterproof canvas.

tarragon n. an aromatic herb.

tarsier n. a small monkey-like animal of the East Indies.

tarsus n. (pl. **tarsi**) the set of small bones forming the ankle. □ **tarsal** adj.

tart n. 1 a pie or flan with a sweet filling. 2 (slang) a prostitute. ● adj. acid in taste or manner. □ **tart up** (colloquial) dress gaudily; smarten up. □ **tartly** adv., **tartness** n.

tartan n. a checked pattern (originally of a Scottish clan); cloth with this.

tartar n. 1 a hard deposit forming on teeth. 2 a deposit formed by fermentation in a wine cask. 3 (**Tartar**) a member of a group of central Asian peoples. 4 a bad-tempered or difficult person.

tartare sauce n. a cold sauce of mayonnaise, chopped gherkins, etc.

tartaric adj. of or derived from tartar.

tartlet n. a small tart.

tartrazine n. a yellow dye from tartaric acid, used as food colouring.

task n. a piece of work to be done. □ **take to task** rebuke.

task force n. a group organized for a special task.

taskmaster n. a person who makes others work hard.

tassel n. an ornamental bunch of hanging threads. □ **tasselled** adj.

taste n. 1 the sensation caused in the tongue by things placed on it; the ability to perceive this. 2 a small quantity (of food or drink); a slight experience. 3 a liking. 4 the ability to perceive what is beautiful or fitting. ● v. 1 discover or test the flavour of; have a certain flavour. 2 experience.

tasteful adj. showing good taste. □ **tastefully** adv.

tasteless adj. 1 having no flavour. 2 showing poor taste. □ **tastelessly** adv., **tastelessness** n.

taster n. a person who judges teas, wines, etc. by tasting them. 2 a small sample.

tasty adj. (**tastier**) having a strong flavour, appetizing.

tat n. (colloquial) tattiness; tasteless or fussily ornate things.

tattered adj. ragged.

tatters n.pl. torn pieces.

tatting n. lace made by hand with a small shuttle.

tattle v. chatter idly, reveal information in this way. ●n. idle chatter.

tattoo v. mark (skin) by puncturing it and inserting pigments; make (a pattern) in this way. ●n. **1** a tattooed pattern. **2** a military display or pageant. **3** a tapping sound.

tatty adj. (**tattier**) (colloquial) ragged, shabby and untidy. □ **tattily** adv., **tattiness** n.

taught see **teach**.

taunt v. jeer at provocatively. ●n. a taunting remark.

taupe (tohp) n. pale greyish brown.

taut adj. stretched firmly, not slack.

tauten v. make or become taut.

tautology n. pointless repetition, esp. using a word or phrase of the same grammatical function (e.g. free, gratis, and for nothing). □ **tautological** adj., **tautologous** adj.

tavern n. (old use) an inn, a pub.

taverna n. a Greek restaurant.

tawdry adj. (**tawdrier**) showy but without real value. □ **tawdrily** adv., **tawdriness** n.

tawny adj. orange-brown.

tax n. **1** money to be paid to a government. **2** something that makes a heavy demand. ●v. **1** impose a tax on. **2** make heavy demands on. □ **taxable** adj., **taxation** n.

tax evasion n. illegal non-payment of tax.

taxi n. (also **taxicab**) a car with a driver which may be hired. ●v. (**taxied, taxiing**) (of an aircraft) move along the ground under its own power.

taxidermy n. the process of preparing, stuffing, and mounting the skins of animals in life-like form. □ **taxidermist** n.

taxonomy n. a scientific classification of organisms. □ **taxonomical** adj., **taxonomist** n.

taxpayer n. a person who pays tax (esp. income tax).

tax return n. a form declaring income and expenditure for a particular year, used for tax assessment.

TB abbr. tuberculosis.

t.b.a. abbr. to be announced.

T-bone n. a T-shaped bone, esp. in a steak from the thin end of the loin.

tbsp abbr. tablespoonful.

Tc symb. technetium.

Te symb. tellurium.

te n. (Music) the seventh note of a major scale, or the note B.

tea n. **1** the dried leaves of a tropical evergreen shrub; a hot drink made by infusing these (or other substances) in boiling water. **2** an afternoon or evening meal at which tea is drunk.

tea bag n. a small porous bag holding tea for infusion.

teacake n. a flat sweet bread-bun served toasted and buttered.

teach v. (**taught, teaching**) impart information or skill to (a person) or about (a subject). □ **teacher** n.

tea chest n. a large wooden box in which tea is exported, also used as a container when moving house etc.

tea cloth n. a tea towel.

teacup n. a cup from which tea etc. is drunk.

teak n. strong heavy wood of an Asian evergreen tree; this tree.

teal n. (pl. **teal** or **teals**) a small duck.

tea leaf n. a dried leaf of tea, esp. after infusion.

team n. a set of players; a set of people or animals working together. ● v. combine into a team or set.

teamster n. **1** a driver of a team. **2** (Amer.) a lorry driver.

teamwork n. organized co-operation.

teapot n. a vessel with a spout, in which tea is made.

tear¹ (tair) v. (**tore, torn, tearing**) **1** pull forcibly apart or away or to pieces; make (a hole etc.) in this way; become torn. **2** move or travel hurriedly. ● n. a hole etc. torn.

tear² (teer) n. a drop of liquid forming in and falling from the eye. □ **in tears** crying.

tearaway n. an unruly young person.

tearful adj. crying or about to cry. □ **tearfully** adv.

tear gas n. a gas causing severe irritation of the eyes.

tearoom n. = tea shop.

tea rose n. a rose with a scent like tea.

tease v. **1** try to provoke in a playful or unkind way. **2** pick into separate strands. ● n. a person fond of teasing others.

teasel n. a plant with bristly heads.

teaset n. a set of cups and plates etc. for serving tea.

tea shop n. a small restaurant serving tea and light refreshments.

teaspoon n. a small spoon for stirring tea etc.; the amount held by this. □ **teaspoonful** n.

tea towel n. a towel for drying washed crockery etc.

teat n. a nipple on a milk-secreting organ; a device of rubber etc. on a feeding bottle, through which the contents are sucked.

technetium (tek-nee-shŭm) n. an artificial radioactive element (symbol Tc).

technical adj. **1** of the mechanical arts and applied sciences; of a particular subject or craft etc.; using technical terms. **2** in a strict legal sense. □ **technically** adv., **technicality** n.

technician n. **1** an expert in the techniques of a subject or craft. **2** a person employed to look after technical equipment.

Technicolor n. (trade mark) a process of producing films in colour; vivid colour.

technique n. a method of doing something.

technocracy n. government by technical experts. □ **technocrat** n.

technology n. the study of mechanical arts and applied sciences; these subjects; their application in industry etc. □ **technological** adj., **technologically** adv., **technologist** n.

teddy n. (in full **teddy bear**) a soft toy bear.

Teddy boy n. (colloquial) a youth of the 1950s wearing Edwardian-style clothes.

tedious adj. tiresome because of length, slowness, or dullness. □ **tediously** adv., **tediousness** n., **tedium** n.

tee n. a cleared space from which a golf ball is driven at the start of play; a small heap of sand or piece of wood for supporting this ball. ● v. (**teed, teeing**) place (a ball) on a tee. □ **tee off** make the first stroke in golf.

teem v. be full of; be present in large numbers; (of water or rain) pour.

teenager n. a person in his or her teens. □ **teenage** adj. **teenaged** adj.

teens n.pl. the years of age from 13 to 19.

teeny adj. (**teenier**) (colloquial) tiny.

teepee var. of **tepee**.

tee shirt n. = **T-shirt**.

teeter v. stand or move unsteadily.

teeth see **tooth**.

teethe v. (of a baby) have its first teeth appear through the gums.

teething troubles n.pl. problems in the early stages of an enterprise.

teetotal adj. abstaining completely from alcohol. □ **teetotaller** n.

TEFL abbr. teaching of English as a foreign language.

Teflon n. (trade mark) a non-stick coating for saucepans etc.

tele- comb. form **1** at a distance. **2** by telephone. **3** television.

telecommunications n.pl. communication by telephone, radio, cable, etc.

telegram n. a message sent by telegraph.

telegraph n. a system or apparatus for sending messages, esp. by electrical impulses along wires. ● v. communicate thus.

telegraphist n. a person employed in telegraphy.

telegraphy n. communication by telegraph. □ **telegraphic** adj., **telegraphically** adv.

telekinesis n. the supposed process of moving things without touching them.

telemarketing n. selling by unsolicited telephone calls.

telemessage n. a message sent by telephone or telex, delivered in printed form.

telemeter n. an apparatus for recording the readings of an instrument and transmitting them by radio. □ **telemetry** n.

telepathy n. communication between minds other than by the senses. □ **telepathic** adj.

telephone n. a device for transmitting speech by wire or radio. ● v. speak to (a person) by telephone; send (a message) by telephone. □ **telephonic** adj., **telephonically** adv., **telephony** n.

telephonist n. an operator of a telephone switchboard.

telephoto lens n. a photographic lens producing a large image of a distant object.

teleprinter n. a device for transmitting, receiving, and printing telegraph messages.

teleprompter n. a device that unrolls a television broadcaster's script out of sight of the viewers.

telesales n.pl. selling by telephone.

telescope n. an optical instrument for making distant objects appear larger. ● v. make or become shorter by sliding each section inside the next; compress or become compressed forcibly. □ **telescopic** adj., **telescopically** adv.

teletext n. a service transmitting written information to television screens.

telethon n. a long television programme broadcast to raise money for charity.

televise v. transmit by television.

television n. a system for reproducing on a screen a view of scenes etc. by radio transmission; televised programmes; (in full **television set**) an apparatus for receiving these. □ **televisual** adj.

telex n. (also **Telex**) a system of telegraphy using teleprinters and public transmission lines. ● v. send a message to (a person) by telex.

tell v. (**told**, **telling**) **1** make known in words; give information to; reveal a secret. **2** decide or determine; distinguish. **3** produce an effect. **4** direct, order. □ **tell off** (*colloquial*) reprimand. **tell tales** reveal secrets.

teller *n.* **1** a narrator. **2** a person appointed to count votes. **3** a bank cashier.

telling *adj.* having a noticeable effect.

tell-tale *n.* a person who tells tales.

tellurium *n.* an element (symbol Te) used in semiconductors.

telly *n.* (*colloquial*) a television.

temerity *n.* audacity, rashness.

temp *n.* (*colloquial*) a temporary employee.

temper *n.* **1** a state of mind as regards calmness or anger. **2** a fit of anger. **3** calmness under provocation. ● *v.* **1** bring (metal or clay) to the required hardness or consistency. **2** moderate the effects of.

tempera *n.* a method of painting using colours mixed with egg.

temperament *n.* a person's nature as it controls his or her behaviour.

temperamental *adj.* of or relating to temperament; excitable or moody. □ **temperamentally** *adv.*

temperance *n.* **1** self-restraint, moderation. **2** total abstinence from alcohol.

temperate *adj.* self-restrained, moderate; (of a climate) without extremes. □ **temperately** *adv.*

temperature *n.* the degree of heat or cold; a body temperature above normal.

tempest *n.* a violent storm.

tempestuous *adj.* stormy.

template *n.* a pattern or gauge, esp. for cutting shapes.

temple *n.* **1** the flat part between the forehead and the ear. **2** a building dedicated to the worship of a god or gods.

tempo *n.* (*pl.* **tempos** or **tempi**) the speed of a piece of music; the rate of motion or activity.

temporal *adj.* **1** secular. **2** of or denoting time. **3** of the temple(s) of the head.

temporary *adj.* lasting for a limited time. □ **temporarily** *adv.*

temporize *v.* (also **-ise**) avoid committing oneself in order to gain time. □ **temporization** *n.*

tempt *v.* persuade or try to persuade by the prospect of pleasure or advantage; arouse a desire in. □ **temptation** *n.,* **tempter** *n.,* **temptress** *n.*

ten *adj. & n.* one more than nine (10, X).

tenable *adj.* able to be defended or held. □ **tenability** *n.*

tenacious *adj.* holding or sticking firmly. □ **tenaciously** *adv.,* **tenacity** *n.*

tenancy *n.* the use of land or a building as a tenant.

tenant *n.* a person who rents land or a building from a landlord.

tench *n.* (*pl.* **tench**) a freshwater fish of the carp family.

tend *v.* **1** take care of. **2** have a specified tendency.

tendency *n.* the way a person or thing is likely to be or behave; an inclination.

tendentious *adj.* biased, not impartial. □ **tendentiously** *adv.*

tender *adj.* **1** not tough or hard; delicate; painful when touched. **2** sensitive; loving, gentle. ● *n.* **1** a formal offer to supply goods or carry out work at a stated price. **2** a vessel or vehicle conveying goods or passengers to and from a larger one. **3** a truck attached to a steam locomotive and carrying fuel and water etc. ● *v.* offer formally; make a tender for. □ **legal tender** currency that must, by law, be accepted in payment. □ **tenderly** adv., **tenderness** n.

tendon n. a strip of strong tissue connecting a muscle to a bone etc.

tendril n. a threadlike part by which a climbing plant clings; a slender curl of hair etc.

tenement n. a large house let in portions to tenants.

tenet n. a firm belief or principle.

tenfold adj. & adv. ten times as much or as many.

tenner n. (colloquial) a ten-pound note.

tennis n. a ball game played with rackets over a net, with a soft ball on an open court (**lawn tennis**), or with a hard ball in a walled court (**real tennis**).

tenon n. a projection shaped to fit into a mortise.

tenor n. **1** general meaning. **2** the highest ordinary male singing voice. ● adj. of tenor pitch.

tense adj. stretched tightly; nervous, anxious. ● v. make or become tense. ● n. (Grammar) any of the forms of a verb that indicate the time of the action. □ **tensely** adv., **tenseness** n.

tensile adj. of tension; capable of being stretched.

tension n. **1** stretching. **2** tenseness of feelings; mental strain. **3** an effect produced by forces pulling against each other. **4** electromagnetic force.

tent n. a portable shelter or dwelling made of canvas etc.

tentacle n. a slender flexible part of certain animals, used for feeling or grasping.

tentative adj. **1** hesitant. **2** done as a trial, experimental. □ **tentatively** adv.

tenterhooks n.pl. □ **on tenterhooks** in suspense because of uncertainty.

tenth adj. & n. next after ninth. □ **tenthly** adv.

tenuous adj. very thin; very slight. □ **tenuousness** n.

tenure n. the holding of an office or of land or accommodation etc.

tepee (tee-pee) n. (also **teepee**) a conical tent used by North American Indians.

tepid adj. slightly warm, lukewarm.

tequila n. a Mexican liquor made from the sap of an agave plant.

terbium n. a metallic element (symbol Tb).

tercentenary n. a 300th anniversary.

tergiversate v. change one's party or principles; make conflicting or evasive statements.

term n. **1** a fixed or limited period; a period of weeks during which a school etc. is open or in which a law court holds sessions. **2** a word or phrase; each quantity or expression in a mathematical series or ratio etc. **3** (**terms**) conditions offered or accepted; relations between people. □ **come to terms with** reconcile oneself to (a difficulty etc.).

termagant n. a bullying woman.

terminal adj. **1** of or forming an end. **2** of or undergoing the last stage of a fatal disease. ● n. **1** a terminus. **2** a building where air passengers arrive and depart. **3** a point of connection in an electric circuit. **4** an apparatus with a VDU and keyboard connected to a large computer etc. □ **terminally** adv.

terminate v. end. □ **terminator** n.

termination n. **1** terminating or being terminated. **2** an induced abortion.

terminology n. the technical terms of a subject. □ **terminological** adj.

terminus n. (pl. **termini** or **terminuses**) the end; the last stopping place on a rail or bus route.

termite n. a small insect that is destructive to timber.

tern n. a seabird similar to a gull.

ternary adj. composed of three parts.

terrace n. 1 a raised level place; a paved area beside a house. 2 a row of houses joined by party walls.

terracotta n. brownish-red unglazed pottery; its colour.

terra firma n. dry land, the ground.

terrain n. land with regard to its natural features.

terrapin n. a freshwater tortoise.

terrarium n. (pl. **terrariums** or **terraria**) 1 a place for keeping small land animals. 2 a sealed glass container with growing plants inside.

terrestrial adj. of the earth; of or living on land.

terrible adj. appalling, distressing; (colloquial) very bad. □ **terribly** adv.

terrier n. a small active dog.

terrific adj. (colloquial) of great size; excellent. □ **terrifically** adv.

terrify v. fill with terror.

terrine n. a kind of pâté; an earthenware dish for this.

territorial adj. of territory.

Territorial Army n. a volunteer reserve force.

territory n. 1 land under the control of a person, state, city, etc. 2 a sphere of action or thought.

terror n. extreme fear; a terrifying person or thing; (colloquial) a troublesome person or thing.

terrorism n. the use of violence and intimidation for political purposes. □ **terrorist** n.

terrorize v. (also **-ise**) fill with terror; coerce by terrorism.

terry adj. a looped cotton fabric used esp. for towels.

terse adj. concise, curt. □ **tersely** adv., **terseness** n.

tertiary (ter-sher-i) adj. next after secondary.

Terylene n. (trade mark) a synthetic fibre.

TESSA abbr. tax exempt special savings account.

tessellated adj. resembling mosaic.

test n. 1 something done to discover a person's or thing's qualities or abilities etc.; an examination (esp. in a school) on a limited subject. 2 a test match. ● v. subject to a test. □ **tester** n.

testament n. 1 a will. 2 evidence, proof. □ **Old Testament** the books of the Bible telling the history and beliefs of the Jews. **New Testament** the books of the Bible telling the life and teachings of Christ.

testate adj. having left a valid will at death. □ **testacy** n.

testator n. a person who has made a will.

testatrix n. a woman who has made a will.

testes see **testis**.

testicle n. a male organ that secretes sperm-bearing fluid.

testify v. bear witness to; give evidence; be evidence of.

testimonial n. 1 a formal statement testifying to character, abilities, etc. 2 a gift showing appreciation.

testimony n. a declaration (esp. under oath); supporting evidence.

testis n. (pl. **testes**) a testicle.

test match n. one of a series of international cricket or rugby matches.

testosterone n. a male sex hormone.

test tube n. a tube of thin glass with one end closed, used in laboratories.

test-tube baby n. (colloquial) a baby conceived by in vitro fertilization.

testy adj. (**testier**) irritable. □ **testily** adv.

tetanus n. a bacterial disease causing painful muscular spasms.

tête-à-tête (tet-a-tet) n. a private conversation, esp. between two people. ● adj. & adv. together in private.

tether n. a rope or chain for tying an animal to a spot. ● v. fasten with a tether. □ **at the end of one's tether** having reached the limit of one's endurance.

tetrahedron n. (pl. **tetrahedra** or **tetrahedrons**) a solid with four sides.

Teutonic adj. of Germanic peoples or their languages.

text n. **1** the main body of a book as distinct from illustrations used. **2** a passage from Scripture used as the subject of a sermon. □ **textual** adj.

textbook n. a book of information for use in studying a subject.

textile n. a woven or machine-knitted fabric. ● adj. of textiles.

texture n. the way a fabric etc. feels to the touch. □ **textural** adj.

textured adj. having a noticeable texture; (of a yarn or fabric) crimped, curled, or looped.

Th symb. thorium.

thalidomide n. a sedative drug found to have caused malformation of babies whose mothers took it during pregnancy.

thallium n. a toxic metallic element (symbol Tl).

than conj. & prep. used to introduce the second element in a comparison.

thank v. express gratitude to. ● n.pl. (**thanks**) expressions of gratitude; (colloquial) thank you. □ **thank you** a polite expression of thanks.

thankful adj. feeling or expressing gratitude.

thankfully adv. in a thankful way; let us be thankful that.

■ **Usage** The use of thankfully to mean 'let us be thankful that' is common, but it is considered incorrect by some people.

thankless adj. not likely to win thanks. □ **thanklessness** n.

thanksgiving n. an expression of gratitude, esp. to God.

that adj. & pron. (pl. **those**) the (person or thing) referred to; the further or less obvious (one) of two. ● adv. to such an extent. ● rel.pron. used to introduce a defining relative clause. ● conj. introducing a dependent clause.

thatch n. a roof made of straw or reeds etc. ● v. cover (a roof) with thatch. □ **thatcher** n.

thaw v. make or become unfrozen; become less cool or less formal in manner. ● n. thawing, weather that thaws ice etc.

the adj. (called the definite article) applied to a noun standing for a specific person or thing, or one or all of a kind, or used to emphasize excellence or importance.

theatre n. (Amer. **theater**) **1** a place for the performance of plays etc.; plays and acting. **2** a lecture hall with seats in tiers. **3** a room where surgical operations are performed.

theatrical adj. of or for the theatre; exaggerated for effect. ● n.pl. (**theatricals**) theatrical (esp. amateur) performances. □ **theatrically** adv., **theatricality** n.

thee pron. (old use) the objective case of thou.

theft *n.* stealing.

their *adj.* of or belonging to them.

theirs *poss.pron.* belonging to them.

theism (th'ee-izm) *n.* belief that the universe was created by a god. □ **theist** *n.*, **theistic** *adj.*

them *pron.* the objective case of *they.*

theme *n.* **1** the subject being discussed. **2** a melody which is repeated. □ **thematic** *adj.*

theme park *n.* a park with amusements organized round one theme.

themselves *pron.* an emphatic and reflexive form of *they* and *them.*

then *adv.* **1** at that time. **2** next, afterwards. **3** in that case. ● *adj.* & *n.* (of) that time.

thence *adv.* (*old use*) from that place or source.

thenceforth *adv.* (*old use*) from then on.

theocracy *n.* a form of government by a divine being or by priests. □ **theocratic** *adj.*

theodolite *n.* a surveying instrument for measuring angles.

theology *n.* the study or a system of religion. □ **theological** *adj.*, **theologian** *n.*

theorem *n.* a mathematical statement to be proved by reasoning.

theoretical *adj.* based on theory only. □ **theoretically** *adv.*

theorist *n.* a person who theorizes.

theorize *v.* (also **-ise**) form theories.

theory *n.* a set of ideas formulated to explain something; an opinion, a supposition; a statement of the principles of a subject.

theosophy *n.* a system of philosophy that aims at direct intuitive knowledge of God. □ **theosophical** *adj.*

therapeutic (the-rǎ-pew-tik) *adj.* contributing to the relief or curing of a disease etc. □ **therapeutically** *adv.*

therapist *n.* a specialist in therapy.

therapy *n.* a healing treatment.

there *adv.* in, at, or to that place; at that point; in that matter. ● *n.* that place. ● *int.* an exclamation of satisfaction or consolation.

thereabouts *adv.* near there.

thereafter *adv.* after that.

thereby *adv.* by that means.

therefore *adv.* for that reason.

therein *adv.* (*formal*) in that place.

thereof *adv.* (*formal*) of that.

thereto *adv.* (*formal*) to that.

thereupon *adv.* in consequence of that, because of that.

thermal *adj.* of or using heat; warm, hot. ● *n.* a rising current of hot air.

thermocouple *n.* a device for measuring temperature by means of the thermoelectric voltage developing between two pieces of wire of different metals joined at each end.

thermodynamics *n.* the science of the relationship between heat and other forms of energy.

thermoelectric *adj.* producing electricity by a difference of temperatures.

thermometer *n.* an instrument for measuring temperature.

thermonuclear *adj.* of or using nuclear reactions that occur only at very high temperatures.

thermoplastic *adj.* & *n.* (a substance) becoming soft when heated and hardening when cooled.

thermos *n.* (in full **thermos flask**) (*trade mark*) a vacuum flask.

thermosetting *adj.* (of plastics) setting permanently when heated.

thermostat n. a device that regulates temperature automatically. □ **thermostatic** adj., **thermostatically** adv.

thesaurus n. (pl. **thesauri** or **thesauruses**) a dictionary of synonyms.

these see **this**.

thesis n. (pl. **theses**) 1 a theory put forward and supported by reasoning. 2 a lengthy written essay submitted for a university degree.

thespian adj. of the theatre. ● n. an actor or actress.

theta n. the eighth letter of the Greek alphabet (Θ, θ).

they pron. people or things mentioned or unspecified.

thiamine (also **thiamin**) a vitamin of the B complex found in unrefined cereals.

thick adj. 1 of a great or specified distance between opposite surfaces. 2 dense; fairly stiff in consistency. 3 (colloquial) stupid. 4 (colloquial) friendly, intimate. ● adv. thickly. ● n. (**the thick**) the busiest or most intense part. □ **thickly** adv., **thickness** n.

thicken v. make or become thicker.

thicket n. a close group of shrubs or small trees.

thickset adj. 1 stocky, burly. 2 set or growing close together.

thick-skinned adj. not sensitive to criticism or snubs.

thief n. (pl. **thieves**) a person who steals. □ **thievish** adj.

thieve v. be a thief; steal. □ **thievery** n.

thigh n. the upper part of the leg, between the hip and the knee.

thimble n. a hard cap worn to protect the end of the finger in sewing.

thin adj. (**thinner**) 1 not thick; lean, not plump. 2 lacking substance, weak. ● adv. thinly. ● v. (**thinned**) make or become thinner. □ **thinly** adv., **thinness** n.

thine adj. & poss.pron. (old use) belonging to thee.

thing n. 1 whatever is or may be perceived, known, or thought about; an act, fact, idea, task, etc.; an item; an inanimate object. 2 (**things**) belongings; utensils, equipment; circumstances.

think v. (**thought**, **thinking**) exercise the mind, form ideas; form or have as an idea, opinion, or plan. ● n. (colloquial) an act of thinking. □ **think better of it** change one's mind after thought. □ **thinker** n.

think-tank n. a group providing ideas and advice on national or commercial problems.

thinner n. a substance for thinning paint.

third adj. next after second. ● n. the third thing, class, etc.; one of three equal parts. □ **thirdly** adv.

third degree n. long and severe questioning.

third-degree burn n. a burn of the most severe kind.

third party n. another person etc. besides the two principals.

third-party insurance n. insurance only giving protection against liability for damage or injury to any other person.

third-rate adj. very inferior in quality.

Third World n. the developing countries of Asia, Africa, and Latin America.

thirst n. the feeling caused by a desire to drink; any strong desire. ● v. feel a thirst. □ **thirsty** adj., **thirstily** adv.

thirteen adj. & n. one more than twelve (13, XIII). □ **thirteenth** adj. & n.

thirty adj. three times ten (30, XXX). □ **thirtieth** adj. & n.

this adj. & pron. (pl. **these**) the person or thing near or present or mentioned.

thistle n. a prickly plant. □ **thistly** adj.

thistledown n. the very light fluff on thistle seeds.

thither adv. (old use) to or towards that place.

thong n. a strip of leather used as a fastening or lash etc.

thorax n. (pl. **thoraces** or **thoraxes**) the part of the body between the neck and the abdomen. □ **thoracic** adj.

thorium n. a radioactive metallic element (symbol Th).

thorn n. a small sharp projection on a plant; a thorn-bearing tree or shrub. □ **thorny** adj.

thorough adj. complete in every way; detailed; methodical. □ **thoroughly** adv., **thoroughness** n.

thoroughbred n. & adj. (a horse etc.) bred of pure or pedigree stock.

thoroughfare n. a public way open at both ends.

those see **that**.

thou pron. (old use) you.

though conj. in spite of the fact that, even supposing. ● adv. (colloquial) however.

thought see **think**. n. the process or power of thinking; a way of thinking; an idea or intention; consideration.

thoughtful adj. 1 thinking deeply; thought out carefully. 2 considerate. □ **thoughtfully** adv., **thoughtfulness** n.

thoughtless adj. careless; inconsiderate. □ **thoughtlessly** adv., **thoughtlessness** n.

thousand adj. & n. ten hundred (1000, M). □ **thousandth** adj. & n.

thrall n. □ **in thrall** in bondage; enslaved.

thrash v. beat, esp. with a stick or whip; defeat thoroughly;

thresh; make flailing movements. □ **thrash out** discuss thoroughly.

thread n. 1 a thin length of spun cotton or wool etc.; something compared to this. 2 the spiral ridge of a screw. ● v. pass a thread through; pass (a strip or thread) through or round something; make (one's way) through a crowd etc. □ **threader** n.

threadbare adj. (of cloth) with the nap worn off and threads visible; shabbily dressed.

threadworm n. a small threadlike parasitic worm.

threat n. an expression of intention to punish, hurt, or harm; a person or thing thought likely to bring harm or danger.

threaten v. make or be a threat (to).

three adj. & n. one more than two (3, III).

three-dimensional adj. having or appearing to have length, breadth, and depth.

threefold adj. & adv. three times as much or as many.

threesome n. three together, a trio.

thresh v. beat out (grain) from husks of corn; make flailing movements.

threshold n. 1 a piece of wood or stone forming the bottom of a doorway; a point of entry. 2 the lowest limit at which a stimulus is perceptible.

threw see **throw**.

thrice adv. (old use) three times.

thrift n. 1 economical management of resources. 2 a plant with pink flowers. □ **thrifty** adj., **thriftily** adv.

thrill n. a wave of feeling or excitement. ● v. feel or cause to feel a thrill.

thriller n. an exciting story or play etc., esp. involving crime.

thrive v. (**throve** or **thrived**, **thriven** or **thrived**, **thriving**) grow or develop well; prosper.

throat n. the front of the neck; the passage from the mouth to the oesophagus or lungs.

throaty adj. (**throatier**) uttered deep in the throat; hoarse. □ **throatily** adv.

throb v. (**throbbed**) vibrate or sound with a persistent rhythm; (of the heart or pulse) beat with more than usual force. ● n. a throbbing beat or sound.

throes n.pl. severe pangs of pain. □ **in the throes of** struggling with the task of.

thrombosis n. (pl. **thromboses**) the formation of a clot of blood in a blood vessel or organ of the body.

throne n. a ceremonial seat for a monarch, bishop, etc.; sovereign power.

throng n. a crowded mass of people. ● v. move or press in a throng; fill with a throng.

throttle n. a valve controlling the flow of fuel or steam etc. to an engine; the lever controlling this. ● v. strangle.

through prep. 1 from end to end or side to side of; from beginning to end (of). 2 by the agency, means, or fault of. ● adv. 1 from end to end or side to side; from beginning to end; entering at one point and coming out at another. 2 so as to have finished, so as to have passed (an examination). 3 so as to be connected by telephone. ● adj. (of journeys etc.) going through; passing without stopping.

throughout prep. & adv. right through, from beginning to end (of).

throughput n. the amount of material processed.

throve see **thrive**.

throw v. (**threw**, **thrown**, **throwing**) 1 send with some force through the air; cause to fall; put (clothes etc.) on or off hastily. 2 cause to be in a certain state; have (a fit or tantrum); (colloquial) disconcert. 3 operate (a switch or lever). 4 (colloquial) give (a party). 5 shape (pottery) on a wheel. ● n. an act of throwing; the distance something is thrown. □ **throw away** part with as useless or unwanted; fail to make use of. **throw in the towel** admit defeat or failure. **throw out** discard; reject. **throw up** vomit; bring to notice. □ **thrower** n.

throwback n. □ an animal etc. showing characteristics of an earlier ancestor.

thru (Amer.) = **through**.

thrum v. (**thrummed**) strum, sound monotonously.

thrush n. 1 a songbird, esp. one with a speckled breast. 2 a fungal infection of the mouth, throat, or vagina.

thrust v. (**thrust**, **thrusting**) push forcibly; make a forward stroke with a sword etc. ● n. a thrusting movement or force.

thud n. a dull low sound like that of a blow. ● v. (**thudded**) make or fall with a thud.

thug n. a vicious ruffian. □ **thuggery** n.

thulium n. a metallic element (symbol Tm).

thumb n. the short thick finger set apart from the other four. ● v. 1 touch or turn (pages etc.) with the thumbs. 2 request (a lift) by signalling with one's thumb. □ **under the thumb of** completely under the influence of.

thumbnail n. the nail of the thumb. ● adj. brief, concise.

thump v. strike or knock heavily (esp. with the fist), thud. ● n. a heavy blow; a sound of thumping.

thunder n. the loud noise that accompanies lightning; any similar sound. ● v. sound with or like thunder; utter loudly or angrily. □ **steal a person's thunder** forestall him or her. □ **thundery** adj.

thunderbolt n. **1** a lightning flash. **2** a sudden unexpected event or piece of news.

thunderclap n. a crash of thunder.

thundering adj. (colloquial) very big.

thunderous adj. like thunder.

thunderstorm n. a storm accompanied by thunder.

thunderstruck adj. amazed.

Thur. abbr. (also **Thurs.**) Thursday.

Thursday n. the day after Wednesday.

thus adv. (formal) in this way; as a result of this; to this extent.

thwack v. strike with a heavy blow. ● n. this blow or sound.

thwart v. prevent from doing what is intended; frustrate. ● n. a rower's bench across a boat.

thy adj. (old use) belonging to thee.

thyme (tym) n. a herb with fragrant leaves.

thymus n. (pl. **thymi**) the ductless gland near the base of the neck.

thyroid adj. & n. (in full **thyroid gland**) a large ductless gland in the neck, secreting a growth hormone.

thyself pron. an emphatic and reflexive form of thou and thee.

Ti symb. titanium.

tiara n. a woman's jewelled semicircular headdress.

tibia n. (pl. **tibiae**) the shin bone.

tic n. an involuntary muscular twitch.

tick n. **1** a regular clicking sound, esp. made by a clock or watch. **2** (colloquial) a moment. **3** a small mark placed against

an item in a list etc., esp. to show that it is correct. **4** a blood-sucking mite or parasitic insect. **5** (colloquial) financial credit. ● v. **1** (of a clock etc.) make a series of ticks. **2** mark with a tick. □ **tick off** (colloquial) reprimand. **tick over** (of an engine) idle.

ticket n. **1** a marked piece of card or paper entitling the holder to a certain right (e.g. to travel by train etc.). **2** a label. **3** notification of a traffic offence. **4** a certificate of qualification as a ship's master or pilot etc. **5** a list of candidates for office. **6** (**the ticket**) (colloquial) the correct or desirable thing. ● v. (**ticketed**) put a ticket on.

ticking n. strong fabric used for covering mattresses, pillows, etc.

tickle v. touch or stroke lightly so as to cause a slight tingling sensation; feel this sensation; amuse, please. ● n. the act or sensation of tickling.

ticklish adj. **1** sensitive to tickling. **2** (of a problem) requiring careful handling.

tick-tack n. semaphore signalling used by racecourse bookmakers.

tidal adj. of or affected by tides.

tidbit n. (Amer.) a titbit.

tiddler n. (colloquial) a small fish, esp. a stickleback or minnow.

tiddly adj. (**tiddlier**) (colloquial) **1** very small. **2** slightly drunk.

tiddlywinks n. a game involving flicking small counters into a cup.

tide n. **1** the sea's regular rise and fall. **2** a trend of feeling or events etc. □ **tide over** help temporarily.

tidemark n. a mark made by the tide at high water; a line left round a bath by dirty water.

tidings n.pl. (literary) news.

tidy *adj.* (**tidier**) neat and orderly. ● *v.* make tidy. □ **tidily** *adv.*, **tidiness** *n.*

tie *v.* (**tied, tying**) **1** attach or fasten with cord etc.; form into a knot or bow; unite. **2** make the same score as another competitor. **3** restrict, limit. ● *n.* **1** a cord etc. used for tying something. **2** a strip of material worn below the collar and knotted at the front of the neck. **3** something that unites or restricts. **4** an equal score between competitors. **5** a sports match between two of a set of teams or players. □ **tie in** link or (of information etc.) be connected with something else. **tie up** fasten with cord etc.; make (money etc.) not readily available for use; occupy fully.

tie-break *n.* a means of deciding the winner when competitors have tied.

tied *adj.* **1** (of a public house) bound to supply a particular brewer's beer. **2** (of a house) for occupation only by a person working for its owner.

tiepin *n.* an ornamental pin for holding a necktie in place.

tier (teer) *n.* any of a series of rows, ranks, or units of a structure placed one above the other.

tie-up *n.* a connection, a link.

tiff *n.* a petty quarrel.

tiger *n.* a large striped animal of the cat family.

tight *adj.* **1** held or fastened firmly, fitting closely; tense, not slack. **2** (of money etc.) severely restricted; (*colloquial*) stingy. **3** (*colloquial*) drunk. ● *adv.* tightly. □ **tightly** *adv.*, **tightness** *n.*

tight corner *n.* a difficult situation.

tighten *v.* make or become tighter.

tight-fisted *adj.* stingy.

tightrope *n.* a tightly stretched rope on which acrobats perform.

tights *n.pl.* a garment closely covering the legs and lower part of the body.

tigress *n.* a female tiger.

tike var. of **tyke**.

tilde *n.* a mark (˜) put over a letter to change its pronunciation.

tile *n.* a thin slab of baked clay etc. used for covering roofs, walls, or floors. ● *v.* cover with tiles.

till¹ *prep.* & *conj.* up to (a specified time).

■ **Usage** Till can be replaced by *until*, which is more formal in style.

till² *n.* a drawer for money in a shop etc. ● *v.* prepare and use (land) for growing crops.

tiller *n.* a bar by which the rudder of a boat is turned.

tilt *v.* move into a sloping position. ● *n.* a sloping position. □ **at full tilt** at full speed or force.

timber *n.* wood prepared for use in building or carpentry; trees suitable for this; a wooden beam used in constructing a house or ship.

timbered *adj.* constructed of timber or with a timber framework; (of land) wooded.

timbre (tambr) *n.* the characteristic quality of the sound of a voice or instrument.

time *n.* **1** all the years of the past, present, and future; a point or portion of this; an occasion, an instance; (**times**) contemporary circumstances. **2** rhythm in music. **3** (**times**) expressing multiplication. ● *v.* choose the time for; measure the time taken by. □ **behind the times** out of date. **for the time being** until another arrangement is made. **from time to time** at intervals.

in time not late; eventually. **on time** punctually.

time-and-motion adj. concerned with measuring efficiency of effort.

time bomb n. a bomb that can be set to explode at a particular time.

time-honoured adj. respected because of antiquity; traditional.

time lag n. an interval between two connected events.

timeless adj. not affected by the passage of time. □ **timelessness** n.

timely adj. (**timelier**) occurring at just the right time. □ **timeliness** n.

timepiece n. a clock or watch.

timeshare n. a share in a property that allows use by several joint owners at agreed different times.

time switch n. a switch operating automatically at a set time.

timetable n. a list showing the times at which certain events take place.

time zone n. a region (between parallels of longitude) where a common standard time is used.

timid adj. easily alarmed, not bold, shy. □ **timidly** adv., **timidity** n.

timing n. the way something is timed; the control of the opening and closing of valves in an engine.

timorous adj. timid. □ **timorously** adv., **timorousness** n.

timpani n.pl. (also **tympani**) kettledrums. □ **timpanist** n.

tin n. 1 a metallic element (symbol Sn), a silvery-white metal. 2 a metal box or other container, one in which food is sealed for preservation. ● v. (**tinned**) 1 seal (food) in a tin. 2 coat with tin.

tincture n. 1 a solution of a medicinal substance in alcohol. 2 a slight tinge.

tinder n. any dry substance that catches fire easily.

tine n. a prong or point of a fork, harrow, or antler.

tinge v. (**tinged, tingeing**) colour slightly; give a slight trace of an element or quality to. ● n. a slight colouring or trace.

tingle v. have a slight pricking or stinging sensation. ● n. this sensation.

tinker n. 1 a travelling mender of pots and pans. 2 (colloquial) a mischievous person or animal. ● v. work at something casually, trying to repair or improve it.

tinkle n. a series of short light ringing sounds. ● v. make or cause to make a tinkle.

tinnitus n. repeated ringing or other sounds in the ears.

tinny adj. (**tinnier**) (of metal objects) flimsy; (of sound) thin and metallic.

tinpot adj. worthless, inferior.

tinsel n. glittering decorative metallic strips or threads.

tint n. a variety or slight trace of a colour. ● v. colour slightly.

tiny adj. (**tinier**) very small.

tip n. 1 tilt, topple; pour out (a thing's contents) by tilting. 2 make a small present of money to, esp. in acknowledgement of services. 3 name as a likely winner. 4 put a substance on the end of (something small or tapering). ● n. 1 a small money present. 2 a useful piece of advice. 3 the end of something small or tapering. 4 a place where rubbish etc. is tipped. □ **tip off** give a warning or hint to. □ **tipper** n.

tip-off n. a warning or hint.

tippet n. a small cape or collar of fur with hanging ends.

tipple v. drink (wine or spirits etc.) repeatedly. ● n. (colloquial) an alcoholic drink.

tipster n. a person who gives tips about racehorses etc.

tipsy *adj.* (**tipsier**) slightly drunk.

tiptoe *v.* (**tiptoed, tiptoeing**) walk very quietly or carefully.

tiptop *adj.* (*colloquial*) first-rate.

tirade *n.* a long angry speech.

tire *v.* make or become tired. ● *n.* Amer. sp. of **tyre**.

tired *adj.* feeling a desire to sleep or rest. □ **tired of** having had enough of; bored or impatient with.

tireless *adj.* not tiring easily. □ **tirelessly** *adv.*, **tirelessness** *n.*

tiresome *adj.* annoying; tedious.

tiro var. of **tyro**.

tissue *n.* 1 a substance forming an animal or plant body. 2 tissue paper; a piece of soft absorbent paper used as a handkerchief etc.

tissue paper *n.* thin soft paper used for packing things.

tit *n.* 1 any of several small birds. 2 (*slang*) a breast. □ **tit for tat** blow for blow.

titanic *adj.* gigantic.

titanium *n.* a grey metallic element (symbol Ti).

titbit *n.* (*Amer.* **tidbit**) a choice bit of food or item of information.

tithe *n.* one-tenth of income or produce formerly paid to the Church.

titillate *v.* excite or stimulate pleasantly. □ **titillation** *n.*

titivate *v.* (*colloquial*) smarten up, put finishing touches to. □ **titivation** *n.*

title *n.* 1 the name of a book, poem, picture, etc. 2 a word denoting rank or office, or used in speaking of or to the holder. 3 a championship or sport. 4 the legal right to ownership of property.

titled *adj.* having a title of nobility.

title deed *n.* a legal document proving a person's right to a property.

title role *n.* the part in a play etc. from which the title is taken.

titmouse *n.* (*pl.* **titmice**) a small active tit.

titter *n.* a high-pitched giggle. ● *v.* give a titter.

tittle-tattle *v.* & *n.* gossip.

titular *adj.* of a title; having the title of ruler etc. but no real authority.

tizzy *n.* (*colloquial*) a state of nervous agitation or confusion.

T-junction *n.* a junction where one road meets another at right angles but does not cross it.

Ti *symb.* titanium.

Tm *symb.* thulium.

TNT *abbr.* trinitrotoluene, a powerful explosive.

to *prep.* 1 towards; as far as. 2 as compared with, in respect of. 3 for (a person or thing) to hold or possess or be affected by. 4 (with a verb) forming an infinitive, or expressing purpose or consequence etc.; used alone when the infinitive is understood. ● *adv.* 1 to a closed position. 2 into a state of consciousness or activity. □ **to and fro** backwards and forwards.

toad *n.* a froglike animal living chiefly on land.

toad-in-the-hole *n.* sausages baked in batter.

toadstool *n.* a fungus (usu. poisonous) with a round top on a stalk.

toady *n.* a person who flatters in order to gain advantage. ● *v.* behave in this way.

toast *n.* 1 toasted bread. 2 a person or thing in whose honour a company is requested to drink; this request or instance of drinking. ● *v.* 1 brown (bread etc.) by heating. 2 express good wishes to by drinking.

toaster *n.* an electrical device for toasting bread.

tobacco *n.* a plant with leaves that are used for smoking or snuff; its prepared leaves.

tobacconist *n.* a shopkeeper who sells cigarettes etc.

toboggan *n.* a small sledge used for sliding downhill. □ **tobogganing** *n.*

toby jug *n.* a mug or jug in the form of a stout old man.

tocsin *n.* an alarm bell or signal.

today *adv. & n.* (on) this present day; (at) the present time.

toddle *v.* (of a young child) walk with short unsteady steps.

toddler *n.* a child who has only recently learnt to walk.

toddy *n.* a sweetened drink of spirits and hot water.

to-do *n.* a fuss or commotion.

toe *n.* any of the divisions (five in humans) of the front part of the foot; part of a shoe or stocking covering the toes. ● *v.* touch with the toe(s). □ **be on one's toes** be alert or eager. **toe the line** conform; obey orders.

toehold *n.* **1** a slight foothold. **2** a small beginning or advantage.

toff *n.* (*slang*) an upper-class or well-dressed person.

toffee *n.* a sweet made with heated butter and sugar.

toffee apple *n.* a toffee-coated apple on a stick.

tofu *n.* soft white soya bean curd.

tog *n.* **1** a unit for measuring the warmth of duvets or clothing. **2** (**togs**) (*colloquial*) clothes. □ **tog out** or **up** (**togged**) (*colloquial*) dress.

toga *n.* a loose outer garment worn by men in ancient Rome.

together *adv.* in or into company or conjunction; towards each other; simultaneously.

toggle *n.* **1** a short piece of wood etc. passed through a loop as a fastening device. **2** a switch that turns a function on and off alternately.

toil *v.* work or move laboriously. ● *n.* laborious work. □ **toilsome** *adj.*

toilet *n.* **1** a lavatory. **2** (also **toilette**) the process of dressing and grooming oneself.

toiletries *n.pl.* articles used in washing and grooming oneself.

toilet water *n.* a light perfume.

token *n.* **1** a sign, a symbol. **2** a voucher that can be exchanged for goods. **3** a disc used as money in a slot machine etc. **4** a keepsake. ● *adj.* nominal; symbolic.

tokenism *n.* the granting of only small concessions esp. to a minority group.

told *see* **tell**. □ **all told** counting everything or everyone.

tolerable *adj.* endurable; passable. □ **tolerably** *adv.*

tolerance *n.* **1** willingness to tolerate. **2** a permitted variation. □ **tolerant** *adj.*, **tolerantly** *adv.*

tolerate *v.* permit without protest or interference; endure. □ **toleration** *n.*

toll *n.* **1** a tax paid for the use of a public road etc. **2** loss or damage caused by a disaster. **3** a stroke of a tolling bell. ● *v.* ring with slow strokes, esp. to mark a death.

toll gate *n.* a barrier preventing passage on a road etc. until a toll is paid.

tom *n.* (in full **tom-cat**) a male cat.

tomahawk *n.* a light axe used by North American Indians.

tomato *n.* (*pl.* **tomatoes**) a red fruit used as a vegetable.

tomb *n.* a grave or other place of burial.

tombola *n.* a lottery with tickets drawn for immediate prizes.

tomboy *n.* a girl who enjoys rough and noisy activities.

tombstone *n.* a memorial stone set up over a grave.

tome *n.* a large book.

tomfoolery *n.* foolish behaviour.

tommyrot *n.* (*slang*) nonsense.

tomorrow *adv. & n.* (on) the day after today; (in) the near future.

tom-tom *n.* a drum beaten with the hands.

ton *n.* a measure of weight, either 2,240 lb (**long ton**) or 2,000 lb (**short ton**) or 1,000 kg (**metric ton**); a unit of volume in shipping.

tone *n.* **1** a musical or vocal sound, esp. with reference to its pitch, quality, and strength. **2** an interval of a major second in music (e.g. between C and D). **3** proper firmness of muscles etc. **4** a shade of colour. **5** the spirit or character of something. ● *v.* **1** give tone to. **2** harmonize in colour. □ **tone down** make less intense. □ **tonal** *adj.*, **tonally** *adv.*, **tonality** *n.*

tone-deaf *adj.* unable to perceive differences of musical pitch.

toneless *adj.* without positive tone, not expressive. □ **tonelessly** *adv.*

tone poem *n.* an orchestral composition illustrating a poetic idea.

toner *n.* an ink-like substance used in a photocopier, printer, etc.

tongs *n.pl.* an instrument with two arms used for grasping things.

tongue *n.* **1** the muscular organ in the mouth, used in tasting and speaking; the tongue of an ox etc. as food. **2** a language. **3** a projecting strip; a tapering jet of flame.

tongue-in-cheek *adj.* ironic, insincere.

tongue-tied *adj.* silent from shyness etc.

tongue-twister *n.* a sequence of words difficult to pronounce quickly and correctly.

tonic *n.* **1** a medicine with an invigorating effect. **2** tonic water. **3** a keynote in music. ● *adj.* invigorating.

tonic sol-fa *n.* the system of syllables *doh*, *ray*, *me*, etc. representing notes of a musical scale.

tonic water *n.* a carbonated soft drink flavoured with quinine.

tonight *adv. & n.* (on) the present evening or night, or that of today.

tonnage *n.* a ship's carrying capacity expressed in tons; the charge per ton for carrying cargo.

tonne *n.* a metric ton, 1000 kg.

tonsil *n.* either of two small organs near the root of the tongue.

tonsillitis *n.* inflammation of the tonsils.

tonsorial *adj.* of a barber.

tonsure *n.* shaving the top or all of the head as a religious symbol; this shaven area. □ **tonsured** *adj.*

too *adv.* **1** to a greater extent than is desirable; (*colloquial*) very. **2** also.

took *see* **take**.

tool *n.* a thing used for working on something; a person used by another for his or her own purposes. ● *v.* shape or ornament with a tool; equip with tools.

toot *n.* a short sound produced by a horn or whistle. ● *v.* make or cause to make a toot.

tooth *n.* (*pl.* **teeth**) each of the white bony structures in the jaws, used in biting and chewing; a toothlike part or projection. □ **toothed** *adj.*

toothpaste *n.* paste for cleaning the teeth.

toothpick n. a small pointed instrument for removing food from between the teeth.

toothy adj. (**toothier**) having many or large teeth.

top n. **1** the highest point, part, or position; the upper surface; something forming the upper part or covering. **2** a garment for the upper part of the body. **3** the utmost degree or intensity. **4** a toy that spins on its point when set in motion. ● adj. highest in position or rank etc. ● v. (**topped**) **1** provide or be a top for; reach the top of; be higher than. **2** add as a final thing. □ **on top of** in addition to. **top up** fill up (something half empty).

topaz n. a semiprecious stone of various colours, esp. yellow.

topcoat n. **1** an overcoat. **2** a final coat of paint etc.

top dog n. (colloquial) the master or victor.

top dress v. apply fertilizer on the top of (soil).

top hat n. a man's tall hat worn with formal dress.

top-heavy adj. heavy at the top and liable to fall over.

topiary n. the art of clipping shrubs etc. into ornamental shapes. □ **topiarist** n.

topic n. the subject of a discussion or written work.

topical adj. having reference to current events. □ **topically** adv., **topicality** n.

topknot n. a tuft, crest, or bow on top of the head.

topless adj. wearing nothing on top, having the breasts bare.

topmost adj. highest.

top-notch adj. (colloquial) first-rate.

topology n. the study of geometrical properties unaffected by changes of shape or size.

topography n. local geography, the position of the rivers, roads,
buildings, etc., of a place or district. □ **topographical** adj.

topper n. (colloquial) a top hat.

topple v. be unsteady and fall; cause to do this.

top secret adj. of the highest category of secrecy.

topside n. beef from the upper part of the haunch.

topsoil n. the top layer of the soil.

topspin n. a spinning motion given to a ball by hitting it forward and upward.

topsy-turvy adv. & adj. upside down; in or into great disorder.

tor n. a hill or rocky peak.

torch n. a small hand-held electric lamp; a burning piece of wood etc. carried as a light. □ **torchlight** n.

tore see **tear¹**.

toreador n. a fighter (esp. on horseback) in a bullfight.

torment n. (tor-ment) severe suffering; a cause of this. ● v. (tor-ment) subject to torment; tease, annoy. □ **tormentor** n.

torn see **tear¹**.

tornado n. (pl. **tornadoes** or **tornados**) a violent destructive whirlwind.

torpedo n. (pl. **torpedoes**) an explosive underwater missile. ● v. attack or destroy with a torpedo.

torpid adj. sluggish and inactive. □ **torpidly** adv., **torpidity** n.

torpor n. a sluggish condition.

torque n. a force that produces rotation.

torr n. (pl. **torr**) a unit of pressure.

torrent n. a rushing stream or flow; a downpour. □ **torrential** adj.

torrid adj. intensely hot; passionate.

torsion n. twisting, a spiral twist.

torso n. (pl. **torsos**) the trunk of the human body.

tort *n.* (*Law*) any private or civil wrong (other than breach of contract) for which damages may be claimed.

tortilla *n.* a Mexican flat maize cake eaten hot.

tortoise *n.* a slow-moving reptile with a hard shell.

tortoiseshell *n.* the mottled yellowish-brown shell of certain turtles, used for making combs etc.

tortoiseshell cat *n.* a cat with mottled colouring.

tortuous *adj.* full of twists and turns. □ **tortuously** *adv.*

torture *n.* severe pain; the infliction of pain as a punishment or means of coercion. ● *v.* inflict torture upon. □ **torturer** *n.*

Tory *n.* (*colloquial*) a member or supporter of the Conservative Party.

toss *v.* throw lightly; throw up (a coin) to settle a question by the way it falls; roll about from side to side; coat (food) by gently shaking it in dressing etc. ● *n.* tossing. □ **toss off** drink rapidly; compose or finish rapidly. **toss up** toss a coin to decide a choice etc.

toss-up *n.* **1** the tossing of a coin. **2** an even chance.

tot *n.* **1** a small child. **2** a small quantity of spirits. ● *v.* (**totted**) (*colloquial*) add up. □ **tot up**

total *adj.* including everything or everyone; complete. ● *n.* a total amount. ● *v.* (**totalled**; *Amer.* **totaled**) reckon the total of; amount to. □ **totally** *adv.*, **totality** *n.*

totalitarian *adj.* of a regime in which no rival parties or loyalties are permitted. □ **totalitarianism** *n.*

totalizator *n.* (also **totalisator**, **totalizer**) a device that automatically registers bets, so that

the total amount can be divided among the winners.

tote *n.* (*colloquial*) a totalizator. ● *v.* (*Amer.*) carry.

totem *n.* a tribal emblem among North American Indians.

totem pole *n.* a pole decorated with totems.

totter *v.* walk or rock unsteadily. ● *n.* a tottering walk or movement. □ **tottery** *adj.*

toucan *n.* a tropical American bird with an immense beak.

touch *v.* **1** be, come, or bring into contact; feel or stroke; press or strike lightly. **2** reach; affect; rouse sympathy in. **3** (*slang*) persuade to give or lend money. ● *n.* **1** an act, fact, or manner of touching; the ability to perceive things through touching them. **2** a style of workmanship. **3** a detail; a slight trace. □ **touch down** touch the ball on the ground behind the goal line in rugby; (of an aircraft) land. **touch off** cause to explode; start (a process). **touch on** mention briefly. **touch up** improve by making small additions; (*slang*) molest; touch sexually.

touch-and-go *adj.* uncertain as regards the result.

touché (too-*shay*) *int.* an acknowledgement of a hit in fencing, or of a valid criticism.

touching *adj.* rousing kindly feelings or pity. ● *prep.* concerning.

touchline *n.* the side limit of a football field.

touchstone *n.* a standard or criterion.

touchy *adj.* (**touchier**) easily offended. □ **touchiness** *n.*

tough *adj.* hard to break, cut, or chew; hardy; unyielding, resolute; difficult; severe. ● *n.* a rough violent person. □ **toughness** *n.*

toughen *v.* make or become tough or tougher.

toupee (too-pay) n. a small wig.

tour n. a journey, visiting things of interest or giving performances. ● v. make a tour (of). □ **on tour** touring.

tour de force n. (pl. **tours de force**) a feat of strength or skill.

tourism n. the organization of holidays and services for tourists.

tourist n. a person visiting a place for recreation.

tourmaline n. a mineral possessing unusual electric properties and used as a gem.

tournament n. a contest of skill involving a series of matches.

tourniquet n. a strip of material pulled tightly round a limb to stop the flow of blood from an artery.

tousle v. make (hair etc.) untidy by ruffling.

tout v. 1 pester people to buy tickets etc. 2 spy on racehorses in training. ● n. a person who touts; a person who sells tickets for popular events at high prices.

tow v. pull along behind. ● n. 1 the act of towing. 2 coarse fibres of flax or hemp.

towards prep. (also **toward**) in the direction of; in relation to; as a contribution to; near, approaching.

towel n. a piece of absorbent material for drying things. ● v. (**towelled**; Amer. **toweled**) rub with a towel.

towelling n. (Amer. **toweling**) fabric for towels.

tower n. a tall structure, esp. as part of a church or castle etc. ● v. be very tall.

tower block n. a tall building with many storeys.

tower of strength n. a person who gives strong reliable support.

town n. a collection of buildings (larger than a village); its inhabitants; a central business and shopping area. □ **go to town** (colloquial) do something lavishly or with enthusiasm. □ **townsman** n., **townswoman** n.

town hall n. a building containing local government offices etc.

township n. a small town; (S. Afr. hist.) an urban area set aside for black people.

towpath n. a path beside a canal or river, originally for horses towing barges.

toxaemia (tok-see-miă) n. (Amer. **toxemia**) 1 blood poisoning. 2 abnormally high blood pressure in pregnancy.

toxic adj. of or caused by poison; poisonous. □ **toxicity** n.

toxicology n. the study of poisons. □ **toxicologist** n.

toxin n. a poison produced by a living organism, esp. one formed in the body.

toy n. a thing to play with. ● adj. (of a dog) of a small variety. □ **toy with** handle idly; deal with (a thing) without seriousness.

toyboy n. (colloquial) a woman's much younger male lover.

trace n. 1 a track or mark left behind; a sign of what has existed or occurred. 2 a very small quantity. 3 each of the two sidestraps by which a horse draws a vehicle. ● v. 1 follow or discover by observing marks or other evidence. 2 mark out; copy by using tracing paper or carbon paper. □ **kick over the traces** become insubordinate or reckless. □ **traceable** adj., **tracer** n.

trace element n. a chemical element occurring or required only in minute amounts.

tracery n. a decorative pattern of interlacing lines, esp. in stone.

trachea (tră-kee-ă) n. (pl. **tracheae** or **tracheas**) the windpipe.

tracheotomy | trailer

tracheotomy (tra-ki-ot-ŏmi) *n.* a surgical opening made in the trachea.

tracing *n.* a copy of a map or drawing etc. made by tracing it.

tracing paper *n.* transparent paper used for making tracings.

track *n.* **1** marks left by a moving person or thing. **2** a course; a path, a rough road; a railway line. **3** a section on a CD, tape, etc. **4** a continuous band round the wheels of a tank, tractor etc. ● *v.* follow the track of, find or observe in this way. □ **keep or lose track of** keep or fail to keep oneself informed about. □ **tracker** *n.*

tracksuit *n.* a loose warm suit worn for exercise etc.

tract *n.* **1** a stretch of land. **2** a system of connected parts of the body, along which something passes. **3** a pamphlet with a short essay, esp. on a religious subject.

tractable *adj.* easy to deal with or control, docile. □ **tractability** *n.*

traction *n.* **1** pulling. **2** the grip of wheels on the ground. **3** the use of weights etc. to exert a steady pull on an injured limb etc.

tractor *n.* a powerful vehicle for pulling farm equipment etc.

trad *adj.* (colloquial) traditional.

trade *n.* the exchange of goods for money or other goods; business of a particular kind, people engaged in this; trading. ● *v.* engage in trade, buy and sell; exchange (goods) in trading. □ **trade in** give (a used article) as partial payment for a new one. **trade off** exchange as a compromise. □ **trader** *n.*

trade mark *n.* (also **trademark**) a company's registered emblem or name etc. used to identify its goods.

tradesman *n.* (pl. **-men**) a person engaged in trade; a shopkeeper, a supplier.

trade union *n.* an organized association of employees formed to protect and promote their common interests.

trade unionist *n.* a member of a trade union.

trade wind *n.* a constant wind blowing towards the equator.

tradition *n.* a belief or custom handed down from one generation to another; a long-established procedure. □ **traditional** *adj.*, **traditionally** *adv.*

traditionalist *n.* a person who upholds traditional beliefs etc. □ **traditionalism** *n.*

traduce *v.* misrepresent in an unfavourable way. □ **traducement** *n.*

traffic *n.* **1** vehicles, ships, or aircraft moving along a route. **2** trading. ● *v.* (**trafficked, trafficking**) trade. □ **trafficker** *n.*

traffic warden *n.* an official who controls the movement and parking of road vehicles.

tragedian *n.* a writer of tragedies; an actor in tragedy.

tragedienne *n.* an actress in tragedy.

tragedy *n.* a serious drama with unhappy events or a sad ending; an event causing great sadness.

tragic *adj.* of or in tragedy; causing great sadness. □ **tragically** *adv.*

tragicomedy *n.* a drama of mixed tragic and comic events.

trail *v.* **1** drag behind, esp. on the ground; hang loosely; move wearily; lag, straggle. **2** track. ● *n.* **1** a thing that trails; a line of people or things following something. **2** a track, a trace; a beaten path.

trailer *n.* **1** a truck etc. designed to be hauled by a vehicle. **2** a short extract from a film etc., shown in advance to advertise it.

train n. **1** a railway engine with linked carriages or trucks. **2** a retinue. **3** a sequence of things. **4** part of a long robe that trails behind the wearer. ● v. **1** bring or come to a desired standard of efficiency, condition, or behaviour by instruction and practice. **2** cause (a plant) to grow in the required direction. **3** aim (a gun etc.). □ **in train** in preparation.

trainee n. a person being trained.

trainer n. **1** a person who trains horses or athletes etc. **2** (**trainers**) soft sports or running shoes.

traipse v. (colloquial) trudge.

trait n. a characteristic.

traitor n. a person who behaves disloyally, esp. to his or her country. □ **traitorous** adj.

trajectory n. the path of a projectile.

tram n. a public passenger vehicle running on rails laid in the road.

tramcar n. a tram.

tramlines n.pl. **1** rails on which a tram runs. **2** (colloquial) a pair of parallel sidelines in tennis etc.

trammel v. (**trammelled**; Amer. **trammeled**) hamper, restrain.

tramp v. **1** walk with heavy footsteps; go on foot across (an area); trample. ● n. **1** the sound of heavy footsteps; a long walk. **2** a vagrant. **3** a cargo boat that does not travel a regular route. **4** (slang) a promiscuous woman.

trample v. tread repeatedly, crush or harm by treading.

trampoline n. a sheet of canvas attached by springs to a frame, used for jumping on in acrobatic leaps. ● v. use a trampoline.

trance n. a sleeplike or dreamy state.

tranquil adj. calm and undisturbed. □ **tranquilly** adv., **tranquillity** n.

tranquillize v. (also **-ise**; Amer. **tranquilize**) make calm.

tranquillizer n. (also **-iser**; Amer. **tranquilizer**) a drug used to relieve anxiety and tension.

transact v. perform or carry out (business etc.). □ **transaction** n.

transatlantic adj. on or from the other side of the Atlantic; crossing the Atlantic.

transceiver n. a radio transmitter and receiver.

transcend v. go beyond the range of (experience, belief, etc.); surpass. □ **transcendent** adj., **transcendence** n.

transcendental adj. transcendent; abstract, obscure, visionary.

transcontinental adj. crossing or extending across a continent.

transcribe v. copy in writing; produce in written form; arrange (music) for a different instrument etc. □ **transcription** n.

transcript n. a written copy.

transducer n. a device that receives waves or other variations from one system and conveys related ones to another.

transept n. a part lying at right angles to the nave in a church.

transexual var. of **transsexual**.

transfer v. (trans-**fer**; -**ferred**) convey from one place, person, or application to another. ● n. (**trans**-fer) transferring; a document transferring property or a right; a design for transferring from one surface to another. □ **transferable** adj., **transference** n.

transfigure v. change in appearance to something nobler or more beautiful. □ **transfiguration** n.

transfix v. **1** pierce through, impale. **2** make motionless with fear or astonishment.

transform | transubstantiation

transform v. **1** change greatly in appearance or character. **2** change the voltage of (electric current). □ **transformation** n.

transformer n. an apparatus for changing the voltage of an alternating current.

transfuse v. **1** give a transfusion of or to. **2** permeate; imbue.

transfusion n. an injection of blood or other fluid into a blood vessel.

transgress v. break (a rule or law). □ **transgression** n., **transgressor** n.

transient adj. passing away quickly. □ **transience** n.

transistor n. a very small semiconductor device which controls the flow of an electric current; a portable radio set using transistors. □ **transistorized** adj.

transit n. the process of going or conveying across, over, or through.

transition n. the process of changing from one state or style etc. to another. □ **transitional** adj.

transitive adj. (Grammar) (of a verb) used with a direct object.

transitory adj. lasting only briefly.

translate v. express in another language or other words; be able to be translated; transfer. □ **translation** n., **translator** n.

transliterate v. convert to the letters of another alphabet. □ **transliteration** n.

translucent adj. allowing light to pass through but not transparent. □ **translucence** n.

transmigrate v. (of the soul) pass into another body after a person's death. □ **transmigration** n.

transmission n. **1** transmitting; a broadcast. **2** the gear transmitting power from engine to axle in a motor vehicle.

transmit v. (transmitted) **1** pass on from one person, place, or thing to another. **2** send out (a signal or programme etc.) by cable or radio waves. □ **transmitter** n., **transmissible**, **transmittable** adj.

transmogrify v. transform in a surprising or magical way.

transmute v. change in form or substance. □ **transmutation** n.

transom n. a horizontal bar across the top of a door or window.

transparency n. **1** being transparent. **2** a photographic slide.

transparent adj. **1** able to be seen through. **2** easily understood, obvious. □ **transparently** adv.

transpire v. **1** become known. **2** (of plants) give off vapour from leaves etc. □ **transpiration** n.

transplant v. (trans-**plaht**) remove and replant or establish elsewhere; transfer (living tissue). ● n. (trans-**plaht**) transplanting of tissue; something transplanted. □ **transplantation** n.

transport v. (trans-**port**) convey from one place to another. ● n. (trans-**port**) **1** the process of transporting; a means of conveyance. **2** (transports) strong emotion; transports of joy. □ **transportation** n., **transporter** n.

transpose v. cause (two or more things) to change places; change the position of. **2** put (music) into a different key. □ **transposition** n.

transsexual n. (also **transsexual**) a person who feels himself or herself to be a member of the opposite sex; a person who has had a sex change.

transubstantiation n. the doctrine that the bread and wine in the Eucharist are converted by consecration into the body and blood of Christ.

transuranic adj. having atoms heavier than those of uranium.

transverse adj. crosswise.

transvestism n. dressing in clothing of the opposite sex. □ **transvestite** n.

trap n. **1** a device for capturing an animal; a scheme for tricking or catching a person. **2** a curved section of a pipe holding liquid to prevent gases from coming upwards. **3** a compartment from which a dog is released in racing; a trapdoor. **4** (slang) the mouth. **5** a two-wheeled horse-drawn carriage. ●v. (**trapped**) catch or hold in a trap.

trapdoor n. a door in a floor, ceiling, or roof.

trapeze n. a kind of swing on which acrobatics are performed.

trapezium n. (pl. **trapezia** or **trapeziums**) **1** a quadrilateral with only one opposite sides parallel. **2** (Amer.) a trapezoid.

trapezoid n. **1** a quadrilateral with no sides parallel. **2** (Amer.) a trapezium.

trapper n. a person who traps animals, esp. for furs.

trappings n.pl. accessories; symbols of status.

trash n. worthless stuff. □ **trashy** adj.

trash can n. (Amer.) a dustbin.

trattoria n. an Italian restaurant.

trauma n. a wound, an injury; an emotional shock having a lasting effect. □ **traumatic** adj., **traumatize** v.

travail n. & v. labour.

travel v. (**travelled**; Amer. **traveled**) go from one place to another; journey along or through. ●n. travelling, esp. abroad.

traveller n. (Amer. **traveler**) a person who travels; a person living the life of a gypsy.

traveller's cheque n. a cheque for a fixed amount, able to be cashed in other countries.

travelogue n. a film or lecture about travel.

traverse v. travel, lie, or extend across. ●n. something that lies across another; a lateral movement.

travesty n. an absurd or inferior imitation. ●v. make or be a travesty of.

trawl n. a large wide-mouthed fishing net. ●v. fish with a trawl.

trawler n. a boat used in trawling.

tray n. a board with a rim for carrying small articles; an open receptacle for office correspondence.

treacherous adj. showing treachery; not to be relied on, deceptive. □ **treacherously** adv.

treachery n. betrayal of a person or cause; an act of disloyalty.

treacle n. a thick sticky liquid produced when sugar is refined. □ **treacly** adj.

tread v. (trod, trodden, treading) walk, step; walk on, press or crush with the feet. ●n. **1** the manner or sound of walking. **2** a horizontal surface of a stair. **3** the part of a tyre that touches the ground. □ **tread water** keep upright in water by making treading movements.

treadle n. a lever worked by the foot to drive a wheel.

treadmill n. **1** a mill-wheel formerly turned by people treading on steps round its edge. **2** monotonous routine work.

treason n. treachery towards one's country.

treasonable adj. involving treason.

treasure n. a collection of precious metals or gems; a highly

valued object or person. ●v. value highly; store as precious.

treasurer n. a person in charge of the funds of an institution.

treasure trove n. treasure of unknown ownership, found hidden.

treasury n. **1** the department managing a country's revenue. **2** a place where treasure is kept.

treat v. **1** act or behave towards or deal with in a specified way; give medical treatment to; subject to a chemical or other process. **2** buy something for (a person) in order to give pleasure. ●n. something special that gives pleasure.

treatise n. a written work dealing with one subject.

treatment n. **1** a manner of dealing with a person or thing. **2** something done to relieve illness etc.

treaty n. a formal agreement, esp. between countries.

treble adj. **1** three times as much or as many. **2** (of a voice) high-pitched, soprano. ●n. **1** a treble quantity or thing. **2** a treble voice, a person with this. □ **trebly** adv.

tree n. a perennial plant with a single thick woody stem.

trefoil n. a plant with three leaflets (e.g. clover); something shaped like this.

trek n. a long arduous journey. ●v. (**trekked**) make a trek.

trellis n. a light framework of crossing strips of wood etc.

tremble v. shake, quiver; feel very anxious. ●n. a trembling movement. □ **trembly** adj.

tremendous adj. immense; (colloquial) excellent. □ **tremendously** adv.

tremolo n. (pl. **tremolos**) a trembling effect in music.

tremor n. a slight trembling movement; a thrill of fear etc.

tremulous adj. trembling, quivering. □ **tremulously** adv.

trench n. a deep ditch.

trenchant adj. (of comments, policies, etc.) strong and effective.

trench coat n. a belted double-breasted raincoat.

trend n. a general tendency.

trendsetter n. a person who leads the way in fashion etc.

trendy adj. (**trendier**) (colloquial) following the latest fashion. □ **trendily** adv., **trendiness** n.

trepidation n. nervousness.

trespass v. enter land or property unlawfully; intrude. ●n. the act of trespassing. □ **trespasser** n.

tress n. a lock of hair.

trestle n. one of a set of supports on which a board is rested to form a table. □ **trestle-table** n.

trews n.pl. close-fitting usu. tartan trousers.

tri- comb. form three times, triple.

triad n. **1** a group of three. **2** a Chinese secret society.

trial n. **1** an examination in a law court by a judge. **2** the process of testing qualities or performance. **3** a person or thing that tries one's patience. □ **on trial** undergoing a trial.

triangle n. a geometric figure with three sides and three angles; a triangular steel rod used as a percussion instrument.

triangular adj. **1** shaped like a triangle. **2** involving three people.

triangulation n. the measurement or mapping of an area by means of a network of triangles.

tribe n. a related group of families living as a community. □ **tribal** adj., **tribesman** n.

tribulation n. great affliction.

tribunal n. a board of officials appointed to pass judgement on a particular problem.

tributary n. & adj. (a stream) flowing into a larger stream or a lake.

tribute n. **1** something said or done as a mark of respect. **2** a payment that one country or ruler was formerly obliged to pay to another.

trice n. □ **in a trice** in an instant.

triceratops n. a dinosaur with three horns on the forehead and a bony frill above the neck.

trichology n. the study of hair and its diseases. □ **trichologist** n.

trick n. **1** something done to deceive or outwit someone; a mischievous act. **2** a technique, a knack. **3** a mannerism. ● v. deceive or persuade by a trick. □ **do the trick** (colloquial) achieve what is required.

trickery n. the use of tricks, deception.

trickle v. (cause to) flow in a thin stream; come or go gradually. ● n. a trickling flow.

tricky adj. (**trickier**) **1** crafty, deceitful. **2** requiring careful handling, difficult, awkward. □ **trickiness** n.

tricolour n. (Amer. **tricolor**) a flag with three colours in stripes.

tricycle n. a three-wheeled pedal-driven vehicle.

trident n. a three-pronged spear.

triennial adj. happening every third year; lasting three years.

trier n. a person who tries hard.

trifle n. **1** something of only slight value or importance; a very small amount. **2** a sweet dish of sponge cake and jelly etc. topped with custard and cream. □ **trifle with** toy with. □ **trifler** n.

trifling adj. trivial.

trigger n. a lever for releasing a spring, esp. to fire a gun. ● v. (also **trigger off**) set in action, cause.

trigger-happy adj. apt to shoot on slight provocation.

trigonometry n. a branch of mathematics dealing with the relationship of sides and angles of triangles.

trike n. (colloquial) a tricycle.

trilateral adj. having three sides or three participants.

trilby n. a man's soft felt hat.

trill n. a vibrating sound, esp. in music or singing. ● v. sound or sing with a trill.

trillion n. a million million; (formerly, esp. Brit.) a million million million.

trilobite n. a marine creature now found only as a fossil.

trilogy n. a group of three related books, plays, etc.

trim adj. (**trimmer**) neat and orderly. ● v. (**trimmed**) **1** reduce or neaten by cutting. **2** decorate. **3** make (a boat or aircraft) evenly balanced by distributing its load. ● n. **1** ornamentation; the colour or type of upholstery etc. in a car. **2** trimming of hair etc. □ **trimly** adv., **trimness** n.

trimaran n. a boat like a catamaran, with three hulls.

trimming n. **1** something added as a decoration. **2** (**trimmings**) pieces cut off when something is trimmed. **3** (**trimmings**) the usual accompaniments, extras.

trinity n. a group of three; (**the Trinity**) the three members of the Christian deity (Father, Son, Holy Spirit) as constituting one God.

trinket n. a small fancy article or piece of jewellery.

trio n. (pl. **trios**) a group or set of three; music for three instruments or voices.

trip v. (**tripped**) **1** go lightly and quickly. **2** (cause to) catch one's foot and lose balance; (cause to) make a blunder. **3** release (a switch etc.) so as to operate a mechanism. ● n. **1** a journey or excursion, esp. for pleasure. **2** (colloquial) a visionary experience caused by a drug. **3** a stumble. **4** a device for tripping a mechanism.

tripartite adj. consisting of three parts.

tripe n. **1** the stomach of an ox etc. as food. **2** (colloquial) nonsense.

triple adj. having three parts or members; three times as much or as many. ● v. increase by three times its amount.

triplet n. one of three children born at one birth; a set of three.

triplex adj. triple, threefold.

triplicate adj. existing in three examples. ● n. each of a set of three copies. □ **in triplicate** as three identical copies.

tripod n. a three-legged stand.

tripos n. the final examination for the BA degree at Cambridge University.

tripper n. a person who goes on a pleasure trip.

triptych (trip-tik) n. a picture or carving with three panels fixed or hinged side by side.

trite adj. commonplace, hackneyed.

tritium n. a radioactive isotope of hydrogen (symbol T).

triumph n. great success; victory; joy at this. ● v. be successful or victorious; rejoice at this.

triumphal adj. celebrating or commemorating a triumph.

■ Usage Triumphal, as in triumphal arch, should not be confused with triumphant.

triumphant adj. victorious, successful; exultant. □ **triumphantly** adv.

■ Usage See note at triumphal.

triumvirate n. government or control by a board of three.

trivet n. a metal stand for a kettle or hot dish etc.

trivia n.pl. trivial things, trifles.

trivial adj. of only small value or importance. □ **trivially** adv., **triviality** n.

trod, trodden see tread.

troglodyte n. a cave dweller.

troika n. **1** a Russian vehicle with a team of three horses. **2** an administrative group of three people.

troll n. a giant or dwarf in Scandinavian folklore.

trolley n. (pl. **trolleys**) a platform on wheels for transporting goods; a small cart; a small table on wheels.

trollop n. a promiscuous or disreputable woman.

trombone n. a large brass wind instrument with a sliding tube.

trompe l'oeil n. a painting on a wall designed to give an illusion of reality.

troop n. a company of people or animals; a cavalry or artillery unit. ● v. go as a troop or in great numbers.

trooper n. a soldier in a cavalry or armoured unit; (Amer. & Austral.) a mounted or state police officer.

trophy n. something taken in war or hunting etc. as a souvenir of success; an object awarded as a prize.

tropic n. a line of latitude 23° 27′ north or south of the equator; (**the tropics**) the region between these, with a hot climate. □ **tropical** adj.

troposphere n. the layer of the atmosphere between the earth's surface and the stratosphere.

trot n. the running action of a horse etc.; a moderate running pace. ● v. (**trotted**) go or cause to go at a trot. □ **on the trot** (colloquial) continually busy; in succession. **trot out** (colloquial) produce.

troth n. a promise; fidelity.

trotter n. an animal's foot as food.

troubadour n. a medieval romantic poet.

trouble n. difficulty, distress, misfortune; a cause of this; conflict; inconvenience, exertion. ● v. cause trouble to; make or be worried; exert oneself.

troubleshooter n. a person employed to deal with faults or problems.

troublesome adj. causing trouble.

trough n. **1** a long open receptacle, esp. for animals' food or water. **2** a depression between two waves or ridges; a region of low atmospheric pressure.

trounce v. defeat heavily.

troupe n. a company of actors or other performers.

trouper n. a member of a troupe; a staunch colleague.

trousers n.pl. a two-legged outer garment reaching from the waist usu. to the ankles.

trousseau (troo-soh) n. (pl. **trousseaux** or **trousseaus**) a bride's collection of clothing etc. for her marriage.

trout n. (pl. **trout** or **trouts**) a freshwater fish valued as food.

trowel n. a small garden tool for digging; a similar tool for spreading mortar etc.

troy weight n. a system of weights used for precious metals and gems.

truant n. a pupil who stays away from school without leave. □ **play truant** stay away as a truant. □ **truancy** n.

truce n. an agreement to cease hostilities temporarily.

truck n. an open container on wheels for transporting loads; an open railway wagon; a lorry.

trucker n. a lorry driver.

truculent adj. defiant and aggressive. □ **truculently** adv., **truculence** n.

trudge v. walk laboriously. ● n. a laborious walk.

true adj. **1** in accordance with fact, genuine; exact, accurate. **2** loyal, faithful. ● adv. truly, accurately.

truffle n. **1** a rich-flavoured underground fungus valued as a delicacy. **2** a soft chocolate sweet.

trug n. a shallow wooden basket for gardening.

truism n. a statement that is obviously true, esp. a hackneyed one.

truly adv. truthfully; genuinely; faithfully.

trump n. **1** a playing card of a suit temporarily ranking above others. **2** (colloquial) a helpful or excellent person. □ **trump up** invent fraudulently. **turn up trumps** (colloquial) be successful or helpful.

trumpet n. a metal wind instrument with a flared tube; something shaped like this. ● v. (**trumpeted**) proclaim loudly; (of an elephant) make a loud sound with its trunk. □ **trumpeter** n.

truncate v. shorten by cutting off the end. □ **truncation** n.

truncheon n. a short thick stick carried as a weapon.

trundle v. roll along, move along heavily on wheels.

trunk n. **1** a tree's main stem. **2** the body apart from the head and limbs. **3** a large box with a hinged lid, for transporting or storing clothes etc. **4** an elephant's long flexible nose. **5**

(*Amer.*) the boot of a car. **6** (**trunks**) men's shorts for swimming etc.

trunk road *n.* an important main road.

truss *n.* **1** a cluster of flowers or fruit. **2** a framework supporting a roof etc. **3** a device worn to support a hernia. ● *v.* tie up securely.

trust *n.* **1** a firm belief in the reliability or truth or strength etc. of a person or thing; a confident expectation. **2** responsibility, care. **3** property legally entrusted to someone. **4** an organization formed to promote or protect something; an association of business firms, formed to defeat competition. ● *v.* **1** have or place trust in; entrust. **2** hope earnestly. **in trust** held as a trust. **on trust** accepted without investigation. □ **trustful** *adj.*, **trustfully** *adv.*

trustee *n.* a person who administers property held as a trust; one of a group managing the business affairs of an institution.

trustworthy *adj.* worthy of trust. □ **trustworthiness** *n.*

trusty *adj.* (**trustier**) trustworthy.

truth *n.* the quality of being true; something that is true.

truthful *adj.* habitually telling the truth; true. □ **truthfully** *adv.*, **truthfulness** *n.*

try *v.* (**tries, tried, trying**) **1** attempt; test, esp. by use. **2** be a strain on. **3** hold a trial of. ● *n.* **1** an attempt. **2** a touchdown in rugby, entitling the player's side to a kick at goal. □ **try on** put on (a garment) to see if it fits. **try out** test by use.

trying *adj.* annoying.

tsar *n.* (also **czar**) the title of the former emperors of Russia.

tsetse *n.* an African fly that transmits disease by its bite.

T-shirt *n.* (also **tee shirt**) a short-sleeved casual cotton top.

tsp *abbr.* teaspoonful.

T-square *n.* a large T-shaped ruler used in technical drawing.

tsunami *n.* a series of long high sea waves; a tidal wave.

tub *n.* an open usu. round container.

tuba *n.* a large low-pitched brass wind instrument.

tubby *adj.* (**tubbier**) short and fat. □ **tubbiness** *n.*

tube *n.* **1** a long hollow cylinder; something shaped like this. **2** a cathode ray tube esp. in a television set.

tuber *n.* a short thick rounded root or underground stem from which shoots will grow.

tubercle *n.* a small rounded swelling.

tubercular *adj.* of or affected with tuberculosis.

tuberculosis *n.* an infectious wasting disease, esp. affecting lungs.

tuberous *adj.* of or like a tuber; bearing tubers.

tubing *n.* tubes; a length of tube.

tubular *adj.* tube-shaped.

TUC *abbr.* Trades Union Congress.

tuck *n.* a flat fold stitched in a garment etc. ● *v.* put into or under something so as to be concealed or held in place; cover or put away compactly. □ **tuck in** (*colloquial*) eat heartily.

tuck shop *n.* a shop selling cakes and sweets etc. to schoolchildren.

Tues. *abbr.* (also **Tue.**) Tuesday.

Tuesday *n.* the day after Monday.

tufa *n.* porous rock formed round springs of mineral water.

tuft *n.* a bunch of threads, grass, hair, etc., held or growing together at the base. □ **tufted** *adj.*

tug *v.* (**tugged**) pull vigorously; tow. ● *n.* **1** a vigorous pull. **2** a

tug-of-war *n.* a contest of strength with two teams pulling opposite ways on a rope.

tuition *n.* teaching or instruction, esp. of an individual or small group.

tulip *n.* a garden plant with a cup-shaped flower.

tulle (tewl) *n.* a fine silky net fabric.

tumble *v.* (cause to) fall; roll, push, or move in a disorderly way; perform somersaults etc.; rumple. ● *n.* a fall; an untidy state. □ **tumble to** (*colloquial*) grasp the meaning of.

tumbledown *adj.* dilapidated.

tumble-dryer *n.* (also **tumble-drier**) a machine for drying washing in a heated rotating drum.

tumbler *n.* **1** a drinking glass with no handle or foot. **2** a pivoted piece in a lock etc. **3** an acrobat.

tumbril *n.* (also **tumbrel**) (*old use*) an open cart, esp. used to take condemned people to the guillotine.

tumescent *adj.* swelling. □ **tumescence** *n.*

tummy *n.* (*colloquial*) the stomach.

tumour *n.* (*Amer.* **tumor**) an abnormal mass of new tissue growing in or on the body.

tumult *n.* uproar; a conflict of emotions.

tumultuous *adj.* making an uproar.

tun *n.* a large cask; a fermenting vat.

tuna *n.* (*pl.* **tuna** or **tunas**) a large edible sea fish.

tundra *n.* vast level Arctic regions where the subsoil is frozen.

tune *n.* a melody. ● *v.* **1** put (a musical instrument) in tune. **2** set (a radio) to the desired wavelength. **3** adjust (an engine) to run smoothly. □ **in tune** playing or singing at the correct musical pitch; harmonious. **out of tune** not in tune.

tuneful *adj.* melodious.

tuner *n.* **1** a person who tunes pianos etc. **2** a radio receiver as part of a hi-fi system.

tungsten *n.* a heavy grey metallic element (symbol W).

tunic *n.* a close-fitting jacket worn as part of a uniform; a loose garment reaching to the knees.

tunnel *n.* an underground passage. ● *v.* (**tunnelled**; *Amer.* **tunneled**) make a tunnel (through), make (one's way) in this manner.

tunny *n.* = **tuna.**

tup *n.* a male sheep.

turban *n.* a Muslim or Sikh man's headdress of cloth wound round the head; a woman's hat resembling this.

turbid *adj.* **1** (of liquids) muddy, not clear. **2** confused, disordered. □ **turbidity** *n.*

■ Usage □ *Turbid* is sometimes confused with *turgid*.

turbine *n.* a machine or motor driven by a wheel that is turned by a flow of water or gas.

turbo- *comb. form* using a turbine; driven by such engines.

turbot *n.* a large flat edible sea fish.

turbulent *adj.* in a state of commotion or unrest; moving unevenly. □ **turbulently** *adv.*, **turbulence** *n.*

turd *n.* (*vulgar slang*) a lump of excrement.

tureen *n.* a deep covered dish from which soup is served.

turf *n.* (*pl.* **turfs** or **turves**) **1** short grass and the soil just below it; a piece of this. **2** (**the turf**) horse racing; a racecourse.

● *v.* lay (ground) with turf. □ **turf out** (*colloquial*) throw out.

turf accountant *n.* a bookmaker.

turgid *adj.* **1** swollen, inflated. **2** (of language) pompous. □ **turgidity** *n.*

■ **Usage** *Turgid* is sometimes confused with *turbid*.

Turk *n.* a native of Turkey.

turkey *n.* (*pl.* **turkeys**) a large bird reared for its flesh; this as food.

Turkish *adj.* & *n.* (the language) of Turkey.

Turkish bath *n.* a hot-air or steam bath followed by massage etc.

Turkish delight *n.* a sweet of flavoured gelatin coated in powdered sugar.

turmeric *n.* a bright yellow spice from the root of an Asian plant.

turmoil *n.* a state of great disturbance or confusion.

turn *v.* **1** move round a point or axis; take or give a new direction (to); aim; pass round (a point). **2** pass (a certain hour or age). **3** change in form or appearance etc.; make or become sour; shape in a lathe; give an elegant form to. ● *n.* **1** the process of turning; a change of direction or condition etc.; a bend in a road. **2** an opportunity or obligation coming in succession. **3** a service of a specified kind. **4** a short performance in an entertainment. **5** (*colloquial*) an attack of illness; a momentary nervous shock. □ **in turn** in succession. **out of turn** before or after one's proper turn; indiscreetly, presumptuously. **to a turn** so as to be cooked perfectly. **turn against** make or become hostile to. **turn down** reduce the volume or flow of; reject. **turn in** hand in; (*colloquial*) go to bed. **turn off** stop the operation of; (*colloquial*) cause to lose interest. **turn on** start the operation of; (*colloquial*) excite, esp. sexually. **turn out** expel; turn off (a light etc.); equip, dress; produce by work; empty and search or clean; (*colloquial*) prove to be the case. **turn the tables** reverse a situation and put oneself in a superior position. **turn up** discover; be found; make one's appearance.

turncoat *n.* a person who changes sides in a dispute etc.

turner *n.* a person who works with a lathe.

turning *n.* a place where one road meets another, forming a corner.

turning point *n.* a point at which a decisive change takes place.

turnip *n.* a plant with a round white root used as a vegetable.

turnout *n.* **1** the process of turning out a room etc. **2** the number of people attending a gathering. **3** an outfit.

turnover *n.* **1** the amount of money taken in a business. **2** the rate of replacement. **3** a small pie of pastry folded over a filling.

turnpike *n.* (*old use*) a toll gate, a toll road.

turnstile *n.* a revolving barrier for admitting people one at a time.

turntable *n.* a circular revolving platform.

turn-up *n.* a turned-up part, esp. at the lower end of a trouser leg.

turpentine *n.* oil used for thinning paint and as a solvent.

turpitude *n.* wickedness.

turps *n.* (*colloquial*) turpentine.

turquoise *n.* a blue-green precious stone; its colour.

turret *n.* a small tower; a revolving tower for a gun on a warship or tank. □ **turreted** *adj.*

turtle *n.* a sea creature like a tortoise. □ **turn turtle** capsize.

turtle-dove *n.* a wild dove noted for its soft cooing.

turtleneck *n.* a high round close-fitting neckline.

tusk *n.* a long pointed tooth outside the mouth of an elephant, walrus, etc.

tussle *n.* a struggle, a conflict.

tussock *n.* a tuft or clump of grass.

tutelage *n.* guardianship; tuition.

tutti *adv.* (*Music*) with all voices or instruments together.

tutor *n.* a private or university teacher. ● *v.* act as tutor (to).

tutorial *adj.* of a tutor. ● *n.* a student's session with a tutor.

tut-tut *int.* an exclamation of annoyance, impatience, or rebuke.

tutu *n.* a dancer's short skirt made of layers of frills.

tuxedo (tuk-see-doh) *n.* (*pl.* **tuxedos** or **tuxedoes**) (*Amer.*) a dinner jacket.

TV *abbr.* television.

twaddle *n.* nonsense.

twang *n.* **1** a sharp ringing sound like that made by a tense wire when plucked. **2** a nasal intonation. ● *v.* make or cause to make a twang.

tweak *v.* pinch, twist, or pull with a sharp jerk. ● *n.* an instance of tweaking.

twee *adj.* affectedly dainty or quaint.

tweed *n.* a thick woollen fabric; (also **tweeds**) clothes made of tweed. □ **tweedy** *adj.*

tweet *v.* chirp.

tweeter *n.* a small loudspeaker for reproducing high-frequency signals.

tweezers *n.pl.* small pincers for handling very small things.

twelve *adj.* & *n.* one more than eleven (12, XII). □ **twelfth** *adj.* & *n.*

twenty *adj.* & *n.* twice ten. □ **twentieth** *adj.* & *n.*

twerp *n.* (also **twirp**) (*slang*) a stupid person.

twice *adv.* two times; in double amount or degree.

twiddle *v.* twist idly about. ● *n.* an act of twiddling. □ **twiddle one's thumbs** have nothing to do.

twig *n.* a small shoot growing from a branch or stem. ● *v.* (**twigged**) (*colloquial*) realize, grasp the meaning of.

twilight *n.* light from the sky after sunset; the period of this.

twill *n.* fabric woven so that parallel diagonal lines are produced. □ **twilled** *adj.*

twin *n.* one of two children or animals born at one birth; one of a pair that are exactly alike. ● *adj.* being a twin or twins. ● *v.* (**twinned**) combine as a pair; link with (another town) for cultural and social exchange.

twine *n.* strong thread or string. ● *v.* twist; wind or coil.

twinge *n.* a slight or brief pang.

twinkle *v.* shine with a flickering light. ● *n.* a twinkling light.

twin towns *n.pl.* two towns that establish special social and cultural links.

twirl *v.* twist lightly or rapidly. ● *n.* a twirling movement; a twirled mark. □ **twirly** *adj.*

twirp var. of **twerp.**

twist *v.* **1** (strands etc.) round each other, esp. to form a single cord; bend round; rotate. **2** distort, pervert. ● *n.* **1** the process of twisting; something formed by twisting. **2** an unexpected development, esp. in a story.

twister *n.* (*colloquial*) a swindler.

twit *n.* (*colloquial*) a foolish person. ● *v.* (**twitted**) taunt.

twitch *v.* pull with a light jerk; quiver or contract spasmodically. ● *n.* a twitching movement.

twitcher n. (slang) a birdwatcher who tries to see as many species as possible.

twitter v. make light chirping sounds; talk rapidly in an anxious or nervous way. ● n. twittering.

two adj. & n. one more than one (2, II).

two-dimensional adj. having or appearing to have length and breadth but no depth.

two-faced adj. insincere, deceitful.

twofold adj. & adv. twice as much or as many.

twosome n. two together, a pair.

two-time v. be unfaithful to (a lover); deceive.

tycoon n. a wealthy influential industrialist.

tying see **tie.**

tyke n. (also **tike**) **1** a coarse or unpleasant person. **2** a small child.

tympani var. of **timpani.**

tympanum n. (pl. **tympana** or **tympanums**) **1** the eardrum. **2** a recessed triangular area above a door.

type n. **1** a kind, a class; a typical example or instance; (colloquial) a person of specified character. **2** a set of characters used in printing. ● v. **1** classify according to type. **2** write with a typewriter.

typecast v. (**typecast, typecasting**) cast (an actor or actress) repeatedly in the same type of role.

typeface n. a set of printing types in one design.

typescript n. a typewritten document.

typesetter n. a person or machine that sets type for printing.

typewriter n. a machine for producing printlike characters on paper by pressing keys. □ **typewritten** adj.

typhoid n. (in full **typhoid fever**) a serious infectious feverish disease.

typhoon n. a violent hurricane.

typhus n. an infectious feverish disease transmitted by parasites.

typical adj. having the distinctive qualities of a particular type of person or thing. □ **typically** adv.

typify v. be a representative specimen of.

typist n. a person who types.

typography n. the art or style of printing. □ **typographical** adj.

tyrannize v. (also **-ise**) rule as or like a tyrant.

tyrannosaurus n. (also **tyrannosaur**) a large flesh-eating dinosaur that walked on its hind legs.

tyranny n. government by a tyrant; tyrannical use of power. □ **tyrannical** adj., **tyrannically** adv., **tyrannous** adj.

tyrant n. a ruler or other person who uses power in a harsh or oppressive way.

tyre n. (Amer. **tire**) a covering round the rim of a wheel to absorb shocks.

tyro n. (also **tiro**) (pl. **tyros**) a beginner.

tzatziki n. a Greek dish of yoghurt and cucumber.

Uu

U symb. uranium. ● abbr. universal; a film classification indicating suitability for all ages.

UAE abbr. United Arab Emirates.

UB40 abbr. a card issued to people claiming unemployment benefit.

ubiquitous adj. found everywhere. □ **ubiquity** n.

UCCA *abbr.* Universities Central Council on Admissions.

udder *n.* a bag-like milk-secreting organ of a cow, goat, etc.

UDI *abbr.* unilateral declaration of independence.

UEFA *abbr.* Union of European Football Associations.

UFO *abbr.* unidentified flying object.

ugly *adj.* (**uglier**) unpleasant to look at or hear; threatening, hostile. □ **ugliness** *n.*

UHF *abbr.* ultra-high frequency.

UHT *abbr.* ultra heat treated (esp. of milk).

UK *abbr.* United Kingdom.

ukulele (yoo-kă-lay-li) *n.* a small four-stringed guitar.

ulcer *n.* an open sore. □ **ulcerous** *adj.*

ulcerated *adj.* affected with an ulcer. □ **ulceration** *n.*

ulna *n.* (*pl.* **ulnae** or **ulnas**) the thinner long bone of the forearm. □ **ulnar** *adj.*

ult. *abbr.* ultimo.

ulterior *adj.* beyond what is obvious or admitted.

ultimate *adj.* last, final; fundamental. □ **ultimately** *adv.*

ultimatum *n.* (*pl.* **ultimatums** or **ultimata**) a final demand, with a threat of hostile action if this is rejected.

ultimo *adj.* of last month.

ultra- *pref.* beyond; extremely.

ultra-high *adj.* (of a frequency) between 300 and 3000 megahertz.

ultramarine *adj.* & *n.* brilliant deep blue.

ultrasonic *adj.* above the range of normal human hearing.

ultrasound *n.* ultrasonic waves.

ultraviolet *adj.* of or using radiation with a wavelength shorter than that of visible light rays.

ululate *v.* howl, wail. □ **ululation** *n.*

umbel *n.* a flower cluster like an umbrella.

umber *n.* a natural brownish colouring matter.

umbilical *adj.* of the navel.

umbilical cord *n.* a flexible tube connecting the placenta to the navel of a foetus.

umbra *n.* (*pl.* **umbrae** or **umbras**) an area of total shadow cast by the moon or earth in an eclipse.

umbrage *n.* a feeling of being offended. □ **take umbrage** take offence.

umbrella *n.* a circle of fabric on a folding framework of spokes on a central stick, used as a protection against rain.

umlaut *n.* a mark (") over a vowel indicating a change in pronunciation, used esp. in Germanic languages.

umpire *n.* a person appointed to supervise a contest and see that rules are observed. ● *v.* act as umpire in.

umpteen *adj.* (*colloquial*) very many. □ **umpteenth** *adj.*

UN *abbr.* United Nations.

un- *pref.* not; reversing the action indicated by a verb, e.g. *unlock.* (*The number of words with this prefix is almost unlimited and many of those whose meaning is obvious are not listed below.*)

unaccountable *adj.* unable to be accounted for; not having to account for one's actions etc. □ **unaccountably** *adv.*

unadulterated *adj.* pure, complete.

unalloyed *adj.* pure.

unanimous *adj.* all agreeing; agreed by all. □ **unanimously** *adv.*, **unanimity** *n.*

unarmed *adj.* without weapons.

unassuming *adj.* not arrogant, unpretentious.

unattended *adj.* (of a vehicle etc.) having no person in charge of it.

573

unavoidable adj. unable to be avoided. □ **unavoidably** adv.

unawares adv. unexpectedly; without noticing.

unbalanced adj. not balanced; biased; mentally unsound.

unbeknown adj. (also **unbeknownst**) unknown.

unbend v. (**unbent, unbending**) **1** change from a bent position. **2** become relaxed or affable.

unbending adj. inflexible, refusing to alter one's demands.

unbidden adj. not commanded or invited.

unblock v. remove an obstruction from.

unbolt v. release (a door) by drawing back the bolt.

unborn adj. not yet born.

unbounded adj. without limits.

unbridled adj. unrestrained.

unburden v. □ **unburden oneself** reveal one's thoughts and feelings.

uncalled for adj. given or done impertinently or unjustifiably.

uncanny adj. (**uncannier**) strange and rather frightening; extraordinary. □ **uncannily** adv.

uncared-for adj. neglected.

unceasing adj. not ceasing.

unceremonious adj. without proper formality or dignity.

uncertain adj. not known or not knowing certainly; not to be depended on. □ **uncertainly** adv., **uncertainty** n.

uncle n. a brother or brother-in-law of one's father or mother.

unclean adj. not clean; ritually impure.

uncoil v. unwind.

uncommon adj. not common, unusual.

uncompromising adj. not allowing or not seeking compromise, inflexible.

unconcern n. lack of concern.

unconditional adj. not subject to conditions. □ **unconditionally** adv.

unconscionable adj. unscrupulous; contrary to what one's conscience feels is right.

unconscious adj. not conscious; not aware; done without conscious intention. □ **unconsciously** adv., **unconsciousness** n.

unconsidered adj. disregarded.

uncork v. pull the cork from.

uncouple v. disconnect (things joined by a coupling).

uncouth adj. awkward in manner, boorish.

uncover v. remove a covering from; reveal, expose.

unction n. **1** anointing with oil, esp. as a religious rite. **2** excessive politeness.

unctuous adj. having an oily manner; smugly virtuous. □ **unctuously** adv., **unctuousness** n.

undeceive v. free (a person) from a misconception.

undecided adj. not yet certain; not yet having made up one's mind.

undeniable adj. undoubtedly true. □ **undeniably** adv.

under prep. **1** in or to a position or rank etc. lower than; less than. **2** governed or controlled by. **3** subjected to. **4** in accordance with; designated by. ● adv. **1** in or to a lower position or subordinate condition. **2** in or into a state of unconsciousness. **3** below a certain quantity, rank, age, etc. □ **under age** not old enough, esp. for some legal right; not yet of adult status. **under way** making progress.

under- pref. **1** below; lower, subordinate. **2** insufficiently.

underarm adj. & adv. **1** in the armpit. **2** with the hand brought forwards and upwards.

underbelly n. the undersurface of an animal, vulnerable to attack.

undercarriage n. an aircraft's landing wheels and their supports; the supporting framework of a vehicle.

underclass n. a social class below mainstream society.

undercliff n. a terrace or lower cliff formed by a landslip.

underclothes n.pl. (also **underclothing**) underwear.

undercoat n. a layer of paint used under a finishing coat.

undercover adj. done or doing things secretly.

undercurrent n. a current flowing below a surface; an underlying feeling, influence, or trend.

undercut v. (**undercut, undercutting**) 1 cut away the part below. 2 sell or work for a lower price than.

underdog n. a person etc. in an inferior or subordinate position.

underdone adj. not thoroughly cooked.

underestimate v. make too low an estimate (of).

underfelt n. felt for laying under a carpet.

underfoot adv. on the ground; under one's feet.

underfunded adj. provided with insufficient funds.

undergo v. (**undergoes, underwent, undergone, undergoing**) experience; be subjected to.

undergraduate n. a university student who has not yet taken a degree.

underground adj. under the surface of the ground; secret. ● n. an underground railway.

undergrowth n. thick growth of shrubs and bushes under trees.

underhand adj. 1 done or doing things slyly or secretly. 2 underarm.

underlay n. material laid under another as a support.

underlie v. (**underlay, underlain, underlying**) be the cause or basis of. □ **underlying** adj.

underline v. 1 draw a line under. 2 emphasize.

underling n. a subordinate.

undermanned adj. having too few staff or crew etc.

undermine v. weaken gradually; weaken the foundations of.

underneath prep. & adv. below or on the inside of (a thing).

underpants n.pl. an undergarment covering the lower part of the body and part of the legs.

underpass n. a road passing under another.

underpay v. (**underpaid, underpaying**) pay (a person) too little.

underpin v. (**underpinned**) support, strengthen from beneath.

underprivileged adj. not having the normal standard of living or rights.

underrate v. underestimate.

underscore v. underline.

underseal v. coat the lower surface of (a vehicle) with a protective layer. ● n. this coating.

undersell v. (**undersold, underselling**) sell at a lower price than.

undershoot v. (**undershot, undershooting**) land short of (a runway etc.).

undersigned adj. who has or have signed this document.

underskirt n. a petticoat.

understand v. (**understood, understanding**) see the meaning or importance of; know the ways or workings of; know the explanation; infer; take for granted. □ **understandable** adj., **understandably** adv.

understanding adj. showing insight or sympathy. ● n. 1 ability to understand; sympathetic insight. **2** an agreement; a thing agreed.

understate v. express in restrained terms; represent as less than it really is. □ **understatement** n.

understudy n. an actor who studies another's part in order to be able to take his or her place if necessary. ● v. be an understudy for.

undertake v. (**undertook, undertaken, undertaking**) agree or promise (to do something).

undertaker n. a person whose business is to organize funerals.

undertaking n. **1** work etc. undertaken. **2** a promise, a guarantee.

undertone n. **1** a low or subdued tone. **2** an underlying quality or feeling.

undertow n. an undercurrent moving in the opposite direction to the surface current.

underwear n. garments worn under indoor clothing.

underwent see **undergo**.

underworld n. **1** part of society habitually involved in crime. **2** (in mythology) the abode of spirits of the dead, under the earth.

underwrite v. (**underwrote, underwritten, underwriting**) accept liability under (an insurance policy); undertake to finance. □ **underwriter** n.

undesirable adj. not desirable, objectionable. □ **undesirably** adv.

undies n.pl. (colloquial) women's underwear.

undo v. (**undoes, undid, undone, undoing**) **1** unfasten, unwrap. **2** cancel the effect of, ruin.

undone adj. unfastened; not done.

undoubted adj. not disputed. □ **undoubtedly** adv.

undreamed adj. (also **undreamt**) not imagined, not thought to be possible.

undress v. take clothes off.

undue adj. excessive.

undulate v. have or cause to have a wavy movement or appearance. □ **undulation** n.

unduly adv. excessively.

undying adj. everlasting.

unearned income n. income from interest on investments, rent from tenants, etc.

unearth v. uncover or bring out from the ground; find by searching.

unearthly adj. **1** supernatural; mysterious. **2** (colloquial) absurdly early or inconvenient.

uneasy adj. (**uneasier**) not comfortable; not confident; anxious. □ **uneasily** adv., **uneasiness** n., **unease** n.

uneatable adj. not fit to be eaten.

uneconomic adj. not profitable.

unemployable adj. not fit for paid employment.

unemployed adj. **1** without a paid job. **2** not in use. □ **unemployment** n.

unending adj. endless.

unequalled adj. (Amer. **unequaled**) without an equal; supreme.

unequivocal adj. clear and not ambiguous. □ **unequivocally** adv.

unerring adj. making no mistake.

UNESCO abbr. (also **Unesco**) United Nations Educational, Scientific, and Cultural Organization.

uneven adj. not level, not smooth; not uniform. □ **unevenly** adv., **unevenness** n.

unexampled adj. without precedent.

unexceptionable *adj.* entirely satisfactory.

■ **Usage** *Unexceptionable* is sometimes confused with *unexceptional.*

unexceptional *adj.* normal, ordinary.

■ **Usage** *Unexceptional* is sometimes confused with *unexceptionable.*

unexpected *adj.* not expected; surprising. □ **unexpectedly** *adv.*

unfailing *adj.* constant, reliable. □ **unfailingly** *adv.*

unfair *adj.* not impartial, not in accordance with justice. □ **unfairly** *adv.*, **unfairness** *n.*

unfaithful *adj.* not loyal; having committed adultery. □ **unfaithfully** *adv.*, **unfaithfulness** *n.*

unfeeling *adj.* lacking sensitivity; callous. □ **unfeelingly** *adv.*

unfit *adj.* **1** unsuitable. **2** not in perfect physical condition.

unflappable *adj.* (*colloquial*) remaining calm in a crisis.

unfold *v.* **1** open, spread out. **2** become known.

unforeseen *adj.* not foreseen.

unforgettable *adj.* impossible to forget. □ **unforgettably** *adv.*

unfortunate *adj.* having bad luck; regrettable. □ **unfortunately** *adv.*

unfounded *adj.* with no basis.

unfrock *v.* dismiss (a priest) from the priesthood.

unfurl *v.* unroll; spread out.

ungainly *adv.* awkward-looking, not graceful. □ **ungainliness** *n.*

ungodly *adj.* **1** irreligious; wicked. **2** (*colloquial*) outrageous. □ **ungodliness** *n.*

ungovernable *adj.* uncontrollable.

ungracious *adj.* not courteous or kindly. □ **ungraciously** *adv.*

unguarded *adj.* **1** not guarded. **2** incautious.

unguent (ung-wěnt) *n.* ointment, a lubricant.

ungulate *n.* a hoofed animal.

unhand *v.* (*literary*) let go of.

unhappy *adj.* (**unhappier**) **1** not happy, sad. **2** unfortunate. **3** unsuitable. □ **unhappily** *adv.*, **unhappiness** *n.*

unhealthy *adj.* (**unhealthier**) not healthy; harmful to health. □ **unhealthily** *adv.*

unheard-of *adj.* not previously known or done.

unhinge *v.* cause to become mentally unbalanced.

unholy *adj.* (**unholier**) **1** wicked, irreverent. **2** (*colloquial*) dreadful, outrageous.

unicorn *n.* a mythical horselike animal with one straight horn on its forehead.

uniform *n.* distinctive clothing identifying the wearer as a member of an organization or group. ● *adj.* always the same. □ **uniformly** *adv.*, **uniformity** *n.*

unify *v.* unite. □ **unification** *n.*

unilateral *adj.* done by or affecting only one person or group. □ **unilaterally** *adv.*

unimpeachable *adj.* completely trustworthy.

uninterested *adj.* not interested; showing no concern.

uninviting *adj.* unattractive, repellent.

union *n.* uniting, being united; a whole formed by uniting parts; an association; a trade union.

Union Jack *n.* the national flag of the UK.

unionist *n.* **1** a member of a trade union; a supporter of trade unions. **2** one who favours union.

unionize *v.* (also **-ise**) organize into or cause to join a union. □ **unionization** *n.*

unique | unobtrusive

unique *adj.* **1** being the only one of its kind; unequalled. **2** remarkable. □ **uniquely** *adv.*

■ **Usage** The use of *unique* to mean 'remarkable' is considered incorrect by some people.

unisex *adj.* suitable for people of either sex.

unison *n.* □ **in unison** all together; sounding or singing together.

unit *n.* **1** an individual thing, person, or group, esp. as part of a complex whole. **2** a fixed quantity used as a standard of measurement.

Unitarian *n.* a person who believes that God is not a Trinity but one person.

unitary *adj.* of a unit or units.

unite *v.* join together, make or become one; act together, cooperate.

United Kingdom *n.* Great Britain and Northern Ireland.

unit trust *n.* an investment company paying dividends based on the average return from the various securities which they hold.

unity *n.* **1** the state of being one or a unit; a complex whole. **2** agreement.

universal *adj.* of, for, or done by all. □ **universally** *adv.*

universe *n.* all existing things, including the earth and its creatures and all the heavenly bodies.

university *n.* an educational institution for advanced learning and research.

unkempt *adj.* looking untidy or neglected.

unkind *adj.* cruel, harsh, hurtful. □ **unkindly** *adv.*, **unkindness** *n.*

unknown *adj.* not known. ● *n.* an unknown person, thing, or place.

unleaded *adj.* (of petrol etc.) without added lead.

unleash *v.* release, let loose.

unleavened *adj.* (of bread) made without yeast or other raising agent.

unless *conj.* except when; except on condition that.

unlettered *adj.* illiterate.

unlike *adj.* not like. ● *prep.* differently from.

unlikely *adj.* (**unlikelier**) not likely to happen or be true or be successful.

unlimited *adj.* not limited; very great in number.

unlisted *adj.* not included in a (published) list.

unload *v.* remove a load or cargo (from); get rid of; remove the ammunition from (a gun).

unlock *v.* release the lock of (a door etc.); release by unlocking.

unlooked-for *adj.* unexpected.

unmanned *adj.* operated without a crew.

unmask *v.* remove a mask (from); expose the true character of.

unmentionable *adj.* so bad that it is not fit to be spoken of.

unmistakable *adj.* clear, not able to be doubted or mistaken for another. □ **unmistakably** *adv.*

unmitigated *adj.* not modified, absolute.

unmoved *adj.* not moved, not persuaded, not affected by emotion.

unnatural *adj.* not natural; not normal. □ **unnaturally** *adv.*

unnecessary *adj.* not necessary; more than is necessary. □ **unnecessarily** *adv.*

unnerve *v.* cause to lose courage or determination.

unnumbered *adj.* **1** not marked with a number. **2** countless.

unobtrusive *adj.* not making oneself or itself noticed. □ **unobtrusively** *adv.*

unpack v. open and remove the contents of (a suitcase etc.); take out from its packaging.

unparalleled adj. never yet equalled.

unpick v. undo the stitching of.

unplaced adj. not placed as one of the first three in a race etc.

unpleasant adj. not pleasant. □ **unpleasantly** adv., **unpleasantness** n.

unplumbed adj. not fully investigated or understood.

unpopular adj. not popular; disliked. □ **unpopularity** n.

unprecedented adj. for which there is no precedent; unparalleled.

unprepared adj. not prepared beforehand; not ready or not equipped to do something.

unprepossessing adj. unattractive, not making a good impression.

unpretentious adj. not pretentious, not showy or pompous.

unprincipled adj. without good moral principles, unscrupulous.

unprintable adj. too indecent or libellous etc. to be printed.

unprofessional adj. contrary to the standards of behaviour for members of a profession. □ **unprofessionally** adv.

unprofitable adj. not profitable, useless. □ **unprofitably** adv.

unprompted adj. not prompted, spontaneous.

unqualified adj. **1** not qualified. **2** not restricted or modified.

unquestionable adj. too clear to be doubted. □ **unquestionably** adv.

unravel v. (**unravelled**; Amer. **unraveled**) disentangle, undo (knitted fabric); become disentangled.

unready adj. **1** not ready; not prompt in action. **2** (old use) rash, lacking good advice.

unreal adj. imaginary, not real; (slang) incredible, amazing.

unreasonable adj. not reasonable; excessive, unjust. □ **unreasonably** adv.

unrelenting adj. not becoming less in intensity.

unremitting adj. not ceasing.

unrequited adj. (of love) not returned or rewarded.

unreservedly adv. without reservation, completely.

unrest n. disturbance; dissatisfaction.

unrivalled adj. (Amer. **unrivaled**) having no equal, incomparable.

unroll v. open out (something rolled).

unruly adj. (**unrulier**) not easy to control, disorderly. □ **unruliness** n.

unsaid adj. not spoken or expressed.

unsaturated adj. (of fat or oil) capable of further reaction by combining with hydrogen.

unsavoury adj. (Amer. **unsavory**) disagreeable to the taste or smell; morally disgusting.

unscathed adj. without suffering any injury.

unscramble v. sort out; make (a scrambled transmission) intelligible.

unscrew v. loosen (a screw etc.); unfasten by removing screw(s).

unscripted adj. without a prepared script.

unscrupulous adj. not prevented by scruples of conscience.

unseasonable adj. not seasonable; untimely.

unseat v. dislodge (a rider); remove from a position of authority.

unseen adj. not seen, invisible; (of translation) done without preparation.

unselfish adj. not selfish; considering others' needs before one's own. □ **unselfishly** adv., **unselfishness** n.

unsettle v. make uneasy, disturb.

unsettled adj. uneasy; (of weather) changeable.

unshakeable adj. (also **unshakable**) firm.

unsightly adj. not pleasant to look at, ugly. □ **unsightliness** n.

unskilled adj. not having or needing skill or special training.

unsociable adj. disliking company.

■ **Usage** Unsociable is sometimes confused with unsocial.

unsocial adj. not social; not conforming to normal social practices; antisocial.

■ **Usage** Unsocial is sometimes confused with unsociable.

unsolicited adj. not requested.

unsophisticated adj. simple and natural or naive.

unsound adj. not sound or strong, not free from defects. □ **of unsound mind** insane.

unsparing adj. giving lavishly.

unspeakable adj. too bad to be described in words.

unstable adj. not stable; mentally or emotionally unbalanced.

unstick v. (**unstuck, unsticking**) detach (what is stuck). □ **come unstuck** (colloquial) suffer disaster, fail.

unstinting adj. given freely and generously.

unstudied adj. natural in manner, spontaneous.

unsung adj. not acknowledged or honoured.

unsuspecting adj. feeling no suspicion.

unswerving adj. not turning aside; unchanging.

untenable adj. (of a theory or position) not valid because of strong arguments against it.

unthinkable adj. too bad or too unlikely to be thought about.

unthinking adj. thoughtless.

untidy adj. (**untidier**) not tidy. □ **untidily** adv., **untidiness** n.

untie v. (**untied, untying**) unfasten; release from being tied up.

until prep. & conj. up to (a specified time), till.

■ **Usage** Until, as opposed to till, is used especially at the beginning of a sentence and in formal style, as in Until you told me, I had no idea or He resided there until his decease.

untimely adj. happening at an unsuitable time; premature. □ **untimeliness** n.

unto prep. (old use) to.

untold adj. **1** not told. **2** too much or too many to be counted.

untouchable adj. not able or not allowed to be touched. ● n. a member of the lowest Hindu social group.

untoward adj. unexpected and inconvenient.

untried adj. not yet tried or tested.

untruth n. an untrue statement, a lie. □ **untruthful** adj., **untruthfully** adv.

unusual adj. not usual; remarkable, rare. □ **unusually** adv.

unutterable adj. too great to be expressed in words. □ **unutterably** adv.

unvarnished adj. not varnished; plain and straightforward.

unveil v. remove a veil (from); remove concealing drapery from; disclose, make publicly known.

unwaged adj. unemployed.

unwarranted adj. unjustified, unauthorized.

unwary adj. not cautious.

unwell adj. not in good health.

unwieldy adj. (**unwieldier**) awkward to move or control because of its size, shape, or weight.

unwilling adj. not willing, reluctant.

unwind v. (**unwound, unwinding**) draw out or become drawn out from being wound; (colloquial) relax from work or tension.

unwise adj. not wise, foolish. □ **unwisely** adv.

unwitting adj. unaware; unintentional. □ **unwittingly** adv.

unwonted (un-**wohn**-tid) adj. not customary, not usual.

unworldly adj. spiritually-minded, not materialistic. □ **unworldliness** n.

unworthy adj. (**unworthier**) worthless; not deserving; unsuitable to the character (of a person or thing).

unwrap v. (**unwrapped**) remove the wrapping from; open, unfold.

unwritten adj. not written down; based on custom not statute.

unzip v. (**unzipped**) open by the undoing of a zip fastener.

up adv. **1** to, in, or at a higher place or state etc.; to a vertical position; to a larger size. **2** as far as a stated place, time, or amount. **3** out of bed. **4** (colloquial) amiss, happening. ● prep. upwards along or through or into; at a higher part of. ● adj. **1** directed upwards. **2** travelling towards a central place. ● v. (**upped**) (colloquial) raise; get up (and do something). □ **up against** (colloquial) confronted with (a problem etc.). **ups and downs** alternate good and bad fortune. **up to** occupied with, doing; required as a duty or obligation from; capable of. **what's up?** the matter?

upbeat n. (Music) an unaccented beat. ● adj. (colloquial) optimistic, cheerful.

upbraid v. reproach.

upbringing n. training and education during childhood.

update v. bring up to date.

upend v. set or rise up on end.

upgrade v. raise to a higher grade; improve (equipment etc.).

upheaval n. a sudden heaving upwards; a violent disturbance.

uphill adj. & adv. going or sloping upwards.

uphold v. (**upheld, upholding**) support.

upholster v. put a fabric covering, padding, etc. on (furniture).

upholstery n. upholstering; material used in this.

upkeep n. keeping (a thing) in good condition and repair; the cost of this.

upland n. & adj. (of) the higher or inland parts of a country.

uplift v. raise. ● n. being raised; a mentally elevating influence.

upmarket adj. & adv. of or towards the more expensive end of the market.

upon prep. on.

■ Usage Upon is usually more formal than on, but it is standard in once upon a time and upon my word.

upper adj. higher in place, position, or rank. ● n. part of a shoe above the sole. □ **the upper hand** mastery, dominance.

upper case n. capital letters in printing or typing.

upper class n. the highest social class. □ **upper-class** adj.

upper crust n. (colloquial) the aristocracy.

uppermost adj. & adv. in, on, or to the top or most prominent position.

uppish adj. (also **uppity**) (colloquial) self-assertive; arrogant.

upright adj. **1** in a vertical position. **2** strictly honest or honourable. ● n. a vertical part or support.

uprising n. a rebellion.

uproar n. an outburst of noise and excitement or anger.

uproarious adj. noisy; with loud laughter. □ **uproariously** adv.

uproot v. pull out of the ground together with its roots; force to leave an established place.

upset v. (up-set) (**upset, upsetting**) overturn; disrupt; distress; disturb the temper or digestion of. ● n. (up-set) upsetting, being upset.

upshot n. an outcome.

upside down adv. & adj. **1** with the upper part underneath. **2** in great disorder.

upstage adv. & adj. nearer the back of a theatre stage. ● v. divert attention from; outshine.

upstairs adv. & adj. to or on a higher floor.

upstanding adj. strong and healthy, well set up.

upstart n. a person newly risen to a high position, esp. one who behaves arrogantly.

upstream adj. & adv. in the direction from which a stream flows.

upsurge n. an upward surge, a rise.

upswing n. an upward movement or trend.

uptake n. □ **quick on the uptake** (colloquial) quick to understand.

uptight adj. (colloquial) nervously tense; annoyed.

up to date adj. modern, fashionable.

upturn v. (up-tern) turn up, upwards, or upside down. ● n. (up-tern) an upheaval; an upward trend; an improvement.

upward adj. moving or tending up. ● adv. (also **upwards**) towards a higher level etc.

upwind adj. & adv. in the direction from which the wind is blowing.

uranium n. a metallic element (symbol U), a heavy grey metal used as a source of nuclear energy.

urban adj. of a city or town.

urbane adj. courteous; having elegant manners. □ **urbanely** adv.; **urbanity** n.

urbanize v. (also **-ise**) change (a place) into an urban area. □ **urbanization** n.

urchin n. a mischievous child.

Urdu n. a language related to Hindi.

ureter (yoo-ree-ter) n. the duct from the kidney to the bladder.

urethra (yoo-ree-thrā) n. the duct which carries urine from the body.

urge v. press, encourage to proceed; recommend strongly or earnestly. ● n. a feeling or desire urging a person to do something.

urgent adj. needing or calling for immediate attention or action. □ **urgently** adv.; **urgency** n.

urinal n. a receptacle or structure for receiving urine.

urinate v. discharge urine from the body. □ **urination** n.

urine n. waste liquid which collects in the bladder and is discharged from the body. □ **urinary** adj.

urn n. **1** a vase, esp. for holding a cremated person's ashes. **2** a large metal container with a tap, for keeping water etc. hot.

ursine adj. of or like a bear.

us pron. the objective case of we.

US, USA abbr. United States (of America).

usable adj. able or fit to be used.

usage n. the manner of using or treating something; customary practice.

use v. (yooz) cause to act or to serve for a purpose; treat; exploit selfishly. ● n. (yooss) using, being used; the power of using; a purpose for which a thing is used.

used (yoozd) *adj.* second-hand.

used to (yoost) *v.* was accustomed to (do). ●*adj.* familiar with by practice or habit.

■**Usage** The usual negative and question forms of *used to* are, for example, *You didn't use to go there* and *Did you use to go there?* Both are, however, rather informal, so it is better in formal language to use *You used not to go there* and a different expression such as *Were you in the habit of going there?* or *Did you go there when you lived in London?*

useful *adj.* fit for a practical purpose; able to produce good results. □ **usefully** *adv.*, **usefulness** *n.*

useless *adj.* not usable, not useful. □ **uselessly** *adv.*, **uselessness** *n.*

user *n.* a person who uses something.

user-friendly *adj.* easy for a user to understand and operate.

usher *n.* a person who shows people to their seats in a public hall etc. ● *v.* lead, escort.

usherette *n.* a woman who ushers people to seats in a cinema etc.

USSR *abbr.* (*hist.*) Union of Soviet Socialist Republics.

usual *adj.* such as happens or is done or used etc. in many or most instances. □ **usually** *adv.*

usurp *v.* take (power, a position, or right) wrongfully or by force. □ **usurpation** *n.*, **usurper** *n.*

usury *n.* the lending of money at excessively high rates of interest. □ **usurer** *n.*

utensil *n.* an instrument or container, esp. for domestic use.

uterus *n.* the womb. □ **uterine** *adj.*

utilitarian *adj.* useful rather than decorative or luxurious.

utilitarianism *n.* the theory that actions are justified if they benefit the majority.

utility *n.* **1** usefulness; a useful thing. **2** (also **public utility**) a company supplying water, gas, electricity, etc. to the community. ●*adj.* severely practical.

utility room *n.* a room for large domestic appliances.

utilize *v.* (also **-ise**) use, find a use for. □ **utilization** *n.*

utmost *adj.* furthest, greatest, extreme. ● *n.* the furthest point or degree etc.

Utopia *n.* an imaginary place or state where all is perfect. □ **Utopian** *adj.*

utter *adj.* complete, absolute. ● *v.* make (a sound or words) with the mouth or voice; speak. □ **utterance** *n.*, **utterly** *adv.*

uttermost *adj.* & *n.* = **utmost**.

U-turn *n.* the driving of a vehicle in a U-shaped course to reverse its direction; a reversal of policy or opinion.

UV *abbr.* ultraviolet.

uvula (yoov-yoo-lă) *n.* (*pl.* **uvulae**) the small fleshy projection hanging at the back of the throat. □ **uvular** *adj.*

uxorious *adj.* excessively fond of one's wife.

Vv

V *n.* (as a Roman numeral) 5. ●*abbr.* volt(s). ●*symb.* vanadium.

v. *abbr.* **1** versus. **2** very.

vac *n.* (*colloquial*) **1** a vacation. **2** a vacuum cleaner.

vacancy *n.* the state of being vacant; a vacant place or position etc.

vacant *adj.* **1** unoccupied. **2** showing no interest. □ **vacantly** *adv.*

vacate *v.* cease to occupy.

vacation *n.* **1** an interval between terms in universities and law courts; (*Amer.*) a holiday. **2** the vacating of a place etc.

vaccinate *n.* inoculate with a vaccine. □ **vaccination** *n.*

vaccine (vak-seen) *n.* a preparation that gives immunity from an infection.

vacillate *v.* keep changing one's mind. □ **vacillation** *n.*

vacuous *adj.* unintelligent, inane; expressionless. □ **vacuously** *adv.*, **vacuity** *n.*

vacuum *n.* (*pl.* **vacuums** or **vacua**) a space from which air has been removed. ● *v.* clean with a vacuum cleaner.

vacuum cleaner *n.* an electrical apparatus that takes up dust by suction.

vacuum flask *n.* a container for keeping liquids hot or cold.

vacuum-packed *adj.* sealed after removal of air.

vagabond *n.* a wanderer; a vagrant.

vagary *n.* a capricious act or idea; a fluctuation.

vagina *n.* the passage leading from the vulva to the womb. □ **vaginal** *adj.*

vagrant *n.* a person without a settled home. □ **vagrancy** *n.*

vague *adj.* not clearly explained or perceived; not expressing oneself clearly. □ **vaguely** *adv.*, **vagueness** *n.*

vain *adj.* **1** conceited. **2** useless, futile. ● **in vain** uselessly, without success. □ **vainly** *adv.*

vainglory *n.* great vanity. □ **vainglorious** *adj.*

valance *n.* a short curtain or hanging frill.

vale *n.* a valley.

valediction *n.* a farewell. □ **valedictory** *adj.*

valence *n.* (also **valency**) the combining power of an atom as compared with that of the hydrogen atom.

valentine *n.* a romantic greetings card sent, often anonymously, on St Valentine's Day (14 Feb.); a person to whom one sends this.

valet *n.* a man's personal attendant. ● *v.* (**valeted**) **1** act as valet to. **2** clean (a car) thoroughly.

valetudinarian *n.* a person of poor health or unduly anxious about health.

valiant *adj.* brave. □ **valiantly** *adv.*

valid *adj.* **1** having legal force, legally acceptable. **2** logical. □ **validity** *n.*

validate *v.* make valid, confirm. □ **validation** *n.*

valise *n.* (*Amer.*) a small suitcase.

valley *n.* (*pl.* **valleys**) a low area between hills.

valour *n.* (*Amer.* **valor**) bravery.

valuable *adj.* of great value or worth. ● *n.pl.* (**valuables**) valuable things.

valuation *n.* estimation or an estimate of a thing's worth.

value *n.* **1** the amount of money or other commodity etc. considered equivalent to something else; usefulness, importance. **2** (**values**) moral principles. ● *v.* estimate the value of; consider to be of great worth.

value added tax *n.* a tax on the amount by which a thing's value has been increased at each stage of its production.

value judgement *n.* a subjective estimate of quality etc.

valuer *n.* a person who estimates values professionally.

valve *n.* **1** a device controlling flow through a pipe; a structure allowing blood to flow in one direction only. **2** each half of the hinged shell of an oyster etc. □ **valvular** *adj.*

vamoose v. (Amer. slang) depart hurriedly.

vamp n. the upper front part of a boot or shoe. ● v. improvise (esp. a musical accompaniment).

vampire n. a ghost or reanimated body supposed to suck blood.

van n. **1** a covered vehicle for transporting goods etc.; a railway carriage for luggage or goods. **2** the vanguard, the forefront.

vanadium n. a hard grey metallic element (symbol V).

vandal n. a person who damages things wilfully. □ **vandalism** n.

vandalize v. (also **-ise**) damage wilfully.

vane n. **1** a weathervane. **2** the blade of a propeller, sail of a windmill, etc.

vanguard n. the foremost part of an advancing army etc.

vanilla n. a flavouring, esp. obtained from the pods of a tropical orchid.

vanish v. disappear completely.

vanity n. **1** conceit. **2** futility.

vanity case n. a small case for carrying cosmetics etc.

vanquish v. conquer.

vantage n. (in tennis) advantage.

vantage point n. a position giving a good view.

vapid adj. insipid, uninteresting. □ **vapidly** adv., **vapidity** n.

vaporize v. (also **-ise**) convert or be converted into vapour. □ **vaporization** n.

vapour n. (Amer. **vapor**) moisture suspended in air, into which certain liquids or solids are converted by heating. □ **vaporous** adj.

variable adj. varying. ● n. something that varies. □ **variability** n.

variance n. □ **at variance** in disagreement.

variant adj. differing. ● n. a variant form, spelling, etc.

variation n. **1** varying, the extent of this; a variant. **2** a repetition of a melody in a different form.

varicose adj. (of veins) permanently swollen. □ **varicosity** n.

varied adj. of different sorts.

variegated adj. having irregular patches of colours. □ **variegation** n.

variety n. **1** the quality of not being the same; a quantity of different things. **2** a sort or kind. **3** light entertainment made up of a series of short unrelated performances.

various adj. of several kinds; several. □ **variously** adv.

■ **Usage** Various (unlike several) is not a pronoun and therefore cannot be used with of, as (wrongly) in Various of the guests arrived late.

varlet n. (old use) **1** a menial servant. **2** a rascal.

varnish n. a liquid that dries to form a shiny transparent coating. ● v. coat with varnish.

vary v. make or be become different.

vascular adj. of vessels or ducts for conveying blood or sap.

vase n. a decorative jar for holding cut flowers.

vasectomy n. a surgical removal of part of the ducts that carry semen from the testicles, esp. as a method of birth control.

Vaseline n. (trade mark) a thick oily cream made from petroleum, used as an ointment or lubricant.

vassal n. a humble subordinate.

vast adj. very great in area or size. □ **vastly** adv., **vastness** n.

VAT abbr. value added tax.

vat n. a large tank for liquids.

vault n. **1** an arched roof; a cellar used as a storage place. **2** a burial chamber. **3** an act of vaulting. ● v. jump, esp. with the help of the hands or a pole. □ **vaulted** adj.

vaunt v. boast, brag. ● n. a boast.

VCR abbr. video cassette recorder.

VD abbr. venereal disease.

VDU abbr. visual display unit.

veal n. calf's flesh as food.

vector n. **1** a thing (e.g. velocity) that has both magnitude and direction. **2** the carrier of an infection.

veer v. change direction.

vegan (vee-găn) n. a person who eats no meat or animal products.

vegetable n. a plant grown for food. ● adj. of or from plants.

vegetarian n. a person who eats no meat or fish. □ **vegetarianism** n.

vegetate v. live an uneventful life.

vegetation n. plants collectively.

veggie n. (colloquial) = **vegetarian**.

vehement adj. showing strong feeling. □ **vehemently** adv. **vehemence** n.

vehicle n. **1** a conveyance for transporting passengers or goods on land or in space. **2** a means by which something is expressed or displayed. □ **vehicular** adj.

veil n. a piece of fine net or other fabric worn to protect or conceal the face. ● v. cover with or as if with a veil.

vein n. **1** any of the blood vessels conveying blood towards the heart. **2** a threadlike structure. **3** a narrow layer in rock etc. **4** a mood, a manner. □ **veined** adj.

Velcro n. (trade mark) a fastener consisting of two strips of fabric which cling together when pressed.

veld n. (also **veldt**) open grassland in South Africa.

vellum n. fine parchment; smooth writing paper.

velocity n. speed.

velour (vĕ-loor) n. (also **velours**) a plush fabric resembling velvet.

velvet n. a woven fabric with a thick short pile on one side. □ **velvety** adj.

velveteen n. cotton velvet.

velvet glove n. outward gentleness concealing strength or inflexibility.

Ven. abbr. Venerable.

venal adj. able to be bribed; influenced by bribery. □ **venality** n.

vend v. sell, offer for sale.

vendetta n. a feud.

vending machine n. a slot machine where small articles are obtained.

vendor n. a seller.

veneer n. a thin covering layer of fine wood; a superficial show of a quality. ● v. cover with a veneer.

venerable adj. worthy of great respect; (**Venerable**) the title of an archdeacon.

venerate v. respect deeply. □ **veneration** n.

venereal adj. (of infections) contracted by sexual intercourse with an infected person.

Venetian adj. & n. (a native) of Venice.

venetian blind n. a window-blind of adjustable horizontal slats.

vengeance n. retaliation. □ **with a vengeance** in an extreme degree.

vengeful adj. seeking vengeance.

venial adj. (of a sin) pardonable, not serious. □ **veniality** n.

venison n. deer's flesh as food.

Venn diagram n. one using overlapping circles etc. to show relationships between sets.

venom n. **1** a poisonous fluid secreted by snakes etc. **2** bitter feeling or language. □ **venomous** adj.

venous adj. of veins.

vent n. **1** an opening allowing gas or liquid to pass through. **2** a slit at the lower edge of a coat. ● v. give vent to. □ **give vent to** give an outlet to (feelings).

ventilate v. **1** cause air to enter or circulate freely in. **2** express publicly. □ **ventilation** n.

ventilator n. **1** a device for ventilating a room etc. **2** a respirator.

ventral adj. of or on the abdomen.

ventricle n. a cavity, esp. in the heart or brain. □ **ventricular** adj.

ventriloquist n. an entertainer who can produce voice sounds so that they seem to come from a puppet etc. □ **ventriloquism** n.

venture n. an undertaking that involves risk. ● v. dare; dare to go or utter. □ **venturesome** adj.

venue n. an appointed place for a meeting, concert, etc.

veracious adj. truthful; true. □ **veraciously** adv., **veracity** n.

veranda n. a roofed terrace.

verb n. a word indicating an action, state, or occurrence.

verbal adj. **1** of or in words; spoken. **2** of a verb. □ **verbally** adv.

verbalize v. (also **-ise**) express in words; be verbose. □ **verbalization** n.

verbatim (ver-bay-tim) adv. & adj. in exactly the same words.

verbiage n. an excessive number of words.

verbose adj. using more words than are needed. □ **verbosely** adv., **verbosity** n.

verdant adj. (of grass etc.) green.

verdict n. a decision reached by a jury; a decision or opinion reached after testing something.

verdigris n. a green deposit forming on copper or brass.

verdure n. green vegetation.

verge n. the extreme edge, the brink; a grass edging of a road etc. □ **verge on** border on.

verger n. a church caretaker.

verify v. check the truth or correctness of. □ **verification** n.

verisimilitude n. the appearance of being true.

veritable adj. real, rightly named.

vermicelli n. pasta made in slender threads.

vermiform adj. worm-shaped.

vermilion adj. & n. bright red.

vermin n. (pl. **vermin**) an animal or insect regarded as a pest. □ **verminous** adj.

vermouth n. fortified wine flavoured with herbs.

vernacular n. the ordinary language of a country or district.

vernal adj. of or occurring in spring.

veronica n. a flowering herb or shrub.

verruca (vě-roo-kǎ) n. (pl. **verrucas** or **verrucae**) a wart, esp. on the foot.

versatile adj. able to do or be used for many different things. □ **versatility** n.

verse n. poetry, a poem; a group of lines forming a unit in a poem or hymn; a numbered division of a Bible chapter.

versed adj. □ **versed in** experienced in.

versify v. express in verse; compose verse. □ **versification** n.

version n. a particular account of a matter; a translation; a special or variant form.

verso n. (pl. **versos**) the left-hand page of a book; the back of a leaf of manuscript.

versus prep. against.

vertebra n. (pl. **vertebrae**) a segment of the backbone. □ **vertebral** adj.

vertebrate n. & adj. (an animal) having a backbone.

vertex n. (pl. **vertices** or **vertexes**) the highest point of a hill etc.; an apex.

vertical adj. perpendicular to the horizontal, upright. ● n. a vertical line or position. □ **vertically** adv.

vertiginous adj. causing vertigo.

vertigo n. dizziness.

verve n. enthusiasm, vigour.

very adv. in a high degree, extremely; exactly. ● adj. actual, truly such; extreme. □ **very well** an expression of consent.

vesicle n. a sac, esp. containing liquid; a blister.

vessel n. **1** a receptacle, esp. for liquid. **2** a ship, a boat. **3** a tubelike structure conveying blood or other fluid in the body of an animal or plant.

vest n. an undergarment covering the trunk; (Amer.) a waistcoat. ● v. confer (power) as a firm or legal right.

vested interest n. a personal interest in a state of affairs, usu. with an expectation of gain.

vestibule n. an entrance hall; a porch.

vestige n. a small amount or trace. □ **vestigial** adj.

vestment n. a ceremonial garment, esp. of clergy or a church choir.

vet n. **1** a veterinary surgeon. **2** (Amer.) a military veteran. ● v. (**vetted**) examine critically for faults etc.

vetch n. a plant of the pea family used as fodder.

veteran n. a person with long experience, esp. in the armed forces.

veterinarian n. a veterinary surgeon.

veterinary adj. of or for the treatment of diseases and disorders of animals.

veterinary surgeon n. a person qualified to treat animal diseases and disorders.

veto n. (pl. **vetoes**) an authoritative rejection of something proposed; the right to make this. ● v. reject by a veto.

vex v. annoy. □ **vexation** n., **vexatious** adj.

vexed question n. a problem that is much discussed.

VHF abbr. very high frequency, (in radio) 30–300 MHz.

via prep. by way of, through.

viable adj. capable of living or surviving; practicable. □ **viability** n.

viaduct n. a long bridge over a valley.

vial n. a small bottle.

viands n.pl. (old use) articles of food.

vibes n.pl. (colloquial) **1** a vibraphone. **2** mental or emotional vibrations; a feeling communicated.

vibrant adj. vibrating, resonant, thrilling with energy.

vibraphone n. a percussion instrument like a xylophone but with a vibrating effect.

vibrate v. move rapidly and continuously to and fro; sound with rapid slight variation of pitch. □ **vibrator** n., **vibratory** adj.

vibration n. **1** vibrating. **2** (**vibrations**) mental influences; an atmosphere or feeling communicated.

vibrato n. (in music) a rapid slight fluctuation in the pitch of a note.

vicar n. a member of the clergy in charge of a parish.

vicarage n. the house of a vicar.

vicarious adj. felt through sharing imaginatively in the feelings or activities etc. of another person; acting or done etc. for another. □ **vicariously** adv.

vice n. **1** great wickedness; criminal and immoral practices. **2** (Amer. **vise**) an instrument with two jaws for holding things firmly.

vice- comb. form a substitute or deputy for; next in rank to.

vice-chancellor n. the chief administrator of a university.

viceroy n. a person governing a colony etc. as the sovereign's representative. □ **viceregal** adj.

vice versa adj. with terms the other way round.

vichyssoise n. a creamy leek and potato soup, often served chilled.

vicinity n. the surrounding district. □ **in the vicinity (of)** near.

vicious adj. brutal, strongly spiteful; savage and dangerous. □ **viciously** adv.

vicious circle n. a bad situation producing effects that intensify its original cause.

vicissitude n. a change of circumstances or luck.

victim n. a person injured or killed or made to suffer.

victimize v. (also **-ise**) single out to suffer ill treatment. □ **victimization** n.

victor n. a winner.

Victorian n. & adj. (a person) of the reign of Queen Victoria (1837-1901).

victorious adj. having gained victory.

victory n. success achieved by winning a battle or contest.

victualler (vit-ler) n. (Amer. **victualer**) a person who sells food.

victuals n.pl. food, provisions.

vicuña (vik-oon-yā) n. a South American animal related to the llama; soft cloth made from its wool.

video n. (pl. **videos**) a recording or broadcasting of pictures; an apparatus for this; a videotape. ● v. make a video of.

videotape n. magnetic tape suitable for recording television pictures and sound. ● v. record on this.

videotex n. (also **videotext**) teletext or viewdata.

vie v. (**vied**, **vying**) carry on a rivalry, compete.

view n. **1** a range of vision; things within this. **2** fine scenery. **3** a mental attitude, an opinion. ● v. **1** look at, inspect; watch television. **2** consider. □ **in view of** having regard to. **on view** displayed for inspection. **with a view to** with the hope or intention of. □ **viewer** n.

viewdata n. a news and information service to which a television screen is connected by a telephone link.

viewfinder n. a device on a camera showing the extent of the area being photographed.

viewpoint n. a point of view; a place from which there is a good view.

vigil n. a period of staying awake to keep watch or pray.

vigilant adj. watchful. □ **vigilantly** adv., **vigilance** n.

vigilante (vi-ji-lan-ti) n. a member of a self-appointed group trying to prevent crime etc.

vignette (veen-yet) n. a short written description.

vigour n. (Amer. **vigor**) active physical or mental strength; forcefulness. □ **vigorous** adj., **vigorously** adv., **vigorousness** n.

Viking | visa

Viking *n.* an ancient Scandinavian trader and pirate.

vile *adj.* extremely disgusting or wicked. □ **vilely** *adv.*, **vileness** *n.*

vilify *v.* say evil things about. □ **vilification** *n.*, **vilifier** *n.*

villa *n.* a rented holiday home, esp. abroad; a house in a suburban district.

village *n.* a collection of houses and other buildings in a country district. □ **villager** *n.*

villain *n.* a wicked person. □ **villainous** *adj.*, **villainy** *n.*

villein *n.* (*hist.*) a feudal tenant subject to a lord.

vim *n.* (*colloquial*) vigour.

vinaigrette *n.* a salad dressing of oil and vinegar.

vindicate *v.* clear of blame; justify. □ **vindication** *n.*

vindictive *adj.* showing a desire for vengeance. □ **vindictively** *adv.*, **vindictiveness** *n.*

vine *n.* a climbing plant whose fruit is the grape.

vinegar *n.* a sour liquid made from wine, malt, etc., by fermentation. □ **vinegary** *adj.*

vineyard *n.* a plantation of vines for wine-making.

vintage *n.* **1** wine from a season's grapes. **2** a date of origin or existence. ● *adj.* of high quality, esp. from a past period.

vintner *n.* a wine-merchant.

vinyl (vy-nil) *n.* a kind of plastic.

viola¹ (vi-oh-lă) *n.* an instrument like a violin but of lower pitch.

viola² (vy-ŏ-lă) *n.* a plant of the genus to which violets and pansies belong.

violate *v.* **1** break (an oath, treaty, etc.). **2** treat (a sacred place) irreverently; disturb. **3** rape. □ **violation** *n.*, **violator** *n.*

violent *adj.* involving great force or intensity; using excessive physical force. □ **violently** *adv.*, **violence** *n.*

violet *n.* a small plant, often with purple flowers; a bluish-purple colour. ● *adj.* bluish purple.

violin *n.* a musical instrument with four strings of treble pitch, played with a bow. □ **violinist** *n.*

violoncello *n.* (*pl.* **violoncellos**) a cello.

VIP *abbr.* very important person.

viper *n.* a small poisonous snake.

virago *n.* (*pl.* **viragos**) an aggressive woman.

viral (vy-răl) *adj.* of a virus.

virgin *n.* **1** a person (esp. a woman) who has never had sexual intercourse. **2** (**the Virgin**) Mary, mother of Christ. ● *adj.* **1** never having had sexual intercourse. **2** untouched, not yet used. □ **virginal** *adj.*, **virginity** *n.*

virile *adj.* having masculine strength or procreative power. □ **virility** *n.*

virology *n.* the study of viruses. □ **virologist** *n.*

virtual *adj.* being so in effect though not in name. □ **virtually** *adv.*

virtual reality *n.* a computer-generated simulation of reality.

virtue *n.* moral excellence, goodness; chastity; a good characteristic. □ **by** or **in virtue of** because of, on the strength of.

virtuoso *n.* (*pl.* **virtuosos** or **virtuosi**) an expert performer. □ **virtuosity** *n.*

virtuous *adj.* morally good. □ **virtuously** *adv.*, **virtuousness** *n.*

virulent *adj.* (of poison or disease) extremely strong or virulent; bitterly hostile. □ **virulently** *adv.*, **virulence** *n.*

virus *n.* **1** a minute organism capable of causing disease. **2** a destructive code hidden in a computer program.

visa *n.* an official mark on a passport, permitting the holder to enter a specified country.

visage n. (*literary*) a person's face.

vis-à-vis (veez-ah-vee) prep. with regard to; as compared with.

viscera (viss-er-ă) n.pl. the internal organs of the body. □ **visceral** adj.

viscid (vi-sid) adj. thick and sticky. □ **viscidity** n.

viscose n. viscous cellulose; a fabric made from this.

viscount (vy-kownt) n. a nobleman ranking between earl and baron.

viscountess n. a woman holding the rank of viscount; a viscount's wife or widow.

viscous adj. thick and gluey. □ **viscosity** n.

vise Amer. sp. of **vice** (sense 2).

visibility n. the state of being visible; the range of vision.

visible adj. able to be seen or noticed. □ **visibly** adv.

vision n. 1 ability to see, sight; foresight. 2 something seen in the imagination or a dream. 3 a person of unusual beauty.

visionary adj. fanciful; not practical. ● n. a person with visionary ideas.

visit v. go or come to see; stay temporarily with or at. ● n. an act of visiting. □ **visitor** n.

visitation n. 1 an official visit or inspection. 2 trouble regarded as divine punishment.

visor (vy-zer) n. (also **vizor**) a movable front part of a helmet, covering the face; a shading device at the top of a vehicle's windscreen.

vista n. an extensive view, esp. seen through a long opening.

visual adj. of or used in seeing. □ **visually** adv.

visual display unit n. a device displaying a computer output or input on a screen.

visualize v. (also **-ise**) form a mental picture of. □ **visualization** n.

vital adj. 1 essential to life; essential to a thing's existence or success. 2 full of vitality. ● n.pl. (**vitals**) the vital organs of the body, e.g. the heart and brain. □ **vitally** adv.

vitality n. liveliness, persistent energy.

vital statistics n.pl. 1 statistics of births, deaths, and marriages. 2 (*colloquial*) the measurements of a woman's bust, waist, and hips.

vitamin n. any of the organic substances present in food and essential to nutrition.

vitiate (vish-i-ayt) v. make imperfect or ineffective. □ **vitiation** n.

viticulture n. vine-growing.

vitreous adj. like glass in texture, finish, etc.

vitrify v. change into a glassy substance. □ **vitrifaction** n.

vitriol n. 1 sulphuric acid. 2 savagely hostile remarks. □ **vitriolic** adj.

vituperate v. use abusive language. □ **vituperation** n., **vituperative** adj.

viva¹ n. (*colloquial*) a viva voce examination.

viva² int. long live (a person or thing).

vivace (vi-vah-chi) adv. in a lively manner.

vivacious adj. lively, high-spirited. □ **vivaciously** adv., **vivacity** n.

vivarium n. (pl. **vivaria**) a place for keeping living animals etc. in natural conditions.

viva voce (vy-vă voh-chi) n. an oral university examination.

vivid adj. bright and strong; clear; (of imagination) lively. □ **vividly** adv., **vividness** n.

vivify v. put life into.

viviparous adj. bringing forth young alive, not egg-laying.

vivisection n. performance of experiments on living animals.

vixen n. a female fox.

viz. adv. namely.

vizor var. of **visor**.

V-neck n. a V-shaped neckline on a pullover etc.

vocabulary n. a list of words with their meanings; words known or used by a person or group.

vocal adj. of, for, or uttered by the voice. ● n. a piece of sung music. □ **vocally** adv.

vocalist n. a singer.

vocalize v. (also **-ise**) utter. □ **vocalization** n.

vocation n. a strong desire or feeling of fitness for a certain career; a trade or profession. □ **vocational** adj.

vociferate v. say loudly, shout. □ **vociferation** n.

vociferous adj. making a great outcry. □ **vociferously** adv.

vodka n. an alcoholic spirit distilled chiefly from rye.

vogue n. current fashion; popularity. □ **in vogue** in fashion.

voice n. sounds formed in the larynx and uttered by the mouth; an expressed opinion, the right to express an opinion. ● v. express; utter.

voice-over n. a narration in a film etc. without a picture of the speaker.

void adj. **1** empty. **2** not valid. ● n. an empty space, emptiness. ● v. make void; excrete.

voile n. a very thin dress fabric.

volatile adj. **1** evaporating rapidly. **2** lively, changing quickly in mood. □ **volatility** n.

vol-au-vent (vol-oh-von) n. a small puff pastry case filled with a savoury mixture.

volcano n. (pl. **volcanoes**) a mountain with a vent through which lava is expelled. □ **volcanic** adj.

vole n. a small rodent.

volition n. the exercise of one's will.

volley n. (pl. **volleys**) **1** a simultaneous discharge of missiles etc.; an outburst of questions or other words. **2** a return of the ball in tennis etc. before it touches the ground. ● v. send in a volley.

volleyball n. a game for two teams of six people sending a large ball by hand over a net.

volt n. a unit of electromotive force.

voltage n. electromotive force expressed in volts.

voltameter n. an instrument measuring an electrical charge.

volte-face (volt-fass) n. a complete change of attitude to something.

voltmeter n. an instrument measuring electrical potential in volts.

voluble adj. speaking or spoken with a great flow of words. □ **volubly** adv., **volubility** n.

volume n. **1** a book. **2** an amount of space occupied or contained; the quantity of something. **3** strength of sound.

voluminous adj. having great volume, bulky; copious.

voluntary adj. done, given, or acting by choice; working or done without payment; maintained by voluntary contributions. □ **voluntarily** n.

volunteer n. **1** a person who offers to do something. **2** a person who enlists voluntarily for military service. ● v. undertake or offer voluntarily; be a volunteer.

voluptuary n. a person fond of luxury and sensual pleasure.

voluptuous adj. **1** full of or fond of sensual pleasure. **2** (of a woman) having a full attractive figure. □ **voluptuously** adv., **voluptuousness** n.

vomit v. (**vomited**) eject (matter) from the stomach through the mouth. ● n. vomited matter.

voodoo n. a form of religion based on witchcraft. □ **voodooism** n.

voracious adj. greedy; ravenous; insatiable. □ **voraciously** adv., **voracity** n.

vortex n. (pl. **vortexes** or **vortices**) a whirlpool or whirlwind.

vote n. a formal expression of one's opinion or choice on a matter under discussion; a choice etc. expressed in this way; the right to vote. ● v. express, decide, or support etc. by a vote. □ **voter** n.

votive adj. given to fulfil a vow.

vouch v. **vouch for** guarantee the accuracy or reliability etc. of.

voucher n. a document exchangeable for certain goods or services; a receipt.

vouchsafe v. give or grant.

vow n. a solemn promise, esp. to a deity or saint. ● v. make a vow.

vowel n. a speech sound made without audible stopping of the breath; the letter representing this.

voyage n. a journey made by water or in space. ● v. make a voyage. □ **voyager** n.

voyeur (vwah-yer) n. a person who gets sexual pleasure from watching others having sex or undressing.

vox pop n. popular opinion represented by informal comments from the public.

vs. abbr. versus.

V-sign n. a sign of abuse made with the first two fingers pointing up and the back of the hand facing out; a similar sign with the palm facing outwards as a symbol of victory.

VSO abbr. voluntary service overseas.

vulcanite n. hard black vulcanized rubber.

vulcanize v. (also **-ise**) strengthen (rubber) by treating with sulphur. □ **vulcanization** n.

vulgar adj. lacking refinement or good taste. □ **vulgarity** n.

vulgarly adv.

vulgar fraction n. a fraction represented by numbers above and below a line (rather than decimally).

vulgarian n. a vulgar (esp. rich) person.

vulgarism n. a coarse word or expression.

vulnerable adj. able to be hurt or injured; exposed to danger or criticism. □ **vulnerability** n.

vulture n. a large bird of prey that lives on the flesh of dead animals.

vulva n. the external parts of the female genital organs.

vying see **vie**.

Ww

W. abbr. **1** west; western. **2** watt(s). ● symb. tungsten.

wacky adj. (**wackier**) (slang) crazy.

wad n. **1** a pad of soft material. **2** a bunch of papers or banknotes. ● v. (**wadded**) pad.

wadding n. padding.

waddle v. walk with short steps and a swaying movement. ● n. a waddling gait.

wade v. walk through water or mud; proceed slowly and laboriously (through whole etc.).

wader n. **1** a long-legged waterbird. **2** (**waders**) high waterproof boots worn in fishing etc.

wadi n. a rocky watercourse, dry except in the rainy season.

wafer n. a thin light biscuit; a small thin slice.

waffle n. **1** (*colloquial*) vague wordy talk or writing. **2** a small cake of batter eaten hot, cooked in a **waffle-iron.** ● v. (*colloquial*) talk or write waffle.

waft v. carry or travel lightly through air or over water. ● n. a wafted odour.

wag v. (**wagged**) shake briskly to and fro. ● n. **1** a wagging movement. **2** a humorous person.

wage n. (also **wages**) regular payment to an employee for his or her work. ● v. engage in (war).

waged adj. employed, working.

wager n. a bet. ● v. bet.

waggle v. wag. ● n. a waggling motion. □ **waggly** adj.

wagon n. (also **waggon**) a four-wheeled vehicle for heavy loads; an open railway truck. □ **on the wagon** (*colloquial*) abstaining from alcohol.

wagtail n. a small bird with a long tail that wags up and down.

waif n. a homeless child.

wail v. & n. (utter) a long sad cry.

wainscot n. (also **wainscoting**) wooden panelling in a room.

waist n. the part of the human body between ribs and hips; a narrow middle part.

waistcoat n. a close-fitting waist-length sleeveless jacket.

waistline n. the outline or size of the waist.

wait v. **1** postpone an action until a specified time or event occurs; be postponed; pause. **2** wait on people at a meal. ● n. an act or period of waiting. □ **wait on** serve food and drink to (a person) at a meal; fetch and carry things for.

waiter n. a man employed to serve customers in a restaurant etc.

waitress n. a woman employed to serve customers in a restaurant etc.

waive v. refrain from using (a right etc.). □ **waiver** n.

wake v. (**woke** or **waked**, **woken** or **waked**, **waking**) **1** cease to sleep; cause to cease sleeping. **2** evoke. ● n. **1** the track left on water's surface by a ship etc. **2** a watch by a corpse before burial; lamentations and merrymaking accompanying this. □ **wake up** wake; make or become alert. **in the wake of** behind; following.

wakeful adj. unable to sleep; sleepless. □ **wakefulness** n.

waken v. wake.

walk v. progress by setting down one foot and then lifting the other(s) in turn; travel (over) in this way; accompany in walking. ● n. a journey on foot; a manner or style of walking; a place or route for walking. □ **walk of life** social rank; occupation. **walk out** depart suddenly and angrily; go on strike suddenly. **walk out on** desert. □ **walker** n.

walkabout n. an informal stroll among a crowd by royalty etc.

walkie-talkie n. a small portable radio transmitter and receiver.

walking stick n. a stick carried or used as a support when walking.

walkout n. a sudden angry departure, esp. as a strike.

walkover n. an easy victory.

wall n. a continuous upright structure forming one side of a building or room or area; something like this in form or function. ● v. surround or enclose with a wall.

wallaby n. a marsupial like a small kangaroo.

wallet n. a small folding case for banknotes or documents.

wallflower n. **1** a garden plant with clusters of fragrant flowers.

2 (*colloquial*) a woman sitting out dances for lack of partners.

wallop (*slang*) *v.* (**walloped**) thrash, hit hard. ●*n.* a heavy blow.

wallow *v.* roll in mud or water etc. ●*n.* an act of wallowing. □ **wallow in** take unrestrained pleasure in.

wallpaper *n.* paper for covering the interior walls of rooms.

wally *n.* (*slang*) a stupid person.

walnut *n.* a nut containing a wrinkled edible kernel; the tree bearing this; its wood.

walrus *n.* a large seal-like Arctic animal with long tusks.

waltz *n.* a ballroom dance; the music for this. ●*v.* dance a waltz; (*colloquial*) move gaily or casually.

wan *adj.* pale, pallid. □ **wanly** *adv.*, **wanness** *n.*

wand *n.* a slender rod, esp. associated with the working of magic.

wander *v.* go from place to place with no settled route or purpose; stray; digress. ●*n.* an act of wandering. □ **wanderer** *n.*

wanderlust *n.* a strong desire to travel.

wane *v.* decrease in vigour or importance; (of the moon) show a decreasing bright area after being full. □ **on the wane** waning.

wangle *v.* (*slang*) obtain or arrange by trickery or scheming.

want *v.* desire; need; lack; fall short of. ●*n.* a desire; a need; a lack.

wanted *adj.* (of a suspected criminal) sought by the police.

wanting *adj.* lacking, deficient.

wanton *adj.* irresponsible, lacking proper restraint.

wapiti (wo-pi-ti) *n.* a large North American deer.

war *n.* (a period of) fighting (esp. between countries); open hostil-

ity; conflict. ●*v.* (**warred**) make war.

warble *v.* sing, esp. with a gentle trilling note. ●*n.* a warbling sound.

ward *n.* **1** a room with beds for patients in a hospital. **2** a division of a city or town, electing a councillor to represent it. **3** a child under the care of a guardian or law court. □ **ward off** keep at a distance; repel.

warden *n.* an official with supervisory duties.

warder *n.* a prison officer.

wardrobe *n.* a large cupboard for storing hanging clothes; a stock of clothes or costumes.

ware *n.* manufactured goods of the kind specified; (**wares**) articles offered for sale.

warehouse *n.* a building for storing goods or furniture.

warfare *n.* making war, fighting.

warhead *n.* the explosive head of a missile.

warlike *adj.* fond of making war, aggressive; of or for war.

warm *adj.* **1** moderately hot; providing warmth. **2** enthusiastic, hearty. **3** kindly and affectionate. ●*v.* make or become warm. □ **warm to** become cordial towards (a person) or more animated about (a task). **warm up** warm; reheat; prepare for exercise etc. by practice beforehand; make or become more lively. □ **warmly** *adv.*, **warmness** *n.*

warm-blooded *adj.* having blood that remains warm permanently.

warmonger *n.* a person who seeks to bring about war.

warmth *n.* warmness.

warn *v.* inform about a present or future danger or difficulty etc., advise about action in this. □ **warn off** tell (a person) to keep away or to avoid (a thing).

arning *n.* something that serves to warn a person.

arp *v.* make or become bent by uneven shrinkage or expansion; distort, pervert. ● *n.* **1** a warped condition. **2** the lengthwise threads in a loom.

arrant *n.* a written authorization; a voucher; proof, a guarantee. ● *v.* justify; prove, guarantee.

arranty *n.* a guarantee.

arren *n.* a series of burrows where rabbits live; a building or district with many winding passages.

arrior *n.* a person who fights in a battle.

art *n.* a small hard abnormal growth.

arthog *n.* an African wild pig with wart-like lumps on its face.

ary *adj.* (**warier**) cautious, looking out for possible danger or difficulty. □ **warily** *adv.*, **wariness** *n.*

ash *v.* **1** cleanse with water or other liquid; wash oneself or clothes etc.; be washable. **2** flow past, against, or over; carry by flowing. **3** coat thinly with paint. **4** (*colloquial*) of (reasoning) be valid. ● *n.* **1** the process of washing or being washed; clothes etc. to be washed. **2** disturbed water or air behind a moving ship or aircraft etc. **3** a thin coating of paint. □ **wash one's hands of** refuse to take responsibility for. **wash out** make (a sport) impossible by heavy rainfall; (*colloquial*) cancel. **wash up** wash (dishes etc.) after use; cast up on the shore.

ashable *adj.* able to be washed without suffering damage.

ashbasin *n.* a bowl (usu. fixed to a wall) for washing one's hands and face.

ashed out *adj.* faded; pallid.

ashed up *adj.* (*slang*) defeated, having failed.

washer *n.* a ring of rubber or metal etc. placed between two surfaces to give tightness.

washing *n.* clothes etc. to be washed.

washing-up *n.* dishes etc. for washing after use; the process of washing these.

washroom *n.* (*Amer.*) a room with a lavatory.

wash-out *n.* (*colloquial*) a complete failure.

washy *adj.* (**washier**) too watery or weak; lacking vigour.

Wasp *n.* (also **WASP**) a middle-class American white Protestant.

wasp *n.* a stinging insect with a black and yellow striped body.

waspish *adj.* making sharp or irritable comments. □ **waspishly** *adv.*

wassail *n.* a festive occasion; merrymaking with a lot of drinking.

wastage *n.* loss or diminution by waste; loss of employees by retirement or resignation.

waste *v.* use or be used extravagantly or without adequate result; fail to use; make or become gradually weaker. ● *adj.* left or thrown away because not wanted; (of land) unfit for use. ● *n.* the process of wasting; waste material or food etc.; wasteland; a waste pipe.

wasteful *adj.* extravagant. □ **wastefully** *adv.*, **wastefulness** *n.*

waste pipe *n.* a pipe carrying off used or superfluous water or steam.

waster *n.* a wasteful person; a good-for-nothing person.

watch *v.* keep under observation; wait alertly, take heed; exercise protective care. ● *n.* **1** the act of watching, constant observation or attention; a sailor's period of duty, those on duty in this. **2** a small portable device indicating the time. □ **on watch** waiting alertly. **watch out** be on one's guard. □ **watcher** n.

watchdog n. **1** a dog kept to guard property. **2** a guardian of people's rights etc.

watchful adj. watching closely. □ **watchfully** adv., **watchfulness** n.

watchmaker n. a person who makes or repairs watches.

watchman n. (pl. **-men**) a person employed to guard a building etc.

watchtower n. a tower from which observation can be kept.

watchword n. a word or phrase expressing a group's principles.

water n. **1** a colourless odourless tasteless liquid that is a compound of hydrogen and oxygen; this as supplied for domestic use; a lake, the sea; the level of the tide. **2** a watery secretion; urine. ● v. **1** sprinkle, supply, or dilute with water. **2** secrete tears or saliva. □ **by water** in a boat etc. **water down** dilute; make less forceful.

waterbed n. a mattress of rubber etc. filled with water.

water biscuit n. a thin unsweetened biscuit.

water-butt n. a barrel used to catch rainwater.

water-cannon n. a device giving a powerful jet of water to dispel a crowd etc.

water chestnut n. the edible corm from a sedge.

water closet n. a lavatory flushed by water.

watercolour n. (Amer. **watercolor**) artists' paint mixed with water (not oil); a painting done with this.

watercourse n. a stream, brook, or artificial waterway; its channel.

watercress n. a kind of cress that grows in streams and ponds.

waterfall n. a stream that falls from a height.

waterfront n. part of a town that borders on a river, lake, or sea.

water ice n. frozen flavoured water.

watering can n. a container with a spout for watering plants.

watering place n. a pool where animals drink; a spa, a seaside resort.

water lily n. a plant with broad floating leaves.

waterline n. the line along which the surface of water touches a ship's side.

waterlogged adj. saturated with water.

water main n. a main pipe in a water supply system.

watermark n. a manufacturer's design in paper, visible when the paper is held against light.

water-meadow n. a meadow that is flooded periodically by a stream.

watermelon n. a melon with green skin, red pulp, and watery juice.

watermill n. a mill worked by a waterwheel.

water pistol n. a toy pistol that shoots a jet of water.

water polo n. a ball game played by teams of swimmers.

water power n. power obtained from flowing or falling water.

waterproof adj. unable to be penetrated by water. ● n. a waterproof garment. ● v. make waterproof.

water rat n. a small rodent living beside a lake or stream.

watershed n. **1** a line of high land separating two river systems. **2** a turning point in the course of events.

water-skiing n. the sport of skimming over water on skis while towed by a motor boat.

waterspout n. a column of water between sea and cloud, formed by a whirlwind.

water-table n. the level below which the ground is saturated with water.

watertight adj. **1** made or fastened so that water cannot get in or out. **2** impossible to disprove.

waterway n. a navigable channel, a canal.

waterwheel n. a wheel turned by a flow of water to work machinery.

water wings n.pl. floats worn on the shoulders by a person learning to swim.

waterworks n. an establishment with machinery etc. for supplying water to a district.

watery adj. **1** of or like water; containing too much water. **2** (of colour) pale.

watt n. a unit of electric power.

wattage n. an amount of electric power expressed in watts.

wattle n. **1** interwoven sticks used as material for fences, walls, etc. **2** a fold of skin hanging from the neck of a turkey etc.

wave n. **1** a moving ridge of water. **2** a wave-like curve, e.g. in hair. **3** an advancing group. **4** a temporary increase of an influence or condition. **5** an act of waving. **6** a wave-like motion by which heat, light, sound, or electricity etc. is spread; a single curve in this. ● v. **1** move loosely to and fro or up and down; move (one's arm etc.) in this way as a signal. **2** give or have a wavy course or appearance.

waveband n. a range of wavelengths.

wavelength n. the distance between corresponding points in a sound wave or electromagnetic wave.

wavelet n. a small wave.

waver v. be or become unsteady; show hesitation or uncertainty. □ **waverer** n.

wavy adj. (**wavier**) having waves or curves.

wax n. beeswax; any of various similar soft substances; polish containing this. ● v. **1** coat, polish, or treat with wax. **2** remove unwanted hair from (legs etc.) using wax. **3** increase in vigour or importance; (of the moon) show an increasing bright area until becoming full. □ **waxy** adj.

waxwing n. a small bird with red tips on some of its wing-feathers.

waxwork n. a wax model, esp. of a person.

way n. **1** a line of communication between two places; a route; a space free of obstacles so that people etc. can pass; progress; a specified direction or aspect. **2** a method, style, or manner; a chosen or desired course of action; (**ways**) habits. ● adv. (colloquial) far. □ **by the way** incidentally, as an irrelevant comment. **by way of** by going through; serving as. **in a way** to a limited extent; in some respects. **in the way** forming an obstacle or hindrance. **on one's way** in the process of travelling or approaching. **on the way** on one's way; (of a baby) conceived but not yet born.

waybill n. a list of the passengers or goods being carried in a vehicle.

wayfarer n. a traveller.

waylay v. (**waylaid, waylaying**) lie in wait for.

waymarker n. an arrow showing the direction of a footpath.

wayside n. the side of a road or path.

wayward adj. childishly self-willed, hard to control. □ **waywardness** n.

WC abbr. water closet.

we pron. used by a person referring to himself or herself and another or others; used instead of 'I' in newspaper editorials and by a royal person in formal proclamations.

weak adj. lacking strength or power; not convincing; much diluted. □ **weakly** adv.

weaken v. make or become weaker.

weakling n. a feeble person or animal.

weakness n. **1** the state of being weak; a weak point, a fault. **2** a self-indulgent liking.

weal n. a ridge raised on flesh esp. by the stroke of a rod or whip.

wealth n. money and valuable possessions; possession of these; a great quantity.

wealthy adj. (**wealthier**) having wealth, rich.

wean v. accustom (a baby) to take food other than milk; cause to give up something gradually.

weapon n. a thing designed or used for inflicting harm or damage; a means of coercing someone.

wear v. (**wore, worn, wearing**) **1** have on the body as clothing or ornament. **2** damage or become damaged by prolonged use. **3** endure continued use. ● n. **1** wearing, being worn. **2** clothing. **3** the capacity to endure being used. □ **wear down** overcome (opposition etc.) by persistence. **wear off** pass off gradually. **wear out** use or be used until no longer usable; tire or become tired out. □ **wearer** n., **wearable** adj.

wearisome adj. causing weariness.

weary adj. (**wearier**) very tired, tiring, tedious. ● v. make or become weary. □ **wearily** adv. **weariness** n.

weasel n. a small wild animal with a slender body and reddish-brown fur.

weather n. the state of the atmosphere with reference to sunshine, rain, wind, etc. ● v. **1** (cause to) be changed by exposure to the weather. **2** come safely through (a storm). □ **under the weather** feeling unwell or depressed.

weather-beaten adj. bronzed or worn by exposure to the weather.

weatherboard n. a sloping board for keeping out rain.

weathervane n. (also **weathercock**) a revolving pointer to show the direction of the wind.

weave v. (**wove, woven, weaving**) **1** make (fabric etc.) by passing crosswise threads or strips under and over lengthwise ones; form (thread etc.) into fabric in this way. **2** compose (a story etc.). **3** move in an intricate course. ● n. a style or pattern of weaving.

weaver n. **1** a person who weaves. **2** a tropical bird that builds an intricately woven nest.

web n. **1** a network of fine strands made by a spider etc. **2** skin filling the spaces between the toes of ducks, frogs, etc.

webbing n. a strong band of woven fabric used in upholstery.

Wed. abbr. (also **Weds.**) Wednesday.

wed v. (**wedded**) marry; unite.

wedding n. a marriage ceremony and festivities.

wedge n. a piece of solid substance thick at one end and tapering to a thin edge at the other. ● v. force apart or fix firmly with a wedge; crowd tightly.

wedlock *n.* the married state.

Wednesday *n.* the day after Tuesday.

wee *adj.* (Scot.) little; (colloquial) tiny. ● *n.* & *v.* (colloquial) = **wee-wee.**

weed *n.* **1** a wild plant growing where it is not wanted. **2** a thin weak-looking person. ● *v.* uproot and remove weeds (from). □ **weed out** remove as inferior or undesirable. □ **weedy** *adj.*

week *n.* a period of seven successive days, esp. from Monday to Sunday or Sunday to Saturday; the weekdays of this; the working period during a week.

weekday *n.* a day other than Sunday or other than the weekend.

weekend *n.* Saturday and Sunday.

weekly *adj.* & *adv.* (produced or occurring) once a week. ● *n.* a weekly periodical.

weeny *adj.* (**weenier**) (colloquial) tiny.

weep *v.* (**wept, weeping**) shed (tears); shed or ooze (moisture) in drops. ● *n.* a spell of weeping. □ **weeper** *n.*, **weepy** *adj.*

weeping *adj.* (of a tree) having drooping branches.

weevil *n.* a small beetle that feeds on grain etc.

wee-wee (colloquial) *v.* urinate. ● *n.* urine, the act of urinating.

weft *n.* the crosswise threads in weaving.

weigh *v.* **1** measure the weight of; have a specified weight. **2** consider the relative importance of. **3** have influence. **4** be burdensome. □ **weigh anchor** raise the anchor and start a voyage. **weigh down** bring or keep down by its weight; depress, oppress. **weigh up** form an estimate of.

weighbridge *n.* a weighing machine with a plate set in a road etc. for weighing vehicles.

weight *n.* **1** an object's mass numerically expressed using a recognized scale of units; a unit or system of units used for this; a piece of metal of known weight used in weighing; a heavy object; heaviness; a load. **2** influence. ● *v.* **1** attach a weight to; hold down with a weight; burden. **2** bias. □ **weightless** *adj.*, **weightlessness** *n.*

weighting *n.* extra pay or allowances given in special cases.

weighty *adj.* (**weightier**) **1** heavy. **2** showing or deserving earnest thought. **3** influential.

weir *n.* a small dam built so that some of a stream's water flows over it; the waterfall formed from this.

weird *adj.* uncanny, bizarre. □ **weirdly** *adv.*, **weirdness** *n.*

welch var. of **welsh.**

welcome *adj.* **1** received with pleasure. **2** freely permitted. ● *int.* a greeting expressing pleasure. ● *n.* a kindly greeting or reception. ● *v.* **1** give a welcome to. **2** receive gladly.

weld *v.* unite or fuse (pieces of metal) by heating or pressure; unite into a whole. ● *n.* a welded joint. □ **welder** *n.*

welfare *n.* well-being; organized efforts to ensure people's well-being.

welfare state *n.* a system attempting to ensure the welfare of all citizens by means of state operated social services.

well[1] *n.* a shaft sunk to obtain water, oil, etc.; an enclosed shaft-like space. ● *v.* rise or spring.

well[2] *adv.* (**better, best**) in a good manner or style; rightly; thoroughly; favourably, kindly; with good reason; easily; probably. ● *adj.* in good health; satisfactory. ● *int.* expressing surprise, relief, or resignation etc., or said when one is hesitating. □ **as well** in addition; desirable; desirably. **as well as** in addition to.

well-appointed *adj.* well-equipped.

well-being *n.* good health, happiness, and prosperity.

well-disposed *adj.* having kindly or favourable feelings.

well-heeled *adj.* (*colloquial*) wealthy.

wellington *n.* a boot of rubber or other waterproof material.

well-meaning *adj.* (also **well-meant**) acting or done with good intentions.

well-nigh *adv.* almost.

well off *adj.* in a satisfactory or good situation; fairly rich.

well-read *adj.* having read much literature.

Welsh *adj.* & *n.* (the language) of Wales. □ **Welshman** *n.*, **Welshwoman** *n.*

welsh *v.* (also **welch**) avoid paying one's debts; break an agreement. □ **welsher** *n.*

Welsh rabbit *n.* (also **Welsh rarebit**) melted cheese on toast.

well-spoken *adj.* speaking in a polite and correct way.

welt *n.* **1** a leather rim attaching the top of a boot or shoe to the sole. **2** a ribbed or strengthened border of a knitted garment. **3** a weal. ● *v.* **1** provide with a welt. **2** raise weals on; thrash.

welter *n.* a turmoil; a disorderly mixture. ● *v.* wallow; lie soaked in blood etc.

welterweight *n.* a boxing weight between lightweight and middleweight.

well-to-do *adj.* fairly rich.

wen *n.* a benign tumour on the skin.

wend *v.* □ **wend one's way** go.

went *see* **go**.

wept *see* **weep**.

werewolf *n.* (*pl.* **werewolves**) (in myths) a person who at times turns into a wolf.

west *n.* the point on the horizon where the sun sets; the direction in which this lies; a western part. ● *adj.* in the west; (of wind) from the west. ● *adv.* towards the west. □ **go west** (*slang*) be destroyed, lost, or killed.

westerly *adj.* towards or blowing from the west.

western *adj.* of or in the west. ● *n.* a film or novel about cowboys in western North America.

westerner *n.* a native or inhabitant of the west.

westernize *v.* (also **-ise**) make (a person or country) more like a western one in ideas and institutions. □ **westernization** *n.*

westernmost *adj.* furthest west.

westward *adj.* towards the west. □ **westwards** *adv.*

wet *adj.* (**wetter**) **1** soaked or covered with water or other liquid; rainy; not dry. **2** (*colloquial*) lacking vitality. ● *v.* (**wetted**) make wet. ● *n.* moisture; water; wet weather. □ **wetly** *adv.*, **wetness** *n.*

wet blanket *n.* a miserable person.

wether *n.* a castrated ram.

wet-nurse *n.* a woman employed to suckle another's child. ● *v.* act as wet-nurse to; coddle as if helpless.

wetsuit *n.* a rubber garment worn by divers, windsurfers, etc.

whack (*colloquial*) *v.* strike with a sharp blow. ● *n.* **1** a sharp blow. **2** an attempt. □ **do one's whack** (*slang*) do one's share.

whacked *adj.* (*colloquial*) tired out.

whale *n.* a very large sea mammal. □ **a whale of a** (*col-

loquial) an exceedingly great or good (thing).

whalebone *n.* a horny substance from the upper jaw of whales, formerly used as stiffening.

whaler *n.* a whaling ship; a seaman hunting whales.

whaling *n.* hunting whales.

wham *int.* & *n.* the sound of a forcible impact.

wharf *n.* (*pl.* **wharfs** or **wharves**) a landing stage where ships load and unload.

what *adj.* asking for a statement of amount, number, or kind; how great or remarkable; the or any that. ● *pron.* what thing(s); what did you say? ● *adv.* to what extent or degree. ● *int.* an exclamation of surprise. □ **what for?** why?

whatever *pron.* anything or everything; no matter what. ● *adj.* of any kind or number.

whatnot *n.* something trivial or indefinite.

whatsoever *adj.* & *pron.* whatever.

wheat *n.* grain from which flour is made; the plant producing this.

wheatear *n.* a small bird.

wheaten *adj.* made from wheat.

wheatmeal *n.* wholemeal wheat flour.

wheedle *v.* coax.

wheel *n.* a disc or circular frame that revolves on a shaft passing through its centre; a wheel-like motion. ● *v.* **1** push or pull (a cart or bicycle etc.) along. **2** turn; move in circles or curves. □ **at the wheel** driving a vehicle, directing a ship; in control of affairs. **wheel and deal** engage in scheming to exert influence.

wheelbarrow *n.* an open container for moving small loads, with a wheel at one end.

wheelbase *n.* the distance between a vehicle's front and rear axles.

wheelchair *n.* a chair on wheels for a person who cannot walk.

wheel-clamp *v.* fix a clamp on (an illegally parked car etc.).

wheelwright *n.* a maker and repairer of wooden wheels.

wheeze *v.* breathe with a hoarse whistling sound. ● *n.* this sound. □ **wheezy** *adj.*

whelk *n.* a shellfish with a spiral shell.

whelp *n.* a young dog, a pup. ● *v.* give birth to (whelps).

when *adv.* at what time, on what occasion; at which time, ● *conj.* at the time that; whenever; as soon as; although. ● *pron.* what or which time.

whence *adv.* & *conj.* (*formal*) from where; from which.

whenever *conj.* & *adv.* at whatever time; every time that.

where *adv.* & *conj.* at or in which place or circumstances; in what respect; from what place or source; to what place. ● *pron.* what place.

whereabouts *adv.* in or near what place. ● *n.* a person's or thing's approximate location.

whereas *conj.* since it is that; but in contrast.

whereby *conj.* by which.

whereupon *conj.* after (and then.)

wherever *adv.* & *co* whatever place.

wherewithal *n.* t money) needed f

wherry *n.* a lig large light ba

whet *v.* (**whe** rubbing a stimulate

whether *co* alterna

whets stone us

whey *n.* the watery liquid left when milk forms curds.

which *adj.* & *pron.* what particular one(s) of a set. ● *rel.pron.* the thing or animal referred to.

whichever *adj.* & *pron.* any which, no matter which.

whiff *n.* a puff of air or odour.

Whig *n.* (*hist.*) a member of the political party in the 17th–19th centuries opposed to the Tories.

while *n.* a period of time; time spent in doing something. ● *conj.* **1** during the time that, as long as. **2** although; on the other hand. □ **while away** pass (time) in a leisurely or interesting way.

whilst *conj.* while.

whim *n.* a sudden fancy.

whimper *v.* make feeble crying sounds. ● *n.* a whimpering sound.

whimsical *adj.* impulsive and playful; fanciful, quaint. □ **whimsically** *adv.*, **whimsicality** *n.*

whinchat *n.* a small songbird.

whine *v.* make a long high complaining cry or a similar shrill sound; complain or utter with a whine. ● *n.* a whining sound or complaint. □ **whiner** *n.*

whinge *v.* (*colloquial*) whine, complain.

whinny *n.* a gentle or joyful neigh. ● *v.* utter a whinny.

whip *n.* **1** a cord or strip of leather on a handle, used for striking a person or animal. **2** a dessert made with whipped cream etc. ● *v.* (**whipped**) **1** strike or urge on with a whip. **2** beat into a froth. **3** move or take suddenly. □ **have the whip hand** have control. **whip up**

whipcord *n.* a cord of tightly twisted strands; twilled fabric with prominent ridges.

whiplash *n.* the lash of a whip; a sudden jerk.

whippet *n.* a small dog resembling a greyhound, used for racing.

whipping boy *n.* a scapegoat.

whippy *adj.* flexible, springy.

whip-round *n.* (*colloquial*) a collection of money from a group.

whirl *v.* **1** swing or spin round and round. **2** convey or go rapidly in a vehicle. ● *n.* **1** a whirling movement. **2** a confused state. **3** bustling activity.

whirlpool *n.* a current of water whirling in a circle.

whirlwind *n.* a mass of air whirling rapidly about a central point.

whirr *n.* a continuous buzzing or vibrating sound. ● *v.* make this sound.

whisk *v.* **1** convey or go rapidly; brush away lightly. **2** beat into a froth. ● *n.* **1** a whisking movement. **2** an instrument for beating eggs etc.; a bunch of bristles etc. for brushing or flicking things.

whisker *n.* a long hairlike bristle on the face of a cat etc.; (**whiskers**) hairs growing on a man's cheek. □ **whiskered** *adj.*, **whiskery** *adj.*

whisky *n.* (*Irish* & *Amer.* **whiskey**) a spirit distilled from malted grain (esp. barley).

whisper *v.* speak or utter softly, not using the vocal cords; rustle. ● *n.* a whispering sound or remark; a rumour.

whist *n.* a card game usu. for two pairs of players.

whistle *n.* a shrill sound made by blowing through a narrow opening between the lips; a similar sound; an instrument for producing this. ● *v.* make this sound; signal or produce (a tune) in this way. □ **whistler** *n.*

whistle-stop *n.* a brief stop made during a tour.

Whit *adj.* of or close to Whit Sunday, the seventh Sunday after Easter.

white *adj.* **1** of the colour of snow or common salt; having a light-coloured skin; pale from illness, fear, etc. **2** (of coffee or tea) served with milk. ●*n.* **1** a white colour or thing; a member of the race with light-coloured skin. **2** the transparent substance round egg yolk. □ **whiteness** *n.*

white ant *n.* a termite.

whitebait *n.* (*pl.* **whitebait**) a small silvery-white fish.

whiteboard *n.* a white board which can be written on with coloured pens and wiped clean.

white-collar worker *n.* one not engaged in manual labour.

white elephant *n.* a useless possession.

white gold *n.* gold mixed with platinum.

whiten *v.* make or become white or whiter.

white hope *n.* a person expected to achieve much.

white horses *n.pl.* white-crested waves on the sea.

white-hot *adj.* (of metal) glowing white after heating.

white lie *n.* a harmless lie.

white noise *n.* noise containing many frequencies with equal intensities, a harsh hissing sound.

White Paper *n.* a government report giving information.

white sale *n.* a sale of household linen.

white spirit *n.* light petroleum used as a solvent.

whitewash *n.* **1** a liquid containing quicklime or powdered chalk, used for painting walls or ceilings etc. **2** a means of glossing over mistakes. ●*v.* **1** paint with whitewash. **2** gloss over mistakes in.

whither *adv.* (*old use*) to what place.

whiting *n.* (*pl.* **whiting**) a small sea fish used as food.

whitlow *n.* an inflammation near a fingernail or toenail.

Whitsun *n.* Whit Sunday and the days close to it. □ **Whitsuntide** *n.*

whittle *v.* trim (wood) by cutting thin slices from the surface; reduce by removing various amounts.

whizz *v.* (also **whiz**) (**whizzed**) make a sound like something moving at great speed through air; move very quickly. ●*n.* a whizzing sound.

whizz-kid *n.* (also **whiz-kid**) (*colloquial*) a brilliant or successful young person.

who *pron.* what or which person(s)?; the particular person(s).

whodunnit *n.* (*Amer.* **whodunit**) (*colloquial*) a detective story or play.

whoever *pron.* any or every person who, no matter who.

whole *adj.* with no part removed or left out; not injured or broken. ●*n.* the full amount, all parts or members; a complete system made up of parts. □ **on the whole** considering everything; in general.

wholefood *n.* food which has not been unnecessarily processed.

wholehearted *adj.* without doubts or reservations.

whole number *n.* a number consisting of one or more units with no fractions.

wholemeal *adj.* made from the whole grain of wheat etc.

wholesale *n.* selling of goods in large quantities to be retailed by others. ●*adj. & adv.* **1** in the wholesale trade. **2** on a large scale. □ **wholesaler** *n.*

wholesome adj. good for health or well-being. □ **wholesomeness** n.

wholism var. of **holism**.

wholly adv. entirely.

whom pron. the objective case of *who*.

whoop v. utter a loud cry of excitement. ● n. this cry.

whooping cough n. an infectious disease esp. of children, with a violent convulsive cough.

whopper n. (slang) something very large; a great lie.

whore n. a prostitute.

whorl n. a coiled form, one turn of a spiral; a circle of ridges in a fingerprint; a ring of leaves or petals.

who's who is, who has.

■ **Usage** Because it has an apostrophe, *who's* is easily confused with *whose*. They are each correctly used in *Who's there?* (= *Who is there?*), *Who's taken my pen?* (= *Who has taken my pen?*), and *Whose book is this?* (= *Who does this book belong to?*).

whose pron. & adj. of whom; of which.

whosoever pron. whoever.

why adv. for what reason or purpose; on account of which. ● int. an exclamation of surprised discovery or recognition.

wick n. a length of thread in a candle or lamp etc., by which the flame is kept supplied with melted grease or fuel.

wicked adj. 1 morally bad, offending against what is right. 2 formidable, severe. 3 mischievous. 4 (slang) excellent. □ **wickedly** adv., **wickedness** n.

wicker n. osiers or thin canes interwoven to make furniture, baskets, etc. □ **wickerwork** n.

wicket n. a set of three stumps and two bails used in cricket; the part of a cricket ground between or near the two wickets.

wide adj. measuring much from side to side; having a specified width; extending far; fully opened; far from the target. ● adv. widely. □ **wide awake** fully awake or (colloquial) alert. □ **widely** adv., **wideness** n.

widen v. make or become wider.

widespread adj. found or distributed over a wide area.

widgeon n. (also **wigeon**) a wild duck.

widow n. a woman whose husband has died and who has not remarried. □ **widowhood** n.

widowed adj. made a widow or widower.

widower n. a man whose wife has died and who has not remarried.

width n. wideness; the distance from side to side; a piece of material of full width as woven.

wield v. hold and use (a tool etc.); have and use (power).

wife n. (pl. **wives**) a married woman in relation to her husband. □ **wifely** adj.

wig n. a covering of hair worn on the head.

wigeon var. of **widgeon**.

wiggle v. move repeatedly from side to side, wriggle. ● n. an act of wiggling. □ **wiggly** adj.

wigwam n. a conical tent as formerly used by North American Indians.

wild adj. 1 not domesticated, tame, or cultivated; not civilized; disorderly. 2 stormy; full of strong unrestrained feeling. 3 extremely foolish. 4 random. ● adv. in a wild manner. □ (**the wilds**) a desolate uninhabited place. □ **wildly** adv., **wildness** n.

wildcat n. reckless; (of strikes) unofficial and irresponsible.

wildebeest n. (pl. **wildebeest** or **wildebeests**) a gnu.

wilderness *n.* a wild uncultivated area.

wildfire *n.* □ **spread like wildfire** spread very fast.

wildfowl *n.pl.* birds hunted as game.

wild-goose chase *n.* a useless quest.

wildlife *n.* wild animals and plants.

wile *n.* a piece of trickery.

wilful *adj.* (*Amer.* **willful**) intentional, not accidental; self-willed. □ **wilfully** *adv.*, **wilfulness** *n.*

will¹ *v.aux.* used with *I* and *we* to express promises or obligations, and with other words to express a future tense.

will² *n.* **1** the mental faculty by which a person decides upon and controls his or her actions; determination; a person's attitude in wishing good or bad to others. **2** written directions made by people for disposal of their property after their death. ● *v.* **1** exercise one's will-power, influence by doing this. **2** bequeath by a will. □ **at will** whenever one pleases. **have one's will** get what one desires.

willie var. of **willy**.

willing *adj.* desiring to do what is required, not objecting; given or done readily. ● *n.* willingness. □ **willingly** *adv.*, **willingness** *n.*

will-o'-the-wisp *n.* **1** a phosphorescent light seen on marshy ground. **2** a hope or aim that can never be fulfilled.

willow *n.* a tree or shrub with flexible branches; its wood.

willowy *adj.* slender and supple.

will-power *n.* control exercised by one's will.

willy *n.* (also **willie**) (*slang*) the penis.

willy-nilly *adv.* whether one desires it or not.

wilt *v.* lose or cause to lose freshness and droop; become limp from exhaustion.

wily *adj.* (**wilier**) full of wiles, cunning. □ **wiliness** *n.*

wimp *n.* (*colloquial*) a feeble or ineffective person.

win *v.* (**won**, **winning**) be victorious (in); obtain as the result of a contest etc., or by effort; gain the favour or support of. ● *n.* a victory, esp. in a game.

wince *v.* make a slight movement from pain or embarrassment etc. ● *n.* this movement.

winceyette *n.* cotton fabric with a soft downy surface.

winch *n.* a machine for hoisting or pulling things by a cable that winds round a revolving drum. ● *v.* hoist or pull with a winch.

wind¹ (wind) *n.* **1** a current of air; gas in the stomach or intestines; breath as needed in exertion or speech etc. **2** an orchestra's wind instruments. **3** useless or boastful talk. ● *v.* detect by a smell; cause to be out of breath. □ **get wind of** hear a hint or rumour of. **in the wind** happening or about to happen. **put the wind up** (*colloquial*) frighten. **take the wind out of a person's sails** take away an advantage, frustrate by anticipating him or her.

wind² (wynd) *v.* (**wound**, **winding**) move or go in a curving or spiral course; wrap closely around something or round upon itself; move by turning a windlass or handle etc.; wind up (a clock etc.). □ **wind up** set or keep (a clock etc.) going by tightening its spring; bring or come to an end; settle the affairs of and close (a business company); (*colloquial*) provoke or tease. □ **winder** *n.*

windbag *n.* (*colloquial*) a person who talks a lot.

windbreak n. a screen shielding something from the wind.

wind-chill n. the cooling effect of the wind.

windfall n. **1** fruit blown off a tree by the wind. **2** an unexpected gain, esp. a sum of money.

wind instrument n. a musical instrument sounded by a current of air, esp. by the player's breath.

windlass n. a winch-like device using a rope or chain that winds round a horizontal roller.

windmill n. a mill worked by the action of wind on projecting parts that radiate from a shaft.

window n. an opening in a wall etc. to admit light and air, usu. filled with glass; this glass; a space for display of goods behind the window of a shop.

window box n. a trough fixed outside a window, for growing flowers etc.

window dressing n. **1** arranging a display of goods in a shop window. **2** presentation of facts so as to give a favourable impression.

window-shopping n. looking at goods displayed in shop windows without buying.

windpipe n. the air passage from the throat to the bronchial tubes.

windscreen n. (Amer. **windshield**) the glass in the window at the front of a vehicle.

windsock n. a canvas cylinder flown at an airfield to show the direction of the wind.

windsurfing n. the sport of surfing on a board to which a sail is fixed. □ **windsurfer** n.

windswept adj. exposed to strong winds.

wind tunnel n. an enclosed tunnel in which winds can be created for testing things.

windward adj. situated in the direction from which the wind blows. ● n. this side or region.

wine n. **1** fermented grape juice as an alcoholic drink; a fermented drink made from other fruits or plants. **2** dark red. ● v. drink wine; entertain with wine.

wine bar n. a bar or small restaurant serving wine as the main drink.

wing n. each of a pair of projecting parts by which a bird or insect etc. is able to fly; a winglike part of an aircraft; a projecting part; the bodywork above the wheel of a car; either end of a battle array; a player at either end of the forward line in football or hockey etc., the side part of the playing area in these games; an extreme section of a political party; (**wings**) the sides of a theatre stage. ● v. fly, travel by wings; wound in the wing or arm. □ **on the wing** flying. **take wing** fly away. **under one's wing** under one's protection.

wing collar n. a high stiff collar with turned-down corners.

winged adj. having wings.

winger n. a wing player in football etc.

wingspan n. the measurement across wings from one tip to the other.

wink v. blink one eye as a signal; shine with a light that flashes or twinkles. ● n. an act of winking. □ **not a wink** no sleep at all.

winker n. a flashing indicator.

winkle n. an edible sea snail. □ **winkle out** extract, prise out.

winner n. a person or thing that wins; something successful.

winning adj. charming, persuasive. ● n.pl. (**winnings**) money won in betting etc.

winnow v. fan or toss (grain) to free it of chaff.

wino n. (pl. **winos**) (slang) an alcoholic.

winsome adj. charming.

winter n. the coldest season of the year. ● v. spend the winter. □ **wintry** adj.

wipe v. clean, dry, or remove by rubbing; spread thinly on a surface. ● n. the act of wiping. □ **wipe out** cancel; destroy completely.

wiper n. a device that automatically wipes rain etc. from a windscreen.

wire n. a strand of metal; a length of this used for fencing, conducting electric current, etc. ● v. provide, fasten, or strengthen with wire(s).

wiring n. a system of electric wires in a building, vehicle, etc.

wiry adj. (**wirier**) like wire; lean but strong. □ **wiriness** n.

wisdom n. being wise, soundness of judgement; wise sayings.

wisdom tooth n. a hindmost molar tooth, not usu. cut before the age of 20.

wise adj. showing soundness of judgement; having knowledge. □ **wisely** adv.

wiseacre n. a person who pretends to have great wisdom.

wisecrack (colloquial) n. a witty remark. ● v. make a wisecrack.

wish n. a desire, a mental aim; an expression of desire. ● v. have or express as a wish; hope or express hope about another person's welfare.

wishbone n. a forked bone between a bird's neck and breast.

wishful adj. desiring.

wishy-washy adj. weak in colour, character, etc.

wisp n. a small separate bunch; a small streak of smoke etc. □ **wispy** adj., **wispiness** n.

wisteria n. (also **wistaria**) a climbing shrub with hanging clusters of flowers.

wistful adj. full of sad or vague longing. □ **wistfully** adv., **wistfulness** n.

wit n. amusing ingenuity in expressing words or ideas; a person who has this; intelligence. □ **at one's wits' end** worried and not knowing what to do.

witch n. a person (esp. a woman) who practises witchcraft; a bewitching woman.

witchcraft n. the practice of magic.

witch doctor n. a tribal magician.

witch hazel n. a North American shrub; an astringent lotion made from its leaves and bark.

witch-hunt n. persecution of people thought to be holders of unpopular views.

with prep. in the company of, among; having, characterized by; by means of; of the same opinion as; at the same time as; because of; under the conditions of; by addition or possession of; in regard to, towards.

withdraw v. (**withdrew, withdrawn, withdrawing**) **1** take back; remove (deposited money) from a bank etc.; cancel (a statement). **2** go away from a place or from company. □ **withdrawal** n.

withdrawn adj. (of a person) unsociable.

wither v. **1** shrivel, lose freshness or vitality. **2** subdue by scorn.

withhold v. (**withheld, withholding**) refuse to give; restrain.

within prep. inside; not beyond the limit or scope of; in a time no longer than. ● adv. inside.

without prep. not having; in the absence of; with no action of. ● adv. outside.

withstand v. (**withstood, withstanding**) endure successfully.

witless adj. foolish.

witness n. a person who sees or hears something; one who gives evidence in a law court; one who confirms another's signature; something that serves as evidence. ● v. be a witness of.

witter v. (*colloquial*) speak lengthily about something trivial.

witticism n. a witty remark.

witty adj. (**wittier**) full of wit. □ **wittily** adv., **wittiness** n.

wives see **wife**.

wizard n. a male witch, a magician; a person with amazing abilities. □ **wizardry** n.

wizened (wiz-ĕnd) adj. full of wrinkles, shrivelled with age.

woad n. a blue dye obtained from a plant; this plant.

wobble v. stand or move unsteadily; quiver. ● n. a wobbling movement; a quiver.

wobbly adj. (**wobblier**) unsteady; quivering. □ **throw a wobbly** (*colloquial*) have a fit of annoyance or panic.

wodge n. (*colloquial*) a chunk or lump.

woe n. sorrow, distress; trouble causing this, misfortune. □ **woeful** adj., **woefully** adv., **woefulness** n.

woebegone adj. looking unhappy.

wog n. (*offensive slang*) a foreigner, esp. a non-white one.

wok n. a large bowl-shaped frying pan used esp. in Chinese cookery.

woke, woken see **wake**.

wold n. (esp. **wolds**) an area of open upland country.

wolf n. (pl. **wolves**) a wild animal of the dog family. ● v. eat quickly and greedily. □ **cry wolf** raise false alarms. □ **wolfish** adj.

wolfram n. tungsten (ore).

wolf whistle n. a man's admiring whistle at a woman.

wolverine n. a North American animal of the weasel family.

woman n. (pl. **women**) an adult female person; women in general.

womanhood n. the state of being a woman.

womanize v. (also **-ise**) (of a man) chase after women □ **womanizer** n.

womankind n. women in general.

womanly adj. having qualities considered characteristic of a woman. □ **womanliness** n.

womb n. the hollow organ in female mammals in which the young develop before birth.

wombat n. a burrowing Australian marsupial like a small bear.

women see **woman**.

womenfolk n. women in general; the women of one's family.

won see **win**.

wonder n. a feeling of surprise and admiration or curiosity or bewilderment; a remarkable thing. ● v. **1** feel wonder or surprise. **2** desire to know; try to decide.

wonderful adj. arousing admiration, excellent. □ **wonderfully** adv.

wonderland n. a place full of wonderful things.

wonderment n. a feeling of wonder.

wonky adj. (**wonkier**) (*slang*) unsteady; crooked.

wont (wohnt) adj. (*old use*) accustomed.

woo v. (*old use*) court; try to achieve, obtain, or coax.

wood n. the tough fibrous substance of a tree; this cut for use; (also **woods**) trees growing fairly densely over an area of ground. □ **out of the wood** clear of danger or difficulty.

woodbine n. wild honeysuckle.

woodcock n. a game bird.

woodcut n. an engraving made on wood; a picture made from this.

wooded adj. covered with trees.

wooden adj. **1** made of wood. **2** showing no expression. □ **woodenly** adv.

woodland n. wooded country.

woodlouse n. (pl. **woodlice**) a small wingless creature with many legs, living in decaying wood etc.

woodpecker n. a bird that taps tree trunks with its beak to discover insects.

wood pigeon n. a large pigeon.

woodwind n. wind instruments made (or formerly made) of wood, e.g. the clarinet.

woodwork n. the art or practice of making things from wood; wooden things or fittings.

woodworm n. the larva of a kind of beetle that bores in wood.

woody adj. (**woodier**) like or consisting of wood; full of woods.

woof n. a dog's gruff bark. ● v. make this sound.

woofer n. a loudspeaker for reproducing low-frequency signals.

wool n. the soft hair from sheep or goats etc.; yarn or fabric made from this. □ **pull the wool over someone's eyes** deceive him or her.

woollen adj. (Amer. **woolen**) made of wool. ● n.pl. (**woollens**) woollen cloth or clothing.

woolly adj. (**woollier**) **1** covered with wool; like wool, woollen. **2** vague. ● n. (colloquial) a woollen garment. □ **woolliness** n.

woozy adj. (**woozier**) (colloquial) dizzy, dazed.

word n. sound(s) expressing a meaning independently and forming a basic element of speech; this represented by letters or symbols; something said; a message, news; a promise; a command. ● v. express in words. □ **word of mouth** spoken (not written) words.

wording n. the way a thing is worded.

word-perfect adj. having memorized every word perfectly.

word processor n. a computer program designed for producing and altering text and documents; a computer and printer designed for this purpose.

wordy adj. (**wordier**) using too many words.

wore see **wear**.

work n. **1** the use of bodily or mental power in order to do or make something; employment; something to be done; something done or produced by work; a literary or musical composition. **2** ornamentation of a certain kind, articles with this; things made of certain materials or with certain tools. **3** (**works**) operations of building etc.; the operative parts of a machine; a factory; a defensive structure. **4** (**the works**) (colloquial) everything, all that is available. ● v. **1** perform work; make efforts; be employed. **2** operate, do this effectively. **3** bring about, accomplish. **4** shape, knead, or hammer etc. into a desired form or consistency. **5** make (a way) or pass gradually by effort; become (loose etc.) through repeated stress or pressure; be in motion; ferment. □ **work off** get rid of by activity. **work out** find or solve by calculation; plan the details of; have a specified result; take exercise. **work to rule** cause delay by over-strict observance of rules, as a form of protest. **work up** bring gradually to a more developed state; excite progressively; advance (to a climax).

workable adj. able to be done or used successfully.

workaday adj. ordinary, everyday; practical.

workbook n. a student's book with exercises.

worker n. a person who works; a member of the working class; a neuter bee or ant etc. that does the work of the hive or colony.

workhouse n. a former public institution where people unable to support themselves were housed.

working adj. engaged in work, esp. manual labour; working-class. ● n. excavations made in mining, tunnelling, etc.; a mine or quarry.

working class n. the class of people who are employed for wages, esp. in manual or industrial work. □ **working-class** adj.

working knowledge n. knowledge adequate to work with.

workman n. (pl. -men) a person employed to do manual labour.

workmanlike adj. characteristic of a good worker, practical.

workmanship n. skill in working or in a thing produced.

workout n. a session of physical exercise or training.

workshop n. a room or building in which manual work or manufacture etc. is carried on.

workstation n. 1 a computer terminal and keyboard; a desk with this. 2 the location of a stage in a manufacturing process.

worktop n. a flat surface for working on, esp. in a kitchen.

world n. 1 the earth. 2 the people or things belonging to a certain class or sphere of activity; everything, all people. 3 a very great amount.

worldly adj. of or concerned with earthly life or material gains, not spiritual. □ **worldliness** n.

worldwide adj. extending through the whole world.

worm n. 1 a creature with a long soft body and no backbone or limbs; (worms) internal parasites. 2 an insignificant or contemptible person. 3 the spiral part of a screw. ● v. 1 make one's way with twisting movements; insinuate oneself; obtain by crafty persistence. 2 rid of parasitic worms. □ **wormy** adj.

worm-cast n. a pile of earth cast up by an earthworm.

worm-eaten adj. full of holes made by insect larvae.

wormwood n. a woody plant with a bitter flavour.

worn see **wear**. adj. damaged or altered by use or wear; looking exhausted.

worn out adj. exhausted; extremely worn.

worried adj. feeling or showing worry.

worry v. 1 be troublesome to; give way to anxiety. 2 seize with the teeth and shake or pull about. ● n. a worried state, mental uneasiness; something causing this. □ **worrier** n.

worse adj. & adv. more bad or badly, more evil or ill. ● n. something worse.

worsen v. make or become worse.

worship n. reverence and respect paid to a god; adoration or devotion to a person or thing. ● v. (**worshipped**; Amer. **worshiped**) honour as a god; take part in an act of worship; idolize, treat with adoration. □ **worshipper** n.

worst adj. & adv. most bad or badly. ● n. the worst part, feature, event, etc. ● v. defeat, outdo. □ **get the worst of** be defeated in.

worsted (wuu-stid) n. a smooth woollen yarn or fabric.

worth adj. having a specified value; deserving; possessing as wealth. ● n. value, merit, usefulness; the amount that a specified sum will buy. □ **for all one is worth** (*colloquial*) with all one's energy. **worth one's while** worth the time or effort needed.

worthless *adj.* having no value or merit. □ **worthlessness** *n.*

worthwhile *adj.* worth the time or effort spent.

worthy *adj.* (**worthier**) having great merit; deserving. ● *n.* a worthy person. □ **worthily** *adv.*, **worthiness** *n.*

would *v.aux.* used in senses corresponding to *will*¹ in the past tense, conditional statements, questions, and polite requests and statements, and to express probability or something that happens from time to time.

would-be *adj.* desiring or pretending to be.

wound¹ (woond) *n.* an injury done to tissue by violence; injury to feelings. ● *v.* inflict a wound upon.

wound² (wownd) *see* **wind**².

wove, woven *see* **weave**.

wow *int.* an exclamation of astonishment. ● *n.* (*slang*) a sensational success.

WP *abbr.* word processor.

WPC *abbr.* woman police constable.

w.p.m. *abbr.* words per minute.

WRAC *abbr.* Women's Royal Army Corps.

wrack *n.* seaweed.

WRAF *abbr.* Women's Royal Air Force.

wraith *n.* a ghost, a spectral apparition of a living person.

wrangle *v.* argue or quarrel noisily. ● *n.* a noisy argument.

wrap *v.* (**wrapped**) arrange (a soft or flexible covering) round (a person or thing). ● *n.* a shawl. □ **be wrapped up in** have one's attention deeply occupied by.

wrapper *n.* a cover of paper etc. wrapped round something.

wrapping *n.* material for wrapping things.

wrasse (rass) *n.* a brightly coloured sea fish.

wrath *n.* anger, indignation. □ **wrathful** *adj.*, **wrathfully** *adv.*

wreak *v.* inflict (vengeance etc.).

wreath (reeth) *n.* flowers or leaves etc. fastened into a ring, used as a decoration or placed on a grave etc.

wreathe (reeth) *v.* encircle; twist into a wreath; wind, curve.

wreck *n.* the destruction esp. of a ship by storms or accident; a ship that has suffered this; something ruined or dilapidated; a person whose health or spirits have been destroyed. ● *v.* cause the wreck of; involve in shipwreck. □ **wrecker** *n.*

wreckage *n.* the remains of something wrecked.

Wren *n.* (*hist.*) a member of the WRNS.

wren *n.* a very small bird.

wrench *v.* twist or pull violently round; damage or pull by twisting. ● *n.* **1** a violent twisting pull. **2** pain caused by parting. **3** an adjustable spanner-like tool.

wrest *v.* wrench away; obtain by force or effort; twist, distort.

wrestle *v.* fight (esp. as a sport) by grappling with and trying to throw an opponent to the ground; struggle to deal with.

wretch *n.* a wretched or despicable person; a rascal.

wretched *adj.* miserable, unhappy; worthless; contemptible. □ **wretchedly** *adv.*, **wretchedness** *n.*

wriggle *v.* move with short twisting movements; escape (out of a difficulty etc.) cunningly. ● *n.* a wriggling movement.

wring *v.* (**wrung, wringing**) twist and squeeze; press to remove liquid; squeeze firmly or forcibly; obtain with effort or difficulty.

wrinkle n. **1** a small crease; a small ridge or furrow in skin. **2** (colloquial) a useful hint. ● v. form wrinkles (in). □ **wrinkly** adj.

wrist n. the joint connecting the hand and forearm; part of a garment covering this.

writ n. a formal written authoritative command.

write v. (**wrote, written, writing**) make letters or other symbols on a surface, esp. with a pen or pencil; compose in written form; be an author; write and send a letter. □ **write off** recognize as lost. **write up** write an account of; write entries in.

write-off n. something written off as lost; a vehicle too damaged to be worth repairing.

writer n. a person who writes; an author.

writer's cramp n. cramp in the muscles of the hand.

write-up n. (colloquial) a published account of something, a review.

writhe v. twist one's body about, as in pain; wriggle; suffer because of embarrassment.

writing n. handwriting; literary work. □ **in writing** in written form.

writing paper n. paper for writing (esp. letters) on.

written see **write**.

WRNS abbr. (hist.) Women's Royal Naval Service.

wrong adj. incorrect, not true; morally bad; contrary to justice; not what is required or desirable; not in a satisfactory condition. ● adv. in a wrong manner or direction, mistakenly. ● n. what is wrong, a wrong action etc.; an injustice. ● v. treat unjustly. □ **in the wrong** not

having truth or justice on one's side. □ **wrongly** adv.

wrongdoer n. a person who behaves illegally or immorally. □ **wrongdoing** n.

wrongful adj. contrary to what is right or legal. □ **wrongfully** adv.

wrote see **write**.

wrought (rawt) adj. (of metals) shaped by hammering.

wrought iron n. a pure form of iron used for decorative work.

wrung see **wring**.

WRVS abbr. Women's Royal Voluntary Service.

wry adj. **1** (of the face) contorted in disgust or disappointment. **2** (of humour) dry, mocking. □ **wryly** adv., **wryness** n.

wych elm n. an elm with broad leaves and spreading branches.

WYSIWYG adj. (Computing) indicating that the text is displayed on the screen as it will appear in print (from what you see is what you get).

Xx

X n. (as a Roman numeral) ten. ● symb. (of films) suitable for adults only.

X chromosome n. a sex chromosome, of which female cells have twice as many as males.

Xe symb. xenon.

xenon (zen-on) a chemical element (symbol Xe), a colourless, odourless gas.

xenophobia (zen-ŏ-foh-biă) n. a strong dislike or distrust of foreigners.

Xerox (zeer-oks) n. (trade mark) a machine for producing photocopies; a photocopy. ● v. photocopy.

Xmas n. (colloquial) Christmas.

X-ray *n.* a photograph or examination made by electromagnetic radiation (**X-rays**) that can penetrate solids. ● *v.* photograph, examine, or treat by X-rays.

xylophone *n.* a musical instrument with flat wooden bars struck with small hammers.

Yy

Y *symb.* yttrium.

yacht (yot) *n.* a light sailing vessel for racing; a vessel used for private pleasure excursions. □ **yachting** *n.*, **yachtsman** *n.*, **yachtswoman** *n.*

yak *n.* a long-haired Asian ox.

yam *n.* a tropical climbing plant; its edible tuber; a sweet potato.

yang *n.* the active male principle in Chinese philosophy.

yank (*colloquial*) *v.* pull sharply. ● *n.* **1** a sharp pull. **2** (**Yank**) an American.

yap *n.* a shrill bark. ● *v.* (**yapped**) bark shrilly.

yard *n.* **1** a measure of length, = 3 feet (0.9144 metre). **2** a piece of enclosed ground, esp. attached to a building. **3** a pole slung from a mast to support a sail.

yardage *n.* length measured in yards.

yardstick *n.* a standard of comparison.

yarmulke (yar-mul-kă) *n.* (also **yarmulka**) a skullcap worn by Jewish men.

yarn *n.* **1** any spun thread. **2** (*colloquial*) a tale.

yarrow *n.* a plant with feathery leaves and strong-smelling flowers.

yashmak *n.* a veil worn by Muslim women in certain countries.

yaw *v.* (of a ship or aircraft) fail to hold a straight course.

yawl *n.* a fishing or sailing boat.

yawn *v.* open the mouth wide and draw in breath, as when sleepy or bored; have a wide opening. ● *n.* the act of yawning.

yaws *n.* a tropical skin disease.

Yb *symb.* ytterbium.

Y chromosome *n.* a sex chromosome occurring only in males.

yd *abbr.* yard.

year *n.* the time taken by the earth to orbit the sun (about 365¼ days); (also **calendar year**) the period from 1 Jan. to 31 Dec. inclusive; any consecutive period of twelve months.

yearbook *n.* an annual publication of events or aspects of the previous year.

yearling *n.* an animal between 1 and 2 years old.

yearly *adj.* happening, published, or payable once a year. ● *adv.* annually.

yearn *v.* feel great longing.

yeast *n.* a fungus used to cause fermentation in making beer and wine and as a raising agent.

yell *n.* a shout or scream. ● *v.* shout, scream.

yellow *adj.* of the colour of buttercups and ripe lemons; (*colloquial*) cowardly. ● *n.* a yellow colour or thing. ● *v.* turn yellow.

yellowhammer *n.* a bird of the finch family with a yellow head, neck, and breast.

yellowish *adj.* rather yellow.

yelp *n.* a shrill yell or bark. ● *v.* utter a yelp.

yen *n.* **1** (*pl.* **yen**) a unit of money in Japan. **2** (*colloquial*) a longing, a yearning.

yeoman *n.* (*pl.* **-men**) (*hist.*) a man who owned and farmed a small estate.

yes *int.* & *n.* an expression of agreement or consent, or of reply to a summons etc.

yes-man n. (pl. **-men**) a person who is always ready to agree with a superior.

yesterday adv. & n. (on) the day before today; (in) the recent past.

yet adv. **1** up to this or that time; still; besides; eventually. **2** even; nevertheless. ◆ conj. nevertheless, in spite of that.

yeti n. a large manlike or bearlike animal said to exist in the Himalayas.

yew n. an evergreen tree with dark needle-like leaves; its wood.

Y-fronts n.pl. men's briefs with a Y-shaped seam at the front.

YHA abbr. Youth Hostels Association.

Yiddish n. the language used by Jews from eastern Europe.

yield v. **1** give as a fruit, gain, or result. **2** surrender; allow (victory, right of way, etc.) to another; be able to be forced out of the natural shape. ◆ n. an amount yielded or produced.

yin n. the passive female principle in Chinese philosophy.

yippee int. an exclamation of delight or excitement.

YMCA abbr. Young Men's Christian Association.

yob n. (slang) a lout, a hooligan.

yodel v. (**yodelled**; Amer. **yodeled**) sing with a quickly alternating change of pitch. ◆ n. a yodelling cry. □ **yodeller** n.

yoga n. a Hindu system of meditation and self-control; exercises used in this.

yogurt n. (also **yoghurt**) food made of milk that has been thickened by the action of certain bacteria.

yoke n. **1** a wooden crosspiece fastened over the necks of two oxen pulling a plough etc.; a piece of wood shaped to fit a person's shoulders and have a load slung from each end. **2** the

top part of a garment. **3** oppression. ◆ v. **1** harness with a yoke. **2** unite.

yokel n. a country person; a bumpkin.

yolk n. the round yellow internal part of an egg.

yonder adj. & adv. over there.

yonks n.pl. (slang) a long time.

yore n. □ **of yore** long ago.

Yorkshire pudding n. a baked batter pudding eaten with gravy or meat.

you pron. the person(s) addressed; one, anyone, everyone.

young adj. having lived or existed for only a short time; youthful; having little experience. ◆ n. the offspring of animals.

youngster n. a young person, a child.

your adj. of or belonging to you.

yours poss.pron. belonging to you.

yourself pron. (pl. **yourselves**) the emphatic and reflexive form of you.

youth n. **1** the state or period of being young. **2** a young man; young people.

youth club n. a club providing leisure activities for young people.

youth hostel n. a hostel providing cheap accommodation for walkers or holidaymakers.

youthful adj. young; characteristic of young people. □ **youthfulness** n.

yowl v. & n. (make) a loud wailing cry.

yo-yo n. (pl. **yo-yos**) (trade mark) a disc-shaped toy that can be made to rise and fall on a string that winds round it in a groove.

YTS abbr. Youth Training Scheme.

ytterbium (it-er-biùm) n. a metallic element (symbol Yb).

yttrium (it-riùm) n. a metallic element (symbol Y).

yuan n. (pl. **yuan**) the chief monetary unit in China.

yucca n. a tall plant with white flowers and spiky leaves.

yuck int. (also **yuk**) (slang) an expression of disgust.

Yuletide n. (old use) the Christmas festival.

yummy adj. (**yummier**) (colloquial) delicious.

yuppie n. (also **yuppy**) (colloquial) a young middle-class professional person working in a city.

YWCA abbr. Young Women's Christian Association.

....

Zz

....

zabaglione (za-ba-lyoh-ni) n. a dessert of whipped egg yolks, sugar, and wine.

zany adj. (**zanier**) crazily funny. ● n. a zany person.

zap v. (**zapped**) (slang) **1** hit; kill. **2** move quickly.

zeal n. enthusiasm, hearty and persistent effort.

zealot (zel-ŏt) n. a zealous person, a fanatic.

zealous (zel-ŭs) adj. full of zeal. □ **zealously** adv.

zebra n. an African horse-like animal with black and white stripes.

zebra crossing n. a pedestrian crossing where the road is marked with broad white stripes.

zebu n. an ox with a humped back.

Zen n. a form of Buddhism.

zenith n. the part of the sky that is directly overhead; the highest point.

zephyr n. (literary) a soft gentle wind.

zero n. (pl. **zeros**) nought, the figure 0; nil; a point marked 0 on

a graduated scale, a temperature corresponding to this. □ **zero in on** take aim at; focus attention on.

zero hour n. the hour at which something is timed to begin.

zest n. **1** keen enjoyment or interest. **2** orange or lemon peel as flavouring. □ **zestful** adj., **zestfully** adv.

zigzag n. a line or course turning right and left alternately at sharp angles. ● adj. & adv. as or in a zigzag. ● v. (**zigzagged**) move in a zigzag.

zilch n. (slang) nothing.

zinc n. a white metallic element (symbol Zn).

zing (colloquial) n. vigour. ● v. move swiftly or with a shrill sound.

Zionism n. a movement that campaigned for a Jewish homeland in Palestine. □ **Zionist** n.

zip n. **1** a short sharp sound. **2** (in full **zip fastener**) a fastening device with teeth that interlock when brought together by a sliding tab. **3** vigour, liveliness. ● v. (**zipped**) **1** fasten with a zip fastener. **2** move with vigour or at high speed.

Zip code n. (Amer.) a postal code.

zipper n. a zip fastener.

zircon n. a bluish-white gem cut from translucent mineral.

zirconium n. grey metallic element (symbol Zr).

zit n. (slang) a pimple.

zither n. a stringed instrument played with the fingers.

zloty n. the unit of money in Poland.

Zn symb. zinc.

zodiac n. (in astrology) a band of the sky divided into twelve equal parts (**signs of the zodiac**) each named from a constellation. □ **zodiacal** adj.

zombie n. **1** (in voodoo) a corpse said to have been revived by witchcraft. **2** (*colloquial*) a person who seems to have no mind or will.

zone *n.* an area with particular characteristics, purpose, or use. ● *v.* divide into zones. □ **zonal** *adj.*

zoo *n.* a place where wild animals are kept for exhibition, conservation, and study.

zoology *n.* the study of animals. □ **zoological** *adj.*, **zoologist** *n.*

zoom *v.* **1** move quickly, esp. with a buzzing sound; rise quickly. **2** (in photography) make a distant object appear gradually closer by means of a **zoom lens**.

zoophyte *n.* a plantlike animal, esp. a coral, sea anemone, or sponge.

Zr *symb.* zirconium.

zucchini (zoo-kee-ni) *n.* (*pl.* **zucchini** or **zucchinis**) (esp. *Amer.*) a courgette.

Zulu *n.* a member of a Bantu people of South Africa; their language.

zygote *n.* a cell formed by the union of two gametes.

Word-games Supplement

Players of word-games are often at an advantage if they have ready access to a supply of short words that can be regarded as valid for the purposes of the game. Particularly useful are words of only two letters, words with a *q* not followed by *u*, and words beginning with *x*. Many of these words are excluded from a small dictionary (which concentrates on current usage) because they are rare, obsolete, or occur only in dialects.

All the words in the following lists are attested in one of the great historical dictionaries (such as the twenty-volume *Oxford English Dictionary* and the *English Dialect Dictionary*) or are included in a major American dictionary.

Words marked with an asterisk (*) are obsolete but are subject to modern rules of inflection. Those marked with a dagger (†) are obsolete and were not in use after the Middle English period (ending in 1500); these words cannot be assumed to form plurals and verbal inflections in the modern style (for example, the plural of *ac* is *aec*, not 'acs').

Excluded from these lists are names of people and places etc. and abbreviations (such as Dr, Mr) which are not pronounced as they are spelt, and suffixes and other elements which have never been current as independent words; most word-games do not regard these as valid items. An arbitrary limit of six letters has been imposed throughout.

Two-letter words

Note This list does not include plurals of the names of letters of the alphabet (*bs, ds, ms, ts,* etc.). These are correct formations but are not always regarded by word-game players as acceptable.

aa *n.* rough cindery lava.

ab *v.* (*dialect*) hinder. ● *n.* (*dialect*) hindrance.

†ac *n.* (*pl.* **aec**) oak.

ad *n.* (*colloquial*) advertisement.

ae *adj.* (*Scot.*) one.

af *prep.* (*dialect*) of; off.

ah *int.* expressing surprise.

ai *n.* three-toed sloth.

ak *n.* (*dialect*) oak.

†al *adj. & n.* all.

Word-games Supplement

am *present tense* of be.

an *indefinite article* one.

ar *n.* letter r.

as *adv.* & *conj.* similarly.

at *prep.* having as position etc.

†au *n.* awe.

aw *n.* water-wheel board.

ax *n.* & *v.* axe.

ay *int.* ah.

ba *n.* (*Egyptian myth*) soul.

be *v.* exist.

bi *n.* & *adj.* (*slang*) bisexual (person).

bo *n.* a kind of fig-tree.

bu *n.* former Japanese coin.

by *prep.* & *adv.* beside.

ca *n.* (*pl.* **caas** or **cais**) (*Scot.*) calf.

ce *n.* letter c.

†co *n.* jackdaw.

†cu *n.* cow.

†cy *n.pl.* cows.

da *n.* Indian fibre plant.

de *prep.* of; from.

di *n.* note in music-scale.

do *v.* perform. ● *n.* performance.

*du *v.* (*Scot.*) do.

*dw *v.* (*Scot.*) do.

*dy *n.* (*pl.* **dyce** or **dys**) gaming-die.

ea *n.* (*dialect*) river; stream.

*eb *n.* ebb.

†ec *adv.* also, too.

*ed *adj.* distinguished.

ee *n.* (*pl.* **een**) (*Scot.*) eye.

ef *n.* letter f.

*eg *n.* egg.

eh *int.* & *v.* expressing surprise.

†ei *adj.* & *pron.* any.

*ek *adv.* & *v.* eke.

el *n.* letter l.

em *n.* unit of print measure.

en *n.* half an em.

†eo *pron.* you.

er *int.* & *v.* expressing hesitation.

es *n.* (*pl.* **esses**) letter s.

†et *prep.* at.

†eu *n.* yew.

ew *v.* (*dialect*) owe.

ex *n.* (*pl.* **exes**) former spouse etc.

ey *n.* (*dialect*) water.

fa *n.* note in music-scale.

fe *n.* (*old use*) note in music scale.

fo *n.* (*dialect*) area measure.

fu *n.* (*pl.* **fu**) Chinese district.

fy *int.* fie.

ga *n.* (*old use*) note in music-scale.

ge *n.* (*old use*) note in music-scale.

go *v.* move. ● *n.* (*pl.* **goes**) energy; turn.

gu *v.* (*dialect*) go.

gy *n.* (*Scot.*) guide-rope. ●*v.t.* guide.

ha *int.* & *v.* expressing surprise.

he *pron.* male mentioned.

hi *int.* attracting attention.

hm *int.* expressing doubt.

ho *int.* & *v.* expressing surprise.

hu *n.* Chinese liquid measure.

†hv *adv.* how.

†hw *n.* yew.

hy *v.* (*Scot.*) hie.

*ia *n.* (*Scot.*) jay.

†ic *pron.* I.

id *n.* mind's impulses.

ie *n.* Pacific islands tree.

if *conj.* & *n.* (on) condition (that).

†ig *pron.* I.

†ih *pron.* I.

†ik *pron.* I.

†il *n.* hedgehog.

†im *pron.* him.

in *prep.* & *adv.* within. ● *n.* passage in.

io *n.* Hawaiian hawk.

*ir *n.* ire.

is *present tense* of be.

it *pron.* thing mentioned.

iv *prep.* (*dialect*) in; of.

†iw *n.* yew.

ja *v.* & *n.* (*dialect*) jaw, talk.

*je *adv.* yea.

jo *n.* (*pl.* **joes**) (*Scot.*) darling.

ka *n.* (*Egyptian myth*) spirit.

*ke *n.* (*Scot.*) jackdaw.

ki n. liliaceous plant.

ko n. (pl. **ko**) Chinese liquid measure.

ku n. (dialect) ulcer in the eye.

ky n.pl. (Scot.) cows.

la n. note in music-scale.

le n. = **li**.

li n. (pl. **li**) Chinese unit.

lo int. expressing surprise.

lu v. (Orkney) listen.

ly n. = **li**.

ma n. (colloquial) mother.

me pron. objective case of **I**.

mi n. note in music-scale.

mo n. (colloquial) moment.

mu n. Greek letter m.

my adj. belonging to me.

na adv. (Scot.) no.

ne adv. & conj. (old use) not.

***ni** n. = **ny**.

no adj. not any.

nu n. Greek letter n.

†nv adv. & conj. now.

†nw adv. & conj. now.

***ny** n. brood of pheasants.

***ob** n. wizard.

†oc conj. but.

od n. hypnotic force.

oe n. small island.

of prep. belonging to.

oh int., n., & v. (give) cry of pain.

oi int. attracting attention.

†ok n. oak.

ol n. hydroxyl atom group.

om n. mantra syllable.

on prep. & adv. supported by; covering. ● n. one side of a cricket field.

oo n. (Scot.) wool.

op n. (colloquial) operation.

or conj. as an alternative.

os[1] n. (pl. **ora**) orifice.

os[2] n. (pl. **osar** or **osars**) geological ridge.

os[3] n. (pl. **ossa**) bone.

ot n. (dialect) urchin.

ou int. (Scot.) oh.

†ov pron. you.

ow int. expressing pain.

ox n. (pl. **oxen** or **oxes**) a kind of animal.

oy n. (Scot.) grandchild.

pa n. (colloquial) father.

pe n. Hebrew letter p.

pi n. Greek letter p.

po n. (pl. **pos**) (colloquial) chamber pot.

pu n. (pl. **pu**) Chinese measure of distance.

***py** n. pie.

qi n. (Chinese philosophy) life-force.

***qu** n. half-farthing.

ra n. Arabic letter r.

re n. note in music-scale.

ri n. (pl. **ri**) Japanese measure of distance.

***ro** n. & v. (Scot.) repose.

†ru v. rue.

***ry** n. rye.

sa adv. & conj. (dialect) so.

se n. Japanese measure of area.

sh int. command to silence.

si n. note in music-scale.

so[1] adv. & conj. therefore.

so[2] n. note in music-scale.

st int. attracting attention.

su pron. (dialect) she.

sy n. (dialect) scythe.

ta n. Arabic letter t.

te n. = **ti**.

ti n. note in music-scale.

to prep. & adv. towards.

tu n. 250 li.

***ty** n. & v. tie.

†ua n. woe.

ug v. & n. (dialect) dread.

uh int. inarticulate sound.

um int. hesitation in speech.

un pron. (dialect) one; him.

†uo n. foe.

up adv. & prep. towards. ● v. raise. ● n. upward direction.

us pron. objective case of **we**.

ut n. music note C.

†uu n. yew.

†uv n. yew.

uz pron. (dialect) us.

va n. (Scot.) woe.

Word-games Supplement

vg v. = ug.

vi n. Polynesian fruit.

vo n. size of book.

*vp adv. up.

*vs pron. us.

*vy v. vie.

wa n. Siamese measure.

we pron. self and others.

*wg v. (Scot.) = ug.

†**wi** n. battle; conflict.

wo int. recalling a hawk.

*wp adv. (Scot.) up.

†**wr** pron. our.

*ws pron. (Scot.) us.

†**wu** adv. how.

wy n. (Scot.) heifer.

*xa n. shah.

xi n. Greek letter x.

xu n. (pl. **xu**) Vietnamese coin.

ya n. Arabic letter y.

†**yd** pron. it.

ye pron. (old use) you.

*yf conj. if.

yi adv. (dialect) yes.

*yk pron. I.

*yl n. isle.

†**yn** n. inn.

yo int. expressing effort.

*yr n. ire.

*ys pron. his.

*yt pron. it.

yu n. Chinese wine-vessel.

*yw pron. you.

za n. Arabic letter z.

†**ze** adj. the.

zi adv. & conj. (dialect) so.

zo n. (pl. **zos**) hybrid yak.

†**zy** adj. the.

Words with a q not followed by u

The spelling *qw* was a frequent variant of *qu* and *wh* in Middle English (*c.* 1150–1500), especially in Scotland and northern England. In the words listed below, most of such forms are attested in the *Oxford English Dictionary*; several hundred others are to be found in the *Dictionary of the Older Scottish Tongue* but are excluded from the list through lack of space.

cinq n. number 5 on a die.

eqwal n. (dialect) green woodpecker.

faqih n. (pl. **faqihs** or **fuqaha**) fakir.

faqir n. fakir.

fiqh n. Islamic jurisprudence.

*liqor n. liquor.

miqra n. Hebrew biblical text.

qabab n. kebab.

qadhi n. = qadi.

qadi n. Muslim civil judge.

qaf n. letter of the Arabic alphabet.

qaid n. = qadi.

qanet n. irrigation tunnel.

qaneh n. ancient Hebrew measure (= 6 ells).

qanon n. dulcimer-like instrument.

qantar n. Middle Eastern unit of weight.

qasab n. (pl. **qasab**) ancient Mesopotamian measure of length.

qasaba n. (pl. **qasaba**) ancient Arabian measure of area.

qasida n. Arabic or Persian poem.

qat n. Ethiopian bush.

qazi n. = qadi.

qere n. marginal word in Hebrew Bible.

qeri n. = qere.

Word-games Supplement

qhat *adj. & pron.* what.

†**qheche** *adj. & pron.* which.

†**qhete** *n. (Scot.)* wheat.

†**qhom** *pron.* whom.

†**qhwom** *pron.* whom.

qi *n. (Chinese philosophy)* life-force.

qibla *n.* direction towards Mecca.

qiblah *n.* = **qibla**.

qibli *n.* sirocco.

qindar *n. (pl.* **qindarka**) Albanian coin.

qintar *n.* Albanian coin.

***qirk** *n.* quirk.

qirsh *n. (pl.* **qurush**) Saudi Arabian coin.

qiviut *n.* belly-wool of the musk-ox.

qiyas *n.* Islamic judgement.

qoph *n.* Hebrew letter q.

qre *n.* = **qere**.

***qvair** *n.* quire.

***qvan** *adv. & conj.* when.

qvare *n.* quire.

†**qvarte** *n.* quart.

†**qvayr** *n.* quire.

†**qveise** *v.* quease (= squeeze).

†**qvele** *n.* wheel.

†**qvene** *n.* queen.

†**qverel** *n.* quarrel.

†**qveyse** *v.* quease (= squeeze).

†**qvyk** *v.* quicken.

†**qvylte** *n.* quilt.

†**qwa** *pron.* who.

†**qwaint** *adj.* quaint.

***qwaire** *n.* quire.

†**qwal** *n. (Scot.)* whale.

†**qwalke** *n.* whelk, pimple.

†**qwall** *n. (Scot.)* whale.

†**qwalle** *n. (Scot.)* whale.

†**qwappe** *v.* quap (= quiver).

†**qwar** *adv. & conj. (dialect)* where.

†**qware** *adv. & conj. (dialect)* where.

†**qwarte** *adj. & n.* = **qwert**.

†**qwarto** *adv.* whereto.

†**qwartt** *adj. & n.* qwert.

†**qwasse** *v.* quash.

qwat *v. (dialect)* squash flat.

†**qwate** *n.* divination.

†**qwatte** *v. (dialect)* = **qwat**.

***qway** *n.* whey.

†**qwaylle** *n. (Scot.)* whale.

†**qwayer** *n.* quire.

†**qwaynt** *adj.* quaint.

†**qwe** *n. (pl.* **qwes**) *(Scot.)* musical instrument (pipe).

†**qwech** *adj. & pron.* which.

qweche *adj. & pron.* which.

†**qwed** *adj. & n.* evil.

†**qwede** *n.* will or bequest.

†**qwedyr** *n. (dialect)* quiver.

†**qweed** *adj. & n.* evil.

†**qweer** *n.* choir.

†**qwel** *adj. & pron.* which.

***qwele** *n. (dialect)* wheel.

†**qwelke** *n.* whelk.

†**qwell** *n. (dialect)* wheel.

†**qwelp** *n. (Scot.)* whelp.

†**qwelpe** *n. (Scot.* whelp.

†**qwem** *v.* please.

†**qweme** *v.* please.

†**qwen** *n.* queen.

†**qwench** *v.* quench.

†**qwene** *n.* queen.

†**qwenne** *adv. & conj.* when.

†**qwens** *adv. & conj.* whence.

***qwent** *adj.* quenched.

†**qwer** *n.* choir.

†**qwere** *n.* choir.

†**qwerf** *n.* wharf.

qwerk *n. (dialect)* twist; bend.

†**qwerle** *n.* whirl.

***qwern** *n.* quern.

†**qwerne** *n.* piece of ice.

†**qwert** *n.* health. ● *adj.* healthy.

†**qwerte** *n. & adj.* = **qwert**.

qwerty *n.* standard layout of typewriter keyboard.

†**qweryn** *n.* piece of ice.

†**qwest** *n.* quest.

†**qwesye** *adj.* queasy.

†**qwet** *n. (Scot.)* wheat.

†**qwete** *n. (Scot.)* wheat.

***qwey** *n.* whey.

†**qweyll** *n. (dialect)* wheel.

†**qweynt** *adj.* quaint.

*qwha *pron.* (*Scot.*) who.

†qwhar *adv.* & *conj.* where.

†qwhare *adv.* & *conj.* where.

†qwheet *n.* (*Scot.*) wheat.

†qwheit *n.* (*Scot.*) wheat.

†qwhele *n.* (*dialect*) wheel.

†qwhen *adv.* & *conj.* when.

†qwhene *n.* queen.

†qwher *adv.* & *conj.* where.

†qwhete *n.* (*Scot.*) wheat.

†qwheyn *adv.* & *conj.* (*Scot.*) when.

*qwhil *n.* (*Scot.*) while.

†qwhile *n.* (*Scot.*) while.

†qwhill *n.* (*Scot.*) while.

†qwhit *adj.* (*Scot.*) white.

qwhite *adj.* (*Scot.*) white.

†qwhois *adj.* & *pron.* (*Scot.*) which.

†qwhom *pron.* whom.

†qwhome *pron.* whom.

*qwhos *pron.* (*Scot.*) whose.

†qwhy *adv.* (*Scot.*) why.

†qwhyet *adj.* (*Scot.*) white.

*qwhyl *n.* (*Scot.*) while.

†qwhyt *adj.* white.

†qwhyte *adj.* (*Scot.*) white.

†qwi *adv.* why.

†qwiche *adj.* & *pron.* which.

†qwike *adj.* quick.

†qwikk *adj.* quick.

†qwil *n.* quill.

†qwile *n.* (*Scot.*) while.

†qwilk *adj.* & *pron.* which.

†qwill *n.* (*Scot.*) while.

†qwine *n.* quince.

qwine *n.* (*dialect*) money; corner.

qwirk *n.* (*dialect*) twist, bend.

†qwitte *v.* quit.

†qwo *n.* who.

†qwom *pron.* whom.

†qwome *pron.* whom.

†qwon *adv.* & *conj.* (*Scot.*) when.

qwop *v.* (*dialect*) throb with pain.

†qworle *n.* whorl.

†qwose *pron.* (*Scot.*) whose.

qwot *v.* (*dialect*) = qwat.

†qwy *n.* (*Scot.*) heifer.

†qwyce *n.* gorse.

†qwych *adj.* & *pron.* (*Scot.*) which.

†qwyche *adj.* & *pron.* (*Scot.*) which.

†qwye *n.* (*Scot.*) heifer.

†qwyet *n.* (*Scot.*) wheat.

†qwyght *adj.* white.

*qwyk *adj.* quick.

†qwyken *v.* quicken.

†qwykyr *n.* wicker.

†qwykyn *v.* quicken.

*qwyl *n.* (*dialect*) wheel.

†qwyle *n.* (*Scot.*) while.

†qwylte *n.* quilt.

†qwylum *adv.* (*Scot.*) whilom (= while).

†qwylys *adv.* & *n.* (*Scot.*) whiles (= while).

†qwynce *n.* quince.

†qwyne *adv.* (*dialect*) whence.

†qwynne *n.* whin.

†qwynse *n.* quinsy.

†qwype *n.* (*Scot.*) whip.

*qwyt *adj.* white.

*qwyte *adj.* white.

†qwyuer *n.* quiver.

*qwyver *n.* quiver.

†qwytt *v.* quit.

shoq *n.* Indian tree.

suq *n.* Arab market place.

tariqa *n.* Muslim ascetics' spiritual development.

tariqah *n.* = tariqa.

Words beginning with *x*

*****xa** *n.* shah.

†**xal** *v.* shall.

†**xall** *v.* shall.

†**xalle** *v.* shall.

*****xaraf** *n.* Oriental money-changer.

*****xaroff** *n.* = xaraf.

xebec *n.* sailing boat.

xebeck *n.* = xebec.

†**xel** *v.* shall.

xeme *n.* fork-tailed gull.

xenia *n.* (*pl.* **xenias**) foreign pollen effect.

xenial *adj.* of hospitality.

xenium *n.* (*pl.* **xenia**) gift to a guest.

xenon *n.* heavy inert gas.

xeque *n.* sheikh.

xeric *adj.* having little moisture.

xeriff *n.* Muslim title.

xeroma *n.* abnormal bodily dryness.

xerox *v.* photocopy.

xi *n.* Greek letter x.

*****xiph** *n.* swordfish.

*****xisti** *pl.* of **xystus**.

xoanon *n.* (*pl.* **xoana**) carved image.

†**xowyn** *v.* shove.

xu *n.* (*pl.* **xu**) Vietnamese coin.

†**xul** *v.* shall.

†**xuld** *v.* shall.

†**xulde** *v.* shall.

†**xwld** *v.* shall.

xylan *n.* carbohydrate in plants.

xylary *adj.* of xylem.

xylate *n.* salt of xylic acid.

xylem *n.* plant tissue.

xylene *n.* hydrocarbon from wood-spirit.

xylic *adj.* of a kind of acid.

xylo *n.* (*colloquial*) Xylonite (a kind of celluloid).

xylol *n.* xylene.

xylose *n.* substance obtained from xylan.

xylyl *n.* derivative of xylene.

xyrid *n.* sedge-like herb.

xyst *n.* xystus.

xysta *pl.* of **xystum**.

xyster *n.* surgeon's instrument.

xysti *pl.* of **xystus**.

xyston *n.* ancient Greek spear.

xystos *n.* xystus.

xystum *n.* (*pl.* **xysta**) xystus.

xystus *n.* (*pl.* **xysti**) covered portico.

Appendix I
Roman numerals

I	=	1	XX	=	20
II	=	2	XXX	=	30
III	=	3	XL	=	40
IV	=	4	L	=	50
V	=	5	LX	=	60
VI	=	6	LXX	=	70
VII	=	7	LXXX	=	80
VIII	=	8	XC	=	90
IX	=	9	C	=	100
X	=	10	CC	=	200
XI	=	11	CCC	=	300
XII	=	12	CD	=	400
XIII	=	13	D	=	500
XIV	=	14	DC	=	600
XV	=	15	DCC	=	700
XVI	=	16	DCCC	=	800
XVII	=	17	CM	=	900
XVIII	=	18	M	=	1000
XIX	=	19	MM	=	2000

MCMXCV = 1995

Appendix II
The Greek Alphabet

Capital	Lower case	English transliteration	Name
A	α	a	alpha
B	β	b	beta
Γ	γ	g	gamma
Δ	δ	d	delta
E	ϵ	e	epsilon
Z	ζ	z	zeta
H	η	ē	eta
Θ	θ	th	theta
I	ι	i	iota
K	κ	k	kappa
Λ	λ	l	lambda
M	μ	m	mu
N	ν	n	nu
Ξ	ξ	x	xi
O	o	o	omicron
Π	π	p	pi
P	ρ	r	rho
Σ	σ (at end of word ς)	s	sigma
T	τ	t	tau
Y	υ	u	upsilon
Φ	ϕ	ph	phi
X	χ	kh	chi
Ψ	ψ	ps	psi
Ω	ω	ō	omega

' (rough breathing) over vowel = prefixed h ($\dot{\alpha}$ = ha)

over rho = suffixed h ($\dot{\rho}$ = rh)

' (smooth breathing) over vowel or rho: not transliterated

. (iota subscript) under vowel = suffixed i (α = ai)

Appendix III
The metric system of weights and measures

Linear Measure

1 millimetre	= 0.039 inch
1 centimetre = 10 mm	= 0.394 inch
1 decimetre = 10 cm	= 3.94 inches
1 metre = 100 cm	= 1.094 yards
1 decametre = 10 m	= 10.94 yards
1 hectometre = 100 m	= 109.4 yards
1 kilometre = 1,000 m	= 0.6214 mile

Square Measure

1 square centimetre	= 0.155 sq. inch
1 square metre	= 1.196 sq. yards
1 are = 100 sq. metres	= 119.6 sq. yards
1 hectare = 100 ares	= 2.471 acres
1 square kilometre	= 0.386 sq. mile

Cubic Measure

1 cubic centimetre	= 0.061 cu. inch
1 cubic metre	= 1.308 cu. yards

Capacity Measure

1 millilitre	= 0.002 pint (British)
1 centilitre = 10 ml	= 0.018 pint
1 decilitre = 10 cl	= 0.176 pint
1 litre = 100 cl	= 1.76 pints
1 decalitre = 10 litres	= 2.20 gallons

Note 1 litre is almost exactly equivalent to 1,000 cubic centimetres.

Weight

1 milligram	= 0.015 grain
1 centigram = 10 mg	= 0.154 grain
1 decigram = 10 cg	= 1.543 grains
1 gram = 10 dg	= 15.43 grains
1 decagram = 10 g	= 5.64 drams

Weights and measures

1 hectogram = 100 g	= 3.527 ounces
1 kilogram = 1,000 g	= 2.205 pounds
1 tonne (metric ton) = 1,000 kg	= 0.984 (long) ton

. . . .

Appendix IV

Temperatures: Celsius (centigrade) and Fahrenheit

. . . .

Celsius	Fahrenheit
−17.8°	0°
−10°	14°
0°	32°
10°	50°
20°	68°
30°	86°
40°	104°
50°	122°
60°	140°
70°	158°
80°	176°
90°	194°
100°	212°

To convert Celsius into Fahrenheit: multiply by 9, divide by 5, and add 32.

To convert Fahrenheit into Celsius: subtract 32, multiply by 5, and divide by 9.

Appendix V
Rules of English Spelling and Punctuation

SPELLING

British spelling was largely standardized by the mid 18th century, and American variants established by the early 19th, but many conventions were fixed by printers as early as 1500, and as pronunciation has changed since then, present-day pronunciation and spelling are often at variance. Also, the 'neutral' vowel sound of unstressed syllables gives no guidance as to spelling, which is usually determined by the origin of the word, and care must be taken with words containing unstressed syllables such as *de-*, *di-*, *en-*, *in-*, *-par-*, *-per-*, and pairs such as *affect* and *effect*, *complement* and *compliment*. These notes cover a few of the more common difficulties: for other individual points of uncertainty, the main part of the dictionary should be consulted.

i before e For words pronounced with an 'ee' sound, the traditional rule '*i* before *e* except after *c*' is fairly reliable. Exceptions include *seize* and *heinous*, *either* and *neither* (if you pronounce them that way), *species*, and words in which a stem ending in *-e-* is followed by a suffix beginning with *-i-*, e.g. *caffeine*, *plebeian*, *protein*. Note the appearance of *-feit* in *counterfeit*, *forfeit*, *surfeit*, and that *mischief* is spelt like *chief*.

Words pronounced with an 'ay' or long 'i' (as in 'eye') sound generally have *-ei-*: e.g. *beige*, *reign*, *veil*; *eiderdown*, *height*, *kaleidoscope*. Words with other sounds follow no rules and must simply become familiar to the eye, e.g. *foreign*, *friend*, *heifer*, *leisure*, *Madeira*, *sieve*, *sovereign*, *their*, *view*, *weir*, *weird*.

Doubling consonants When a suffix beginning with a vowel (such as *-able*, *-ed*, *-er*, *-ing*, or *-ish*) is added to a word ending in a consonant, the consonant is usually doubled if it is a single consonant preceded by a single vowel, and comes at the end of a stressed syllable. So *controlled*, *dropped*, *permitted*, *transferred*, *trekked*, *bigger*, *reddish*, but *sweated*, *sweeter*, *appealing*, *greenish* (more than one vowel), *planting* (more than one consonant), *balloted*,

happened, profited, targeted (not ending a stressed syllable). A secondary stress is often sufficient to give a double consonant, e.g.
caravanning, formatted, programmed, zigzagged, and (in British
use) *kidnapped, worshipped,* though note (in British use) *benefited*.
Verbs ending in a vowel followed by *-c* generally form inflections
in *-cked, -cking,* e.g. *bivouac, mimic, picnic.*

Derivative verbs formed by the addition of prefixes follow the pattern of the root verb, as in *inputting, leapfrogged.*

In British English, the letter *l* is doubled if it follows a single
vowel, regardless of stress, e.g. *labelled, travelled, jeweller,* but
heeled, airmailed, cooler (more than one vowel). In American
English the double *l* occurs only if ending a stressed syllable, e.g.
labeled, traveling, jeweler in American use, but *dispelled* in both
British and American use (the double *l* may be retained in the present tense in American use, e.g. *appall, enthrall*). Exceptions
retaining single *l: paralleled, devilish;* exceptions having double *l*
(especially in British use): *woollen, woolly;* note variability of
cruel(l)er, cruel(l)est.

The letter *s* is not usually doubled before the suffix *-es,* either in
plural nouns, e.g. *focuses, gases, pluses, yeses,* or in the present
tense of verbs, e.g. *focuses, gases.* However, verbal forms in *-s(s)ed,
-s(s)ing* are variable, and doubling after stressed syllables is usually preferable, as in *gassed,* though with an unstressed syllable,
single *-s-,* as in *biased* and *focused,* is perhaps better. The consonants *w, x,* and *y* are never doubled; nor are silent consonants
(e.g. *crocheted, ricocheted*).

Dropping silent e A final silent *e* is usually dropped when adding
a suffix beginning with a vowel, e.g. *bluish, bravest, continuous,
queued, refusal, writing.* However, it may be retained in certain
words to preserve the sound of the preceding consonant.

The *e* is retained in *dyeing, singeing, swingeing,* to distinguish
them from *dying, singing, swinging.* It is usually retained in *ageing, cueing, whingeing,* and sometimes in *glu(e)ing, hing(e)ing,
ru(e)ing, spong(e)ing, ting(e)ing.* It is also retained in *-ee, -oe, -ye,*
e.g. *canoeing, eyeing, fleeing, shoeing.* Otherwise it is dropped:
charging, icing, staging, etc.

The *e* is retained to preserve the sound of the consonant in words
such as *advantageous, noticeable, manageable, peaceable.*
However, the dropping of *e* before *-able* is very unpredictable, and
the first (or only) spelling given in the main part of the dictionary

should be preferred. The *e* is more often dropped in American English.

The *e* is usually dropped before *-age*: *cleavage*, *dosage*, *wastage*. Exceptions: *acreage* (always), *mil(e)age* (optional; note also *lineage*). It is also usually dropped before *-y*, as in *bony*, *icy*, *grimy*. Exceptions: *cagey*, *dicey*, *gluey*. The *e* is retained in *holey* to distinguish it from *holy*, and an extra *e* is added to separate two *y*s, e.g. *clayey*.

Forming plurals Regular plurals are formed by adding *s*, or after *s*, *sh*, *ss*, *z*, *x*, *ch* (unless pronounced 'hard') adding *es*: *books*, *boxes*, *pizzas*, *queues*, *arches*, *stomachs*. An apostrophe should not be used. Nouns ending in *-y* preceded by a consonant (or *-quy*) form plurals ending in *-ies*, e.g. *rubies*, *soliloquies*, but *boys*, *monkeys*. Exceptions: *laybys*, *stand-bys*. Nouns ending in *-f* or *-fe* (but not *-ff*, *-ffe*) may form plurals in *-ves*, either always (e.g. *halves*, *leaves*) or optionally (e.g. *hooves*, *scarves*), or they may always have regular plurals (e.g. *beliefs*, *chiefs*). Nouns ending in *-o* or *-i* should be checked; a number of words have only plurals in *-oes* (e.g. *heroes*, *potatoes*, *tomatoes*) but plurals in *-os* are common, especially among words which are less naturalized (e.g. *arpeggios*), or are formed by abbreviation (e.g. *kilos*), or have a vowel preceding the *-o* (e.g. *radios*). Nouns ending in *-ful* form regular plurals in *-fuls*. Only the letter *z* is regularly doubled in forming plurals: *fezzes*, *quizzes*. Nouns ending in *-man* form plurals in *-men*, e.g. *chairmen*, *postmen*, *spokeswomen*, but note *caymans* and *talismans*. Other irregular plurals are noted in the main text of the dictionary.

Most compound nouns pluralize the last element: *break-ins*, *forget-me-nots*, *major generals*. Exceptions include *daughters-in-law*, *ladies-in-waiting*, *passers-by*, *runners-up*.

Words adopted into English generally form regular English plurals, and though words not fully naturalized may form the plural as in the language of origin, e.g. *bureaux*, *cherubim*, *formulae*, *indices*, *lire*, *virtuosi*, many nouns regularly form only English plurals, e.g. *censuses*, *octopuses*, *omnibuses*. Care should be taken with words ending in *-a*, e.g. *addenda*, *bacteria*, *criteria*, *phenomena*, and *strata* are plural, but *vertebra* is singular.

Common suffixes Several common suffixes occur in different forms which may cause spelling difficulties: users of the dictionary should be careful to check if unsure of accepted usage. The most frequent sources of uncertainty are *-able/-ible*, *-ance/-ence*, *-ant/-ent*, *-cede/-ceed*, *-er/-or*, *-er/-re*, *-ice/-ise*, *-our/-or*.

The verbal ending *-ize* has been in general use since the 16th century; it is favoured in American English and in much British writing, and in the current house style of Oxford University Press. However, the alternative spelling *-ise* is now widespread (partly under the influence of French), especially in Britain, and may be adopted provided that its use is consistent. A number of verbs always end in *-ise* in British use, notably *advertise*, *chastise*, *despise*, *disguise*, *franchise*, *merchandise*, *surmise*, and all verbs ending in *-cise*, *-prise*, *-vise* (including *comprise*, *excise*, *prise* (open), *supervise*, *surprise*, *televise*, etc.), but *-ize* is always used in *prize* (= value), *capsize*, *size*. Spellings with *-yze* (*analyze*, *paralyze*) are acceptable only in American use.

PUNCTUATION

apostrophe 1 Used to indicate the possessive case:

singular	*a boy's book*; *a day's work*; *the boss's chair*
plural with *s*	*a girls' school*; *two weeks' holiday*; *the bosses' chairs*
plural without *s*	*children's books*; *women's liberation*
names: singular	*Bill's book*; *Thomas's coat*
	Barnabas' (or *Barnabas's*) *book*; *Nicholas'* (or *Nicholas's*) *coat*

    names ending in *-es* pronounced /-iz/ are treated like plurals: *Bridges' poems*; *Moses' mother*

    before the word *sake*: *for God's sake*; *for goodness' sake*; *for Charles's sake*

    business names often omit the apostrophe: *Debenhams*; *Barclays Bank*

2 Used to mark an omission of one or more letters:
e'er (= ever); *he's* (= he is or he has); *we'll* (= we shall or we will); *'88* (= 1988)

- Incorrect uses: (i) the apostrophe must not be used with a plural where there is no possessive sense, as in *tea's are served here*; (ii) there is no such word as *her's*, *our's*, *their's*, *your's*

- Confusions: *it's* = it is or it has (not 'belonging to it'); correct uses are *it's here* (= it is here); *it's gone* (= it has gone); but *the dog wagged its tail* (no apostrophe) *who's* = who is or who has; correct uses are *who's there?*; *who's taken my pen?*; but *whose book is this? (whose* = belonging to whom)

colon 1 Used to introduce an example or a list:

Please send the following items: passport, two photographs, the correct fee.

2 Used to introduce an interpretation or description of what precedes it:

There is one thing we need: money.
I have news for you: we have won!

3 Used to introduce speech in a play or in a newspaper report where quotation marks are omitted:

Defence lawyer: Objection!
Judge: Objection overruled.

comma The comma marks a slight break between words or phrases etc. Among its specific uses are the following:

1 to separate items in a list:

red, white, and blue
bread, butter, jam, and cake

2 to separate main clauses:

Cars will park here, coaches will turn left.

3 after (or before and after) a vocative or a clause etc. with no finite verb:

Reader, I married him.
Well, Mr Jones, we meet again.
Having had lunch, we went back to work.

4 to separate phrases etc. in order to clarify meaning:

In the valley below, the villages looked very small.
In 1995, 1918 seems a long time ago.

5 following words that introduce direct speech, or after direct speech where there is no question mark or exclamation mark:

They answered, 'Here we are'.
'Here we are,' they answered.

6 after *Dear Sir, Dear John,* etc. in letters, and after *Yours faithfully, Yours sincerely,* etc.; after a vocative such as *My Lord.*

7 to separate a parenthetical word, phrase, or clause:

I am sure, however, that it will not happen.
Autumn, the season of mists, is here again.

No comma is needed between month and year in dates (e.g. *in December 1992*) or between number and road in addresses (e.g. *17 Belsyre Court*).

dash 1 Used to mark the beginning and end of an interruption in the structure of a sentence:

My son—where has he gone?—would like to meet you.

2 In print, a line slightly longer than a hyphen is used to join pairs or groups of words where it is often equivalent to *to* or *versus*:

the 1914–18 war; *the London–Horsham–Brighton route*; *the Marxist–Trotskyite split*

(See also **hyphen**.)

exclamation mark Used after an exclamatory word, phrase, or sentence, or an interjection:

Well! If it isn't John!
Order! Order!

full stop 1 Used at the end of all sentences that are not questions or exclamations.

2 Used after abbreviations:

H. G. Wells; *B.Litt.*; *Sun.* (= Sunday); *Jan.* (= January); *p. 7* (= page 7); *e.g.*; *etc.*; *a.m.*; *p.m.*

- A full stop should not be used with the numerical abbreviations *1st*, *2nd*, *3rd*, etc., nor with acronyms such as *Aslef*, *NAAFI*, nor with words that are colloquial abbreviations (e.g. *Co-op*, *demo*, *recap*, *vac*).

- Full stops are not essential in abbreviations consisting entirely of capitals (e.g. *BBC*, *NNE*, *AD*, *BC*, *PLC*), nor with *C* (= Celsius), *F* (= Fahrenheit), chemical symbols, and measures of length, weight, time, etc. (except for *in.* = inch), nor for *Dr*, *Revd*, *Mr*, *Mrs*, *Ms*, *Mme*, *Mlle*, *St*, *Hants*, *Northants*, *p* (= penny or pence).

hyphen 1 Used to join two or more words so as to form a single expression:

father-in-law; *happy-go-lucky*; *non-stick*; *self-control*

2 Used to join words in an attributive compound:

a well-known man (but 'the man is well known')
an out-of-date list (but 'the list is out of date')

3 Used to joint a prefix etc. to a proper name:

anti-Darwinian; *half-Italian*; *non-German*

4 Used to prevent misconceptions, by linking words:
a poor-rate collection; a poor rate-collection
or by separating a prefix:
re-cover/recover; re-present/represent; re-sign/resign

5 Used to separate two similar consonant or vowel sounds, as a
help to understanding and pronunciation:
pre-empt; pre-exist; Ross-shire

6 Used to represent a common second element in the items of a
list:
two-, three-, or fourfold

7 Used at the end of a line of print to show that a word not usually
hyphenated has had to be divided.

question mark 1 Used after every question that expects a separate answer:
Why is he here? Who invited him?

2 Placed before a word or date etc. whose accuracy is doubted:
T. Tallis ?1505–85

• It is not used in indirect questions, e.g. *We asked why he was
there and who had invited him.*

quotation marks Used round a direct quotation:
'That is nonsense,' he said.
The commas stand outside the quotation marks when *he said*
interrupts the quotation:
'That', he said, 'is nonsense.'

semicolon Used to separate those parts of a sentence between
which there is a more distinct break than would be called for by a
comma but which are too closely connected to be made into separate sentences:
To err is human; to forgive, divine.